Cell Biology

T. DEVASENA
Assistant Professor
Centre for Nanoscience and Technology
College of Engineering Guindy
Anna University

OXFORD
UNIVERSITY PRESS

OXFORD
UNIVERSITY PRESS

Oxford University Press is a department of the University of Oxford. It furthers the University's objective of excellence in research, scholarship, and education by publishing worldwide. Oxford is a registered trade mark of Oxford University Press in the UK and in certain other countries.

Published in India by
Oxford University Press
YMCA Library Building, 1 Jai Singh Road, New Delhi 110001, India

First published in 2012

ISBN-13: 978-0-19-807551-6
ISBN-10: 0-19-807551-0

Typeset in Times New Roman
by Anvi Composers, New Delhi 110063
Printed in India by Rajkamal Electric Press, Kundli, Haryana

Preface

Everything you'll ever need to know is within you;
The secrets of the universe are imprinted on the cells of your body;
But you haven't learned how to read the wisdom of the body.

—Dan Millman

Cells are the fundamental structural units of all living organisms, providing site for all the biochemical reactions essential for life. They produce copies of themselves inheriting all their properties, protect themselves through their membranes, perform specific functions, accommodate biomolecules, organize their own functions via subcellular organelles, communicate with each other by signalling, and they even commit suicide! Cells mimic every vital function of an organism and could be termed as the microcosm of a living being.

Features of cells are studied under a distinct branch of life science called *cell biology*. Cell biology deals with the morphology, ultrastructure, and functions of cells. It is the most vital branch of life science as it lies at the core of all other branches of life sciences, including biochemistry, biophysics, biotechnology, bioinformatics, developmental biology, genetics, genomics, immunology, microbiology, molecular biology, neurobiology, pharmacology, physiology, and plant science. Hence, knowledge of cell biology is a requisite for scientists in diverse disciplines such as life sciences, medical sciences, molecular biology, cloning, and biomedical engineering to keep pace with the rapid advances in technology.

Cell biology is an integral part of the curriculum for students pursuing various disciplines in life sciences, such as biochemistry, biotechnology, nanobiotechnology, genetics, genomics, and microbiology, both at undergraduate and postgraduate levels.

About the Book

This book—*Cell Biology*—offers a simplified and comprehensive approach to the subject with self-explanatory line diagrams, and flow charts, which are very essential for the students to acquire clear insights about this field of study. The book will also be useful for preparing for competitive examinations such as GATE and NET. The principal objective of this book is therefore, to help students gain an extensive knowledge in cell biology. The book presents streamlined information on the basic concepts and applications of cell biology in a coherent and lucid manner through its simplified presentation.

The book, in fourteen chapters, discusses every integral part of a cell along with its functional significance. Continuity is maintained across various chapters to illustrate the 'irreducible complexity' characteristic of the cell. Being a field of study which often renews itself, the book also presents contemporary information on topics such as artificial cells, cell nanofactories, and liver cell imaging, among others.

Chapters 1–13 reveal that information about the cells are mostly obtained with the help of various instrumentation and biochemical techniques, which were discovered from time to time—the most important being the microscopic technique. Since biochemical techniques are used to characterize the cell components, a separate chapter (Chapter 14) has been devoted for discussing 'the techniques used in cell biology'.

The book is meant for undergraduate engineering students of biotechnology, postgraduate students of biotechnology, and also for students pursuing other allied disciplines, which offer a complete course on this subject.

The key features of this book could be summarized as follows:

- Self-explanatory line diagrams and flow charts for easy understanding of the process.
- Key terms explained at the end of every chapter for reinforcing important topics learnt in that chapter.
- Tabular presentation of facts throughout the book for a quicker grasp of information.
- Multiple-choice questions (MCQs) at the end of every chapter to help students in scoring well both in academic and competitive examinations such as GATE, and NET.
- Discussion of recent topics such as stem cell biology, brain cell imaging through gold nanoparticles, cell nanofactories.

Content and Structure

The book comprises 14 chapters. The following is a brief description of the contents of each chapter.

Chapter 1 introduces the concept of cell biology. The chapter starts with the discovery of cells and its historical perspectives. It deals with the types, structure, and functions of cells. It also discusses the mechanics of cell division.

Chapter 2 deals with membrane transport, focusing on different types of transport, and roles of proteins and channels in membrane transport. It also throws light on transportation across intracellular organelles.

Chapter 3 discusses the cells of the immune system. It includes the structure of lymphoid organs, as well as the functions and mechanism of action of the immune cells.

Chapter 4 delves into biomolecues, an important constituent of cells that include nucleic acids, proteins, carbohydrates, lipids, pigments, and water. The chapter elaborates upon the structure, types, and functions of each of these biomolecules.

Chapter 5 elaborates upon the various stages of haematopoiesis, and the significance of haematopoietic growth factors. It also highlights the structure, and functions of erythrocytes, before discussing artificial erythrocytes and nanorobots.

Chapter 6 presents genetic mechanisms in cell—a rapidly growing field of cell biology. This chapter delves on prokaryotic and eukaryotic genomes, human genome project, and mechanism of replication, transcription, and translation. It also throws light on circadian clock and gene expression.

Chapter 7 discusses the metabolic pathways of the biomolecules, and their significance. It also highlights the various diseases that result due to abnormal metabolism.

Chapter 8 throws light on germ cells and discusses germ cell lineage, germ cell cycle, formation of germ cells, types of fertilization, and tumours of germ cells.

Chapter 9 focuses on cytoskeleton, and describes different types of cytoskeleton, its formation and properties. The chapter also discusses the various drugs influencing the function of cytoskeleton.

Chapter 10 discusses the types, structure, and functions of cell junctions and extracellular matrix, which enable communication between cells thereby coordinating the functions of cells. The chapter also elaborates upon the structure and functions of plant cell wall.

Chapter 11 is on cell signalling—the most exciting area of cell biology. This chapter describes the types of cell signalling, functions of cell signalling molecules, cell signalling networks, regulatory mechanisms of cell signalling, and signal transduction in plant cells.

Chapter 12 explains the mechanism of cell differentiation in unicellular and multicellular organisms, its determinants, causes, prevention, diagnosis, and treatment of cancer, stages in carcinogenesis, and host–tumour interaction.

Chapter 13 provides contemporary information on stem cell biology and thereby discusses the types and characteristics, derivation and applications of embryonic stem cells and adult stem cells, functions of amniotic fluid stem cells and cord blood stem cells, cardiovascular stem cells, gut stem cells, liver stem cells, and bone marrow transplantation.

Chapter 14 discusses a very significant topic in this book—techniques in cell biology. This chapter informs about the principles and applications of various biochemical and imaging techniques. Some of these include centrifugation, chromatography, electrophoresis, radioactive techniques, and electrochemical techniques, spectroscopic techniques, polymerase chain reaction, RFLP, and DNA footprinting applied to cell biology. The chapter also elaborates upon Imaging of cells using nanoparticles—a topic of interest among researchers.

Appendix includes topics such as types of cells in the human body, cancer-related genes, biochemical reference values, Pi values of amino acids, molecular weight of proteins, specific rotation of carbohydrates, absorption maxima, coefficient of some biomolecules, and glossary of a few important terms.

Acknowledgements

I thank Dr P. Mannar Jawahar, Vice Chancellor, Anna University, Chennai, for encouraging and supporting me in writing this book. I acknowledge Dr R. Jayavel, Director, Centre for Nanoscience and Technology, Anna University, Chennai, for his help and encouragement throughout the preparation of this book. I am indebted to Dr S. Sundaram, Professor, Department of Indian History, University of Madras, Chennai, for igniting the spark in me to write. I am grateful to Dr N. Sushma, HOD, Department of Biochemistry, Vel's University, Chennai, for her kind help in collecting manuscripts and reference books.

I have taken utmost care and sincere effort in ensuring that this book is free of errors. Readers are welcome to send any suggestions for improvement of the book at tdevasenabio@sify.com.

T. Devasena

Features of the Book

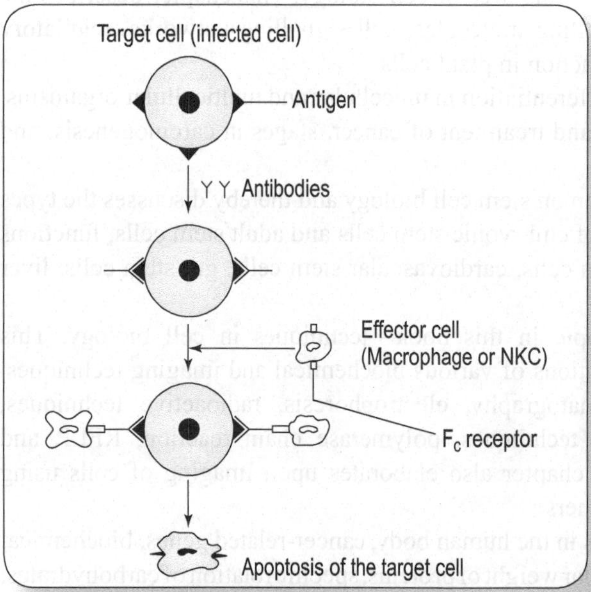

Illustrative Presentation

The book is packed with illustrations to clarify concepts, explain processes, and depict the structures of cells.

Step 1 Macrophage, being an APC, phagocytoses the antigen and expresses it on the cleft of the MHC-II complex on its surface. The macrophage then presents the antigen to the T_H cells (cell-mediated immunity). T_H cells are then activated.

Step 2 Activated T_H cells undergo mitosis and generate more T_H cells at the site of antigen presentation. Some of these cells become *memory T cells*. Simultaneously, active T_H cells secrete cytokines (interleukins).

Step 3 Interleukins function as chemo-attractants to attract and activate B cells at the site of antigen presentation.

Step 4 Activated B cells undergo mitosis and differentiation. Mitosis results in a large number of B cells, which become *memory B cells*. Differentiation results in the conversion of B cells to plasma cells.

Step 5 B cells have antibodies as insoluble receptors expressed on their surface which is specific for the antigens presented to the T cells. Plasma cells secrete immunoglobulins (soluble form of the receptor) at a frequency of several thousand per

Step-by-step Explanation

The book follows a step-by-step approach to explain concepts. This will help students visualize every step and understand better.

14.9.2 Cell Imaging by Nanoparticles

Cell imaging is defined as the process of locating a diseased cell or cancerous cell for further treatment. Cell imaging is the best method for early detection of cancer or tumour. Recently, nanoparticles have proved to be a good tool for imaging cancer cells. Quantum dots (QD) are nanocrystals with 2 to 8 nm diameter. They are made of atoms from groups II and VI elements (e.g., CdSe and CdTe) or groups III and V elements (e.g., InP and

14.9.12 Cells as Nanofactories

DNA can be manipulated into two- and three-dimensional nanostructures, which in turn can be used as template to engineer other nanostructures as well as a scaffold for organizing nanoparticles and proteins. Hao Yan from Biodesign Institute at Arizona State University has used cells as factories producing DNA nanostructures. He used the four basic building blocks of DNA—A, G, T, and C. Bacterial cells were manipulated into a nanofactory using phages and such factory can produce DNA nanostructures.

In this technique, a single-stranded DNA of 1.7 kb is folded into an octahedron (nanostructure) in the presence of helper DNA strands. Initially, the DNA is introduced into a phagemid vector and the vector is transferred into E. coli (host). Following

Contemporary Discussions

In addition to conventional topics, the book discusses contemporary topics on this subject such as brain cell imaging using gold nanoparticles, artificial cells, cell nanofactories.

SUMMARY

Genome of an organism is its complete set of DNA, bearing all the information required to build and maintain that organism. The human genome was studied by the Human Genome Project. The genome undergoes several molecular processes such as replication, transcription, and translation, collectively called genetic mechanisms.

The DNA of the genome is replicated by the process of replication, catalyzed by various enzymes and proteins, including helicase, single strand stabilizing protein, primase, ligase, and DNA polymerase. The semiconservative mode of replication was proved by Meselson Stahl experiment. The rate of DNA replication varies between prokaryotes and eukaryotes. In prokaryotes, the replication rate is 1000 nucleotides per second, which contributes to a higher error transcription. Three types of RNAs are produced by transcription: messenger RNA (mRNA), ribosomal RNA (rRNA), and transfer RNA (tRNA). Transcription is catalyzed by RNA polymerase and transcription factors.

The nascent RNA undergoes post-transcriptional modification to become biologically active and functional. Removal of spacers, trimming of tRNA and addition of CCA sequence are common types of post transcriptional modification in prokaryotes. 5' capping, 3' polyadenylation, and splicing are three types of post- transcriptional modification in eukaryotes. The information encoded in the m-RNA is converted into proteins by enzymes, ribosomes, and tRNA in a process called translation. The proteins undergo post-translational modification, such as

Chapter-end Summary

It provides a quick recap of the detailed discussions within a chapter.

TERMS TO REMEMBER

Adherens junctions Anchoring junctions of cadherin family, which interlink actin filaments of two cells.

Anchoring junctions Transmembrane proteins which connect cytoskeletal filaments of a cell to the ECM or to the cytoskeleton of the neighbouring cells.

Basal laminae A sheet-like layer of ECM with thickness of 60 to 120 nm, secreted by the epithelial cells.

Cadherin A family of glycoproteins, often called cadherin super family, crucial for cell adhesion in embryogenesis.

Cell junctions Specialized cell–cell or cell–matrix junctions which usually appear like discrete patches of protein strands.

Cell wall The extracellular matrix of plant cells composed of a pectin matrix over which a network of cellulose with glycan crosslinks is embedded.

intermediate filaments of the cells with the ECM.

Heterophilic adhesion Adhesion between two cells mediated by CAMs of different types.

Heterotypic adhesion CAM-mediated adhesion between cells of different types.

Homophilic adhesion Adhesion between two cells due to CAMs of same type.

Homotypic adhesion CAM-mediated adhesion between cells of same type.

Ig superfamily Protein family containing one or more Ig-like structural domains. They can function as calcium-independent homophilic CAMs or heterophilic CAMs.

Integrin A transmembrane glycoprotein that plays a role in cell signalling and in the final step of leucocyte migration.

Laminin The major multiadhesive protein of basal lamina,

Terms to Remember

This is an important feature which aids in remembering the key terminologies discussed within the chapter.

REVIEW QUESTIONS

1. Define LB broth and LB and media.
2. Explain the protocol for plant cell culture.
3. What is the role of sodium azide, glycerol, and DNAse in tissue homogenization?
4. Name the dyes used for staining the DNA and intracellular structures for microscopic examination.
5. Explain the principle and working of TEM.
6. Describe the principle and applications of AFM.
7. How are dynabeads used in cell sorting?
8. How will you count the blood cells using haemocytometer?
9. Write a note on cryopreservation.
10. Derive the formula for relative centrifugal field
14. Mention the disadvantages or heat generated during electrophoresis.
15. Agarose gel is preferred for DNA separation, whereas polyacrylamide gel is preferred for protein separation. Justify this statement.
16. Write the role of glycerol in agarose gel electrophoresis.
17. Explain the protocol for resolving a mixture of different molecular weights.
18. Write about Craig's countercurrent apparatus and its applications.
19. Define silanization.
20. Define electroendosmosis.
21. What is PFGE?

Chapter-end Review Questions

Every chapter ends with a set of review questions which will help students test their understanding of the chapter.

MULTIPLE-CHOICE QUESTIONS

1. The combination of a ribose and a base is called
   (a) ribonucleotide
   (b) ribonucleoside
   (c) deoxyribonucleotide
   (d) deoxyribonucleoside
2. Double helical DNA is also called
   (a) Watson and Crick model
   (b) clover leaf model
   (c) zinc finger model
   (d) all the above
3. Enzyme that joins the Okazaki fragment is
   (a) endonuclease
   (b) exonuclease
   (c) helicase
   (d) ligase

   (c) aspergillus niger
   (d) staphylococcus aureus
7. Which of the following is the enzyme used in translation?
   (a) transformylase
   (b) deformylase
   (c) helicase
   (d) topoisomerase
8. A short stretch of RNA used to initiate replication is termed
   (a) promoter
   (b) primer
   (c) primase
   (d) replisome
9. Messelson Stahl's experiment proved

Multiple-choice Questions

Multiple choice questions at the end of every chapter will help students both in academic and competitive examinations.

Brief Contents

Detailed Contents

In the year 1665, Robert Hooke discovered cells.

CHAPTER 1

Introduction to Cell Biology

After reading this chapter, students should be able to understand and appreciate the following:

- Discovery of cells
- Cell theory and cell complexity
- Prokaryotes and eukaryotes
- Phases and regulation of cell cycle
- Historical perspectives in cell biology
- Basic properties and chemical structure of cells
- Mechanism of cell division

1.1 ROBERT HOOKE

Robert Hooke, born on 18th July 1635 in Freshwater on the Isle of Wight, England, discovered cells. His thirst for knowledge was so great that once he stung himself with nettles, a wild plant, to understand the mechanism of transmission of poison into his hand. In 1660, he designed a microscopic lens that could illuminate objects by sunlight. In 1665, Robert Hooke made known the discovery of cells in the Royal Society of London. He was observing a thin sheet of cork and found a honeycomb-like structure, with spaces that resembled small rooms. He used the term 'cell', meaning a little room in Latin. His discovery strongly suggested that living organisms are made of fundamental functional units called cells and they are the building blocks of all organisms. Hooke also calculated the number of cells in a cubic inch to be 12,597,120,002. This finding suggested that each cell is a tiny compartment. He, was thus, credited for the discovery of the cell. This discovery is considered as a milestone in scientific history, which later opened doors for his successors to study cells at greater depth.

1.2 CELL BIOLOGY—HISTORICAL PERSPECTIVES

'Those who cannot remember the past are condemned to repeat it', said the Spanish philosopher and essayist, George Santayana (1863–1952). Therefore, it becomes necessary to introduce the readers to the history of cell biology.

In 1682, Antonie van Leeuwenhoek observed live cells, including bacteria and nucleus under the microscope. In 1831, Robert Brown named the nucleus and described its role in fertilization. In 1838–39, Theodor Schwann proposed that all living organisms are entirely made of cells. In 1855, Robert Remak isolated the membrane and proposed the role of membrane in cell division. In 1858, Rudolf Virchow discovered cell division and its role in pathology. Major historical episodes in cell biology are summarized in Table 1.1. Each discovery provided an impetus to further developments in this discipline, which revealed more and more details about cells. These discoveries brought to light the following aspects of cells:

- Morphology of cells
- Structural architecture of cells and cell organelles
- Functions of cells and cellular biomolecules
- Mechanics of cell division
- Significance of apoptosis and proliferation process in cells
- Role of cells in eliciting immune response (immunity)
- Role of thrombocytes (blood cells) in blood clotting
- Genetic mechanisms occurring in cells
- Cellular metabolic pathways
- Energy production required for driving a cell's functions
- Structure of germ cells and their role in fertilization
- Maintenance of cell structure by cytoskeletal organization
- Mechanism of cell–cell adhesion and cell–extracellular matrix adhesion and compounds involved in these processes
- Cell signalling mechanism responsible for specific functions of the cell and specific processes, such as proliferation, survival, and apoptosis, occurring in the cell
- Formation of cancer and genes responsible for cancer
- Origin, lineage, and differentiation of stem cells

Discussions on the aspects listed above form the basis for subsequent chapters of this book (Chapters 1–13). Table 1.1 and Chapters 1–13, reveal that the information about the cells are mostly obtained with the help of various instrumentation techniques and biochemical techniques, which were discovered from time to time, the most important being the microscopic technique. In fact, the cell itself was discovered by Robert Hooke using a microscopic lens. Even the nanoscale images were provided with high resolution by atomic force microscopy. Biochemical techniques are used to characterize cell components. Therefore, a separate chapter (Chapter 14) is allotted for discussing 'the techniques used in cell biology'.

TABLE 1.1 Major discoveries in the field of cell biology

Year	Scientist	Discovery
1665	Robert Hooke	Discovered cell using a microscope
1674	Antonie van Leeuwenhoek	Observed the cell of bacteria and protozoa
1695	Antonie van Leeuwenhoek	Discovered red blood cells
1830	Joseph Jackson Lister	Invented a prototype of compound microscope
1838	Matthias Jakob Schleiden and Theodor Schwann	Proposed cell theory
1840	Albrecht von Roelliker	Proposed that sperms and ova are germ cells

(Contd)

TABLE 1.1 (Contd)

Year	Scientist	Discovery
1848	Rudolf Virchow	First to describe the association between gastric cancer and the signal node
1856	N. Pringsheim	Observed the penetration of sperm cell into the egg cell
1857	Albert von Kölliker	Discovered mitochondria
1858	Rudolf Virchow	Discovered that cells come from pre-existing cells
1877	Wilhelm Friedrich Kühn	Coined the term 'enzyme' to describe catalytic proteins
1879	Walther Flemming	Discovered chromosomal behaviour during mitosis
1885	Wilhelm Roux	First to establish the principle of cell/tissue culture
1898	Camillo Golgi	Discovered Golgi apparatus
1903	Richard Zsigmondy	Developed ultramicroscope for high resolution
1923	Theodor Svedberg	Developed ultracentrifuge for molecular characterization
1925	Arne Wilhelm Kaurin Tiselius	Developed electrophoresis
1931	Ernst Ruska and Max Knoll	Invented electron microscope
1932	Frits Zernike	Invented the phase-contrast microscope
1938	Martin Behrens	Used differential centrifugation to separate nuclei from cytoplasm
1938	Albert Szent-Györgyi von Nagyrápolt	Identified the presence of the actin cytoskeleton in muscle
1950	Archer John Porter Martin and Richard	Laurence Millington Synge developed chromatography technique
1952	George Otto Gey and co-workers	Established a continuous human cell line
1955	Harry Eagle	Defined a medium for culturing animal cells
1957	Meselson, Stahl, and Jerome Vinograd	Developed density gradient centrifugation in caesium chloride solutions for separating nucleic acids
1969	Gerald M. Edelman	Proposed the complete structure of antibodies
1980	Sir Martin J. Evans	Developed embryonic stem cell culture method from the blastocyst
1981	Gerd Binnig and Heinrich Rohrer	Invented scanning tunnelling microscope, using which bacteriorhodopsin was characterized
1984	Masatoshi Takeichi	Discovered the first cadherin molecule, responsible for cell–cell adhesion
1986	Gerd Binnig, Cal Quate, and Christoph Gerber	Invented the first atomic force microscope, for nanoscale imaging
1995	Arne Wilhelm Kaurin Tiselius	Identified mutant green fluorescent protein with enhanced spectral properties
2005	Michael J. Berridge	Received Nobel Prize for the discovery of the role of inositol-3-phosphate in calcium-mediated cell signalling

1.3 CELL THEORY

Cell theory refers to the hypothesis that states: Cells are the fundamental units of any living organism. Cell theory was proposed by Theodor Schwann (Fig. 1.1), Matthias Jakob Schleiden, and Rudolf Virchow. After the discovery of cells by Robert Hooke, Ludolph Christian Treviranus and Johann Jacob Paul Moldenhawer suggested that cells can be separated as individual units. In 1839, Schwann and Schleiden proposed that cells were the basic unit of life. Schleiden devised a method called 'free cell formation' for spontaneous crystallization. Later, Rudolf Virchow proposed that new cells are produced from already existing cells. With the development of microscope, several other hypotheses evolved, all of which led to the modern cell theory, which states the following:

FIGURE 1.1 Theodor Schwann, founder of cell theory

1. The cell is the basic structural and functional unit of all living organisms.
2. All organisms are either unicellular or multicellular.
3. In multicellular organisms, cells are separable individual units.
4. Cells are produced from pre-existing cells through cell division.
5. Cell is the compartment for all the biochemical processes and metabolic pathways.
6. Cell encloses the heredity material which is inherited by the next generation during cell division.
7. Organisms from similar species are made up of cells of identical types and composition.
8. The function of the entire organism is determined by the activity of individual cells.

The above hypothesis supports the fact that cell is the basic structural and functional unit of all organisms.

1.4 CELL COMPLEXITY

The cell theory suggests that cell is the functional unit of all organisms. This means that the cell, which the human eye cannot even see, has an irreducible complexity. The cell complexity is more in eukaryotic cells. The term irreducible complexity means that the cells, though tiny, possess many interacting parts and components contributing to their functions, and removal of any part would impair the function of the cells. Each chapter in this book focuses on one or more of the functional components of the cell, thus, proving the cell complexity. To summarize:

- The small hair-like thread called flagella helps in locomotion via mechanical movement of structural proteins (see further discussions in this chapter and Chapter 9).
- Barriers, called membranes, surrounding the cells transport the molecules selectively, into and outside cells (see Chapter 2).

- Immune cells perform the complex function of killing and processing the foreign substances (see Chapter 3).
- Cellular biomolecules, such as proteins, lipids, carbohydrates, and even water, are very important in maintaining the structure and function of the cells (see Chapter 4).
- Blood cells supply oxygen to the cells (see Chapter 5).
- The nucleic acids inherit the genetic information and encode proteins (see Chapter 6).
- Compounds in the cells are metabolized to produce energy in the form of ATP (see Chapter 7).
- Germ cells are involved in the production of new organisms via fertilization (see Chapter 8).
- An intricate network of protein called cytoskeleton is responsible for the cell division and mechanical properties of the cell (see Chapter 9).
- Functions of the cells are coordinated by communication with each other (see Chapter 10).
- Cells perform their function in a regulated manner, depending on the signals they receive from the signalling molecules (see Chapter 11).
- A single type of cell called the stem cell can be differentiated into many cell types (see Chapter 13).

1.5 BASIC PROPERTIES OF CELLS—CELL SHAPE, SIZE, AND ARCHITECTURE

All living organisms have cells that vary in size and shape. The following factors determine the shape of a cell:

- Surface tension of the cytoplasm
- Viscosity of the cytoplasm
- Cell–cell communication distance
- Fluidity and rigidity of the cell membrane
- Function of the cell

The shape of bacterial cells depends on the species and they may be spherical, spiral, or rod-shaped (Section 1.7.1). The average size of a bacterial cell is 2 micrometres (μm) in length and 0.5 μm in diameter. A bacterium called *Mycoplasma gallisepticum* is just 0.14 μm in size, and it is the smallest cell observed so far.

Viruses are also of different shapes. They may be spherical, rod-like, bullet-like, or icosahedral. Many viruses have diameter between 10 and 300 nanometres (nm). However, filoviruses are 1400 nm in length and about 80 nm in diameter.

Animal cells are larger than those of prokaryotes. The typical size of an animal cell is 10–100 μm. The largest cell known is the egg cell of the ostrich which is about 75 mm in diameter. Animal cells are nearly spherical, but cells of different shapes are also distributed in different parts of the body. For example, brain cells called neurons are branched, epithelial cells are elongated or columnar, and erythrocytes are discoidal. Cells found in the lining or layers are flat. Most of the animal cells, for example, muscle cells, brain cells, erythrocytes, and epithelial cells possess a stable shape. Sometimes, a single cell may vary in its shape, for example, leukocytes and macrophages, which circulate in the blood and migrate to the site of infection or injury can be spherical or amoeboid (Chapters 3, 5, and 10). Plant cells possess a rectangular shape (Section 1.7.2 and Chapter 10). The average size of a plant cell is 10–100 μm. The architecture of the cell varies between the two major classes of organisms called prokaryotes and eukaryotes. This is discussed in Section 1.7.2.

1.6 CHEMISTRY OF A CELL—IMPORTANCE OF CARBON AND WATER

Water is the primary component of cell, that constitutes about 70% of the weight. Two hydrogen atoms are linked to a single oxygen atom forming a water molecule. Because of its abundance, water forms an ideal aqueous environment in the intracellular compartment, where most of the biochemical reactions and metabolic pathways occur. The structure, chemical bonding, properties, and functions of water in a cell are discussed in detail in Chapter 4.

Most compounds other than water contain carbon. This is because carbon atoms have an outstanding ability to form huge organic polymers. Carbon has four valence electrons with which it can form highly covalent bonds with other atoms or other carbon atoms and form larger organic molecules. Polymers in turn are the biological macromolecules responsible for the regulation of structure and functions of the cell. Cells have four major organic polymers:

1. Proteins
2. Carbohydrates
3. Lipids
4. Nucleic acids

All these polymers are synthesized from covalent bonding of their monomers. The polymers can in turn be catabolised into their monomers, depending on the cellular need and cellular signalling. Each and every reaction of the compound in the cell is mediated by cell signalling. The synthesis and degradation of biomolecules are discussed in Chapter 7. Cell signalling is discussed in Chapter 11.

Proteins are formed from amino acid monomers. Cells have twenty types of amino acids, which combine in different ways to form different types of proteins (Chapter 4). Proteins are synthesized by a process called translation (Chapter 6). They perform various functions, such as oxygen transport, catalysis, hormonal regulation, act as nutrient, mechanical support, and many more. The role of proteins as a cytoskeleton is discussed in Chapter 9. Proteins are also responsible for cell–cell adhesion and this is discussed in Chapter 10. Carbohydrates are organic polymers made of sugar residues. They are the used as energy sources. They can be stored in plant and animal cells as energy reserve. Lipids are polymers of fatty acids, which form the major constituent of cell membrane. Some of them are stored in adipose tissues as energy reserve. Nucleic acids are the storehouse of genetic codes. The structure and functions of all the four macromolecules are discussed in detail in Chapter 4. The genetic mechanisms associated with nucleic acids, namely replication, transcription, and translation are discussed in Chapter 6. All the four macromolecules are catabolised to enter into the Kreb's cycle and finally converted into energy via the electron transport chain and oxidative phosphorylation cycle (refer Chapter 7). Cellular biomolecules can be separated and characterized by various biochemical techniques as discussed in Chapter 14. Apart from the four primary organic polymers, cells contain other smaller molecules too. They are inorganic ions and precursors of amino acids, precursors of nucleotides, and precursors of fatty acids. The chemical composition of a prokaryotic (bacterial) cell and a eukaryotic (mammalian) cell is given in Table 1.2.

TABLE 1.2 Chemical composition of cell

Chemical component	Bacterial cell (% by total cell weight)	Mammalian cell (% by total cell weight)
Water	70	70
Proteins	15	18
Nucleic acids	7	2
Polysaccharides	2	2
Phospholipids	2	3
Other lipids	Nil	2
Total cell volume	2×10^{-12} cm^3	4×10^{-9} cm^3
Relative cell volume	1	2000

1.7 DIFFERENT CLASSES OF CELLS

On the basis of morphology, structure, and reproduction, cells are broadly classified into two types: prokaryotes (e.g., bacteria) and eukaryotes (animals, plants, and fungi). The primary differences between prokaryotes and eukaryotes are listed in Table 1.3. This section discusses prokaryotic and eukaryotic cells along with another class of organisms called virus.

TABLE 1.3 Differences between prokaryotes and eukaryotes

Feature	Prokaryotes	Eukaryotes
Subcellular organelles	Absent	Present
DNA	DNA is naked, e.g., plasmid	DNA is enclosed in a membrane-bound nucleus
Cell division	Takes place by binary fission	Takes place by mitosis and meiosis
Genetic mechanism	No separate compartments for transcription	Transcription occurs in nucleus; translation occurs in cytoplasm
Metabolism	Aerobic or anaerobic	Aerobic
Respiratory enzymes	Embedded in the plasma membrane	Integrated with the mitochondrial membrane
Cell wall	Present	Absent in animal cells Present only in plant cells
Vesicular transport	Does not occur	Occurs by exocytosis and endocytosis

1.8 PROKARYOTIC CELL

Prokaryotic cells do not contain membrane-bound organelles and are mostly unicellular, for example, bacteria. This section discusses the structure, function, and reproduction in prokaryotes in terms of bacteria.

1.8.1 Structure

Bacteria range in size from 1.1 to 1.5 μm (in width) and 2 to 6 μm (in length). They are diverse in their shapes and are accordingly divided into cocci, bacilli, vibrios, and spirilla (Table 1.4 and Fig. 1.2.). Cocci (singular: coccus) are spherical bacterial cells; examples of cocci include *Streptococcus pyogenes*, *Streptococcus mutans*, and *Lactococcus lactis*. Bacilli (singular: bacillus) are a group of rod-shaped bacteria. An example of bacillus is *Bacillus megatarium*. Vibrios are comma-shaped or curve-shaped bacteria; an example is *Vibrio cholerae*. Some bacteria, such as *Rhodospirillum rubrum*, are spiral shaped and are called spirilla.

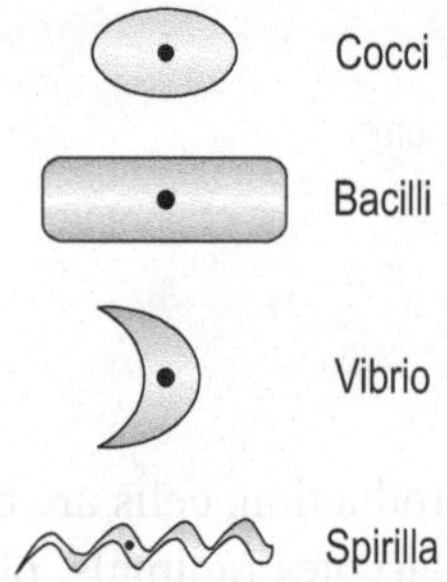

FIGURE 1.2 Shapes of bacteria

TABLE 1.4 Types of bacteria based on shape

Types	Shape	Example
Cocci	Spherical	*Staphylococcus aureus*
Bacilli	Rod	*Bacillus thuringiensis*
Vibrio	Comma or curve	*Vibrio fischeri*
Spirilla	Spiral	*Spirillum volutans*

The structure of a bacterial cell is shown in Fig. 1.3. The functions of the structural component of bacteria are summarized in Table 1.5.

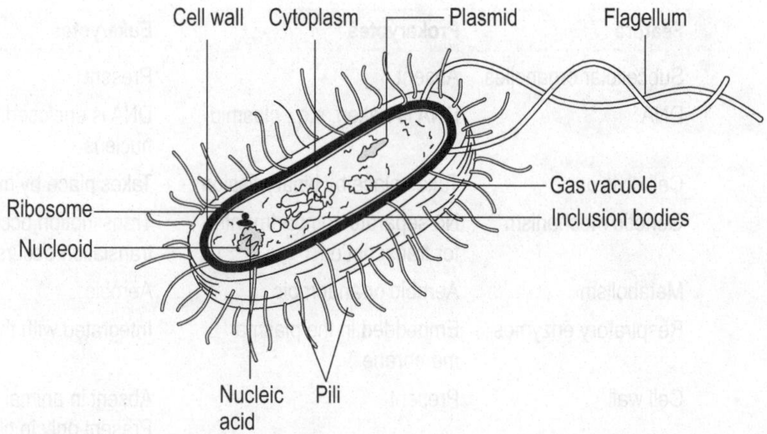

FIGURE 1.3 Structure of a bacterial cell

TABLE 1.5 Bacterial cell structures and functions

Structure	Function
Cell wall	
Lipopolysaccharide (LPS) layer	Antigen (Gram-negative bacteria)
Peptidoglycan layer	Rigidity and shape of the cell
Plasma membrane	Molecular transport and permeability barrier
Plasmid DNA	Genetic material
Gas vacuole	Buoyancy for floating
Ribosomes	Protein synthesis
Inclusion bodies	Storage site
S-layer	Protection of cells
Cell adhesion	Stabilization of cell membrane
Flagellum	Locomotion
Pili	Reproduction
Capsule	Resistance against phagocytosis and desiccation
Slime layer	Cell adhesion and formation of biofilm
Magnetosomes	Orientation of bacterial cell in magnetic field

The cell wall forms the outer layer of the bacterial cell. It maintains the rigidity and integrity of the cell and also functions as a selective barrier for the transport of molecules across the cell. Bacterial cells have higher intraconcentrations of molecules, such as proteins, when compared to their external environment. This results in a turgor pressure. The cell wall is crucial in protecting the cells against turgor pressure. On the basis of the composition of the cell wall, bacteria are divided into two types: Gram positive and Gram negative (Fig. 1.4). Gram-positive bacterial cell wall has an inner layer called the plasma membrane and an outer thick layer of peptidoglycan (refer Chapter 4), both separated by a periplasmic space. Peptidoglycan is an important component, forming around 90% of the cell wall. Gram-negative bacterial cell wall has an inner plasma membrane, surrounded externally by a thin layer of peptidoglycan, both separated by periplasmic space. The peptidoglycan in turn is surrounded by an outer layer of lipopolysaccharide (LPS). A second periplasmic space exists between the LPS and peptidoglycan layers. The peptidoglycan layer is porous and it maintains the shape and the rigidity of the cells. The LPS layer imparts a high negative charge to the surface of the Gram-negative cells. In most of the bacteria, LPS functions as the antigenic component. The membrane is made of phospholipids of various fatty acid types, including hydroxyl acids and cyclic fatty acids. It contains hopanoids (Fig. 1.5) contributing to the fluidity of the membrane. The membrane helps in the transport of molecules and acts as a permeability barrier. The periplasmic space is made of peptidoglycan and is responsible for cell signalling.

A hopanoid serves as a mechanical barrier, site for respiration, photosynthesis, chemotaxis, and also aids in the transport of materials across the cell. The inner layer of the membrane is invaginated and folded into vesicles called *mesosomes*. Mesosomes vary in shape and are accordingly known as spherical, lamellar (leaf-like), tubular, and flat mesosomes (Fig. 1.6.). Studies suggest that they are produced during the chemical fixation of bacterial cells for microscopic observation. They are considered as artifacts.

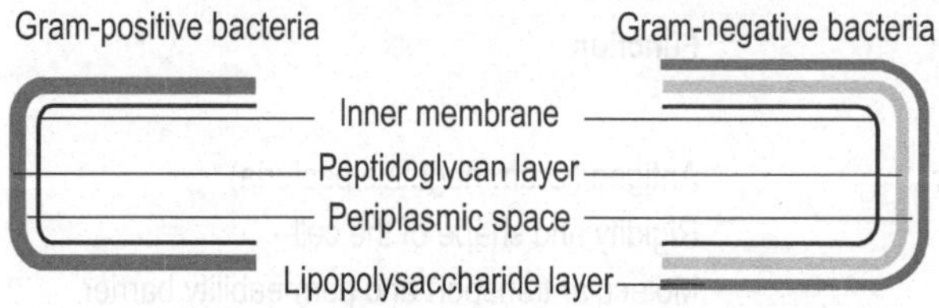

FIGURE 1.4 Composition of the cell wall in Gram-positive and Gram-negative bacteria

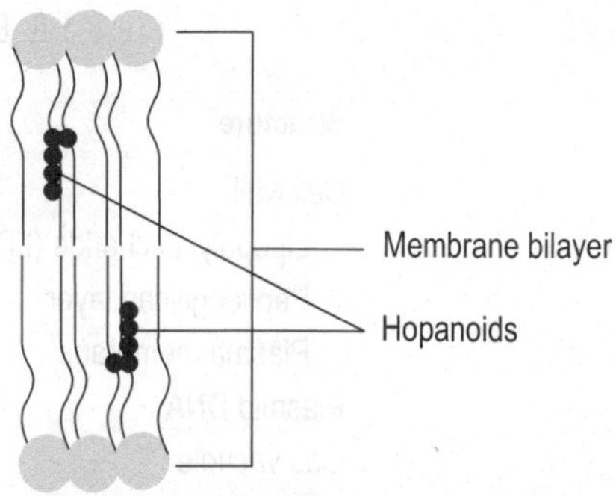

FIGURE 1.5 Hopanoids in bacterial cell membrane

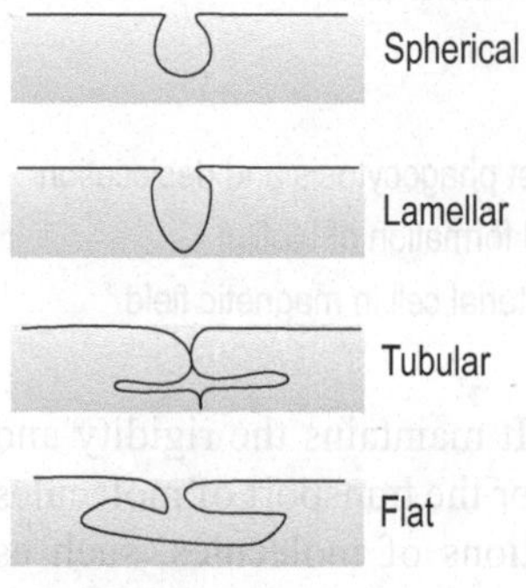

FIGURE 1.6 Mesosomes of bacterial cell membrane

In some cases, a strong, rigid layer of polysaccharide called capsule is present on the surface of the bacteria, exterior to the cell wall. Capsules may contain proteins also. Capsule is an impermeable layer and helps in protecting the bacterial cell from phagocytosis and desiccation. If the layer is unorganized, forming a diffused zone-like structure, it is called slime layer. Slime layer is used in cell adhesion and formation of biofilms. It is also used as food reserve.

Some of the bacteria contain a two-dimensional, porous surface layer called S-layer. The S-layer is 5 to 25 nm in thickness, and its pores ranges between 2 and 8 nm in diameter. It is made of proteins and glycoproteins and has the following functions:

- It protects the cell against abnormal pH, phagocytosis, and bacteriophages.
- It functions as a permeability barrier against high-molecular-weight molecules.
- It facilitates cell adhesion.
- It stabilizes the membrane.

1.8.2 Cellular Organization

The matrix present within the bacterial membrane is called cytoplasm. It is the major site of metabolism. About 80% of the cytoplasm is composed of water. Organelles, including nucleic acids (DNA and RNA), ribosomes, proteins, enzymes, amino acids, carbohydrates, lipids, inorganic ions, and many low-molecular-weight compounds are distributed in the cytoplasm. In some bacteria, cytoplasmic inclusion bodies are also found. A few vacuoles are also found in the cytoplasm; these are called *gas vacuoles* and they provide buoyancy when cells float. Some small vesicle-like structures are also found in the bacterial cell; these are called *inclusion bodies*. They act as storage vesicles in which various compounds, such as sodium, potassium, and starch are stored. Endospores are small spherical structures which offer survival capacity to the cells under harsh environmental conditions. Nearly in the centre of the cytoplasm a discrete region called nucleoid is found, where the DNA is stored. Nucleoid contains the extrachromosomal circular DNA called *plasmid* or *cytoplasmic* DNA.

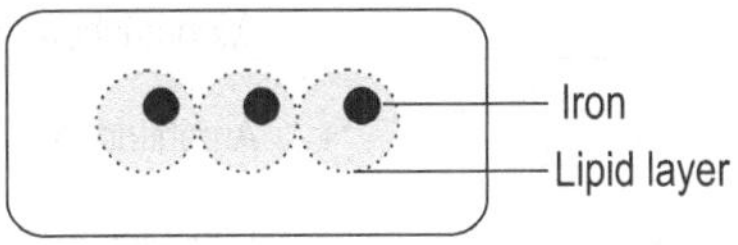

FIGURE 1.7 Structure of a magnetosome

Magnetosomes

Magnetosomes are specialized inclusion bodies found in the cytoplasm of magnetotactic bacteria, those capable of orienting in the magnetic field line of earth's magnetic field. *Aquaspirillum magnetotacticum* is an example of a magnetotactic bacterium. The structure of a magnetosome is shown in Fig. 1.7. Magnetosomes are spheres with diameters ranging from 35 to 125 nm. It has an outer lipid layer coat and an inner matrix. The matrix contains a chain of iron spheres. Each sphere functions as a tiny magnet. Iron exists in the following three forms:

1. Magnetite (Fe_3O_4)
2. Gregite (Fe_3S_4)
3. Pyrite (FeS_2)

When bacteria are exposed to magnetic field, magnetosomes help in giving a wave-like motion.

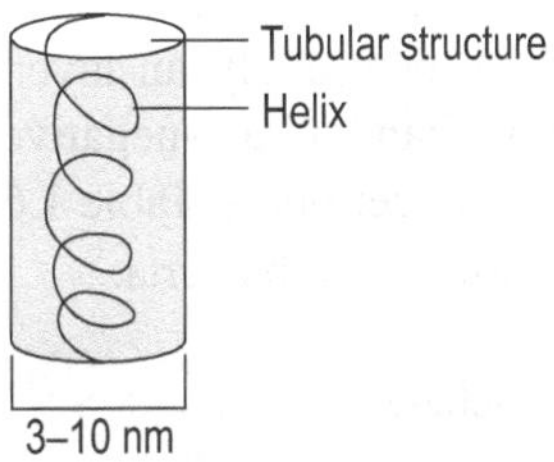

FIGURE 1.8 Fimbriae

Appendages

Appendages are outgrowths protruding from the surface of a bacterial cell. Fimbriae, pili, and flagella are the bacterial appendages.

Fimbriae Fimbriae are very small slender tubular structures of 3–10 nm diameter, widely distributed in Gram-negative bacteria. They are made up of helically arranged proteins (proteins folded into secondary structure) (Fig. 1.8). Each bacterium has approximately 1000 fimbriae/cell. Fimbriae help in cell attachment, that is, in anchoring the cell to the host cell, thus, helping in virulence and cell movements such as gliding and jerking.

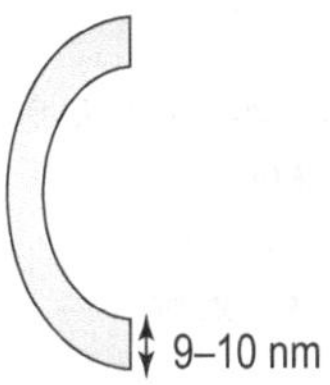

FIGURE 1.9 Pili

Pili Pili are slender hair-like structures larger than fimbriae. They are 9–10 nm in diameter and have a wavy appearance. They are responsible for conjugation (reproduction) and attachment of virus to the cell surface (Fig. 1.9). Pili are made of oligomeric protein called pilin. An enzyme called sortase is also present in pili. It functions as a protease and transpeptidase.

Flagella Flagellum (singular) is a rigid thin outgrowth of about 15–20 μm long and 20 nm diameter. It moves like a whip or thread and facilitates locomotion. The structure of a flagellum is shown in Fig. 1.10. It protrudes from the deep interior of the cell via a basal body. The body of the flagellum is called flagellar filament and is made up of protein subunits called flagellin. The filament is linked to the basal body through a flagellar hook. A protein called capping protein is present on the tip of the flagellum. The number and the distribution of flagella vary among bacteria, giving rise to different types of bacteria (Fig. 1.11). Bacteria with a single flagellum at one pole are called monotrichous; those with flagella at both ends are called amphitrichous; bacteria with a tuft of flagella at one end are called lophotrichous; and bacteria with flagella distributed throughout the outer surface of the cell are called peritrichous. Flagellar motion is aided by microtubules which can be moved mechanically by motor proteins, as discussed in Chapter 9.

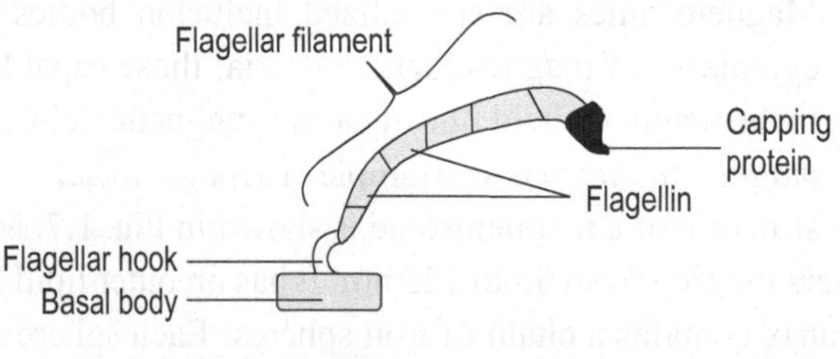

FIGURE 1.10 Flagellum

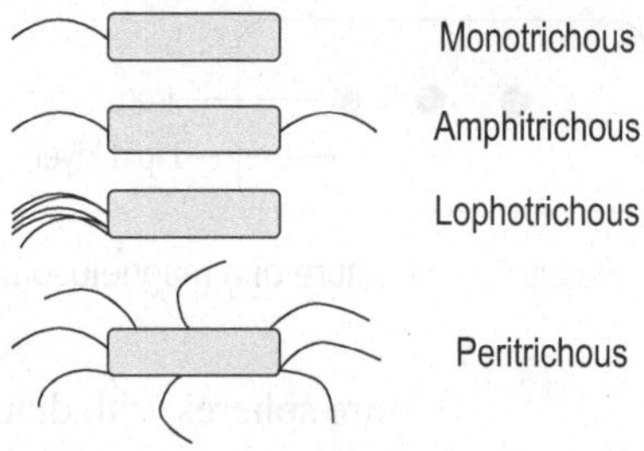

FIGURE 1.11 Types of bacteria according to the presence of flagella

1.8.3 Gram-positive and Gram-negative Bacteria

As discussed earlier, bacteria may also be classified into Gram positive and Gram negative depending on the nature of their cell walls (Fig. 1.4). This difference in the cell wall was first identified by Hans Christian Gram, using the staining technique; hence, the name 'Gram'. The absorption and exclusion of dye (stain) varies between the two types of the cell wall.

Gram-positive bacterial cell wall has a highly cross-linked thick peptidoglycan layer. This layer possesses less number of pores of small size. On the other hand, Gram-negative bacteria have thin cell wall with less cross-links and more number of larger pores. Table 1.6 and Fig.1.4 show the differences between Gram-positive and Gram-negative bacteria.

TABLE 1.6 Differences between Gram-positive and Gram-negative bacteria

Gram-positive bacteria	Gram-negative bacteria
The cell wall has a thick layer of peptidoglycan.	The cell wall has a thin layer of peptidoglycan.
Cell wall is not covered by outer membrane.	Cell wall is covered by outer membrane.
Teichoic acid is present in the peptidoglycan layer.	Teichoic acid is not present in the peptidoglycan layer.
Surface proteins are attached to peptidoglycan via sortase enzyme.	Surface proteins and sortase are absent.
Brauns lipoprotein is absent.	Brauns lipoprotein is present in the membrane.

Crystal violet dye is used to stain the bacterial cells to identify the nature of the cell wall. Crystal violet is a basic dye possessing positive charge in aqueous solution, thus, behaving as a cation. When the cells are treated with the dye, the dye molecules penetrate the peptidoglycan layer and the underlying membrane, making their way into the cell matrix or the cytoplasm. After staining with the dye, Gram's iodine (mixture of iodine and potassium iodide) is added to form an insoluble dye-iodine complex. The complex is much larger than the Gram's iodine or the dye and is very well retained in the cytoplasm. Then, a decolourizing agent, usually a mixture of ethanol and acetone, is added. Ethanol replaces the water present in the peptidoglycan layer. This leads to shrinkage and tightening of the peptidoglycan layer in Gram-positive bacteria. As a result, the larger complex (dye-iodine complex) cannot exit via the tightened peptidoglycan layer and is therefore, trapped within the cell. Then, safranin counter-stain (red/pink in colour) is added. Safranin penetrates the cell, combines with the crystal violet and forms a purple-coloured complex. Under the microscope, the cells now appear purple.

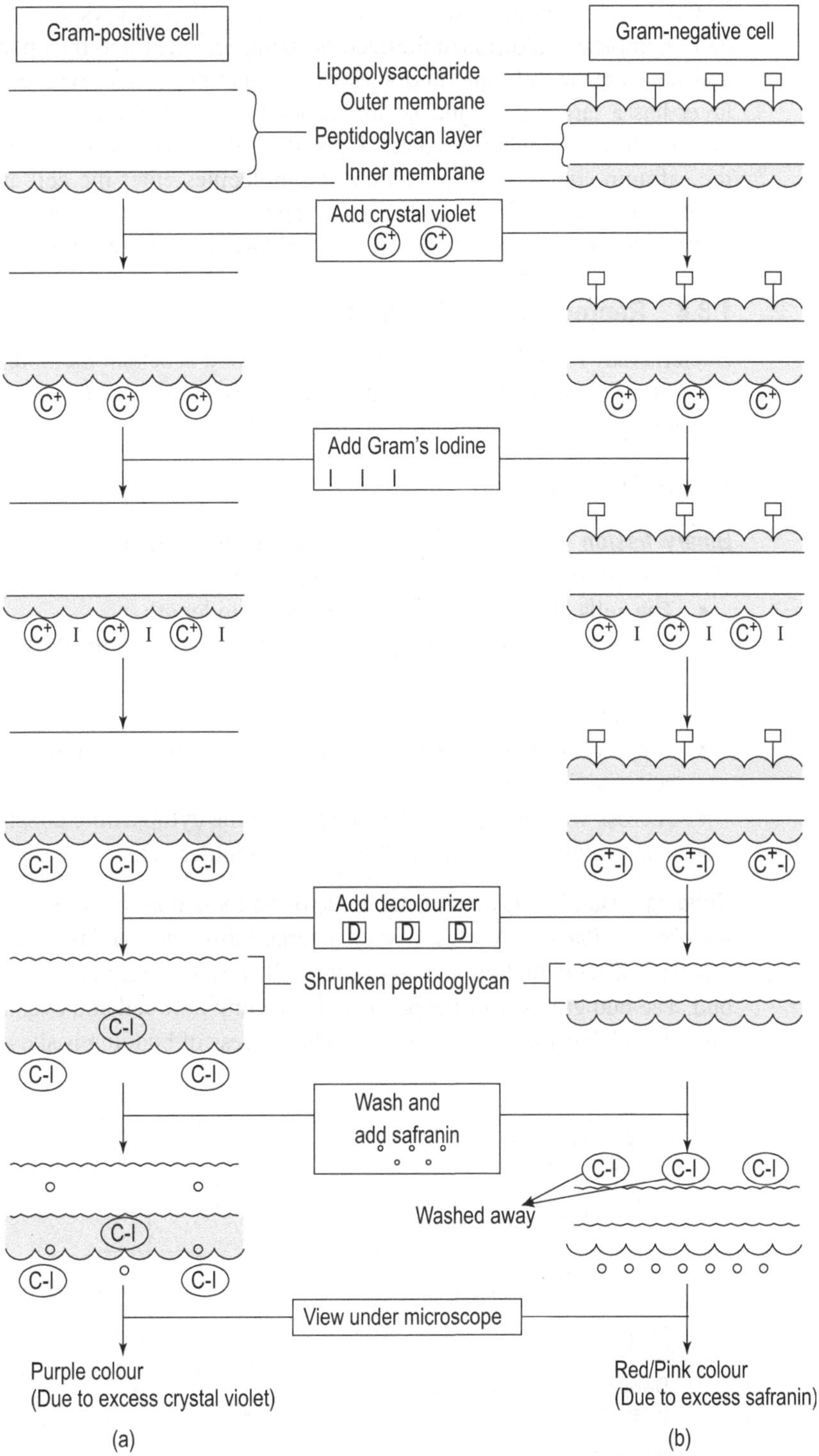

Figure 1.12 Gram staining (a) Gram-positive bacteria (b) Gram-negative bacteria

On the other hand, in Gram-negative bacteria, though crystal violet and iodine enter the cell's cytoplasm, addition of the decolourizing agent extracts the lipopolysaccharide (LPS), and the outer membrane, and shrinks the peptidoglycan layer. Still, the peptidoglycan layer has a larger pore and is thin enough to allow the escape of crystal violet in the reverse direction, when washed. Thus, the cells will appear colourless. However, when the safranin counter-stain is added, the molecules enter the cell and possess the colour of the safranin itself, that is, the cells appear pink or red under microscope. Figure 1.12 represents the event of Gram-positive and Gram-negative staining.

1.8.4 Reproduction in Bacteria

Bacteria may reproduce by either asexual method or sexual method. Asexual reproduction takes place by binary fission, budding, or fragmentation, while sexual reproduction occurs by conjugation, transformation, or transduction.

Asexual reproduction

Binary fission This is the most common mode of asexual reproduction in bacteria (Fig. 1.13 a). It occurs via the following events:

- The cells elongate, showing an increase in length.
- Chromosomes undergo replication (refer Chapter 6) resulting in doubling of nucleic acids.
- The nucleoid and nucleic acids segregate.
- A cytoplasmic constriction occurs across along the width of the cell, marking the cell-division path.
- A cross wall or septum is formed along the cytoplasmic constriction.
- The septum divides the cell into two daughter cells.

Budding Budding is an unusual mode of reproduction in bacteria. This is observed in the members of Planctomycetes, Cyanobacteria, Firmicutes, and prosthecate Proteobacteria. In this type of reproduction, the end of the cell shows a slight bulging, which develops into a bud. The bud grows over the parent cell as a tiny new cell and eventually pinches off from the cell and develops into a new cell. The process of budding is shown in Fig. 1.13b.

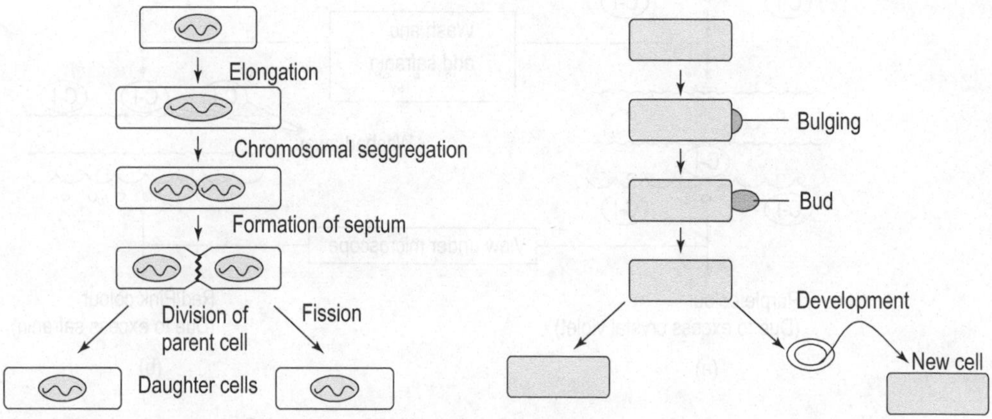

FIGURE 1.13 Asexual reproduction of bacterial cell by (a) binary fission and (b) budding

Fragmentation Fragmentation occurs in photosynthetic bacteria. During fragmentation, a portion of the parent cell is fragmented and the fragment develops into a new organism.

Sexual reproduction

Sexual reproduction occurs by the direct exchange of genetic material between two bacterial cells. Sexual reproduction is of the following three types:

Conjugation Conjugation is brought about by a bacterial DNA sequence that is capable of producing sex pilus. Such DNA is known as the F-plasmid, fertility factor, or sex factor. The bacteria with sex pilus are donors, capable of donating their genetic material to a recipient bacterium. The donor possessing the F-plasmid is called F^+ and the recipient that lacks the F-plasmid is called F^-. The F-plasmid contains a *tra* locus containing the pilin gene. Pilin gene forms the pili. The *trb* locus of the F-plasmid produces proteins responsible for attaching the donor to the recipient cell.

Conjugation is initiated by an enzyme called relaxase. Relaxase produces a nick in one of the strands of the donor plasmid. The nicked strand unwinds and separates from the unbroken strand of the plasmid. The donor and the recipient cells come closer and are conjugated via the sex pilus. The unwound strand is transferred to the recipient cell in 5′ to 3′ direction. The strands in the donor and the recipient cells can make their own complementary strands. As a result, each cell will develop a plasmid, capable of producing pili. Thus, the two daughter cells will be F^+. The process of conjugation is shown in Fig. 1.14.

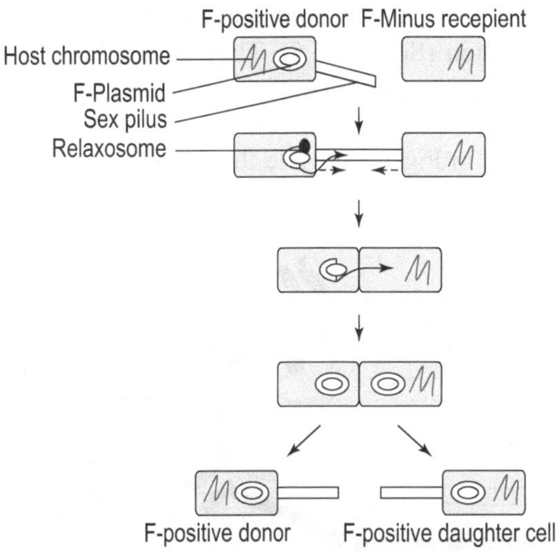

FIGURE 1.14 Sexual reproduction of bacteria by conjugation

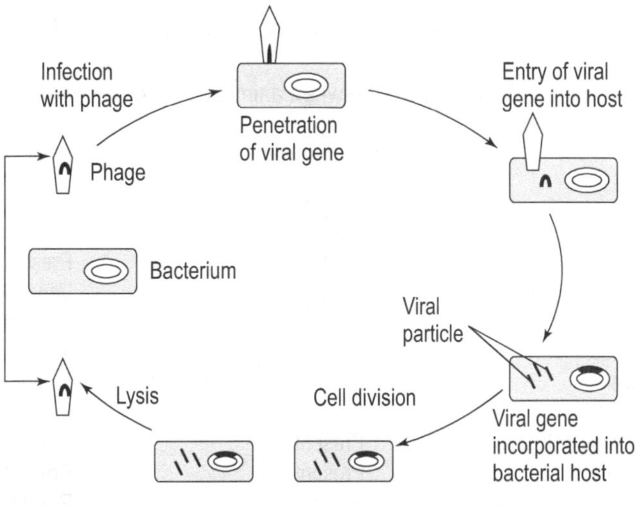

FIGURE 1.15 Transduction

Transformation Transformation refers to the uptake of naked DNA by the bacteria. Bacterial cell capable of taking the DNA is called a competent cell. This method of reproduction is less common, and occurs rarely in nature. However, artificial transformation is possible under stringent laboratory conditions. In natural transformation, the competent cell produces proteins which form a machinery to transfer the exogenous DNA across the cell membrane.

Transduction In the transduction process, the DNA is introduced into the bacteria by a bacteriophage. It can also be done artificially in the laboratory, using viral vectors for gene transfer. The process of bacteriophage-mediated transduction is shown in Fig. 1.15.

1.9 EUKARYOTES

Eukaryotic cells have distinct membrane-bound organelles. Animal cell is the best example of a eukaryotic cell. These cells are almost spherical but may vary in their shape depending on the function and tissue distribution (Section 1.5). They contain an outer membrane and inner cytoplasm. Membrane-bound subcellular organelles are embedded in the cytoplasm. Plant cells possess a thick cell wall that covers the membrane. The structure of plant and animal cells are depicted in Fig.1.16, while the features of animal and plant cells are compared in Table 1.7.

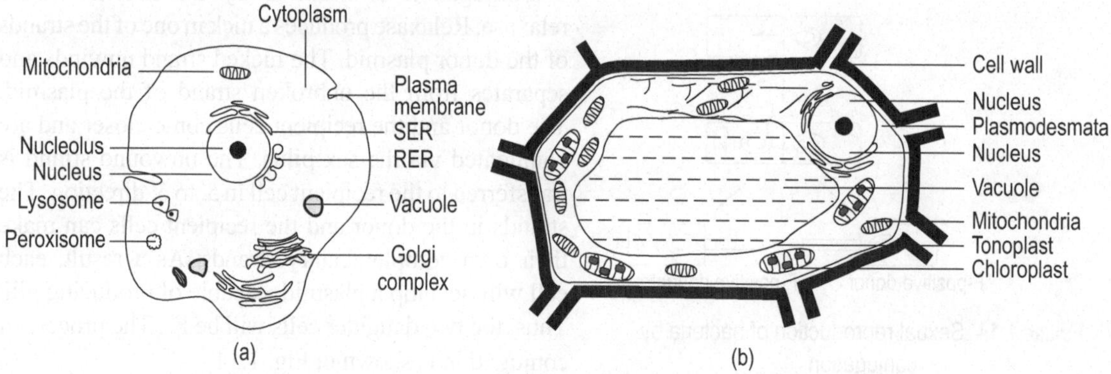

FIGURE 1.16 Structure of (a) animal cell and (b) plant cell

TABLE 1.7 Differences between animal cell and plant cell

Feature	Animal cell	Plant cell
Shape	Spherical or diverse shape	Rectangular
Chloroplast	Absent	Present
Plastid	Absent	Present
Vacuole	Small	Very huge
Cell wall	Absent	Present
Lysosomes	Distinct in cytoplasm	Not distinct
Centriole	Present in all animal cells	Present in lower plant forms only
Cytoplasm	Present	Present
Endoplasmic reticulum	Present	Present
Ribosomes	Present	Present
Mitochondria	Present	Present
Golgi apparatus	Present	Present
Cytoskeleton	Present	Present
Cytokinesis (cytoplasmic division)	Takes place by constriction	Takes place by plate formation

1.9.1 Subcellular Organelles

Membrane-bound subcellular organelles present in eukaryotic cells are discussed as follows:

Cytoplasm

The major portion of the cell interior that forms the cell matrix is called cytoplasm. It is amorphous, translucent, and homogeneous colloidal in nature. It is composed of inorganic constituents (water, sodium salts, potassium salts, and metals), organic constituents (carbohydrates, lipids, proteins, nucleoproteins). It has living structures called organelles and non-living structures called *paraplasm*, *deutoplasm*, or *inclusions* (e.g., pigments). The peripheral part of the cytoplasm which is nearer to the membrane is non-granular and viscous, and it is called ectoplasm or *plasmagel*. The inner part of the cytoplasm which is nearer to the nucleus is granular and less viscous, and this part is called *endoplasm*. The cytoplasm stores many enzymes required for metabolic pathways. It is the site for protein synthesis.

Nucleus

Nucleus is a more or less spherical organelle, located almost in the centre of the cell. The structure of nucleus is given in Fig. 1.17. The nucleus is enveloped by a double membrane structure called nuclear envelope, which is interrupted by small holes called nuclear pores. The nuclear pores accommodate nuclear pore complex which contains protein granules (refer Chapter 2). The inner membrane has a leaf-like structure called nuclear lamina, which is a network of fibres. The nuclear envelope is the dynamic part of the nucleus. The outer membrane of the envelope is anchored to another organelle called rough endoplasmic reticulum, which in turn is attached to the smooth endoplasmic reticulum. A small spherical structure called nucleolus is present in the nucleus, in which the nucleic acids are stored (refer Chapters 4 and 6).

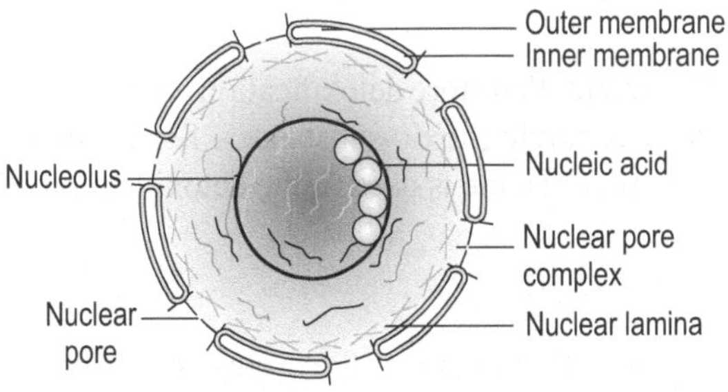

FIGURE 1.17 Structure of nucleus

Endoplasmic reticulum

The endoplasmic reticulum (ER) is a highly convoluted membranous sac formed of interconnected tubules. Almost 50% of the cell membrane is contributed by ER. It is continuous with the outer membrane of the nucleus (Fig. 1.18). ER may be smooth (SER) or rough (RER, studded with ribosomes). Differences between RER and SER are shown in Table 1.8. Role of ER in the intracellular transport is discussed in Chapter 2.

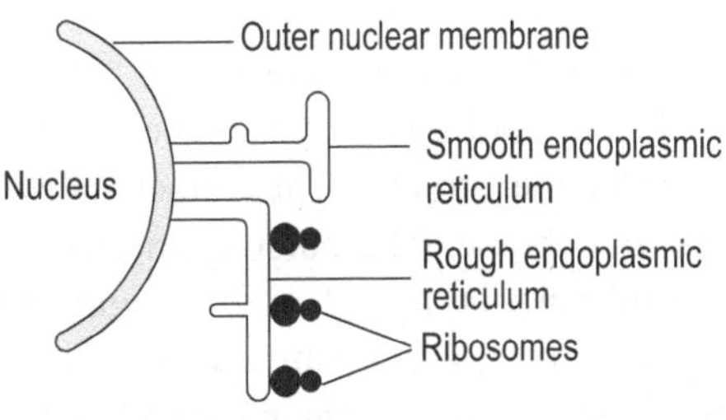

FIGURE 1.18 Structure of endoplasmic reticulum

Golgi complex/Golgi stack

Golgi complex (GC) is a stack of flattened membranous discs. GC is associated with a tiny membranous vesicle called Golgi vesicle. In plant cells, Golgi vesicles are well developed and are called dictyosomes. The function of the GC is the intracellular transport of macromolecules such as lipids, proteins, and carbohydrates (refer Chapter 2).

TABLE 1.8 Differences between rough endoplasmic reticulum (RER) and smooth endoplasmic reticulum (SER)

Rough endoplasmic reticulum	Smooth endoplasmic reticulum
Tubules are highly convoluted	Tubules are fine
Tubules are studded with ribosomes	Ribosomes are absent
Site for the synthesis of proteins	Site for the synthesis of detoxification enzymes, such as catalase
Widely distributed in secretory cells	Widely distributed in cells that actively metabolize lipids, e.g., liver cells

Lysosomes

Lysosomes are tiny irregular or nearly spherical vesicles. A lysosome is the storehouse of hydrolytic enzymes collectively called *acid hydrolase*. These enzymes catalyze the digestion of macromolecules. Some examples of lysosomes include RNAse involved in the digestion of RNA; acid phosphatase involved in digestion of phosphates; DNAse involved in the digestion of DNA; proteases (cathepsin, collagenase, peptidase) involved in digestion of proteins; galactosidase, hexosaminidase, lysozyme, and hyaluronidase involved in the digestion of carbohydrates; and esterase and phospholipase involved in degrading lipids.

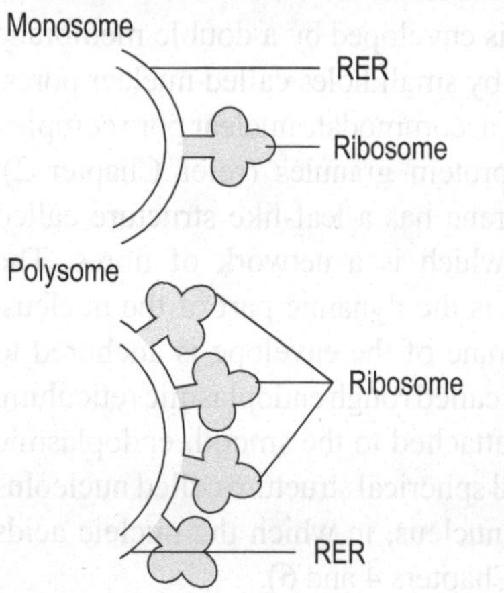

FIGURE 1.19 Monosome and polysome

Peroxisomes

Peroxisomes are small vesicles that store detoxification enzymes and help in detoxification by removing toxic substances. For example, catalase (CAT) stored in peroxisome is used in the detoxification of hydrogen peroxide.

Ribosomes

Ribosomes are the primary constituents of the protein synthesizing machinery (refer Chapter 6). Ribosomes can exist as a single unit associated with ER (monosomes) or multiple units associated with mRNA (polysomes). Monosomes mediate in the synthesis of secretory proteins (proteins secreted out of the cell) and organelle-targeted proteins (proteins transported into specific organelles within the cell). Polysomes mediate in the synthesis of cytosolic proteins (Fig. 1.19).

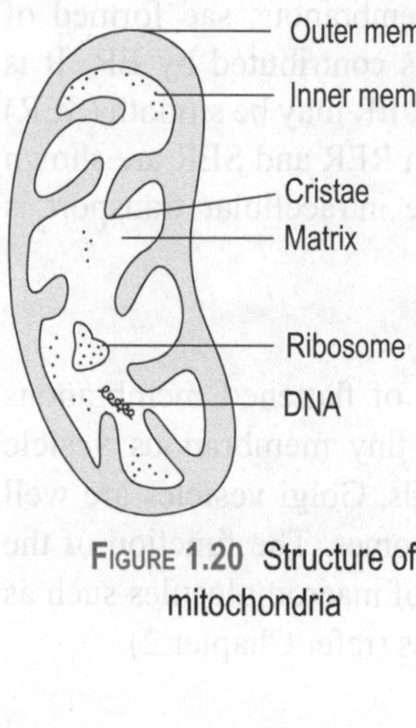

FIGURE 1.20 Structure of mitochondria

Mitochondria

Mitochondria, the power house or the power plant of the cell, is responsible for respiration of cells and energy production (refer Chapter 7). Figure 1.20 reveals the structure of mitochondria. It has an envelope of double membrane. The outer membrane is smooth, while the inner membrane is folded and invaginated into the mitochondrial matrix, forming a structure called cristae. Enzymes responsible for the electron transport chain are embedded in the inner membrane (refer Chapter 7). The mitochondrial matrix contains independently replicating DNA, ribosomes, and enzymes. The genetic code of the mitochondrial DNA is different from that of nuclear DNA.

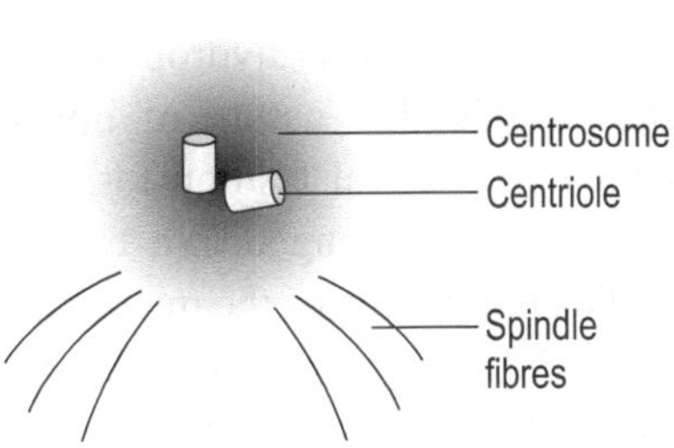

FIGURE 1.21 Centrosome

Centrosome

Centrosome is the dense region of cytoplasm around the nucleus (Fig. 1.21). It plays a major role in cell division (discussed in Section 1.11.1). The centrosome is made of a pair of cytoskeletal structures called centrioles, arranged perpendicular to each other.

Vacuole

Vacuoles are hollow spheres that are spherical or irregular in shape. Plant vacuoles are larger than animal vacuoles and are bound by a well-defined membrane called tonoplast. The space of the vacuole is filled with cell liquids. Vacuole plays a role in the maintenance of the internal pressure of the cell.

Membrane

The cell is covered by a protective barrier called membrane, or the plasma membrane. This membrane is a porous, elastic, ultra-thin, and semipermeable bilayer. As mentioned at the beginning of this section, all intracellular organelles are membrane bound. A membrane is extremely important for a cell since it performs the following major functions:

- It is responsible for the transport of molecules into and outside the cell (refer Chapter 2).
- It promotes intracellular traffic by transporting molecules from one organelle to the other (refer Chapter 2).
- It can also form vesicles and helps in vesicular transport of molecules (refer Chapter 2).
- It serves as a barrier between the intracellular and the extracellular compartment.
- It acts as a mechanical support to the cell and segregates intracellular events, such as genetic mechanisms (refer Chapter 6) and metabolism and energy production (refer Chapter 7).
- It can help in signal transduction (refer Chapter 11).
- It can mediate cell–cell communication and cell–extracellular matrix (ECM) adhesion (refer Chapter 10).
- It can help in phagocytosis (refer Chapter 3).
- It helps in fertilization (refer Chapter 8).

TABLE 1.9 Composition of membrane

Component	Type of the component
Lipids	Phospholipids Glycolipids Cholesterol
Proteins	Integral proteins Peripheral proteins Membrane-bound enzymes Receptors
Carbohydrates	Glycolipids Glycoproteins

1.9.2 Composition of Membrane

The principal component of the membrane is lipid, the other components being proteins and carbohydrates. The composition of cell membrane is given in Table 1.9.

Lipids

Lipids are esters (polymers) of fatty acids. The fatty acid chain of the lipid forms a tail and it is non-polar or hydrophobic. The alcoholic moiety and its derivative

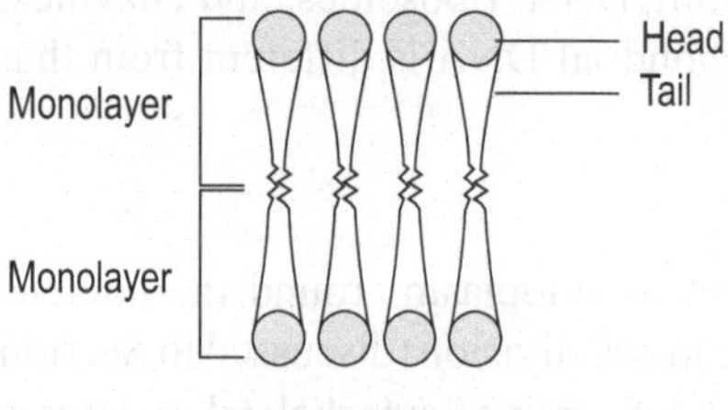

FIGURE 1.22 Lipid bilayer

forms the head of the lipid molecule and it is hydrophilic or polar. This contributes to the amphipathic nature of lipids. The lipid molecules are arranged in a lateral fashion forming a monolayer. Two such monolayers are arranged in a tail-to-tail fashion, forming a bilayer. The bilayer forms the membrane architecture (Fig. 1.22). Lipid bilayer of the membranes undergoes two types of movements: lateral diffusion and translational movement.

There are three types of lipids found in the membrane: phospholipid, glycolipid, and cholesterol. In majority of the cell types, phospholipid is predominant. Phospholipids are amphiphilic molecules, that is, they have both water-soluble and water-insoluble moieties. Each phospholipid is made of a hydrophobic hydrocarbon tail and a hydrophilic polar head. Types of phospholipids include phosphatidylcholine, phosphatidylethanolamine, phosphatidylserine, phosphatidylinositol, and sphingomyelin. Phosphatidylinositol is present in a minor amount. Bacterial cell membrane is made of types of phospholipid only, but no cholesterol. Eukaryotic cells, on the other hand, are made of mixture of different types of phospholipids and a large amount of cholesterol.

Asymmetry of lipid bilayer Asymmetry refers to differences in phospholipid composition of outer and inner monolayers. For example, in erythrocytes, the outer monolayer is composed of phospholipids with choline group—phosphatidylcholine and sphingomyelin—while the inner layer is made of phospholipid with terminal primary amino group (phosphatidylethanolamine and phosphatidylserine). The phosphatidylserine gives a negative charge to the inner layer leading to a charge difference within the bilayer.

Functions of phospholipids The functions of phospholipids are attributed to its asymmetry. The following are functions of some of the phospholipids:

- Negatively charged phosphatidylserine distributed in the inner membrane is the binding site and/or activation site for cytosolic proteins such as enzymes. For example, the enzyme protein kinase C (PKC) binds to the phosphatidylserine in response to an extracellular signal, and it gets activated (refer Chapter 11).
- Phosphatidylinositol, a minor phospholipid present in the cytosolic monolayer of the membrane, is the binding site for lipid phosphorylating enzyme, namely, phosphatidylinositol-3-kinase or PI3 kinase. PI3 kinase adds phosphate group to the inositol ring, which in turn acts as a binding site that recruits specific proteins from the cytosol to the membrane.
- Enzymatically cleaved phosphatidylinositol is used in releasing calcium from the endoplasmic reticulum. In the presence of extracellular signal, PLC (phospholipase C) is activated. Active PLC cleaves inositol phospholipid into two fragments—head fragment and tail fragment. The tail fragment remains in the membrane and activates PKC, while the head fragment gets into the cytosol and stimulates the release of calcium ions from the endoplasmic reticulum. This process is generally called relay signal (refer Chapter 11).
- Phospholipid asymmetry enables one to differentiate live and dead cells. In dead cells, phosphatidylserine is translocated from cytosolic monolayer to the extra-

cellular monolayer and exposed on the cell surface. The exposed phosphatidyl-serine is the signal for the neighbouring macrophages for phagocytosing the dead cells. Translocation is mediated by inactivation of phospholipid translocator (which transports phosphatidylserine from the non-cytosolic monolayer to the cytosolic monolayer) and activation of scramblase, which induces bidirectional transfer.

Cholesterol Apart from phospholipids, membranes contain cholesterol and glycolipids. Eukaryotic membrane has one molecule of cholesterol per phospholipid. The hydroxyl group of cholesterol is close to the polar head group of the phospholipid. This leads to the interaction of rigid steroid ring of the cholesterol with the adjacent hydrocarbon chain of the phospholipid that is close to the polar head. This arrangement leads to the following situations:

- Decrease in fluidity or mobility
- Increase in the degree of immobilization in the membrane structure
- Membrane stiffening
- Decrease in phase transition

The net result is the decrease in permeability of small water-soluble molecules.

Glycolipids Glycolipids are lipids containing sugar residues. They are derived from serine in eukaryotic cells and from glycerol in bacteria. Sugar molecules are added to the lipid molecules within the lumen of the Golgi apparatus. They show extreme asymmetry in their cellular distribution. They are predominant in non-cytosolic monolayer, with the sugar residues exposed to the cell surface. Glycolipids are of two types:

- Galactocerebroside—It is neutral due to the presence of an uncharged head group.
- Ganglioside—It is negatively charged due to the presence of sialic acid residues. It is predominant in nerve cells and scarce in other cell types.

Functions of glycolipids The following are some important functions of glycolipids:

- They prevent the intestinal epithelial cells from harsh conditions, such as low pH and enzyme action.
- They alter the electrical field across membrane and alter the concentration of ions, especially calcium ions, across the membrane.
- They are responsible for cell–cell recognition and cell adhesion (refer Chapter 10).
- Glycolipids are entry points for bacterial toxin, for example, cholera toxin, which increases the cellular cAMP level causing the efflux of sodium ions and water into the intestinal lumen (refer Chapter 11).

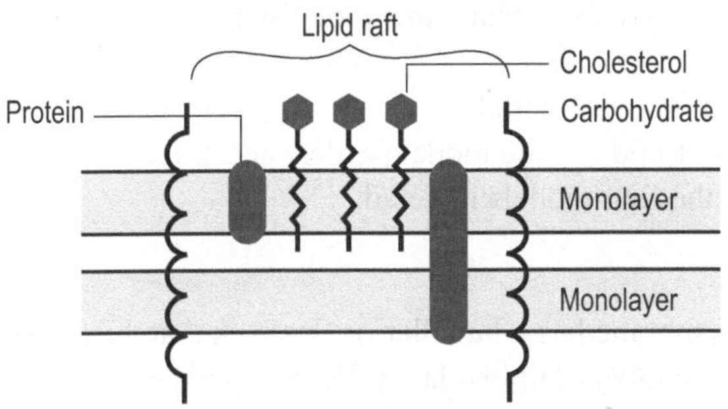

FIGURE 1.23 Lipid raft

Lipid raft Lipid rafts are tiny domains (Fig. 1.23) of approximately 70-nm diameter in animal cell membranes that are enriched with sphingolipids and cholesterol. The hydrocarbon chains of this domain are long, straight, and saturated. Thus, these rafts are thicker than other parts of the membrane bilayer, and therefore, accommodate certain membrane proteins. The hydrocarbon chains of both the monolayers in the raft region communicate with each other and do not move independently unlike other regions where the molecule of one monolayer moves independently of the other.

Following are some functions of lipid rafts:

- Organizing proteins
- Concentrating proteins for transport
- Enabling proteins to function together during signal transduction

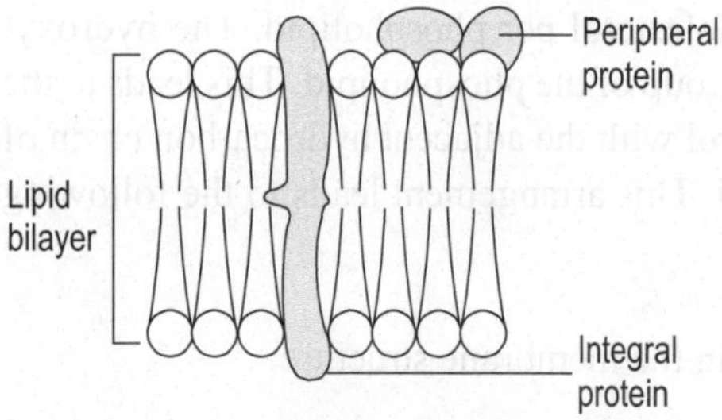

FIGURE 1.24 Integral proteins and peripheral proteins

Proteins

Membranes have two types of proteins—peripheral proteins or extrinsic proteins and integral proteins or intrinsic proteins (Fig. 1.24). Peripheral proteins are loosely bound to the membrane and are easily dissociable. On the other hand, integral proteins are tightly bound to the membrane and span the bilayer. One end of the integral protein forms the extracellular domain and the other end forms the intracellular domain. Most receptors are integral proteins.

Membrane proteins provide mechanical support to the membrane and act as channels or carriers for transport of molecules across the membrane (refer Chapter 2). They also serve as membrane-bound enzymes and receptors. Some of the membrane-bound proteins help in signal transduction (refer Chapter 11).

Carbohydrates

Membranes contain very little amount of carbohydrates when compared to the amount of lipids. The carbohydrate chains in the membrane have sialic acid as the terminal residue and galactose as the penultimate residue. Majority of membrane carbohydrates are in the form of glycolipids. Glycocalyx is the carbohydrate found in the membranes of bacteria and epithelial cells containing microvilli. They perform the following important functions:

- They protect the membrane from injury.
- They help the immune cells to invade the foreign antigens.
- Abnormal glycocalyx are markers of cancers, which is recognized by immune cells for degradation.
- They enable cell–cell adhesion.

1.9.3 Molecular Architecture of Membrane

The following four models were proposed to explain the molecular architecture of membrane (Fig. 1.25):

1. Lamellar model
2. Micelle model
3. Protein crystal model
4. Fluid mosaic model (widely accepted model).

We now proceed to discuss each of the four models in detail.

Lamellar model

The *lamellar model* proposes that a membrane has a lamellar or 'leaf -like' architecture. The leaf-like structure is formed by an assembly of lipid bilayer. The bilayer is sandwiched between protein layers. The protein layers are associated with sialic acid residues.

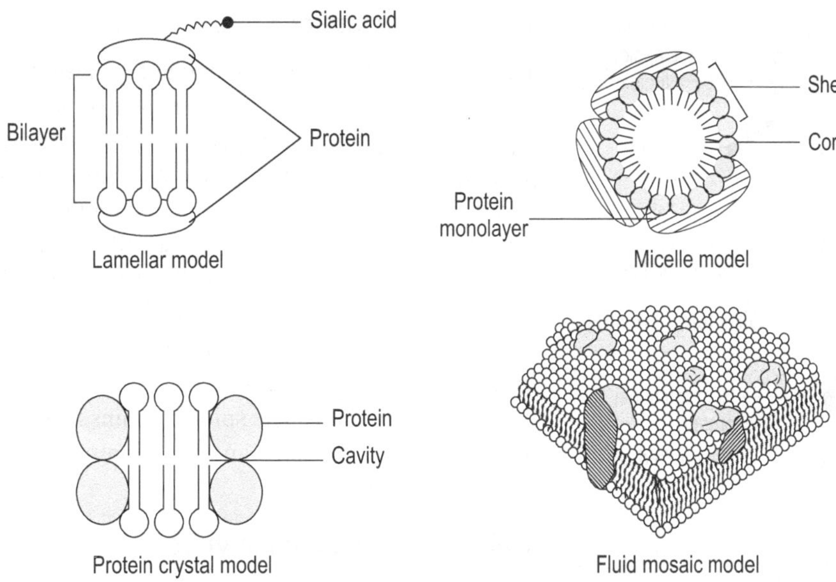

FIGURE 1.25 Models for molecular architecture of membrane

Micelle model

According to the *micelle model*, the membrane layer has a spherical structure, covered by proteins. The sphere contains a hydrophobic core and a hydrophilic shell, thus forming a micelle-like structure.

Protein crystal model

The *protein crystal model* proposes that the lipid bilayer occupies the cavity between protein molecules giving a crystal-like appearance to the membrane.

Fluid mosaic model

The *fluid mosaic model* is a widely accepted model for the study of membrane architecture. This model was proposed by Sanger and Nicholson (1972). According to this model, membranes have a mosaic-like structure made up of both proteins and lipids. Peripheral proteins are anchored to the surface of the bilayer, and integral proteins are tightly associated with the membrane. The bilayer is not rigid, instead it is mobile and exhibits a fluid state. The fluidity of the membrane can, however, be altered by changing the temperature. This process is called *phase transition*. During phase transition, the carbon–carbon bond in the lipid rotates by 120° and assumes a new conformation called 'Gauche' conformation. The exact temperature at which phase transition occurs is called the transition temperature. The fluidity is higher for lipids with unsaturated fatty acids, hence, the transition temperature is low. Lipids with saturated fatty acids have less mobility and fluidity, hence, the transition temperature is high.

1.10 VIRUS

Viruses are a class of microorganism of 10 to 400 nm in size. Viruses stay alive inside a living cell, known as the host cell.

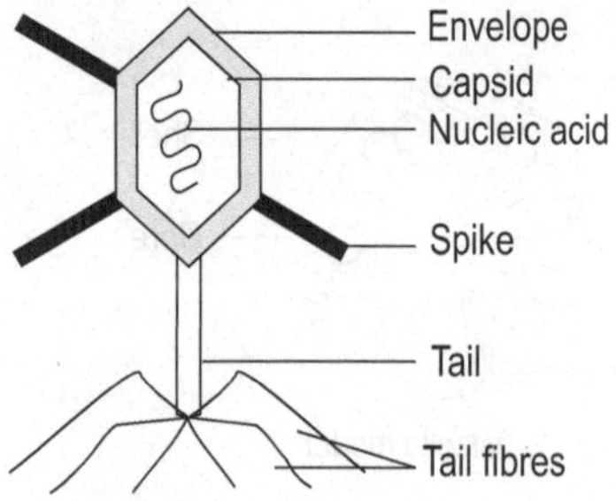

FIGURE 1.26 Structure of a virus

1.10.1 Structure of a Virus

The structure of a virus is shown in Fig. 1.26. It consists of a genetic material (DNA or RNA, refer Chapter 6) encapsulated in a protein coat called capsid. The capsid is surrounded by an envelope. The posterior side of the envelope has a tail made of protein. The genetic material of the virus gets integrated with the host chromosome. The capsid and the genetic material are collectively called nucleocapsid. The proteins in the capsid are called protomeres. The capsid is either helical as in tobacco mosaic virus (TMV), or icosahedral as in Siman virus-40. The envelope is made of lipids. Proteins are present in the lipids and they project out to form a spike (Fig. 1.27), distributed at a distance of 7 to 8 nm. The spike contains hemagglutinin protein and neuraminidase. The spike and the tail are used to attach the virus to the host cell.

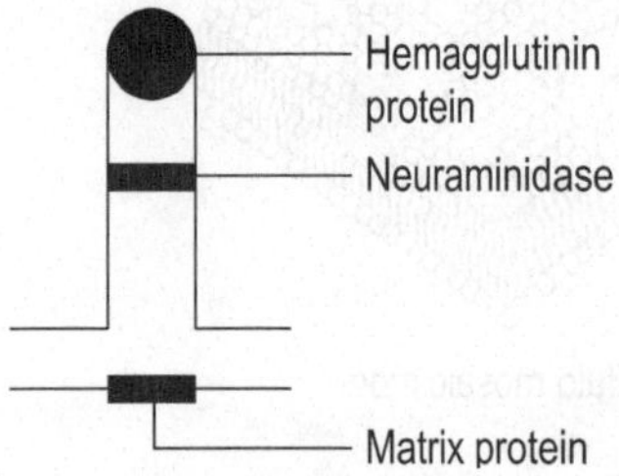

FIGURE 1.27 Structure of the spike

1.10.2 Reproduction in Virus

Virus reproduces in a living cell. The reproductive process has the following six steps (These steps are also depicted in Fig. 1.28).

- The virus attaches to the host cell via the spike and the tail.
- The virus penetrates its genetic material into the host cell.
- The viral genome integrates with the host chromosome and gets expressed.
- Because of gene expression, viral coat and proteins are produced.
- The products formed from the gene expression are assembled into new virus particles called virions.
- The host cell ruptures to release the virions.
- The virions can further infect other host cells and reproduce. The cycle continues.

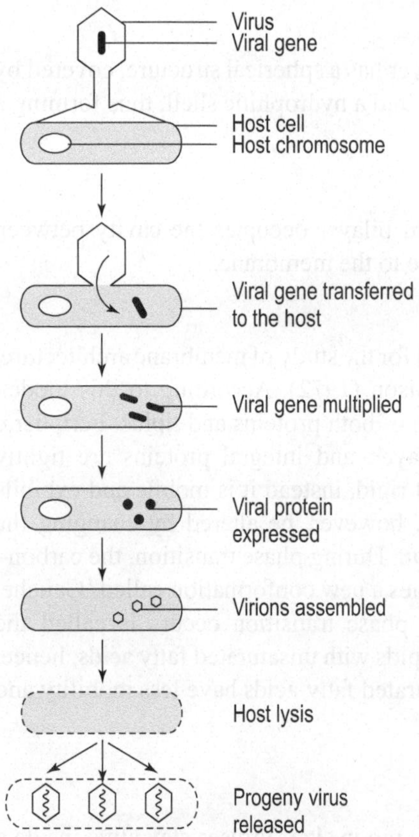

FIGURE 1.28 Reproduction in virus

1.10.3 Virus as a Host for Drug Delivery

The infective capacity of virus makes it an excellent nanotool for drug delivery. Use of virus as a drug carrier is the hot area of research, but is not in practical use due to ethical issues. Stefan Franzen and Steve Lommel used red clover necrotic mosaic virus (RCNMV) for the delivery of an anticancer drug, doxorubicin. RCNMV has an icosahedral capsid of size 38 nm. A 7-nm space within the capsid is used for storing the therapeutics and subsequent drug delivery. The RCNMV infused with the drugs is then surface modified with targeting antibodies. The virus is then used to deliver the drugs in a cell-specific manner. The drug delivery capability of virus was studied in vitro using doxorubicin in Hela cells. It was found that the cancer cells infected with the virus showed apoptosis, confirming the delivery of the drug.

Myongsoo Lee at Yonsei University in Seoul has developed an artificial virus, containing a nanostructure (70 nm) made of β-sheet peptides. The protein is called nanoribbon, which can incorporate RNA, thus, mimicking a virus structure. This structure is, therefore, known as an artificial virus. The gene incorporated into the pleated sheet can be a therapeutic gene and can be used for gene therapy. The artificial virus can also hold hydrophobic drug molecules. This can result in the simultaneous delivery of therapeutic gene and hydrophobic anticancer drug, thus increasing the therapeutic potential.

1.11 CELL DIVISION

Cell division is the process of reproduction in which the cell divides into two equal halves. Two types of cell division are known: mitosis and meiosis. Mitosis occurs in somatic cells (body cells), whereas meiosis occurs in germ cells (sperms and ova, refer Chapter 8). During mitosis, the chromosome or the DNA is duplicated, that is, each daughter cell formed after the division gets the same number of chromosomes possessed by the parent cell. In meiosis, the chromosome is exactly halved so that each daughter cell receives half of the chromosomes from the parent cell. The chromosome number of a somatic cell is denoted as $2n$, that is, the chromosomes are paired, and the cells are called diploid cells. However, chromosome number of a germ cell is denoted as $1n$, that is, the chromosomes are unpaired, and the cells are called haploid cells. We will discuss about the mechanics of mitosis in the following subsection and meiosis will be discussed later in the chapter.

1.11.1 Mitosis

Mitosis takes place through various phases that are listed as follows:

- Prophase
- Prometaphase
- Metaphase
- Anaphase
- Telophase
- Cytokinesis

All the phases are collectively called the mitotic phase or the M phase, where the actual nuclear division (mitosis) and the cytoplasmic division (cytokinesis) occur.

As an overview of mitosis, several changes occur in the nucleus and the cytoplasm of the cell. The nuclear envelope degenerates and reassembles and the chromosomes condense, separate, and decondense. The main cytoplasmic component required for the alignment and separation of chromosome is the spindle of microtubules. Microtubules are cytoskeletal proteins (refer Chapter 9). There are three types of microtubules (Fig. 1.29) formed during different phases of mitosis:

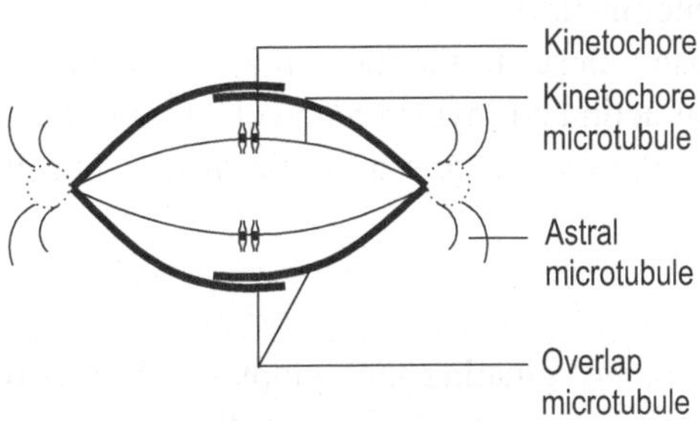

FIGURE 1.29 The three types of microtubules formed during mitosis

- Astral microtubules
- Kinetochore microtubules
- Overlap microtubules

The role of these tubules is mentioned while discussing the respective phases. The following events occur stepwise during each phase of mitosis:

Prophase

1. A matured diploid somatic cell has $2n$ chromosomes in the nucleus (formed after replication, refer Chapter 6), and centrosomes at the pole in the cytoplasm.
2. The replicated chromosomes become paired and condense.
3. An intricate network of protein cytoskeleton called microtubules (refer Chapter 9) assembles in the cytoplasm, forming the mitotic spindle. Some of the microtubules surround the centrosomes and radiate in all directions. These radiating microtubules are called *astral microtubules*.
4. The astral microtubules so formed contribute to the pulling force needed to separate the poles.

Prometaphase

1. The nuclear envelope breaks abruptly.
2. Chromosomes attach to the mitotic spindle via their kinetochores. The microtubules attached to the kinetochores are called *kinetochore microtubules*.
3. Chromosomes align along the spindle and acquire a 'V' shape with a bend near the kinetochore.

Metaphase

The chromosomes are aligned at the equator of the spindle. The equator lies midway between the two poles of the spindle. The microtubules that interdigitate in the equator are called *overlap microtubules*. They maintain the symmetry and the bipolar shape of the spindle (refer Chapter 9 for the discussion on bipolar spindle).

Anaphase

The sister chromatids of the chromosomes of each of the paired chromosome separate synchronously, forming two daughter chromosomes. Separation takes place as the two chromatids are pulled apart towards opposite poles. During this phase, the kinetochores get shorter.

Telophase

1. The two sets of unpaired chromosomes (called daughter chromosomes) align at the two poles, and decondense.
2. A new nuclear envelope assembles around each set of chromosomes. This results in the formation of two nuclei (called daughter nuclei).
3. A contractile ring appears in the cytoplasm, closer to the membrane, between the two poles. The contractile ring is made of actin and myosin filaments. It is formed in the cortex of the cytoplasm. It provides the force required to split the parent cell into two daughter cells.

Cytokinesis

1. The contractile ring moves deeply inward, invaginating the cytoplasm from two sides. This leads to the formation of a cleavage furrow, which marks the division of cytoplasm (the cytokinesis).
2. The nuclear envelope gets completed.

3. The microtubule spindle is disaligned and reformed around each daughter nuclei to restart the interphase.
4. The cell slowly divides along the cleavage furrow and separates into two separate, identical daughter cells. The cell can again replicate and double the chromosomes. This phase is called the synthetic phase or S phase. After this, the cell can have a dormant phase or a gap phase (G phase); it then re-enters into prophase and the cycle continues. This is known as cell cycle.

The phases of mitosis are shown in Fig. 1.30.

The assembly and the movement of the microtubules during mitosis are governed by motor proteins called dyneins and kinesin-related proteins (refer Chapter 9).

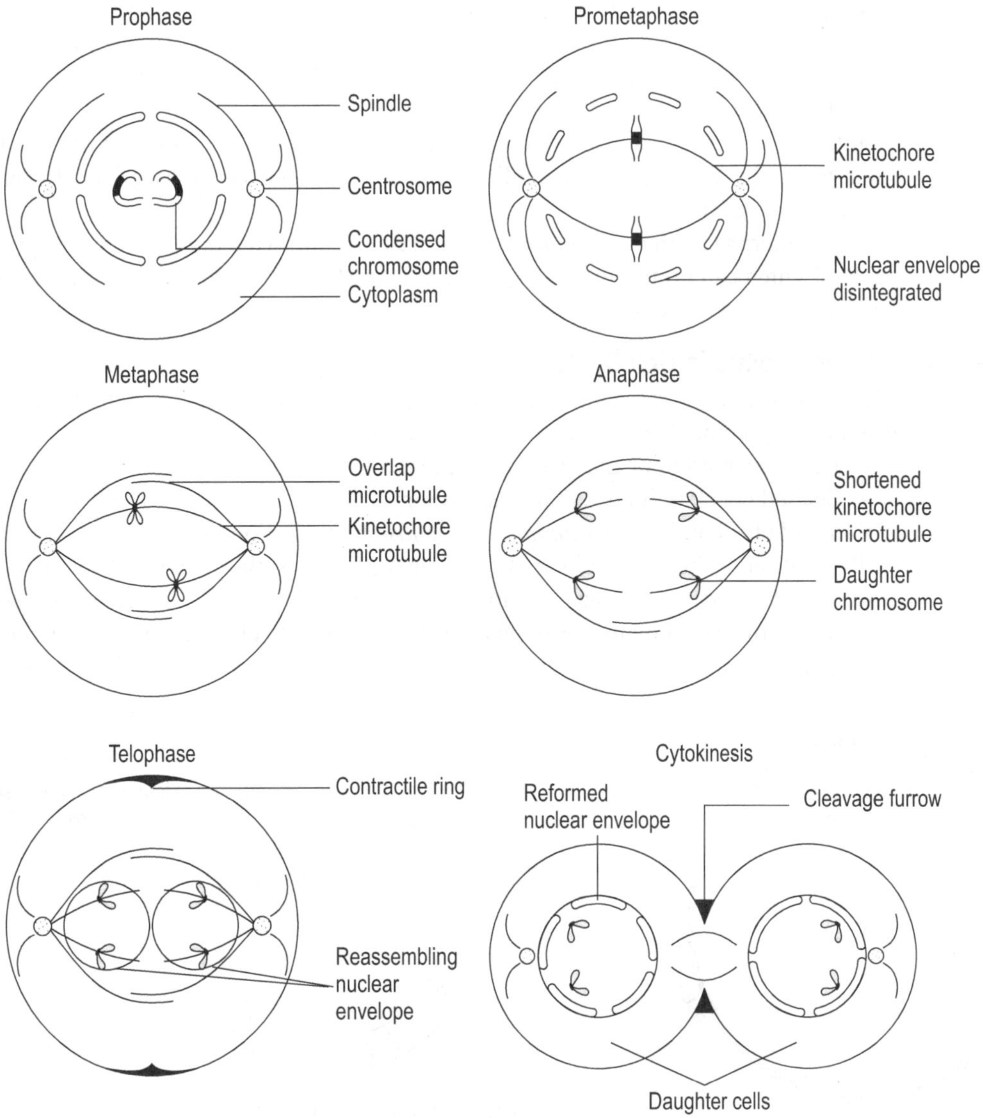

FIGURE 1.30 Phases of mitosis

1.12 CELL CYCLE

Cell cycle is a process where newly formed cells go through the interphase (G1 phase, S phase, and G2 phase) and the M phase, generation after generation (Fig. 1.31).

1.12.1 Phases of Cell Cycle

In the previous section, we discussed that two new cells are formed after the process of cytokinesis in the M phase. The two new daughter cells then undergo the first gap phase (G1 phase). In this phase, size of the cell increases and the mass of the cellular content, such as proteins and organelles, also increase.

Following the G1 phase, the cells replicate their DNA. Because of the synthesis of new DNAs, this phase is called synthesis phase or S phase.

Following the S phase the cells undergo a second gap phase (G2 phase) where the cell monitors the environment and if the environment is found to be suitable, prepares itself for cell division.

Interphase is the term used to collectively refer to the G1, S, and G2 phases. During interphase, the centrosome is duplicated and each centrosome is pulled towards the opposite pole of the cell.

After interphase, the cells enter the M phase. Again two daughter cells are produced, each of which can go through G1 → S → G2 → M phases and the cycle continues. This process is called cell cycle.

1.12.2 Check Points in Cell Cycle

The cell cycle is regulated at three different phases. These phases are called check points or regulatory points (Fig. 1.32). They are listed as follows:

- First check point or Late G1 check point
- Second check point or G2/M check point
- Third check point or *metaphase–anaphase* transition point

Regulation of the cell cycle at check points occurs using regulatory proteins, such as cyclin, cyclin-dependent kinase, anaphase-promoting complex, and many other proteins. The details of the regulatory proteins and their target check points are discussed in the subsequent section.

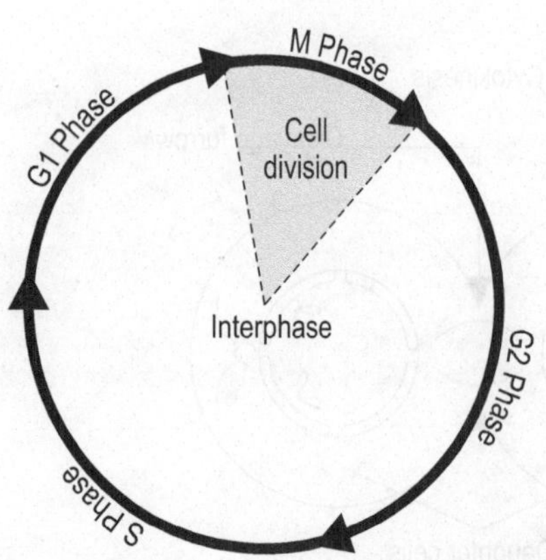

FIGURE 1.31 Cell cycle

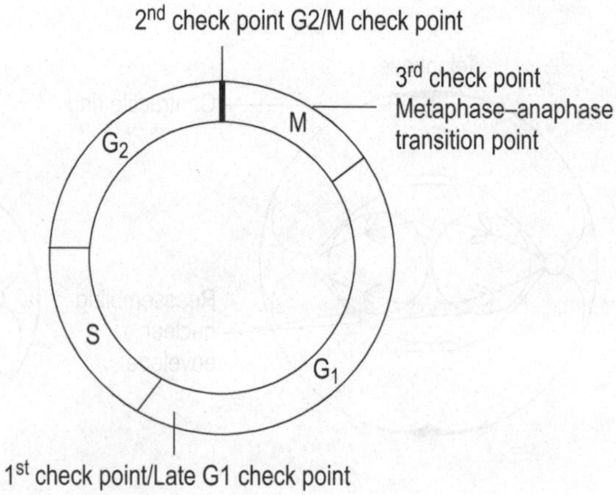

FIGURE 1.32 Check points in a cell cycle

1.12.3 Role of Cyclin-dependent Kinases (CdK) in Cell-cycle Control

Cyclin-dependent kinase (CdK) is the crucial enzyme for the regulation of cell cycle. It is a protein with molecular weight 35–40 k Da. It phosphorylates the serine or threonine residues of its target protein. The target sequence for CdK is represented as [S/T]PX[K/R], where S/T is serine or threonine, P is proline, X is any amino acid, K is lysine, and R is arginine. Phosphorylated proteins in turn result in the initiation or regulation of major cell cycle events such as replication (S phase), mitosis, and cytokinesis (M phase).

The phosphorylation role of CdK, however, is mediated by another protein called cyclin. Cyclin combines with CdK forming an active cyclin-CdK complex (CCC). CCC is functional, that is, CdK is active only in the presence of cyclin. There are three types of cyclin, depending on the phases of cell cycle at which they function:

1. G1/S cyclin, which binds to the CdK at the end of the G1 and functions in the S phase by promoting DNA synthesis;
2. S cyclin, which binds to CdKin the S phase and stimulates DNA replication; and
3. M cyclin, which binds to CdK in the M phase and promotes mitosis.

Regulation of CCC

The activity of CdK is regulated by the formation or dissociation of the CCC. The active site of CdK is a cleft blocked by a T-loop (located in the carboxyl end of the CdK). Binding of cyclin to the CdK dislocates the loop and exposes the active site, forming CCC. Phosphorylation of a threonine residue in the T-loop induces a conformational change in the loop and forms a completely active CdK, capable of phosphorylating the proteins involved in the cell cycle regulation. Activation of CdK is shown in Fig. 1.33.

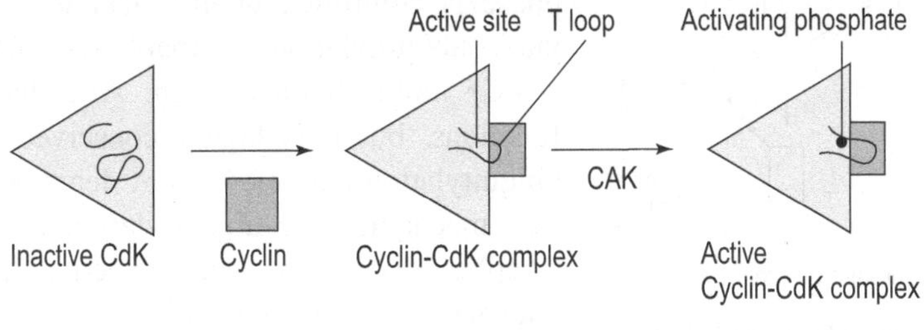

FIGURE 1.33 Activation of cyclin-CdK complex

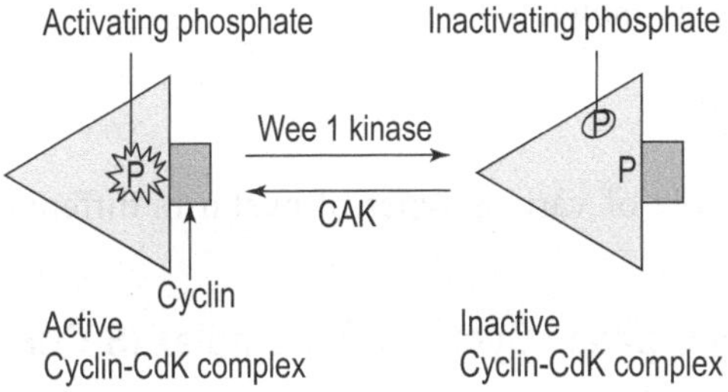

FIGURE 1.34 Inhibition of cyclin-CdK complex

CdK contains a tyrosine residue at the roof of the active site. This is the site of inhibition. An enzyme called wee 1 kinase phosphorylates the tyrosine, leading to the inactivation of the CCC. Phosphorylation by wee 1 kinase can be reversed by another enzyme called CdK-activating kinase (CAK). CAK dephosphorylates the roof of the active site and reactivates the CCC. Reversible inhibition of CCC is shown in Fig. 1.34.

1.12.4 Role of Enzyme Complexes (SCF and APC) in Cell Cycle Regulation

Cyclin-dependent kinase inhibitor protein (CKI or CdKI or CDK1) is an inhibitor of CdK. CdKI arrests the cell cycle at the G1 phase, thus, controlling the progression of cell cycle. Arrest of G1 phase can be controlled by degradation of CdKI via ubiquitylation (process of addition of ubiquitin to the protein to be degraded). A protein complex called SCF complex is responsible for degradation of CdKI by ubiquitylation. SCF is named after its protein components:

- Skp1(S-phase kinase-associated protein 1)
- Cullin
- F-box domain

It is an E3 ubiquitin ligase (refer Chapter 11) which adds ubiquitin to the cyclin (and even to their proteins) destined for degradation in proteosome. The mechanism of action of SCF is shown in Fig. 1.35.

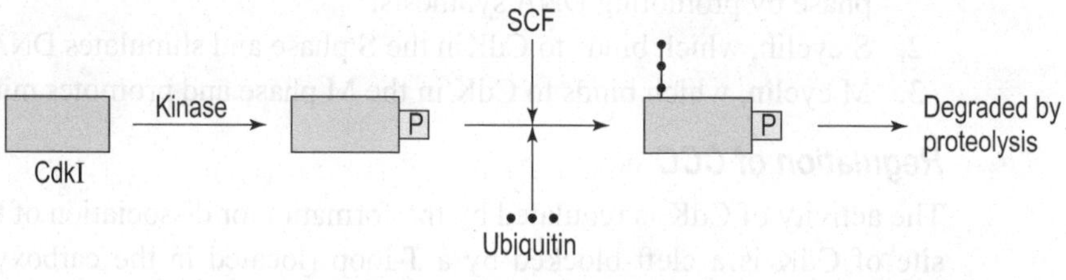

FIGURE 1.35 Mechanism of action of SCF

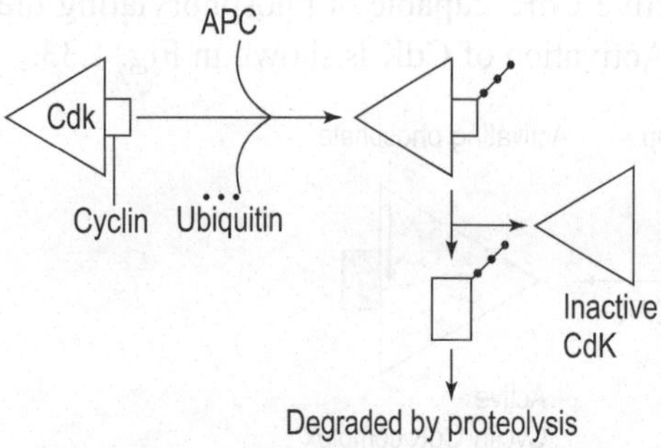

FIGURE 1.36 Mechanism of action of APC

Cyclin, the crucial component for CdK is regulated by degradation via ubiquitylation. The initiation of S phase is controlled by the degradation of G1/S cyclin via ubiquitylation. Anaphase-promoting complex (APC) multisubunit protein. APC has some unknown functions, but it is highly conserved and mediates the ubiquitylation and subsequent degradation of M cyclins. The mechanism of ubiquitylation is well described in Chapter 11. As mentioned in Section 1.12.3, the cyclin is phosphorylated above the active site by wee 1 kinase. The phosphorylated cyclin is the target for APC which adds the ubiquitin residues to cyclin, marking the later for subsequent degradation in the 26S proteosome. The mechanism of action of APC is shown in Fig. 1.36.

1.13 INTRACELLULAR CONTROL OF CELL CYCLE

This section discusses about the regulatory role of various types of cyclin in different phases of the cell cycle.

1.13.1 Stimulation of DNA Synthesis by S-phase Cyclin-CdK Complex (S-CdK)

S cyclin-CdK complex (S-CdK) plays a dual role of initiating as well as inhibiting replication. During the G1 phase, a large multisubunit protein complex called origin-

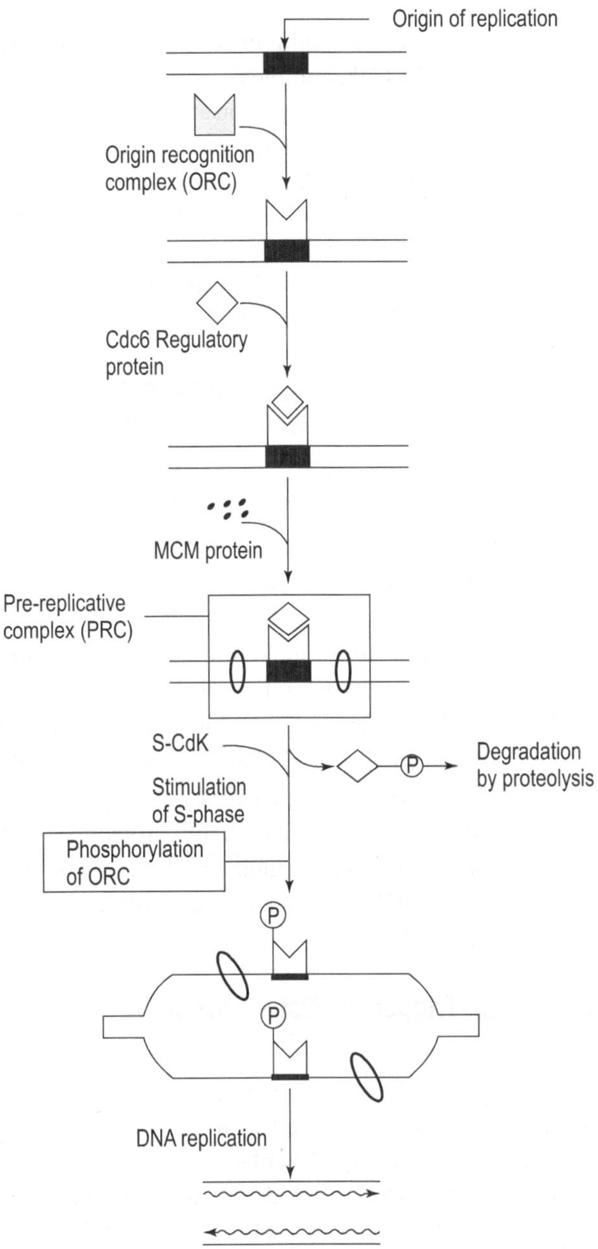

FIGURE 1.37 Regulation of DNA replication by S-CdK

recognition complex (ORC) binds to the origin or replication (refer Chapter 6). ORC remains in the origin of replication throughout the cell cycle. Several regulatory proteins bind to ORC during the cell cycle. In early G1 phase, a regulatory protein called Cdc6 is elevated in the cell. Cdc6 binds to the ORC. Cdc6 enhances the binding of MCM proteins forming MCM ring complex adjacent to the origin of replication. At this stage, S-CdK does the following to promote DNA synthesis in S phase:

- Assembles the DNA polymerase to initiate replication
- Assembles other replicative proteins (refer Chapter 6)
- Activates MCM ring complex to drag along the DNA strands for unwinding the DNA to continue replication.

S-CdK can also prevent re-replication in the following manner:

- It phosphorylates the Cdc6 triggering its dissociation from ORC, and subsequent degradation in proteosomes as mentioned in Section 1.9 and Chapter 11.
- It stimulates the export of MCM proteins from the nucleus.

The role of S-CdK complex in firing replication and preventing re-replication is shown in Fig. 1.37.

1.13.2 Stimulation of Mitosis by M-CdK

M-CdK refers to M-phase cyclin-CdK complex. As the S-CdK triggers DNA replication in the S phase, M-CdK triggers mitosis in the M phase. To trigger mitosis, M-CdK is activated by a positive feedback loop as shown in Fig. 1.38. The mechanism of activation involves the following steps:

- M-cyclin level is elevated gradually in the G2 and M phases. M-cyclin binds to CdK forming the M-CdK complex.
- CAK phosphorylates the activating site of the M-CdK, while the wee 1 kinase phosphorylates on the inhibitory site. The resulting complex is an *inactive M-CdK* complex.
- By the end of G2 phase, a phosphatase (Cdc25 phosphatase) is activated by Polo kinase.

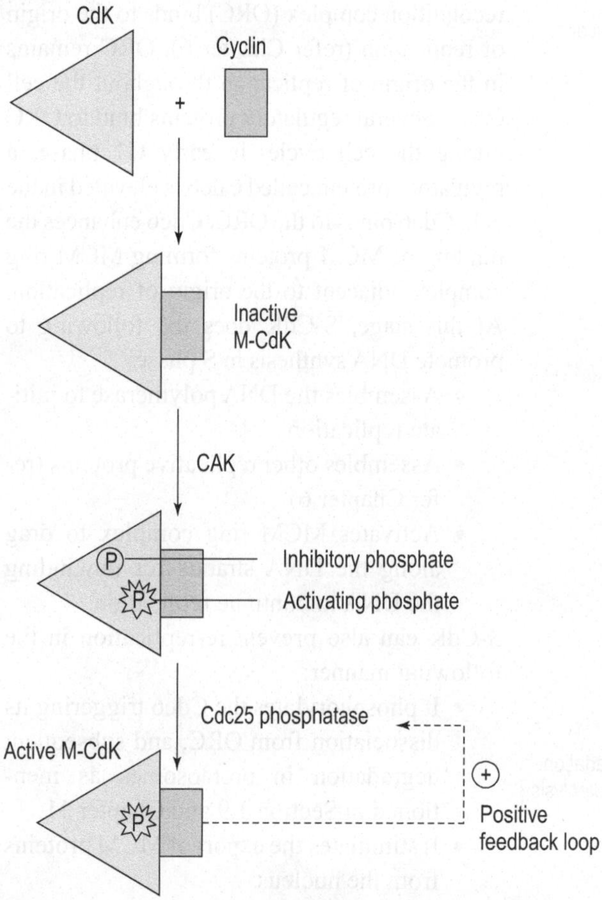

FIGURE 1.38 M-CdK activation required for the progress of mitosis

- Active phosphatase removes phosphate from the inhibitory site of the inactive M-CdK complex.
- Dephosphorylation at the inhibitory site activates M-CdK.
- Active M-CdK triggers mitosis.
- Active M-CdK can also stimulate the Cdc25 activity, forming more molecules of active M-CdK, thus, forming a positive feedback loop.

1.13.3 Blocking of Mitosis by Inactivation of M-CdK

While the activation of M-CdK triggers mitosis, its inactivation blocks mitosis. Once the mitosis has progressed and the chromosomes have segregated, the spindle must be dissembled, and the nuclear envelope must be reformed (refer Section 1.11.1 on mitosis). For all this to occur, any further mitosis must be blocked. This can be established by the inactivation of M-CdK by removing the active-site phosphate and subsequent degradation by proteolysis via ubiquitylation. Ubiquitylation is carried out by the APC complex. As a result, mitosis is arrested.

1.13.4 Role of Securin and Separase in Triggering Sister Chromatid Separation (anaphase)

As mentioned in Section 1.11, once the DNA is duplicated and aligned along the equator, the sister chromatids separate synchronously and are pulled apart towards opposite poles. At this stage the metaphase → anaphase transition takes place. For this process to occur, the attachment between the two sister chromatids should be disturbed. The sister chromatids are held together by a protein called cohesin. A cysteine protease called separase hydrolyzes and degrades the cohesin, triggering anaphase.

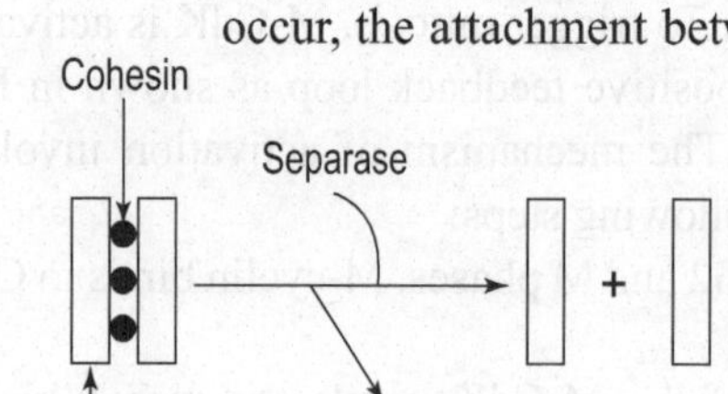

FIGURE 1.39 Role of separase and securin in regulation of mitosis

When progression to anaphase is not needed, the cohesin is maintained intact. This is made possible by two proteins called cyclin-CdK complex and securin. Cyclin-CdK phosphorylates separase. Phosphorylated separase remains bound to securin and is not active to cleave the cohesin. Whenever anaphase has to be triggered, the securin is ubiquitylated and degraded, releasing the separase for

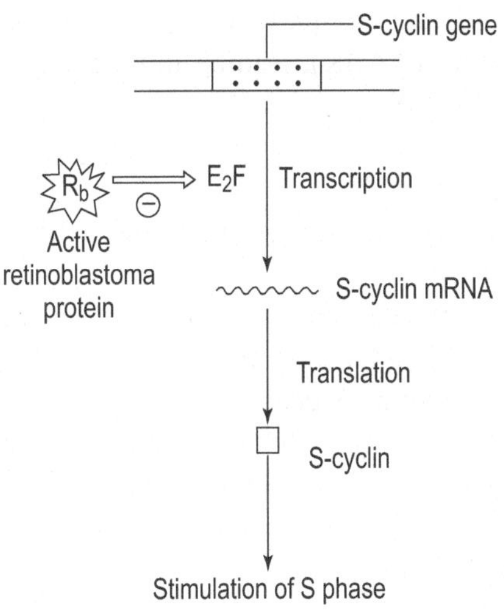

FIGURE 1.40 Role of retinoblastoma protein in cell cycle

dephosphorylation. Dephosphorylated separase can degrade the cohesin and helps in the separation of sister chromatids. The role of separase and securin in regulating mitosis (anaphase) of the cell cycle is shown in Fig. 1.39.

1.13.5 Retinoblastoma Protein (Rb protein) as an Inhibitor of Cell Cycle

Retinoblastoma protein (Rb) is a tumour suppressor protein, capable of preventing the cell cycle by interfering with the G1 → S progression. The target for Rb is the transcription factors of E_2F family. These transcription factors are called E_2F proteins. E_2F protein induces the transcription of S phase genes, including the genes coding for S-cyclin and G1/S cyclin, thus, triggering the S phase. Rb is active in its dephosphorylated form. Active Rb binds to and inactivates the E_2F, thus, preventing the transcription of S phase genes and progression of replication. However, if the cells need to proceed to the S phase for the subsequent cell division, the Rb is phosphorylated by the active G1-CdK. Phosphorylated Rb is inactive. Role of Rb in inhibiting S phase of the cell cycle is shown in Fig. 1.40.

1.14 EXTRACELLULAR CONTROL OF CELL CYCLE

Some of the extracellular molecules can control the cell cycle by acting on the intracellular regulatory points. Following are the three major extracellular regulatory molecules:

1. Mitogens 2. Growth factors 3. Survival factors

Mitogen is a protein. It acts as an extracellular ligand and stimulates the signal transduction leading to the activation of protein kinases (called mitogen-activated protein kinases or MAP kinases). MAP kinases activate a transcription factor that stimulates the transcription of myc gene. This in turn increases the level of a gene regulatory protein called Myc. Myc stimulates S phase of the cell cycle by three overlapping mechanisms that are discussed as follows (Fig. 1.41):

1.14.1 Mechanism 1

1. Myc stimulates the transcription of cyclin D gene. Cyclin D drives the G1/S phase transition.
2. Cyclin D activates the G1-CdK.
3. Active G1-CdK phosphorylates Rb.
4. Phosphorylated Rb increases the activity of E_2F.
5. E_2F stimulates the initiation of S phase.

1.14.2 Mechanism 2

1. Myc stimulates SCF subunit gene.

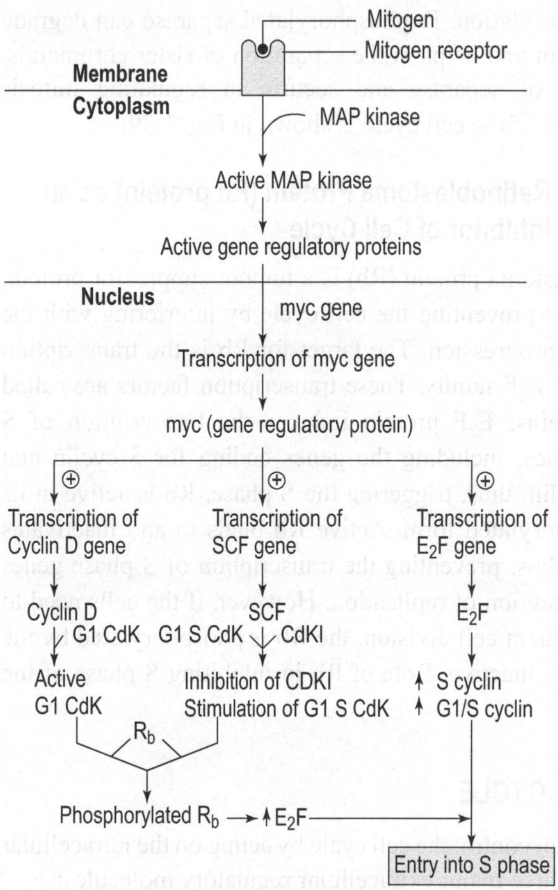

FIGURE 1.41 Role of extracellular mitogen in stimulating S phase of the cell cycle

2. SCF subunit complex degrades the CdKI (inhibitor of CdK), leading to stimulation of G1/S CdK.
3. Active G1/S CdK induces phosphorylation of Rb.
4. Phosphorylated Rb increases the activity of E_2F.
5. E_2F stimulates the transcription of the genes coding for S-cyclin and G1/S cyclin, thus, triggering S phase.

(The last two steps of this mechanism overlap with those of the second mechanism).

1.14.3 Mechanism 3

1. Myc stimulates the transcription of E_2F gene.
2. E_2F transcription factor level rises.
3. E_2F turns on the genes coding for S-cyclin and G1/S cyclin, thus, triggering S phase (this step overlaps with the last step of the second mechanism).

Growth factors (GF) bind to growth factor receptor (GFR), thereby activating the phosphatidylinositol-3-kinase (PI3K). PI3K phosphorylates and activates several proteins kinases (refer Chapter 11), including S6 kinase. S6 kinase phosphorylates the ribosomal protein S6 which is responsible for controlling cell growth and proliferation. Platelet-derived growth factor is another example for the extracellular GF that functions similar to the mitogen.

Survival factors bind to cell surface receptors and activate signalling pathways that lead to prevention of apoptosis (programmed cell death). For example, insulin-like growth factor is known to suppress apoptosis. In mammalian cells and drosophila, the survival factor is known to suppress apoptosis by activating Protein kinase B (PKB) and MAP kinase, respectively (Fig. 1.42).

In mammalian cells, the survival factor binds to its receptor and activates the PKB. PKB acts on a protein of Bcl2 family called Bcl2 associated death (BAD). BAD is a promoter of apoptosis and is bound to Bcl2. PKB phosphorylates BAD. Phosphorylated BAD dissociates from Bcl2, and the Bcl2 in turn suppresses the apoptosis. Another target for PKB is the regulatory protein of fork head family (RPFF). RPFF stimulates the transcription of genes encoding the proteins that promote apoptosis. PKB phoshorylates and inactivates RPFF, leading to suppression of apoptosis.

Head involution defective (HID) protein inhibits another protein called inhibitor of apoptosis (IAP). In drosophila, the survival factor stimulates the MAP kinase activity. MAP kinase phosphorylates and inactivates HID protein. Thus, IAP is free to suppress apoptosis.

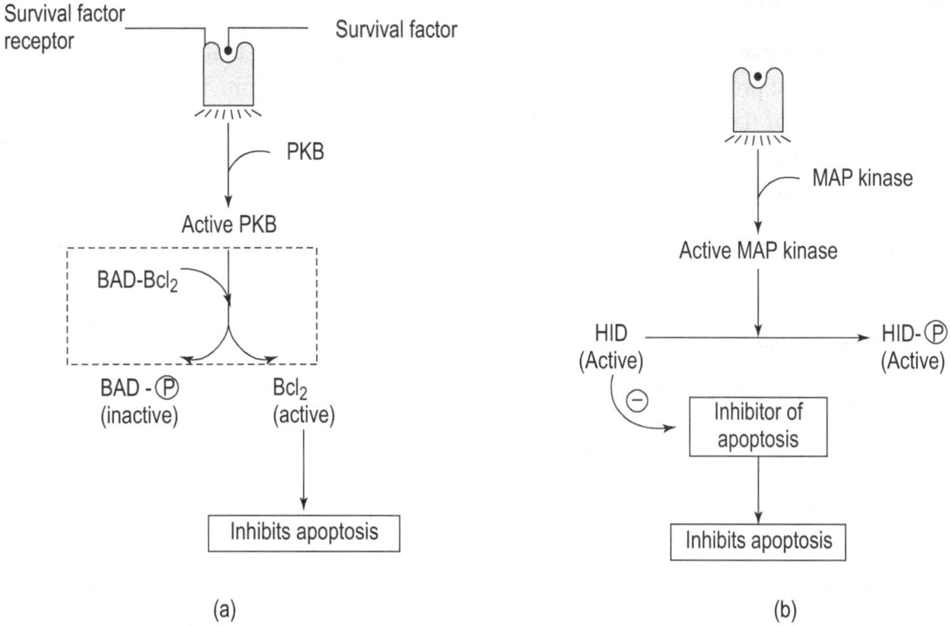

FIGURE 1.42 Mechanism of suppression of apoptosis by survival factor

1.15 CELL CYCLE REGULATORY PROTEINS

Cell cycle regulatory proteins refer collectively to all the proteins, including enzymes and transcription factors, involved in the regulation of the cell cycle. This section summarizes the mechanism of action of cell cycle regulatory proteins which stimulate or inhibit the different phases of the cell cycle (Table 1.10).

TABLE 1.10 Cell cycle regulatory proteins

Regulatory protein	Mechanism of action
CdK-activating kinase (CAK)	Phosphorylates the active site of CdK
Wee 1 kinase	Phosphorylates the inhibitory site of CdK
CdK inhibitory protein (CdKI)	Inhibits CdK
SCF-protein complex	Induces the ubiquitylation and proteolysis of regulatory proteins, including CKIs
Anaphase-promoting complex (APC)	Induces the ubiquitylation and degradation of proteins, including securin and M cyclin
E2F	Promotes the transcription of S phase genes, including the G1/S cyclin gene
Securin	Binds to separase and prevents the separation of sister chromatids
Separase	Degrades the cohesin and separates the sister chromatids to promote anaphase
Mitogen	Stimulates the S phase
Extracellular growth factor	Promotes cell growth
Survival factor	Inhibits apoptosis

1.16 TARGETS OF CELL CYCLE CONTROL SYSTEM

So far, we have discussed about various regulatory proteins that stimulate or inhibit cell cycle. These proteins are considered to form a control system. This section summarizes the targets of these regulatory proteins, that is, the exact phase of the cell cycle which is regulated by a specific regulatory protein. Table 1.11 shows the targets of the cell cycle control system.

TABLE 1.11 Targets of the cell cycle control system

Component of the control system	Target phase in the cell cycle	Effect on the target
S cyclin-CdK complex (SCdK)	S phase	Stimulates replication Inhibits re-replication
M cyclin-CdK complex (MCdK)	M phase	Stimulates mitosis
Anaphase-promoting complex	M phase	Inhibits mitosis
Separase	Anaphase	Stimulates anaphase
Securin	Anaphase	Inhibits anaphase
Retinoblastoma protein	S phase	Inhibits replication
Mitogen	S phase	Stimulates replication
Extracellular growth factor	G1	Stimulates growth
Survival factor	-	Prevents apoptosis

1.17 MEIOSIS

Meiosis is a type of reduction division in which haploid cells are produced from a diploid cell. In this process, the chromosome number is exactly halved, that is, reduced, hence, the name reduction division. Meiosis occurs in germ cells (refer Chapter 8).
The process of meiosis is described as follows:

Meiosis occurs in two stages, meiosis I and meiosis II. The outline of meiosis is shown in Fig. 1.43. In meiosis I, a diploid cell is converted into two haploid cells. In meiosis II, a total of four haploid cells are obtained from the two haploid cells.

Following are the phases of meiosis:

Interphase The gap phase just before the occurrence of meiosis is called interphase. In this phase the genetic material is duplicated by replication.

Meiosis I has the following phases.

Prophase I The duplicated chromatins condense and undergo homologous recombination (crossing over).

Metaphase I Homologous chromosome assembles in the equator of the cell as described in mitosis.

Anaphase I The homologous chromosome pair separates, while the sister chromatids are still together.

Telophase I The cells divide to form two daughter cells, each containing a single chromosome of the homologous pair.

Meiosis II is similar to mitotic division, which finally producces four haploid cells, each with 1n chromosomes.

Chromosomal changes that occur during meiosis are shown in Fig. 1.44.

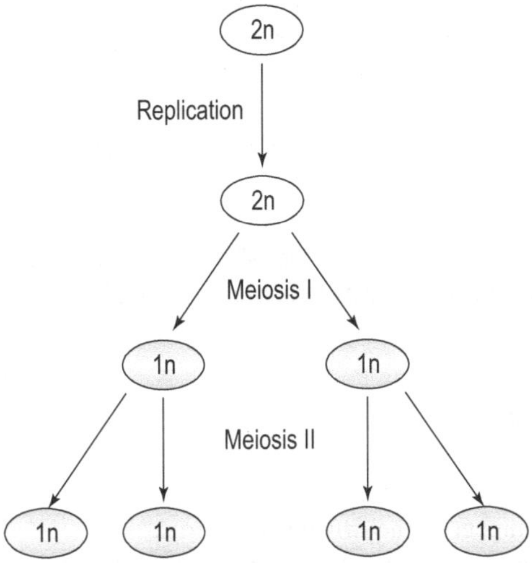

FIGURE 1.43 Outline of meiosis

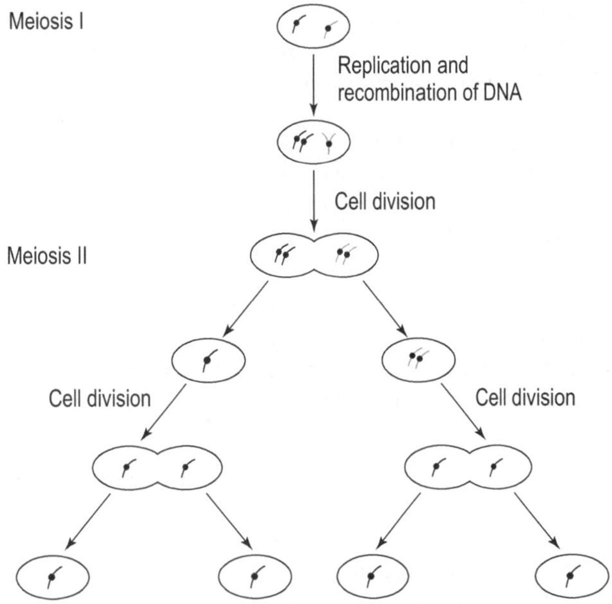

FIGURE 1.44 Process of meiosis

SUMMARY

Robert Hooke discovered cells in 1665. Subsequently, many other discoveries were made about cells. These discoveries revealed a lot of information about the structure and functions of cells. Cell theory was proposed by Theodor Schwann, Matthias Jakob Schleiden, and Rudolf Virchow, which confirmed that cells are the basic units of life. From cell theory it was understood that cell has complexity, which is greater for eukaryotes. This chapter discussed about the general aspects of cells, including the shape, size, and architecture of prokaryotic and eukaryotic cells, and mechanics and the regulation of cell division. Basically, two types of organisms exist: prokaryotes (e.g., bacteria) and eukaryotes (e.g., mammals and plants). Sometimes, viruses are considered to be the third type of organisms. Prokaryotes may be spherical, rod-like, or spiral in shape and range in size from 0.5 to 3 μm. They lack membrane-bound organelles. They reproduce by binary fission, budding, fragmentation, and conjugation.

In eukaryotes, cells may vary in their shape depending on the tissues they reside in and their functions. They possess distinct membrane-bound organelles. Mostly the cells are spherical but some are branched (nerve cells), elongated (epithelial cells), flat (skin cells), biconcave (red blood cells), or amoeboid (phagocytes). The typical size of an animal cell is 10 to 100 μm. The primary compounds of cell are water and carbon compounds. Carbon is the basic compound for the major biomolecules, including proteins, carbohydrates, lipids, and nucleic acids. Cells divide by two processes: mitosis and meiosis. Mitosis occurs in somatic cells or body cells. Meiosis occurs in germ cells such as sperms and ova.

TERMS TO REMEMBER

Capsule A strong rigid layer of polysaccharide present on the surface of the bacteria, which helps in protecting the bacterial cell from phagocytosis and desiccation.

Cell division A process of reproduction in which the cell divides into two equal halves.

Cell theory The proposal or the hypothesis which states that cells are the fundamental units of every living organism.

Cytoplasm An amorphous, translucent, and homogeneous colloidal matrix of the cell.

Eukaryotic cells Cells of higher forms of organisms containing distinct membrane-bound organelles.

Flagellum A rigid thin outgrowth of about 15 to 20 µm length and 20 nm diameter that facilitates locomotion in bacteria.

Hopanoids Pentacyclic compounds that improve the fluidity of the bacterial membrane.

Magnetosomes Specialized inclusion bodies found in the cytoplasm of magnetotactic bacteria, those capable of orienting in the magnetic field line of the earth's magnetic field.

Mesosomes Invagination of inner membrane of bacteria.

Pili A slender hair-like structure of 9 to 10 nm in diameter responsible for conjugation (reproduction) and attachment of virus to the cell surface.

Prokaryotic cells Cells without membrane-bound organelles, mostly unicellular.

Slime layer An unorganized capsule forming a diffused-zone-like structure; it is used in cell adhesion and formation of biofilms.

S layer A two-dimensional, porous surface layer of 5 to 25 nm thickness that protects the cell against abnormal pH, phagocytosis, and bacteriophages, and facilitates cell adhesion.

REVIEW QUESTIONS

1. Write a note on cell theory.
2. Differentiate between prokaryotic and eukaryotic cells.
3. Explain the different methods of reproduction in bacteria.
4. Write a note on the structure and functions of bacterial appendages.
5. Compare the features of an animal cell and a plant cell.
6. Comment on the structure of a virus.
7. How is virus used as a nanocarrier for drugs? Explain with an example.
8. Explain the mechanics of mitosis with a neat, labelled sketch.
9. Write about the different phases of cell cycle.
10. How is cell cycle controlled?

MULTIPLE-CHOICE QUESTIONS

1. Cell was discovered by
   (a) Leeuwonhoek
   (b) Schleiden
   (c) Schwann
   (d) Robert Hooke
2. Which of the following factors does not determine the shape of the cell?
   (a) membrane fluidity
   (b) size of the vacuole
   (c) viscosity of the cytoplasm
   (d) cell–cell distance
3. The primary constituent of a cell is
   (a) carbon
   (b) water
   (c) cytoplasm
   (d) gas vacuole
4. The cell volume in a mammal is approximately
   (a) 4×10^{-9} cm^3
   (b) 3×10^{-9} cm^3
   (c) 4×10^{-8} cm^3
   (d) 3×10^{-8} cm^3
5. Lipopolysaccharide is the crucial component of
   (a) gram-positive bacteria

(b) gram-negative bacteria
(c) virus
(d) plant cell wall

6. Pili is involved in
   (a) magnetic orientation of bacterial cell
   (b) reproduction in bacteria
   (c) infection by virus
   (d) cell division in plants
7. Hopanoids are present in
   (a) mitochondrial matrix
   (b) nuclear envelope
   (c) cell membrane
   (d) cytoplasm
8. Pyrite and gregite are present in
   (a) magnetosomes
   (b) golgi stack
   (c) nucleoid
   (d) periplasm
9. Decolourizing agent used in Gram staining is
   (a) ethanol
   (b) crystal violet
   (c) iodine
   (d) safranin
10. Plant vacuole is termed as
    (a) tonoplast
    (b) magnetosomes
    (c) capsid
    (d) hopanoid
11. Lipid raft is rich in
    (a) cholesterol and sphingolipid
    (b) peripheral proteins
    (c) integral proteins
    (d) phospholipid
12. Retinoblastoma proteins is a
    (a) tumour suppressor protein
    (b) catalytic protein
    (c) mitochondrial protein
    (d) transport protein
13. Chromosomes are held together by
    (a) securin
    (b) separase
    (c) cohesin
    (d) mitogen factor
14. The target for separase is
    (a) securin
    (b) survival factor
    (c) capsid
    (d) cytoskeleton

Answers to Multiple-choice Questions

1. (d)	2. (b)	3. (a)	4. (a)	5. (b)
6. (b)	7. (c)	8. (a)	9. (a)	10. (a)
11. (a)	12. (a)	13. (c)	14. (a)	

Erwin Neher and Bert Sakmann won the Nobel Prize for Physiology/Medicine in 1991 for developing a technique to measure the voltage changes produced by single-ion channels in cell membranes.

CHAPTER 2

Membrane Transport

After reading this chapter, students should be able to understand and appreciate the following:

- Types of transport proteins and their mechanism of action
- Channels involved in membrane transport
- Mechanism of transport mediated by vesicles
- Modes of membrane transport
- Transport across nucleus, mitochondria, chloroplast, peroxisome, and endoplasmic reticulum (ER)
- Proteins involved in membrane transport

The structure and properties of membrane have already been discussed in Chapter 1. The primary function of the membrane is to transport molecules. Membrane transport refers to the movement of small molecules, such as ions, across the plasma membrane (the lipid bilayer) to both sides of the cell. The transport process is mediated by specific transporters in the membrane which are usually proteins. The degree of transport or the permeability of different types of molecules across the membrane is shown in Fig. 2.1.

Degree of permeability	Nature of the molecule	Example
	Hydrophobic	O_2, CO_2, N_2
	Small uncharged polar	Water, urea, glycerol
	Large uncharged polar	Glucose, Sucrose
	Ions	Na^+, K^+, Mg^{2+}, Ca^{2+}, Cl^-, H^-, HCO^{3-}

FIGURE 2.1 Degree of permeability of various cellular molecules across the membrane bilayer

2.1 NEED FOR MEMBRANE TRANSPORT

Before discussing the mechanism, types, proteins, carriers, and channels involved in membrane transport, the need for membrane transport should be understood. Membrane transport is needed:

- for the transport of lipophilic molecules across the membrane of a cell
- to enable the cell to ingest essential nutrients
- to excrete metabolic waste products from the cell
- to regulate the intracellular concentration of molecules such as ions
- to regulate the proper functioning of intracellular organelles
- to maintain the composition and pH of different compartments within the cell
- to promote intracellular traffic

2.2 ENERGY FOR MEMBRANE TRANSPORT

Active transport of molecules across the membrane is driven by the cell's metabolic energy. Two-thirds of the total metabolic energy of the cell is utilized for active membrane transport. Adenosine triphosphate (ATP) hydrolysis is the source of energy for membrane transport. The difference in ion concentration across the membrane is the signal for the transport of ions across the membrane to reach a normal value. This concentration difference creates an electrochemical gradient across the membrane, which stores energy. The energy so produced can also drive membrane transport. Occasionally, the transport can also occur without the input of energy, hence, the name *passive transport.*

2.3 TRANSPORT PROTEINS

Transport proteins or transporters are mobile or immobile conveyors that facilitate membrane transport. Mutation studies in bacteria indicate that membrane transfer proteins are specific for transport of specific molecules. All the membrane transport proteins that have been investigated so far are multipass transmembrane proteins. Three types of transport proteins are found in a cell—carriers (also called permeases), channels, and ionophores.

2.3.1 Mechanism of Action of Transport Proteins

Carriers aid transport by undergoing reversible conformational changes. They bind to the target molecule to be transported. This is similar to the binding of enzyme to its substrate binding site. A carrier has two binding sites, one on each side of the membrane. Binding of the target molecule (molecule to be transported) to one side of the carrier, induces a conformational change in the other side of the carrier structure. Consequent to the conformational change, the other binding site is exposed which results in the transport of molecule to the other side of the membrane. After the molecule is released, the carrier regains its original conformation. The rate of transfer depends on two factors:

1. Rate of flip-flop between two conformations
2. Extent of saturation of binding site

The mechanism of action of carrier is shown in Fig. 2.2. As a result of conformational change, the binding site faces the opposite site of the membrane, releasing the target (shown as dotted arrow). The carrier retrieves its original conformation after transport.

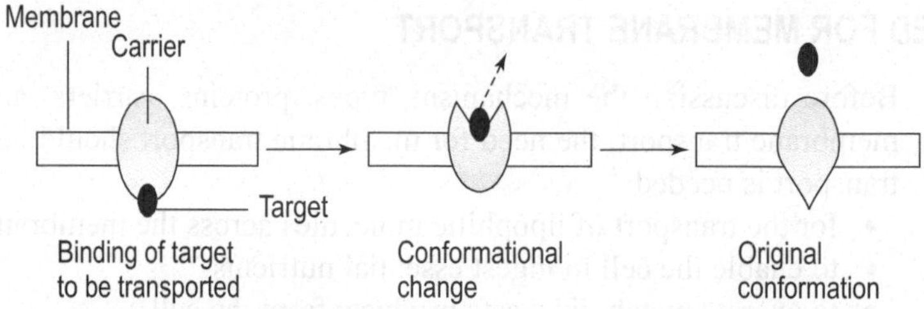

FIGURE 2.2 Mechanism of action of carrier protein

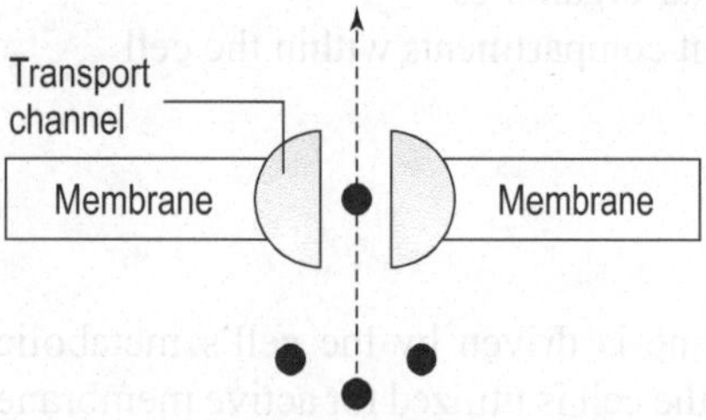

FIGURE 2.3 A transport channel located across the membrane (arrow indicates the direction of transport through the channel)

In addition to carriers, there are two more types of transport protein present in a cell: channels and ionophores. Channels are small openings or pores extending across the membrane (Fig. 2.3) to aid the diffusion of molecules from one side of the cell to the other.

Ionophores are hydrophobic molecules usually produced by bacteria to transport inorganic ions. Ionophores do not depend on ATP or any other energy source for their function. They are of two types, *mobile ionophores*, which bind to the target molecules and release them after undergoing a conformational change, and *immobile ionophores*, which form channels in the membrane. Examples of mobile ionophores are

1. valinomycin, for transfer of potassium ions down the electrochemical gradient,
2. FCCP, for transfer of proton across the proton gradient and dissipation in mitochondrial membrane, and
3. A23187 (Calcimycin), for transport of calcium into the cytosol. Gramicidin A is a channel ionophore that transports monovalent cation to flow down the electrochemical gradient.

2.4 MEMBRANE CHANNELS IN TRANSPORT PROCESS

Channels are membrane-bound proteins that help in transport. They are of two types: porins and gap junctions. Porins are hydrophilic pores present across the membrane, capable of closing and opening for the transport of inorganic ions, hence the name *ion channels*. They are found in microbial animal and plant cells and are predominant in nerve cells. Gap junctions are channels connecting the cytoplasm of two adjacent cells (refer Chapter 11).

2.4.1 Advantages and Features of Channel-mediated Transport

Channels have two major advantages over carriers: the rate of transfer by channels is 105 times greater than the fastest rate of the carrier protein, and channels are energy independent for their function. Ion-channel-mediated transport is always passive; it is also known as downhill transport. It allows fluxing of inorganic ions, primarily sodium, potassium, calcium, and chloride ions, to diffuse down the concentration gradient across the lipid bilayer. Ion channels are used for receiving, conducting, and transmitting electrical signals and are responsible for electricial excitability in muscle cells. They are capable of conformational changes and gating.

2.4.2 Types of Ion Channels

Ion channels are gated, which makes them open briefly and close again. A specific stimulus induces opening of the channel. The function of these channels is illustrated in Fig. 2.4. Depending on the stimulus used for opening, they are classified into different types (Table 2.1).

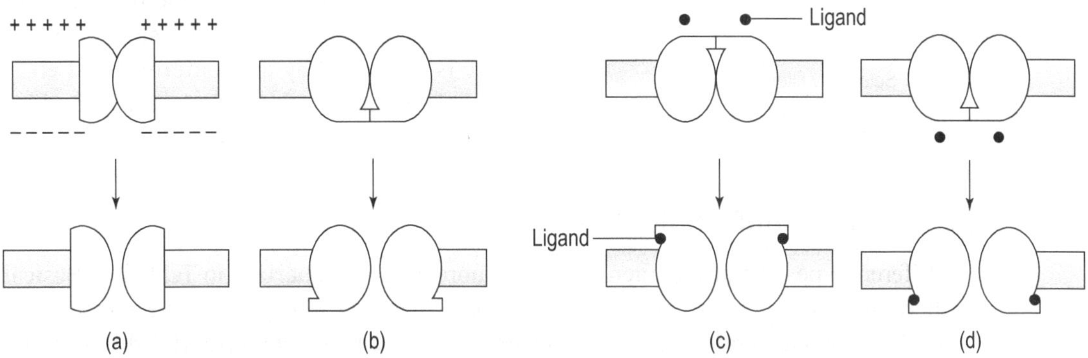

FIGURE 2.4 Types of gated channel illustrating their function

TABLE 2.1 Types of ion channels based on gating stimulus

Name of the channel	Stimulus for opening the gate	Example
Voltage-gated channel	Fluctuation in voltage across the membrane	Channels in neurons
Mechanical-gated channel	Mechanical stress	Channels in the smooth muscle cells
Extracellular-ligand-gated channel	Binding of neurotransmitters	Nicotinic acetylcholine receptor
Intracellular-ligand-gated channel	Binding of ions	Type 1 inositol 1,4,5 trisphosphate receptor
Nucleotide-gated channel	Binding of nucleotides	Cyclic-nucleotide-gated channel alpha found in retinal photoreceptors and olfactory neurons

2.4.3 K^+ Channel

Potassium channel (Fig. 2.5) is a classic example of a gated channel that is well studied using X-ray crystallography. It is predominant in nerve cells. This channel consists of an outer entrance which forms the pore on the extracellular side and an inner entrance which forms the pore on the cytosolic side of the cell. Structurally, it is made of two identical subunits. Each subunit resembles a conical structure with the wider part facing

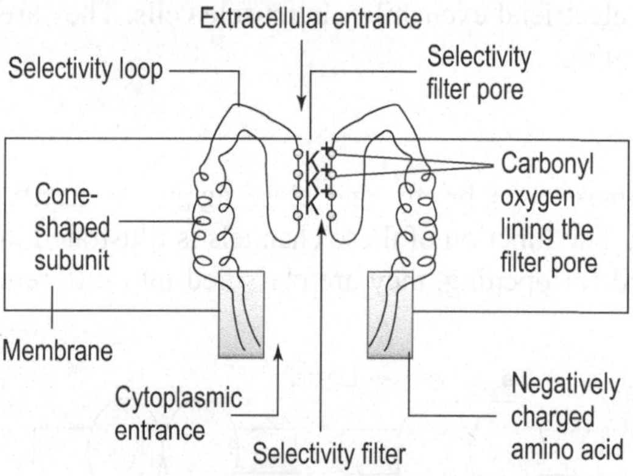

FIGURE 2.5 Structure of K^+ ion channel

the extracellular side. The subunit, which is made of two transmembrane helical motifs, projects its negatively charged amino acids at the cytosolic entrance. The other side of the helical motif connects the two helix via selectivity loop. The selectivity loop of both the subunits lie adjacent to each other facing the extracellular side, forming a narrow rigid pore called 'selectivity filter'. The filter is lined by carbonyl oxygen through which the potassium ions are transported from the cytosolic side. Closing of the pore is facilitated by the conformational change in the helical motif which shuts the pore like a diaphragm in the cytosolic side.

2.5 TYPES OF MEMBRANE TRANSPORT

Different types of transport across cell membrane are summarized in Table 2.2. Basically there are three types of membrane transport—passive transport, active transport, and gradient-mediated transport. The mechanism of these types of transport is discussed with suitable examples in this section.

TABLE 2.2 Types of membrane transport

Mode of transport	Mechanism
Passive transport	
Simple diffusion	Independent passage of molecules through the lipid bilayer, down the electrochemical gradient.
Facilitated diffusion	Channel- or carrier-mediated passage of molecules through the lipid bilayer, down the electrochemical gradient.
Active transport	Carrier-mediated passage of molecules through the lipid bilayer. This process is tightly coupled to metabolic energy such as that produced by ATP hydrolysis.
Gradient-mediated transport	Electrochemical gradient combines the membrane potential and concentration gradient to increase the driving force on ions across the membrane.

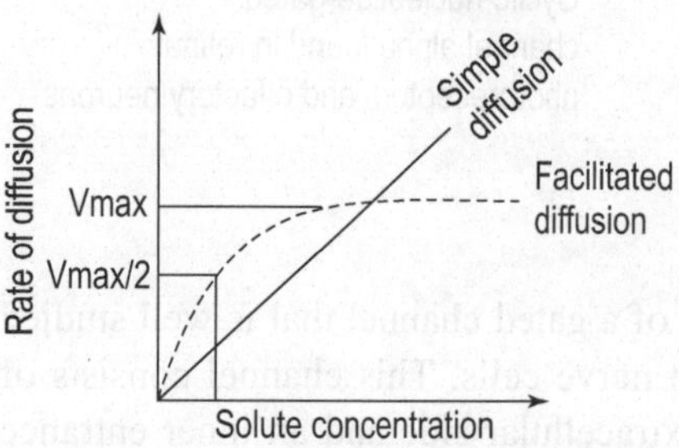

FIGURE 2.6 Kinetics of passive diffusion (solid line indicates the kinetics of simple diffusion and dotted line represents the kinetics of facilitated diffusion)

2.5.1 Passive Transport

Passive transport takes place independent of any energy source or gradient. It is of two types—simple diffusion and facilitated diffusion. Simple diffusion takes place independent of any facilitator, whereas facilitated diffusion occurs via facilitators, such as carriers or channels, discussed in the previous section. The rate of simple diffusion depends on the concentration of the molecules transported (Fig. 2.6), that is, it increases linearly with the concentration of the molecules (of the solute) to be transported. Facilitated diffusion increases with solute concentration initially, but at a certain concentration the rate 'levels off' without further increase. Thus, the kinetics of simple diffusion are linear and that of facilitated diffusion are parabolic. The solute concentration at which the facilitator is half saturated is k_m, the binding constant. Each solute has its own binding constant.

2.5.2 Active Transport

Active transport involves coupling of carrier proteins to energy, in order to pump the solutes uphill against the electrochemical gradient. Cells mediate coupling in two ways—by coupling carriers, and by ATP-driven pump.

Coupling carriers

Coupling carriers couple uphill transport of one molecule and downhill transport of another molecule. The energy is harvested from the electrochemical gradient of one solute, typically ions. The energy derived from movement of ion downhill (i.e., down the electrochemical gradient) is used as the driving force to move another ion uphill (i.e., against the electrochemical gradient). Carriers that transport single solute are called *uniporters*, carriers that transport two types of molecules in the same direction are called *symporters*, and those that transport two types of molecules in opposite directions are called *antiporters*. Mechanism of active transport mediated by coupling transport is shown in Fig. 2.7.

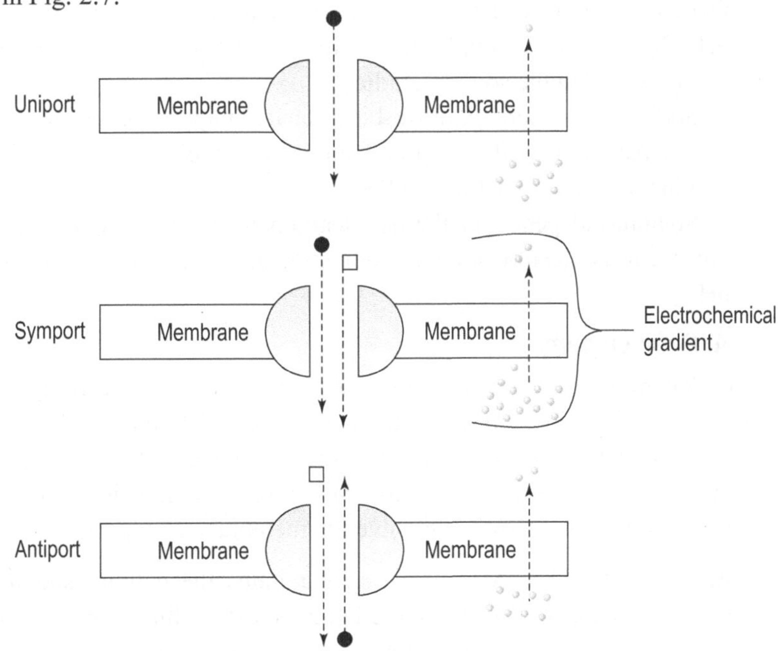

FIGURE 2.7 Active transport by coupling carriers

For the symport process, sodium ion is the usual cotransport ion. Electrochemical gradient of sodium ion provides a large driving force for the transport of the second solute molecule. For example, sodium is used to transport glucose into the cell. Proton gradient is used for transport across membranes in bacteria, yeast, and organelles because of the predominance of proton pump and the virtual absence of sodium pump in these membranes. For example, permease is a symporter that transports lactose across bacterial membrane at the expense of proton gradient. In symport process, the binding of two ligands—the solute to be transported and the accompanying ion (Na^+ or H^+) to the symporter—shows cooperativity, that is, binding of either of the ligand enhances the binding of the other ligand by means of inducing conformational changes in the symporter. The conformational changes will be reversed on release of both the ligands.

Antiport process, mediated by proton gradient, is used in maintaining and regulating the cellular pH. Maintenance of cellular pH is crucial for the maintenance of the structure and function of cellular biomolecules that are concentrated or comparmentalized in different organelles. This is true for proteins. Each protein has its own optimum pH at which it functions maximally. Acid hydrolase, for example, is a member of hydrolases which are compartmentalized in lysosomes, which have pH 5. Cytosolic protein, on the other hand, shows optimal activity at a pH of approximately 7.2. Thus, to maintain the pH in different organelles, the hydrogen ion concentration should be optimal. This is achieved by (i) sodium–hydrogen antiporter, which increases the pH, (ii) sodium-dependent chloride–bicarbonate exchanger, which increases the pH, and (iii) sodium-independent chloride–bicarbonate exchanger, which decreases the pH.

Whenever the acidity of the cell or a compartment is raised (pH decreased) due to acid-forming reactions or acid influx, the sodium–proton exchanger is used to normalize the pH. This exchanger couples the influx of sodium ion to an efflux of proton by using the energy stored in the sodium gradient.

Sodium-dependent chloride–bicarbonate antiporter couples the influx of sodium and bicarbonate to the efflux of proton and chloride ion. Thus, sodium bicarbonate is formed within the cell and HCl is expelled out.

Sodium-independent chloride–bicarbonate antiporter couples the efflux of bicarbonate ion down its electrochemical gradient to the influx of chloride, thereby decreasing the pH.

ATP-driven pump

ATP pumps are a family of enzyme proteins which catalyse the hydrolysis of ATP, forming free phosphate ion. During hydrolysis, energy is liberated (refer Chapter 15). The energy is harnessed to drive the transport process. As the hydrolysis is catalysed by the ATPase enzyme, the pump is also called ATPase transporter. This section discusses the different types of ATP-driven pump involved in membrane transport.

Na^+–K^+ pump The Na^+–K^+ pump maintains the cellular sodium- and potassium-ion concentrations. Sodium is the predominant extracellular cation, whereas potassium is the predominant intracellular cation. Sodium is effluxed and potassium is influxed against their respective concentration gradients, at the expense of the energy derived from ATP hydrolysis. ATP hydrolysis is catalysed by the Na^+–K^+ pump (Fig. 2.8), which is called Na^+–K^+ ATPase because of its function. The Na^+ and K^+ gradients across the membrane are depicted using shaded block arrow. Na^+ and K^+ transport is shown by dotted arrow, the pump is shown as a shaded circle, and the energy derived to drive the pump through ATP hydrolysis is shown with a thin solid arrow.

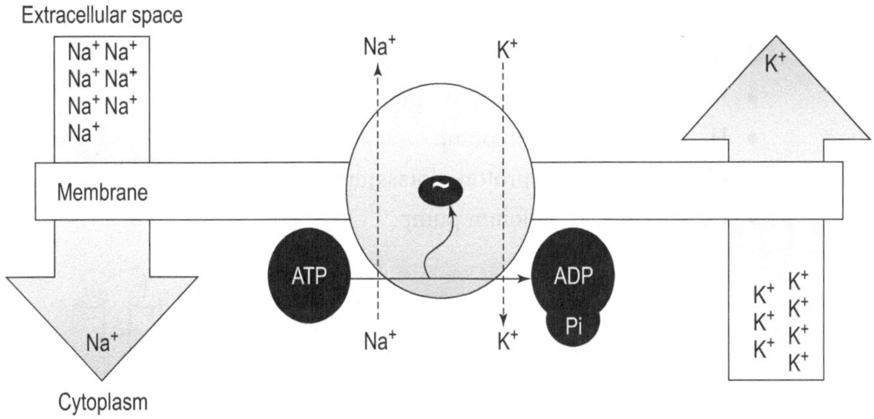

FIGURE 2.8 Na^+–K^+ pump

The transport process is mediated by phosphorylation and dephosphorylation of the pump. The phosphate of the ATP is transferred to the aspartate residue of the protein constituting the pump. The phosphorylated pump then undergoes conformational changes and develops affinity for sodium ion causing the efflux of sodium ions. Later, dephosphorylation transforms the pump to its original conformation, which in turn binds to potassium ions and internalizes into the cell.

The Na^+–K^+ pump regulates the cell volume by regulating osmolarity and controlling the solute concentration inside the cell. This can be explained by placing the RBCs in different saline concentrations. In hypotonic saline, where water is more, the water moves into the cell down its concentration gradient. This leads to bulging and swelling of the cells. In hypertonic saline, where water concentration is less, water moves out of the cell, thus, shrinking the cells. Thus, the Na^+–K^+ pump mediates osmotic water flow.

Calcium pump is another example of an ATP-driven pump which is located in the membrane of sarcoplasmic reticulum. Sarcoplasmic reticulum is a specialized endoplasmic reticulum found in the cytosol of muscle cells. It is the reservoir of intracellular calcium ions. In response to appropriate signal, calcium ions will be released into the cytosol to promote muscle contraction. After the contraction stroke, calcium ions have to be pumped back into the sarcoplasmic reticulum. Calcium pumping is mediated by the Ca^{2+} pump which constitutes 90% of the sarcoplasmic reticulum membrane. The calcium pump is involved in muscle contraction, as explained in Chapter 9, Section 9.11.1.

P-type ATPase or E_1–E_2 ATPase P-type ATPase is a transporter used to pump cations across the membrane and to flip the membrane phospholipids. The structure of a P-type ATPase is shown in Fig. 2.9. It consists of three domains:

1. *P domain* or the phosphorylating domain
2. *N domain*
3. *A domain* or the actuator domain

The P domain is made of seven-stranded, parallel β-sheet structure, the N domain has two helical segments and a seven-stranded, antiparallel β-sheet structure. The A domain is the smallest of these and is made of two short helical segments. The hydrolysis of ATP usually occurs in the P–N interface. The P-type ATPases are called so because they autophosphorylate at their aspartate residues. Following are some examples of P-type ATPAses:

- Na^+–K^+ ATPase or sodium–potassium pump
- H^+ ATPase or proton pump
- H^+–K^+ ATPase or proton–potassium pump
- Ca^{2+} ATPase or calcium pump

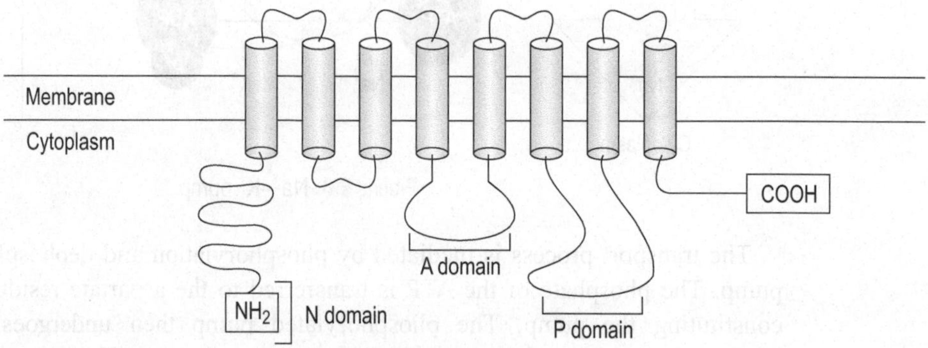

FIGURE 2.9 Structure of a P-type ATPase

The mechanism of action of P-type ATPase has the following steps (described in Fig. 2.10):

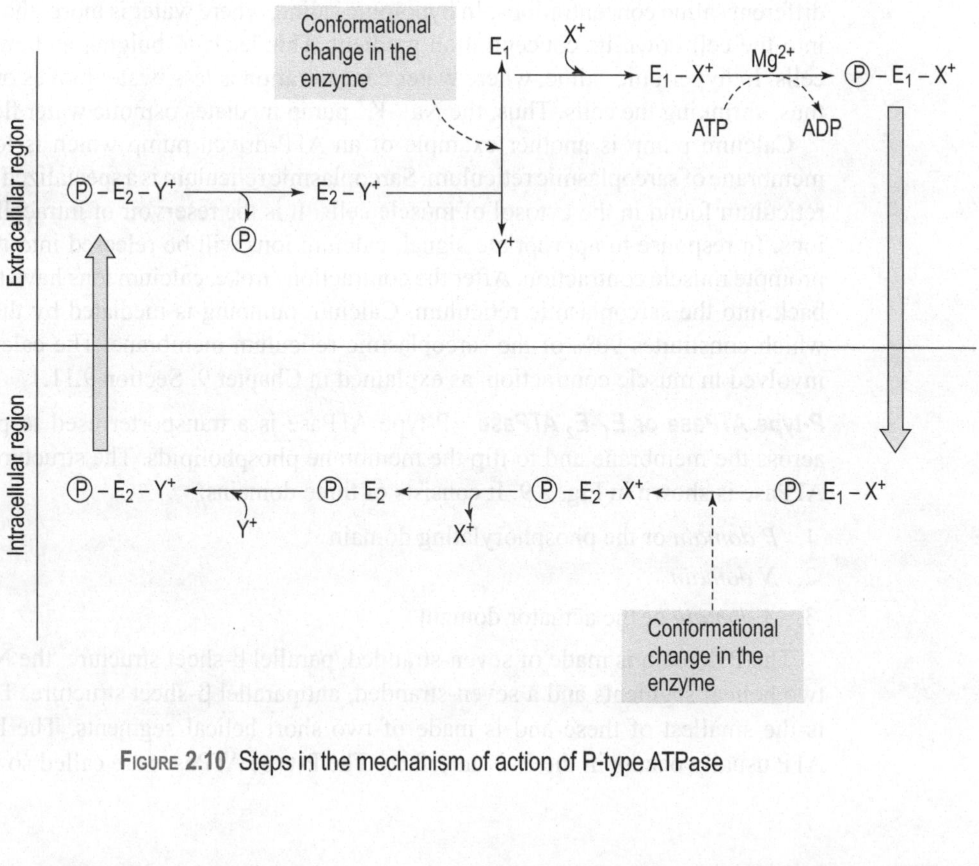

FIGURE 2.10 Steps in the mechanism of action of P-type ATPase

1. In the extracellular region, the cation (X^+) binds to the ATPase enzyme in dephosphorylated form (E_1), which has a high affinity for the ion, thus, forming E_1–X^+.
2. Magnesium-bound ATP (Mg^{2+}–ATP) is hydrolyzed and the phosphate from the ATP is transferred to the aspartate residue of the E_1, forming P–E_1–X^+.
3. P–E_1–X^+ moves to the intracellular compartment. Here the enzyme undergoes a conformational change, forming a low-affinity state. This phosphorylated state, denoted by P–E_2–X^+, has low affinity for the ion X^+. X^+ is, therefore, released and P–E_2 is formed.
4. The second ion, say Y^+, is then bound to P–E_2 with high affinity forming P–E_2–Y^+.
5. P–E_2–Y^+ is transported to the extracellular region, where the aspartate residue is dephosphorylated, forming E_2–Y^+. This dephosphorylated form of the enzyme has low affinity for Y^+; hence, it is released and the enzyme is regenerated in its E_1 form.
6. Thus, the E_1 state of the enzyme transports X^+ into the cell, while the E_2 state of the enzyme transports Y^+ to the extracellular compartment. Because of the different states that the enzyme assumes (E_1 and E_2) during the mechanism, it is also called E_1–E_2 ATPase.

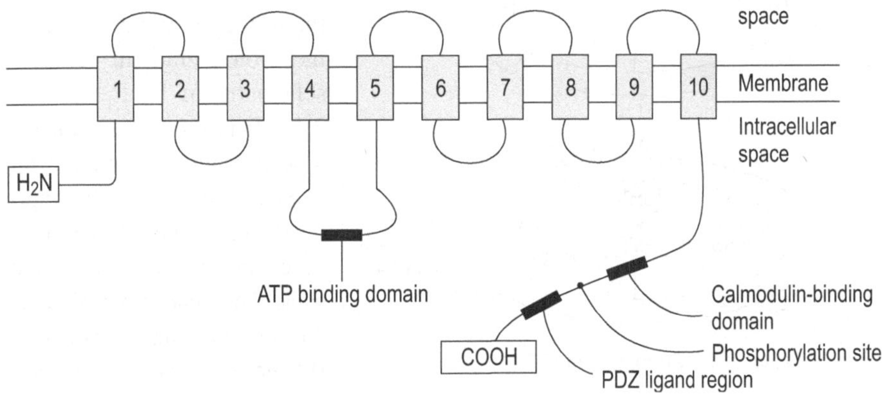

FIGURE 2.11 Structure of calcium ATPase

Calcium ATPase Calcium ATPase (also called plasma membrane Ca^{2+} ATPase, PMCA) is a P-type ATPase pump required for the transport of calcium. It is responsible for the regulation of muscle contraction, heart function, and cellular processes, such as cell proliferation and cell death. The calcium ATPase located in the membrane of sarcoplasmic reticulum of muscle cells is responsible for muscle contraction (refer Chapter 9) and it is called SERCA (Sarco/Endoplasmic Reticulum Calcium ATPase). It pumps the calcium ions from the cytosol to the sarcoplasmic reticulum. PMCA, present in the plasma membrane, removes calcium from the cell by extrusion. This process is crucial for cell signalling (refer Chapter 11). The structure of PMCA is shown in Fig. 2.11. It consists of 10 transmembrane domains interlinked by extracellular and intracellular domains. The ATP-binding

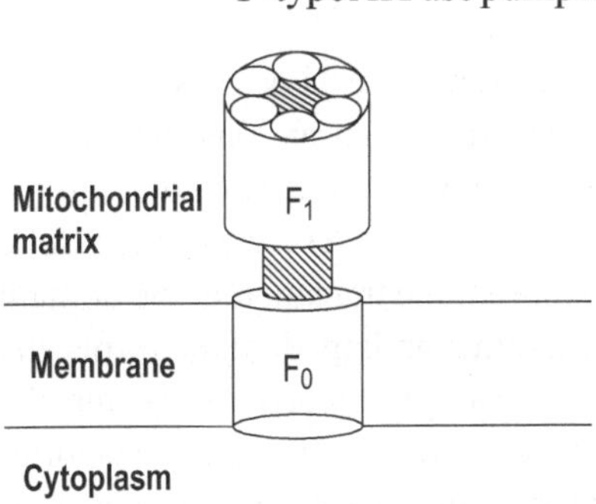

FIGURE 2.12 Structure of F_0–F_1 ATPase

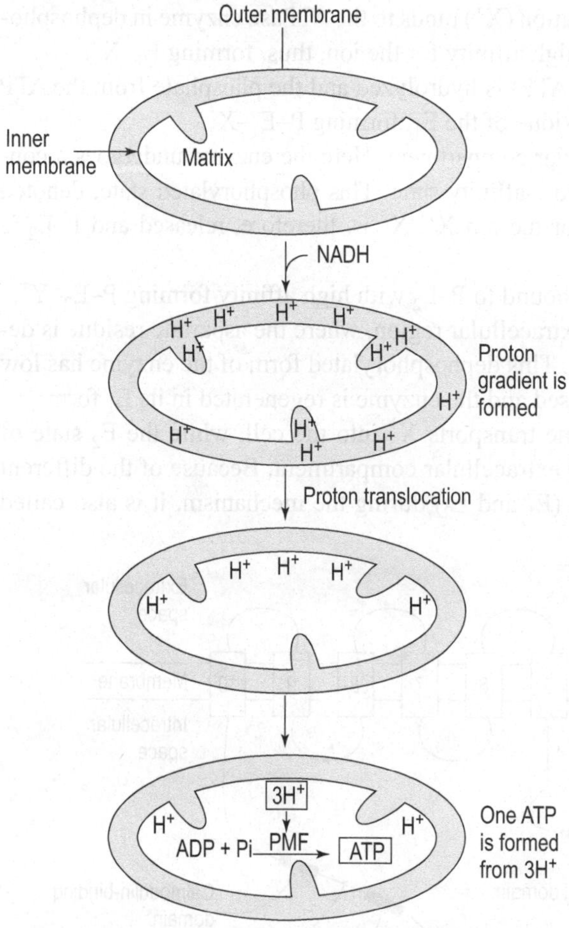

FIGURE 2.13 Formation of ATP

domain is located between the C-terminal and the N-terminal regions, in the intracellular domain. Near the C-terminal region is the PDZ ligand region. Adjacent to the PDZ region is the calmodulin-binding region. This region is responsible for binding the calcium to be transported. Between the PDZ and the calmodulin-binding domain is the phosphorylation site, where the phosphate group of the ATP is added (as explained under *P-type ATPase*). The mechanism of action of PMCA is similar to the general mechanism explained for P-type ATPase.

F_0 and F_1 ATPase F_0–F_1–ATPase is a membrane-bound enzyme found in the inner membrane of the mitochondria. It is an ATP synthase. It reversibly catalyses the formation of ATP in the mitochondria at the expense of energy from the proton gradient across the membrane. The schematic of F_0–F_1 ATPase is shown in Fig. 2.12. It is a multisubunit protein, consisting of a head group or F_1 region facing the matrix and a base plate or the F_0 region embedded in the inner mitochondrial membrane. The head and the base plate are linked by a oligomycin-sensitive stalk. The head group functions as a catalyst and the base plate functions in conducting the protons. Whenever NADH/NADPH is liberated in the oxidation reaction, protons accumulate in the outer mitochondrial membrane forming a proton gradient (refer Chapter 7, Sections 7.11 and 7.12). Each time, three protons are transported from the outer membrane into the matrix via the enzyme complex. Proton translocation creates a protomotive force which is coupled to the synthesis of ATP from ADP and Pi (inorganic phosphate) within the matrix. One ATP is formed by three protons (Fig. 2.13).

ABC transporter ABC transporter (ATP-binding cassette transporter) is a transmembrane protein involved in the active transport of various molecules, such as ions, lipids, and drugs, across membranes. The transporter has two domains: transmembrane domain (or integral-membrane domain or membrane-spanning domain) and the ATP-binding cassette domain (or the nucleotide-binding domain). In ABC exporter, the transmembrane domain forms the N-terminal region and the ATP-binding cassette binding domain forms the C-terminal region. Hemolysin exporter in E. coli is an example. In ABC importer, the orientation is reversed, that is, ABC domain is the N-terminal region. Each transporter has atleast two copies of each domain. The transmembrane domain has variable sequence, thus, representing the ABC transporter for the transport of a wide range of molecules. The

ATP-binding domain, however, has a conserved sequence. When a molecule binds to the transporter, the ATP binds to the ATP-binding domain and it is hydrolyzed. The energy released is used to induce conformational changes in the transmembrane domain. Due to the conformational change, the molecules are transported from one side to the other.

2.5.7 Gradient-driven Pump

Proton gradient is predominant in aerobic bacterial membrane, mitochondrial membrane, and thylakoid membrane of the chloroplast. This is due to the protons generated during the electron-transport chain. Proton gradient is also present in halobacterium where it is generated by light-absorbing rhodopsin. In all these cases, the proton gradient helps ATP synthase to generate ATP. The role of proton gradient is well studied in the mitochondrial membrane during oxidative phosphorylation and ATP production. This concept is described in detail in Chapter 7, Section 7.12.

2.6 TRANSPORT ACROSS INTRACELLULAR MEMBRANES

As discussed in Chapter 1, all eukaryotic cells are characterized by the presence of membrane-bound subcellular organelles in the cytoplasm. Transport across these membranes, that is, movement of molecules between cytosol and the lumen of the subcellular organelles, is mediated by the protein present in the organelle membranes. In fact, the protein itself is transported from the ribosomes in the cytosol to the different organelles by different mechanisms (Table 2.3). The three modes of transport, namely, gated transport, transmembrane transport, and vesicular transport, are guided by the presence of signal amino acids which are recognized by the receptor proteins in the membrane. In gated transport, nuclear pore complex functions as gate for protein transfer into the nucleus. In transmembrane transport, the protein is unfolded on reaching the organelle membrane. Unfolded protein passes through the membrane-bound translocator to enter into the lume of the orgnelles such as mitochondria or ER. In vesicular transport, a vesicle buds from one organelle, which is called the donor organelle. This vesicle functions as a cargo to carry the transportable molecules or solutes. It then detaches and reaches the other organelle, which is the target organelle. Finally, it fuses with the target organelle and releases its content. The transport of soluble proteins from the ER to Golgi bodies takes place by vesicular transport.

TABLE 2.3 Mechanism of transport across intracellular membrane

Mechanism	Target Organelle
Gated transport	Nucleus
Transmembrane transport	Mitochondria
	Endoplasmic reticulum
	Plastid
	Peroxisome
Vesicular transport	Golgi body
	Endosome
	Lysosome
	Cell surface

2.6.1 Transport across Nuclear Membrane

The nuclear membrane is very important in facilitating the transfer of many proteins to and from the nucleus. This sections discusses the nuclear protein signals and various nuclear membrane transporters.

Nuclear localizing signals

Some proteins which function inside the nucleus, for example, DNA polymerase and RNA polymerase, histones, and so on, are called *nuclear proteins*. These proteins are synthesized in the cytosol and thereafter, they are transported to the nucleus. They contain one or two short, conserved sequences of positively charged amino acids, such as lysine and arginine. These sequences are called nuclear localizing signals (NLS). NLS are present only in nuclear proteins, to be transferred to nucleus. Therefore, the specificity of nuclear proteins to pass into nucleus is due to NLS. NLS are organized as loops or patches on the protein surface.

Nuclear import receptor

A group of cytosolic proteins called *nuclear import receptors* (NIR) are used in the transport of nuclear proteins. Short amino acid repeats of phenylalanine and glycine present in NIR are called FG repeats (phenylalanine–glycine repeats) through which the NIR binds to the protein directly (or through an adaptor protein) and forms a complex (Fig. 2.14). The complex is then imported into nucleus in a folded conformation. This is in contrast to other organelles, where the proteins imported from the cytosol should be unfolded for transfer. During transport, the nuclear pore complex diameter is distended to 26 nm. After releasing the protein into the nuclear matrix, the free NIR return backs to the cytosol. Studies on nuclear import receptors, such as importin-11 and kap 120, suggest that NIR and NLS bind directly, however, they are separated by Ran (RAs-related nuclear protein) GTP.

Nuclear pore complex Nuclear pore complex is a protein present in nuclear membrane and is responsible for the transport of various molecules across nuclear membrane. SEM analysis reveals that the complex is made of four subunits which form the building blocks of the complex as shown in Fig. 2.15.

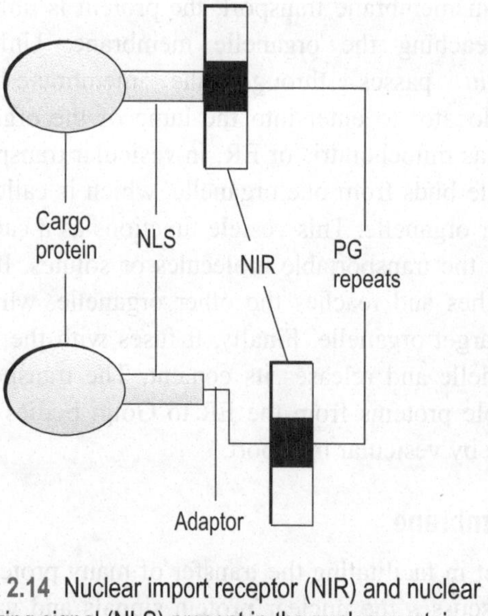

FIGURE 2.14 Nuclear import receptor (NIR) and nuclear localizing signal (NLS) present in the nuclear protein to be transported into the nuclues

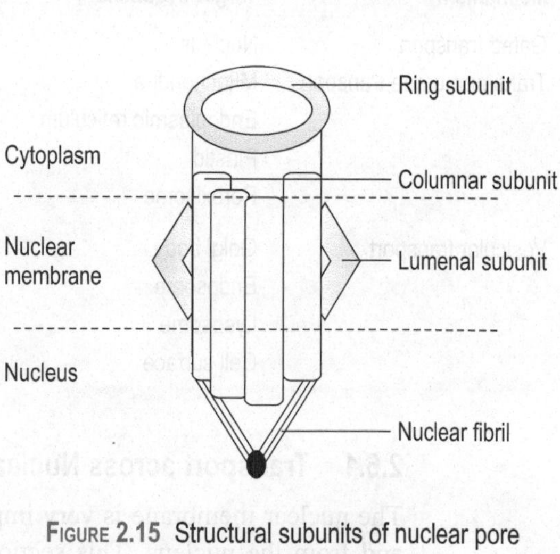

FIGURE 2.15 Structural subunits of nuclear pore complex transporter

The complex has several small motifs arranged as a ring to form a ring subunit facing the cytosolic side of the cell. The ring subunit is extended lengthwise by the column subunit, which in turn has annular subunits on its surface. The column subunits have lateral subunits called lumenal subunits. From the end of the column, numerous fibrous structures extend into the nucleoplasm and are called nuclear fibrils. Nuclear pore complex can form channels of 9 nm diameter and 15 nm length. The channel allows passive diffusion of small molecules with molecular weight less than 5,000 Da and impermeable to large molecules such as DNA polymerase, RNA polymerase, and ribosomal subunits.

Karyopherins

Karyopherin is a protein used for the transport of cargo across the nuclear membrane via the nuclear pore complex. It functions as a nuclear receptor. There are two types of nuclear receptors: (i) nuclear import receptor (NIR) or importin and (ii) nuclear export receptor (NER) or exportin. NIR transports protein into the nucleus, while NER transports protein from the nucleus to the cytoplasm. Both receptors are from the same gene family, hence, they work in similar fashion but in opposite directions.

tRNA and ribosomal subunit, 5S RNA, pass out of the nucleus by NER. The signal sequence in the molecules to be exported is called nuclear export signal (NES). NER binds to the NES and then to the nucleopore complex to guide the transport of the nucleic acids across the membrane to reach the cytoplasm. After reaching the cytosol, the molecules are liberated and the receptor comes back into the nucleus. The mechanism of coordination of bidirectional flow across the membrane is still unknown. Recent reports have confirmed the role of karyopherins in regulating the assembly of the nuclear pore complex, assembly of the nuclear envelope, mitosis, and replication.

Ran cycle

Ran refers to **RA**s-related **N**uclear protein. Ran is a small, GTP-binding protein with molecular weight of 25 kD. Ran functions with karyopherins to enable the transport of cargo, such as protein and RNA, across the nuclear membrane. It has intrinsic GTPase activity. Ran protein occurs in two forms in different locations, that is, in cytosol and nucleus: GDP-bound form (Ran-GDP) and GTP-bound form (Ran-GTP). Ran's intrinsic GTPase activity is switched 'on' (stimulated) by the cytoplasmic GTPase activating protein (GAP). Thus, the GTP in Ran-GTP is hydrolyzed to form Ran-GDP. Thus, the concentration of Ran-GDP predominates in cytoplasm. Ran-GDP translocates from the cytoplasm to the nucleus by another protein called nuclear transport factor 2 (NTF2). Within the nucleus, the Ran guanine nucleotide exchange factor (GEF) is responsible for exchanging Ran's GDP for GTP, forming Ran-GTP. RCC1 (regulator of chromosome condensation 1) is the GEF used in flipping of Ran. It is predominant in nucleus and is found bound to the chromatin. Thus, the concentration of Ran-GTP is predominant in the nucleus. The concentration of GAP and GEF differs in the cellular compartment. So, Ran-GTP: Ran-GDP ratio is high in the nucleus and low in the cytoplasm. The Ran cycle which flips Ran-GTP and Ran-GDP is explained schematically in Fig. 2.16. In fact, Ran mediates the release of imported karyopherin-bound cargo in the nucleus. In addition, it also facilitates the export of cargo from the nucleus to the cytoplasm (Fig. 2.17). The area of import is shown as the shaded box. The area of export is shown as the unshaded box.

Role of Ran is highlighted with bold dotted arrows. Other processes are shown using thin solid arrows. Both the functions are due to the flipping of Ran between GDP-bound form and GTP-bound form. The cargo protein that is transported into the nucleus is bound to importin (karyopherin). At this stage, the Ran-GTP which predominates in the nucleus, releases the cargo from the importin. On the other hand, cargos which have to go to the cytoplasm will be bound to the exportin and Ran GTP, forming a ternary complex. The complex reaches the cytoplasm, where the Ran-GTP is hydrolyzed by GAP and intrinsic GTPase to dissociate the complex. Finally, the cargo is released.

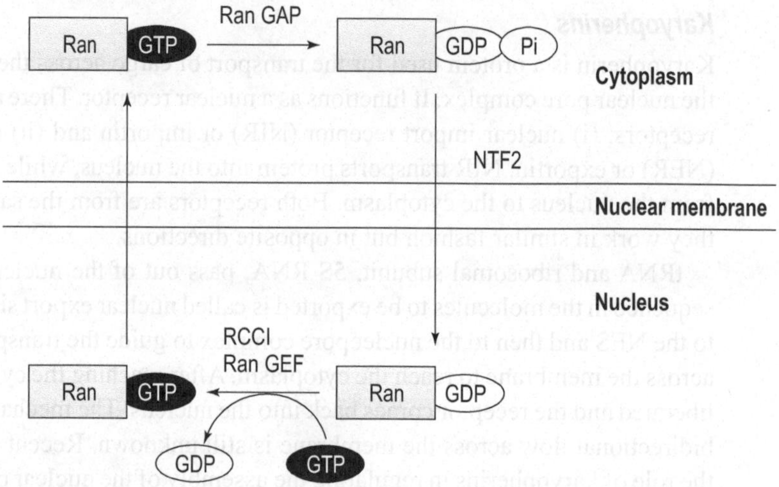

FIGURE 2.16 Ran cycle used in cargo transport across nuclear membrane

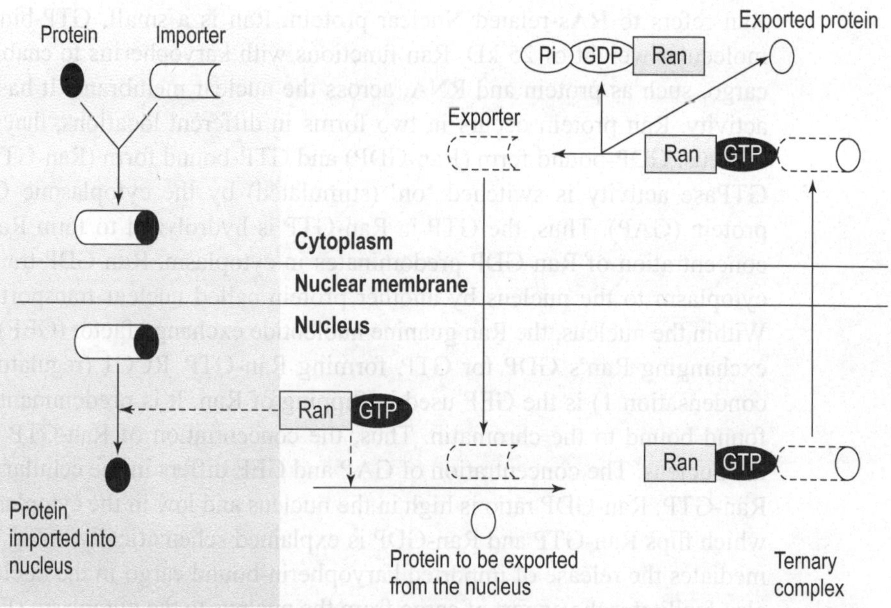

FIGURE 2.17 Import and export of cargo across the nuclear membrane

Transport of gene regulatory protein

Transport across nuclear membrane is very important in immune response. In resting T cell, nuclear factor (NF) is a gene regulatory protein, responsible for regulating the expression of genes coding for cell surface proteins and cytokines—both required for proper immune response. Thus, NF has to actively move across the nuclear membrane. Depending on the stimulus, intracellular calcium level signals the import and export of NF. High calcium-ion concentration in activated T cell signals the binding of calcium to the NF which dephosphorylates the NF, and exposes the import signal. As a result, the NF crosses the membrane and enters the nucleus to activate the transcription of the target genes for all surface proteins and cytokines which in turn elicit immune response. On the other hand, if the intracellular calcium decreases, as in resting T cell, that is, when there is no need for immune response, NF is phosphorylated by active protein kinase and ATP. On phosphorylation, the export signal is exposed causing the NF to re-enter into cytoplasm. Transport of NF across the nuclear membrane is illustrated in Fig. 2.18.

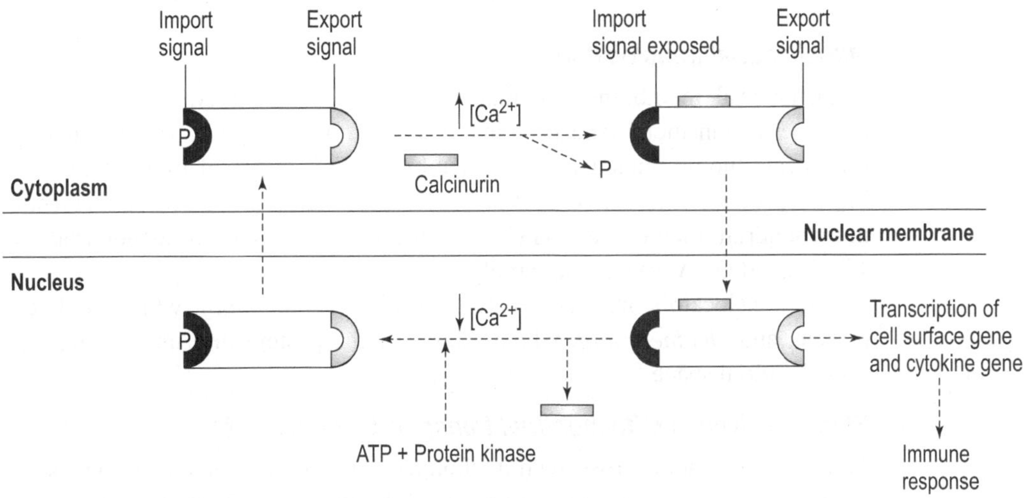

FIGURE 2.18 Transport of gene regulatory protein across the nuclear membrane

2.6.2 Transport across Mitochondrial Membrane

Mitochondria peforms a very crucial function in the cell: the synthesis of energy in the form of ATP. To execute this task, mitochondria makes use of several proteins. These proteins are normally synthesized in the cytosol and then transported through the mitochondrial membrane to reach the mitochondrial matrix or inner membrane. This section describes the transport of proteins across the mitochondrial membranes.

Mitochondrial proteins

Proteins to be transferred to mitochondria are composed of a specific amino acid sequence called signal sequence, at their N-terminus (Fig. 2.19). The signal sequence exists as an amphipathic helix, with the positively charged polar amino acids oriented towards one side of the helix and the neutral non-polar amino acids distributed at the opposite side. The proteins with signal sequence are maintained unfolded due to interaction with

HSP 70 family chaperons (Chapter 4). Mitochondrial proteins with signal peptide and the chaperon are called mitochondrial precursor protein (MPP).

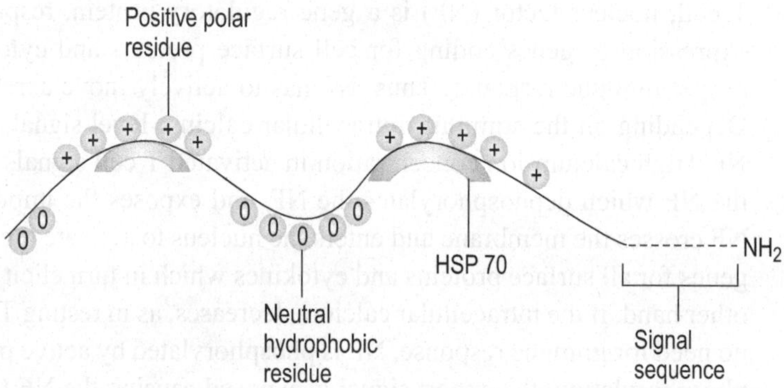

FIGURE 2.19 Signal sequence of transportable mitochondrial proteins

Mitochondrial translocators

Mitochondrial membrane contains specific receptors called translocators, which are multimeric transmembrane proteins. Translocators recognize the signal sequence and translocate the proteins across the membrane. Four types of translocators are found to import proteins into mitochondria (Table 2.4). They are TOM (translocator of outer mitochondrial membrane), TIM23 (translocator of inner mitochondrial membrane), TIM22, and OXA (oxidase assembly).

The translocators are made of two motifs: *receptor motif*, which binds to the cargo protein, and *channel motif*, which transports the protein into intermembrance space or mitochondrial space.

Steps involved in mitochondrial transport by translocators

Transport of protein across the mitochondrial membrane is an energy-dependent process. Energy is derived from two sources: ATP and H^+ gradient. Import of proteins into mitochondria occurs in the following six steps:

1. Recognition of protein by TOM complex and removal of HSP 70 by ATP hydrolysis
2. Insertion of the protein into inner mitochondrial membrane by electrochemical H^+ gradient

TABLE 2.4 Translocators of mitochondrial membrane

Translocator	Function	Location
TOM	Imports the outer membrane proteins	Outer mitochondrial membrane
TIM23	Imports the inner membrane proteins and matrix proteins	Spans inner and outer membrane
TIM22	Imports proteins that are carriers for ATP, ADP, and Pi	Inner mitochondrial membrane
OXA	Exports proteins from mitochondrial matrix to the inner membrane	Inner mitochondrial membrane

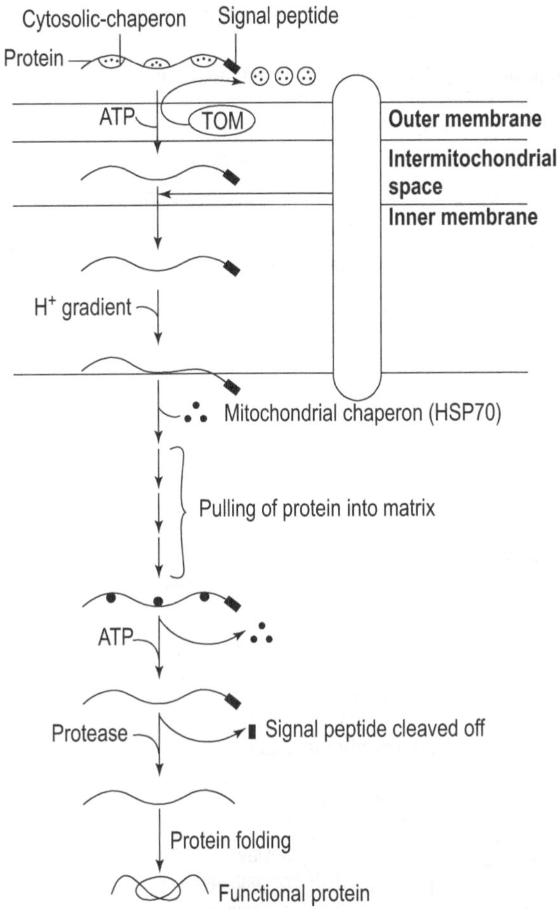

FIGURE 2.20 Mitochondrial transport

3. Pulling of protein into the matrix by mitochondrial HSP 70
4. Removal of HSP 70 by ATP hydrolysis
5. Removal of signal peptide by protease
6. Folding of protein to become functional

The two membranes of the mitochondria are held together by TIM23 complex (as it spans both the membranes). Protein translocation into mitochondria occurs at special contact sites. The signal sequence of MPP is recognized by the receptor motif of the TOM complex. This causes the binding of MPP to TOM complex. After this, the chaperon is stripped-off from the protein at the expense of ATP hydrolysis. The signal sequence enters the translocation channel leading the protein into intermembrane space. Later, the protein is recognized by TIM to be inserted into inner membrane. An electrochemical proton gradient is created across the inner membrane by pumping of proton from the matrix to the intermitochondrial space. This gradient transports the protein along the TIM complex and across the inner membrane allowing it to span both the membranes and drive the positive signal sequence protruding towards the matrix. The protein is then pulled into the matrix by binding of mitochondrial HSP 70. After reaching the matrix, the signal sequence in the protein is degraded by protease enzyme and the HSP 70 is degraded by the ATP hydrolysis, allowing the protein to fold as well as become functional. Mitochondrial transport is illustrated in Fig. 2.20.

Role of OXA complex

OXA is a multisubunit protein located in the inner mitochondrial membrane, where it translocates proteins from the mitochondrial matrix to inner membrane. It also associates with TIM to help insert proteins into the inner memberane. Proteins to be incorporated into the inner membrane or intermembrane space follow an alternate pathway that depends on OXA complex. These proteins have hydrophobic sequence after the N-terminal signal sequence. This hydrophobic sequence signals the protein to reach the matrix, where the signal peptide is cleaved by matrix signal peptidase. Consequently, the hydrophobic sequence becomes the new N-terminus, which signals the protein to reach the inner membrane or intermitochondrial space using OXA complex.

2.6.3 Transport across Chloroplast Membrane (Thylakoid Membrane)

Recall the structure of chloroplast discussed in Chapter 1. Comparing the transport process across mitochondria and chloroplast (Table 2.5) leads to easy understanding of this section.

TABLE 2.5 Comparison of transport across mitochondrial and chloroplast membranes

Factor	Mitochondrial membrane	Chloroplast membrane
Translocator complex	Separate complex for each membrane	Separate complex for each membrane
Transporting site	Special contact site	Special contact site
Protein recognition guide	N-terminal positively charged signal sequences	N-terminal positively charged signal sequences
Energy dependence for transport	Dependent	Dependent
Energy source	H^+ gradient across the inner membrane and ATP hydrolysis	Hydrolysis of GTP and ATP
Signal sequence	Matrix signal sequence	Stromal and thylakoid signal sequence
Removal of signal sequence	By matrix signal peptidase	By stromal signal peptidase

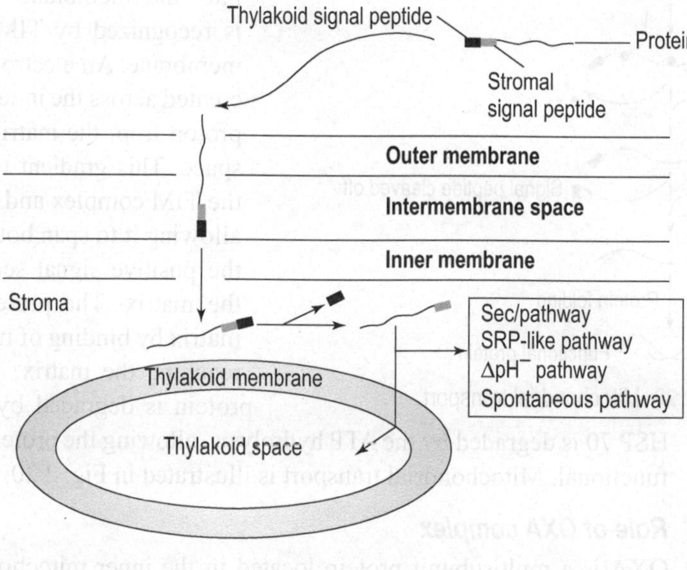

FIGURE 2.21 Transport across chloroplast membrane

Information given in Table 2.5 reveals that the transport across the chloroplast membrane is more or less similar to that in mitochondrial membrane. Proteins targeted to chloroplast have two adjacent signal peptides at the N-terminus. They are stromal signal sequence and thylakoid signal sequence. Specific receptors on the chloroplast membrane recognize the stromal signal sequence. At the expense of ATP or GTP hydrolysis, the protein then crosses the outer and the inner membranes at the 'contact sites' and reaches the stroma or matrix. The signal sequence is then cleaved-off by the stromal matrix peptidase, subsequently exposing the thylakoid sequence. The protein is then translocated to the thylakoid space (another subcompartment) via thylakoid membrane (Fig. 2.21). This translocation may take place by any one of the four pathways:

- Secretory pathway (Sec pathway)
- Signal recognition particle (SRP) like pathway
- ΔpH pathway
- Spontaneous pathway

These pathways are illustrated in Fig. 2.21.

Sec pathways use Sec proteins as carrier and ATP for transport. Sec proteins are membrane-bound proteins, which include SecY and SecE. This pathway is used for the transport of proteins, such as plastocyanin and OE33, into the chloroplast lumen. SRP-like pathway uses signal-recognition particle and ATP for translocation. This pathway can function as a post-translational process. Mechanism of SRP-like pathway is still not well understood, but the proteins cpSRP (chloroplast signal recognition particle). ΔpH, is the change in pH. This means that there is a pH gradient giving rise to a proton gradient across the chloroplast membrane. The energy from this gradient is used to power the transport of molecules across the chloroplast membrane. (This is similar to the proton gradient across the mitochondrial membrane, discussed in Chapter 7). In spontaneous pathway, the protein is spontaneously inserted into the membrane without depending on proton gradient or carriers or translocators. The reason for the dependency on multiple pathways for translocation across the chloroplast is yet to be elucidated.

2.6.4 Transport across Peroxisomal Membrane

The structure and function of peroxisome has been dealt with in Chapter 1. Unlike mitochondria and chloroplast, peroxisome has a single membrane and has no genome of its own. Hence, the protein required for the function of peroxisome is imported from the cytosol. Peroxisomal proteins have a signal sequence of three amino acid residues at their C-terminus. Soluble proteins import receptors called peroxins (encoded by PEX5 gene) that recognize the signal, dock the protein to the cytosolic side of the peroxisomal membrane. The protein–receptor complex is then transported by energy obtained by ATP hydrolysis. PEX5 releases the protein into the peroxisomal lumen and then moves back to the cytosol.

2.6.5 Cotranslational Transport across Endoplasmic Reticulum

Transport of endoplasmic reticulum (ER) proteins takes place without the protein being released into the cytosol. That is, as soon as the ER protein is synthesized, but before translation terminates, the ribosomes come in contact with the ER membrane to release the protein into the ER lumen.

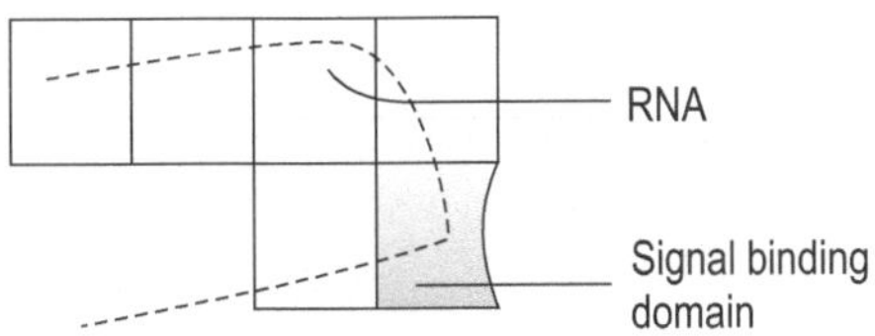

Figure 2.22 Structure of SRP

The following components are involved in the transport of proteins across ER membrane:

- Signal sequence present in the protein to be transported
- SRP (signal recognition protein), which shuttles between cytosol and ER membrane
- SRP receptor in the ER membrane
- Translocator complex containing aqueous pore

SRP is a hexamer with six subunits and one RNA molecule (Fig. 2.22).

One of the terminal subunits is called the signal-binding domain which is made of a pocket-like structure with hydrophobic amino acid residues rich in methionine. These residues bind to the signal sequence of the protein which is still bound to ribosome, pulling the entire 'protein synthesizing assembly' (Chapter 6) towards the ER membrane. The SRP-bound assembly is recognized by the SRP receptor in the ER membrane. On binding to the receptor, the protein chain is extended further with the simultaneous detachment of SRP. The ribosome then makes a tight contact with ER membrane. The signal peptide functions as 'start transfer signal' or 'start transfer peptide' to open the aqueous pore in the membrane-bound translocator. The protein then enters into the ER lumen through the pore. Thus, the protein transport into ER occurs cotranslationally, in contrast to mitochondrial and chloroplast transports which occur post-translationally after the proteins are liberated into the cytosol. After the protein is transported, 'start signal' sequence is cleaved-off by the signal peptidase. Transport across ER membrane is illustrated in Fig. 2.23.

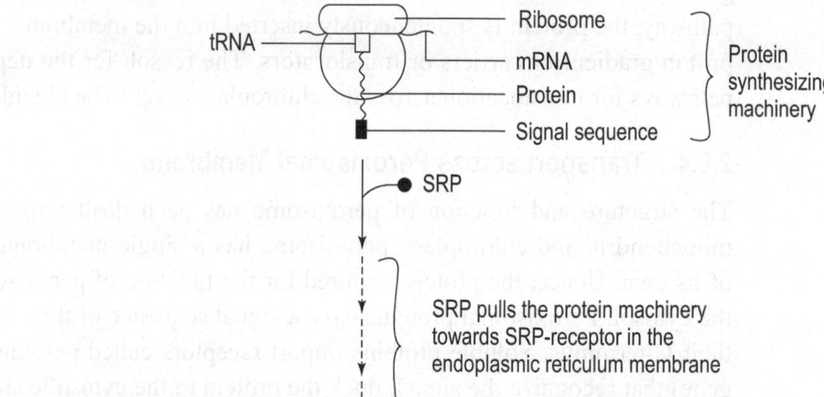

FIGURE 2.23 Transport across endoplasmic reticulum membrane

Integration of transmembrane protein into ER membrane

In addition to the 'start signal' sequence, the peptide possesses a segment of hydrophobic sequence, which is called 'stop transfer' signal. This sequence enters the translocator, discharging the protein laterally into the lipid bilayer, thereby integrating the protein with the membrane.

ER resident proteins

Proteins that are transported into the ER lumen and subsequently retained there are called ER resident proteins. Most of the proteins are transported to other destinations. ER resident proteins have a signal sequence of three amino acids at their C-terminus. This is called ER retention signal, which makes the proteins to be retained in the ER. ER proteins are glycosylated, by the addition of oligosaccharide (usually containing maltose, glucose, N-acetyl glucosamine) to the amino group of asparagine residue in the Asp-X-Ser sequence or Asp-X-Thr sequence (where X = any amino acid). These oligosaccharides are initially linked to a special membrane lipid called dolichol via pyrophosphate bond. Addition of oligosaccharides is catalysed by a membrane-bound enzyme called oligosaccharide transferase which has its active site facing the ER lumen. Improperly folded proteins are dislocated from the ER lumen to the cytosol through the same translocator (which aided the import of the same protein). In the cytosol, the oligosaccharides are cleaved-off by the enzyme glycanase and then ubiquitylated (refer Chapter 11) and finally degraded.

2.7 VESICULAR TRANSPORT

Vesicles are supramolecular assembly of lipid molecules or membranes to form a capsule. Some proteins are transported after being encapsulated in a vesicle. The vesicle is usually formed in a donor membrane, from where it is transported to yet another membrane, the target membrane, to release the protein. The entire process of vesicular transport can be divided into different stages as discussed in the following text.

2.7.1 Steps Involved in Vesicular Transport

Vesicular transport is used for the transport across different cellular compartment, and for the transport in the cell surface. It forms a major mechanism of molecular traffic in the cell. In this process, the molecule to be transported is encapsulated in a membrane vesicle which is then pinched-off into the compartment. Vesicular transport occurs by four steps which are discussed below.

1. Formation of membranous vesicle

The cargo protein to be transported is bound to another protein, adaptin, which aids the fusion of the protein with the membrane, to form a membranous vesicle.

2. Coat assembly

A protective coat is assembled around the vesicle. The coating molecules are recruited and assembled by 'coat recruitment GTPases' which are members of GTPase family. ADP ribosylation factor (ARF) protein is responsible for the assembly of COPI coat and clathrin coat in the Golgi membrane. SarI protein assembles COPII coat in the ER membrane. (COPI and COPII stands for coat protein I and coat protein II). Coat recruitment GTPases exist in two forms:

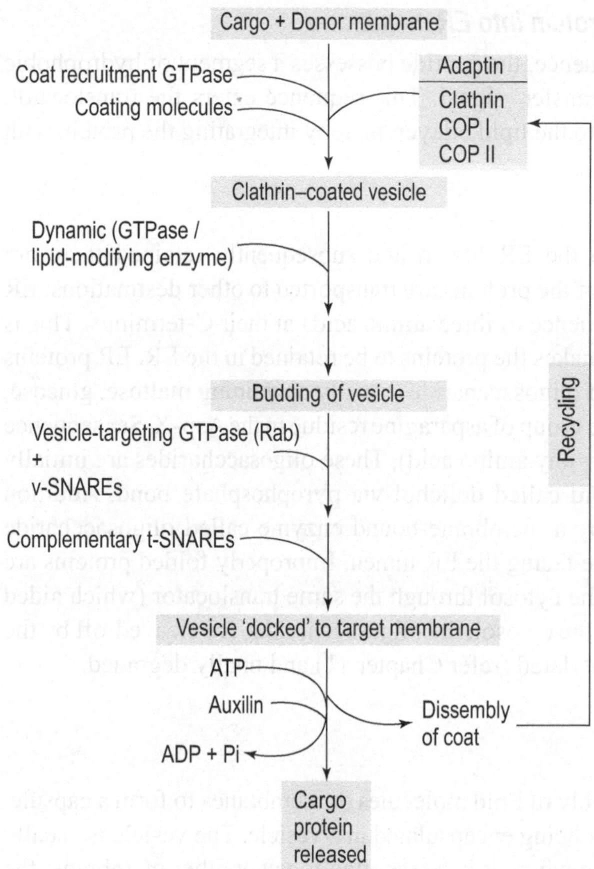

FIGURE 2.24 Vesicular transport

(i) Inactive form or GDP-bound form which is concentrated in cytosol
(ii) Active form or GTP-bound form which is concentrated in the membrane

Whenever the cell has to form the coat around the protein for vesicular transport, the inactive form is acted upon by GEF (guanine nucleotide exchange factor). This results in exchange of GDP for GTP (present in the cytosol), thereby flipping the inactive form into active form which in turn inserts into the membrane.

3. Budding and targeting

The coated vesicle buds-off from the membrane. Budding is accelerated by a protein dynamin. The bud is transported to the target membrane by vesicle targeting GTPase (Rab protein) and v-SNAREs. SNARE refers to **S**oluble **N**SF **A**ttachment Protein **RE**ceptors. They mediate vesicle fusion. They are of two types: v-SNARE (v for vesicle), which functions during budding process, and t-SNARE (t for target compartment) present in the target membrane.

4. Docking and disassembly

On reaching the target membrane, the vesicle docks the membrane and the coat is disassembled to release the protein. Auxilin is responsible for the disassembly of coat. It is an ATPase which hydrolyses ATP and uses the energy to disassemble the coating molecules. During disassembly, GTP in the active GTPase is hydrolyzed to form GDP. Thus, GDP-bound form is regenerated. This causes disassembly. The disassembled coating molecules are recycled for further vesicular coating of other proteins. Vesicular transport is illustrated in Fig. 2.24.

2.8 TRANSPORT ACROSS MEMBRANES OF VACUOLE, LYSOSOMES, AND TRANS-GOLGI NETWORK

In animal cells, vacuolar membrane plays a minor role in transport process, as the vacuoles function as a storage organelle. On the other hand, in plant cells, vacuolar membrane plays a major role in water transport. Vacuolar membrane in plant cells, helps to maintain the pH of the cytosol by allowing the transport of protons into the vacuole. The vacuolar membrane of plant cells is called totonplast. Tonoplast contains 'tonoplast integral protein'(TIP). TIP is an aquoporin, belonging to the family of membrane intrinsic protein (MIP). It is responsible for the passive transport of water molecules. They are impermeable to ions and non-polar solutes. Vacuolar transport of water (rehydration) is regulated by phosphorylation of TIP. A calcium-dependent protein kinase (CDPK) phosphorylates TIP to activate the later. This results in transport of water, causing rehydration. This process in more predominant during seed germination. Some of the small plant vacuolar membranes are known to facilitate the transport of solutes by proton antiport mechanism. This transport process is mediated by adenosine triphosphatase (ATPase) and proton pump and proton phosphatase (PPase) proton pumps.

Lysosomal membrane is responsible for engulfing macromolecules for hydrolysis (endocytosis), cell organelles (autophagy), and antigens (phagocytosis, Chapter 3). Macromolecules from the extracellular fluid is surrounded by the plasmamembrane and the membrane invaginates inside the cell releasing the content into the cytosol (endocytosis). The molecules are transported into irregular vesicles called early endosomes, from where they are transported into late endosomes (immature lysosomes). Late endosomes receive hydrolase enzymes from the endoplasmic reticulum via golgi apparatus. Late endosomes gradually mature into lysosomes. Membrane of the lysosomes facilitates proton transport in order to maintain the acidic pH within the lysosomes. Lysosomal membranes possess proton pump and chloride ion channels to drive the proton transport. Using the energy derived from the ATP hydrolysis, protons are transported into lysosomes to maintain acidic pH. The macromolecules are hydrolysed within the lysosomes, forming metabolic end products such as amino acids, sugars, and nucleotides. These products are transported across the lysosomal membrane into the cytosol for excretion or reutilization. Sometimes, the worn-out cellular organelles (e.g., mitochondria in the liver cells) are surrounded by membrane of unknown source. Thereafter, they are taken up by the lysosomal membrane through endocytosis and are digested. This process is called autophagy. The role of lysosomal membrane in the process of phagocytosis is explained in Chapter 3.

Secretory proteins (proteins to be secreted by the cell into the extracellular fluid) move from endoplasmic reticulum to enter the cis face of the cisternae of the Golgi apparatus and exit through the trans face. The cis and trans faces of the Golgi apparatus is made of tubular, reticular and vesicular networks called cis-Golgi network and trans-Golgi network (TGN), respectively. The TGN is responsible for sorting the secretory proteins. The proteins are encapsulated into vesicles called secretory vesicles (refer section 2.7). In response to the extracellular signal (refer Chapter 11), the secretory vesicles move to the plasma membrane and fuse with the plasma membrane and get secreted into the extracellular fluid (a process called exocytosis). Proteins destined for lysosomes (hydrolase enzymes) are tagged with mannose-6-phosphate and transported from the TGN via endosomes.

SUMMARY

Transport of cellular molecules across plasma membrane and intracellular membrane is a vital process in cell biology. The reason being the ability of this process to enable the cells to ingest essential nutrients, to excrete metabolic waste products, to regulate the intracellular concentration of molecules, such as ions, to regulate the proper functioning of intracellular organelles, and to maintain different compartments within the cell, in order to promote intracellular traffic. Membrane transport is mediated by transporters, such as carriers, channels, and ionophores. Carriers undergo conformational changes in the presence of energy source, whereas channels undergo conformational changes independent of energy. Ionophores are produced in microbial membranes. Channels function by gated mechanisms which are operated by voltage changes, mechanical stress, ligands, and nucleotides. Potassium channel is a classic example of an ion channel. Membrane transport occurs by three modes: passive transport (independent or carrier dependent), active transport (carrier dependent and energy dependent), and gradient-mediated transport (dependent on energy stored in electrochemical gradient).

Apart from plasma-membrane transport process, the cells have a network of intracellular traffic across the membranes of organelles, notably across the membranes of nucleus, mitochondria, peroxisome, and endoplasmic reticulum. Nuclear proteins containing nuclear localizing signals cross the nuclear membrane with the help of nuclear pore complex, karyopherins (importins and exportins), and Ran cycle. Mitochondrial precursor proteins (MPP) contain target signals and chaperons. They are recognized by translocators, such as TIM, TOM, and OXA and inserted into the inner mitochondrial membrane at the expense of the energy provided by the electrochemical proton gradient, further pulled into the matrix by matrix chaperone. The protein is finally made functional by the hydrolysis of chaperone and signal peptide.

In plant cells, the proteins targeted to chloroplasts are recognized by stromal signal and thylakoid signal, by which they cross the two respective membranes to reach the thylakoid space. Entry in thylakoid takes place by Sec pathway, SRP-like pathway, ΔpH pathway, or spontaneous pathway. Peroxisomal proteins which possess C-terminal signal of three amino acid residues are recognized by peroxisomal import receptor peroxin and transported into the lumen via ATP hydrolysis. Moreover, transport is also possible between one membrane to the other within the same cell. This process takes place by vesicles, the vesicular transport. Transport of endoplasmic reticulum (ER) proteins takes place without the protein being released into the cytosol. That is, as soon as the ER protein is synthesized, but before translation terminates, the ribosomes come into contact with the ER membrane to release the protein into the ER lumen. Vesicles are supramolecular assembly of lipid molecules or membranes to form a capsule. Some proteins are transported after being encapsulated in a vesicle. The vesicle is usually formed in a donor membrane, from where it is transported to yet another membrane, the target membrane, to release the protein. Vesicular transport takes place by assembly of coat around the cargo, formation of vesicle, budding-off of the vesicle, transport of the vesicle, and docking into the target membrane to release the cargo. Coat recruitment GTPase, SNAREs, and auxilin are key components of vesicular transport.

TERMS TO REMEMBER

Channels Small openings or pores extending across the membrane to aid the diffusion of molecules from one side of the cell to the other.

Extracellular-ligand-gated channel Membrane channel through which transport is mediated due to binding of neurotransmitters.

Facilitated diffusion A type of passive transport in which molecules are transported down the electrochemical gradient by the influence of carriers.

Gradient-mediated transport A mode of membrane transport in which electrochemical gradient combines the membrane potential and concentration gradient to increase the driving force on ion across the membrane.

Intracellular-ligand-gated channel Membrane channel through which transport is mediated due to binding of intracelluar or extracellular ligands.

Ionophores Hydrophobic molecules usually produced by bacteria to transport inorganic ions.

Karyopherin Protein used for the transport of cargos across the nuclear membrane via the nuclear pore complex. It functions as nuclear receptors— either importins or exportins.

Mechanical-gated channel Membrane channel through which transport is mediated due to mechanical stress across the membrane.

Membrane transport Movement of molecules, such as ions and other small molecules, across the plasma membrane (the lipid bilayer) and organelle membranes.

Mitochondrial precursor protein Mitochondrial proteins with signal peptide and the chaperon.

Mitochondrial translocators Specific multimeric transmembrane receptors which recognize the signal sequence and translocate the mitochondrial precursor proteins across the membrane.

Nuclear import receptors A group of cytosolic proteins that bind to the cargo proteins via their own short amino acid repeats of phenylalanine and glycine present called FG repeats, and transport the cargo.

Nuclear localizing signals Nuclear proteins contain one or two short, conserved sequences of positively charged amino acids, such as lysine and arginine. These sequences are called nuclear localizing signals (NLS) which are used as recognition elements by receptors for subsequent transport.

Nuclear pore complex Multisubunit protein present in nuclear membrane, and it is responsible for the transport of various molecules across nuclear membrane.

Nucleotide-gated channel Membrane channel through which transport is mediated due to binding of nucleotides.

OXA A translocator located in the inner mitochondrial membrane to export proteins from mitochondrial matrix to the outer membrane.

Peroxin A soluble import receptor which recognizes and binds to the signal sequence of three amino acid residues present at the C-terminus of the peroxisomal proteins to aid transport into peroxisomal lumen.

Porins Porins (ion channels) are hydrophilic pores present across the membrane, capable of closing and opening for the transport of inorganic ions.

Ran GTP-binding protein which functions with karyopherins to enable the transport of cargos, such as protein and RNA, across the nuclear membrane.

Simple diffusion A type of passive transport in which the molecules independently cross the membrane down the electrochemical gradient.

TIM A translocator located in the inner mitochondrial membrane to import the inner membrane proteins into mitochondrial matrix.

TOM A translocator located in the outer mitochondrial membrane to import the outer membrane proteins into mitochondrial matrix.

Transport proteins Mobile or immobile conveyors that facilitate membrane transport.

Vesicles Supramolecular assembly of lipid molecules or membranes that form a capsule across the cargo for subsequent transport.

Voltage-gated channel Membrane channel across which transport is mediated due to fluctuation in voltage across the membrane.

REVIEW QUESTIONS

1. Schematize the degree of permeability of different kinds of biomolecules across the cell membrane.
2. What are the parameters that influence the transport rate of carriers?
3. Give reasons for the predominance of proton-pump-mediated transport in organelle membrane.
4. Draw and explain the kinetics of passive diffusion process.
5. Comment on the following fact: 'Ion gradients cannot be maintained across the outer mitochondrial membrane'.
6. Explain nuclear pore complex as a nuclear membrane transporter.
7. State the reason for the difference in Ran-GTP:Ran-GDP ratio between the cytosol and the nucleus.
8. Tabulate the function and location of mitochondrial translocators.
9. Recapitulate the steps involved in the transport of proteins through mitochondrial membrane.
10. Describe the consequence of defective calcinurin in membrane transport.
11. Confirm the role of protease, ATPase, and signal peptidase in mitochondrial membrane transport.
12. Compare the transport process occurring in chloroplast membrane and mitochondrial membrane.
13. Discuss the pathways used in the transport of proteins across the thylakoid membrane.
14. Write a note on the role of PEX5 in membrane transport.
15. List out the enzymes involved in membrane transport.
16. Protein transport across endoplasmic reticulum membrane differs from that occurring in other organelles. Justify this statement.
17. Write the target sequences for the enzyme oligosaccharide transferase within the endoplasmic reticulum lumen.
18. Reiterate the processes leading to vesicular transport of proteins.

MULTIPLE-CHOICE QUESTIONS

1. The energy source for the ΔpH pathway of chloroplast is
(a) ATP
(b) GTP
(c) proton gradient
(d) electrochemical gradient
2. FG repeat refers to
(a) functional gradient repeat
(b) phenylalanine glycine repeat
(c) fusogenic glycine repeat
(d) formyl glycine repeat
3. The degree of transport across cell membrane is
(a) highest for hydrophobic molecules
(b) lowest for hydrophobic molecules
(c) highest for ions
(d) highest for polar molecules
4. Permeases are
(a) protein channels
(b) protein carriers
(c) lipid vesicles
(d) cell-surface carbohydrates
5. OXA complex means
(a) oxygen affinity complex in chloroplast
(b) oxidase assembly complex in chloroplast
(c) oxygen affinity complex in mitochondria
(d) oxidase assembly complex in mitochondria
6. Ran is a
(a) nuclear protein
(b) ribosomal anchor protein
(c) GTP-binding protein
(d) ATP-binding protein
7. The conserved sequence in the NLS is
(a) glycine and phenylalanine
(b) lysine and glycine
(c) lysine and arginine
(d) glycine and arginine
8. SNAREs are used in
(a) mitochondrial transport
(b) nuclear transport
(c) vesicular transport
(d) endoplasmic reticulum transport
9. HSP 70 is a
(a) tumour suppressor protein
(b) chaperon protein
(c) membrane protein
(d) nuclear transport protein
10. Valinomycin is an example for
(a) ionophore
(b) lipid raft
(c) nuclear localizing signal
(d) OXA complex

Answers to Multiple-choice Questions

1. (c)	2. (b)	3. (a)	4. (b)	5. (d)
6. (a)	7. (a)	8. (c)	9. (b)	10. (a)

CHAPTER 3

Cells and Tissues of the Immune System

Ilya Ilyich Mechnikov and Paul Ehrlich, jointly received the Nobel Prize in Physiology/Medicine (1908) in recognition of their work on immunity.

After reading this chapter, students should be able to understand and appreciate the following:

- Haematopoietic stem cells as origin of immune cells
- Structure and function of T cells
- Structure and function of B cells
- Production of memory cells, effector cells, during immune response
- Primary lymphoid organs
- Secondary lymphoid organs
- Structure and functions of macrophages, natural killer cells, and granulocytes
- Role of complement system in phagocytosis
- Cytokines produced by immune cells

The immune system is a collection of organs and cells which elicit defence in response to foreign antigens invading our body. The defence is called immunity or immune response. The immune system consists of primary and secondary lymphoid organs. This chapter first discusses immunity and then proceeds to explain the types, properties of immune cells and their mechanism of action in eliciting immunity.

3.1 IMMUNITY

Immunity is the term used to define the defensive reaction of a body to a foreign antigen. The antigen may be a microorganism or molecule such as a protein or a polysaccharide. Depending on the nature of immune response, there are two types of immunity: natural immunity (innate immunity) and acquired immunity (adaptive immunity/specific immunity). Innate immunity is the natural or inborn response that is elicited within the body when an antigen enters it. Phagocytes [macrophages and natural killer cells (NKCs)], epithelial surface, and complement proteins contribute to innate immunity. Innate immunity helps in protecting the gastrointestinal tract, urinogenital tract, respiratory tract, and skin. It does not require a prior exposure to a specific antigen. On the other hand, adaptive immunity is the response that is elicited after the second exposure to an antigen. In other words, adaptive immunity to an antigen requires prior exposure to that antigen. Effector cells, such as macrophages, neutrophils, and NKCs, contribute to adaptive

TABLE 3.1 Types of immunity

Type	Subtype	Contributor
Innate immunity	-	Epithelial barrier
		Phagocytes
Acquired immunity	Passive	Placental transfer
		Colostrum transfer
	Active	Vaccines
		Antibodies
		T cells

immunity. Acquired immunity is further divided into two types: passive immunity and active immunity. Acquired passive immunity occurs in foetus due to the entry of maternal antibodies via placenta or colostrum. Acquired active immunity occurs through vaccination, production of antibodies (antibody-mediated immunity or humoral immunity), or the action of T cells (cell-mediated immunity). The different types of immunity are summarized in Table 3.1.

3.2 HAEMATOPOIETIC STEM CELL

Haematopoietic stem cell (HSC) is a type of adult stem cell present in the bone marrow. HSCs express specific cell-surface markers, which are proteins. Examples include clusters of differentiation (CD clusters) and major histocompatibility class I (MHC-I) clusters. HSCs are capable of self-renewal and differentiation, depending on signals such as cytokines and growth factors (refer Section 3.11). Development and differentiation of HSCs into immune cells and blood cells are shown in Fig. 3.1. HSCs can differentiate into two types of stem cells or progenitor cells. They are lymphoid stem cells (or lymphoid progenitors) and myeloid stem cells (or myeloid progenitors). Lymphoid stem cells develop into small lymphocytes (B lymphocytes and T lymphocytes) and large granular lymphocytes (natural killer cells or NKCs). Lymphoid lineage forms the major immune cell population. On the other hand, myeloid stem cells develop into blood cells—red blood cells, white blood cells, platelets, monocytes, and mast cells. This chapter focuses on the lymphoid lineage; myeloid lineage is discussed in Chapter 5. B lymphocytes form and get matured with in the bone marrow in association with bone marrow stromal cells. T lymphocytes, on the other hand, develop in the thymus after the migration of the progenitor cells to the thymus.

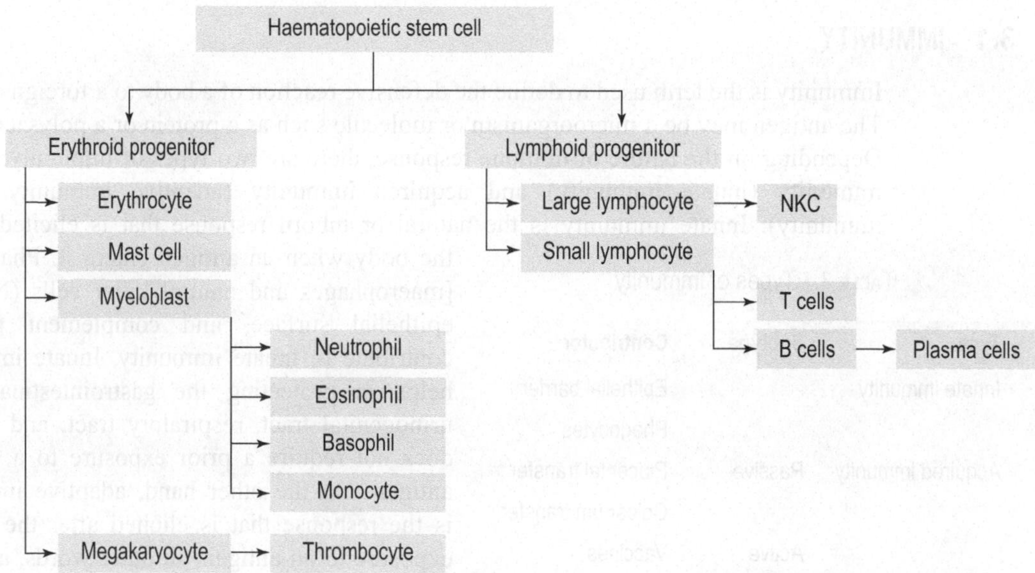

FIGURE 3.1 Development of haematopoietic stem cells into immune cells and blood cells

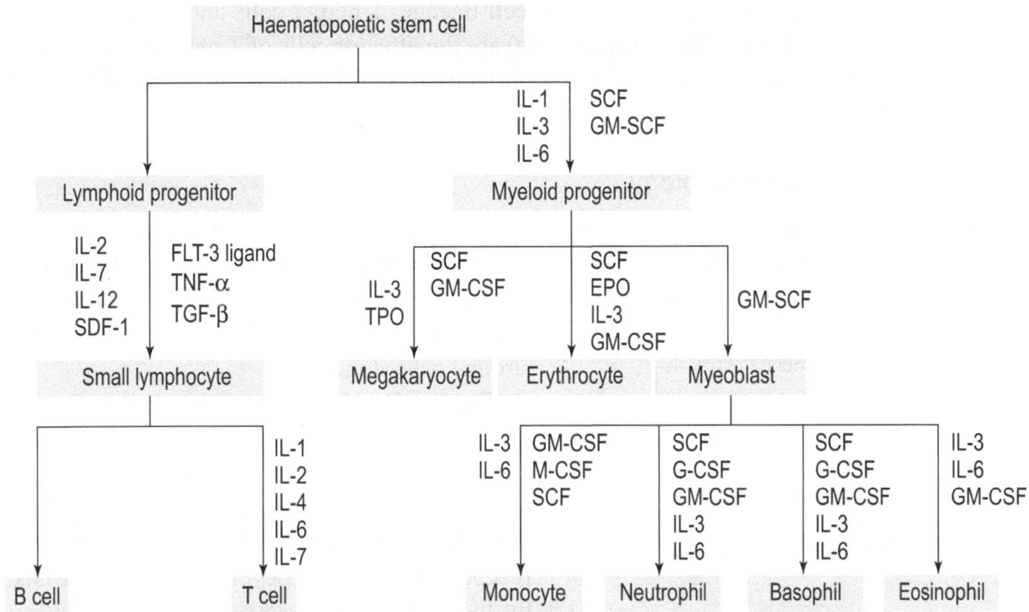

FIGURE 3.2 Role of cytokines in the differentiation of haematopoietic stem cells

Differentiation of HSCs into their lineage is regulated by cytokines and intracellular signalling molecules (discussed in detail in Section 3.11 and Chapter 14). For example, interleukin-7 (IL-7) is required for T-cell and B-cell formation. Deficiency of IL-7 results in severe combined immunodeficiency syndrome (SCID). The receptor for IL-7 (IL-7R) has an γ-chain domain, while a cytokine receptor has a γ-chain (γc) domain. Mutation in the γ chain and γc gene impairs the production of T cells and NKCs. The role of cytokines in the growth and differentiation of HSCs is illustrated in Fig. 3.2. These cytokines are discussed in detail in Section 3.11.

As HSCs are the precursors of immune cells, including blood cells, they are very important in the development of immune system and renewal of blood cells in vivo. In this chapter, the role of HSC as a source for immune cells is described. Other aspects of HSCs, including their cell-surface markers, cytokine stimulants, hallmarks, and applications are discussed in Chapter 14.

3.3 LYMPHOCYTES

Lymphocytes are a category of white blood cells, the other category being granulocytes. They are divided into two types: small lymphocytes and large granular lymphocytes. Small lymphocytes include B cells and T cells, whereas large granular lymphocytes include NK cells. All three types of lymphocytes (B cells, T cells, and NK cells) are non-phagocytic cells. All of them are motile so they can migrate freely in the lymph and blood. As long as there is no antigen, the lymphocytes remain tiny, short-lived, and immature cells (and are known as naive cells). Antigen binding activates the lymphocytes to undergo proliferation in the presence of cytokines to form large cells called lymphoblasts. Lymphoblasts differentiate into effector cells or memory cells. Plasma cells (refer Sections 3.3.1 and

3.4) are the effector cells of B-cell lineage. T helper cells and cytotoxic T lymphocytes (T_H cells and CTLs, respectively) are the effector cells of T-cell lineage. During different stages of development and proliferation, cells express different types of membrane molecules, usually protein clusters called clusters of differentiation (CD).

3.3.1 B-cell Biology

Lymphoid progenitors (lymphoid stem cells) in the bone marrow mature into B cells (B stands for bone marrow in mammals, and bursa of Fabricius in birds). Maturation of progenitor B cell (pro-B cell) into a mature B cell, and then into plasma cell (effector cell), initiates in the presence of IL-7 derived from bone marrow stromal cells (Fig. 3.3).

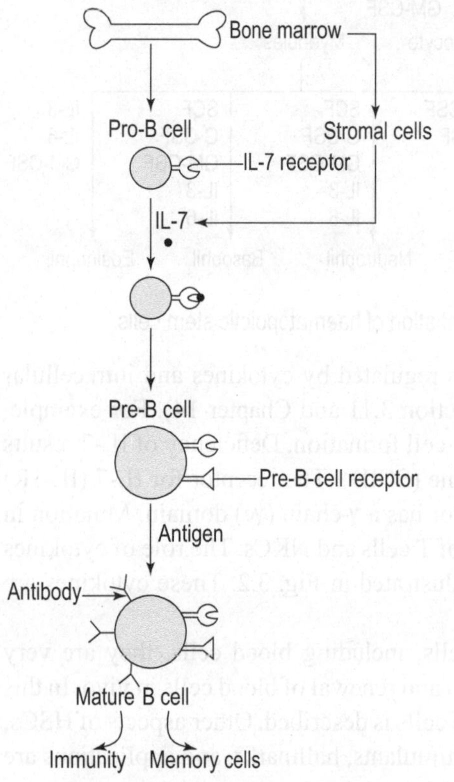

FIGURE 3.3 Maturation of B cell

The pro-B cell possesses interleukin-7 receptors (IL-7R). When the pro-B cell comes in contact with the stromal cells, IL-7 is released, which subsequently gets bound to its receptor in the pro-B cell. As a result, the pro-B cell is converted to pre-B cell which expresses pre-B cell receptor. Pre-B cells develop into B cells characterized by the surface expression of immunoglobulin receptors (IgM or IgD receptors): Igα and Igβ. These three surface molecules are together called B-cell antigen–receptor complex (*BCARC*). BCARC is the site for antigen binding to B cell. The intracellular domain of the Igα and Igβ bears a tyrosine-rich domain called immunoreceptor tyrosine-based activation motif (ITAM). When a B cell recognizes an antigen through BCARC, the ITAM is activated. As a result, ITAM is phosphorylated at its tyrosine residues, and the signal is relayed intracellularly and ends up in the expression of the genes required for the further differentiation of B cell. B cell differentiates into two types of cells—memory B cells and effector B cells (plasma cells). Memory B cells have a long life span and they retain their surface antibodies. In contrast, effector B cells have a short life span of just a few days, and do not retain their surface antibodies. However, effector cells are rich in cytoplasm, and subcellular organelles of protein synthesis, namely, RER and Golgi vesicles. Therefore, they synthesize and secrete a large number of antibodies at a rate of 2,000 Abs/s. These cells are called plasma cells, and the production of antibody as an immune response is called humoral immune response.

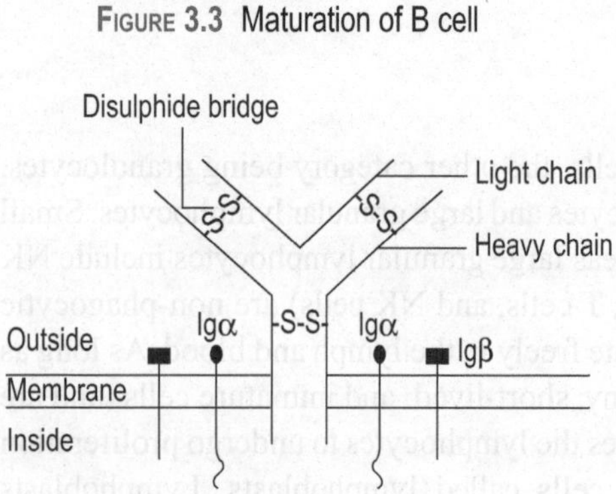

FIGURE 3.4 Structure of a B-cell receptor

B-cell receptor

B-cell receptors (BCRs) are transmembrane protein complex composed of a pair of Igα/Igβ heterodimer separated by an IgM molecule (Fig. 3.4). Each Igα/Igβ

TABLE 3.2 Clusters of differentiation (CD) expressed in B cells

CD type	Functions
CD5	Unknown
CD11a/CD18	Cell adhesion Binding site for ICAM-1 and ICAM-2 (refer Chapter 10)
CD21 (CD2)	Receptor for complement C3d
CD32	Receptor for Fc domain of IgG
CD40	Signal transduction
CD45	Signal transduction
CD54	Cell adhesion

dimer is linked by disulphide bridges and located on either side of the IgG. The cytoplasmic domain of Igα and Igβ is made of a tail of 61 and 44 amino acid residues, respectively. The tails are responsible for signal transduction and activation of B cells.

CD45 cluster is the first marker of B-cell lineage, expressed during the maturation of pre-B cells. CD45 functions in signal transduction and is retained at the surface throughout the life span of B cells. Other types of B cell-specific CD molecules and their functions are listed in Table 3.2.

Coreceptors of B cells

Coreceptors are surface proteins which function along with the BCRs and regulate the B-cell function and signalling. The coreceptors of B cells are the following:

- CD19
- FCγRIIB
- CD22
- CD40–CD40L (CD154)
- CD72
- CD134L
- BMCA and TACI
- CD27
- CD30–CD153

CD19 receptor is a glycoprotein with a molecular weight of 95 kD. CD19, in ligation with an antigen receptor, mediates the signals associated with phosphatidylinositol-3-kinase, and Vav protein (a family of guanine nucleotide exchange factors involved in cell signalling). CD19, in association with CD21, mediates T cell-dependent antigen responses.

FCγRIIB is a coreceptor which inhibits B-cell proliferation mediated by antigen–receptor interaction. The cytoplasmic domain of the FCγRIIB bears a 13-residue motif called phosphorylated immunoreceptor tyrosine inhibitory motif (pITIM). Phosphotyrosine phosphatase (PTP) with Src homology domain (SHP-1) is involved in the negative signalling of pITIM. pITIM binds to and activates the SHP-1 at its carboxy terminal, leading to inhibition of B-cell proliferation.

CD22 is another coreceptor population of B cell. It belongs to sialo-adhesion family of protein, that is, it binds to sialic acid molecule via its Ig-like domain. It is the coreceptor for mIg. Cross-linking of CD22 results in cross-phosphorylation of its tyrosine residues. Phosphorylated CD22 in turn recruits signalling molecules. CD22 inhibits the activity of BCR. Thus, it functions as a regulatory molecule and prevents the overactivation of the immune system.

CD40–CD40L (CD154) is a protein of the receptor–ligand family. Receptor–ligand family is a protein family composed of a receptor–ligand pair. Other members of this family include the following pairs:

- BMCA (B-cell maturation antigen) and TACI (transmembrane activator and CAML-interactor) CD27–CD20
- CD30–CD30L
- Fas (CD95)–FasL

CD40 ligation activates protein tyrosine kinases and mediates phosphorylation of tyrosine residues of various proteins, including phosphatidylinositol-3-kinase and phospholipase C-γ2 (PLC-γ2). Phosphorylated enzymes, in turn, hydrolyze the phosphatidylinositol-4, 5-biphosphate and release diacylalycerol (DAG) and inositol-1, 4, 5-triphosphate (IP3). In addition, CD40 ligation also activates serine/threonine protein kinases, including stress-activated protein kinases (SAPKs) and mitogen-activated protein kinases, (MAPKs). All the activation mediated by CD40 ligation relays the signal and finally activates several transcription factors, including nuclear factor kappa B (NFκB), NFκB-like molecules, and nuclear factor of activated T cells (NFAT). CD40 signals the B cells to function as effective antigen-presenting cells for T cells and activator of T cells. This process is mediated by the upregulation of surface molecules involved in the antigen presentation. CD40 is found to enhance the proliferation, antibody synthesis, and cytokine production in B cells.

CD72 receptor is a member of semaphonin family (proteins involved in neural growth). CD72, on ligation with CD100, enhances the CD40-mediated B-cell activation and antigen presentation. It is also involved in inflammatory response.

CD134L promotes proliferation and differentiation in activated B cells.

BMCA (B-cell maturation antigen) and TACI (transmembrane activator and CAML-interactor) are receptors of tumour necrosis factor receptor (TNFR) family. These receptors recognize the TNF-related proteins such as APRIL (a proliferation-inducing ligand) and TALL-1 (TNF- and ApoL-related leucocyte expressed ligand-1). Both of these ligands can induce B-cell growth and differentiation. APRIL–TALL-I and BCMA–TACI function as a two ligands–two receptors pathway for stimulating B- and T-cell functions.

CD27 receptor is a member of the TNFR family. It is expressed in B cells and is a marker for memory cells. The ligand for CD27 is CD70, which is secreted by activated T cells. CD27–CD70 ligation results in signals that activate the transcription factor NFκB, leading to terminal differentiation of B cells into antibodies producing plasma cells.

CD30–CD153 pair is a negative regulator of B-cell immune response. The signal mediated by this receptor is not yct fully understood.

3.3.2 T-cell Biology

Lymphoid progenitor cells mature in the thymus to form T cells (T for thymus). T cells are characterized by T-cell receptor (TCR) on their surface. TCR is specific for the antigens presented by a specific protein complex called major histocompatibility complex (MHC) or MHC molecules, which are expressed on the surface of the antigen-presenting cells (APC), viral-infected cells, or cancer cells (Fig. 3.5). T cells express various cell-surface molecules such as Thy-1 (CD90), CD3, CD28, and CD45. Thy-1 is the marker for the T-cell lineage. It develops during the maturation of T cells and is retained by them throughout their life span. 'Thy' stands for thymus, where it was originally discovered. CD3 is located along with TCR and helps in mediating signals.

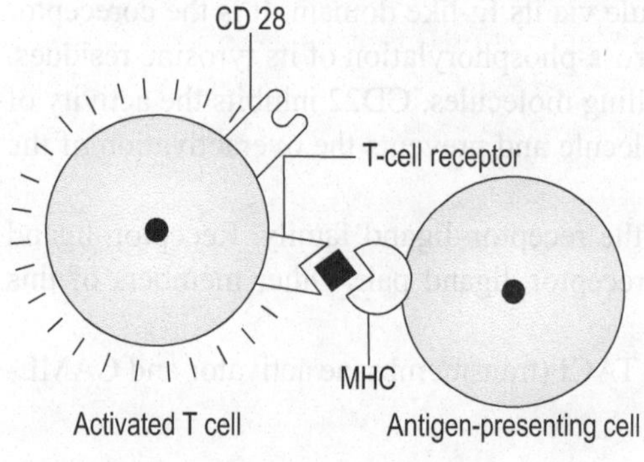

FIGURE 3.5 T-cell receptor and its ligand MHC

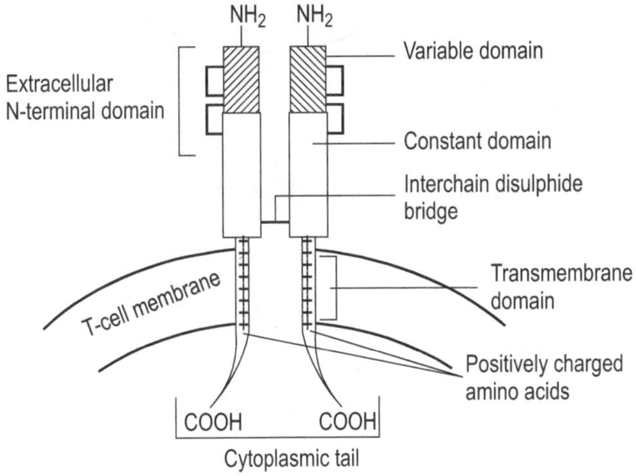

FIGURE 3.6 Structure of a T-cell receptor

CD28 is the target for the B7 molecules of the APCs. It helps in recognition of self and non-self antigens. CD45 is used in signal transduction.

T-cell receptors

T-cell receptors (TCR) are proteins of Ig superfamily having Ig-like domains. They are expressed on the T-cell membrane in association with CD3-signal-transducing complex. They are specific for free antigens as well as MHC-bound antigens. The structure of a TCR is shown in Fig. 3.6. Each TCR is a heterodimer made of either α and β chains or γ and δ chains. Each chain is about 65–70 amino acids long, having an extracellular domain, transmembrane domain, and an intracellular domain. The amino terminal end projects outside the cell as the extracellular domain, while the carboxyl terminal is projected at the intracellular region, as the intracellular domain. Each chain has several intrachain disulphide bonds, and both the chains are linked near the transmembrane domain by a single interchain disulphide bond. The sequence of the N-terminus is variable with three hypervariable regions, whereas the C-terminal domain is conserved. These domains are analogous to the VC domain of Ig. The transmembrane domain is rich in positively charged amino acid sequences, which facilitate the chains to associate with the CD3-signal-transducing complex.

MHC complex and human leucocyte antigens

MHC is the surface complex present on the T cells in mice. The MHC-equivalent in humans is called human leucocyte antigen (HLA). HLA is structurally and functionally analogous to mice MHC. The structure and functions of MHC are studied in mice which are equally applicable to HLA.

There are two types of MHC molecules: class I MHC (MHC-I) and class II MHC (MHC-II). MHC-I is expressed in all nucleated cells and recognized by T_C cells. MHC-II is expressed by antigen-presenting cells, which include dendritic cells, macrophages, and B cells, and are recognized by T_H cells. Structure of MHCs and their interaction with the corresponding T-cell receptors are illustrated in Fig. 3.7.

MHC-I is a heterodimer. One of the monomer is an α chain with three domains: α_1, α_2, and α_3. MHC gene locus encodes the α-chain monomer. The other monomer is β_2 microglobulin, which is not encoded by MHC gene locus. The α_1 and α_2 domains are organized to form a cleft (antigen-binding cleft). The α_3 domain is the binding site for the Ig-like domain of the CD8 receptors expressed on the surface of T_C cells. APCs recognize the antigens, internalize them by phagocytosis, cleave them into antigenic peptide fragments, and bind them to the MHC complex. The processed antigen is exposed on the cleft of the MHC. The T_C cell receptor recognizes the antigen exposed on the MHC cleft, while the CD8 receptor functions as a coreceptor, and activates the function of T_C cell.

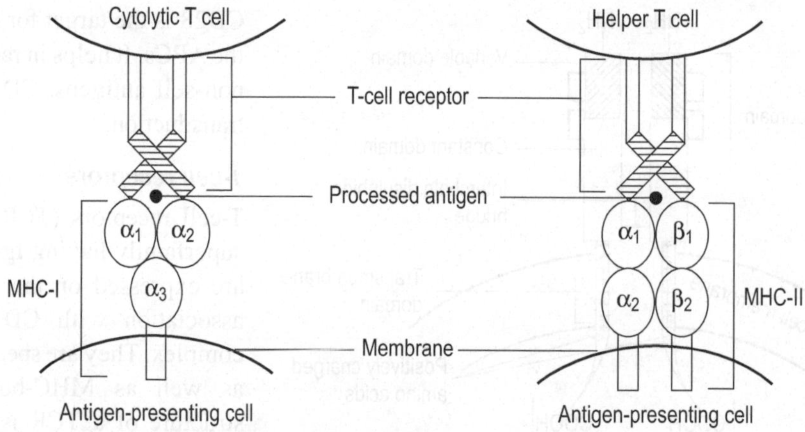

FIGURE 3.7 Structure of MHCs and their interaction with T-cell receptors

MHC-II is also heterodimer. One of the monomer is an α chain with two domains—α_1 and α_2—while the other monomer is a β chain with β_1 and β_2 domains. MHC gene locus encodes both the chains. The α_1 and β_1 domains form a cleft (antigen-binding cleft). The α_2 and β_2 domains are attached to the cell membrane. β_2 domain is the binding site for the Ig-like domain of the CD4 receptors expressed on the surface of T_H cells. The processed antigen is exposed on the cleft of the MHC-II. The T_H cell receptor recognizes the antigen exposed on the MHC cleft, while the CD4 receptor functions as a co-receptor and activates the function of T_H cell. Thus, the immune response produced by the T cells is called cell-mediated immunity.

Maturation of T cells

Naive T cells from the bone marrow reach the thymus and infiltrate the subcapsular cortical region of the thymus, where they still do not express any surface molecules (CD cluster or TCR). This population is called *double-negative cells*. The cells then undergo maturation by rearrangement of genes encoding the TCR and by the expression of CD cluster coreceptors. Double-negative cells interact with the thymic epithelial cells and undergo proliferation, differentiation, and gene rearrangement resulting in the expression of TCR, CD4, and CD8 clusters on the same cell, giving rise to double-positive cells. Double-positive cells are present in the cortex of the thymus, and they may or may not recognize the self-MHC. Cells that can recognize self-MHC undergo maturation and express the TCR and any one of the CD clusters (either CD 4 or CD8) forming T_H cell and T_C cell. These cells are called single-positive cells. Double-positive cells that are unable to recognize self-MHC do not mature and get discarded.

T_H1 and T_H2 responses

T_H cells express CD4 glycoprotein as their markers, and hence, are called CD4+ cells. T_H cells bind to the MHC-II-bound antigen presented by APCs. This results in the activation and proliferation of T_H cells by mitosis to produce numerous daughter cells of the same antigen specificity called effector cells. These cells become T memory cells.

In case of infection with intracellular pathogens, for example, virus, bacteria, or single-celled parasites, the effector cells produce interferon-γ (IFNγ), which in turn activates macrophages to produce immune response. This process is called T_H1 response. Therefore, we observe that T_H1 response involves macrophages.

In case of infection with helminthic intestinal parasites, the effector cells produce cytokines, such as interleukin-3 (IL-3) and interleukin-4 (IL-4), which in turn activate plasma cells to produce IgE, thereby inducing the responses of mast cells, basophils, and eosinophils. This process is called T_H2 response. Therefore, we observe that T_H2 response involves B cells.

TC expresses CD8 clusters as their markers, hence, called $CD8^+$ cells. T_C cells bind to the MHC-I-bound antigen presented by the cells that are altered due to transformation or viral infection. This leads to activation of T_C into effector cells called cytotoxic T lymphocytes (CTLs).

3.4 FORMATION OF MEMORY CELLS AND PLASMA CELLS DURING T_H CELL-MEDIATED IMMUNE RESPONSE

The mechanism of immune response elicited by T_H cells is illustrated in Fig. 3.8. During this response, memory cells and plasma cells [antibody-producing B cells (refer Section 3.3.1)] are produced. The immunity occurs via the following five steps:

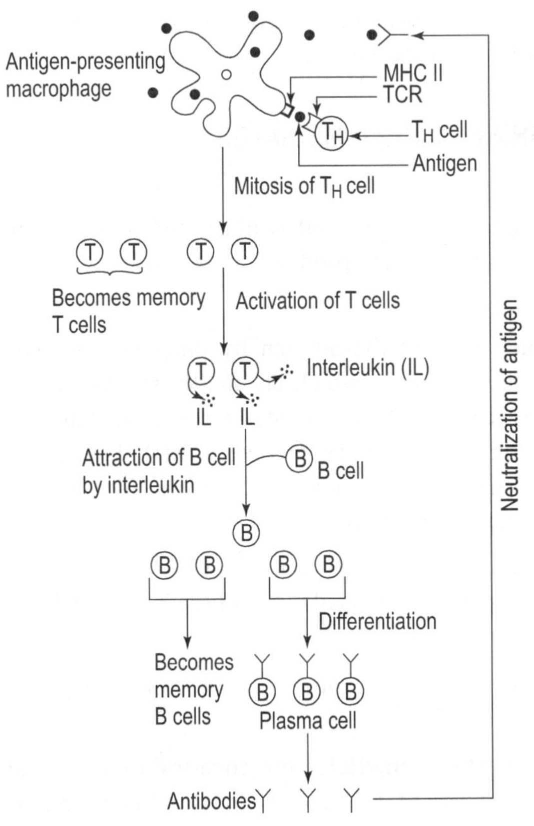

FIGURE 3.8 Mechanism of action of T_H cells

Step 1 Macrophage, being an APC, phagocytoses the antigen and expresses it on the cleft of the MHC-II complex on its surface. The macrophage then presents the antigen to the T_H cells (cell-mediated immunity). T_H cells are then activated.

Step 2 Activated T_H cells undergo mitosis and generate more T_H cells at the site of antigen presentation. Some of these cells become *memory T cells*. Simultaneously, active T_H cells secrete cytokines (interleukins).

Step 3 Interleukins function as chemo-attractants to attract and activate B cells at the site of antigen presentation.

Step 4 Activated B cells undergo mitosis and differentiation. Mitosis results in a large number of B cells, which become *memory B cells*. Differentiation results in the conversion of B cells to plasma cells.

Step 5 B cells have antibodies as insoluble receptors expressed on their surface which is specific for the antigens presented to the T cells. Plasma cells secrete immunoglobulins (soluble form of the receptor) at a frequency of several thousand per

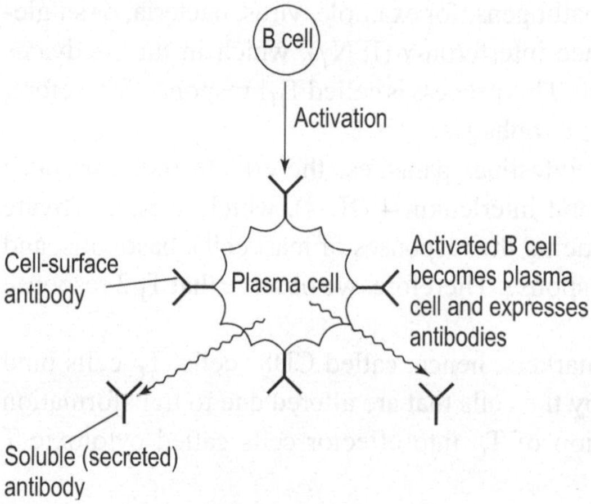

FIGURE 3.9 B cell and plasma cell

second. Immunoglobulins diffuse to the antigen site and bind to and neutralize the antigen (humoral immunity). Plasma cells with antibody receptor also migrate to the antigen site for neutralization.

Memory cells

In steps 2 and 4, memory T cells and memory B cells are produced, respectively. Memory cells can survive in their inactive form and circulate for many years. These cells retain the capacity to recognize and kill the same antigen with higher magnitude if the antigen enters the body for the second time in future. This is called secondary immune response. The secondary responses elicited by memory cells are rapid and of higher magnitude.

Plasma cells

In step 4, plasma cells are formed. Actually, plasma cells are activated B cells. On encountering an antigen, B cells get activated to plasma cells and secrete immunoglobulins specific to the antigen (Fig. 3.9). Plasma cells are active for 4–5 days. In humans, five types of immunoglobulins are produced by plasma cells designated as IgG, IgA, IgM, IgD, and IgE. Immunoglobulins can diffuse freely to reach the antigen site for neutralization.

3.5 FORMATION OF CYTOLYTIC T CELLS DURING T_C-CELL-MEDIATED IMMUNE RESPONSE

The mechanism of immune response elicited by the T_C cell is illustrated in Fig. 3.10. During this response, memory cells and plasma cells are produced. The immunity occurs via the following five steps:

Step 1 As macrophages are APCs, they process the antigen by phagocytosis. The processed antigen is then expressed on their surfaces which are finally recognized by the T_H cells (cell-mediated immunity). T_C cells are then activated by two mechanisms: paracrine signalling and autocrine signalling. APCs produce interleukin-1 (IL-1) which in turn activates T_C cells (paracrine mechanism). APCs also secrete a cytokine, interleukin-2 (IL-2), which activate themselves (autocrine mechanism).

Step 2 Activated T_C cells undergo mitosis and generate more T_C cells at the site of antigen presentation. The activated T_C cell population is called cytolytic T cells (CLTs), as they lyse the infected cells in the subsequent stage.

Step 3 CLTs destruct the target cells by one or more of the following three mechanisms.

Mechanism 1 Target cell destruction by perforin-mediated membrane damage The CLTs secrete a protein called perforin. Granules of $CD8^+$ T cells contain this protein. It is a cytolytic protein, which is released by degranulation due to the activation of T cells. Through its special domain called 'membrane attack complex domain', it induces pores in the target cell membrane. This results in an imbalance in the cell membrane permeability and transport mechanism, allowing water and salts to enter into the cells inducing cell burst and lysis. However, there is an equal chance for the degradation of the membrane of CLTs too. This may lead to self-degradation. To prevent this, CLTs secrete yet another protein called protectin. Protectin is a membrane-bound glycoprotein. It is also called CD59 protein or MIRL (membrane inhibitor of reactive lysis). It coats the surface of CLTs and prevents the action of perforin.

Mechanism 2 Target cell destruction by Fas:Fas ligand-mediated apoptosis Fas ligand is a transmembrane protein of TNF family. On recognizing an antigen, CLTs express Fas ligand on their surface. Fas ligand recognizes the Fas receptors expressed on the surface of the APCs. Fas receptors are death receptors that induce apoptosis. Fas:Fas ligand interaction activates procaspases in the APCs, which lead to signal activation that results in apoptosis.

Mechanism 3 Target-cell destruction by CAM-mediated adhesion The CLTs adhere to the target cell (the APC) via integrin and cell adhesion molecule (refer Chapter 10), and induce a signal to destroy the cell.

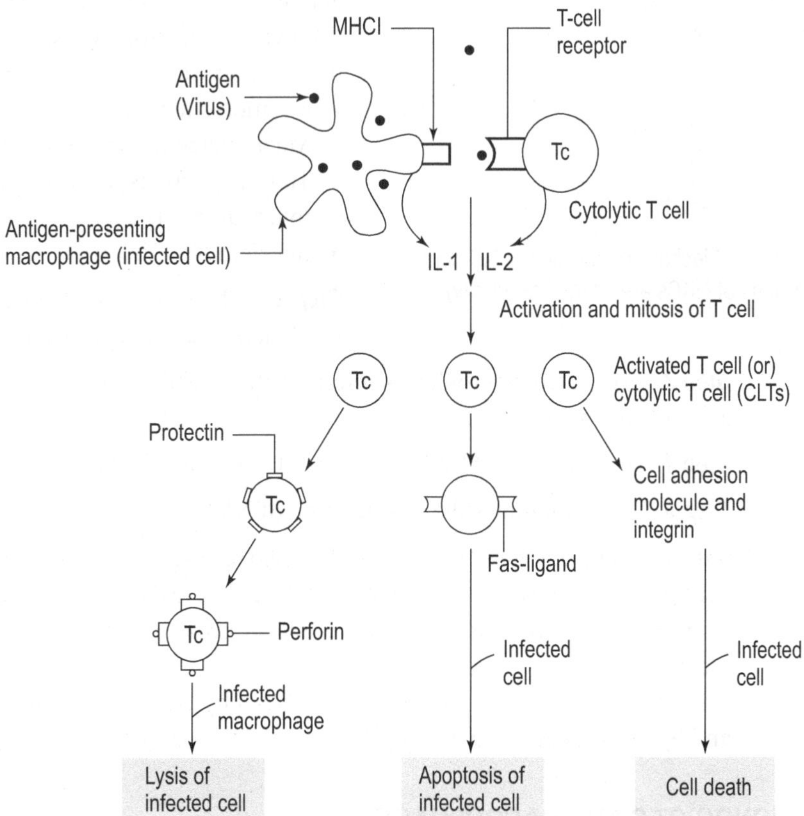

FIGURE 3.10 Mechanism of action of T_C cells

3.6 NATURAL KILLER CELLS

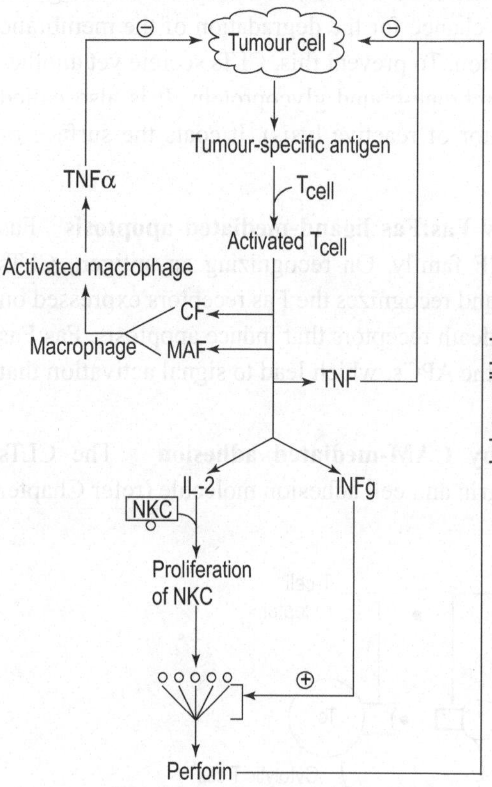

FIGURE 3.11 Mechanism of action of NK cells (actions of NKCs are shaded for clarity)

Natural killer cells (NKCs) are large, granulated, naturally occurring lymphocytes, capable of killing tumour cells by cytotoxicity. They do not express receptors of T- or B-cell lineage, hence, they come under the category of *null cells*. However, the cytotoxic action of NKCs is mediated by T cells. In fact, T cell initiates the mechanism on encountering a tumour cell, while NKCs link in the subsequent steps. The T cell-dependent cytotoxicity induced by NKCs is shown in Fig. 3.11. The granules of NKCs have perforin and granzyme, which help in killing the target cells.

The cytotoxic mechanism has the following steps:

Step 1 Tumour cells express specific cell-surface antigens. The antigens activate T cells to produce the following five cytokines, which mediate cytotoxicity by different ways:

- Chemotactic factor (CF)
- Macrophage activating factor (MAF)
- Tumour necrosis factor (TNF)
- Interleukin-2 (IL-2)
- Interferon-γ (IFN γ)

Step 2 CF attracts macrophages to the tumour site. MAF activates the macrophages.

Step 3 Activated macrophages release TNF-α. TNF-α inhibits tumour cell proliferation and kills the tumour.

Step 4 TNF, produced by T cells, directly destroys the tumour cells.

Step 5 IL-2 stimulates the proliferation of NKCs.

Step 6 IFN γ activates NKCs to produce and release perforin. Perforin induces tumour cell death by degrading the cell membrane as described in the previous section. Downregulation of NKC activity is possible by the action of prostaglandin E_2 released by the activated macrophages.

Thus, it is clear that tumour cells are lysed by NKCs via perforin, by T cells via TNF, and by macrophages via TNF-α.

3.7 MONOCYTES AND MACROPHAGES

Monocytes are tiny, circulating white blood cells (leucocytes) with a single nucleus. They migrate in the circulation and accumulate in tissue spaces and get differentiated

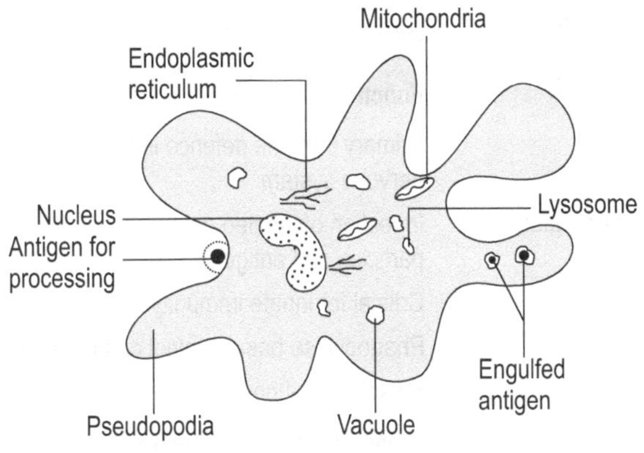

FIGURE 3.12 Structure of a macrophage

into macrophages. During migration and differentiation, following changes take place in monocytes to form macrophages:

- 5- to 10-fold increase in the cell size
- Increase in the number of intracellular organelles
- Increase in the complexity of the cell organelles
- Increase in the phagocytic ability (ability to internalize and process the antigen)
- Increased expression of hydrolytic enzymes in lysosomes
- Increased expression of cytokines

After these changes, monocytes are converted into macrophages. Structure of a typical macrophage is shown in Fig. 3.12. It has an extension called pseudopodia. Some macrophages are motile and exhibit amoeboid movement and are, hence, called wandering macrophages. Some other macrophages permanently reside in the tissues and are called fixed macrophages. Fixed macrophages are specific for each tissue and are called by specific names according to the tissues (Table 3.3). Formation, activation, and functions of macrophages are summarized in Fig. 3.13.

To elicit their function (i.e., to elicit the immune response), the macrophages should be activated. They are activated in the following ways:

- By the bacterial cell wall component (antigenic component)
- By inflammatory mediators such as cytokines
- By immunoglobulin bound to the F_c receptor of the macrophages
- By the C3b complement protein

Activated macrophages perform the following functions, all related to immune response in one way or the other:

- Phagocytosis
- Presentation of antigens for neutralization or cytotoxicity
- Secretion of cytokines to activate other immune cells or to kill tumour cells
- Antibody-dependent cell-mediated cytotoxicity (ADCC)

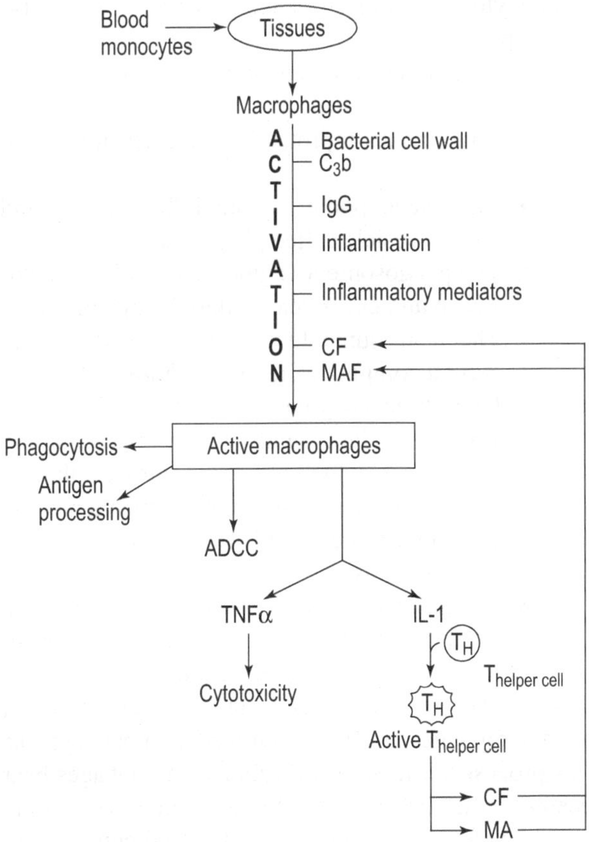

FIGURE 3.13 Formation, activation, and functions of macrophages

TABLE 3.3 Fixed macrophages of specific organs

Organ	Macrophages	Function
Brain	Microglial cells	Primary immune defence in the central nervous system
Lung	Alveolar macrophages or dust cells	Ingestion of inhaled dust particles and antigens
Liver	Kupffer cells	Critical for innate immunity
Kidney	Mesangial cells	Phagocytose basal laminal components
Bone	Osteoclasts	Dissolution of bone cells
Connective tissue	Histocytes	Opsonization and phagocytosis

3.7.1 Phagocytosis

Phagocytosis is a process by which the macrophages ingest, digest, and process antigens. Antigens, such as whole microorganisms, microbial components, cell debris, injured and dead cells, insoluble particles, and activated clotting factors, are reported to be phagocytosed by macrophages. Because of their phagocytic ability, macrophages are also called phagocytes. The process of phagocytosis by a macrophage is shown in Fig. 3.14. Phagocytosis occurs via the following steps:

- Macrophages interact with the antigens or antigen–antibody complex via *chemotaxis*.
- The membrane surface of the macrophage makes a protrusion or extension called *pseudopodium*.
- The pseudopodia surround the antigen and form a vesicle called phagosome.
- The phagosome invaginates towards the cytoplasm and gets released into the cytoplasm.
- The phagosome fuses with the lysosome to form a complex called *phagolysosome*.
- Hydrolytic enzymes stored in the lysosome lyse the antigens into peptide fragments and tiny pieces, which are drawn through the cytoplasm and covered by the plasma membrane to form a vesicle, which is then expelled out by *exocytosis*.
- The antigens are expressed on the cleft of the MHC molecule on the cell surface for further action by T cells (refer Section 3.3.2.)

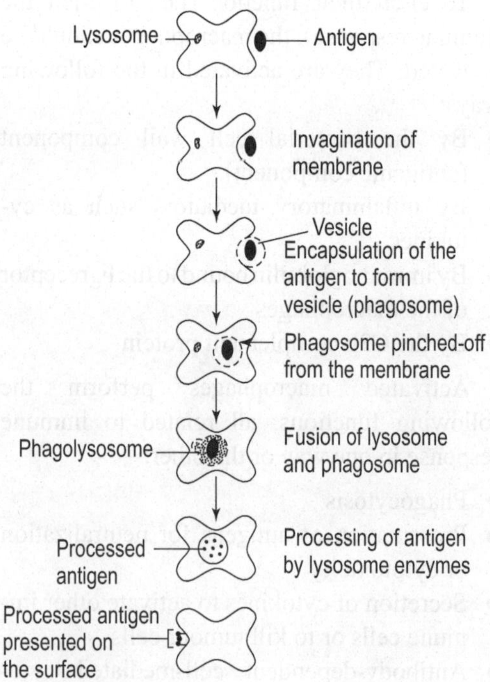

FIGURE 3.14 Phagocytosis by a macrophage

Sometimes, the receptors located on the surface of the macrophages are also involved in phagocytosis. This process is called opsonization. Macrophages bear receptors that are specific for antibodies or certain complement components. Antibodies and complement proteins bound to their antigens bind readily to their respective receptors and get phagocytosed. This process is called opsonization. The antibody and complement components are called opsonins. Details of complement system and opsonization are discussed separately in Section 3.9.

3.7.2 Presentation of Antigens to T cells

Macrophages express the phagocytosed antigens on its MHC cluster and present them to T cells. This process is called antigen presentation. Antigen presentation may be considered as the continuation of phagocytosis. If the antigen is presented to T_H cells, it results in the production of antibodies by B cells. The antibodies neutralize the antigen. If the antigen is presented to T_C cells, it results in the killing of the infected phagocytes (cytoxicity). Both these processes have been explained earlier in Section 3.3.2.

3.7.3 Secretion of Cytokines to Kill Tumour Cells or to Activate Other Immune Cells

Activated macrophages secrete various cytokines to mediate immune response. They secrete TNF-α, which in turn inhibits proliferation of tumour cells and kills them. Details of macrophage activation and macrophage-induced cell killing are given in Sections 3.3.4 to 3.3.6. Macrophages secrete interleukin-1 (IL-1) to mediate the activation of T_H cells. T_H cells in turn produce MAF for activating more macrophages.

3.7.4 Antibody-dependent Cell-mediated Cytotoxicity

Antibody-dependent cell-mediated cytotoxicity (ADCC) is an adaptive immune process in which antibodies are assembled around the target cells to render them susceptible to attack by immune cells. Cytotoxic cells, such as macrophages, NKCs, eosinophils, and neutrophils, are involved in ADCC (Fig. 3.15). These cells express F_c receptors on their surface. These receptors recognize the F_c domain of the antibody bound to a target cells and subsequently cause lysis of the target cell. Majority of ADCC is mediated by NKCs while neutrophils and eosinophils also play a role. ADCC is illustrated in Fig. 3.16.

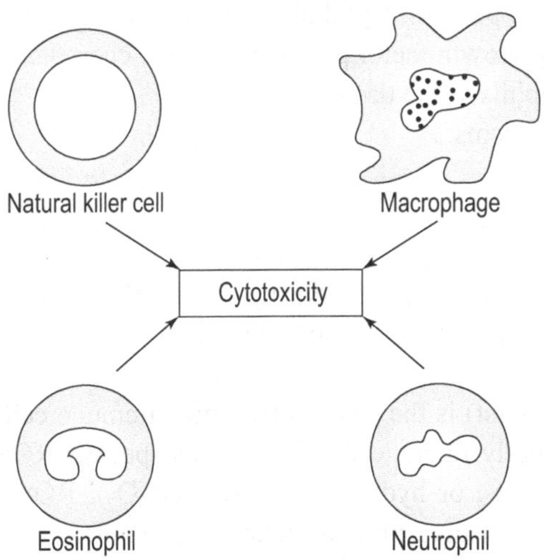

FIGURE 3.15 Cells involved in ADCC

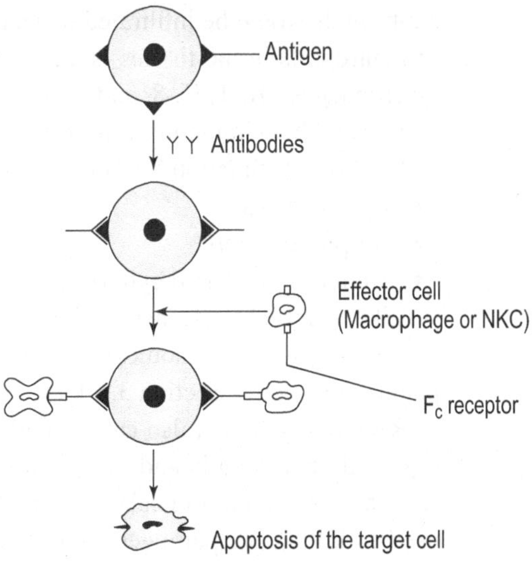

FIGURE 3.16 Antibody-dependent cell-mediated cytotoxicity

3.8 GRANULOCYTES

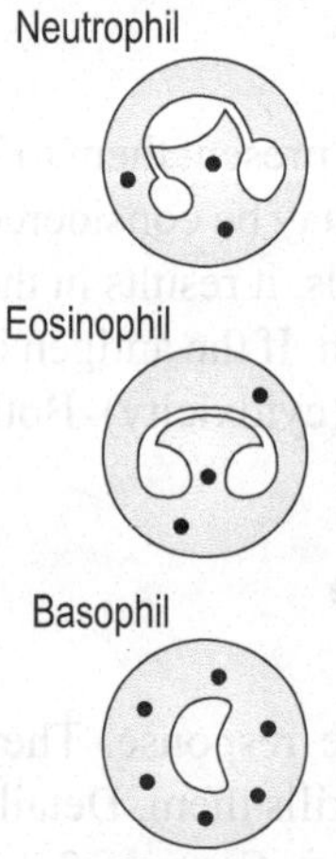

FIGURE 3.17 Types of granulocytes

Granulocytes are a type of leucocytes (white blood cells). They are named so because of the presence of granules in their cytoplasm. They are also called polymorphonuclear leucocytes because of the different morphology and lobed structure of their nucleus. Granulocytes are classified into three types: neutrophils, eosinophils, and basophils. Although all these three types of cells are granulated, they differ in their cytoplasmic staining property and nuclear characteristics. Figure 3.17 shows the types of granulocytes.

3.8.1 Neutrophils

Neutrophils are the most abundant leucocytes. They are formed from the myeloid progenitor-derived myeloblast in the bone marrow and then released into the peripheral blood. They have granulated cytoplasm with a large multilobed nucleus. The cytoplasmic granules stain with both acidic and basic dyes. Two types of granules are distributed in the cytoplasm: primary granules (azurophilic granules) and secondary granules. Azurophilic granules are larger and denser and they store lytic enzymes. Secondary granules have lactoferrin, collagenase, and lysozymes.

Neutrophils circulate in the blood for around 8 h after which they migrate into the tissues and live for around 3 days. During infection, bone marrow synthesizes large numbers of neutrophils, which are released into the blood. Within the blood vessels, neutrophils adhere to the endothelial walls, roll along the wall, escape through the gaps between the adjacent endothelial cell lining, and penetrate through the basement membrane reaching the tissue spaces. The illustrated mechanism of this migration, and the role of cell adhesion molecules in mediating the migration are given in Chapter 10. They penetrate the blood vessels by extravasation to reach the site of inflammation in tissues, always being the first cells to reach the site of inflammation. After reaching the inflammatory tissues, the neutrophils would be infiltrated or accumulated to elicit immune response. Accumulation of neutrophils in the tissues is a chemotactic process. Cytokines produced by activated macrophages, (IL-1, IL-8, and transforming growth factor β) and T cells are chemotactic factors and help in the infiltration of neutrophils in the tissues.

Neutrophils function by following mechanisms:

- Phagocytosis
- Respiratory burst
- Antibody-mediated cytotoxicity

Like macrophages, neutrophils undergo phagocytosis during which the granules are fused with phagosomes (whereas in macrophages, lysosomes are fused with the phagosomes; refer Section 3.7.1).

Respiratory burst (also called oxidative burst) is the process by which immune cells, especially neutrophils and monocytes, abruptly liberate reactive oxygen species (ROS) such as superoxide radicals, hydroxyl radicals, or hydrogen peroxide (H_2O_2). ROS is used in killing the pathogen. NADPH oxidase, present as a membrane complex in the neutrophils, converts oxygen to superoxide anion radical. NADPH can form NAD and proton in the extracellular compartment. The protons enter the membrane proton channel to reach the intracellular compartment and react with the superoxide to form molecular oxygen and H_2O_2. Myeloperoxidase acts upon H_2O_2 and combines it with chloride ion forming hypochlorite. Hypochlorite kills the pathogen. The process of respiratory burst is shown in Fig. 3.18.

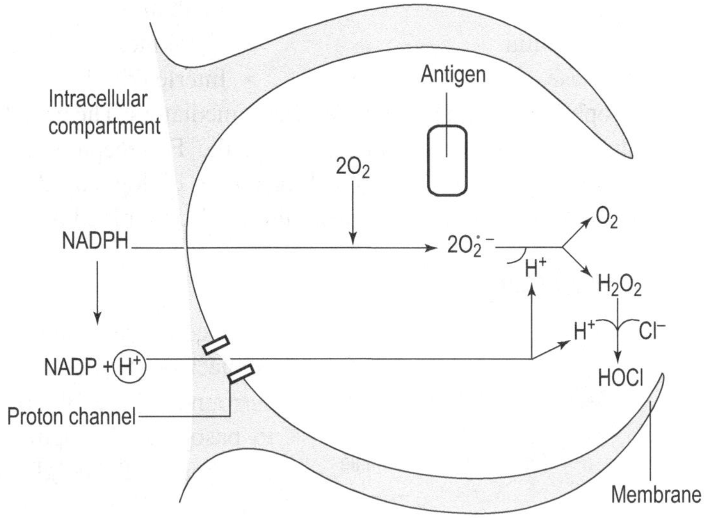

FIGURE 3.18 Respiratory burst and the reactions

Neutrophils express receptors with specificity for the F_c domain of the IgE antibody. They are involved in the antibody-mediated cytotoxicity against the parasitic cells. When the body encounters a parasitic infection, mast cells are sensitized to produce neutrophil chemotactic factor (NCF). In response to NCF, neutrophils induce cytotoxicity.

3.8.2 Eosinophils

Eosinophils are formed from the myeloid progenitor-derived myeloblast in the bone marrow and then released into the peripheral blood. They have granulated cytoplasm with a large multilobed nucleus. The cytoplasmic granules are called eosinophilic granules. They store basic proteins, sulphotransferases, and sulphatases. Like neutrophils, eosinophils undergo extravasation and phagocytosis. They also resemble neutrophils in possessing the receptor for IgE, and responding to chemotactic factor for inducing ADCC. Parasitic antigens are main targets for eosinophils.

When an antigen invades the body, eosinophils undergo degranulation and release eosinophil cationic protein and eosinophil-derived neurotoxin, both of which degrade the RNA virus. Eosinophil peroxidase is also produced by degranulation. It forms ROS and induces the death of the target cells.

Eosinophils perform their immunologic role by releasing growth factors, including TGFβ, VEGF, and PDGF, and cytokines such as IL-1, IL-2, IL-4, IL-5, IL-6, IL-8, IL-13, and TNFα. The functions of these compounds are discussed in Section 3.11 on cytokines.

3.8.3 Basophils

While neutrophils are the most common granulocytes, basophils are the least common ones. As the name indicates, basophils stain with basic dyes. Basophils have bilobed nucleus and cytoplasmic granules containing the following pharmacological mediators:

- Histamine
- Heparin
- Chondroitin
- Elastase
- Lysophospholipase
- Leukotrienes
- Cytokines
- Interleukin-4

Basophils function by secreting these mediators. The cells have F_c receptors. Antigen-sensitized IgE molecules interact with the F_c receptors and stimulate the release of pharmacological mediators, which results in biological effects such as vasodilation and smooth muscle contraction. Basophils are also involved in allergic reactions.

3.8.4 Mast Cells

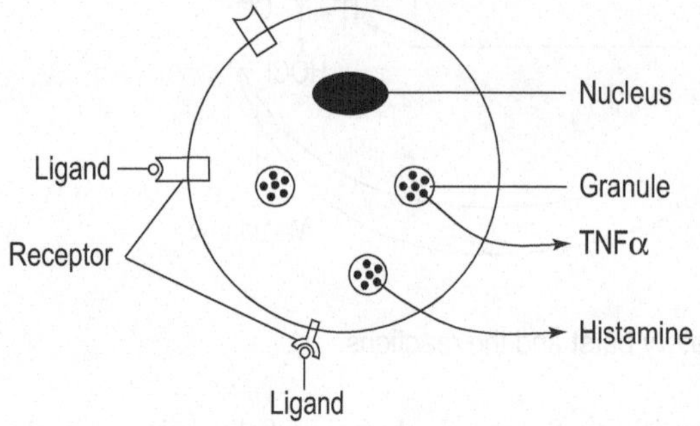

FIGURE 3.19 A mast cell

Mast cells are immune cells (Fig. 3.19) found in connective tissues and mucosa, predominantly surrounding the blood vessels. They are similar to basophils. Basophils express receptors for the F_c domain of the IgE and play a role in allergic reactions. When a system is exposed to an allergen for the first time, B cells are activated to produce IgE antibodies. These antibodies remain *irreversibly* bound to the basophils. On further encounter with the same allergen, the IgE-primed basophils degranulate and secrete mediators, such as histamine and cytokines, which produce allergic reactions (Fig. 3.20).

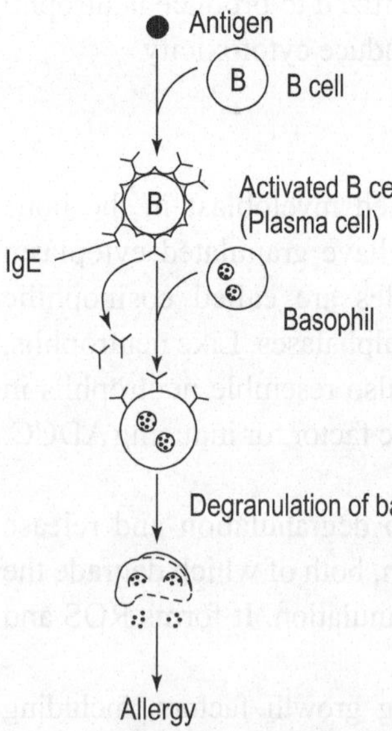

FIGURE 3.20 Role of basophil in allergic reaction

Dendritic cells

Dendritic cells (DCs) are immune cells that differ from other cells discussed so far in that they have projections similar to dendrites of nerve cells. They are found in skin and mucosal lining of nose, lung, stomach, and the intestine. They are enriched with intracellular endosomes, lysosomes, and Birbeck granules of Langerhans cells (LC) of the epidermis for effective antigen processing and antigen presentation. They are unique in that they have large amount of class-II MHC clusters, CD80, CD86, and cell adhesion molecules or CAMs on their surface (refer Chapter 10 for CAMs). These molecules are expressed during the maturation stage of the cells. To perform their functions, DCs should be activated by IFNα, IFNγ, IL-1β, or directly by the virus or bacterial antigens.

Functions The functions of DCs are listed as follows:

- Stimulating the proliferation of allogenic and autologous T cells

- Presentation of antigen to T cells
- Taking up the antigens via macrophage mannose receptor
- Processing of antigen by cross priming
- Delivery of antigens via class-I pathway by utilizing HSP96 (refer Chapter 4 for discussion on HSP96)
- Sensitization of immune system in response to viral infection by activation of CTLs

Types of dendritic cells On the basis of tissue location, DCs are classified as follows:

(i) Langerhans cells (LCs) present in the epidermal layer of the skin

(ii) Interdigitating dendritic cells (IDDCs) located in the secondary lymphoid tissue and the medulla of the thymus

(iii) Interstitial dendritic cells (ISDCs) distributed in many organs, including intestine, kidney, lungs, heart, and liver

(iv) Blood dendritic cells (BDCs) which circulate in the blood

(v) Plasmacytoid dendritic cells (PDCs) found in circulation and peripheral lymphoid organs

(vi) Follicular dendritic cells (FDCs) present in the lymph follicles that are loaded with B cells

3.9 ROLE OF COMPLEMENT SYSTEM IN PHAGOCYTOSIS

Complement system refers to a set of plasma proteins synthesized in the liver. In association with phagocytes and antibodies, they play a role in immunity against extracellular pathogens such as bacteria. They function as opsonins (refer Section 3.7.1) and coat the surface of the pathogen. Receptors for complement components are located on the surface of the phagocytes. Therefore, complement components coated on the surface of the pathogen get bound to the phagocytes (example,macrophages) and elicit immune response. Thus, the main function of the complement system is to enable the phagocytes to recognize the antigen. Alternatively, complement components can also induce direct lysis of the pathogen cell.

3.9.1 Nomenclature

Complement components are named with the first letter 'C' followed by the numerical, say C_1, C_2, …, etc. Most of the complement circulate in inactive form, and are activated during immune response by proteolytic cleavage. As a result of proteolytic cleavage, two fragments will be formed, one small and one large. The smaller fragment is denoted by the lowercase alphabet 'a' and the larger one by 'b'. For example, the complement component C_2 can be cleaved to C_2a and C_2b. The 'a' fragments usually function as chemokines, whereas the 'b' fragments function as opsonins that adhere to the target pathogen.

3.9.2 Complement Cascade

Complement cascade is the mechanism by which the complement components carry out opsonization in a stepwise manner and enable phagocytosis. The crucial component responsible for initiating the classical pathway is 'C_1q'. The structure of C_1q is shown

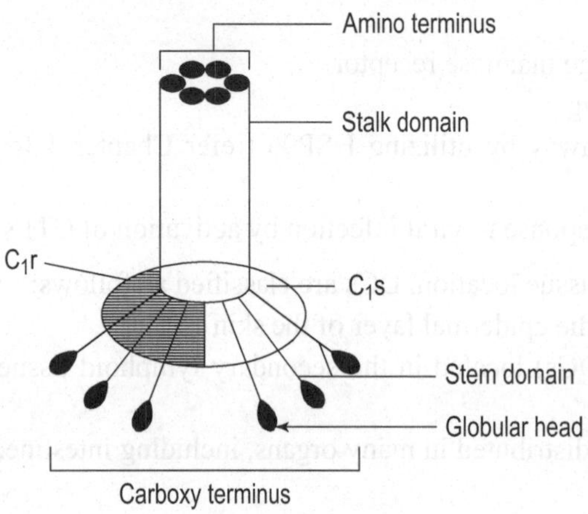

FIGURE 3.21 Structure of C_1q

in Fig. 3.21. It is a protein with molecular weight of approximately 462 kD. It has a hexameric stalk domain which forms the amino terminus. Each stalk domain is extended with a bend to form the stem domain. Each stem domain terminates in a globular domain, which forms the carboxyl terminal. At the junction of the stalk and the stem domain, two components, namely C_1r and C_1s, are located closely adjacent to each other. C_1r is a proenzyme, which activates C_1s.

The steps involved in the complement cascade are shown in Fig. 3.22 and are also described as follows:

(i) When there is an infection by the microbial cell, say bacteria, immunoglobulins (Igs) are raised. Igs coat the surface of the pathogen (the antigen).

(ii) The globular domain of C1q attaches to the Fc domain of the Igs.

(iii) Interaction between the globular domain and the F_c domain activates the C_1r component of C_1q. Activated C_1r in turn activates C_1s component into a serine protease.

(iv) Serine protease cleaves the C_4 component into C_4a and C_4b. C_4b alone attaches to the surface of the pathogen.

(v) It also cleaves the next complement component C_2 to form C_2a and C_2b.

(vi) C_2b binds to the pathogen-bound C_4b, forming C_4b–C_2b complex. This complex is called C_3 convertase.

(vii) C_3 convertase acts upon C_3, cleaving it into C_3a and C_3b. Each C_3 convertase can form about 1,000 C_3b molecules.

(viii) Some molecules of C_3b associate with C_3 convertase to form C_5 convertase. C_5 convertase is literally C_4b–C_2b–C_3b complex. Some molecules of C_3b bind directly to the surface of the pathogen. C_3b is the primary opsonin of the complement system.

(ix) C_5 component is the final complement to complete the opsonization process. C_5 binds to the C_5 convertase at its C_3b component, and gets cleaved into C_5a and C_5b. At this step, *opsonized pathogen* is formed.

(x) The complement component present in the surface of the opsonized pathogen is recognized by the 'complement receptors' present on the surface of the phagocytes.

(xi) The pathogen is processed by the phagocytosis process explained in Section 3.7.1.

(xii) Complement components C_6, C_7, C_8, and C_9 are unique in that they induce direct lysis of certain pathogens.

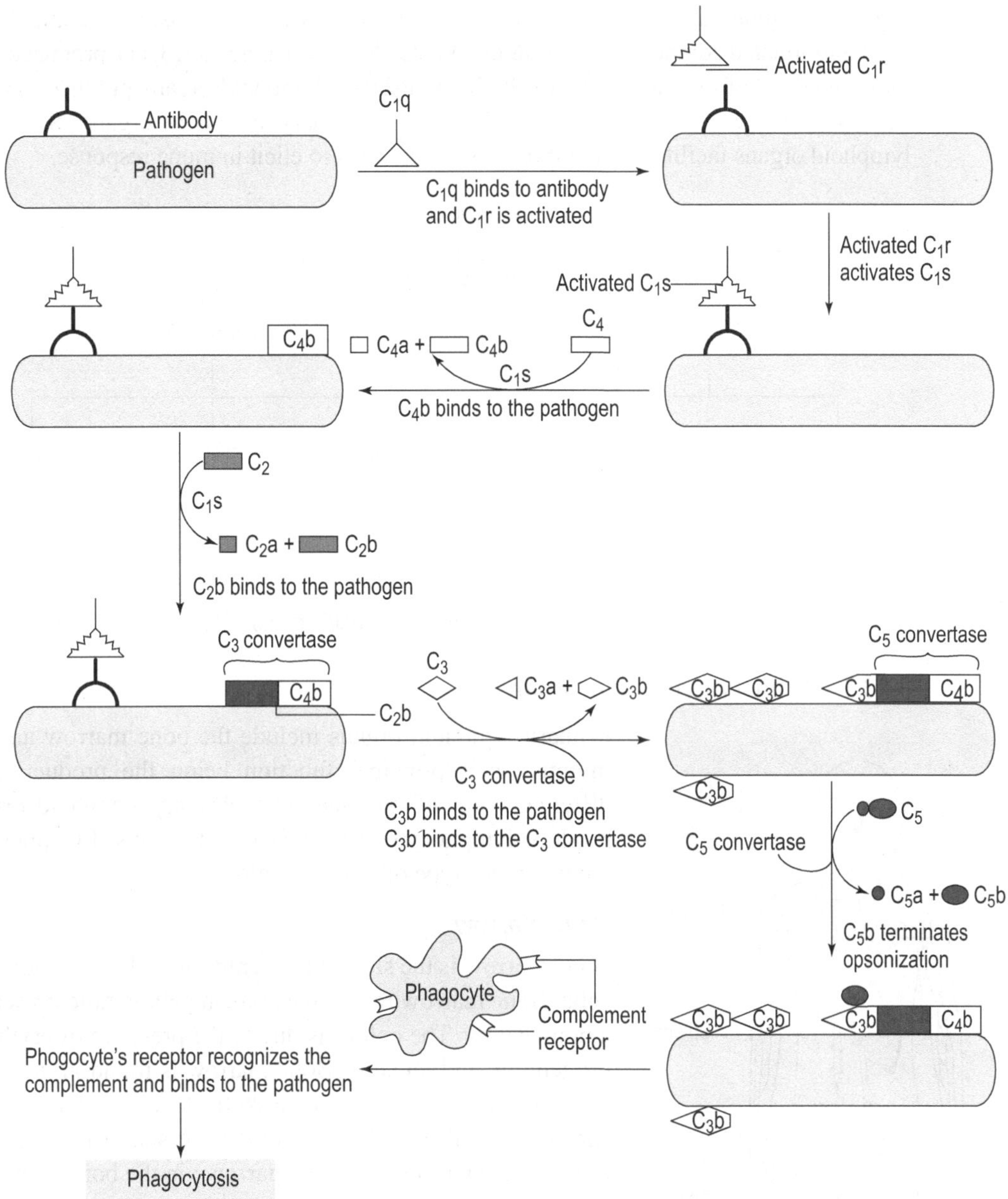

FIGURE 3.22 Function of complement system in enabling phagocytosis

3.10 LYMPHOID ORGANS

So far, the role of different types of cells in various stages of immune response has been discussed. Organs involved in the production, maturation, activation, and function of those immune cells are called lymphoid organs (Fig. 3.23). Basically, two lymphoid organs are present in the immune system—primary and secondary—both enriched with blood vessels and lymphatics. The location of the lymphoid organs is shown in

Fig. 3.24. Lymphoid organs enable supplement of nutrition through transfer of cells and other immunity-related components, such as cytokines (discussed in Section 3.11), produced from these organs to different regions of the body. Primary lymphoid organs produce immune cells. Immune cells migrate via circulation to secondary lymphoid organs. Secondary lymphoid organs facilitate maturation of immune cells to elicit immune response.

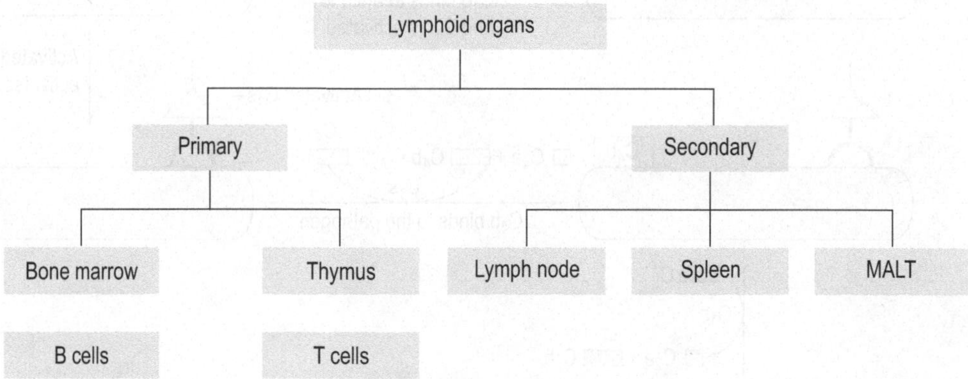

FIGURE 3.23 Types of lymphoid organs

3.10.1 Primary Lymphoid Organs

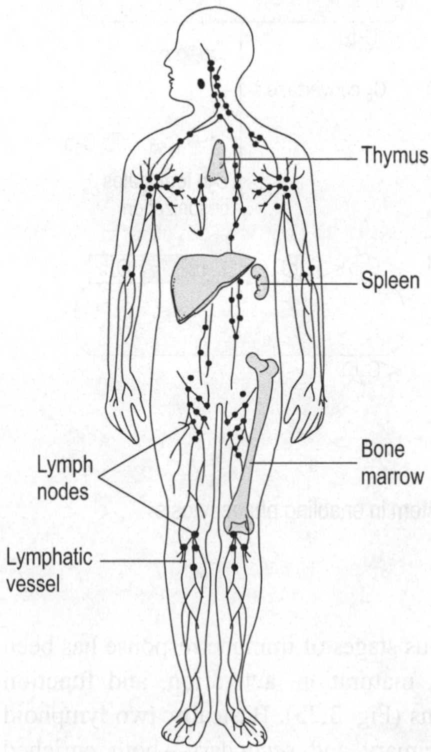

FIGURE 3.24 Location of lymphoid organs

Primary lymphoid organs include the bone marrow and the thymus, their principal function being the production of different types of immune cells. Primary lymphoid organs produce lymphocytes and their effector cells. Lymphocytes are the major type of immune cells.

Bone marrow

Bone marrow is the site for the synthesis and maturation of B cells. Bone marrow refers to a reddish yellow fluid present in a bony cavity. The colour is due to the presence of erythroid progenitors and fat cells. Bone marrow is found in the joints of limb bones. It is also found in skull, clavicles, ribs, sternum, and pelvis. The following blood vessels (illustrated in Fig. 3.25) enter into the bone marrow via the bony shell:

- Central artery running longitudinally along the bone marrow cavity
- Nutrient artery entering the midshaft and joining with the central artery
- Central longitudinal vein running parallel with the central artery
- Venous sinuses (group of tiny vessels running perpendicular to the axis and crossing the cavity), connecting to both the central artery and the central vein. Mature blood cells pass through the venous sinuses.

The bone marrow cavity is partitioned into two compartments, stromal compartment and haematopoietic cell compartment. The stromal compartment is made of several types of cells: endothelial cells, stromal cells, fibroblasts, macrophages, granulocyte progeny, erythroid progeny, adipose cells, osteoblasts, and osteoclasts. Distribution of these cells is shown in Fig. 3.26, while the functions of these cells are listed in Table 3.4. Endothelial cells are interconnected to form a layer lining of the blood vessels. Stromal cells occur as branching network (also called reticular cells) under the endothelial lining. They are mediated by colony-stimulating factors to secrete haematopoietic short-range regulatory molecules. Fibroblasts and adipose cells are distributed in the marrow. Erythroid progeny is found nearer to the epithelial lining. Macrophages appear near the erythroid progeny. Granulocyte progeny exists adjacent to the venous sinusoids. Osteoblasts and osteoclasts are found in the bone.

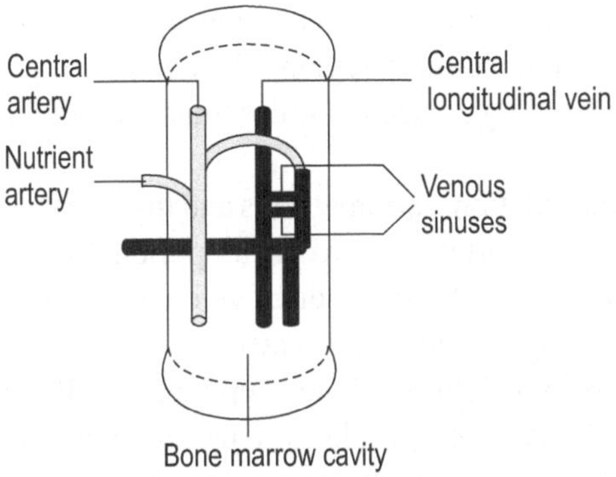

FIGURE 3.25 Blood vessels supplying the bone marrow (arteries are shaded grey and veins are shaded black)

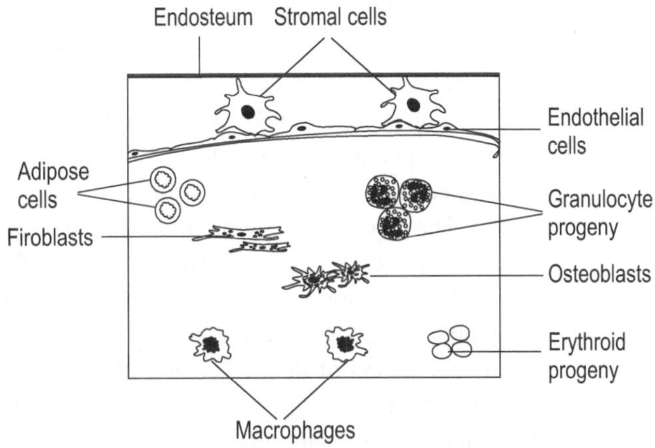

FIGURE 3.26 Distribution of different cell types in the stromal compartment of the bone marrow

TABLE 3.4 Cells forming the stromal cavity of the bone marrow

Cell type	Function
Vascular endothelial cells	Barrier to allow the exit of matured blood cells alone from the bone marrow Prevention of immature blood cells from escaping out of the bone marrow Production of haematopoietic growth factors and cytokines Regulation of blood cell production
Stromal cells	Production of haematopoietic growth factors and cytokines Regulation of blood cell production
Fibroblasts	Production of haematopoietic growth factors and cytokines Regulation of blood cell production
Macrophages	Removal of apoptotic cells
Granulocyte progeny	Differentiation into neutrophils, eosinophils, and basophils
Erythroid progeny	Differentiation into erythrocytes
Adipose cells	Local energy source Synthesis of growth factors
Osteoblasts and osteoclasts	Remodelling of cancerous bone which shells the bone marrow cavity

The haematopoietic cell compartment is vastly supplied with blood vessels branching from the nutrient artery and the sinusoids. This compartment is essential for the maturation and transport of the haematopoietic cells. Sinusoids have an endothelial lining which can secrete growth factors which mediate proliferation and differentiation of haematopoietic cells. Mature haematopoietic cells migrate into the sinuses, and subsequently, enter the circulation via the central vein. This process is called transendothelial migration. Naive blood cells cannot undergo transendothelial migration and remain in contact with the endothelial lining of the blood vessel.

Thymus

Thymus is a bilobed organ located above the heart. The structure of thymus is shown in Fig. 3.27. Structurally, it is made up of many lobules encapsulated in capsules, which extend down along the space between the adjacent lobule, forming a strand called trabecula. Each trabecula is supplied with trabecular venule and trabecular arteriole. Each lobule is made of two regions: outer cortex and inner medulla, separated by a cortico-medullary junction. Trabecula extends till the corticomedullary region. The cortex of each lobule is independent, whereas the medulla is shared by adjacent lobules, thus, giving the appearance of incomplete lobules.

The lobes of the thymus are made of two cell populations: stromal cells and thymocytes. The region of contact between the capsule and the cortex is lined with epithelial cells, and this region is called subcapsular epithelial cell lining. The cortex is enriched with epithelial cells of ectodermal origin and macrophages from bone marrow. The medulla is composed of dendritic cells of the bone marrow origin, medullary epithelial cells of endodermal origin, and Hassall's corpuscle. Hassall's corpuscle is a *whorl* of tightly packed epithelial cells. All the cells of the cortex and the medulla mentioned above are collectively called stromal cells.

The cortical epithelial cells have MHC-I and MHC-II on their surface. Thymocytes include immature T cells (double-negative T cells) in the subcapsular region, double-positive T cells in the outer cortex, and single positive T cells in the inner cortex (refer Section 3.3.2). Thus, around 80–85% of T cells occupy the cortex, while the remaining

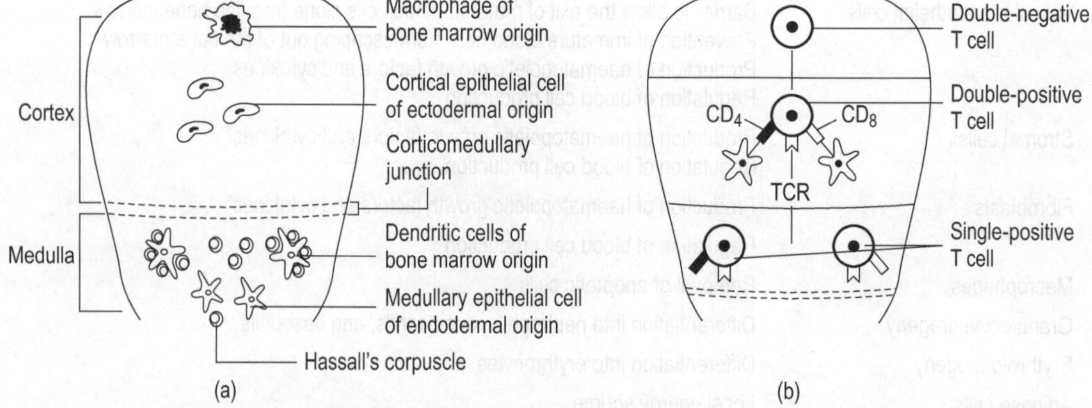

FIGURE 3.27 (a) Structure of the thymus (b) T-cell distribution in the cortex of the thymus, during maturation (arrow indicates the migration of T cells during the process of maturation.)

15–20% occupy the medulla. This leads to denser staining of the cortex. In the outer cortex, the double-positive T cells are in contact with the MHC clusters present on the surface of the epithelial cells. The T-cell distribution profile in the cortex reveals that immature T cells enter the subcapsular region and move to the outer cortex to acquire the two CD markers, CD4 and CD8. After moving to the inner cortex, the cells retain only one CD marker (either CD4 or CD8). These cells are single-positive T cells and are considered as fully matured T cells, which leave the cortex via corticomedullary postcapillary venules. T cells in medulla are fully matured ones.

3.10.2 Secondary Lymphoid Organs

Lymph node and spleen are the major secondary lymphoid organs. Mucosa-associated lymphoid tissues (MALT) are less-organized secondary lymphoid organs. These organs are discussed in detail in this section.

Lymph node

Lymph node is a tiny spherical or bean-shaped structure, widely distributed in many areas of the body, including the armpit, neck, and groins. Lymph nodes function as a filter for the antigens. Lymph nodes are connected by vessels called lymphatics or lymph vessels. Fluid circulating in lymphatics is called lymph. The structure of a lymph node is shown in Fig. 3.28. It consists of the following structures:

- Capsule
- Outer cortex with germinal centres (follicles)
- Inner cortex with $CD4^+$ T cells and endothelial venules
- Medulla containing medullary cords
- Afferent lymphatic vessels with valves
- Subcapsular sinuses under the capsule
- Medullary sinuses
- Efferent lymphatic vessels

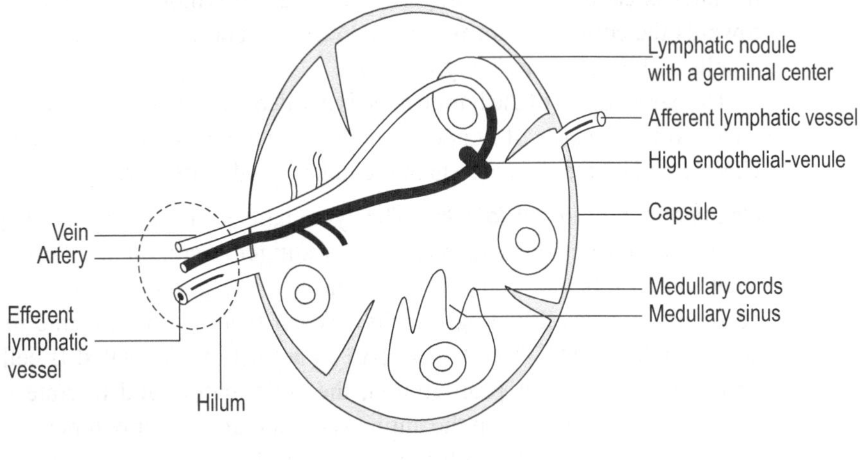

FIGURE 3.28 Structure of a lymph node

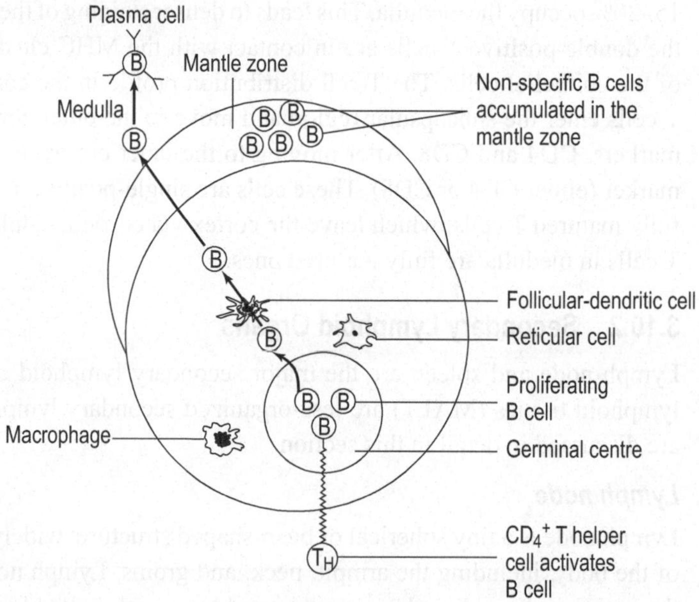

FIGURE 3.29 Structure of a lymphatic nodule (follicle)

Capsule forms the outer layer of the lymph node; it is made of connective tissues and adipose tissues. Afferent lymphatic vessels pierce through the capsule and expand around the capsule as the subcapsular sinus that is located below the capsule. The subcapsular sinus extends down as a paratrabecular sinus and penetrates into the cortex and extends further into the medulla, where it branches into medullary sinuses. Medullary sinuses are lined with endothelial cells and they pierce the capsule on the other side of the lymph node and join with the efferent lymphatic vessel which appears as an outlet. The efferent lymphatic vessel is in close association with the artery and vein supplying the node. The region where the artery enters, and the vein and efferent lymphatic vessel leave the medulla is called hilum. The vein entering the lymph node extends from the medulla towards the cortex and form high endothelial venule (HEV). HEV is the site of entry for T cells and B cells into the lymph node.

The parenchyma of the lymph node is divided into outer cortex, inner cortex, and medulla. Outer cortex consists of lymphatic nodules or lymphatic follicles. Inner cortex consists of $CD4^+$ T cells. Medulla consists of medullary cords surrounding the medullary sinuses.

Germinal centre in the follicle The lymphatic follicles (Fig. 3.29) possess a germinal centre (which develops in response to antigenic stimulant). Follicular dendritic cells (FDCs) and reticular cells are found around the germinal centre. The germinal centre accommodates proliferating B cells. When there is an antigenic invasion, B cells get activated. The $CD4^+$ T cells take part in activation. Activated B cells proliferate in the germinal centre. After proliferation, the cells mature and migrate out of the germinal centre. Follicular cells trap the antibody-bound antigen or complement-bound antigen on their surface. Matured B cell recognizes the FDC-bound antigen. The antigen interacts with specific B cells having high-affinity surface immunoglobulins, and these B cells migrate to the medullary cords and differentiate into plasma cells and leave the lymph node. B cells with low-affinity surface immunoglobulins undergo apoptosis. Apoptotic B cells are engulfed by the macrophages surrounding the germinal centre. Non-specific B cells bypass the follicular cells and accumulate in the mantle zone, surrounding the outer region of the follicle.

Spleen

Spleen is the largest lymphatic organ in the body. Its structure and the vascularization are diagrammatically illustrated in Fig. 3.30. Spleen is highly vascularized and covered by an external covering called capsule. The capsule is made of dense connective tissue with a network of elastic and smooth muscle fibres. The spleen consists of two regions—the outer white pulp and the inner red pulp—with a marginal sinus zone in between.

The white pulp contains the immune cells, which are similar to those present in the lymph node. It also has macrophages and APCs. Trabecular artery enters the spleen via the capsule and extends as central artery. The central artery is surrounded by a sheath of T cells forming a structure called periarteriolar lymphoid sheath (PALS). The central artery extends radially into the marginal zone as radial arterioles. The central artery exits the white pulp as the *penicillary artery*.

The terminal branches of the penicillary artery entering the red pulp are surrounded by macrophages to form *macrophage sheathed capillaries* (MSC). The macrophages engulf the antibody- or complement-bound antigens. The red pulp functions in filtering the aged and damaged red blood cells. The MSCs have two ways to drain: They drain either into the spleenic sinusoid (closed circulation) or directly into the stroma of the red pulp as opened vessels (open circulation). Spleenic sinusoids drain into trabecular vein. These spleenic sinusoids are interconnected networks of vascular spaces, separated by cords of billroth (spleenic cords). The spleenic cords contain stroma formed of reticular cells and fibres. Plasma cells, macrophages, and blood cells are distributed in the stroma.

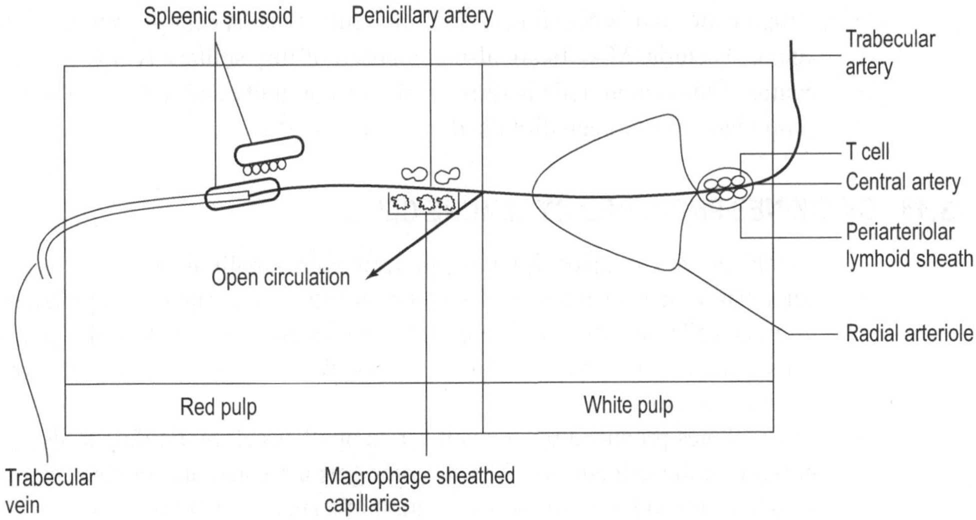

FIGURE 3.30 Vascularization of spleen

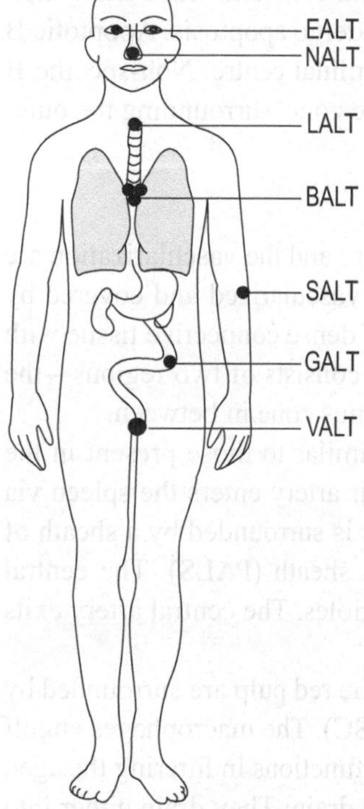

FIGURE 3.31 Mucosa-associated lymphoid tissue (MALT)

Mucosa-associated lymphoid tissue

Mucosa is the name given to the sticky tissues lining various organs, tracts, and cavities of the body. Tracts of digestive system, respiratory system, and urinogenital systems are rich in mucosa. Mucosa is the main site of entry of disease-causing microorganisms. The mucosa, therefore, is very active in defending the antigen and eliciting immune response. Mucosa consists of a group of scattered and widespread lymphoid tissues, approximately 400 m^2. These lymphoid tissues are called mucosa-associated lymphoid tissues (MALTs). Tonsils in throat and Peyer's patches on small intestine are examples of MALT. Depending on the hosting organ, MALT is classified as follows:

- Gut-associated lymphoid tissue (GALT)
- Bronchial/tracheal-associated lymphoid tissue (BALT)
- Nose-associated lymphoid tissue (NALT)
- Vulvovaginal-associated lymphoid tissue (VALT).
- Larynx-associated lymphoid tissue (LALT)
- Skin-associated lymphoid tissue (SALT)
- Eye-associated lymphoid tissue (EALT)

Different types of MALT are shown in Fig. 3.31.

MALTs have larger population of plasma cells compared to the bone marrow, spleen, and lymph node. The mucosa is wrapped by epithelial barrier. The epithelial surface of the mucosa has specialized membranous cells with microfolds. These cells are called M cells. The subepithelial region of the mucosa possesses $CD4^+$ T_H cells and IgA^+/IgM^+ B cells. M cells function as APCs, presenting antigens to the subepithelial lymphoid cells. Under the M cells, follicular-dendritic cells are also present, which bear germinal centre similar to lymph node, and hence, function in immunity. Plasma cells secrete IgA and IgM to fight against bacteria. MALTs are also composed of tiny solitary lymph nodes. The germinal centre of the lymph node is responsible for immunity and defence. The role of germinal centre has already been discussed in Section 3.10.2.

3.11 CYTOKINES PRODUCED BY IMMUNE CELLS

Cytokines are a group of regulatory molecules or signalling molecules secreted by immune cells. They help in intercellular communication. Examples of cytokines produced by immune cells include interferons (IFN) and interleukins (IL). In the context of immune cells, cytokines may be called immunomodulators. Cytokines may be proteins, peptides, or glycoproteins.

Cytokines produced by one cell act on another cell by binding to its specific receptor present on the cell surface. The receptor gets activated and initiates signalling cascade, which results in inhibition or stimulation of a specific cell function or cell process.

Cytokines can be classified depending on their functions into the following:

- Colony-stimulating factors (CSF, stimulate production of blood cells)
- Growth factors, including interleukins, which act on leucocytes

- Chemokines (enable chemotaxis)
- Interferons
- Lymphokines

The role of various cytokines in the immune response is shown in Fig. 3.32.

Colony-stimulating factors (CSF) are soluble glycoproteins. They act upon haematopoietic stem cells to induce their differentiation into specific cell types, usually white-blood-cell lineage. The differentiated cells aggregate into colonies, hence, the name colony-stimulating factor. They function by influencing the signalling pathways via autocrine or paracrine or sometimes endocrine signalling. Depending on the target cells, they are classified into the following:

- Type 1 CSF (CSF1) or macrophage colony-stimulating factor (M-CSF)
- Type 2 CSF (CSF2) or granulocyte macrophage colony-stimulating factor (GM-CSF)
- Type 3 CSF (CSF3) or granulocyte colony-stimulating factor (G-CSF)

M-CSF functions by binding to the CSF1 receptor and signals the haematopoietic cells to differentiate into macrophages for combating the infection of intercellular pathogens. Non-immunologic function of M-CSF includes the development of placenta.

GM-CSF is produced by macrophages, T cells, mast cells, endothelial cells, and fibroblasts. Active GM-CSF is a glycoprotein. It stimulates haematopoietic stem cells to differentiate into granulocytes and monocytes. Monocytes migrate to tissue spaces and mature into macrophages.

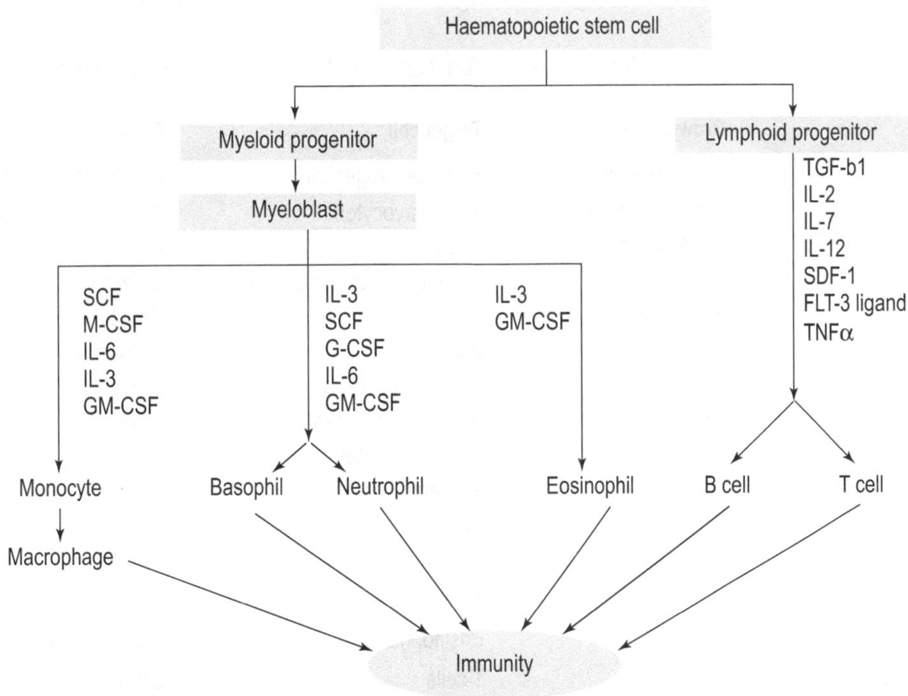

FIGURE 3.32 Cytokines in immune response

G-CSF is a glycoprotein produced by endothelium and macrophages. They are recognized by receptors on the bone marrow haematopoietic stem cells. The receptors are activated (endocrine signalling) to induce the differentiation of stem cells into granulocytes, neutrophils, eosinophils, and basophils. Although haematopoietic cells are the main targets for G-CSF, they also act on other cells, such as neurons, to induce neurogenesis. They are also used as signalling molecules in Janus kinase pathway/signal transducer and activator of transcription (STAT) pathway, Ras/mitogen-activated protein kinase (MAPK), and phosphatidylinositol 3-kinase (PI3K)/protein kinase B pathways (refer Chapter 11). The effect of GCSF is cell survival, cell differentiation, and proliferation.

Growth factors are proteins or steroid hormones capable of promoting cell growth, maturation, and differentiation. As the colony-stimulating factors described earlier are also responsible for the growth and differentiation of haematopoietic cells, sometimes they are also included under growth factors. Some of the growth factors and their target cells are shown in Table 3.5.

Chemokines are very small proteins with molecular weight of 8 to 10 kDa. They are also called chemotactic cytokines. They function in recruiting immune cells, such as leucocytes, monocytes, and neutrophils, to the site of infection by signalling through G-protein-linked transmembrane receptor or chemokine receptor (refer Chapter 11).

Interferons (IFNs) are glycoproteins secreted by lymphocytes. They have the capacity to interfere with the replication of viral pathogens, hence, the name. They are, therefore, used in the eradication of viral infection. They can produce immune response against bacteria and parasites, and can also kill tumour cells by upregulating the antigen presentation and by activating macrophages and natural killer cells.

TABLE 3.5 Target cells and functions of growth factors in immune response

Growth factor	Target cell	Role
Erythropoietin	Erythroid progenitors	Growth and differentiation
Thrombopoietin	Megakaryocytes	Growth and differentiation
Interleukin-1	T cells	Activation
	Interleukin-3 and Interleukin-6	Cofactor
Interleukin-2	T cells	Growth
	B cells	Activation
	Natural killer cells	Activation
Interleukin-3	Pluripotent precursor cells	Stimulation
	Megakaryocytes	
Interleukin-4	Activated B cells	Growth
	Resting T cells	Growth
	Mast cells	Growth
Interleukin-5	B cells	Differentiation
	Eosinophils	Differentiation
Interleukin-6	T cells	Growth
	Plasma cells	Synthesis of Igs
	Monocytes	Maturation, differentiation
Interleukin-7	Lymphoid stem cells	Growth

Lymphokines are cytokines produced by lymphocytes, typically by T cells. Tumour necrosis factor (TNF) is a lymphokine produced by T cells. They enhance the binding of phagocytes to vascular endothelial cells. They mediate tumour killing. They are also thought to be involved in the development of thymus.

SUMMARY

The immune system is a collection of organs and cells which elicit defence in response to foreign antigens invading our body. The defence is called immunity or immune response. Immunity is the defensive reaction of a body to a foreign antigen. The antigen may be a microorganism or molecules such as proteins or polysaccharides. Depending on the nature of immune response, there are two types of immunity: natural immunity (innate immunity) and acquired immunity (adaptive immunity/specific immunity).

Haematopoietic stem cells (HSC), a type of adult stem cells present in the bone marrow, are capable of development and differentiation into immune cells and blood cells. HSCs can differentiate into two types of stem cells or progenitor cells. They are lymphoid progenitors and myeloid progenitors. Lymphoid stem cells develop into B lymphocytes, T lymphocytes, and natural killer cells. B cells possess B-cell receptors (BCRs). Maturation of pre-B cells into B-cell lineage is identified by the CD45 cluster as the first marker. CD45 functions in signal transduction and it is retained on the surface, throughout the life span of B cells. T cells express T-cell receptors (TCRs) on their surface. TCR is specific for the antigens presented by a specific protein complex called major histocompatibility complex (MHC) or MHC molecules, which are expressed on the surface of the antigen-presenting cells (APC) or viral infected cells or cancer cells. There are two types of MHC molecules: MHC class I and MHC class II. MHC-I is expressed by all nucleated cells and recognized by T_C cells. MHC-II is expressed by antigen-presenting cells, which include dendritic cells, macrophages, and B cells, and recognized by T_H cells.

T_H cells express CD4 glycoprotein as their markers, and are, hence, called $CD4^+$ cells. T_H cells bind to the MHC-II-bound antigen presented by APCs. This results in the activation and proliferation of T_H cells by mitosis to produce numerous daughter cells of the same antigen specificity, called effector cells. These cells become T memory cells. In case of infection with intracellular pathogens, for example, virus, bacteria, or single-celled parasites, the effector cells produce interferon-γ (IFNγ), which in turn activates macrophages to produce immune response. This process is called T_H1 response. In case of infection with helminthic intestinal parasites, the effector cells produce cytokines, such as interleukin-3 (IL-3) and interleukin-4 (IL-4), which in turn activate plasma cells to produce IgE, thereby inducing the responses of mast cells, basophils, and eosinophils. This process is called T_H2 response. T_C cells express CD8 clusters as their markers, and are, hence, called $CD8^+$ cells. T_C cells bind to the MHC-I-bound antigen presented by the cells that are altered due to transformation or viral infection. This leads to activation of T_C into effector cells called cytotoxic T lymphocytes (CTLs). Natural killer cells (NKCs) are large, granulated, naturally occurring lymphocytes, capable of killing tumor cells by cytotoxicity. They do not express receptors of T- or B-cell lineage, hence, they come under the cateogory of *null cells*. Monocytes are tiny, circulating white blood cells (leucocytes) with a single nucleus. Monocytes migrate in the circulation and accumulate in the tissue spaces and get differentiated into macrophages. Macrophages process the antigen by phagocytossis.

Granulocytes are classified into three types: neutrophils, eosinophils, and basophils. Though all these three types of cells are granulated, they differ in their cytoplasmic staining property and nuclear characteristics. Neutrophils are the most predominant leucocytes and they function by phagocytosis, respiratory burst, or antibody-mediated cytotoxicity. Basophils are also involved in allergic reactions. Dendritic cells are immune cells that differ from other cells in that they have projections similar to dendrites of nerve cells. They function in antigen presentation to T cells and sensitization of immune system in response to viral infection by activation CTLs. Complement system refers to a set of plasma proteins synthesized in the liver. In association with phagocytes and antibodies, they play a role in immunity against extracellular pathogen such as bacteria. They function as opsonins and coat the surface of the pathogen. Receptors for complement components are located on the surface of the phagocytes. Therefore, complement components coated on the surface of the pathogen get bound to the phagocytes (example, macrophages) and elicit immune response. Thus, the main function of the complement system is to enable the phagocytes to recognize the antigen. Alternatively, complement components can also induce direct lysis of the pathogen cell. Organs involved in the production, maturation, activation, and function of those immune cells are called lymphoid organs. Basically, two lymphoid organs are present in the immune system. They are primary and secondary lymphoid organs, both enriched with blood vessels and lymphatics. Primary lymphoid organs produce immune cells. Immune cells migrate via circulation to secondary lymphoid organs. Secondary lymphoid organs facilitate maturation of immune cells to elicit immune response. Primary lymphoid organs include the bone marrow and the thymus, their principal function being the production of different types of immune cells. Bone marrow is the site for the synthesis and maturation of B cells. Thymus is a bilobed organ located above the heart, responsible for the maturation of T cells.

Lymph node and spleen are the major secondary lymphoid organs. Mucosa-associated lymphoid tissue (MALT) and cutaneous-associated lymphoid tissue (CALT) are less-organized secondary lymphoid organs. Cytokines are group of regulatory molecules or signalling molecules secreted by immune cells. They help in intercellular communication. Examples of cytokines produced by immune cells include interferons (IFN) and interleukins (IL). In the context of immune cells, cytokines may be called immunomodulators. Cytokines may be proteins, peptides, or glycoproteins. Cytokines produced by one cell act on another cell by binding to its specific receptors present on the cell surface. The receptor gets activated and initiates signalling cascade, which results in inhibition or stimulation of a specific cell function or cell process.

TERMS TO REMEMBER

Adaptive immunity The immune response that is elicited after the second exposure to an antigen.

B cell A lymphocyte population, formed and matured in bone marrow to elicit humoral immunity.

Basophil A granulocyte containing pharmacological mediators and involved in allergic reactions.

B-cell receptor A transmembrane protein complex in B cells, composed of a pair of Igα/Igβ heterodimer, separated by an mIg molecule.

CD4+ cells TH cells express CD4 glycoprotein as their markers, in order to recognize MHC class-II molecules.

CD45 cluster The first marker of B-cell lineage, expressed during the maturation process, functions in signal transduction.

CD8+ cells TH cells express CD8 glycoprotein as their markers, in order to recognize MHC class-I molecules.

Complement cascade The mechanism by which the complement components carry out opsonization in a stepwise manner and enable phagocytosis.

Complement system A set of plasma proteins, synthesized in the liver, which associates with phagocytes and antibodies to play a role in immunity against extracellular pathogen such as bacteria.

Cytolytic T cells Activated TC cell population, which destructs the tumour cells by secreting perforin.

Dendritic cells Immune cells with nerve cell appearance, involved in antigen presentation.

Double-negative cells T cells without any surface molecules such as CD cluster or T-cell receptors.

Double-positive cells T cells with both CD4 and CD8 clusters, present in the cortex of the thymus.

Eosinophil A granulocyte with large multilobed nucleus; it induces immune response by phagocytosis.

Granulocytes Granulated immune cells (leucocytes) derived from myeloid progenitor cells.

Haemopoietic stem cell Stem cell of bone marrow origin, capable of both self-renewal and differentiation into blood cell types.

Histocytes Macrophages present in the connective tissues.

Human leucocyte antigens The MHC equivalent in humans, which is structurally and functionally similar to mice MHC.

Immune system A collection of organs and cells which elicit defence in response to foreign antigens invading our body.

Immunity The defensive reaction of a body against a foreign antigen.

Innate immunity The natural or inborn immune response that is elicited within the body when an antigen enters the body.

Kupffer cells Macrophages present in the liver.

Lymphocytes Small, motile, non-phagocytic cells derived from haematopoietic stem cells and involved in immunity.

Macrophage-activating factor Cytokine released from T cell, mediates the activation of macrophages for immunity. Activated macrophages release TNFα. TNFα inhibits tumour cell proliferation and kills the tumour.

Macrophages Matured and developed monocytes accumulated in the tissue spaces. Five- to ten-fold increases in the cell size occurs in monocytes to form macrophages during migration and differentiation.

Memory cells Activated lymphocytes which retain the antigenic specificity and circulate in the blood for a long time, in order to elicit immunity against the same antigen for a second time. They are formed during TH cell-mediated immune response.

Mesangial cells Macrophages present in the kidney.

MHC complex A protein complex present on the surface of mice T cells which helps in antigen presentation.

Microglial cells Macrophages present in the brain.

Monocytes Tiny, mononuclear circulating white blood cells (leucocytes) which mature into macrophages.

Naive T cells Immature T cells present in the bone marrow.

Natural killer cells Large, granulated, naturally occurring lymphocytes, capable of killing tumour cells by cytotoxicity.

Neutrophil The most predominant leucocyte with granulated cytoplasm, elicits immunity by phagocytosis, respiratory burst, and antibody-mediated cytotoxicity.

Osteoclasts Macrophages present in the bone.

Perforins Pore-forming proteins produced by CLTs to induce tumour cell death via membrane damage.

Phagocytosis A process by which the macrophages ingest, digest, and process the antigens.

Phagolysosome A complex formed from phagosome and lysosome.

Phagosome Macrophage membrane vesicle encapsulating the antigen.

Plasma cells Activated B cells, capable of secreting antibodies for neutralizing antigens.

Pseudopodia Protrusion of the macrophage membrane used to entrap the antigen during phagocytosis.

Single-positive cells T cells with either CD4 or CD8 cluster, present in the inner cortex of the thymus.

T cells A population of lymphocytes which matures in thymus and elicits cell-mediated immune response.

T-cell receptors Proteins of Ig superfamily having Ig-like domains, with specificity for free antigens and the MHC-bound antigens.

Tumour necrosis factor A cytokine produced by T cells to destroy the tumour cells.

REVIEW QUESTIONS

1. Define and classify immunity.
2. Draw and explain the haematopoietic stem cell lineage.
3. Illustrate the role of cytokines in haematopoiesis.
4. What is the relation between SCID and HSC?
5. Define and classify CD clusters.
6. Write short notes on BCR and TCR.
7. Draw and compare the structure and functions of MHC-I and MHC-II.
8. Define double-positive cells, double-negative cells, and single-positive cells.
9. Diagrammatically explain the role of T_H and T_C cells in immunity.
10. What are memory cells? How are they formed?
11. Write about CLTs.
12. Describe the role of natural killer cells.
13. What is the immunologic role of mast cells?
14. Explain the formation and functions of macrophages.
15. Describe the phagocytosis process with a suitable diagram.
16. Explain the function of dendritic cells.
17. Define complement system. Illustrate the role of complement components in opsonization and phagocytosis with neat, labelled diagrams.
18. What is Hassall's corpuscle?
19. Define hilum.
20. Describe the structure and function of germinal centre.
21. The cortex of the thymus stains darker than the medulla. Give reasons.
22. Write the role of the germinal centre in eliciting immune response.
23. Define Payer's patch.

24. Tabulate the influence of growth factors on immune cells.
25. Describe the structure and function of primary lymphoid organs.
26. Draw and explain the structure of a lymph node
27. Explain spleen vascularization.
28. Write the functions of the following cytokines:
    (a) Chemotactic factor (CF)
    (b) Macrophage activating factor (MAF)
    (c) Tumour necrosis factor (TNF)
    (d) Interleukin-2 (IL-2)
    (e) Interferon-γ (IFN γ)

MULTIPLE-CHOICE QUESTIONS

1. Macrophages accumulated in the liver are called
   (a) T cells
   (b) dendritic cells
   (c) Kupffer cells
   (d) plasma cells
2. The first marker of B-cell lineage is
   (a) MHC-I
   (b) MHC-II
   (c) CD54
   (d) CD45
3. Immune cells with nerve-cell-like appearance are called
   (a) dendritic cells
   (b) demdrimers
   (c) cytolytic cells
   (d) cytotoxic cells
4. Naive cells are
   (a) natural killer cells
   (b) immature T cells
   (c) antigen-presenting cells
   (d) cytokine-producing cells
5. Any compound that targets the antigen for phagocytosis is called
   (a) opsin
   (b) opsonin
   (c) antigen receptor
   (d) B-cell receptor
6. The MHC equivalent in human immune cells is called
   (a) HLA
   (b) NKC
   (c) BCR
   (d) TCR
7. Plasma cells are derived from
   (a) T cells
   (b) B cells
   (c) antigen-presenting cells
   (d) dendritic cells
8. SCID is characterized by the deficiency of
   (a) tumour necrosis factor
   (b) interleukin-7
   (c) gamma interferon
   (d) memory cells
9. Memory B cells have
   (a) shorter life span than effector B cells
   (b) longer life span than effector B cells
   (c) greater antibodies than effector B cells
   (d) lesser antibodies than effector B cells
10. BMCA is a receptor for
    (a) interleukin
    (b) tumour necrosis factor
    (c) mitogen-activating factor
    (d) macrophages
11. Which of the following is the least predominant granulocyte in blood?
    (a) neutrophil
    (b) eosinophil
    (c) basophil
    (d) plasma cell
12. Which of the following is the most predominant granulocyte in blood?
    (a) neutrophil
    (b) eosinophil
    (c) basophil
    (d) natural killer cell

Answers to Multiple-choice Questions

1. (c)	2. (d)	3. (a)	4. (b)	5. (b)
6. (a)	7. (b)	8. (b)	9. (b)	10. (b)
11. (c)	12. (a)			

CHAPTER 4

Biomolecules in Cell

G.N. Ramachandran, proposed the Ramachandran plot to explain the possible conformations for peptide.

After reading this chapter, students should be able to understand and appreciate the following:

- Definition of biomolecules, types of biomolecules, and bonds holding the cellular biomolecules
- Structure of nucleotides and nucleic acids and functions of nucleic acids
- Structure and functions of amino acids and proteins
- Classification, structure, and functions of enzymes and hormones
- Classification, structure, and functions of carbohydrates
- Classification, structure, and functions of lipids
- Role of water and minerals in biological system

4.1 BIOMOLECULES

The term 'biomolecules' is used to define the organic compounds that exist in, or are produced by, living organisms. Biomolecules are also known as biogenic substances. They are of two types: (i) polymers or macromolecules such as nucleic acids, proteins, carbohydrates, and lipids; and (ii) small molecules or micromolecules such as primary metabolites, secondary metabolites, and natural products. In general, biomolecules play a vital role in maintaining the cellular structure, integrity, metabolism, and specific functions such as transport, storage, defence, signalling, adhesion, and communication. This chapter explains the functions of each molecule separately.

The four major elements constituting biomolecules are carbon, hydrogen, oxygen, and nitrogen, while the minor elements are phosphorus and sulphur. These elements together constitute up to 98% of the mass of all living organisms. These elements bind among themselves to constitute biomolecules, often called biological macromolecules. Carbon is a unique element with the remarkable ability to form strong, stable chemical bonds with other atoms forming chains, branched chains, and rings. Atoms of hydrogen and oxygen, and less frequently nitrogen, sulphur, or phosphorus, are bonded to the carbon skeleton to form monomeric biomolecules which in turn are polymerized in vivo to form giant molecules called macromolecules. Four major types of macromolecules are found in living cells—nucleic acids, proteins, carbohydrates, and lipids. Elements and monomers constituting these macromolecules are given in Table 4.1. Functions of macromolecules are related to the structure and properties of their constituent monomers.

TABLE 4.1 Four major macromolecules constituting living cells

Macromolecule	Monomer	Elements constituting the monomer	Examples
Nucleic acids	Nucleotides	Carbon Hydrogen Oxygen Nitrogen Phosphorus	DNA RNA Ribozymes
Proteins	Amino acids	Carbon Hydrogen Oxygen Nitrogen Sulphur	Hormones Enzymes Haemoglobin (pigment)
Carbohydrates	Sugars (Saccharides)	Carbon Hydrogen Oxygen	Starch (in plants) Glycogen (in animals)
Lipids	Fatty acids and alcohols (esters)	Carbon Hydrogen Oxygen	Fat Oil Wax

4.1.1 Bonds That Hold Cellular Biomolecules

The following bonds are present in cellular biomolecules:

- Covalent bond
- Ionic bond
- Hydrogen bond
- van der Waals interaction
- Hydrophobic interaction

Through these bonds, different parts of a single type of biomolecule (intramolecular bonding) or different types of biomolecules (intermolecular bonding) are held together. Bonds are responsible for maintaining the structure of biomolecules.

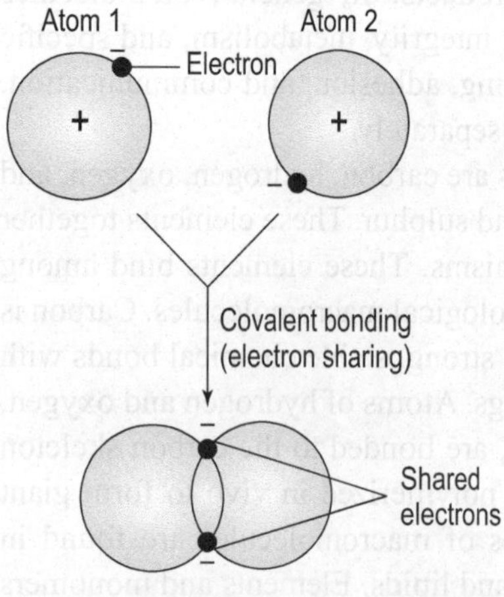

FIGURE 4.1 Covalent bond due to sharing of electrons between two atoms

Covalent bond

Covalent bond is formed between two atoms by sharing of electrons (Fig. 4.1). Carbon is ideal for covalent bond and has a unique role in a cell due to the formation of strong covalent bonds with other carbon atoms. Since the cellular biomolecules listed in Table 4.1 have carbon as the major element, covalent bonds are present in all these molecules.

DNA molecules have two types of covalent bond. The first is between the nucleotides on the same strand of the helix. These *ester bonds* are covalent carbon–oxygen–carbon (C–O–C) bonds. The nucleotides in each DNA strand are joined head to tail by covalent bonds that link the phosphate group on the 5′ carbon of one nucleotide to the 3′ carbon on the next nucleotide. The second type of covalent bond is the hydrogen bond, which is present between the nucleotides on either side of the helix.

Disulphide bonds and *peptide bonds* found in proteins are also covalent bonds. Peptide bond is formed by a dehydration reaction that occurs when two amino acids are positioned such that the carboxyl group of one is adjacent to the amino group of the other.

Amino acid-1–COOH + –H_2N–Amino acid-2 → Amino acid-1–CO–HN–Amino acid-2 + H_2O

Disulphide bond is a weak covalent bond formed between two thiol groups (–SH) possessed by two different cysteine residues. This is the strongest chemical bond present in protein structure. For example, insulin is stabilized by disulphide bridges.

Protein–Cysteine–SH + HS–Cysteine–Protein → Protein–Cysteine–S–S–Cysteine–Protein
(Free thiol) (Disulphide Bridge)

Glycosidic bond is the covalent bond between two monosaccharides to form a disaccharide. Thus, carbohydrates also have covalent bonds. Glycoproteins are proteins that have covalently attached sugar residues. One of the common linkages between the sugar(s) and the protein is an *N*-glycosidic linkage between the –OH group of the sugar at C_1 and the side chain of an asparagine residue of the protein.

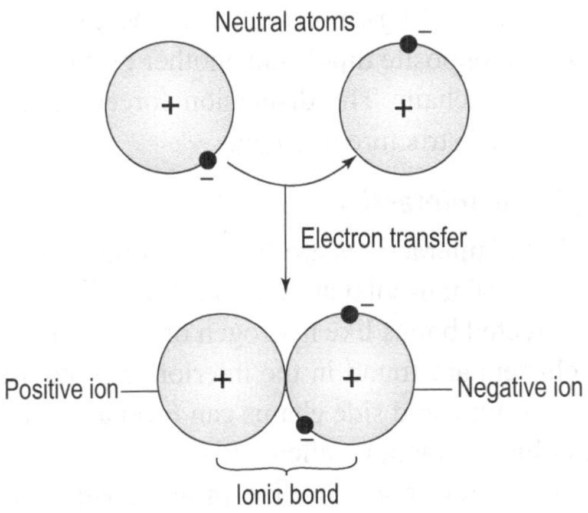

FIGURE 4.2 Formation of ionic bond

Ionic bond

An ionic bond is formed between two atoms bearing opposite charges. The formation of an ionic bond is shown in Fig. 4.2. During the formation of an ionic bond, one atom loses an electron (this atom is called the donor) and the other atom accepts that electron (this atom is called the acceptor). After the transfer of electrons, the donor becomes smaller bearing a positive charge, while the acceptor becomes bigger bearing a negative charge. As a result, a strong electrostatic attraction is established between them which results in the formation of an ionic bond. Compounds bearing ionic bonds are mostly crystals with high melting temperatures, readily soluble in water and polar solvents, and capable of conducting electricity.

Salt bridge Salt bridge is also an example of a weak electrostatic bond or weak ionic bond formed between a positively charged amino acid (arginine/lysine) and a negatively charged amino acid (aspartate/glutamate) in proteins. Salt bridge helps in stabilizing the structure of proteins.

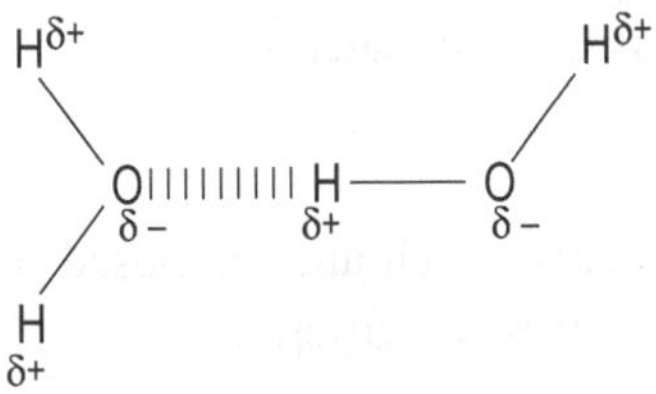

FIGURE 4.3 Representation of hydrogen bond (small vertical lines represent the hydrogen bond)

Hydrogen bond

A hydrogen bond is formed when a hydrogen atom is sandwiched between two electron-attracting (electrophilic) atoms, usually oxygen or nitrogen, as shown in Fig. 4.3. Hydrogen bond is involved in stabilizing the secondary structure of proteins, such as helix and

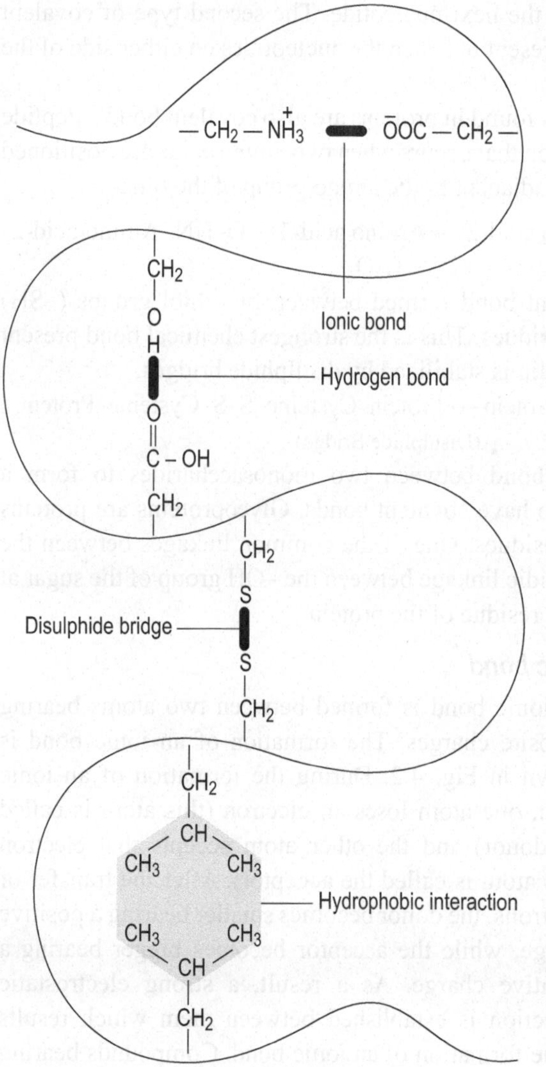

FIGURE 4.4 Bonds stabilizing protein structure

pleated sheets. In DNA helix, base pairs are bonded via hydrogen bonds; two between adenine and thymine and three between guanosine and cytosine.

van der Waals interaction

The van der Waals force (or van der Waals interaction), named after Dutch scientist Johannes Diderik van der Waals, is the attractive or repulsive force between molecules (or between parts of the same molecule) other than those caused due to covalent bonds or electrostatic interaction. This force exists in all compounds. Vibrations of the nucleus induce a temporary dipole, and the consequent dipole–dipole attraction generates van der Waals force. The van der Waals interaction in proteins is due to the presence of a transient, electrical attraction between amino acids owing to the presence of electrical dipole. Several amino acids have quite large hydrocarbon groups in their side chains. Examples of such amino acids include leucine, isoleucine, and phenylalanine. Temporary fluctuating dipoles in one of these groups could induce opposite dipoles in another group on a nearby folded chain. The dispersion forces set up hold the folded structure together.

Hydrophobic interaction

Hydrophobic interactions are found in proteins due to clustering of non-polar amino acid side chains and are not directed bonds like hydrogen or ionic bonds. These clusters are found in the interior of proteins. Aromatic amino acid side chains can form a special hydrophobic interaction called a Pi-stack.

Various bonds that stabilize protein, which is the most important macromolecule, are shown in Fig. 4.4.

4.2 NUCLEIC ACIDS

Nucleic acids are polymers of nucleotides which function as genetic material.

4.2.1 Nucleotides

Nucleotides are monomers or building blocks of nucleic acids which are composed of three structural units: pentose sugar, nitrogenous base, and phosphate group.

Pentose sugar

Two types of pentose (five-carbon) sugars are present in nucleic acids. They are ribose and deoxyribose. If the pentose is ribose, the nucleic acid is called ribonucleic acid (RNA) and

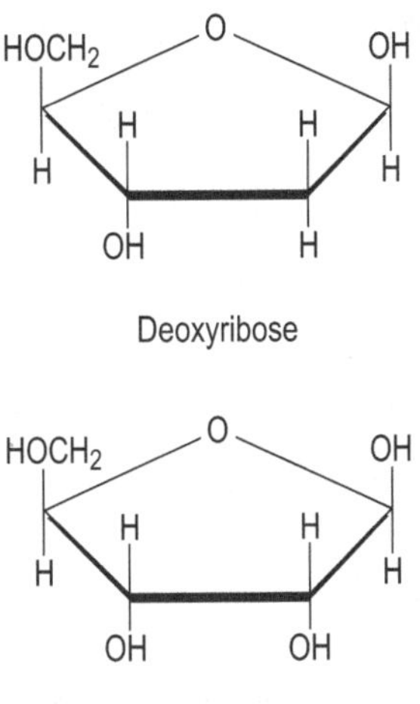

FIGURE 4.5 Structure of sugars found in nucleic acids

if the pentose is deoxyribose, the nucleic acid is called deoxyribonucleic acid (DNA). Structures of these two sugars are shown in Fig. 4.5. Carbon atoms in pentose sugar are denoted as 1′ through 5′.

Nitrogenous base

Two types of nitrogenous bases are found in nucleic acids: purine and pyrimidine. Adenine (A) and guanine (G) are purine bases, whereas cytosine (C), thymine (T), and uracil (U) are pyrimidine bases. Structures of purine and pyrimidine bases are shown in Fig. 4.6. DNA is composed of A, G, C, and T, and RNA is composed of A, G, C, and U. Pentose sugar and nitrogenous base (without the phosphate group) are collectively known as nucleoside.

Phosphate group

Phosphate group is a phosphorus atom surrounded by four oxygen atoms. Phosphate is attached through the oxygen atom of the hydroxyl group of the 5′ carbon of the pentose sugar. Thus, a nucleotide can be named as nucleoside monophosphate. As an example, the structure of a nucleotide (deoxynucleotide and ribonucleotide) with a guanine base is shown in Fig. 4.7. Nomenclatures of various nucleosides found in nucleic acids are given in Table 4.2.

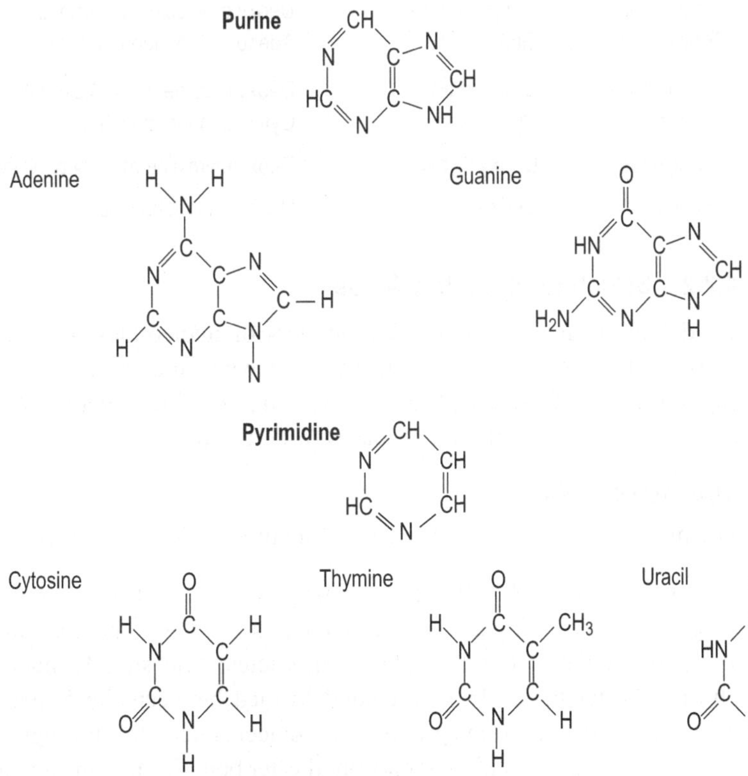

FIGURE 4.6 Structure of purine and pyrimidine bases

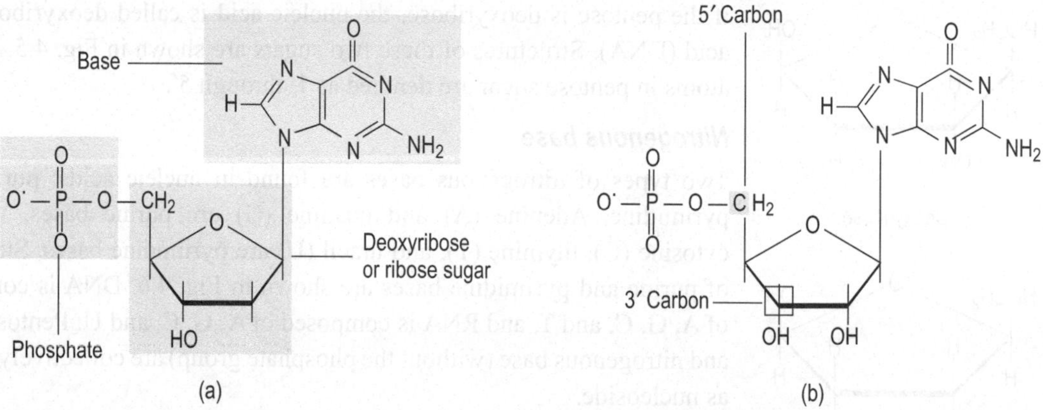

FIGURE 4.7 Structure of a nucleotide (a) Deoxyguanosine monophosphate (b) (Ribo) guanosine monophosphate

TABLE 4.2 Nomenclature of nucleosides and nucleotides

Base	Sugar	Nucleoside	Nucleotide	Occurrence
Adenine	Deoxyribose	Deoxy adenosine	Deoxy adenosine monophosphate	DNA
	Ribose	Adenosine	Adenosine monophosphate	RNA
Guanine	Deoxyribose	Deoxy guanosine	Deoxy guanosine monophosphate	DNA
	Ribose	Guanosine	Adenosine monophosphate	RNA
Cytosine	Deoxyribose	Deoxy cytidine	Deoxy cytidine monophosphate	DNA
	Ribose	Cytidine	Cytidine monophosphate	RNA
Thymine	Deoxyribose	Deoxy thymidine	Deoxy thymidine monophosphate	DNA
Uracil	Ribose	Uridine	Uridine monophosphate	RNA

4.2.2 Structure of Nucleic Acids

By definition, nucleic acids are polymers of nucleotides (discussed in the preceding section). Thus, the structure of DNA is always formed by nucleotides with deoxyribose sugar and one of the four bases A, G, T, and C. The structure of RNA is formed with ribose sugar and one of the four bases A, G, U, and C.

Structure of DNA

The primary, secondary, and tertiary structures of DNA are given in Fig. 4.8.

Primary structure of DNA The primary structure of DNA refers to the nucleotide sequence (determining the order of the bases in a DNA chain is called DNA sequencing) and phosphodiester bond backbone. In nucleic acid strands, nucleotides are linked by phosphodiester bond. This is a bond formed between the 5′-phosphate group of one nucleotide and 3′-hydroxyl group of the adjacent nucleotide through a phosphate molecule. A schematic representation of phosphodiester bond formation and phosphodiester bond is shown in Fig. 4.8a. A single nucleic acid strand formed by phosphodiester bond has two terminals namely 5′ end with phosphate group and 3′ end with hydroxyl group generally represented as: P-5′-----------3′-OH. This terminology refers to the 5′ and 3′ carbons on the sugar. For both DNA and RNA, the 5′ end bears a phosphate, and the 3′ end bears a hydroxyl group. For example, a strand with adenine at 5′ end and thymine at 3′ end and other bases in between can simply be represented as: P-5′A–G–G–C–A–G–C–T–3′–OH. Conventionally, base sequence of DNA is written in the 5′→3′ direction. Primary structure is shown in Fig. 4.8b.

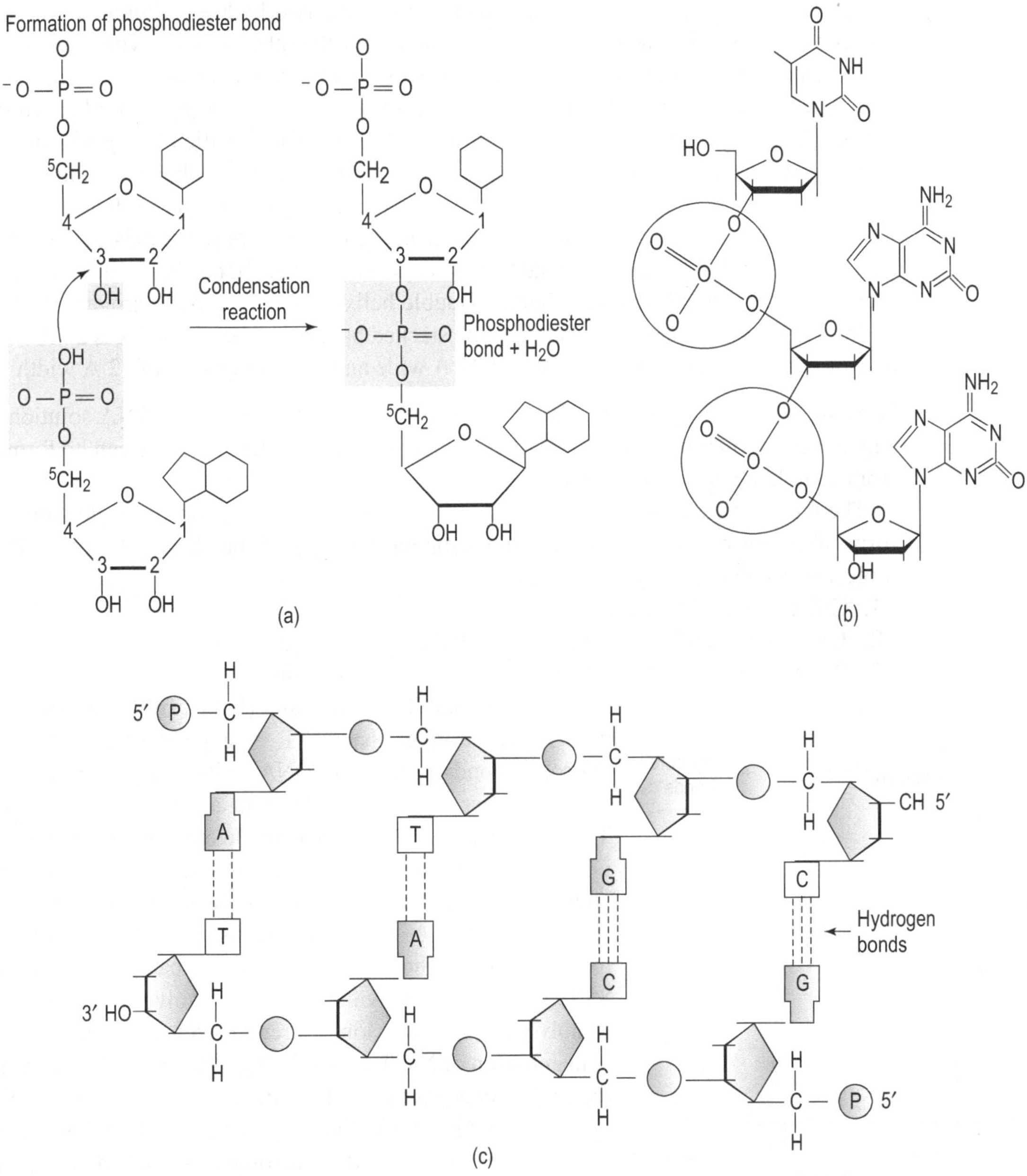

FIGURE 4.8 Structure of DNA: (a) Formation of Phosphodiester bond, (b) Primary structure (Single strand), (c) Secondary structure (Double strands)

Secondary structure of DNA The secondary structure of DNA, studied using X-ray diffraction data is popularly known as the double helical DNA. The double helical DNA model was proposed by James Watson and Francis Crick in the year 1953. This model is also called B-form DNA (Fig. 4.8c). DNA double helix is made up of two DNA strands running in opposite directions, that is, in antiparallel fashion. This implies that the 3′ and 5′ ends of the two strands are in opposite directions and are coiled to form a right-handed helix. The helix has a common axis. The two strands of the helix are held together by complementary base pairing. Here, A pairs with T via two hydrogen bonds, each base serves as H-donor for one bond and as H-acceptor for the other; G pairs with C via three hydrogen bonds, G is an acceptor for one of these, and a donor for two. Therefore, in a double helix, the number of A is equal to T and that of G is equal to C. In other words, DNA will have 1:1 ratio of purine and pyrimidine bases. This is called Chargaff's rule. A helix has a ladder-like appearance, in which alternating sugar and phosphate units form the two sides (outer part) of the ladder with the rungs or steps (inner part of helix) formed by a pair of nucleotide bases (bps). The bases are at right angles to the axis of the helix. The sides of the ladder are held together by weak electrostatic bonds between base pairs; these bonds are called hydrogen bonds. Double helix is stabilized by hydrogen bonding and van der Waals interactions between the stacked base pairs. Double helix is 20 Å wide and 34 Å long. It has a major groove of 22 Å wide and a minor groove of 12 Å width.

Thermal melting Separation of two strands of a helix by heating the DNA solution is called thermal melting (Fig. 4.9). Conversely, a double-stranded molecule can be formed from complementary single stands.

The temperature at which the transition of DNA from helix form to single-stranded form takes place is called the melting temperature (T_m) of that DNA. T_m of a DNA molecule is influenced by three factors:

1. G:C content of the strand
2. Ionic strength of the solution in which DNA is suspended
3. Presence of mismatches (non-complementary base pairs)

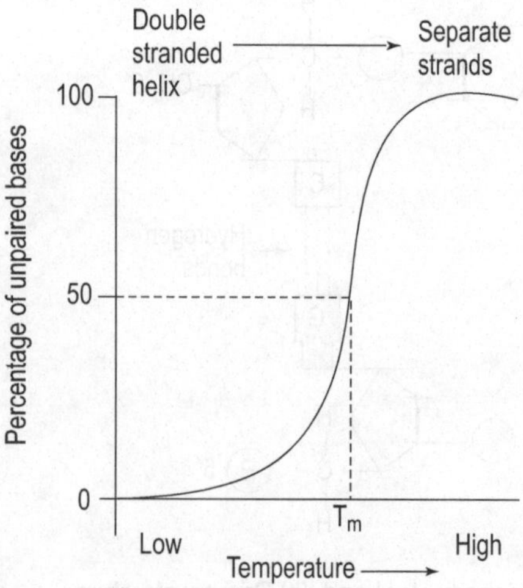

FIGURE 4.9 Thermal melting

We know that G–C base pairs (linked by three hydrogen bonds) are stronger than A–T base pairs (linked by two hydrogen bonds), therefore, DNAs with a high G–C content have a higher T_m than DNAs with a higher A–T content. This can be understood with a simple example where human DNA is compared with bacterial DNA. Human DNA has approximately 50% G–C, whereas *Streptomyces* DNA has approximately 73% G–C content. As a result, human DNA melts at 70°C, whereas DNA from the bacterium melts at 85°C. The T_m of a DNA also depends on solvent composition. A solvent with high ionic strength, for example, a high concentration of NaCl, raises the T_m of a given DNA because the higher concentration of positive sodium ions masks the negative charge of the phosphates in the DNA backbone. DNA double strand made with some mismatched base pairs (non-complementary base pairs) has a lower T_m compared to a completely double-stranded DNA.

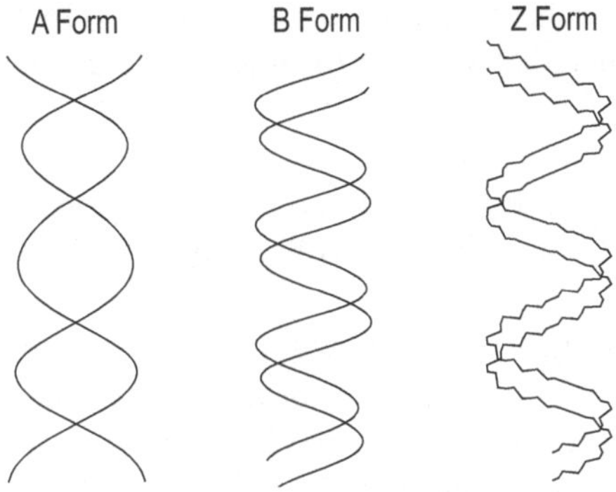

FIGURE 4.10 Three forms of DNA

Hypochromicity Thermal melting of DNA double helix is often detected by the absorbance of ultraviolet light. In the double helix structure, the stacked bases shield each other from light. As a result, the absorbance of UV light (wavelength 260 nanometers, i.e., A_{260}) of the double helix DNA is less than that of the same DNA, whose strands are separated. This effect is called the hypochromicity (less-colour) of the double-helical DNA.

Diversity in DNA helical structure

A form, B form, and Z form DNA helix can occur in three different forms: A form, B form, and Z form. These three forms of DNA are compared in Table 4.3 and shown in Fig. 4.10. B form is the Watson and Crick's double helical model with base pairs at right angles to each other. The B form can undergo partial dehydration and get converted into A form in which the base pairs are not at right angles, but horizontally tilted to 20°. Because of tilting, the A form helix is slightly wider (11 bp) than the B form (10.5 bp), which makes the major groove deeper and the minor groove shallower. The physiological significance of A form is still obscure.

TABLE 4.3 Different forms of DNA

DNA	Helix turn	Diameter of the helix in nm	Number of bp/turn	Degree of rotation /bp	Distance/helical turn in nm	Axial distance between bp
A form	Right-handed	2.6	11	+32.7	2.8	0.25
B form	Right-handed	2.0	10.5	+34.6	3.4	0.33
Z form	Left-handed	1.8	12	–30.0	4.5	0.38

DNA strands that are rich in alternate purine and pyrimidine bases 'CGCGCG' undergo a radical departure from the B form and acquire a zig-zag conformation, which is slender and elongate. These DNAs are called Z form. Stretches of Z-DNA are thought to play a role in gene expression and genetic recombination.

Triplex DNA and Hoogsteen base pairing In addition to B form, the DNA duplex can also take a triplex form, probably due to negative supercoiling stress. This triplex is formed due to additional hydrogen bonding of Watson and Crick base pairs (G≡C and A=T) with protonated bases in the major groove. For example, a cytidine (if protonated) can bond with guanosine of G≡C pair and similarly, a thymidine can bond with adenosine of A=T pair. A triplex is described as H form, because it is stabilized by hydrogen ions (it is a protonated structure). This H form consists of an intramolecular triple helix formed by the pyrimidine strand and half of purine strand, the other half of purine strand being single stranded (Fig. 4.11). Under physiological conditions,

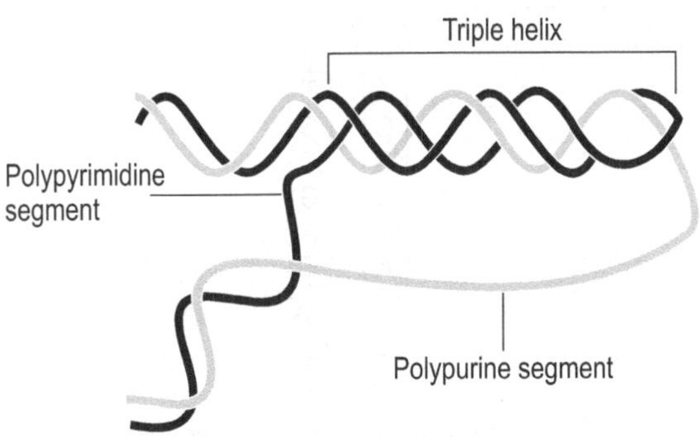

FIGURE 4.11 H-DNA (Triplex form)

H-DNA plays an apparent role in gene expression. One of the repeats in H-DNA, (GAA) n/ (TTC) n, is implicated in the development of a hereditary human disorder—Friedreich's ataxia.

H-form DNA can be formed in vitro using certain chemicals such as osmium tetroxide, chloroacetaldehyde, and psorale. H-DNA sequences are found in promotore regions of genes, where they are bound and activated by transcriptional activators for subsequent regulation of transcription.

Quadruplex DNA DNAs (and RNAs) containing guanine tracts will associate in vitro to form four-stranded, right-handed helices known as quadruplexes or tetraplexes called G-quadruplex (Fig. 4.12). G-quadruplex was later reported to exist in vivo also. Repeated folding of a single polynucleotide chain or intermolecular association of two or four polynucleotide chains results in the G-quadruplex. Structurally, the quadruplex is made of stacked G tetrads (each of which is a square coplanar array) or of four guanine bases each. DNA quadruplex is involved in the regulation of replication and cell proliferation, while RNA quadruplex plays a role in mRNA turnover, translation, and repression.

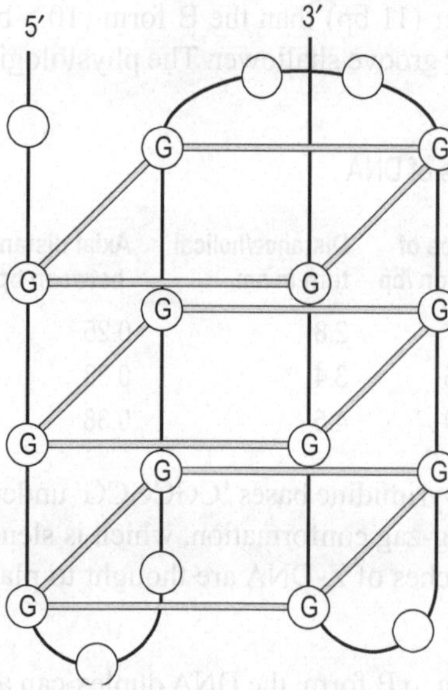

FIGURE 4.12 G-quadruplex DNA

(a) Palindrome (in two strands)
5′-----**TTAGCACGTGCTAA**-----3′
3′-----**AATCGTGCACGATT**-----5′
(b) Mirror repeats (in single strand)
5′-----**TTAGCACCACGATT**-----3′
3′-----AATCGTGGTGCTAA-----5′
(c) Cruciform
5′-----**TGCGAT**ACTC**ATCGCA**-----3′
3′-----**ACGCTA**TGAG**TAGCGT**-----5′

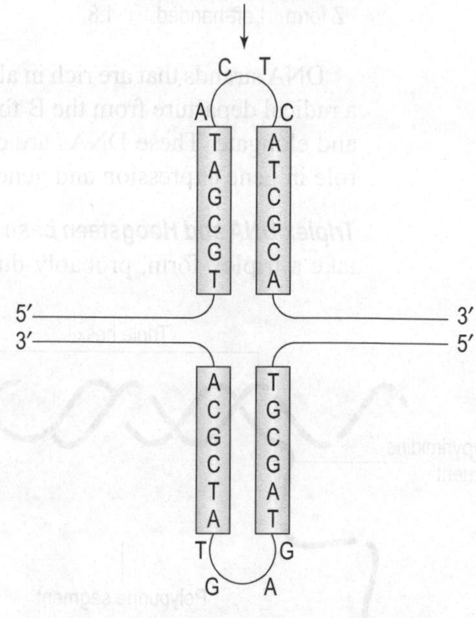

FIGURE 4.13 Palindrome, mirror repeats, and cruciforms in DNA (sequences are shown in bold)

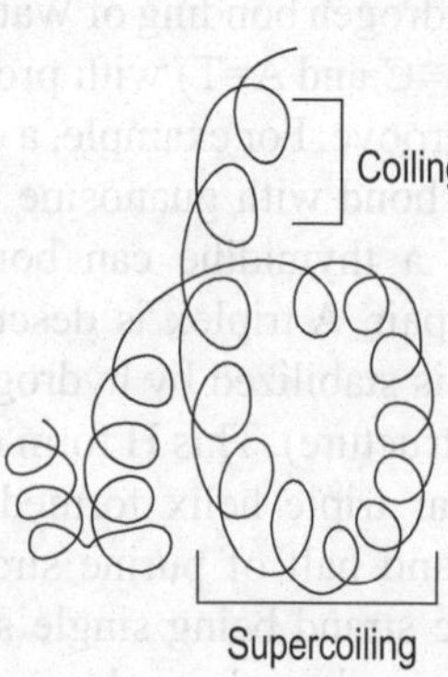

FIGURE 4.14 Supercoiled DNA

Palindrome mirror repeats and cruciforms Palindrome is a word, sentence, or phrase which spells identically in both forward and reverse directions, for example, Malayalam, rotator, nun, etc. In nucleic acids, palindromes are base sequences with two-fold symmetry which are in two separate strands. These palindrome sequences on the two strands read the same when in opposite directions and are called inverted repeats. In order to superimpose one repeat (bold) on the other, it must be rotated 180° about the horizontal axis and then 180° about the vertical axis (Fig. 4.13a). If the inverted repeats exist in a same strand, they are called mirror repeats (Fig. 4.13b). Within each strand, two palindrome sequences are self-complementary, and hence, they tend to form a cruciform structure (Fig. 4.13c).

Tertiary structure of DNA Supercoiled DNA, chromosomes, and chromatin are three types of tertiary structures of DNA. DNA can be twisted and coiled into a compact tertiary structure called supercoiled structure which resembles a coiled telephone wire (Fig. 4.14). Supercoiling may be right handed or left handed.

Linking number The total number of turns in a *covalently closed circular* DNA is called the linking number, or Lk. Linking number is given by the following equation:

$$Lk = Tw + Wr$$

where Tw = number of helical turns in the DNA, which is also called twist and Wr = number of supercoiled turns in the DNA, which is also called writhe. This equation shows that Lk is an integer formed by twist and writhe.

Functions of supercoiled DNA The following are some functions of supercoiled DNA:

1. Opens DNA with ease, during unwinding, for example, during DNA replication.
2. Protects DNA from nuclease attack and other damage.
3. Provides intermediate, condensed structure for higher structure forms, such as, nucleoid structure and chromosome structure.

Topoisomerases DNA topoisomerases are enzymes that alter the Lk of a DNA by bond-breaking and bond-rejoining processes. They are of the following two types:

- Topoisomerase I or nicking-closing enzyme (ATP-independent enzyme)
- Topoisomerase II or DNA gyrase (ATP-dependent enzyme)

Type-I topoisomerase is an ATP-independent enzyme, which converts negatively supercoiled DNA into a relaxed form. This results in an increase in Lk value by increments of one. Type-I enzymes are targets for some anticancer drugs such as camptothecin. Type-II topoisomerases, on the other hand, depend on ATP for energy in order to induce supercoils. This leads to a reduction in Lk in increments of two. Type-II enzymes are targets for the antibiotic naladixic acid. Both the enzymes perform an essential role in DNA replication and transcription.

B-form of DNA is shortened 10,000 fold and folded densely into chromosomes. This occurs during mitosis. DNA also folds into a less dense form called chromatin during the interphase of cell cycle. The DNA helix in the cell is wrapped around by a protein called histone forming a complex called nucleosome. The nucleosome undergoes further coiling during cell division and gets organized into a dense mass of DNA, that is called chromosome (Fig. 4.15), in which the genes are arranged into chromatin. The arrangement of chromatins

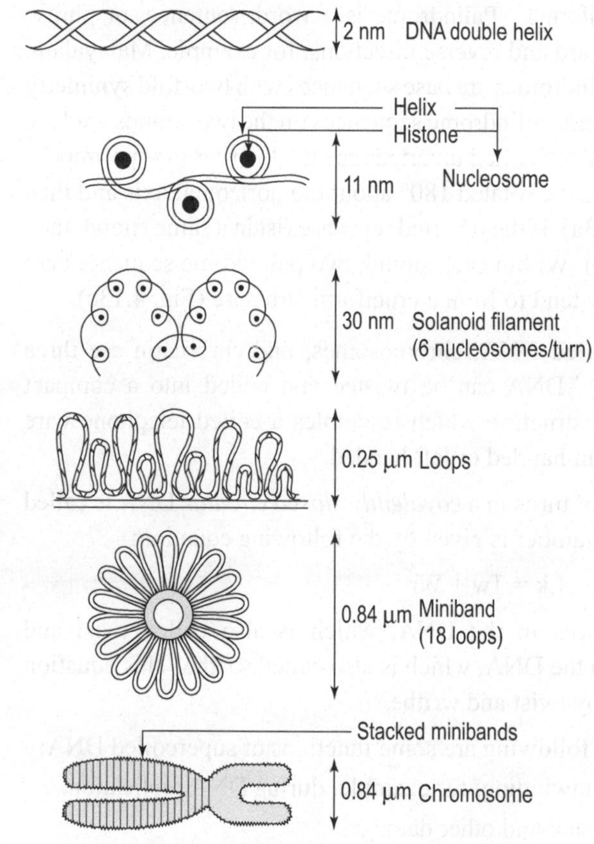

FIGURE 4.15 Organization of chromosomes

and nucleosome resembles beads on a string (Fig. 4.16).

Functions of DNA

The following are some crucial functions of DNA:

Storage of genetic information DNA functions as genetic material. It encodes information to be transcribed into mRNA for further translation into protein (refer Chapter 6).

Energy storage If the phosphate group attached to the pentose of a nucleotide is linked to two more phosphates, nucleoside triphosphate (NTP) is formed (e.g., ATP and GTP). These triphosphates are energy currencies of the cell. They provide energy for driving many cellular biochemical reactions (energy-dependent reactions). ATP (Fig. 4.17) is the widely used energy source in the cell. The three phosphates are named α, β, and γ starting from the pentose end. Pentose-α phosphate bond is an ester bond. α–β bond and β–γ bonds are phosphoanhydride bonds. Cleavage of the ester bond yields 14 kJ/mole of energy and that of the hydride bond yields 30 kJ/mole.

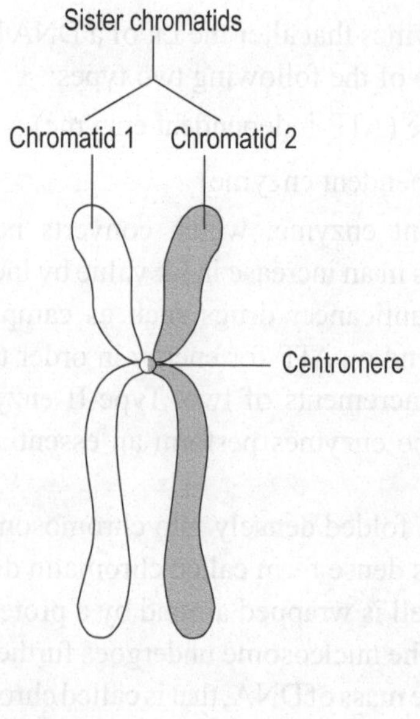

FIGURE 4.16 Structure of chromatid

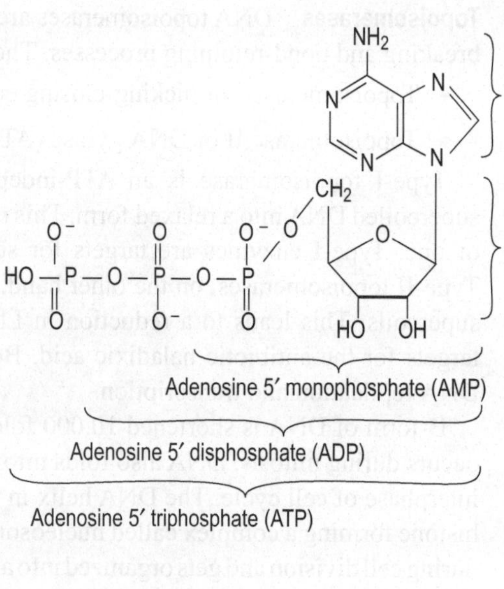

FIGURE 4.17 Structure of ATP

Coenzymes for biochemical reactions Adenosine moiety being a part of various coenzymes and factors is found to enhance the activity of associated enzymes. Removal of adenosine moiety drastically decreases the enzyme activity. Examples of coenzymes include NAD and FAD which catalyse redox reactions.

Regulatory molecules Adenosine 3′, 5′ cyclic monophosphate and guanosine 3′,5′ cyclic monophosphate, commonly called cAMP and cGMP, are produced as second messengers in response to hormones and other signals in the cell and found to have regulatory function (see Chapter 10).

Structure of RNA

Like DNA the structure of RNA also has three levels of organization. Of these, the tertiary structure—commonly called clover-leaf structure—is the most important and involved in protein synthesis.

Primary structure of RNA The primary structure of RNA refers to the sequence of the ribonucleotides and symmetrical phosphodiester bond backbone similar to DNA. The sequence, however, differs among different species and different tissues. In general, the primary structure is mostly single stranded and sometimes double stranded, but unlike DNA, it does not configure into a helix. In 1982, Rong et. al., isolated and determined the primary structure of the 5S RNA of *Philosamia cynthia ricini*.

The primary structure of RNA possesses some functional motifs. Following are a few examples:

- C/D and H/ACA boxes present in small nucleolar RNAs (snoRNAs),
- S_m binding motif of splicesosomal RNA
- Shine–Dalgarno motif and the Kozak consensus motif

Secondary structure of RNA The secondary structure of RNA is called stem loop structure or hairpin loop structure. When two different regions of a single RNA strand are complementary to each other, they attach with one another forming a loop structure (Fig. 4.18a (i) and (ii)). Each loop has at least four bases. At some instances, the bases on one side of the structure cannot form base pairs, this results in a bulge loop as shown in Fig. 4.18b. Another possibility is the interior loop, where bases on both the sides cannot bond as shown in Fig. 4.18c. In some viral genomes, RNA can form a double helix, which differs from DNA helix in that RNA helix is A form in nature. The minor groove of the helix is the target for protein attachment.

Tertiary structure of RNA or tRNA The tertiary structure of mRNA is highly restricted due to the electrostatic repulsion between high negative charges. However, this repulsive force can be counteracted by cations, and folding can be promoted. Millimolar concentration of magnesium stabilizes the RNA tertiary structures by inducing folding. It was also found that the magnesium-induced stabilization of RNA tertiary structure plays a role in catalytic mechanism in ribozymes.

Tertiary structure of RNA is called transfer RNA, tRNA, or clover-leaf structure. This is a very significant structure because it is involved in protein synthesis (translation). RNAs can fold themselves due to complementary base pairing forming a three-looped structure called the clover-leaf structure (Fig. 4.18d). tRNA has four stems/arms and four loops:

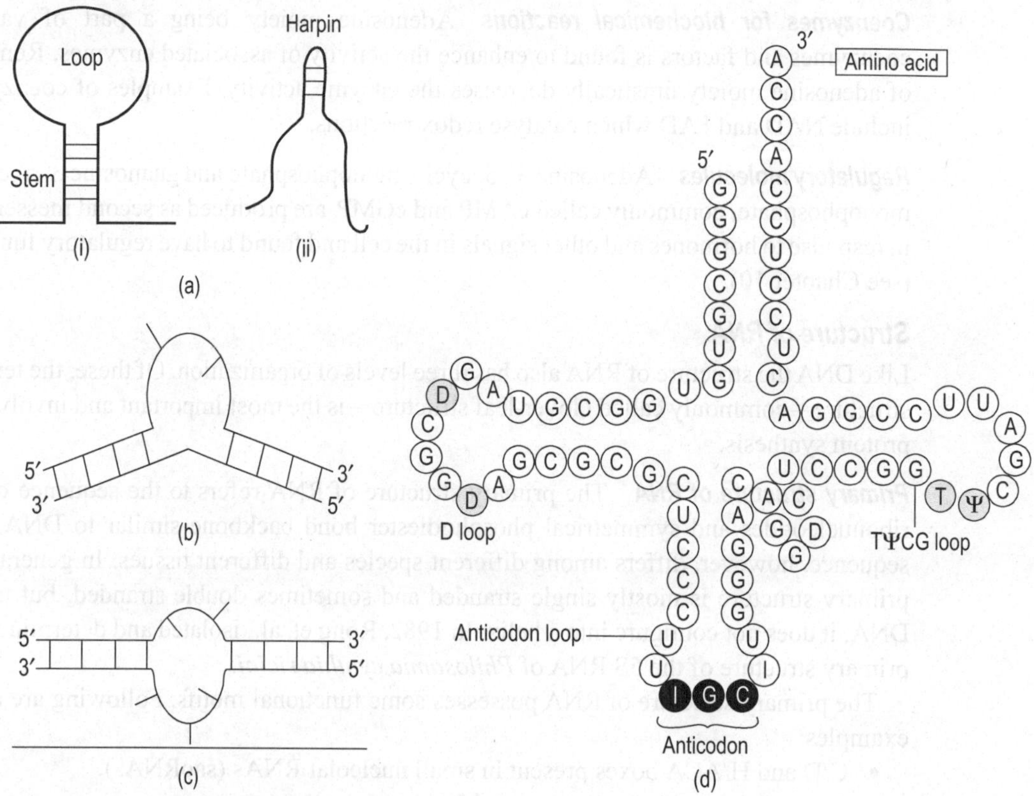

FIGURE 4.18 Structure of RNA: (a) Stem loop, (b) Bulge loop, (c) Interior loop, (d) Clover leaf structure

- Acceptor arm
- Dihydrouridine arm (D arm) with D loop
- Anticodon arm with anticodon loop
- Variable loop (V loop)
- TψC arm with TψC loop

The acceptor arm forms the free 5′ and 3′ end of the RNA molecule. The 3′ end has a conserved (invariable) CCA sequence which is unpaired. This sequence is used as an identity by aminoacyl tRNA synthetase for aminoacylation during translation. This sequence is also used as a primer by tRNA nucleotidyl transferase to begin synthesis of the CCA triplet. The D arm is characterized by modified nucleoside called dihydrouridine. This base is present in the loop of the arm, hence the name D loop. Anticodon arm has a set of three bases called anticodon which base pairs with codon of m-RNA during protein synthesis. Variable loop is a small stretch of RNA with 3–20 nucleotides lying between anticodon arm and acceptor arm. TψC arm has a specific base called thymidine–pseudouridine–cytosine (TψC) present in its loop, hence, the name TψC loop.

Classes of RNA There are three classes of RNA: mRNA, rRNA, and tRNA. These are discussed below.

mRNA is transcribed from DNA, carrying information for protein synthesis. Among the two DNA strands, one is called coding strand and the other is called template strand.

Both are complementary to each other. The sequence of mRNA is complementary to DNA's template strand, and is, therefore, the same as DNA's non-coding strand, except that T is replaced by U. Three consecutive nucleotides in mRNA encode an amino acid or a stop signal for protein synthesis. The trinucleotide is known as a codon. This sequence relationship of DNA, mRNA, and the encoded peptide is shown in Fig. 4.19.

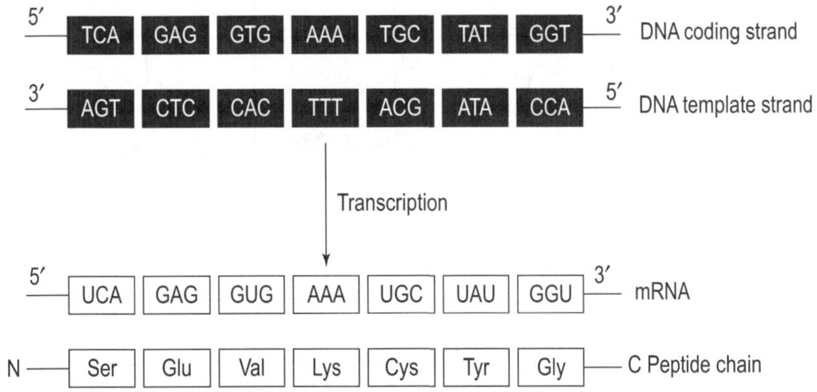

FIGURE 4.19 Sequence relationship of DNA, mRNA, and amino acid

These RNAs have different functions. mRNA is the carrier of genetic information to be translated into proteins. rRNA forms a part of ribosomes to construct protein-synthesizing machinery. tRNA carries amino acids to ribosomes for protein synthesis.

rRNAs are usually expressed in terms of Svedberg units (S) which is a measure of the sedimentation rate. rRNA molecules synthesized in nucleus are transported to the cytoplasm, where they associate with specific proteins to form ribosomal subunits and the subunits then unite to form a structure called ribosome (Table 4.4). In prokaryotes, there are three types of rRNA: 23S, 5S, and 16S. The two prokaryotic ribosomal subunits are 50S and 30S which combine into 70S ribosome. In mammals, there are four types of rRNA: 28S, 5.8S, 5S, and 18S. The two mammalian ribosomal subunits are 60S and 40S which unite into 80S ribosome. Association of rRNAs and proteins to form ribosomes is shown in Fig. 4.20.

tRNA has a central role in biology as the adaptor between mRNA and protein. The tRNA anticodon interacts with the mRNA codon at one end of the L-shaped tRNA allowing an attached amino acid at the other end of the tRNA to condense with the peptide chain on a second tRNA through the action of the ribosome. tRNA is the tertiary structure of RNA and the structure of this RNA has already been described in the previous section and the function will be elaborated under translation (Chapter 6).

TABLE 4.4 Details of prokaryotic and mammalian rRNAs and ribosomal subunits

Components	Prokaryotes			Mammals			
rRNA	23S	5S	16S	28S	5.8S	5S	18S
Ribosomal subunits	50S		30S		60S		40S
Ribosomes	70S						80S

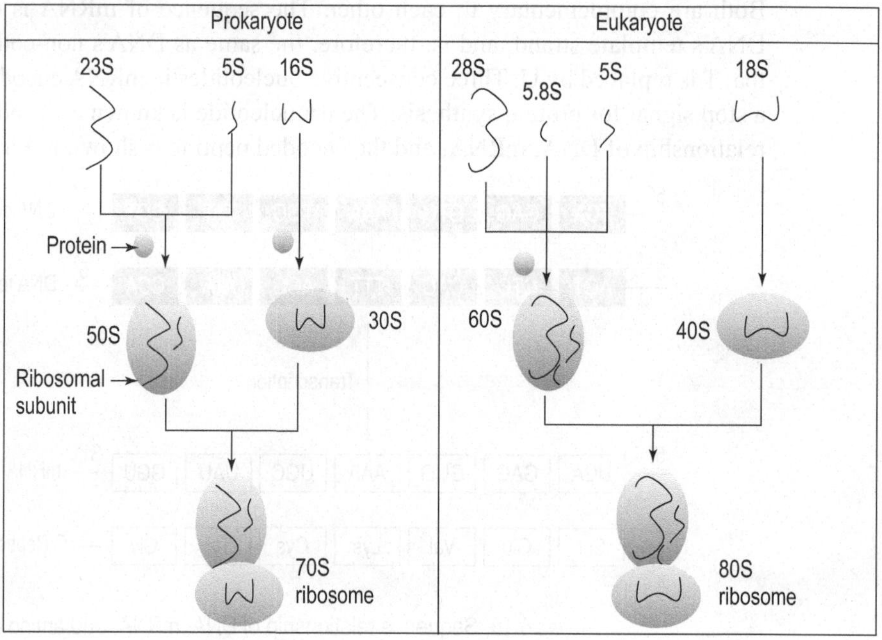

FIGURE 4.20 Formation of prokaryotic and mammalian ribosomes

4.3 AMINO ACIDS AND PROTEINS

Fifty percent of the dry mass of the living system is made of proteins. Hence, proteins are the foremost macromolecules in the biological system. Proteins are synthesized in cytoplasm from RNA by a process called translation. Proteins are linear, unbranched polymers of twenty different α-amino acids that are encoded in the DNA of the genome.

4.3.1 Types of Proteins Based on Function

Proteins have a diverse structure and perform a wide range of functions. Thus, there are two ways of classifying proteins: based on structure and based on function (Table 4.5).

On the basis of their biological function, proteins may be divided into the following eight types:

Enzymes Proteins which catalyse biochemical reactions, for example, urease.

Transport proteins Proteins which transport biological materials; for example, haemoglobin transports oxygen, lipoproteins transport lipids, and transferrin transports iron.

Storage proteins Proteins which store biological compounds, for example, ferritin is the storage form of iron.

Contractile proteins Proteins which perform mechanical work such as muscle contraction, for example, actin and myosin.

Structural proteins Proteins associated with biological structures, for example, collagen in tendon and cartilage, elastin in ligaments, and keratin in hair and nails.

TABLE 4.5 Classes of proteins and their functions

Protein	Example	Function
Enzymes	Urease	Cleavage of urea to form ammonia
	Trypsin	Digestion of proteins
Hormones	Insulin	Regulates blood sugar level
Structural proteins	Keratin	Structural proteins of hair horns, nails, claws and hooves of mammals
	Collagen	Structural proteins of connective tissue
Contractile proteins	Actin and myosin	Involved in muscle contraction
Storage proteins	Ferritin	Stores iron
Immunologic proteins (Defence protein)	Immunoglobulin (Antibodies)	Helps combating antigens and pathogens
Transport proteins	Haemoglobin	Transports oxygen
Nutrient protein	Casein	Provides nutrition

Defence proteins Proteins which provide protection against invading bacteria and other foreign organisms, for example, immunoglobin.

Regulatory proteins or hormones Proteins which regulate metabolic processes, for example, insulin regulates blood glucose.

Nutrient proteins Proteins which provide nutrition, for example, ovalbumin (egg protein) and casein (milk protein).

On the basis of their chemical compositions, proteins may be classified into the following two types—simple proteins and conjugated proteins.

Simple proteins are made of amino acids only. Albumin and globulin are examples of simple proteins. On the other hand, conjugated proteins are made of a non-protein part (called prosthetic group) attached to a simple protein. Depending on the prosthetic group, conjugated proteins are subdivided into the following four types:

- Glycoprotein with carbohydrates as prosthetic group (e.g., immunoglobin)
- Lipoprotein with lipids as prosthetic group [e.g., high-density lipoprotein (HDL)]
- Nucleoprotein with nucleic acid as prosthetic group (e.g., spliceosome, histone)
- Metalloprotein with metals as prosthetic group (e.g., catalase has iron, glutathione peroxidase has selenium)

4.3.2 Structure of Amino Acids

Proteins are polymers made of individual building blocks called amino acids. Amino acids are colourless, crystalline solids. They are water-soluble, high-melting solids and behave like salts. The general structure of an amino acid is represented in Fig. 4.21. It has a central carbon which is called α-carbon. The α-carbon is linked to the following four groups:

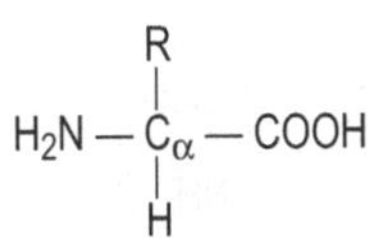

FIGURE 4.21 General structure of amino acid

1. Amino group ($-NH_2$)
2. Carboxyl group (–COOH)
3. Hydrogen atom (–H)
4. Variable group (–R) which is specific for each amino acid

There are twenty possible variable groups giving rise to twenty amino acids. These twenty amino acids can conventionally be grouped under different families. The family, name, and structure of all twenty amino acids are given in Fig. 4.22.

Zwitter ions

Molecules which are electrically neutral with no net charge are called zwitter ions. In an aqueous solution, the carboxyl group of an amino acid is dissociated and negatively charged, while the amino group is protonated and positively charged. Thus, amino acids exist as dipolar molecules; these molecules are called zwitter ions. The characteristic pH at which the net electric charge is zero is called the isoelectric point or isoelectric pH, represented by pI. In acidic and alkaline solutions, amino acids bear net positive or negative charge, respectively. In an acidic solution, that is, at low pH, carboxyl group of the amino acid accepts a proton and becomes uncharged making the molecule positive. In an alkaline solution, that is, at high pH, amino group of the amino acid becomes uncharged making the molecule negative.

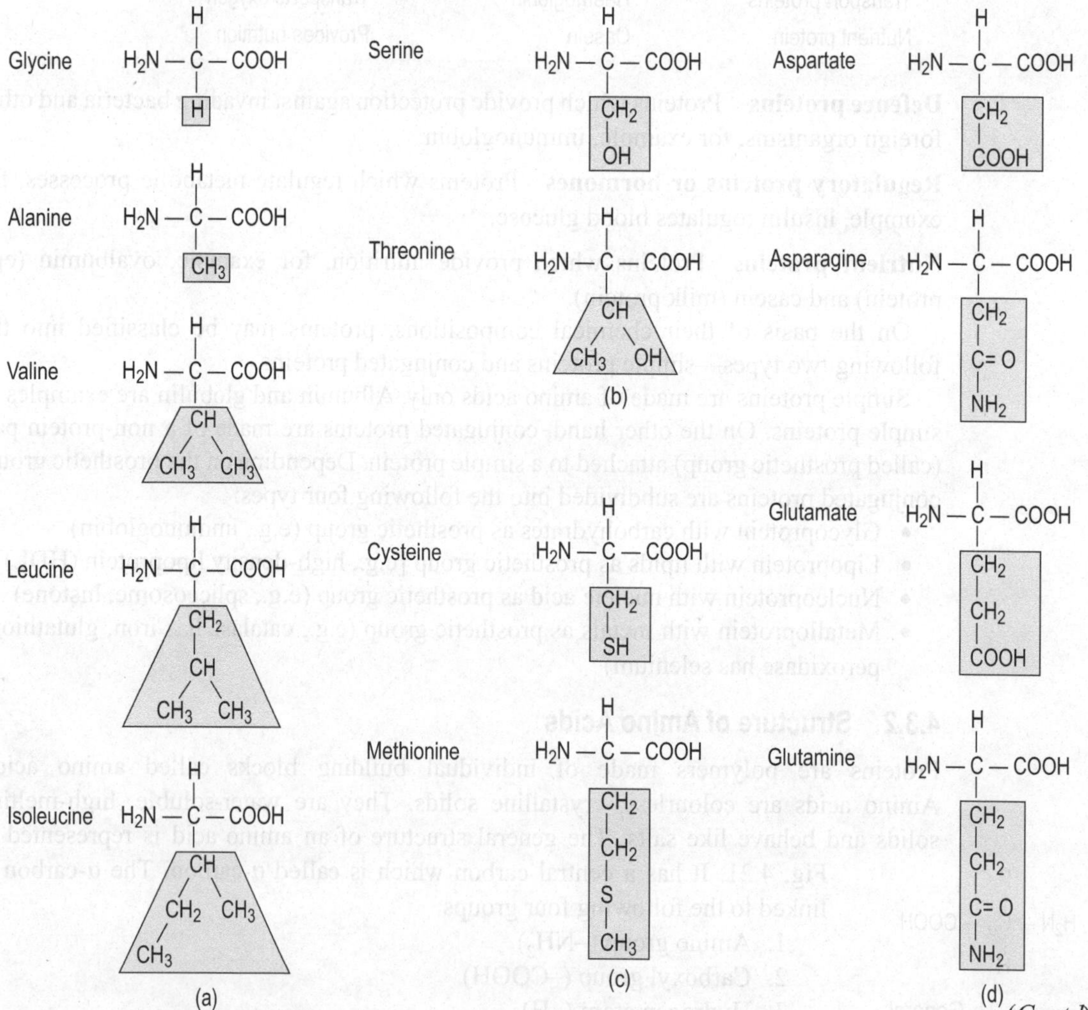

(Contd)

FIGURE 4.22 Family, name, and structure of amino acids: (a) Amino acids with aliphatic side chains–5 (b) Amino acids with hydroxyl group–2 (c) Sulphur containing amino acids–2 (d) Amino acids with acid groups and their corresponding amide group– 2 + 2

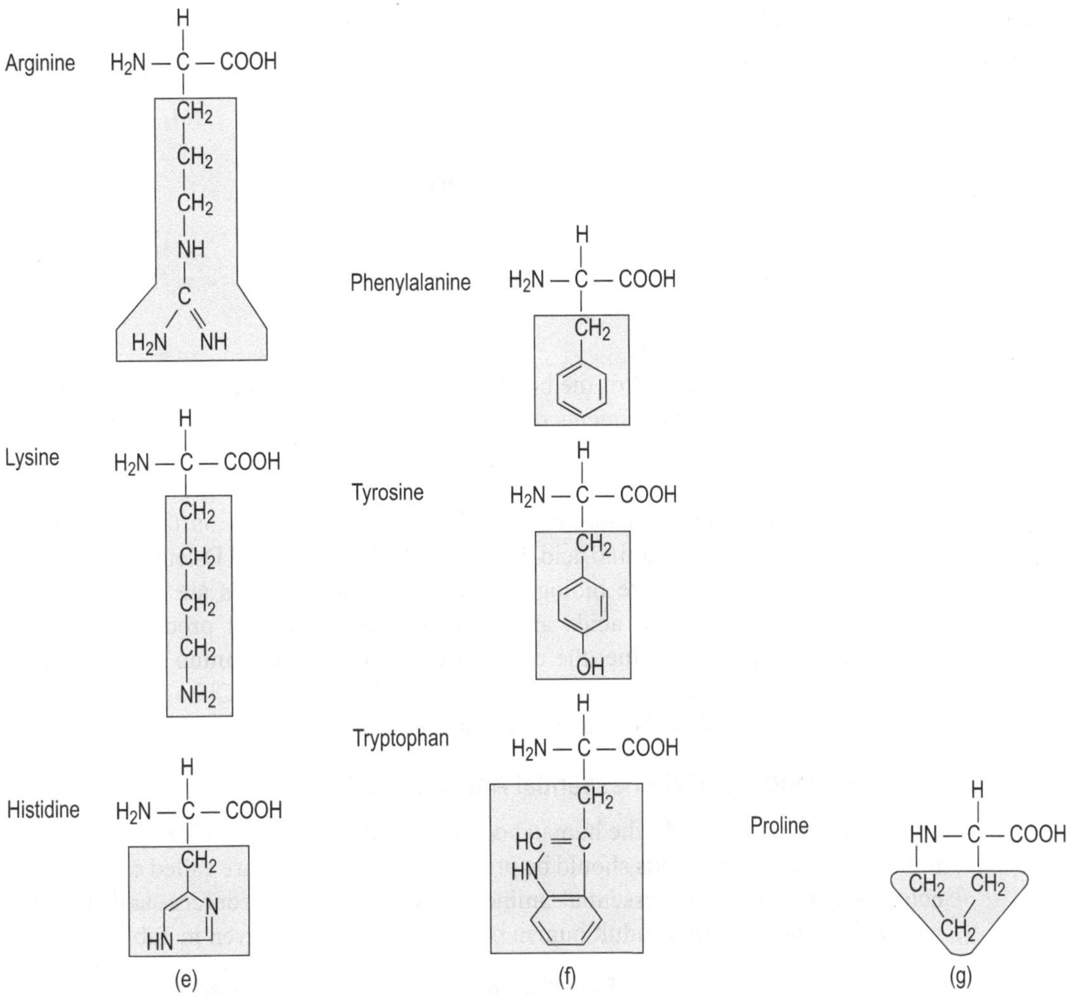

(e) Amino acids with basic group–3 (f) Aromatic amino acids–3 (g) Iminoacid–1

Zwitter ionic form and charged forms of amino acids are shown below:

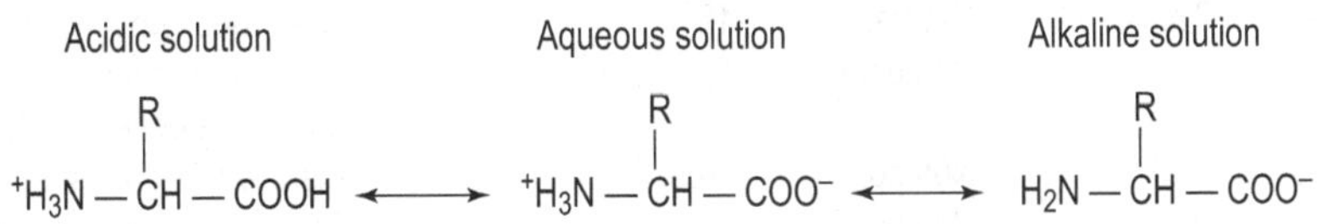

Ampholytes

Substances which have a dual nature, that is, which can act both as an acid (proton donor) and a base (proton acceptor) are called amphoteric substances or ampholytes. Amino acids are amphoteric in nature because the zwitter ionic form of amino acids can donate as well as accept protons shown as follows:

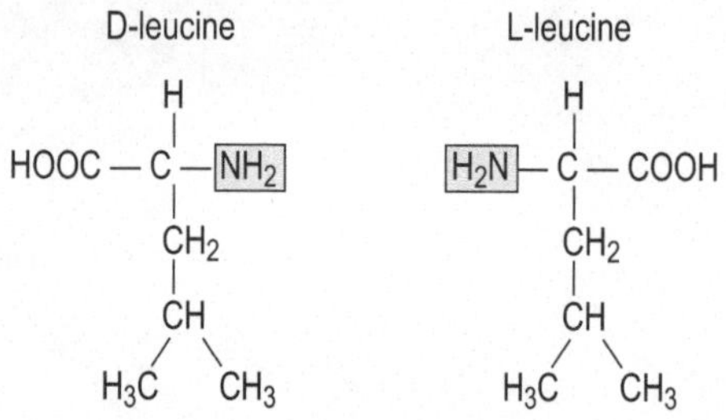

FIGURE 4.23 Structure of D-leucine and L-leucine

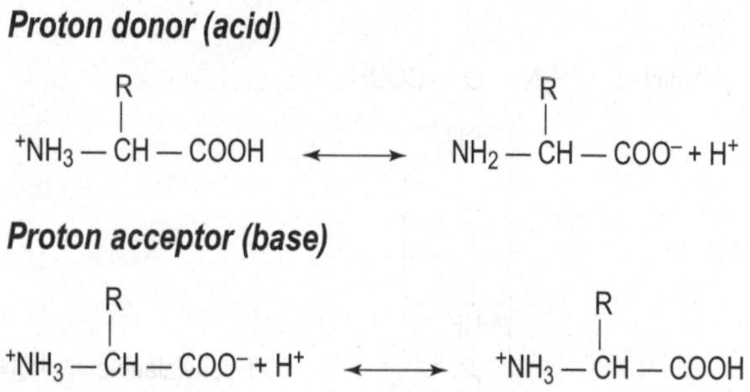

Isomerism in amino acids

The tetrahedral arrangement of the bonding orbital around the alpha carbon atom results in two unique spatial arrangements of the four different groups. Thus, amino acids have two possible stereoisomers which are non-superimposable mirror images of each other. These isomers are called enantiomers. Amino acid with its amino group at the right side of the α-carbon is called a D-amino acid. If the amino group is on the left side of the α-carbon, it is called an L-amino acid. For example, L-leucine and D-leucine are shown in Fig. 4.23. L-amino acids are biologically active. L-amino acids are incorporated into proteins, whereas D-amino acids are found in many bacterial products and peptide antibiotics. Except for glycine, the central carbon of all other amino acids is linked to four different groups. The central carbon is called chiral carbon. Chiral carbon gives rise to optical activity, that is, they can rotate plane polarized light.

4.3.3 Essential and Non-essential Amino Acids

Out of twenty amino acids, the human body can synthesize only twelve amino acids. The remaining eight amino acids should be provided in the diet and are called essential amino acids. A list of the eight essential amino acids, their dietary sources, and the amount required by the body of an adult human being as per WHO is given in Table 4.6.

TABLE 4.6 Essential amino acids (eight out of twenty)

Name of the essential amino acid	Abbreviation	Nature of the amino acid	Recommended amount to be taken in diet (mg/Kg body weight)	Dietary source
Threonine	Thr	Neutral	15	Dairy, beef, poultry, eggs, beans, nuts, and seeds
Tryptophan	Trp	Non polar	4	Chocolate, oats, bananas, dried dates, milk, cottage cheese, meat, fish, turkey, and peanuts
Methionine	Met	Non polar	10.5	Fish, whole grains, and dairy
Isoleucine	Ile	Non polar, branched	20	Eggs, fish, lentils, poultry, beef, seeds, soy, wheat, almonds, and dairy
Leucine	Leu	Non polar, branched	39	Cottage cheese, sesame seeds, peanuts, dry lentils, chicken, and fish
Lysine	Lys	Basic	30	Green beans, lentils, soy bean, spinach, and amaranth
Valine	Val	Non polar, branched	26	Dairy products, grain, meat, mushrooms, peanuts, and soy proteins
Phenylalanine	Phe	Non polar	25	Dairy, almonds, avocados, lima beans, peanuts, and seeds

4.3.4 Non-standard Amino Acids

In addition to the twenty standard amino acids mentioned in Fig. 4.22, other amino acids, called non-standard amino acids, are also found in some proteins. These amino acids are usually modified forms of standard amino acids. 4-hydroxyproline, 5-hydroxylysine, N-methyllysine, γ-carboxyglutamate, desmosine, and selenocysteine are well-known examples of non-standard amino acids. Precursors and occurrence of these non-standard amino acids are given in Table 4.7. The structure of these amino acids is shown in Fig. 4.24.

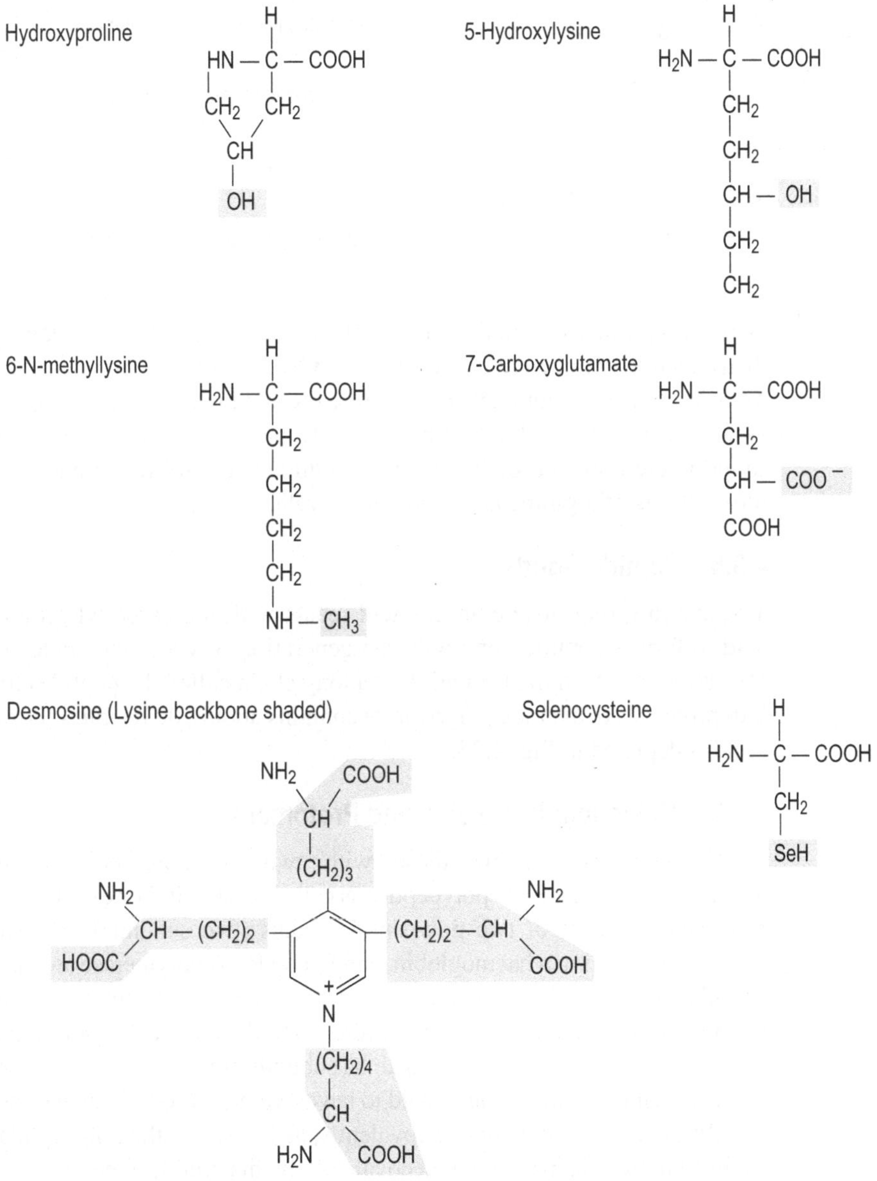

FIGURE 4.24 Structure of non-standard amino acids (modified groups are shaded)

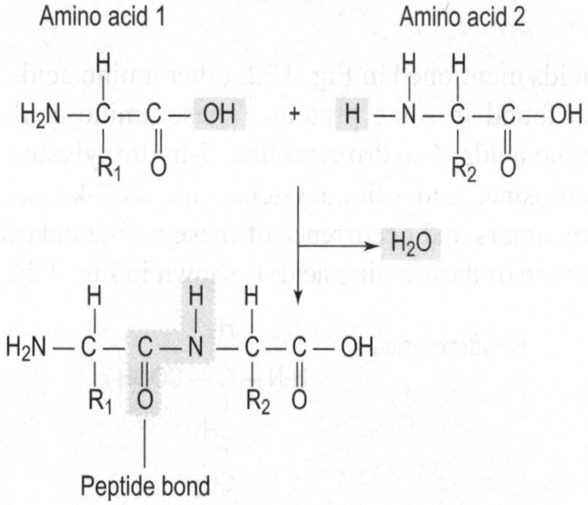

FIGURE 4.25 Formation of peptide bond between two amino acids

TABLE 4.7 Non-standard amino acids

Non-standard amino acids	Precursor	Occurrence
4-hydroxyproline	Proline	Plant cell wall protein
5-hydroxyproline	Proline	Collagen of connective tissue
N-methyllysine	Lysine	Myosin (muscle protein)
γ-carboxyglutamate	Glutamate	Prothrombin (blood clotting protein)
Desmosine	Lysine	Elastin (fibrous protein)
Selenocysteine	Cysteine	Glutathione peroxidase (detoxification enzyme)

4-hydroxyproline is found in plant cell wall proteins, and in collagen, a fibrous protein distributed in connective tissues. N-methyllysine is the component of the muscle protein myosin, γ-carboxyglutamate residues are present in the blood clotting protein prothrombin as well as in calcium-binding proteins. Desmosine, a derivative of four Lys residues, is found in the fibrous protein elastin. Ornithine and citrulline are key intermediates in the biosynthesis of arginine and in the urea cycle.

4.3.5 Peptide Bonds

The amino group of one amino acid reacts with the carboxyl group of another amino acid to form a peptide bond with the generation of water molecule. Thus, peptide-bond formation results in the formation of a long chain called the peptide chain which is folded into proteins. An average polypeptide contains about 300 residues. Formation of a peptide bond is depicted in Fig. 4.25.

4.3.6 Multisubunit Proteins and Protomers

Multisubunit proteins are those with two or more polypeptides associated via non-covalent bonds. Each polypeptide is called a subunit. For example, hexokinase sourced from yeast consists of two polypeptide chains (two subunits), whereas bovine pancreas chymotrypsin, human haemoglobin, and E. coli RNA polymerase are made of four, five, and six subunits, respectively. If at least two subunits of a multisubunit protein are identical then the protein is termed an oligomeric protein and the identical subunits are called protomers. For example, in haemoglobin, two alpha subunits are identical and the two beta subunits are identical. Thus, haemoglobin is said to have two protomers. If the individual polypeptide of a multisubunit protein is linked covalently, as in insulin, the polypeptide is termed a chain. (In insulin, two chains are linked covalently by disulphide bonds.)

Proteins should be purified before the determination of its structure. The techniques used for protein purification are discussed in Chapter 14. The structure of proteins can be described under the following four levels:

TABLE 4.8 Examples of the four levels of structural organization in proteins

Structural level	Example
Primary	Cys-Glu-Gly for glutathione
Secondary	Alpha keratin
	Silk fibroin
	Collagen
Tertiary	Myoglobin
Quaternary	Haemoglobin

1. Primary structure
2. Secondary structure and super secondary structure
3. Tertiary structure
4. Quaternary structure

Examples for these structures and factors stabilizing these structures are summarized in Table 4.8. Diagrammatic representation of the four levels of structural organization in proteins is given in Fig. 4.26.

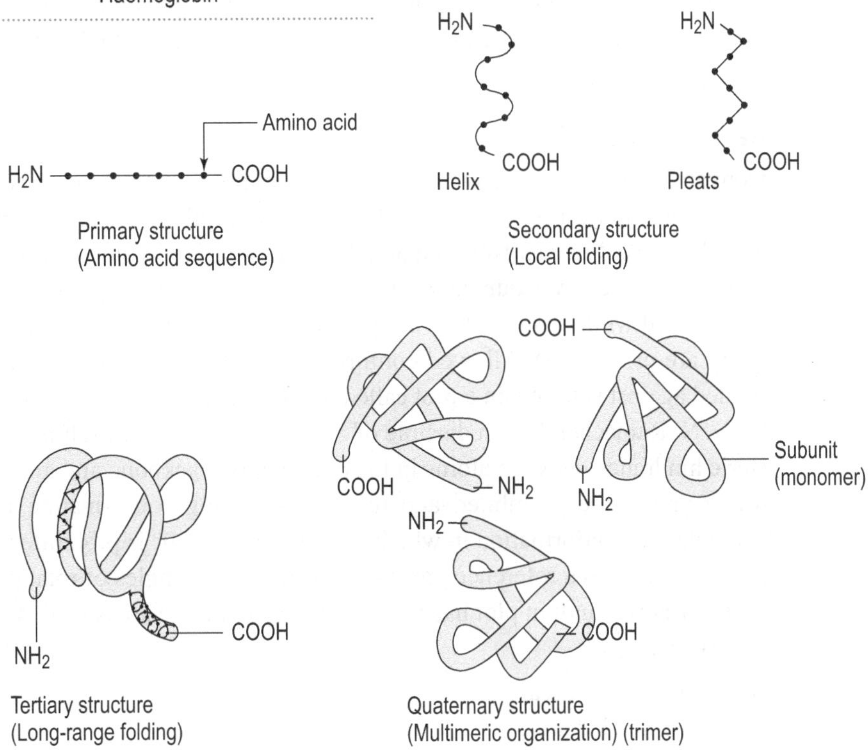

FIGURE 4.26 Structure of proteins

4.3.7 Primary Structure

Peptide bonds link amino acid residues within proteins. The sequence of covalently linked amino acids with amino group at the left end and carboxyl group at the right end is known as the primary structure of a protein. Hence, primary structure deals with amino acid sequencing and peptide-bond backbone.

Peptide bond

The peptide bond backbone of a polypeptide chain is considered a series of rigid planes with consecutive planes sharing a common point of rotation at C_α. Two adjacent C_α are

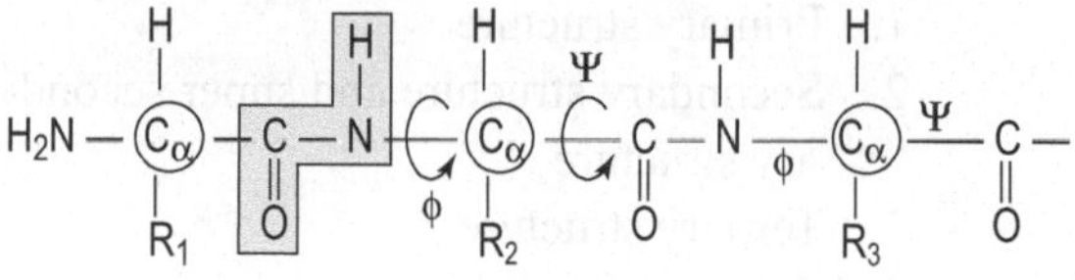

FIGURE 4.27 Peptide bond backbone (peptide bond is shaded)

separated by three covalent bonds denoted as 'C_α–C–N–C_α'', where C–N is the peptide bond. The peptide bond (C–N) is unable to rotate freely with bond angles designated Φ and ψ. Because of their partial double bond character, C_α–C and N–C_α bonds can rotate freely. In peptide bond, the carbon atom is contributed by the carbonyl carbon and the nitrogen is contributed by the amide nitrogen. The atoms associated with the peptide bond are coplanar. There is a resonance or partial sharing of two pairs of electrons between the carbonyl oxygen and the amide nitrogen (Fig. 4.27). The oxygen has a partial negative charge and the nitrogen has a partial positive charge, setting up a small electric dipole. The oxygen atom of the carbonyl group and the hydrogen atom of the amide nitrogen are trans to each other.

Ramachandran plot

Conventionally, the bond angles resulting from rotations at C_α are labelled Φ (phi) for the N–C_α bond and ψ (psi) for the C_α–C bond. For a fully extended polypeptide chain, both Φ and ψ are defined as 180° and all peptide groups are in the same plane. In principle, Φ and ψ can have any value between –180° and +180°. Among these values, some values are allowed while many values are prohibited (i.e., some conformations are allowed and many are prohibited). Allowed conformations (allowed values) are revealed when Φ is plotted against ψ to obtain a plot called the Ramachandran plot. This plot was introduced by G. N. Ramachandran, a disciple of Sir C.V. Raman. The prohibition values are due to steric hindrance between atoms in the polypeptide backbone and amino acid side chains. Many values are prohibited and the corresponding conformation is not formed. For example, the conformation in which both Φ and ψ are 0° is prohibited; this conformation is used merely as a reference point for describing the angles of rotation.

The Ramachandran plot has four quadrants and three areas (Fig. 4.28).

1. White area
2. Dark shaded area
3. Light shaded area

White areas correspond to prohibited conformations. In this conformation, atoms in the polypeptide come closer than the sum of their van der Waals radii. These regions are sterically disallowed for all amino acids, except glycine, which is unique in that it lacks a side chain.

Dark shaded areas correspond to allowed conformations, namely the alpha-helical and beta-sheet conformations. In these conformations, there are no steric clashes.

Light shaded area shows the slightly allowed conformations such as the left-handed alpha-helix. In this conformation, slightly shorter van der Waals radii allow the atoms to come a little closer together.

Glycine has no side chain, and therefore, can adopt Φ and ψ angles in all four quadrants of the Ramachandran plot. Hence, it frequently occurs in turn regions of proteins where any other residue would be sterically hindered.

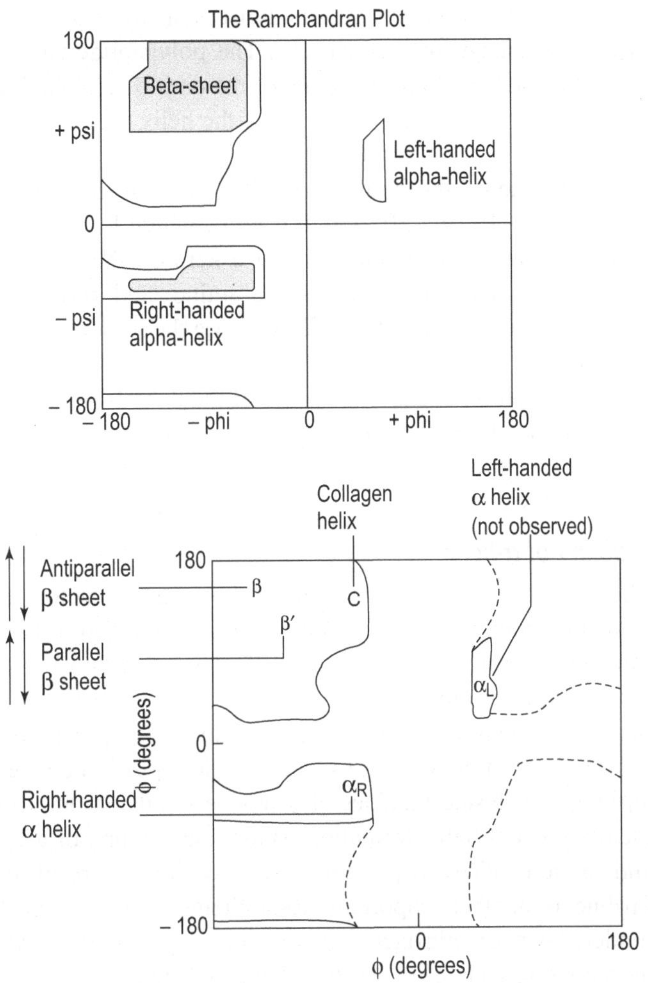

FIGURE 4.28 Ramachandran plot (Two ways of representation)

4.3.8 Secondary Structure

The secondary structure is formed by hydrogen bonding between side chains of different amino acid residues resulting in intricate foldings and bendings of peptide chains (the primary structure). Two types of secondary structures are known—alpha helix, in which the peptide chain is coiled, and β-pleated sheet in which the peptide chain is bent to form zigzag conformation. If a pair of β-pleated sheets has their animo terminal end in opposite direction, they are called antiparallel sheets; if the two amino terminal ends are in the same side, the sheets are called parallel sheets. Helix and pleated sheet structures are stabilized by the hydrogen bond formed between nitrogen atom and oxygen atom of different amino acid residues within the chain. In the secondary structure, all the R groups face outwards.

α-helix is the simplest secondary structure. It was discovered by the scientists Linus Pauling and Robert Corey in the year 1951. This structure is shown to form more readily than other conformations and the twist is always right handed. The α-helix is proposed to have an imaginary helical axis. The polypeptide backbone is tightly wound around the helical axis and the R groups of the amino acid residues protrude outward from the helical backbone. Each single turn of the helix is about 5.4 Å containing 3.6 amino acid residues.

In an α-helix, the amino acids have a conformation contributed by the following angles: $\psi = -45°$ to $-50°$ and $\Phi = -60°$, which leads to a right-handed twist. This type of conformation is predominant in α-keratin, the structural protein of skin and nail. Moreover, about one-fourth of the amino acid residues of any polypeptide possess α-helical conformation. The following are the reasons contributing to the high frequency of α-helix:

- Association of successive turn of the helix by three to four hydrogen bonds.
- Stabilization of the helix by all the existing hydrogen bonds.
- Higher probability of L-amino acids (naturally occurring amino acids) to form α-helical turns.

Stability of α-helix

The stability of the helix is influenced by various factors as listed in Table 4.9. Ion-pair formation and hydrophobic interaction are involved in twisting the helix conformation. Ion pair is formed between positively charged amino acids, which are often found three residues away from negatively charged amino acids. Hydrophobic interaction is formed between two aromatic amino acid residues. The presence of glutamate residues results in negative charge leading to electrostatic repulsion between the side chains. Similarly, arginine and lysine residues also induce repulsion due to positive charges. Both these factors prevent helix formation. Bulky side chains of asparagines, threonine, cysteine, and serine residues, if present close to each other, result in destabilization of the helix. Proline is another important destabilizing factor. In proline, N–C_α bond rotation is hindered because nitrogen is involved in ring formation. As a result, proline introduces a destabilizing kink. Thus, proline, being a destabilizing factor, is rarely found in α-helix.

The side chains of the amino acid residues on the β sheets must be small. Thus, glycine and alanine are found extensively in proteins with β sheets. For example, in silk fibroin, glycine and alanine are frequently found.

TABLE 4.9 Factors stabilizing α-helix structure of proteins

Factor	Influence on the helical structure
Ion-pair formation	Responsible for twisting the helix
Hydrophobic interaction between two aromatic amino acid residues	Responsible for twisting the helix
Glutamate residues	Prevent helix formation due to negative charge repulsion
Arginine or lysine residues	Prevent helix formation due to positive charge repulsion
Asparagine, threonine, cysteine, and serine residues	Destabilize the helix due to bulky side chain
Proline	Destabilizes the helix due to ring structure

β turn

β turn is another form of secondary structure found near the surface of the protein. Each β turn is formed by four amino acid residues. In each turn, the carbonyl oxygen of the first residue is hydrogen bonded to amino group hydrogen of the fourth residue. On the other hand, the second and the third residues are free from bonding. This structure occurs as a connecting element between two adjacent polypeptide chains bearing helical or sheet conformation. Turns linking two antiparallel sheets are more common. β turn is predominant in glycine and proline. Glycine is small in size and highly flexible. Proline gives a tight turn due to the cis configuration of the peptide bond formed by this amino acid.

Left-handed helix

All the helixes have a right-handed twist, for example, α-keratin found as a structural component of hair, nail, wool, horns, and feathers. However, left-handed helix is a distinct secondary structure found in collagen, a structural protein found in bones matrix and tendons. This left-handed α-helix may some times be considered as a separate type of secondary structure. The helix is made of repeating tripeptide sequence Gly–X–Pro or Gly–X–4-Hyp with three residues per turn. Collagen exists as a fibril 3,000 Å long and only 15 Å thick made of three intertwined α-chains (hence, the name triple helix) each having about 1,000 amino acid residues.

4.3.9 Super Secondary Structure

At certain instances, the secondary proteins are orderly arranged or folded into substructures, which are discrete structural domains or functional motifs. Such domains and motifs are called super secondary structures. There are four types of super secondary structure in proteins (Fig. 4.29). They are

- β-hairpin turn—Two antiparallel sheets linked by a hairpin loop
- β-meander—Three antiparallel sheets linked by a hairpin loop
- α-hairpin—Two helixes linked by a hairpin loop
- α-β-motif—Two antiparallel sheets linked by a helix

4.3.10 Tertiary Structure

Tertiary structure represents the spatial arrangement of the secondary structures. Molecular interactions among R groups of different amino acid sequences result in three-

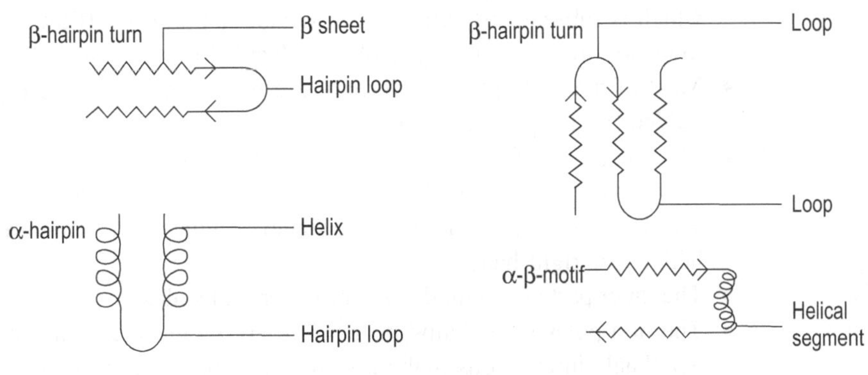

FIGURE 4.29 Super secondary structures in proteins

dimensional or tertiary structure comprising completely folded and compacted single protein molecule. This may have both helical and pleated sheet conformations. In the 3-D structure, hydrophobic amino acid side chains are oriented toward the interior (away from water), whereas hydrophilic side chains are on the surface. Tertiary structure is stabilized by a hydrogen bond, salt bridge, and van der Waals force. Examples of tertiary proteins are myoglobin and dihydrofolate reductase.

The tertiary structure is found in globular proteins, such as enzymes, transport proteins, motor proteins, regulatory proteins, immunoglobulin, (e.g., myoglobin), ribonuclease, cytochrome c, and lysozyme. In a globular protein, different segments of a polypeptide chain (or multiple polypeptide chains) fold back on each other. The folding also provides the structural diversity necessary for proteins to carry out a wide array of biological functions.

Myoglobin

Tertiary structures of proteins were first studied in myoglobin by X-ray diffraction analysis. Studies on myoglobin allowed for the first time to understand the correlation between structure and function of proteins. Myoglobin is an oxygen-carrying protein of muscle cells. It contains a single polypeptide chain of 153 amino acid residues of known sequence and a single iron protoporphyrin, or haeme, group. The polypeptide backbone has α-helical regions. Tertiary structure of myoglobin reveals the following:

- 70% of the poplypeptide has helical structure.
- All the peptide bonds are in the planar trans configuration.
- Three of the four proline residues of myoglobin are found at bends. The fourth proline residue occurs within an α-helix, where it creates a kink necessary for tight helix packing
- The peptide chain is folded into a very compact structure surrounding the haeme group. The haeme group lies in the crevice or pocket in the structure. The iron atom in the centre of the haeme group has two bonding (coordination) positions perpendicular to the plane of the haeme. One of these is bound to the R group of the His residue at position 93, the proximal residue; the other is the site at which an O_2 molecule binds, it is the distal residue. The distal group associates with the dioxygen entity via hydrogen bonds and increases the oxygen-binding constant. Whenever there is oxygen depletion, the oxymyoglobin dissociates from its bound oxygen, which is subsequently utilized for metabolic purposes. If the iron atom is oxidized to ferric form, then the myoglobin molecule loses its capacity to bind to oxygen.
- Most of the hydrophobic R groups are in the interior of the myoglobin molecule, hidden from exposure to water.
- The backbone of the myoglobin molecule is made up of eight relatively straight segments of right-handed α-helix interrupted by bends, some of which are β-turns. The longest α-helix has 23 amino acid residues and the shortest has only 7; all helixes are right-handed.
- The inner part accommodates four water molecules.
- The non-polar side chains in the core are so close together that short-range van der Waals interactions make a significant contribution to stabilizing hydrophobic interactions.

Other examples for tertiary structure

Cytochrome c is a haeme-containing protein. It is composed of a single polypeptide chain with a single haeme group, which is covalently attached to Cys14 and Cys17. Forty per cent of the polypeptide is twisted into a helix, while the remaining portions have β-turns, irregular coils, and extended segments.

Lysozyme is a polypeptide with 129 amino acid residues. Forty per cent of the chains have helical conformation which surrounds a cleft or cavity called the active site. The structure is stabilized by four disulphide bonds.

Ribonuclease (RNAse) has 124 amino acids. The chain has few segments as helix and many segments as β- conformations. The structure is stabilized by four disulphide bridges.

4.3.11 Quaternary Structure

Quaternary structure is the arrangement in space of two or more polypeptide chain which are individually folded and compacted. Each chain in the assembly is called a subunit. The nomenclature of the subunits in quaternary structure is written as follows:

- Monomer—single subunit
- Oligomer—multiple subunits
- Protomer—identical units within the oligomer
- Dimer—two subunits
- Trimer—three subunits
- Tetramer—four subunits

Conventionally, the quaternary proteins are represented in terms of their structural subunits. The subunits are denoted using Greek letters. For example, haemoglobin has four subunits, which can be denoted as $\alpha_2\beta_2$. HIV protease (a retroviral protease with aspartate as the active-site residue, that is crucial for the life-cycle of the human immunodeficiency virus) has two hydrogen-bonded identical subunits, with 180° rotation around a vertical axis, represented as α and β.

The subunits of the quaternary structure exhibit different types of symmetry (Fig. 4.30) so that the individual subunits are superimposed on one another by one or more rotational axes. Different types of symmetry are listed as follows:

1. Rotational symmetry
   (a) Cyclic symmetry
   (b) Dihedral symmetry
   (c) Icosahedral symmetry
2. Helical symmetry

Rotational symmetry

In rotational symmetry, the subunits pack about the rotational axis forming a closed structure. Cyclic symmetry, dihedral symmetry, and icosahedral symmetry are the three types of rotational symmetry.

In cyclic symmetry, subunits are rotated about a single axis. This symmetry is denoted as C_n where C = cyclic and n = number of subunits packed along the axis. The axis is called n-fold rotational axis. In haemoglobin, alpha and beta subunits are related by C_2 symmetry.

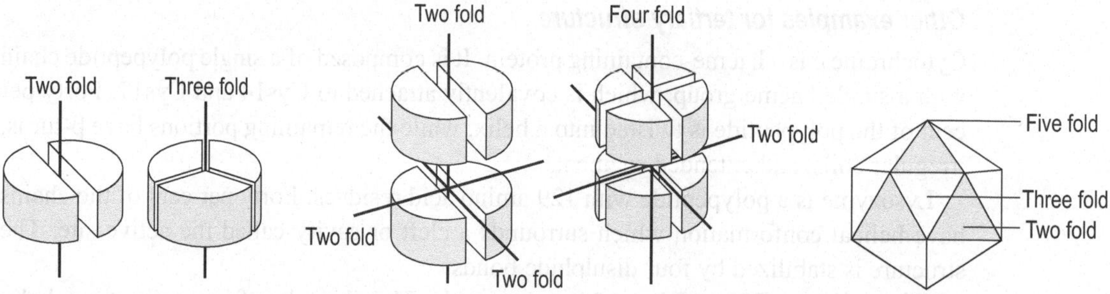

FIGURE 4.30 Symmetry of protein subunits in quaternary structure

In dihedral symmetry, a two-fold rotational axis intersects an n-fold axis at right angles. The symmetry is defined as D_n. A protein with dihedral symmetry has 2^n protomers. Haemoglobin contains four polypeptide chains (hence, called tetramer) and four haeme prosthetic groups, in which the iron atoms are in the ferrous (Fe^{2+}) state. Of these four chains, two are called alpha chains (or subunits) and the other two are called beta chains (or subunits) (hence, the protomers are called αβ protomers). The subunits are arranged in symmetric pairs; each pair having one alpha and one beta subunit.

An icosahedron is a regular 12-cornered polyhedron having 20 equilateral triangular faces. The proteins are supposed to be arranged in the triangular faces. All the faces of the icosahedrons can be related by rotation about one or more of the three separate rotational axes: two-fold, three-fold, and five-fold. Icosahedral symmetry is found in viral capsid coat protein. In human polio virus, three protomers (each made of four polypeptide chains) are packed in each triangular face. Thus, the capsid has a total of 60 protomers.

Helical symmetry

In helical symmetry, subunits are packed in a spiral array with open ends. Helical symmetry is also found in viral capsid. The capsid of tobacco mosaic virus has a helical symmetry because of the spiral arrangement of subunits.

4.3.12 Protein Folding

Protein folding is a physical process by which the polypeptide is folded to achieve its thermodynamically stable, three-dimensional shape to perform its biological functions. Proteins, newly synthesized in cytoplasm are translocated into the ER, where they fold into their three-dimensional structures and become biologically active and functional. Correctly folded proteins are then transported to the Golgi complex and then delivered to the extracellular environment. However, incorrectly folded (misfolded) proteins are detected by a quality-control mechanism and sent along another pathway (the unfolded protein response) in which they are ubiquitinated and then degraded in the cytoplasm by proteasomes. The overview of proteins synthesis, folding, and degradation is summarized in Fig. 4.31.

Molecules assisting protein folding

The following four molecules assist in the process of protein folding:

- Chaperones (e.g., HSP70, DNaK, DNaJ)

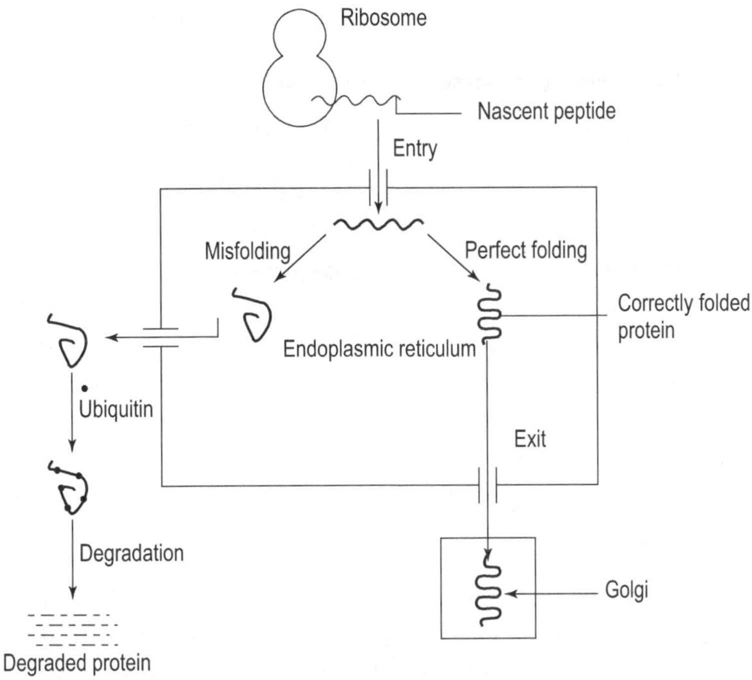

FIGURE 4.31 Overview of protein synthesis and folding

- Chaperonins
- Proteins disulphide isomerase (PDI)
- Protein prolyl cis-trans isomerase (PPI)

Each of these molecules is briefly discussed here.

Chaperones Chaperones are specialized proteins that convert misfolded proteins into perfectly folded proteins. Heat shock protein which has a molecular weight of 70 kDa and is represented as HSP70 is a well-studied chaperone. It acts by the following mechanism:

- By preventing inappropriate aggregation by binding to hydrophobic residues of the polypeptide
- By facilitating quaternary assembly of the protein subunits of oligomers

DnaK and DnaJ are chaperones of E. coli. These chaperones, in the presence of ATP, assist proper folding of proteins which involves the following steps:

- DnaJ first binds to the unfolded or partially folded proteins and then to DnaK which is associated with ATP (DnaK–ATP).
- DnaJ catalyses DnaK-induced hydrolysis of ATP forming DnaK–ADP.
- DnaK–ADP binds tightly to the unfolded proteins.
- ADP is released by nucleotide-stimulating factor–GrpE.
- Another ATP associates with DnaK and the protein is dissociated as a properly folded element.

The other families of chaperones are listed in Table 4.10.

TABLE 4.10 Chaperone family

Neurodegenerative disease	Genes involved	Chaperones
Parkinson's disease	Alpha-synuclein, parkin, UCHL1, PINK1, DJ1, HTRA2, LRRK2	HSP70 , HSP40, HSP90, HSC70, CDC37, TRAP1, 14-3-3, HSPA9, CHIP
Alzheimer's disease	APP, presenilin 1, presenilin 2	HSP72 , HSP28, HSP27 , HSP90
Dementia	Alpha-synuclein	HSP27 , HSP40, HSP60, HSP70, HSP90, HSPA5
Familial amyotropic lateral sclerosis	SOD1	HSC70
Huntington's disease	Huntingtin	HSP40, HSP70
Spinocerebellar ataxis	Ataxins	HSP40, HSP70
Spinal and bulbar muscular atrophy	Androgen receptor	HSP70

Chaperonins Chaperonins are protein complexes. They are responsible for folding proteins which do not fold spontaneously. In E. coli, GroEL/GroES was discovered to assist in protein folding. GroEL is a chaperonin, while GroES is a cochaperonin. Binding and release of GroES in GroEL result in considerable change in GroEL which promotes the folding of the polypeptide. GroEL-induced folding will be driven at the expense of ATP hydrolysis. The mechanism may be outlined as follows:

- GroEL is a heptamer and it has a pocket. GroES is bound to the bottom of the pocket.
- Polypeptide binds to GroEL pocket of GroEL/GroES complex.
- ATP binds to GroEL.
- GroES closes the pocket enclosing the polypeptide.
- Folding occurs inside the closed pocket.
- GroES is then released by the hydrolysis of ATP, to open the pocket.
- The fully folded protein is released.

PDI PDI catalyses interchange and shuffling of disulphide bonds until proper folding of the protein. PDI is an oxidoreductase (thiol-disulphide oxidoreductase). The enzyme itself has a disulphide bridge at its active site. The enzyme recognizes the incorrect disulphide bridge in the misfolded protein and reduces it simultaneously getting oxidized itself. For example, conotoxin is a peptide which is a neurotoxin produced by conus snail venoms. These toxins become biologically active on formation of a disulphide bridge. In the endoplasmic reticulum of eukaryotes, disulphide formation is catalysed by protein disulphide isomerase (PDI); by contrast, prokaryotes produce a family of disulphide bond proteins, which together achieve an equivalent outcome in the bacterial periplasm.

Formation and interchange of a disulphide bond occur in four steps which are outlined as follows. PDI is involved in the third step of the process. Figure 4.32 represents the reaction catalysed by PDI. For convenience, the sulphur atoms involved in the reaction are suffixed with serial numbers as S_1, S_2, S_3, and S_4.

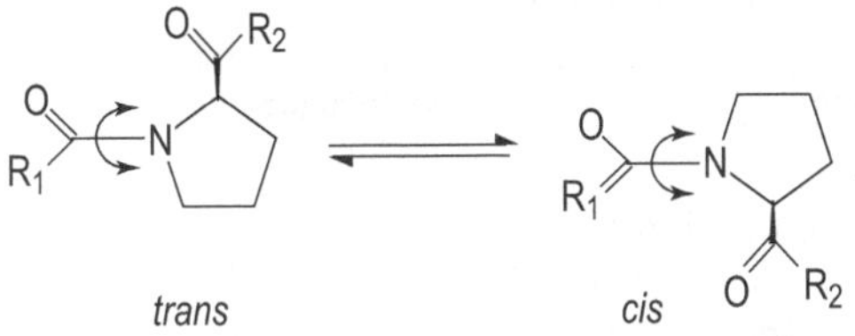

FIGURE 4.32 Reaction catalysed by PDI

Step 1: Reduction of thiol group of a cysteine residue (S_1H) in the presence of glutathione dimer (also called oxidized glutathione, GSSG) to form a mixed disulphide and reduced glutathione (GSH).

Step 2: The mixed disulphide then undergoes an intramolecular exchange reaction resulting in the liberation of reduced glutathione (GSH) and formation of intrachain disulphide bond (S_1–S_2) in the peptide chain.

Step 3: The peptide chain with intrachain disulphide bond is the substrate for PDI. PDI catalyses the isomerization reaction resulting in the exchange of a disulphide bond. Thus, a different sulphur atom, that is, S_3 is involved in the bond formation. The bond is denoted as S_1–S_3. In this process, S_2 (involved in bond formation in the previous step) is again reduced to S_2H.

Step 4: Isomerized chain is acted upon by GSSG to form another disulphide bond between S_2 and S_4 (denoted as S_2–S_4) with the formation of two molecules of GSH.

PPI PPI catalyses the interconversion of cis and trans isomers of proline peptide bonds which can otherwise prevent folding. In protein structures, the planar peptide bond occurs predominantly in the trans conformation. Peptidyl–prolyl cis–trans isomerases (PPI) catalyse cis-trans isomerization of imide bonds (peptidyl–prolyl bonds) in peptides and proteins shown as follows:

The cis–trans isomerization of peptidyl–prolyl bonds is the frequently occurring event in the protein-folding process. This reaction occurs in unfolded polypeptides and (to a variable degree) in partially folded intermediates, but not in native folded proteins. The rate of collagen triple-helix formation, which is limited by proline cis–trans isomerization, is significantly slowed by PPIase inhibition in chicken and human fibroblasts. This evidence suggests that PPI plays a role in triple helix formation. The PPI family includes parvulins, FK506-binding proteins, and cyclophilins.

Disorders associated with protein misfolding

Research suggests that dozens of different human diseases may be caused by protein misfolding.

The following are examples of some diseases caused due to protein misfolding:

- Lysosomal storage disorder
- Neurodegenerative disorder
- Prions disease

Lysosomal storage disorder Lysosomal enzymes are made in the cytoplasm of the cell and are then sent to the lysosome where they break down the substrate. Misfolding in lysosomal enzymes prevents their movement within the cell and consequent entry into lysosomes. As a result, the substrate for the misfolded enzymes accumulates in the lysosomes causing lysosomal storage disorder. Chaperones can be used to cure this disorder.

Neurodegenerative disorder Many neurodegenerative disorders are caused due to aggregation of misfolded proteins. Genes and chaperones involved in these disorders are listed in Table 4.11.

TABLE 4.11 Chaperones involved in neurodegenerative disorder

Chaperone family	Organism	Chaperone	Location	Major function	Associated disease
HSP70	Bacteria	DnaK	Cytosol	Major regulator of heat shock response	
	Mammal	HSC70 (HSP73) HSP70 (HSP72)	Cytosol	Cognate form (HSC70/HSP73) assists constitutive folding and transport of proteins to organelles, such as mitochondria, nucleus, and ER; HSP70/HSP72 is induced upon heat shock and mediates similar functions in response to stress-induced increases in protein misfolding and aggregation	Alzheimer's disease, multiple sclerosis, Parkinson's disease, schizophrenia, mammalian Crohn's disease, and tuberculosis
		HSP110	Cytosol	Stress responsive, prevents protein aggregation	
		HSP70L1	Cytosol	Assists folding of new proteins on ribosome	
		BIP/GRP78	ER	Binds folding and translocation intermediates to prevent aggregation; involved in calcium homeostasis, translocation, folding, transport, and retrotranslocation of polypeptides; regulator of unfolded protein response	
		mtHSP70 (GRP75/ Mortalin)	Mitochondria	Protein folding and translocation in the mitochondria	
HSP40	Bacteria	DnaJ	Cytosol	Modulates ATPase activity of DnaK, association of DnaK with nascent polypeptides, binds unfolded protein	Huntington's disease, Parkinson's disease
	Mammal	HDJ1/2 (HSP40), Auxilin	Cytosol	Modulates ATPase activity and peptide loading of HSC70/HSP70; auxilin coordinates HSC70-mediated uncoating of clathrin vesicles	
HSP90	Bacteria	HTPG	Cytosol	Modulates ATPase activity of DnaK, association of DnaK with nascent polypeptides, and binds unfolded protein	Target of Geldanamycin -derived anti-cancer drugs (e.g., 17-AAG, 17-DMAG) which disrupt HSP90-chaperoned oncogenic signalling pathways; Immunophilin FK506 associated with Leber congenital amaurosis
	Mammal	HSP90 /83/89	Cytosol	Stress responsive protein refolding	
		GRP94	ER	Folding and conformational regulation of signalling proteins, regulation of steroid hormone receptors and kinases Folding and assembly of secretory proteins	

(Contd)

TABLE 4.11 (Contd)

Chaperone family	Organism	Chaperone	Location	Major function	Associated disease
Chaperonin	Bacteria	GroEL	Cytosol	Involved in folding of some cytosolic proteins, especially overproduced proteins; stabilizes proteins in response to stress, assists in protein refolding	
	Mammal	mtHSP60	Mitochondira	Folds newly imported mitochondrial proteins	Impaired chaperonin function associated with McKusick– Kaufman and Bardet –Biedl syndromes, mitochondrial protein folding defects in lactic acidemia and hereditary spastic paraplegia
		TRiC/CCT	Cytosol	Folds about 10% of cytosolic polypeptide chains; downstream of the HSP70 machinery	

(Source: www.enzolifesciences.com)

Prions diseases Prions are defined as *small proteinaceous infectious particles*. Prions diseases are caused due to PRoteinaceous Infections ONly (hence, the name prion) and are a group of progressive conditions that affect the brain and nervous system of many animals, including human beings. Other names for prions disease are:

- Inherited human transmissible spongiform encephalopathies
- Prion-associated disorders
- Prion-induced disorders
- Prion protein diseases
- Transmissible dementias

Transmissible spongiform encephalopathies (TSEs) DNA will have 1:1 ratio of purine and pyrimidine bases.

PRNP gene PRNP (PRioN Protein) is the gene coding for 'prion protein' (PrP), located in chromosome 20. PrP is functional in the brain and numerous other tissues. The exact function of PrP is still unclear, but it is reported to function in the transport into cells, cell signalling, cell protection, and the formation of synaptic junction and cell communication.

Mutations in the PRNP gene cause prion disease. PRNP gene causes cells to produce an abnormal form of the prion protein known as PrPSc. Majority of the PrPsc accumulates as clumps in the brain leading to nerve cell damage. This results in microscopic sponge-like holes in the brain that cause the signs and symptoms of prion disease. Different forms of PrP have been identified in the nervous system.

Specific examples of diseases include the following:

- Scrapie—found in sheep
- TME (transmissible mink encephalopathy)—found in minks
- CWD (chronic wasting disease)—found in mules, deer, and elks

- BSE (bovine spongiform encephalopathy)—found in cows

The familial prion diseases of human beings include the following:

- CJD—Creutzfeld–Jacob disease
- GSS—Gerstmann–Straussler–Scheinker syndrome
- FFI—Fatal familial insomnia
- Kuru
- Alpers syndrome

4.3.13 Protein Structure Determination

The determination of protein structure begins with the determination of the primary structure of proteins which involves the determination of amino acid sequences.

Determination of primary structure

Sequencing of amino acids The sequence of amino acids in the peptide chain is always read from the N-terminus to C-terminus (i.e., from left to right). This structure is stabilized by peptide bond and disulphide bond. Amino acids were first sequenced by Fredrick Sanger in insulin hormone.

Four reagents are used for amino acid sequencing:

1. Sanger's reagent or 1-fluoro-2,4-dinitrobenzene (FDNB)
2. Edman reagent or phenylisothiocyanate
3. Dansyl chloride
4. Dabsyl chloride

The structure of these reagents is given in Fig. 4.33.

The identification of the amino-terminal residue is the first step in sequencing a polypeptide. Sanger's method is used for identifying the amino-terminal residue. FDNB reacts with the amino terminal of the peptide forming 2, 4-dinitrophenyl derivative of the polypeptide. This is hydrolyzed with 6M HCl to form 2, 4-dinitrophenyl derivative of amino-terminal residue which is identified.

Edman degradation procedure discovered by Pehr Edman is used to sequence the entire polypeptide. This is done by Edman reaction using phenylisothiocyanate. The Edman degradation procedure labels and removes only the amino-terminal residue from a peptide, leaving all other peptide bonds intact (Fig. 4.34). The peptide is reacted with phenylisothiocyanate under mildly alkaline conditions, which converts the amino-terminal amino acid to a phenylthiocarbamoyl (PTC) adduct. The peptide bond next to the PTC adduct is then cleaved in a step carried out in anhydrous trifluoroacetic acid, with removal of the amino-terminal amino acid as a nilinothiazolinone derivative. The derivatized amino acid is extracted with organic solvents, converted to the more stable phenylthiohydantoin derivative by treatment with aqueous acid, and then identified. The Edman degradation is carried out on a machine, called a sequenator that mixes reagents in proper proportions, separates the products, identifies them, and records the results.

Dansyl chloride forms a fluorescent N-terminal derivative which is easily detectable. The following reaction occurs:

Steps involved in the determination of the primary structure of proteins are summarized in Fig. 4.35.

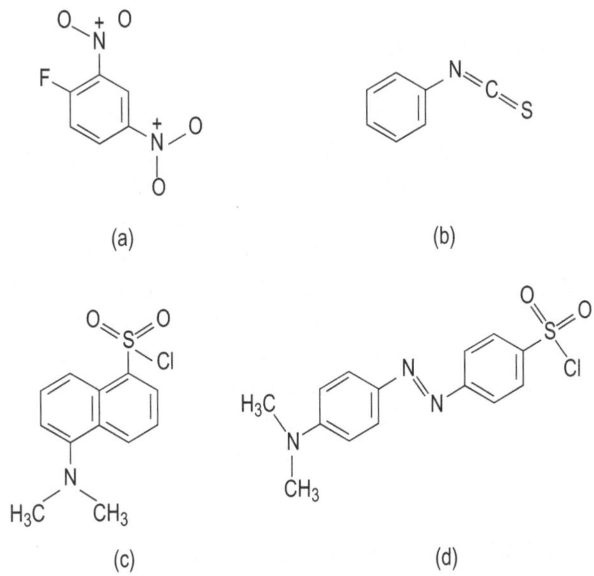

FIGURE 4.33 Structure of chemicals used in amino acid sequencing: (a) Sanger's reagent or 1-fluoro-2, 4-dinitrobenzene (FDNB), (b) Edman reagent or Phenylisothiocyanate, (c) Dansyl chloride, (d) Dabsyl chloride

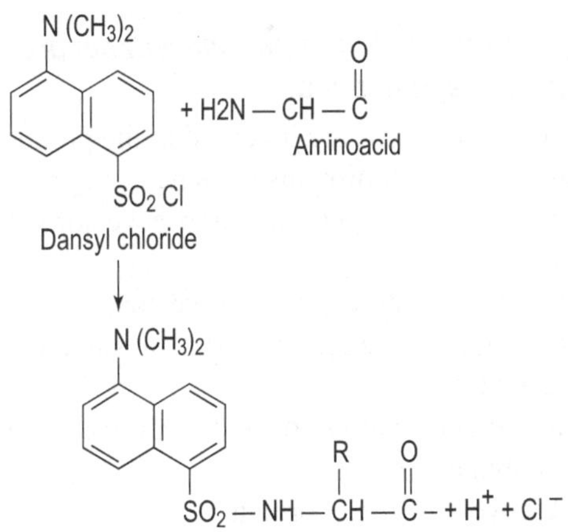

FIGURE 4.34 Edman degradation procedures for amino acid sequencing

Separation of amino acids

Protein sample
6N HCl ↓ 110°C
↓ Keep for one day
Liberation of free amino acids
↓ Cation exchange column
↓ Ion exchange chromatography
Separation of free amino acids
↓ Ninhydrin reagent ↓
Purple colour → a-amino acids
Yellow colour → amino acids (proline)

Identification of N-terminal amino acids

Protein sample
↓ FDNB of Dansyl chloride
N-terminal amino acids derivative
↓
Identification of N-terminal derivative
↓
Identification of N-terminal amino acids

Sequencing of amino acids

Peptide sample with *n* number of amino acids
Phenylisothiocyanate ↓ (Edman reaction)
N-terminal of 1st amino acids/Alkali
↓
Phenylthiocarbamyl derivative
↓
Strong anhydrous acid
↓
Phenylthiohydantoin derivative of the 1st amino acid + remaining peptide with *n*–1 number of amino acids
↓ Chromatography → ↓ N-terminal amino acids identified
↓ Continue the Edman reaction

FIGURE 4.35 Steps involved in the determination of primary structure of proteins/peptides

Determination of secondary and tertiary structures

The secondary and tertiary structures of proteins can be determined using various automated techniques:

- X-ray crystallography
- Nuclear magnetic resonance spectroscopy (NMR)
- Magnetic resonance imaging (MRI)

For X-ray diffraction, protein samples are first crystallized and beamed with X-rays. Various atoms in the protein molecules diffract X-rays differently giving a unique X-ray diffraction pattern. Diffraction pattern can help in determining the structure of protein crystals. NMR, in conjunction with mass spectroscopy, is often the only method for determining structures, conformations, and the nature of interactions between proteins and other molecules. NMR is used to determine the three-dimensional structure of proteins at atomic resolution. Highly relevant molecular structure in solution can be obtained. Protein denaturation can also be studied using NMR. In addition, dynamic features of molecular structure, structural, thermodynamic, and kinetic aspects of interaction between proteins and other macromolecules can be studied. MRI is used to study proteins located deep inside the tissues. Details of techniques used to determine proteins structure are discussed in Chapter 14.

4.3.14 Pharmaceutical Proteins

Proteins which are produced by the process of genetic engineering and used for treatment of diseases are called pharmaceutical proteins. In genetic engineering, the DNA segment coding for a specific protein is expressed in a host where it is transcribed and translated into that specific protein. Such proteins can either be secreted outside the host cell or retained within the host cell. In either case, the protein should be recovered for further use.

Steps in production of pharmaceutical proteins by genetic engineering

The following steps are involved in the production of pharmaceutical proteins by genetic engineering:

- Insertion of target gene into a vector/vehicle gene
- Ligation of target gene and vector
- Insertion of target gene into a suitable host (gene transfer)
- Target gene expression and identification of recombinant host
- Recovery of target protein
- Purification and characterization of target protein

These steps are diagrammatically illustrated in Fig. 4.36.

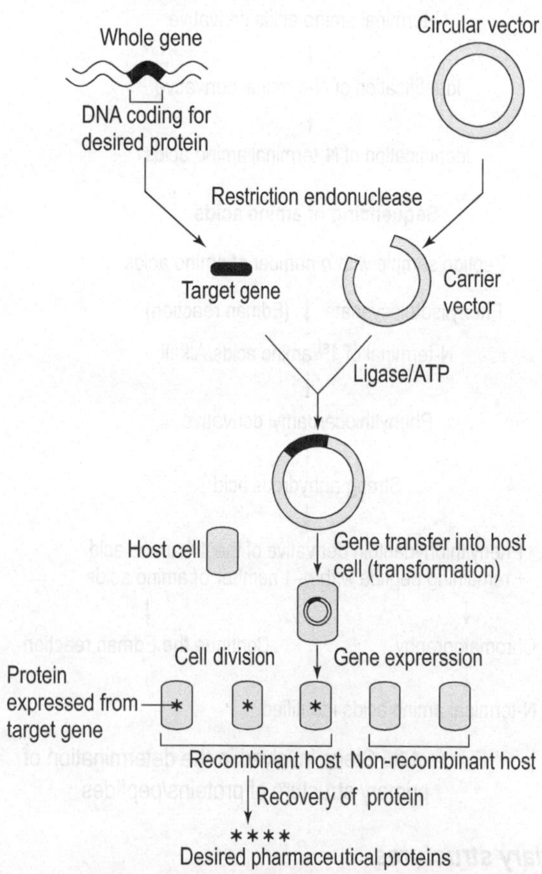

FIGURE 4.36 Production of pharmaceutical proteins by genetic engineering

Identification and isolation of target gene

DNA segment coding for a specific protein of interest is first isolated by cutting the segment using an enzyme called restriction endonuclease (RE). RE is usually obtained from bacteria. For example, EcoR1 is an RE obtained from the bacteria E. coli. The segmented gene is thereafter called the target gene.

Insertion of target gene into a vector vehicle The target gene is inserted into a vector. A vector is also a segment of DNA used to carry the target gene. Plasmid (circular chromosome of bacteria) is the commonly used vector. It is linearized by cutting with an RE. Consequently, two ends of the target gene are attached to the two ends of the linearized plasmid, thus, reforming a circular structure.

Ligation of target gene and vector Ends of target gene are ligated to the ends of vector by the formation of phosphodiester bond. This process is an enzyme-catalysed energy-dependent process. It is catalysed by an enzyme called ligase in the presence of ATP. Thus, ligation results in a stable circular clone.

Insertion of target gene into a suitable host (Gene transfer) The vector-borne target gene is introduced into a host cell to be expressed (transcribed and translated). This process is called gene transfer. Prokaryotic hosts, such as bacteria, can be used in general. In certain cases, eukaryotic host cells, such as yeast cell, animal cell, plant cell, or insect cell can also be used. Gene transfer into a prokaryotic host cell is called transformation, and gene transfer into a eukaryotic host is called transfection. Gene transfer can be done using variety of methods, such as the following:

- Electroporation
- Scrap loading
- Shot gun method
- Liposome mediated transfer
- Viral transfer

Of these methods, electroporation and shot gun method are widely used and are reliable.

Target gene expression (identification of recombinant host) After insertion into the host cell the target gene is duplicated and multiplied by cell division and expressed into functional proteins. Genes cannot be expressed in all the host cells. Cells bearing the functional target proteins are called recombinant host cells and those which are devoid of the target proteins are called non-recombinant hosts. Recombinant hosts can be isolated and multiplied again to get more copies of the target gene.

Recovery of target protein Proteins expressed in the host cell are of two types: (i) secretory proteins or extracellular proteins, which are liberated into the medium of the host cell, and (ii) cytosolic proteins or intracellular proteins, which are retained within the host cell. Such secretory proteins can be recovered by filtration or by centrifugation of the culture medium. Intracellular proteins are recovered by cell disruption in which the cell architecture is disorganized in the presence of a buffer in order to release the cellular content and form a crude extract. Cell disruption can be done by using blenders, homogenizers, mortar and pestle, dynomill, or osmotic shock.

Purification of target protein The target protein which is recovered from the cell should be purified and crystallized for pharmaceutical and medical uses. Purification can be done by ion-exchange chromatography, gel-filtration chromatography, high-pressure liquid chromatography (HPLC), or affinity chromatography. Purified proteins can be characterized for further study. Characterization can be done by using techniques, such as analytical ultracentrifugation, isoelectric focusing, gel-filtration chromatography, SDS-PAGE, and mass spectroscopy.

Examples of pharmaceutical proteins

Various therapeutic enzymes, hormones, interleukins, interferons, clotting factors, and growth factors of pharmaceutical importance are produced by recombinant DNA technology. Table 4.12 lists some important pharmaceutical proteins and the diseases cured by them.

4.4 ENZYMES

Enzymes are catalytic proteins. They are colloidal, high molecular weight, non-dialysable, denaturable structurally diverse group of proteins. Biochemical reactions taking place within the living cells are catalysed by enzymes (hence, the name biocatalysts). Enzymes increase the rate of reaction without undergoing any change themselves and without altering the equilibrium of the reaction.

The speed with which an enzyme catalyses a reaction is known as the velocity or rate of that reaction. It is denoted as V. The substance on which an enzyme acts is called substrate (S) and the new substance that is generated after the reaction is called product (P).

TABLE 4.12 Diseases cured by recombinant pharmaceutical proteins

Proteins	Disease cured
Enzymes	
L-Asparaginase	Cancer
Superoxide dismutase	Oxygen toxicity
Urokinase	Heart attack
DNAse	Cystic fibrosis
Glucocerebrosidase	Gaucher's disease
Galactosidase	Inherited β-galactosidase deficiency
Penicillinase	Penicillin allergy
Papain	Digestive disorder
Hormones	
Insulin	Diabetes
Calcitonin	Osteomalacia
Somatotropin	Growth defect
Erythropoietin	Anaemia
ACTH (adrenocorticotropic hormone)	Rheumatic disease
Chorionic gonadotropin	Infertility
Interleukins and interferons	
Interleukin I	Asthma Cancer
Interferons	Cancer, immune disorders
Clotting factor	
Factor VIII	Haemophilia A
Factor IX	Haemophilia B
Growth factor	
Platelet-derived growth factor	Atherosclerosis
Epidermal growth factor	Burns
Neural growth factor	Brain damage

4.4.1 Ribozymes and Isozymes

Strictly speaking, all enzymes are not proteins because a few RNA molecules also have catalytic properties and are called ribozymes. They are used in the catalysis of RNA splicing, maturation of tRNA, and translation process (peptide-bond formation). RNAse P and RNA component of peptidyl transferase are examples of ribozymes (refer Chapter 6).

At certain instances, a single enzyme can exist in more than one form. They are called isoforms, isozymes, or isoenzymes. They catalyse the same reaction but differ in their amino acid sequence and electrophoretic mobility. Lactate dehydrogenase (LDH) is a classic example for isozymes. LDH exists in five forms denoted as LDH1 to LDH5. Other examples include hexokinase, which exists in four forms, Type 1 to Type 4, and angiotensin converting enzyme, which exists in two forms: somatic form and germinal form. Isoenzymes are used in the diagnosis of diseases related to a specific tissue.

Cytochrome p450 isozymes also exist in isoforms and play a crucial role in metabolism. Cytosolic and mitochondrial malate dehydrogenase and glutamate dehydrogenase (with different molecular weights) are also examples of isozymes. Sometimes, structurally different forms of enzymes are also considered as isoforms, which occur due to allosteric modification of enzymes.

4.4.2 Enzyme Activity

International unit (IU) is an internationally accepted amount of a substance. It is used for the measurement of fat-soluble vitamins (such as vitamins A, D, and E) and certain hormones, enzymes, and biologicals (such as vaccines).

In enzymology, the amount of enzyme that catalyses the formation of 1 mmol of product in 1 min under optimal conditions of pH, temperature, and ionic strength is called 1 International Unit (IU) of an enzyme, and the amount of enzyme catalysing the conversion of 1 mol of substrate to product in 1 s is called 1 Katal (kat).

$$1 \text{ Kat} = 6 \times 107 \text{ IU}$$

4.4.3 Properties of Enzymes

Enzymes possess the following distinctive properties.

Catalytic power

The ratio of the rate of catalysed reaction to the rate of uncatalysed reaction is called catalytic power of an enzyme. For example, the hydration rate of carbon dioxide to form bicarbonate is 1.3×10^{-1} per second in the absence of enzyme. But the rate of the same reaction in the presence of carbonic anhydrase enzyme is 1×10^{6} per second. Thus, the catalytic power of carbonic anhydrase is 7.7×10^{-11} per second.

Specificity

The term specificity refers to the extraordinary ability of an enzyme to recognize a specific substrate to catalyse a specific reaction. For example, urease is specific for urea and hence it reacts only with urea to form ammonia.

$$\text{Urea} \xrightarrow{\text{Urease}} \text{Ammonia} + \text{Carbon dioxide}$$

Reaction conditions

Enzyme catalysis occurs under mild conditions of temperature, atmospheric pressure, and nearly neutral pH. However, there is a certain unique group of enzymes called *extremozymes* which can withstand extreme environmental conditions such as extreme temperatures, very high saline concentration, and desiccation. These enzymes are derived from a group of bacteria called extremophiles. For example, Taq polymerase which is resistant to a temperature of about 100°C is derived from a bacterium called *Thermus aquaticus*. This enzyme is used in polymerase chain reaction (PCR).

Reversibility

Enzymes can catalyse reactions in both directions depending upon the concentration of the substrate and the product. The enzyme usually will not influence the equilibrium of the reaction (i.e., the substrate:product ratio). Addition of enzyme to the substrate solution will lead to formation of the product until the equilibrium ration of S/P is achieved. Addition of the enzyme to the product solution will form the substrate until the same equilibrium ratio is reached. At equilibrium, the rate of S→P reaction will be equal to the rate of P→S reaction. Most of the reactions in the glycolysis pathway are reversible (refer Chapter 7).

Denaturation and solubility

Enzyme denaturation is commonly defined as any noncovalent change in its structure. This change may alter the secondary, tertiary, or quaternary structure of the enzyme molecules.

Enzymes are thermolabile and pH sensitive. Hence, they are denatured by high temperature, strong acids, and strong alkalis. When proteins are exposed to increasing temperature (except hyperthermozyme), interactions responsible for tertiary structure, are broken. A further rise in temperature results in breaking of hydrogen bond increasing the flexibility of the protein leading to denaturation. When enzymes are exposed to very high pH at elevated temperatures it results in alkaline hydrolysis of peptide bonds.

Proteins containing acid-labile groups irreversibly lose their function even with mild acid treatment. This is due to the breaking of specific covalent bonds. Exposure to strong-enough acid at elevated temperatures will first release amide nitrogen from glutamine and asparagine groups and eventually lead to hydrolysis of peptide bonds.

Refolding of the denatured protein to regain its structure and function is called renaturation, i.e., renaturation is the opposite of denaturation. Scientists have proposed that cyclodextran is able to renature proteins. In biological systems, proteins are solubilized in the blood at a pH range of 7.45–7.55. In aqueous buffer, the solubility of the protein is determined by the nature of the amino acids present on the surface. Proteins with a high degree of hydrophobic amino acid on their surface have poor solubility in water and aqueous solvents. On the other hand, proteins with hydrophilic amino acids on their surface have good solubility in aqueous solvents.

Under unoptimal conditions, the tertiary and quaternary structures of enzymes get disorganized due to breakage of hydrogen and ionic bonds. For example, the ordinary polymerase, which catalyses the polymerization reaction at the physiological pH, is replaced by Taq polymerase, which is an extremozyme, for the same reaction in vitro at high temperature at which the DNA pol denatures.

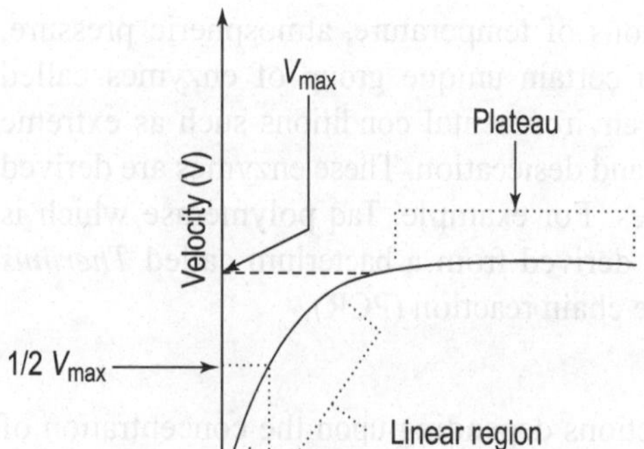

FIGURE 4.37 Effect of substrate concentration on enzyme activity

4.4.4 Factors Influencing Enzyme Activity

Various factors influence the activity of enzymes. These factors include substrate concentration, reaction temperature, and pH of the buffer or reaction medium.

Effect of substrate concentration

The effect of substrate concentration on enzyme activity can be well described by plotting the velocity of the reaction against substrate concentration as shown Fig. 4.37.

At the initial stage of the reaction, the velocity of the enzyme-catalysed reaction increases linearly with an increase in substrate concentration (shown as linear region). This is also termed as initial velocity of the reaction. Generally, it is assumed that the velocity of the enzyme-catalysed reaction is directly proportional to the enzyme concentration in a reaction mixture. As the substrate concentration is raised gradually, the velocity reaches a maximum (V_{max}). Once V_{max} is attained, the velocity is constant even with increase in substrate concentration. This leads to a plateau region. Thus, a hyperbola is obtained.

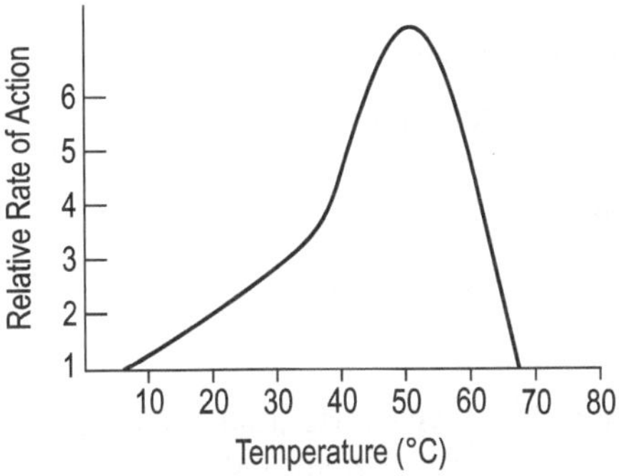

FIGURE 4.38 Effect of temperature on the rate of an enzyme-catalysed reaction

Effect of temperature

The effect of temperature on the rate of an enzyme-catalysed reaction is represented as a plot of velocity versus temperature (Fig. 4.38).

The rate of an enzyme-catalysed reaction increases with an initial rise in temperature until the optimum temperature is reached. This rise is due to the effective collision between reactive groups of the substrate and the enzyme. At optimum temperature, the velocity is maximum. A further increase in temperature leads to a fall in velocity due to disturbance in conformation of active site of the enzyme. This leads to a bell-shaped curve.

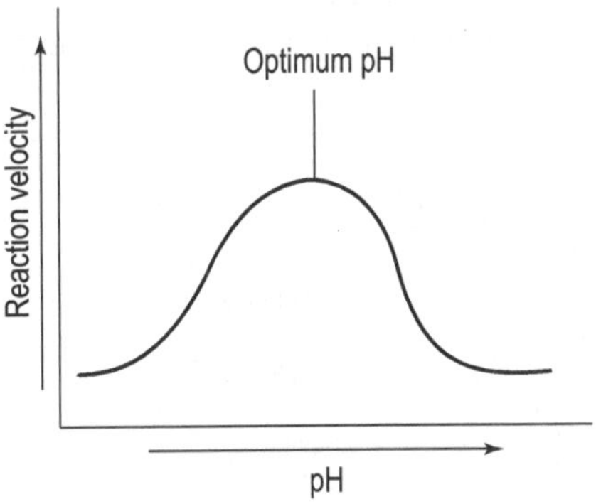

FIGURE 4.39 Effect of pH on velocity of an enzyme-catalysed reaction

Effect of pH

The effect of pH on the velocity of an enzyme-catalysed reaction is represented in Fig. 4.39. The bell-shaped curve of pH versus V is similar to that of T versus V. In the initial stage of the reaction, the velocity increases with increase in pH due to increase in enzyme–substrate binding. At a particular pH, called the optimum pH (usually 6–8), the velocity reaches a maximum. Beyond the optimum pH, the 3-D structure of the enzyme is altered leading to a dissociation of enzyme–substrate (E–S) complex and a fall in velocity.

4.4.5 Structure of Enzymes

Enzymes have a properly folded active conformation. The part of the enzyme that is responsible for catalysis is called the active site. The active site is functionally divided into the following two subsites (Fig. 4.40.):

(i) Specificity subsite (substrate bining site)—the part of the active site where recognition of the substrate takes place.

(ii) Reaction subsite—the part of the active site where chemical reaction occurs.

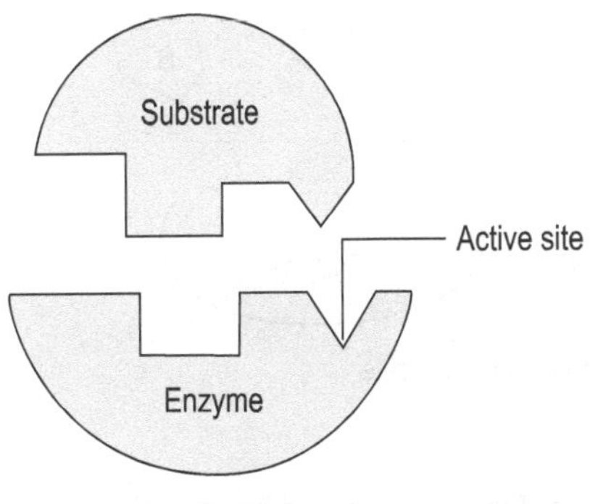

FIGURE 4.40 Structure of enzyme

The substrate (S) binds to the active site of the enzyme (E) forming an ES complex. The ES complex is converted into a high-energy state called the transition state complex ($ES^{\ddagger}$). Subsequently, the transition state is converted into the product (P) and the free enzyme (E) is released as shown in the Fig. 4.41.

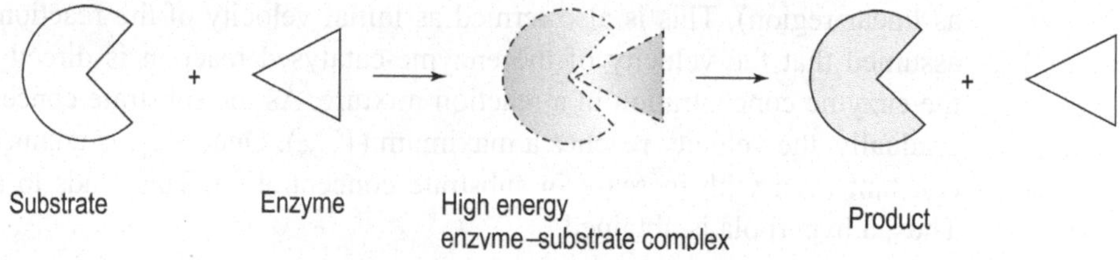

FIGURE 4.41 Enzyme substrate reaction

An enzyme catalysed reaction is represented as: $E + S \rightarrow ES \rightarrow ES^{\ddagger} \rightarrow P + E$

4.4.6 Models for Enzyme–Substrate Reaction

Enzyme–substrate binding can conventionally be explained by the following two models:

Fischer's lock and key model

According to this model (Fig. 4.42a), the active site of the enzyme is rigid and predetermined to be complementary to the substrate. Thus, the substrate fits into the active site as easily as a key fits into a lock.

Koshland's induced fit model

In this model (Fig. 4.42b), the active site of the enzyme is flexible and changes shape on substrate binding, that is, complementarity is induced after the binding of the substrate to the enzyme.

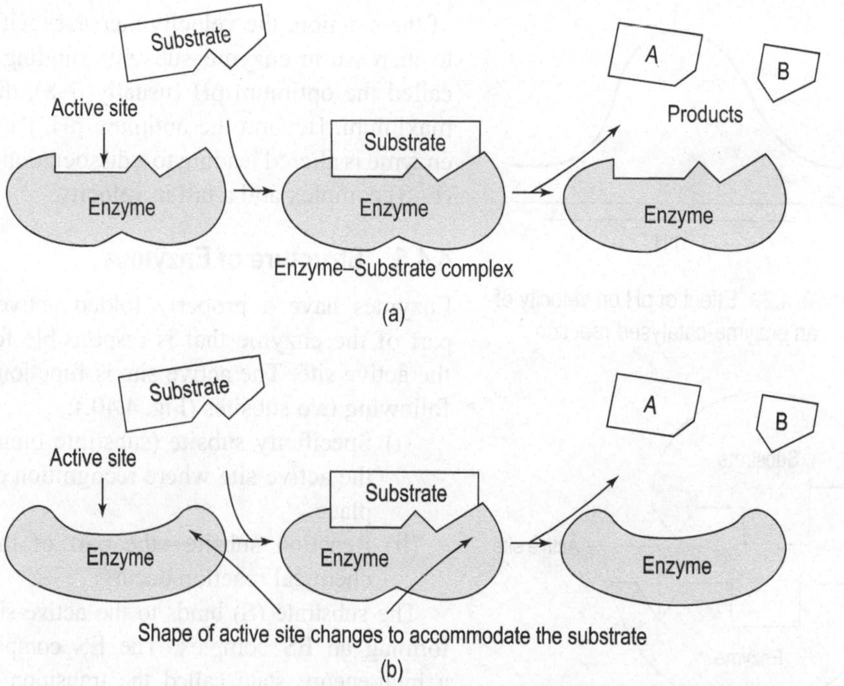

FIGURE 4.42 Two models for enzyme–substrate binding: (a) Fischer's lock and key model and (b) Koshland's induced fit model

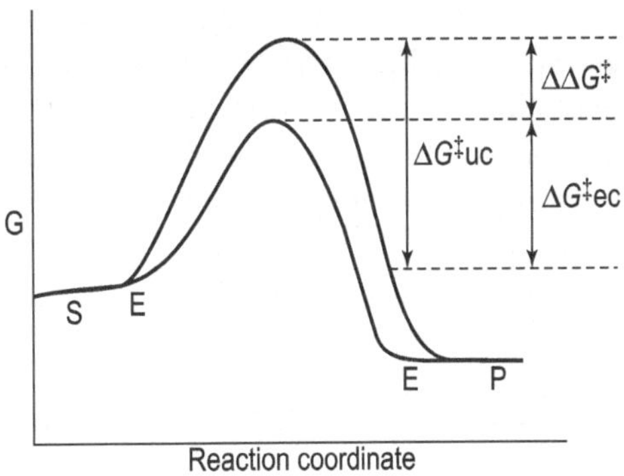

FIGURE 4.43 Transition state diagram showing mechanism of enzyme action

4.4.7 Mechanism of Enzyme Action

The energy required for the formation of the transition state is called activation energy ($\Delta G^{\ddagger}$). Generally, an enzyme increases the reaction rate by lowering the activation energy. This can be explained as follows: After the substrate fits into the active site of the enzyme as per the models described in the preceding section, free energy is released; it is called intrinsic binding energy and is denoted as ΔG_b. When the ES complex is formed, the entropy of S is lost. The entropy loss is denoted as $-T\Delta S$. The amount by which the activation energy is lowered by the enzyme is called catalytic efficiency and is denoted as $\Delta\Delta G^{\ddagger}$. ΔG_b acts as a major source of free energy for the enzyme to lower the activation energy, ΔG_b is equal to the amount by which $\Delta G^{\ddagger}$ is lowered, that is, $\Delta G_b = \Delta\Delta G^{\ddagger}$. This catalytic concept could be described by a transition state diagram (or reaction coordinate diagram) which is a plot of free energy (G) versus progress of the reaction (reaction coordinate) (Fig. 4.43).

G = Free energy
$\Delta G^{\ddagger}$ uc = Activation energy for uncatalysed reaction
$DG^{\ddagger}$ ec = Activation energy for enzyme-catalysed reaction
$\Delta\Delta G^{\ddagger}$ = Amount of activation energy lowered by enzyme (catalytic efficiency)

4.4.8 Rate Equation for Enzyme-catalysed Reaction: Michaelis–Menten Equation

According to Michaelis and Menten, steady-state analysis of an enzyme-catalysed reaction is represented as given in Eqn (4.1):

$$E + S \underset{k_{-1}}{\overset{k_1}{\rightleftharpoons}} ES \xrightarrow{k_2} E + P \tag{4.1}$$

This equation can also be called MM reaction. According to this reaction, the enzyme (E) combines with the substrate (S) forming an ES complex with a rate constant k_1. Now, ES can dissociate into either E+S (with rate constant k_{-1}) or E+P (with rate constant k_2). ES dissociates into P at a velocity V and reaches a maximum velocity. For the MM reaction, k_2 is rate limiting; thus, $k_2 <<< k_{-1}$, hence, it reduces to k_{-1}/k_1 which is defined as the dissociation constant for the ES complex.

During the course of the reaction (Fig. 4.44), the concentration of the free enzyme [E], the substrate-bound enzyme (saturated enzyme) [ES], and the total enzyme [ET] can be correlated by the following equation:

$$[ET] = [E] + [ES] \tag{4.2}$$

From the above postulate, the velocity of an enzyme-catalysed reaction (V) is represented by an equation called MM equation or rate equation. It is given as follows:

$$V = \frac{V_{max}[S]}{K_m + [S]} \tag{4.3}$$

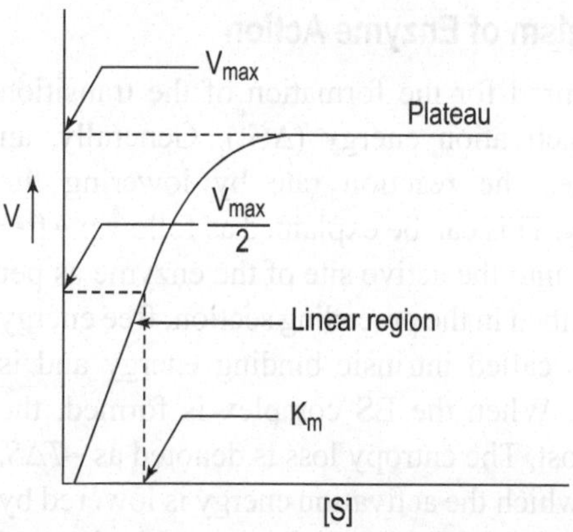

FIGURE 4.44 Course of an enzyme-catalysed reaction

Some important parameters are considered in the derivation of the rate equation; these include concentration of the substrate, concentration of the free enzyme, concentration of the total enzyme, concentration of the enzyme–substrate complex, rate constants for the forward and the reverse reactions, and maximal velocity.

Let us recall the enzyme-catalysed reaction represented in Eqn (4.1):

$$E + S \underset{k_{-1}}{\overset{k_1}{\rightleftarrows}} ES \xrightarrow{k_2} E + P \tag{4.1}$$

E combines with S forming ES with a rate constant k_1. ES can dissociate into either E+S with rate constant k_{-1} or E+P with rate constant k_2.

Rate of ES formation is shown as follows:

$$\frac{d[ES]}{dt}=k_1[E][S]\frac{d[ES]}{dt}=k_1\{[ET]\ [ES]\}[S] \tag{4.4}$$

Rate of ES dissociation is shown as follows:

$$\frac{d[ES]}{dt}=k_{-1}[ES]+k_2[ES] \Rightarrow \frac{d[ES]}{dt}=k_{-1}+k_2)[ES] \tag{4.5}$$

As per steady-state kinetics, rate of formation of ES as given in Eqn (4.4) is equal to the rate of dissociation of ES as given in Eqn (4.5). So, equating Eqn (4.4) and Eqn (4.5), we obtain

$$k_1\{[ET]-[ES]\}\ [S]=(k_{-1}+k_2)\ [ES] \tag{4.6}$$

Bringing the rate constants on one side, we obtain

$$\frac{(k_{-1}+k_2)}{k_1}=\frac{\{[E_T]-[ES]\}[S]}{[ES]} \tag{4.7}$$

In the above equation, $(k_{-1}+k_2)/k_1$, the ratio of rate constants is denoted as K_m. Thus, Eqn (4.7) becomes

$$K_m=\frac{\{[ET]-[ES]\}[S]}{[ES]} \tag{4.8}$$

Taking S as common factor in the above equation, we get

$$Km=\frac{[S][ET]-[S][ES]}{[ES]} \tag{4.9}$$

Cross multiplication of Eqn (4.9) gives

$$[S]\ [ET]-[S]\ [ES]=K_m\ [ES] \tag{4.10}$$

Rearranging Eqn (4.10) we obtain

$$[S]\ [ET]=k_m\ [ES]+[S]\ [ES] \tag{4.11}$$

Taking [ES] as the common factor in Eqn (4.11), we get

$$[S]\ [ET]=[ES]\ \{k_m+[S]\} \tag{4.12}$$

Solving Eqn (4.12) for [S], we get

$$[S] = \frac{[ES]\{k_m + [S]\}}{[ET]} \tag{4.13}$$

In a reaction mixture, [ES] is difficult to measure, but V and V_{max} are easily measurable by variety of methods. So, it is essential to replace [ES] and [ET] in Eqn (4.13) by a measurable factor. This replacement is done on the basis of principle of saturation kinetics. Saturation kinetics states that 'at V_{max}, E is completely saturated with S to form ES. Thus,

$$[E] = 0 \tag{a}$$

Since, [ET] = [E] + [ES] we get

$$[ET] = [ES] \tag{b}$$

Thus,

$$V = k_2\,[ES] \tag{c}$$

and

$$V_{max} = k_2\,[ET] \tag{d}$$

Solving (c) and (d) for k_2, we get

$$k_2 = V/\,[ES] \tag{e}$$

$$k_2 = V_{max}/\,[ET] \tag{f}$$

Combining (e) and (f), we get

$$V/\,[ES] = V_{max}/\,[ET] \tag{g}$$

Solving (g) for velocity terms, we get

$$V/V_{max} = [ES]/\,[ET \tag{h}$$

Using (h), the immeasurable quantity [ES] is related to the measurable quantities V and V_{max}. Hence, (h) is used to solve equation Eqn (4.13).

Thus, substituting (h) in Eqn (4.13), we get

$$[S] = \frac{V}{V_{max}} \times k_m + [S] \tag{4.14}$$

Solving Eqn (4.14) for V, we get

$$V = \frac{V_{max}[S]}{K_m + [S]} \tag{4.15}$$

Equation (4.15) is called *Michaelis–Menten equation*, rate equation, or *velocity equation* where

k_m = Michaelis–Menten constant (MM constant or ratio of rate constant).

4.5 CARBOHYDRATES

Carbohydrates are hydrated carbon molecules with the general formula $C_nH_{2n}O_n$ or $(CH_2O)_n$ where n = number of carbon atoms. The following are a few examples of carbohydrates:

- Glucose = $C_6H_{12}O_6$
- Fructose = $C_6H_{12}O_6$
- Sucrose = $C_{11}H_{22}O_{11}$
- Cellulose = $(C_6H_{10}O_5)_n$
- Starch = $(C_6H_{10}O_5)_n$

The chemical formula for glucose is written as $C_6H_{12}O_6$. It has six carbon, six oxygen (n value), and 12 hydrogen atoms ($2n$ value).

Carbohydrates can exist in two forms: *polyhydroxy ketones* and *polyhydroxy aldehydes*.

4.5.1 Functions of Carbohydrates

Carbohydrates perform the following important functions:

- They provide the majority of the energy requirements of all living organisms.
- They are the structural units of cell wall (e.g., cellulose).
- They are the structural units of cell membranes.
- They serve as metabolic intermediates.
- They comprise large portions of DNA and RNA in the form of deoxyribose and ribose.
- They play a major role in *lubrication* of joints and aid in cell adhesion.
- They play an important role in *immunity* in conjugation with protein (glycoprotein).

4.5.2 Nomenclature of Sugars on the Basis of Number of Carbon Atoms

The building blocks of carbohydrates are called sugar residues (just as amino acid residues are building blocks of proteins). Sugars with three carbon atoms are called triose; those with four carbon atoms are called tetrose, and those with five and six carbons are called pentose and hexose, respectively. Hexose is the most abundant sugar; glucose and fructose are examples of hexose sugar. Triose is the simplest sugar; glyceraldehyde is an example of a triose sugar. Table 4.13 lists different forms of sugar and their examples.

TABLE 4.13 Examples for various types of sugars

Sugar	Number of carbon atoms present	Example
Triose	3	Glyceraldehyde Dihydroxy acetone phosphate
Tetrose	4	Erythrose Erythrulose Threose
Pentose	5	Xylose Ribose Arabionse Lyxose
Hexose	6	Glucose Gulose Mannose Idose Talose Allose Altrose
Heptose	7	Sedoheptulose
Octose	8	Nonose

4.5.3 Classification of Carbohydrates on the Basis of Functional Groups

As mentioned earlier, carbohydrates are polyhydroxy aldehydes or polyhydroxy ketones. This implies that they have either an aldehyde or a ketone as the functional group. On the basis of this, carbohydrates are divided into two types: aldose and ketose. Aldose has an aldehyde as the functional group, for example, glucose, galactose, and maltose. Ketose has a ketone group as the functional group, for example, fructose, ribulose, and xylulose. The structure of some aldoses and ketoses are shown in Fig. 4.45.

4.5.4 Classification of Carbohydrates on the Basis of Number of Sugar Residues

Depending on the number of sugar residues (polyhydroxy aldehyde unit or polyhydroxy ketone unit), carbohydrates can be classified into monosaccharides, disaccharides, trisaccharides, and polysaccharides. Monosaccharides contain only one sugar residue, for example, glucose. Disaccharides contain two sugar residues, for example, maltose is made of two glucose residues, lactose of one glucose residue and one galactose residue, and sucrose of one glucose residue and one fructose residue. Trisaccharides have three sugar residues, for example, raffinose and maltotriose. A short chain of carbohydrates with 4 to 10 sugar residues is called an oligosaccharide. Carbohydrates with more than ten sugar residues are called polysaccharides. Starch and cellulose are examples of polysaccharides.

4.5.5 Structure of Carbohydrates

There are three basic types of structures that comprise carbohydrates:

1. Open-chain form
2. Cyclic form
3. Ring form

Open-chain form

The open-chain form of sugar is linear with its functional group (aldehyde/ketone group) at one end and an alcoholic group at the other end. For example, the open-chain form of glucose is shown in Fig. 4.46. In glucose, carbon atoms are linked one below the other in a linear fashion. The first carbon atom is linked to the aldehyde group (–CHO) and the sixth carbon is linked to the alcoholic group (–CH_2OH). Carbons 2 through 5 have hydrogen on one side and a hydroxyl group on the other side. The second, fourth, and fifth carbon atoms have hydrogen on their left and OH on their right, whereas in the third carbon atom, OH is on the left and hydrogen is on the right.

Presence of a chiral carbon atom gives rise to steroisomerism in monosaccharides All monosaccharides have at least one chiral carbon atom (i.e., a carbon atom with four variable groups) due to which they exist as stereoisomers. Stereoisomers may be defined as different structural forms of the same compound which differ in their spatial arrangements of groups. The possible number of stereoisomers for a monosaccharide depends on the number of chiral carbon atoms and is given by the formula $2n$, where n = number of chiral carbon atoms in the monosaccharide. For example, glyceraldehyde has one chiral carbon atom and hence $2^n = 2^1 = 2$. This implies that there are two stereoisomers for glyceraldehydes. One is L-glyceraldehyde and the other is D-glyceraldehyde. The structures of L-glyceraldehyde and D-glyceraldehyde are shown in Fig. 4.47.

Other sugars are considered to be derived from glyceraldehyde. D-sugar is one that matches the configuration of D-glyceraldehyde around the asymmetric carbon that is farthest from the reactive aldehyde or ketone group. L-sugar correspondingly matches L-glyceraldehyde. In other words, D and L configurations are based on the position of the –OH group on the penultimate carbon. If the –OH group of the penultimate carbon is on the left, the sugar is labelled as L-sugar; and if it is on the right, it is labelled as D-sugar. As an example the structures of L-glucose and D-glucose are shown in Fig. 4.48.

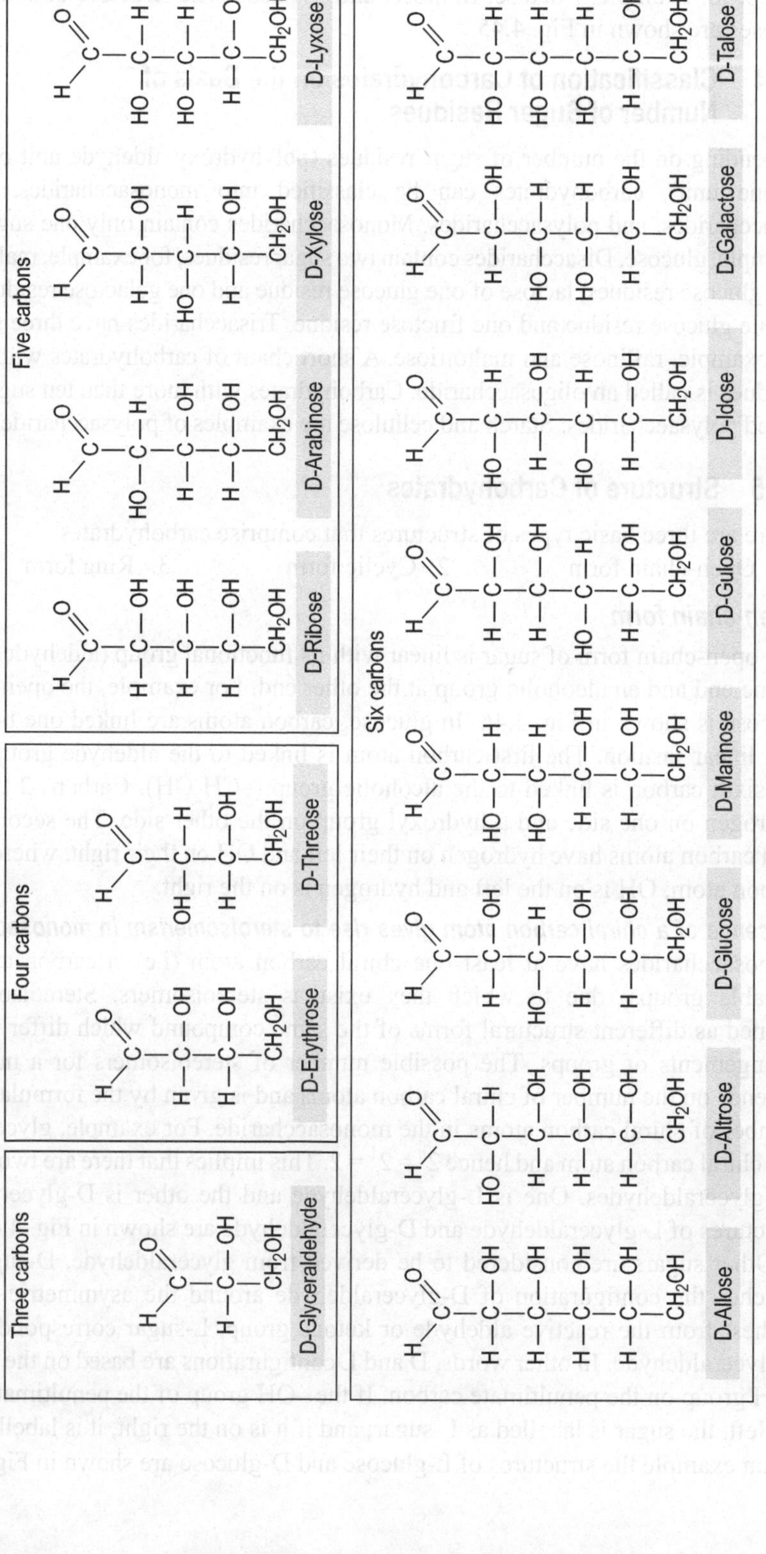

(a)

FIGURE 4.45 (a) Structure of aldoses

(Contd)

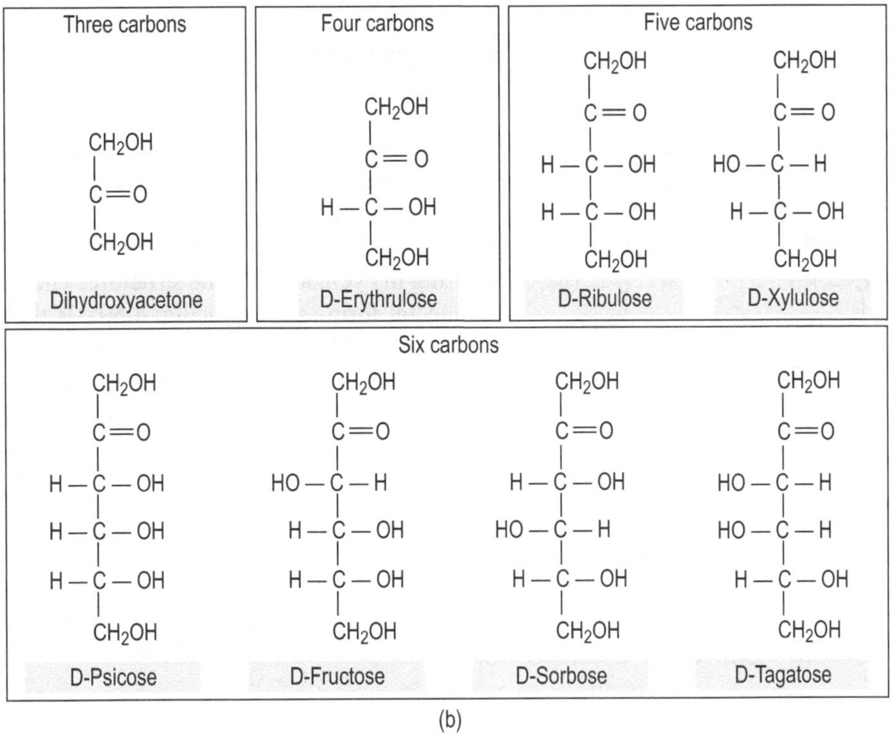

FIGURE 4.45 (b) Structure of ketoses

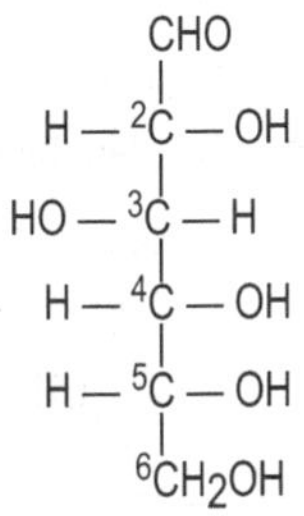

FIGURE 4.46 Structure of open-chain form of glucose

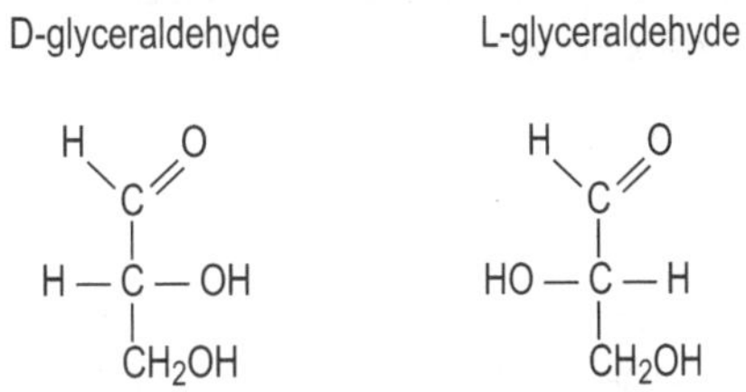

FIGURE 4.47 Structure of D- and L-glyceraldehydes

Enantiomers Compounds that deflect plane polarized light to either right (clockwise) or left (anticlockwise) direction are called enantiomers or optical isomers. Enantiomers are mirror images of each other. Monosaccharides are enantiomers. Monosaccharides that deflect plane polarized light to the right (i.e., in clockwise direction) are called dextrorotatory sugars and are denoted as D (+). On the other hand, sugars that deflect the light to the left (i.e., in anticlockwise direction) are called levorotatory sugars and are denoted as L (–). Glucose is a dextrorotatory sugar, while fructose is a levorotatory sugar.

Epimers Monosaccharides which differ in the position of the –OH group around a single carbon are called epimers. For example, consider glucose, galactose, and mannose

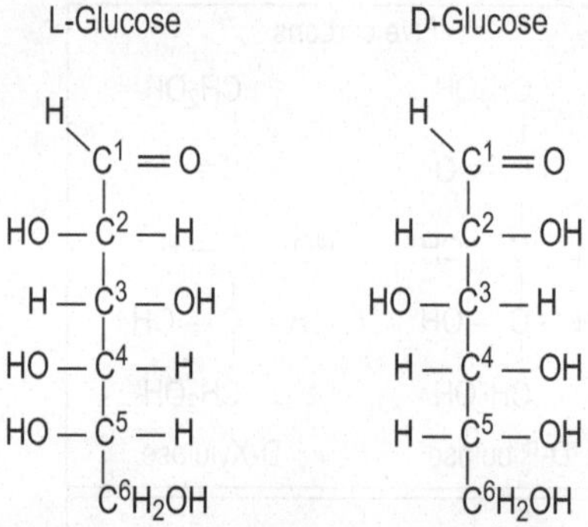

FIGURE 4.48 Structure of L-glucose and D-glucose

as shown in Fig. 4.49. These sugars differ by one carbon which is highlighted in the figure.

Cyclic form

Cyclic form of sugar is formed due to cyclization and bond formation between different carbon atoms of that sugar residue. For example, in glucose, a bond is formed between the carbonyl group of the first carbon and hydroxyl group of the fifth carbon resulting in cyclization of the structure. The cyclic form is called hemiacetal form (shown in Fig. 4.50). If the –OH group of the first carbon of the hemiacetal is on the right side, the sugar is called α-form, and if it is on the left side, the sugar is called β-form. The structures of α-D-glucose and β-D-glucose are shown in Fig. 4.51.

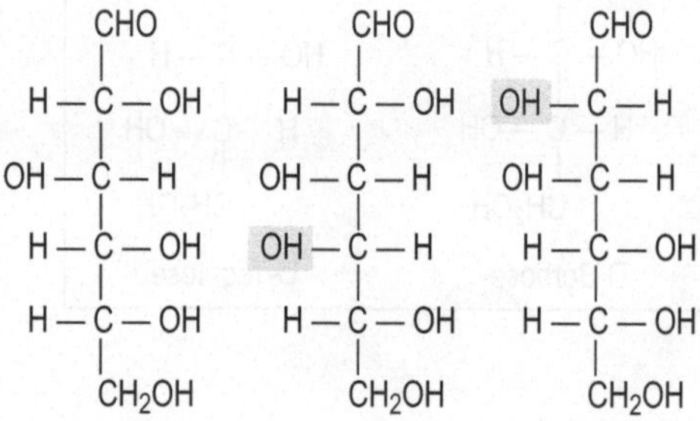

Glucose Galactose Mannose

FIGURE 4.49 Epimers (glucose and galactose differ in the third carbon atom)

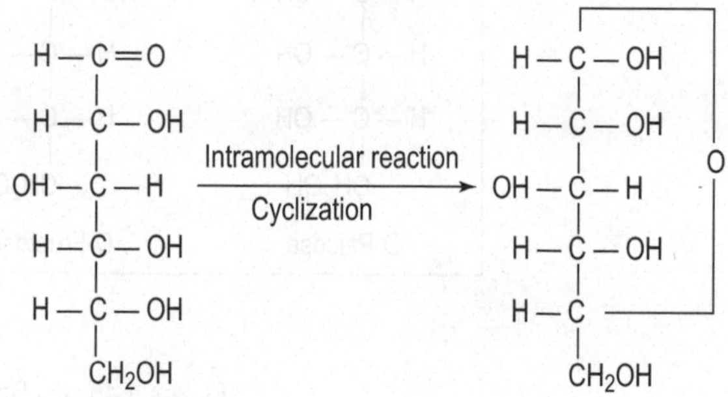

Linear glucose Hemi acetal form of glucose

FIGURE 4.50 Cyclization of linear glucose to hemiacetal form

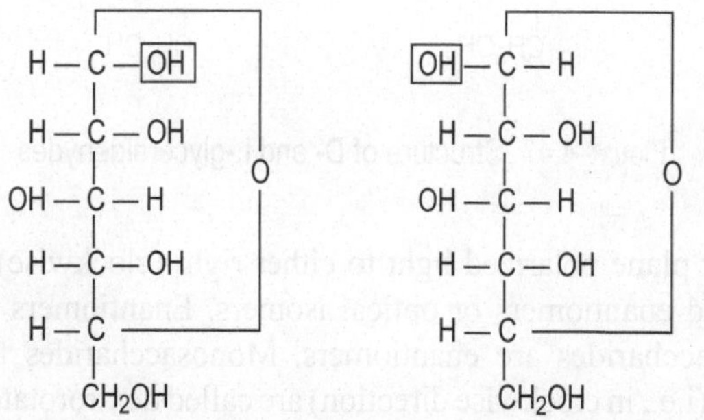

α-D-glucose β-D-glucose

FIGURE 4.51 Structure of α-form and β-form of glucose

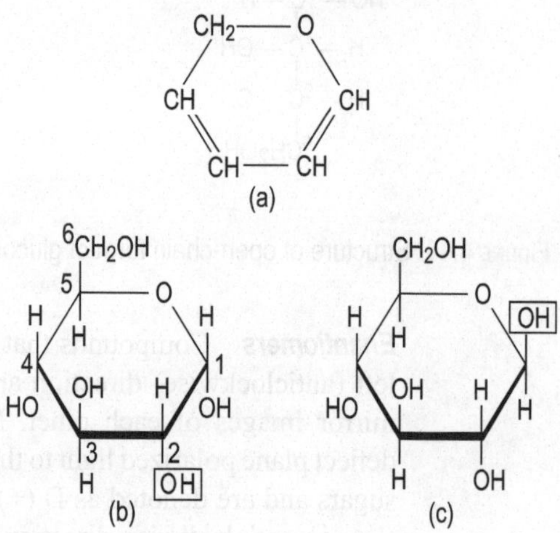

FIGURE 4.52 Structure of (a) pyran ring, (b) α-D-glucose, and (c) β-D-glucose

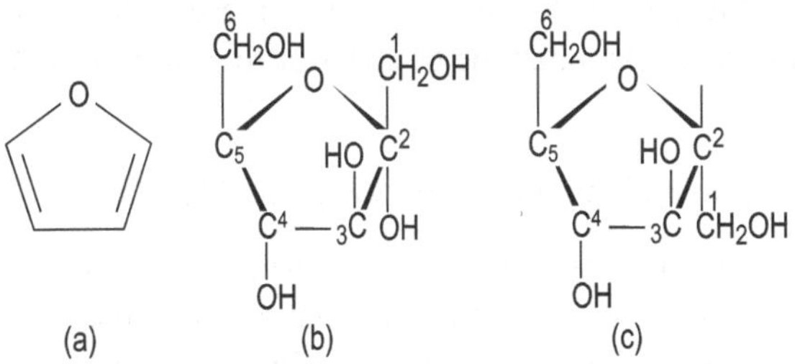

FIGURE 4.53 Structure of (a) furan ring, (b) α-D-fructose, and (c) β-D-fructose

Ring form

The ring form of aldoses resembles the structure of a pyran ring (as shown in Fig. 4.52a). Hence, the ring form of aldoses is also known as the pyranose form. The ring form of ketoses resembles the structure of a furan ring (as shown in Fig. 4.53a). Hence, the ring form of ketoses is also known as the furanose form. If the –OH group of the first carbon of the ring is in the downward direction, that is, below the plane, the sugar is said to be an α-sugar; on the other hand, if the –OH group is in the upward direction, that is, above the plane, it is said to be a β-sugar. Structures of α-D-glucose and β-D-glucose are shown in Fig. 4.52b and c, and those of α-D-fructose and β-D-fructose are shown in Fig. 4.53b and c.

4.5.6 Glycoside Linkage

Sugar residues are linked to one another forming disaccharides, oligosaccharides, and polysaccharides via glycoside linkages. A glycosidic bond is formed as a result of the reaction of an anomeric carbon of one sugar with the hydroxyl group of another sugar. Such sugars involved in glycosidic-bond formation are called non-reducing sugars (e.g., sucrose). If the anomeric carbon of a sugar is not involved in glycosidic-bond formation, that is, if it is free, then it can reduce the ferric and cupric ions. Such sugars are called reducing sugars, for example, glucose. These carbohydrates are usually represented from non-reducing end to reducing end. For example, lactose (milk sugar) is formed from monosaccharides β-D-galactose and β-D-glucose via glycosidic bonds. Maltose is formed as a result of the formation of glycoside bond between the anomeric carbons of the β-D-glucose molecule with the C4 hydroxyl group of (another) β-D-glucose (Fig. 4.54). The bond is called a β (1→4) bond, where β is the configuration of the anomeric carbon, '1' represents the number of the anomeric carbon and '4' represents the number of the carbon of the second sugar involved in the glycosidic bonding. The glucose sugar in the lactose possesses a free anomeric carbon which can reduce ferric ions, thus, glucose forms the reducing end.

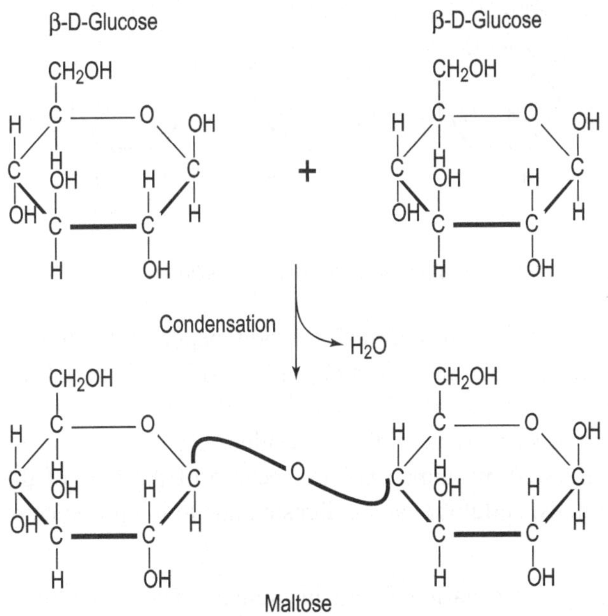

FIGURE 4.54 Formation of glycosidic bond

4.5.7 Polysaccharides

Polysaccharides are polymers of several sugar residues, usually more than ten. There are two types of polysaccharides—homopolysaccharides and heteropolysaccharides. Homopolysaccharides are composed of the same type of sugars, for example, starch, glycogen, and cellulose, which are made of glucose units only. Heteropolysaccharides are composed of more than one type of sugar or their derivatives.

Homopolysaccharides

In this subsection we will briefly discuss some common homopolysaccharides.

Starch Starch is a plant polysaccharide. It is a storage polysaccharide which occurs in the form of granules. It is structurally made of two compounds—amylose and amylopectin (Fig. 4.55). Amylose is made of a linear chain of α-D-glucose units linked by α (1→4) glycosidic linkage. Amylopectin is made of a branched chain of α-D-glucose units, similar to amylose, linked by α (1→6) glycosidic linkage at the branching point. Branching occurs at every 25 to 30 residues. All sugars present in starch are oriented in the same direction.

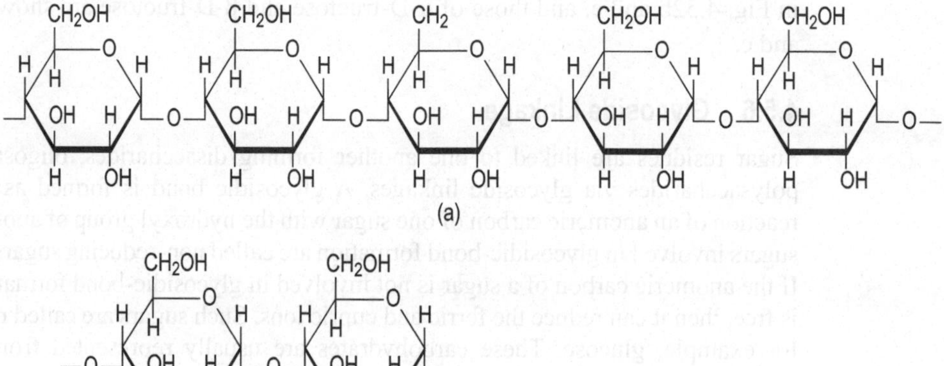

FIGURE 4.55 Structure of starch: (a) amylose and (b) amylopectin

Cellulose Cellulose is a structural polysaccharide found in wood, paper, and cotton. It is formed by polymerization of β-D-glucose units via β (1→4) glycosidic linkage (Fig. 4.56). The orientation of every alternative glucose ring in the cellulose is upside down. This is in contrast to the structure of starch where all the residues are oriented in the same direction. Cellulose is not digested by human beings because of the lack of the enzyme cellulase. However, in animals, cellulose can be digested due to the presence of cellulase.

Glycogen Glycogen is a storage polysaccharide found in animals. The structure of glycogen is similar to the amylopectin component of starch except that the branching occurs more frequently—at every 8 to 12 glucose residues. Unlike starch and cellulose, which are also made of glucose, glycogen has a forked or highly branched glucose chain.

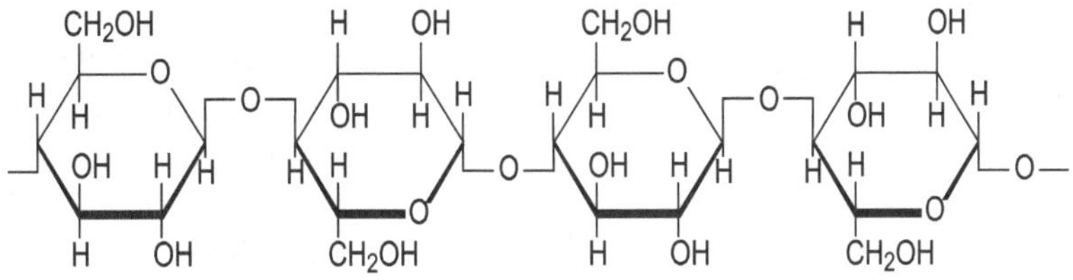

FIGURE 4.56 Structure of cellulose

Chitin Chitin is a homopolysaccharide made up of N-acetylglucosamine units linked by β (1→4) glycosidic linkage. It is hard and tough and found in exoskeleton. It plays a structural role due to its high matrix strength and rigidity. The structure of chitin is given in Fig. 4.57.

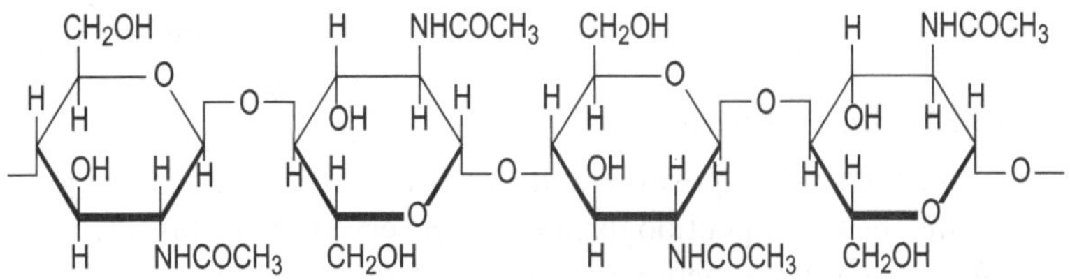

FIGURE 4.57 Structure of chitin

Inulin Inulin is a polymer of fructose units (Fig. 4.58) and is also called polyfructosan. It is found in plants, typically in roots and rhizomes. It serves as a stored energy source.

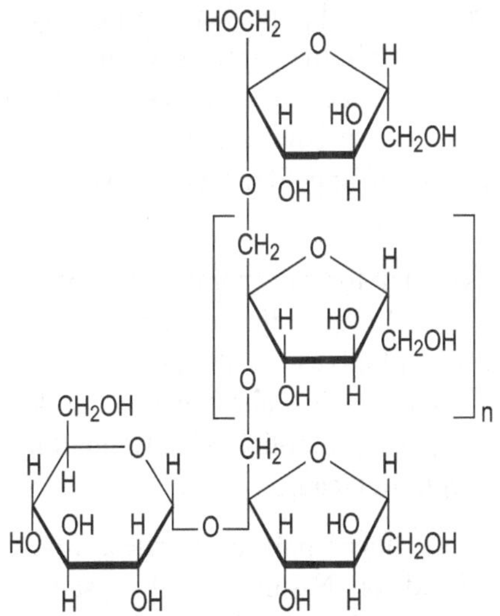

FIGURE 4.58 Structure of inulin

Heteropolysaccharides

Heteropolysaccharides are polymers composed of two or more different types of sugars or sugar derivatives. They are found in bacterial cell wall and extracellular matrix. The following three types of heteropolysaccharides are known:

1. Peptidoglycan
2. Glycosaminoglycan
3. Proteoglycan

Peptidoglycan Peptidoglycans are linear polymers of N-acetyl muramic acid (NAM) and N-acetyl glucosamine (NAG) linked by β(1→4) linkage. NAM residues are attached to tetrapeptide composed of alanine, glutamic acid, and lysine. Tetrapeptides are covalently cross linked by a bridge of pentaglycine residues (Fig. 4.59). The β (1→4) linkage of peptidoglycan is susceptible to attack by an enzyme lysozyme

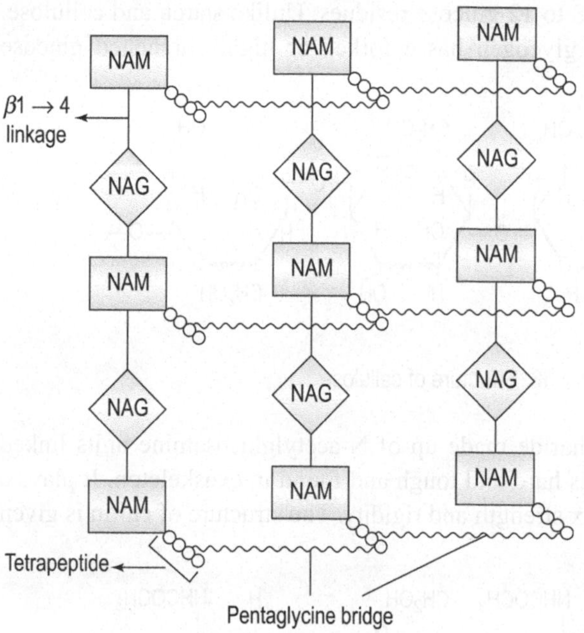

FIGURE 4.59 Structure of peptidoglycan

which results in the degradation of the polymer. This forms the basis for the protection of eyes from bacteria by the lysozyme present in tears. Peptidoglycan is responsible for maintaining the integrity of the bacterial cell. Therefore, peptidoglycan is the target for many antibiotics, which cleave the peptidoglycan leading to osmotic lysis of the cell.

Synthesis of peptidoglycan occurs in three different regions of the cell: cytoplasm, membrane, and cell wall. Initially, the basic subunit of peptidoglycan is synthesized in the cytoplasm, during which the NAM sugar associates with the alanine, glutamic acid, lysine, and alanine precursor. In the next stage, the NAM bonds with the NAG sugar in the membrane. Finally, the peptidoglycan is bonded to the already existing peptidoglycan in the cell wall with subsequent lateral cross-linking via pentaglycine residues.

In both Gram-positive and Gram-negative bacteria, majority of the polysaccharide component of peptidoglycan is conserved, whereas the peptide component varies. Accordingly, the peptide is classified into two types: DAP-type peptide, in which the third amino acid residue is the meso-diaminopimelic acid (DAP), and lys-type peptide, in which the third position is occupied by lysine. DAP-type peptide exists in Gram-negative bacteria and Gram-positive bacilli, whereas lys-type peptide is found in most other Gram-positive bacteria (including Gram-positive cocci). In Gram-positive bacteria, peptidoglycan constitutes half of the cell wall mass, as a thick layer, whereas in Gram-negative bacteria, the peptidoglycan is a thin layer present underneath the lipopolysaccharide layer (discussed in Chapter 1.) When the bacteria infect a host cell, the peptidoglycan and its fragment is considered to be an antigen, which causes enhanced immune response with the production of cytokines and chemokines. The infection produces biological effects such as inflammation, arthritis, fever, drowsiness, loss of appetite, leucocytosis, and thrombocytopaenia.

Functions of peptidoglycan Peptidoglycans are essential for the maintenance of the cellular shape and the architecture. They also play a vital role in the regulation of cell division and diffusion of molecules cross the cell. Peptidoglycan is the anchor for its covalently associated components such as teichoic acid, proteins, and polysaccharides. Counteraction of osmotic pressure of the bacterial protoplast, promotion of cell signalling, and cell–cell communication are other functions of peptidoglycan.

Glycosaminoglycan Glycosaminoglycans (GAGs) are linear polysaccharides, whose disaccharide building blocks consist of N-acetylated or N-sulphated hexosamine and either an uronic acid (glucuronic acid or iduronic acid) or galactose. Due to the combination of carboxyl group of glucuronic acid and sulphates, GAG has a high density of negative charge. Indeed, they are the most anionic molecules produced by animal cells. The negative charges on the surface of the GAG attract osmotically active sodium ions and other cations, resulting in influx of excess of water into the matrix. This generates a swelling pressure or turgor, which facilitates the matrix to endure compressive forces. This type of endurance is seen in the cartilage matrix lining the knee joints. GAGs form porous hydrated gels that fill most of the extracellular space, providing mechanical support to the tissue. GAG forms an interlocking porous meshwork in the extracellular matrix of connective tissues forming a porous pathway for the diffusion of nutrients and oxygen to individual cells.

Hyaluronan, chondroitin sulphate (CS), dermatan sulphate (DS), heparin, and keratan sulphate (KS) are the major types of GAG found in vertebrate cells. The sugar units and functions of these GAGs are listed in Table 4.14.

Hyaluronan lacks sulphate groups, but the rest of the glycosaminoglycans contain sulphates at various positions. DS is distinguished from CS by the presence of iduronic acid. Keratan sulphates lack uronic acids and instead consist of sulphated galactose and N-acetyl glucosamine residues.

Chondroitin sulphate is a linear GAG containing repeating unit of D-galactosamine and D-glucuronic acid. It is involved in the maintenance of structural integrity of tissues. The structure of CS has electrostatic repulsion due to negatively charged sulphate groups, which provide resistance to compression in cartilage.

Dermatan sulphate (DS; also called a mucopolysaccharide) is found mostly in skin, but is also present in blood vessels, heart valves, tendons, and lungs. DS, otherwise called chondroitin sulphate B or CS-B, contains repeated disaccharide units formed by hexosamine and N-acetyl galactosamine (GalNAc) or hexosamine and glucuronic acid (GlcA). The units are linked by β (1→4) or (1→3) linkages, respectively. DS is covalently bonded to serine residues of core proteins to form DSPGs. Secreted matrix proteins, namely decorin (containing one DS chain) and biglycan (containing 1 to 2 DS chains), are the best-studied DSPGs. DS contributes to the pliability of skin. Research reports suggest that dermatan sulphate is less haemorrhagic than heparin at equivalent anti-thrombotic doses, and that the haemorrhagic effect is associated with a glycosaminoglycan-induced platelet defect.

Heparin sulphate is a linear polymer formed by the repeated assembly of disaccharide units with uronic acid (glucuronic or iduronic acid) and glucosamine residues followed by the addition of sulphate groups at different positions. Heparin sulphate is structurally similar to heparin but has a lower density of sulphate esters.

Hyaluronic acid (HA) (also called hyaluronon or hyaluronate) is the simplest glycosaminoglycan composed of two monosaccharides: β (1, 4)-N-acetyl-D-glucosamine and β (1, 3)-D-glucuronic acid. Hyaluronan is secreted by specialized cells such as fibroblasts, keratinocytes, and chondrocytes; it is degraded by hyaluronidase. HA is unique among the GAGs because it does not contain any sulphate and is not found covalently attached to proteins. It forms non-covalently linked complexes with proteoglycans in the extracellular matrix (ECM). It is found in variable amounts in all tissues and fluids in adult animals, and is especially abundant in early embryos. It is a major constituent of the ECM. It provides compression strength, lubrication, and hydration within the ECM. It also regulates cell adhesion and motility and mediates cell proliferation and differentiation,

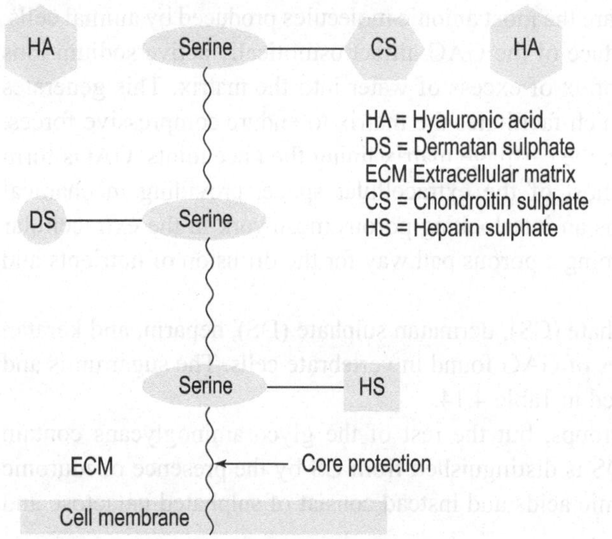

FIGURE 4.60 Glycosaminoglycans

making it not only a structural component of tissues, but also an active, signalling molecule. HA functions as a wound healer and also as a lubricant in joints.

Keratan sulphate (KS) is a proteoglycan in which the core protein is any one of the following: lumican, keratocan, mimecan, fibromodulin, PRELP, osteoadherin, and aggrecan. The KS chain is attached to the asparagine/serine/threonine residues in the core protein. Linkage to asparagines residue is found abundantly in cornea, which maintains tissue hydration making the corneal layer transparent. In cartilage, the linkage is mediated by serine or threonine. Research findings suggest a potential role of KS in the implantation process. Various GAGs are schematically represented in Fig. 4.60.

TABLE 4.14 Glycosaminoglycans

GAG	Individual units	Function	Location
Chondroitin-4-sulphate	N-acetyl galactosamine -4-sulphate D-glucuronic acid	Major structural component of cartilage and other tissues	Cartilage, cornea, bone, skin, arteries
Chondroitin-6-sulphate	N-acetyl galactosamine -6-sulphate D-glucuronic acid	Component of synovium and blood vessels	Cartilage, cornea, bone, skin, arteries
Dermatan sulphate	N-acetyl galactosamine -4-sulphate Iduronic acid	Structural component of skin	Skin, blood vessels, heart, heart valves
Heparin	N-acetyl glucosamine Iduronic acid/glucuronic acid	Anticoagulant liver, immune cells	Arteries, lung
Hyaluronic acid	N-acetyl-D-glucosamine D-glucuronic acid	Lubricant, shock absorber	Various connective tissues, skin, cartilage, synovial fluid
Keratan sulphate	N-acetyl glucosamine Galactose	Maintaining transparency of cornea Implantation process	Cornea, cartilage

Proteoglycans Proteoglycans (POGs) are comprised of a protein core attached to one or more covalently attached glycosaminoglycan chains. POGs synthesized in the cell have one of the following three fates: they may be secreted into the ECM, inserted into the plasma membrane, or stored in secretory granules. Some proteoglycans can be proteolytically cleaved and shed from the cell surface. Syndecan and glypican are membrane-bound POGs. Aggrecan, biglycan, decorin, and prelecan are ECM POGs. Membrane-bound and ECM POGs are represented in Fig. 4.61. POGs interact with other ECM components, such as fibrous proteins that provide tensile strength and elasticity (e.g., various collagens and elastins), adhesive glycoproteins (e.g., fibronectin, laminin, and tenascin) to give a hydrated gel that resists compressive forces. This results in an extracellular environment that regulates cell proliferation and differentiation. In connective tissues, such as cartilages, POGs make noncovalent interactions with other proteoglycans, proteins, and glycosaminoglycans providing strength and resilience.

The carbohydrate part of POGs acts as a binding structure at or near the cell surface. For example, syndecan is a POG which communicates with the intracellular region by linking to the cytoskeleton through its intracellular domain, simultaneously communicating with the extracellular matrix through its association with the protein fibronectin. Thus, the POG functions like a connector between the extracellular and intracellular functions.

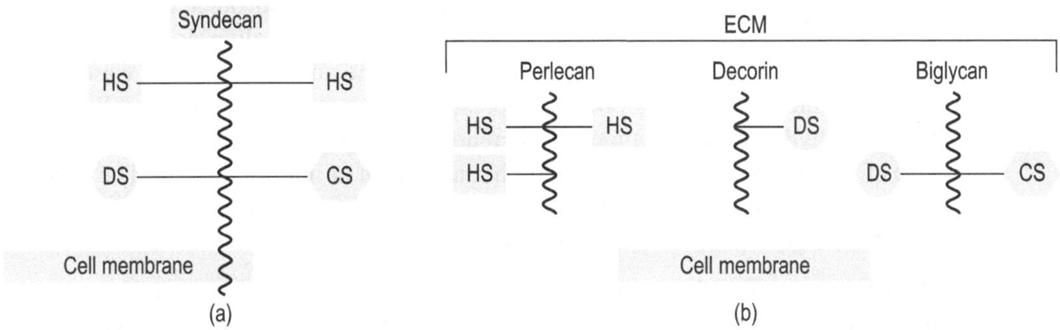

FIGURE 4.61 Proteoglycans (a) membrane-bound proteoglycan and (b) ECM proteoglycan

4.5.8 Glycoproteins

Proteins covalently attached to carbohydrates are called glycoproteins. If the protein moiety is attached to the carbohydrate moiety through the hydroxyl group of serine or threonine residue, it is called an O-linked glycoprotein. If the attachment is via the amino group of asparagine residue, it is called an N-linked glycoprotein. The covalently attached oligosaccharides influence the folding and stability of the proteins, provide critical information about the targeting of newly synthesized proteins, and allow for specific recognition by other proteins.

Functions of glycoproteins

The following are some important functions of glycoproteins:

- They form the structural components of cell membrane and cell wall.
- Mucus is a glycoprotein. It serves as lubricant.
- Immunoglobin is a glycoprotein which functions in immunity.
- Human chorionic gonadotropin (HCG) is a glycoprotein hormone which plays a role in pregnancy.
- Glycoproteins are involved in cell–cell attachment.

Examples of glycoproteins

In this section, we will briefly discuss some examples of glycoproteins to enable greater understanding regarding them.

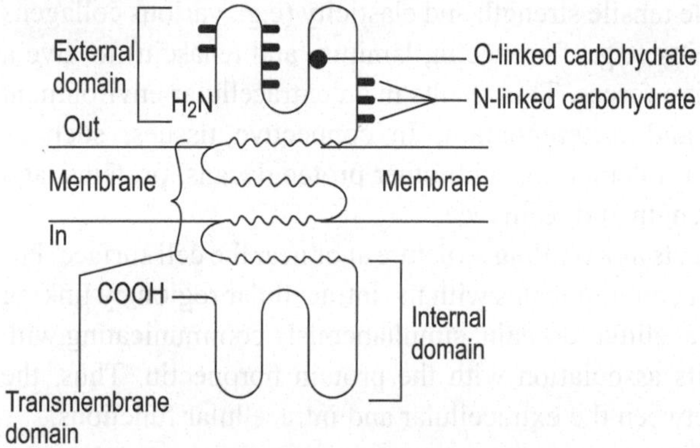

FIGURE 4.62 Glycophorin in red cell membrane

Glycophorin Glycophorin is the best-characterized glycoprotein. It is located in the red blood cell (RBC) membrane. Each RBC has some 500,000 copies of glycophorin embedded in its plasma membrane. About 60% of glycophorin contains 40% protein. Glycophorins have abundant sialic-acid residues that provide red cells with a hydrophilic charged coat. Because of this, the red cells can circulate freely in the lumen of the blood vessels without adhering to the vessel walls or to other cells. Glycophorin also has sixteen oligosaccharide chains, of which fifteen are linked to protein via tyrosine/serine and one is linked via asparagine. The N-terminal end of the protein is present on the outer side of the cell membrane forming a hydrophilic domain or external domain. The C-terminal is located on the inner side of the membrane forming a hydrophobic domain or interior domain. The central part of the protein traverses through the membrane forming the transmembrane domain. The structure of glycophorin is shown in Fig. 4.62.

Ceruloplasmin Ceruloplasmin is a copper-transporting glycoprotein in the blood of human beings and other vertebrates. It also plays a role in iron metabolism. It has oligosaccharide chains ending in sialic acid. Sialic acid determines the life of glycoproteins. If the sialic acid is removed, asialoceruloplasmin (meaning ceruloplasmin without sialic acid) is formed which binds to its specific receptor on the liver cells. Asialoceruloplasmin is degraded in the liver cells by lysosomes.

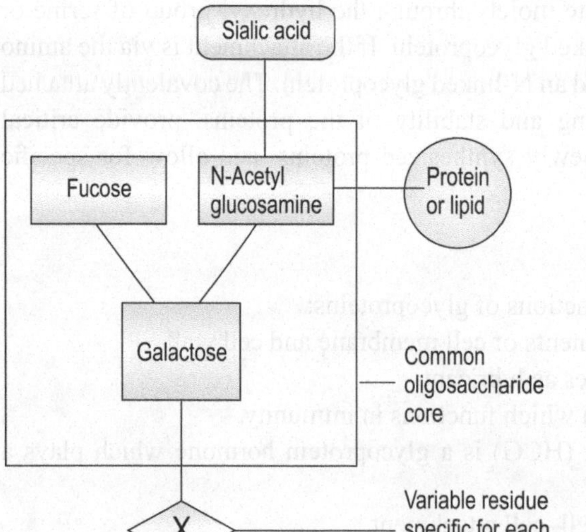

FIGURE 4.63 Structural components of blood group antigen (blood group glycoproteins)

Blood group antigens Blood group antigens which determine the blood grouping of an individual are glycoproteins. Blood group antigens are located on the surface of red blood cells. There are three types of blood group antigens—A-antigen, B-antigen, and O-antigen. All these antigens have a common oligosaccharide core but differ in single sugar moiety attached to the common core. Figure 4.63 illustrates various components present in blood group antigen. The antigen has a core oligosaccharide made of galactose, N-acetyl glucosamine, fucose, and sialic acid. N-acetyl glucosamine is linked to protein (forming glycoprotein), whereas galactose is linked to variable residue denoted as X. X is specific for each type of antigen and it determines the blood group of each individual:

X = N-acetylglucosamine for A-antigen.
X = Galactose for B-antigen.
X = absent for O-antigen.

4.6 LIPIDS

Lipids are organic compounds containing carbon, hydrogen, and oxygen. However, hydrogen and oxygen occur in a much higher ratio in lipids than in water. Lipids form about 3.5% of the total chemical composition of a cell. Carbohydrates and lipids fulfil the majority of the energy requirement of the human body through oxidation. Carbohydrates yield 4 kcal of energy per gram and lipids yield 9 kcal of energy per gram. However, unlike carbohydrates, lipids are reserved energy sources. This is because human system is incapable of storing a large amount of glycogen.

4.6.1 Definition

Lipid is the name given to fatty acids and their derivatives, and substances related biosynthetically or functionally to these compounds. Fatty acids are carboxylic acids with straight chains of carbon atoms (aliphatic chains, commonly C14–C24). They may be saturated or unsaturated, and can contain a variety of substituent groups. In general, fatty acids have the following molecular formula: $CH\text{–}(CH_2)_n\text{–}COOH$. Fat, oil, and wax are examples of lipids.

4.6.2 Constituents of a Lipid Molecule

By definition, two major constituents of lipids are alcohol and fatty acid. The alcohol component may be saturated (e.g., glycerol or cholesterol) or unsaturated (e.g., phytol), while the fatty acid component of lipids may be of the following four types:

1. Saturated fatty acid (without double bond in its structure) (e.g., palmitic acid and stearic acid)
2. Unsaturated fatty acid (with double bond in its structure) (e.g., oleic acid, linoleic acid)
3. Hydroxy fatty acid (e.g., ricinoleic acid, cerebronic acid, and dihydroxy stearic acid)
4. Cyclic fatty acid (e.g., hydnocarpic acid and chaulmoogric acid)

4.6.3 Essential Fatty Acids

Some fatty acids are not present in the body but are essential for normal cell functioning and should be taken through diet. These fatty acids are called essential fatty acids. Examples of essential fatty acids include linoleic acid, linolenic acid, and arachidonic acid.

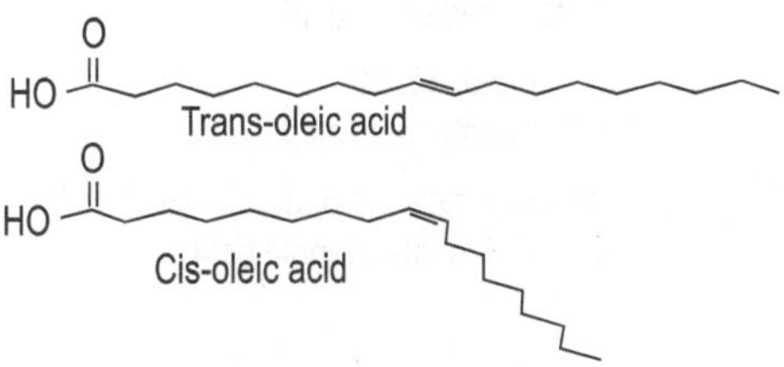

FIGURE 4.64 Geometrical isomerism or cis–trans isomerism in oleic acid

4.6.4 Isomerism due to Unsaturated Fatty Acids

Isomerism is a phenomenon in which a compound has two structural formulae but the same molecular formula. Unsaturated fatty acids exist in two different structural forms having the same molecular formula. They are the cis- form and the trans- form. In cis fatty acids, both parts of the chain adjacent to the double bond are at the same side of the double bond. In trans fatty acids, the two parts of the chain adjacent to the double bond are in opposite side of the double bond. For example, oleic acid is of cis form and elaidic acid is of trans form as shown in Fig. 4.64.

4.6.5 Nomenclature of Fatty Acids

Fatty acid chain has a methyl (CH_3) group at one end and carboxyl group (–COOH) at the other end. The number of carbon atoms in the fatty acid is numbered as C1, C2, etc., starting from the carboxylic end. Beginning from C2, carbon atoms are given Greek alphabetic notation. Thus, C2 is α-carbon, C3 is β-carbon, C4 is γ-carbon, and so on.

$$\begin{array}{ccccccccc} & & & & \varepsilon & \delta & \gamma & \beta & \alpha \\ CH_3– & CH_2– & CH_2– & CH_2– & CH_2– & CH_2– & CH_2– & CH_2– & COOH \\ 9 & 8 & 7 & 6 & 5 & 4 & 3 & 2 & 1 \end{array}$$

Each fatty acid is given a code number with three or four integers. The first integer represents the total number of carbon atoms in the fatty acid chain, the second integer represents the total number of double bonds in the chain. The third and the fourth integers are represented with a delta prefix and they represent the position of the double bond, that is, the position of any double bond is specified by a superscript number (third and fourth number) following D (delta). For example, a 20-carbon fatty acid with one double bond between C9 and C10 and another between C12 and C13 is represented as follows:

$$(CH_3)–(CH_2)_6\text{- }CH{=}CH–CH_2 – CH{=}CH–(CH_2)_7–COOH$$
$$20 \quad 13 \quad 12 \quad 10 \quad 9 \quad 1$$

This fatty acid is designated $20{:}2(\Delta^{9,\,12})$.

Some of the common fatty acids and their nomenclature are listed in Table 4.15.

TABLE 4.15 Some common fatty acids

Common name	Molecular formula	Systemic name	Code
Lauric acid	$CH_3(CH_2)_{10}COOH$	*n*-Dodecanoic acid	12:0
Myristic acid	$CH_3(CH_2)_{12}COOH$	*n*-Tetradecanoic acid	14:0
Palmitic acid	$CH_3(CH_2)_{14}COOH$	*n*-Hexadecanoic acid	16:0
Stearic acid	$CH_3(CH_2)_{16}COOH$	*n*-Octadecanoic acid	18:0
Arachidic acid	$CH_3(CH_2)_{18}COOH$	*n*-Eicosanoic acid	20:0
Lignoceric acid	$CH_3(CH_2)_{22}COOH$	*n*-Tetracosanoic acid	24:0
Palmitoleic acid	$CH_3(CH_2)_5CH{=}CH(CH_2)_7COOH$	cis-9-Hexadecenoic acid	$16{:}1(\Delta^9)$
Oleic acid	$CH_3(CH_2)_7CH{=}CH(CH_2)_7COOH$	cis-9-Octadecenoic acid	$18{:}1(\Delta^9)$
Linoleic acid	$CH_3(CH_2)_4CH{=}CHCH_2CH{=}CH(CH_2)_7COOH$	cis-, cis-9, 12-Octadecadienoic acid	$18{:}2(\Delta^{9,12})$
α- Linolenic acid	$CH_3CH_2CH{=}CHCH_2CH{=}CHCH_2CH{=}CH(CH_2)_7COOH$cis-,	cis-, cis-9, 12, 15-Octadecatrienoic acid	$18{:}3(\Delta^{9,12,15})$
Arachidonic acid	$CH_3(CH_2)_4CH{=}CHCH_2CH{=}CHCH_2CH{=}CHCH_2CH{=}CH(CH_2)_3COOH$	cis-, cis-, cis-, cis-5, 8, 11, 14- Icosatetraenoic acid	$(\Delta^{5,8,11,14})$

4.6.6 Functions of Lipids

The following are some important functions of lipids:

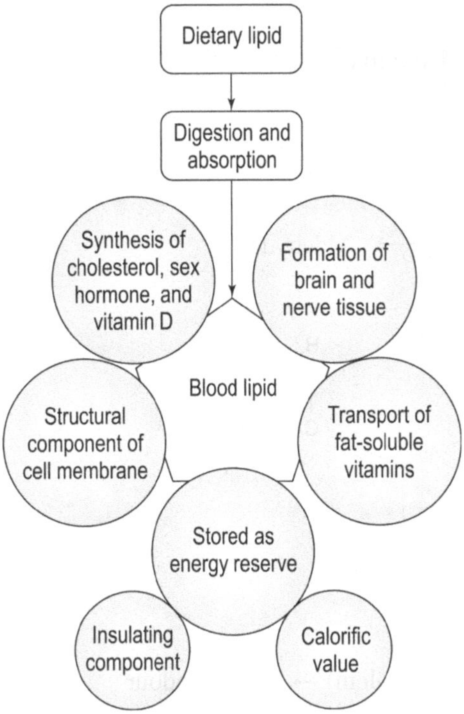

FIGURE 4.65 Functions of lipids

- They are a reserved source of energy.
- They provide high calorific value.
- They are structural components of cell membrane.
- They act as an insulating component.
- They help in absorption and transport of fatty acids.
- They are involved in biosynthesis of cholesterol, sex hormones, and vitamin D.
- They are carriers of fat-soluble vitamins.

The functions of lipids are summarized in Fig. 4.65.

4.6.7 Physical Properties of Lipids

The physical properties of lipids may be listed as follows:

- Lipids are colourless, odourless, and tasteless.
- Lipids are amphipathic in nature, that is, they have polar (–COOH) and non-polar (CH_3) groups.
- Lipids are capable of absorbing odour.
- Lipids are insoluble in water and soluble in organic solvents.
- Lipids have a high melting point. This depends on the degree of unsaturation as melting point is inversely related to unsaturation.
- Lipids have specific gravity less than 1.
- They exhibit geometrical isomerism due to the presence of double bond in unsaturated fatty acid.
- Lipids have high insulating power.
- Lipids exhibit emulsification property. Emulsification is a process by which lipids combine with water or emulsifying agents, such as soaps, gums, or proteins, and form aggregates or droplets called micelles.
- They function as precursors for the synthesis of ketone bodies.

4.6.8 Chemical Properties

The chemical reactions of lipids are listed:

Hydrolysis

$$\text{Lipids} \xrightarrow{\text{Lipase enzyme}} \text{Fatty acid} + \text{Diacyl glylcerol} \xrightarrow{\text{Lipase enzyme}} \text{Fatty acid} + \text{Monoacyl glylcerol}$$

Saponification or soap formation

Fat + alkali (NaOH/KOH) $\longrightarrow$ Glycerol + sodium salt of fatty acid (SOAP)

Sodium salt of higher fatty acid $\longrightarrow$ Hard soap

Potassium salt of higher fatty acid $\longrightarrow$ Soft soap

Hydrogenation

$$\underset{(\text{–CH=CH–})}{\text{Unsaturated fat (vegetable fat)}} \xrightarrow[H_2]{\text{Platinum/palladium}} \underset{(\text{–}CH_2\text{–}CH_2\text{–})}{\text{Saturated fat (solid fat)}}$$

Halogenation

$$\underset{(-CH=CH-)}{\text{Unsaturated fat}} \xrightarrow[\text{Acetic acid/ Methanol}]{\text{Halogen (X)}} \underset{(-CHCl-CHCl-)}{\text{Chloride derivative}}$$

Oxidation

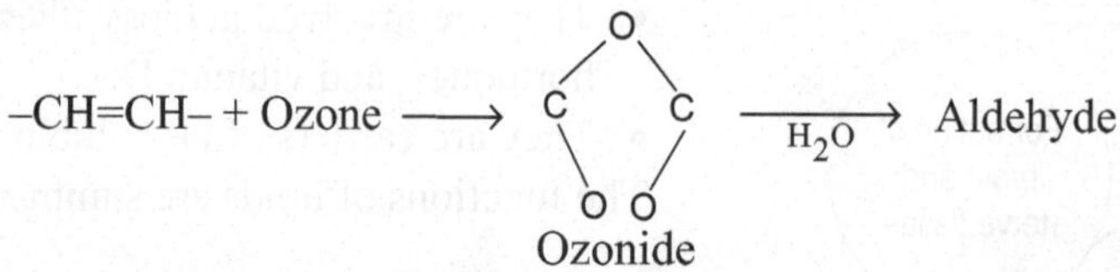

$$\underset{\text{Unsaturated fat}}{-CH=CH-} \xrightarrow{\text{Mild } KMnO_4} \text{Glycol}$$

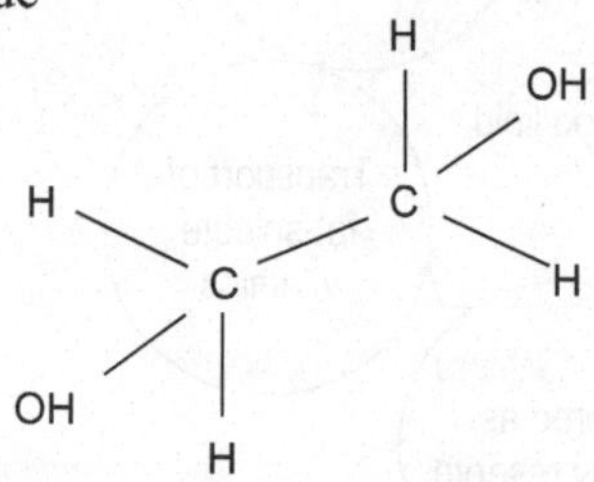

Aldehyde formation

$$\text{Fat} \xrightarrow[\text{Dehydrating agent } NaHSO_4]{\text{Heat}} \text{Unsaturated aldehyde (Acrolein)} \rightarrow \text{Pungent odour}$$

4.6.9 Terminologies Related to Properties of Lipids

Acid number

Acid number is the amount of KOH in milligrams required to neutralize the free fatty acid present in 1 g of fat. This quantity is used to quantify fatty acid in fat.

Saponification number (Sap value/Sap number)

Saponification number is the amount of KOH in milligrams required to saponify 1 g of fat. Sap value is inversely related to fatty acid chain length.

Iodine number

Iodine number is the amount of iodine in grams absorbed by 100 g of fat. Iodine number is directly related to unsaturation.

Reichert–Meissl number (RM number/RM value)

RM number is the amount of 0.1N NaOH in millilitres required to neutralize soluble fatty acid in 5 g fat.

4.6.10 Classification of Lipids

Lipids are classified broadly into simple lipids, compound lipids, and derived lipids. These lipids are subclassified into different types. The classification and subclassification of lipids are outlined in Table 4.16.

TABLE 4.16 Classification of lipids

Simple Lipids	Compound Lipids	Derived Lipids
Fats and oils	**Phospholipids**	**Sterol**
(Triglycerides/ triacylglycerol)	Lecithin	Cholesterol
	Cephalin	
	Inositol	
	Plasmalogen	
Waxes	**Sphingolipids**	**Bile acids**
	Sphingomyelin	
	Sulphatid	**Steroid hormones**
	Cerebroside	Oesterogen
	Ganglioside	Androgen
		Mineralocorticoid
		Glucocorticoid
		Progesterone
		Terpenes
		Monoterpene
		Diterpene
		Triterpene
		Carotenoids
		Lycopene
		Carotene
		Xanthophyll

Simple lipids

Simple lipids are esters of fatty acids with a short-chain alcohol, glycerol. They are also called triacylglycerols or triglycerides (abbreviated as TAG or TG), and are sometimes also called neutral fats. The structure of glycerol and TAG is shown in Fig. 4.66. TGs are named after the types of fatty acids esterified with glycerol. An example for simple lipids is tristearin in which the fatty acid attached to glycerol is stearic acid. Other examples include tripalmitin and triolein named so because of the presence of palmitic acid and oleic acid, respectively. Majority of TGs are stored in specialized cells called adipocytes in animals and human beings and in seeds of plants. Therefore, TGs are called storage lipids. TGs function as stored fuel and insulator against low temperature. TGs provide twice as much energy as carbohydrates because the carbon atoms of fatty acids are more reduced than those of sugar.

Fats and oils Simple lipids containing a large number of saturated fatty acids are called fats and they have a solid consistency. They are also called animal lipids. Examples include butter fat (soft solid) and beef (hard solid). Simple lipids with unsaturated fatty acids are called oils and they have a liquid consistency. They are also called plant lipids. Examples include olive oil, castor oil, etc. Differences between fat (animal lipid) and oil (plant lipid) are listed in Table 4.17.

$$\text{Saturated fatty acid} + \text{glycerol} \rightarrow \text{Fat}$$
$$\text{Unsaturated fatty acid} + \text{glycerol} \rightarrow \text{Oil}$$

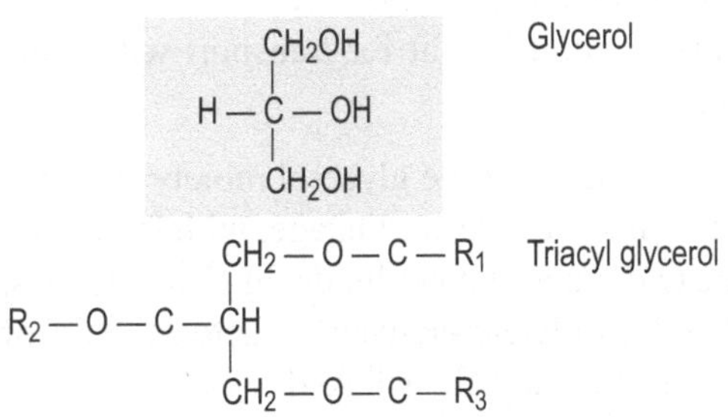

FIGURE 4.66 Structure of glycerol and triacylglycerol (TAG)

Waxes Waxes are hard esters of long-chain saturated fatty acids (containing 14 to 36 carbon atoms) with long-chain monohydroxy alcohols (containing 15 to 30 carbon atoms). Triacontanoylpalmitate is a wax formed due to ester bond formation between palmitic acid with the alcohol triacontanol. Melting points of waxes are higher than those of TGs. Well-known waxes include lanolin wax from lamb wool, bees wax from honey comb, and carnauba wax from Brazilian palm. Juniperic acid and palmitic acid are common long-chain fatty acids found in waxes. Cetyl alcohol, hexacosanol, and myristyl alcohol are common long-chain alcohols found in wax. Waxes are found in cutaneous glands, wool, fur, and some leaves. They are highly resistant to oxidation. Waxes are chief storage form of metabolic fuel in planktons. Waxes function to protect the skin and hair and keep them pliable and water repellent.

TABLE 4.17 Differences between fat and oil (animal lipid and plant lipid)

Fat (animal lipid)	Oil (plant lipid)
Stored in liver and bone marrow	Stored in seeds and fruits
Solid in nature	Liquid in nature
Melting point is high	Melting point is low
Rich in saturated fatty acids	Rich in unsaturated fatty acids
Low iodine value	High iodine value
High RM value	Low RM value
Oxidative rancidity is more frequent	Oxidative rancidity is less frequent

Compound lipids

Compound lipids are also called heterolipids or polar lipids (TGs are neutral lipids). Majority of these lipids are present in membranes, hence, the name membrane lipids. They are subdivided into phospholipids, ether lipids, sphingolipids, galactolipids, and sulpholipids.

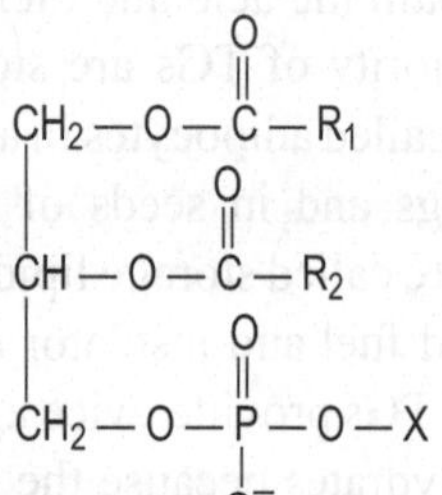

FIGURE 4.67 Structure of phosphatidic acid—the backbone of phospholipids

Phospholipids The common glycerophospholipids are diacylglycerols linked to head-group alcohols through a phosphodiester bond. Phospholipids consist of a glycerol backbone linked to two fatty acids: C16 or C18 saturated fatty acid at C1, and a C18 to C20 unsaturated fatty acid at C2. It also has a head group alcohol at C3 (denoted as X) via a phosphate diester bond. This whole structure is called phosphatidic acid which is shown in Fig. 4.67. Depending upon X (the alcohol group), phospholipids are divided into different types. The name of the phospholipid starts with the term 'phosphatidyl' (because of the phosphatidic acid backbone) and ends with the alcohol group containing compound X. Following are the different types of phospholipids:

- Phosphatidylcholine, if X is choline.
- Phosphatidylethanolamine, if X is ethanolamine.
- Phosphatidylserine, if X is serine.
- Phosphatidylinositol, if X is myoinositol.
- Phosphatidylglycerol, if X is glycerol.
- Cardiolipin, if X is phosphatidylglycerol.

The formula for the substituent X group in the phospholipid is given in Table 4.18.

Phospholipids are the constituents of cell membrane and regulate membrane permeability. They are also used as detergents to emulsify fat for transport within the body. They act as cellular metabolic regulator.

Ether lipids Ether lipids are phospholipids in which C1 of the glycerol moiety is linked to either alkyl or alkenyl chain via ether linkage, instead of ester linkage, with fatty acid. Plasmalogens and platelet-activating factor are two types of ether lipids. In plasmalogens, C1 is linked to an alenyl chain by an ether bond. In platelet-activating factor, the C1 of glycerol is linked to an alkyl chain via an ether linkage and C2 is linked to acetic acid via an ester linkage. Both plasmalogen and platelet-activating factor have choline as the head group alcohol. Structures of ether lipids are shown in Fig. 4.68.

TABLE 4.18 Types of phospholipids on the basis of the nature of alcohol-containing compounds (X) attached to the phosphatidic acid backbone via a phosphodiester bond

Name of the phospholipid (phosphatidic acid – X)	Name of the X group	Formula for X
Phosphatidylcholine	Choline	$OH{-}CH_2{-}CH_2{-}N^+(CH_3)_3$
Phosphatidylethanolamine	Ethanolamine	$OH{-}CH_2{-}CH_2{-}NH_2$
Phosphatidylserine	Serine	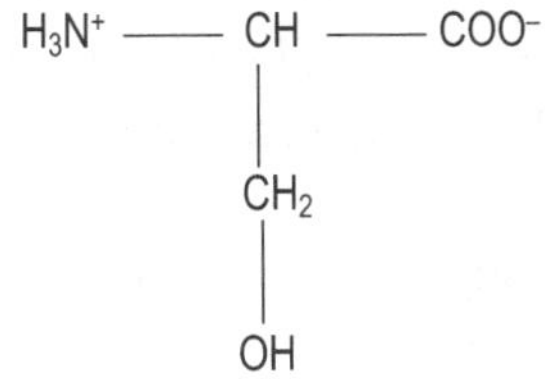
Phosphatidylinositol	Inositol	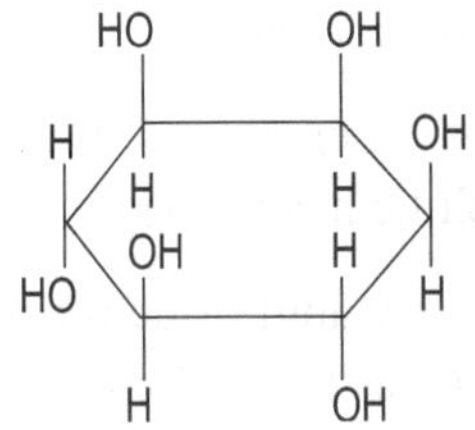
Phosphatidylglycerol	Glycerol	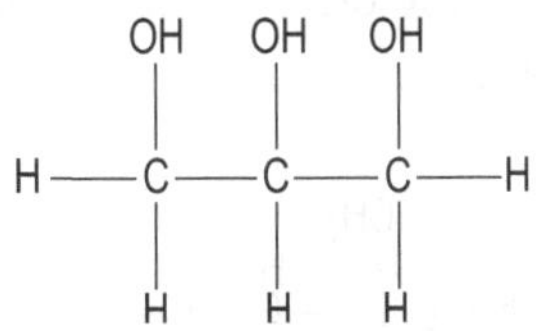
Cardiolipin	Phospatidylglycerol	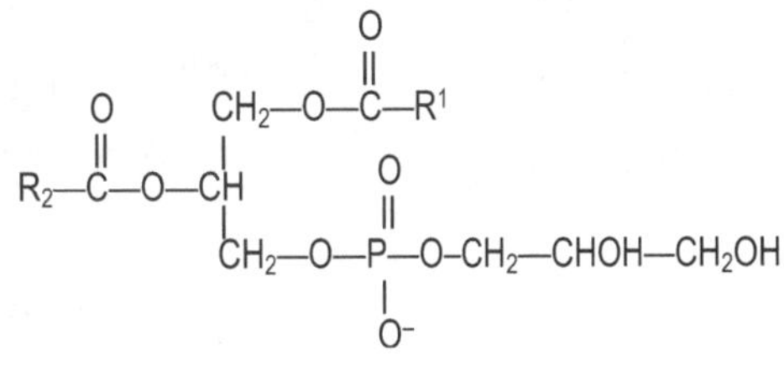

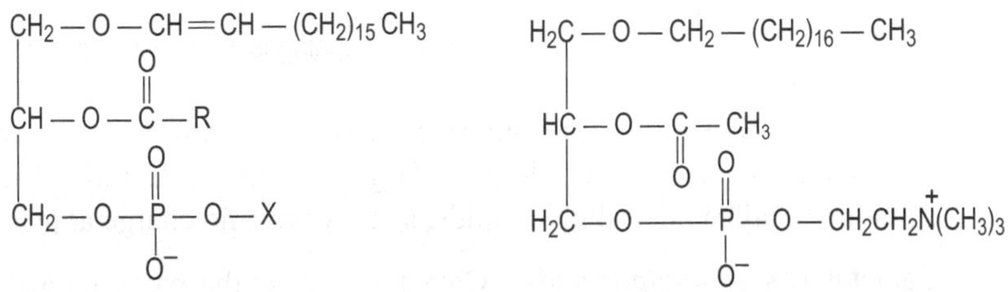

(a) (b)

FIGURE 4.68 Structure of ether lipids (a) Plasmalogen (X = Choline) and (b) Platelet-activating factor

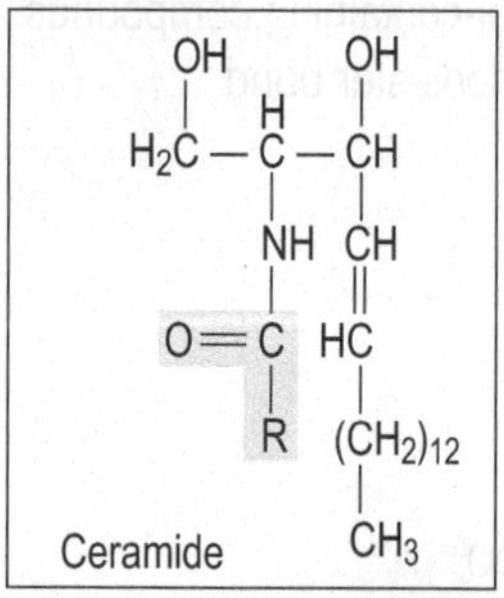

FIGURE 4.69 Structure of ceramide

Sphingolipids Sphingolipids are a large and important class of biomembrane lipids composed of an 18-carbon amino alcohol, sphingosine, which itself is derived from glycerol.

When a fatty acid is attached in amide linkage to the NH_2 on C2, the resulting compound is a ceramide, which is structurally similar to a diacylglycerol (Fig. 4.69). Ceramide is the structural parent of all sphingolipids. There are three subclasses of sphingolipids, all derivatives of ceramide. Depending on the additional groups attached to the sphingosine backbone, sphingolipids are subclassified. Structures of sphingolipids are depicted in Fig. 4.70.

Sphingomyelin Sphingomyelin has a phosphocholine or phosphoethanolamine head group attached to C3. Sphingomyelins are distributed in the plasma membranes of animal cells and are more abundant in myelin sheath or the axon membrane of some neurons.

Cerebroside Cerebroside is a ceramide sphingolipid containing glucose or galactose as the hydrophilic head group.

Ganglioside Ganglioside is another ceramide-containing complex oligosaccharide, such as sialic acid, as the hydrophilic head group.

Cerebrosides and gangliosides, are together referred as glycosphingolipids. They are located in the outer leaflet of the bilayer, with their sugar chains protruding outside the cell surface.

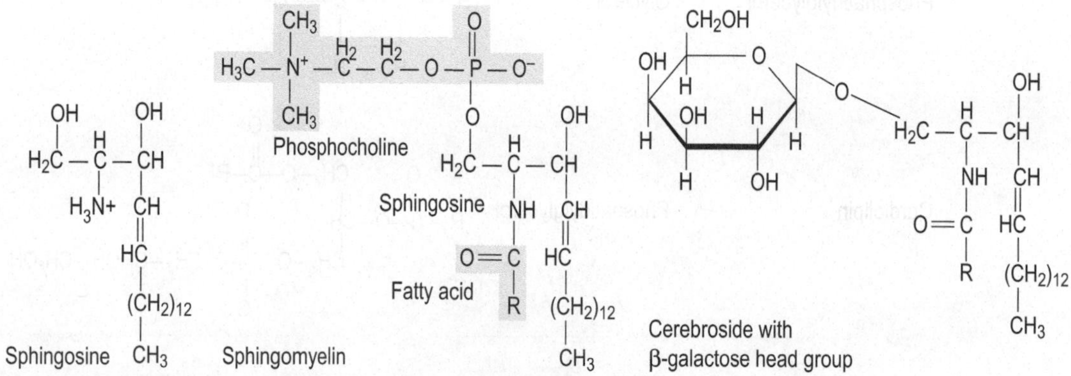

FIGURE 4.70 Structure of sphingosine and sphingolipids

Globosides Globosides are glycosphingolipids with two or more sugars, usually D-glucose, D-galactose, or N-acetyl-D-galactosamine. Cerebrosides and globosides are sometimes called neutral glycolipids, as they have no charge at pH.

Galactolipids and sulpholipids Galactolipids, as the name indicates, consist of one or two galactose attached by a glycosidic linkage to C3 of a 1, 2-diacylglycerol. These lipids are predominant in plant cells where they are found in the thylakoid membrane of chloroplasts. In sulpholipids, sulphated glucose residue is connected by a glycosidic linkage to C3 of a 1, 2-diacylglycerol. This is also predominant in plant cells. Structures of galactolipids and sulpholipids are shown in Fig. 4.71.

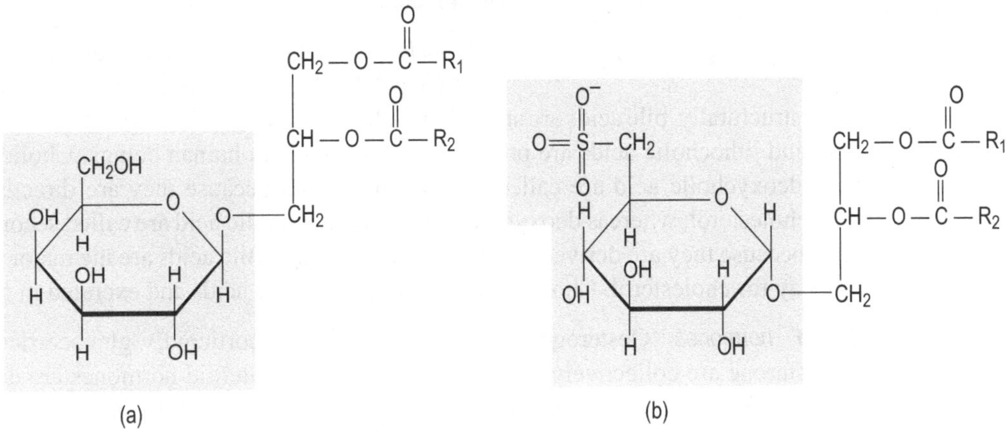

FIGURE 4.71 Structure of (a) galactolipid and (b) sulpholipid

Derived lipids

Derived lipids are obtained by hydrolysis of simple and complex lipids. These lipids contain glycerol and other alcohols. The various types of derived lipids are discussed in the following subsections:

Steroids and sterols Steroids are lipids with four fused carbon rings. The fused ring structure is called steroid nucleus or cyclopentanoperhydrophenanthrene ring. Steroids with a hydroxyl group and an aliphatic chain of at least eight carbons are called sterols, for example, cholesterol. The structures of cyclopentanoperhydrophenanthrene ring and cholesterol are given in Fig. 4.72.

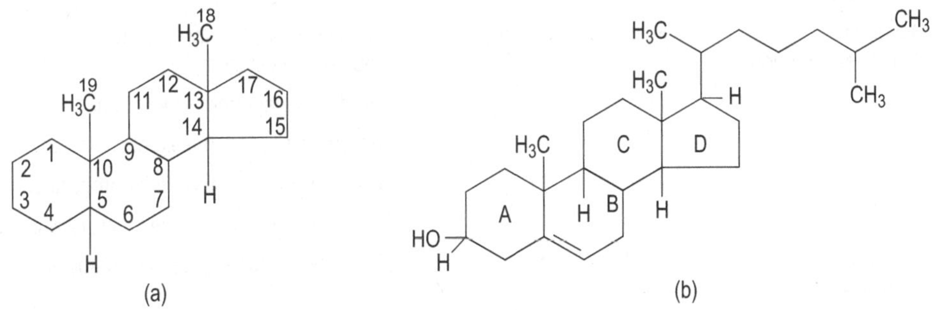

FIGURE 4.72 Structures of (a) cyclopentanoperhydrophenanthrene and (b) cholesterol

Cholesterol Cholesterol is the major sterol in the human system. Structurally, cholesterol is a steroid ring (cyclopentanoperhydrophenanthrene ring with OH group at third carbon, double bond at fifth carbon, and an aliphatic chain at seventeenth carbon) (Fig. 4.72b). Cholesterol is the precursor for bile acids, steroid hormones, and vitamin D3. Cholesterol is the structural component of cell membranes and plasma lipoproteins. Other sterols

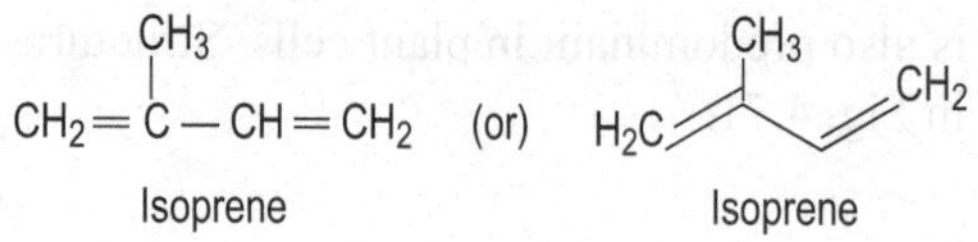

FIGURE 4.73 Structure of isoprene unit

of interest are ergosterol present in yeast and ergot, lanosterol present in wool fat, and spinasterol present in spinach.

Bile acids Bile acids are found in bile juice which is stored in gall bladder. They are derived from cholesterol. Thus, structurally, bile acids are steroids. Cholic acid, chenodeoxycholic acid, deoxycholic acid, and lithocholic acids are predominant bile acids in human beings. Cholic acid and chenodeoxycholic acid are called primary bile acids because they are directly derived from cholesterol, whereas deoxycholic acid and lithocholic acid are called secondary bile acids because they are derived from primary bile acids. Bile acids are the major excretory pathway for cholesterol. Cholesterol is converted to bile acids and excreted in faeces.

Steroid hormones Oesterogen, androgen, mineralocorticoid, glucocorticoid, and progesterone are collectively called steroid hormones. Steroid hormones are derivatives of cholesterol. Occurrence and functions of these hormones are given in Table 4.19.

Terpenes Terpenes are oxygenated derivatives of hydrocarbons. Building blocks of terpenes are called isoprene units (Fig. 4.73).

Depending on the number of carbon atoms, terpenes may be classified as the following:

- Monoterpene—$C_{10}H_{16}$
- Sesquiterpene—$C_{15}H_{24}$
- Diterpene—$C_{20}H_{32}$
- Triterpene—$C_{30}H_{48}$
- Tetraterpene or carotenoids—C_{40}

TABLE 4.19 Steroid hormones

Name of the hormone	Occurrence	Function
Oesterogen (female sex hormone)	Follicle cells of ovary	Regulates the activity of female reproductive system
		Produces menopause symptoms
		Produces secondary sexual characters in females
		Prepares uterine mucosa for progestation hormone action
		Increases skeletal muscle mass, growth, and strength
Androgen (Male sex hormone)	Leydig cells of testes	Produces secondary sexual characters in males
		Regulates the activity of male reproductive system
Progesterone (Pregnancy hormone)	Corpus luteum Placenta	Helps in implantation of fertilized ovum Aids uterine contraction
		Induces maturation of mammary gland
Mineralocorticoid	Cortical cells of kidney	Regulates the metabolism of sodium and potassium
		Regulates water distribution in body
Glucocorticoid	Cortical cells of kidney	Regulates the metabolism of carbohy drates, proteins, and fats

Carotenoids Carotenoids are tetraterpenes with 40 carbon atoms and a high degree of unsaturation. They are also called lipochromes or chromolipids because they are coloured. Examples for carotenoids are carotene present in carrots, lycopene in tomatoes, and xanthophyll present in corn and leaf.

4.7 WATER

4.7.1 Basic Information about Water

Water is the universal solvent. It is present in the body and also supplied to the body directly or by exogenous fluids. Total body water comprises approximately 80% of the mass of most living cells. A normal adult has 49 L of water. Out of this, 35 L is present in intracellular fluid (ICF) and 14 L in extracellular fluid (ECF). Out of 14 L in ECF, 11 L exists in interstitial tissues and 3 L in blood plasma, cerebrospinal fluid, synovial fluid, and aqueous and vitreous humours. Water performs many functions, such as transport of cellular biomolecules, stabilization of cellular structures, lubrication, driving the biochemical reactions and as a partitioning agent. Water is necessary for most of the functions of the body. In spite of its significance, it cannot be stored in the body, demanding regular intake of at least eight glasses per day.

4.7.2 Structure of Water

Atoms in water are arranged in a tetrahedral geometry. The H–O–H bond angle is 104.5°. Since oxygen is more electronegative, electrons are more often on the vicinity of the oxygen atom than hydrogen atoms. Thus, the portion of the water molecule near the oxygen atom is slightly negative (δ–). Portion near the hydrogen atom is slightly is positive (δ+) (Fig. 4.74). Therefore, water molecule is a dipole and it possesses dipole moment.

In the structure, the four valence orbitals of O point to corners of a tetrahedron. The two corners of the tetrahedron are occupied by hydrogen atoms, both of which are covalently bonded to oxygen atom. These corners are weakly electronegative, while the oxygen end is highly electronegative because the shared electrons are more towards the oxygen atom. In liquid state, each water molecule forms hydrogen bonds with an average of 3.40 other water molecules. The continuous motion of water in liquid state indicates rapid formation and breakage of hydrogen bonds.

4.7.3 Hydrogen Bonding

The polar molecules of water are held together by hydrogen bonds. Hydrogen bonds in water are the electrostatic interactions between hydrogen atom and the unshared electron pair of the electronegative atoms of oxygen. The bond is formed between the positively charged H of one water molecule with the negatively charged O of the neighbouring water molecule. Every water molecule can be hydrogen bonded to a maximum of four other neighbouring water molecules. Hydrogen bonding orders water into a higher level of structural organization.

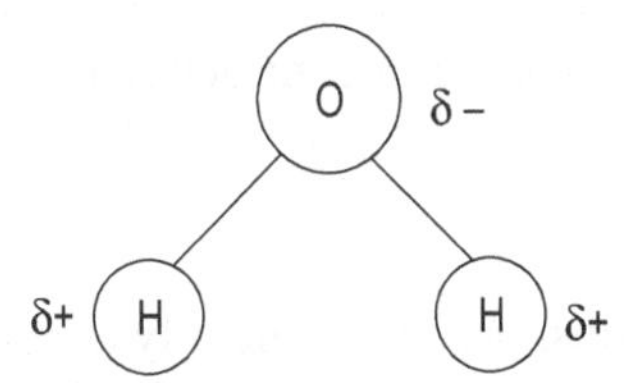

FIGURE 4.74 Structure of water molecule

TABLE 4.20 Properties of water

Name of the property	Meaning	Intensity of the property	Feature contributing to the property
Cohesiveness	Mutual attraction of water molecules	High	Polarity of water
Surface tension	Enhancement of the intermolecular attractive forces at the surface of a liquid	High	More attraction of water molecules to one another and more cohesiveness
Adhesion	Attraction of water to different surfaces	High	Polarity of water
Specific heat	Heat energy required to raise the temperature of a substance	High	Electrostatic attraction between water molecules
Heat of vapourization	Energy required to separate molecules from liquid to gas phase	High	Electrostatic attraction between water molecules
Solvent property	Ability of a liquid to solubilize most of the solutes	High	Small size Polar nature Ability to form water shell structure
Tensile strength	Ability of a substance to resist a pulling force	High	Cohesion of water molecules

4.7.4 Properties of Water

The important properties of water (Table 4.20) are:

- Water molecules exhibit cohesive properties due to which they are held together by hydrogen bonds.
- It has greater surface tension than most liquids.
- It has high specific heat.
- It is a versatile solvent owing to the polarity of the water molecule.
- The dipole movement in water contributes to its solvent property.
- Water molecules being dipoles can interact with positively and negatively charged ions forming a water shell.

4.7.5 Functions of Water

Water molecules play an invaluable role in governing the structure, stability, dynamics, and functions of the major biomolecules—proteins and nucleic acids. Indeed, these biomolecules lack activity in the absence of water. The living world should be thought of, as an equal partnership between proteins, nucleic acids, and water. Experimental information obtained from X-ray diffraction, neutron diffraction, nuclear magnetic resonance (NMR), and femtosecond spectroscopy measurements prove the active role of water in biomolecule structure, stability, and dynamics.

Water stabilizes proteins

Under normal pH, temperature, and ionic strength conditions, proteins generally adopt a preferred three-dimensional (3-D) fold. The hydrophobic effect and hydrogen bonds are also of prime importance in structuring proteins and water is an important constituent in these contributions to protein structure and stability. Globular proteins are folded in such a way that hydrophobic amino acid side chains are buried within the core of the protein. Water assembles to form ordered cages around the hydrophobic groups leading to hydrophobic hydration, resulting in a decrease in the entropy of the system. This is due to hydrophobic effect. However, after the hydrophobic groups are associated with each other, the water molecule is released with a gain in entropy. This is responsible for the free energy of stabilization of the globular protein structure. Thus, water enables protein folding through its capacity of forming hydrophobic interactions.

X-ray and NMR studies have revealed that water molecules serve as an integral part of all folded proteins. They function in the following three ways: (i) by contributing to the structural stability of the proteins via bridging different functional groups by hydrogen bonds, (ii) by lodging themselves within the pockets present in the folded proteins, and (iii) by forming hydration layer on the surface of the proteins. Figure 4.75 schematically represents the way in which water is involved in protein structure and stability.

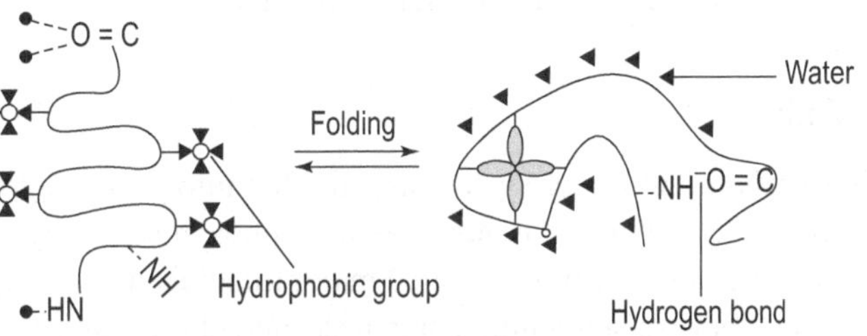

FIGURE 4.75 Role of water in maintaining protein structure and stability

Water stabilizes nucleic acids

Water plays a major role in stabilizing the structure of different forms of nucleic acids. Nucleic acids are highly charged polyanions. X-ray studies have revealed that water plays a major role in the stabilization of nucleic acid structures. In general, the degree of hydration is greater for RNAs than DNAs due to the presence of additional oxygen atom in the ribose ring of RNAs. Water functions along with the base pairing hydrogen bonds and London dispersive forces in stabilizing the structure of nucleic acids. The high dielectric constant of water molecules reduces the phosphate–phosphate electrostatic repulsion in the nucleic acid backbone. The structural forms of DNA are in fact determined by the degree of hydration. High degree of hydration favours and stabilizes B-form DNA, while low hydration level causes a transition from B form to A and C form. B-form DNA helix has water shell wrapping around the surface and spine of hydration in the minor groove, thereby narrowing the minor groove.

Water helps binding of biomolecules

Within the biological system, all biomolecules are either associated with each other or self-assembled. Water plays a vital role in self-assembly of biomolecules. Being able to donate and accept hydrogen atoms, water can exist at the interface between molecules and facilitate multiple hydrogen bonds between atoms and molecules. For example, water is involved in associating proteins and drugs to DNA. Water plays a role in protein–ligand interactions allowing varied ligands to be accommodated in a given binding site.

4.7.6 Biological Significance of Water

Water is needed for:

- maintaining health and integrity of every cell in the body.
- keeping the bloodstream liquid enough to flow through blood vessels.
- eliminating the by-products of the body's metabolism, excess electrolytes, for example, sodium and potassium, and urea (a waste product formed through the processing of dietary protein).
- regulating body temperature through sweating.
- moisturizing the mucous membranes of the lungs and mouth.
- reducing the risk of cystitis by keeping the bladder clear of bacteria.
- assisting digestion and preventing constipation.
- improving the texture of skin.
- transporting nutrients and oxygen to cells.
- functioning as a shock absorber inside the eyes, spinal cord, and in the amniotic sac surrounding the foetus during pregnancy.

4.8 MINERALS

Minerals in the living system are divided into two categories: major minerals and trace minerals. Sodium, potassium, calcium, phosphorus, magnesium, and sulphur are present in larger quantities and are called major minerals. Iron, copper, zinc, molybdenum, cobalt, iodine, fluorine, selenium, manganese, and chromium are trace minerals and are found in small quantities.

4.8.1 Functions of Minerals

Sodium

Sodium is the principal cation of the extracellular fluid. It takes part in fluid and electrolyte balance. Common salt is the principal source of sodium in human diet. Normal blood sodium level is 135–145 $mEqL^{-1}$. The total sodium present in the body of an average adult is 3,700 mmol. Seventy five per cent of this amount is confined to extracellular fluid.

Functions

- Maintenance of acid–base balance in the ECF.
- Maintenance of fluid balance and osmotic pressure.
- Maintenance of normal muscle irritability and permeability of the cell.
- Active transport of sugars and amino acids in the body.

Potassium

Potassium is the principal intracellular cation.

Functions

- Influences cardiac muscle activity.
- Transmission of nerve impulse.
- Cofactor for pyruvate kinase.
- Regulation of acid–base balance and water balance within the cell.

Chloride ion

Chloride ion is involved in the formation of hydrochloric acid in the gastric juice. Chloride is one of the major electrolytes in the living body, necessary for the normal functioning of the cell. In fact, it is the major extracellular anion. The level of chloride ion should be maintained in the body (normal value = 98–108 mmol/L in blood), as an abnormal level leads to deleterious consequences to the body. In kidney diseases and parathyroid hyperactivity, the level of chloride is increased, while the level is decreased during adrenal gland and renal abnormalities as well as after vomiting and excessive sweating. Chloride ion is responsible for the maintenance of osmotic pressure, acid–base homeostasis, muscular activity, and migration of water between fluid compartments.

Magnesium

Magnesium is an important intracellular cation. It is a cofactor for kinases, alkaline phosphatase, and adenylate cyclase. It decreases neuromuscular irritability. It is an inhibitor of renal calcium calculogenesis.

Sulphur

Sulphur forms a part of biomolecules such as proteins, CoA, glycosaminoglycans, thiamin, biotin, lipoic acid, glutathione, and insulin.

Calcium

Calcium is the most abundant cation of the human body. It is involved in the following functions:

- Development of bones and teeth
- Muscle contraction
- Blood clotting
- Nerve transmission
- Activation of enzymes, including pancreatic lipase, ATPase, and succinate dehydrogenase
- Maintenance of membrane integrity and permeability
- Regulation of activity of several enzymes in the form of calcium–calmodulin complex
- Second messenger to epinephrine in hepatic glycogenolysis
- Release of hormones such as insulin, PTH, and calcitonin
- Regulation of microfilament- and microtubule-mediated endocytosis, exocytosis, and cell motility

Phosphorus

Phosphorus occurs in cells as phosphate. Following are the functions of phosphate:

- Development of bone and teeth
- Regulation of acid–base balance
- Structural components of nucleic acids

4.9 PIGMENTS

Pigments are organic compounds distributed in plants and animals. They are usually coloured because they absorb visible light of specific wavelengths and reflect some other wavelengths. For example, green chlorophyll in leaves and brown melanin in animals and human beings. They are also present in some bacteria and in all flowers. Pigments are either organic (containing carbon) or inorganic (without carbon). Some important pigments are discussed in this section.

4.9.1 Plant Pigments

Chlorophyll, carotenoids, lycopene, lutein, and zeaxanthin are plant pigments.

Chlorophyll is a green colour pigment found in leaves and stem cells of plants. They are also found in sepals (green, leaf-like structures that protect the bud prior to opening) of the flower. Chlorophyll pigments, located in the chloroplast of the plant cells reflect light of green wavelength and so they give green colour to the plants. In plants, the chlorophyll pigment is responsible for the process of photosynthesis.

The following three types of chlorophylls are known:

- Chlorophyll 'a' that facilitates photosynthesis in all plants, algae, and cyanobacteria.
- Chlorophyll 'b' that is predominant in green algae and plants.
- Chlorophyll 'c' that exists in photosynthetic aquatic organisms such as chromista and dinoflagellates.

Carotenoids are plant pigments, usually yellow, orange, or red in colour. They function as precursors for vitamin A, and are responsible for antioxidant functions and health maintenance of plants. For example, lycopene is a red colour pigment present in red peppers, pimientos, tomatoes, apples, red onions, pink grapefruits, watermelons, guavas, etc. Lycopene improves memory and prevents prostrate, testicular, cervical, and breast cancers and also heart disorders. It also functions as a detoxifying agent and as an anticholestrolemic agent.

Lutein is a yellow colour pigment (xanthophyll) found in vegetables such as cabbage, lettuce, broccoli, endives, asparagus, spinach, nettles, kiwi fruits, green apples, beans, peas, parsley, pepper grass, green tea, etc., and also in egg yolk. It plays a central role in protecting eyes from stress and in improving eyesight. It is also needed for the proper maintenance of bones and teeth. It also has anticancer properties.

Zeaxanthin is a xanthophyll carotenoid present in corn, saffron, and yellow fruits. It is widely distributed in the central part of the eye, the macula, while the peripheral part is occupied by the lutein. It prevents muscular degeneration that occurs during ageing process.

Betalain is an indole-based yellow colour pigment predominant in plants from the order Caryophyllales, and in higher order fungi of genus Amania. Though the exact function of betalain is unknown, it is found to attract insect for flower pollination and attract birds for the dispersal of seeds.

Anthocyanis are hydrophilic pigments derived from flavonoids. They range in colour from red, blue, and purple, depending on their pH. They function to facilitate pollination and seed dispersion. They also act as antioxidants and plant metabolites.

Phycoerythrin is a red colour protein pigment of red algae and cyanobacteria. They are accessory pigments and responsible for photosynthesis in the hosting organisms.

4.9.2 Animal Pigments

Some important animal pigments are discussed here.

Rhodopsin

Rhodopsin is a purple-coloured pigment present in the rod cells of retina. It is light sensitive and allows vision in black and white and dim light. It contains a protein called opsin that is linked to retinal. In the presence of bright light, rhodopsin cleaves into retinal and opsin, both of which are colourless. In dim light, or in the dark, opsin and retinal associate into rhodopsin.

Melanin

Melanin is the primary pigment in mammals; it gives colour to the hair and fur. Eumelanin and pheomelanin are two types of mammalian melanin that produce diverse colours ranging from black, sandy brown, and sandy red.

Birds also have melanin pigments, occurring as small granules in feathers and skin. Bird melanin reflects black, reddish brown, or yellow colours. Apart from colouring the birds, they also contribute strength and rigidity to the feathers. Thus, melanin-rich feathers are more resistant to wear and tear.

Carotenoid and porphyrin

As discussed earlier, carotenoid is the orange or red pigment of fruits and vegetables. In addition, they are also found in animals and birds. For example, the Northern Cardinal bird appears bright red due to the presence of carotenoid in its feathers. Carotenoid also contributes to the brilliant yellow colour of gold finches and yellow warblers and orange colour of male Blackburnian Warbler. Sometimes, carotenoid merges with melanin to form a different colour, for example, the olive green colour of the female Scarlet Tanager.

Porphyrin is yet another animal pigment that gives colour to owls, pigeons, and turacos.

SUMMARY

Atoms of hydrogen and oxygen and less frequently nitrogen, sulphur, or phosphorus are bonded to the carbon skeleton to form monomeric biomolecules which, in turn, are polymerized in vivo to form giant molecules called macromolecules. Four major types of macromolecules are found in living cells—nucleic acids, proteins, carbohydrates, and lipids.

Nucleic acids are genetic material present in the nucleus of the cell. They are the polymers made of building blocks called nucleotides. Nucleotides are composed of a nitrogenous base (purine or pyrimidine), a ribose sugar, and a phosphate group. Two types of nucleic acids are known: DNA which possesses deoxyribose and RNA which possesses ribose sugar. DNA is converted into RNA which in turn is converted into protein. Thus, the second biomolecule discussed here was protein.

Proteins are the building blocks of amino acids. Amino acid sequence represents the primary structure of proteins; helix and pleated sheets are the secondary structure of proteins. Three-dimensional folding of the amino acid chain (polypeptide chain) leads to tertiary structure of protein. Spatial arrangement of tertiary structure is the quaternary structure of proteins. Proteins have a wide variety of functions. They catalyse biochemical reactions in the cell (enzymes), they regulate cellular activity (hormones), they have a structural role (collagen), they play a role in transport (haemoglobin) or storage (ferritin), or immunity (antibody).

Carbohydrates are the building blocks of sugars (saccharides). Depending on the number of sugar residues carbohydrates are divided into monosaccharides (e.g., glucose), disaccharides (e.g., lactose and maltose), and polysaccharides (e.g., starch and glycogen). They have aldehyde or ketone moieties as functional groups. Homopolysaccharides are made of the same type of sugar (cellulose), while heteropolysaccharides are made of different types of sugar (e.g., GAGs). Carbohydrates serve as energy source, provide lubrication, and form the structural component of cell wall and cell membrane.

Lipids are another class of biomolecules made of fatty acids. Simple lipids, compound lipids, and derived lipids are the three major classes of lipids. Simple lipids function as a reserved fuel, while compound lipids function as membrane component. Derived lipids serve as hormones and pigments.

Apart from these molecules, water and minerals are also considered as biomolecules. Water is of prime importance because it interacts with other molecules, such as proteins and nucleic acids, and helps in their stabilization and coordination of their functions. Minerals, such as sodium, potassium, chloride ion, calcium, magnesium, and phosphorus, play a vital role in the regulation of various biochemical processes and activities.

TERMS TO REMEMBER

Acid number Number of milligrams of KOH required to neutralize free fatty acids in 1 g of fat. This quantity is used to quantify fatty acids in fat.

Ampholytes Substances that have a dual nature, that is, they can act both as acids (proton donors) and as bases (proton acceptors).

Chaperones Chaperones are specialized proteins that convert misfolded proteins into perfectly folded proteins.

Chargaff's rule Chargaff's rule states that DNA strands have 1:1 ratio of purine and pyrimidine bases.

Chitin Hard and tough homopolysaccharide made of N-acetyl glucosamine units linked by β (1→4) glycosidic linkage.

Covalently closed circular DNA A supercoiled DNA which is either circular or coiled upon itself such that there is no nick in the phosphodiester bond to relieve the stress for forming an open circle.

Cyclopentanoperhydrophenanthrene The fused ring structure of steroids and sterols.

Enzymes Catalytic proteins which increase the rate of biochemical reaction in the cells.

Epimers Monosaccharides which differ in the position of –OH group around a single carbon.

Glycogen Storage polysaccharide of animals.

Glycosidic bond The bond formed when the anomeric carbon of one sugar reacts with a hydroxyl group of a second sugar.

Haemoglobin Oxygen-carrying protein of red blood cells which possess quaternary conformation due to spatial arrangement of two alpha subunits and two beta subunits.

Hormones Organic substances, usually a peptide or steroid, produced by tissues of endocrine glands (a series of ductless glands) that produce chemical messages to effect physiological activity.

Hydrogen bond The bond formed when a hydrogen atom is sandwiched between two electron-attracting (electrophilic) atoms, usually oxygen or nitrogen.

Iodine number Number of grams of iodine absorbed by 100 g of fat. Iodine number is directly related to unsaturation.

Melanin Melanin is the main pigment found in mammals—in skin, hair, and feathers.

Myglobin Oxygen-carrying protein of muscle cells which possesses a folded tertiary structure.

Nucleic acids Nucleic acids have purine and pyrimidine bases in the ratio of 1:1.

Palindrome sequence Nucleotide sequence on two DNA strands which reads same when in opposite direction.

Peptide bond Covalent bond formed between two amino acid residues in proteins due to the reaction between the amino group of one amino acid and the carboxyl group of the adjacent amino acid.

Phosphodiester bond Polymers of nucleotides linked by phosphodiester bond formed between 5′ hydroxyl group of one pentose sugar and 3′ hydroxyl group of the adjacent sugar.

Phycoerythrin The red colour pigment present in red algae.

Ramachandran plot Plot of Φ versus ψ which reveals the allowed and prohibited conformations of a polypeptide chain.

Reichert-Meissl number Volume of 0.1N NaOH required for neutralizing soluble fatty acid in 5 g fat.

Rhodopsin A light-sensitive, purple-red organic pigment contained in the rod cells of the retina that allows the eye to see in black and white and in dim light.

Saponification number Number of milligrams of KOH required to saponify 1 g of fat. Sap value is inversely related to fatty acid chain length.

Starch Storage polysaccharide of plants.

Stem loop Secondary structure of nucleic acid formed due to complementary base pairing within a single strand.

Thermal melting Separation of two strands of helix by heating the DNA solution.

van der Waals force Attractive or repulsive force between molecules (or between parts of the same molecule) other than those due to covalent bonds or to electrostatic interaction.

Zeaxanthin A yellow carotenoid pigment which usually occurs in plants together with lutein pigment.

Zwitter ions Amino acids with a net neutral charge due to the presence of positive amino group and negative carboxyl group.

REVIEW QUESTIONS

1. Name the four DNA bases. Which of these are purines and pyrimidines?
2. Which position on purine and pyrimidine bases binds to the ribose?
3. Why is the DNA sugar group known as deoxyribose?
4. Write a reaction to show the formation of phosphodiester bond formation.
5. What force holds complementary base pairs together?
6. Identify the three major groups in the following figure.

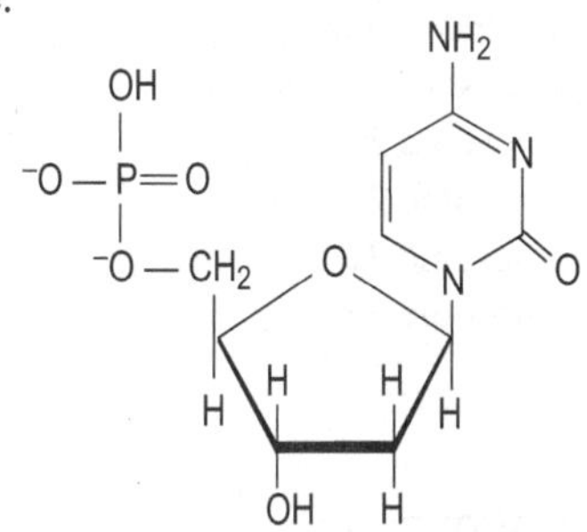

7. How are nucleotides sequenced?
8. Why are the DNA helical strands called antiparallel?
9. What will happen if Chargaff's rule is not obeyed?
10. Explain the structure of clover-leaf form of RNA.
11. Write the equation to explain the formation of a peptide bond.
12. What are the covalent bonds formed in the stabilization of protein structure?
13. Write a note on Ramachandran plot.
14. Why does the alpha helix form more readily than many other possible conformations?
15. What type of helical structure is possessed by the collagen?
16. What is Sanger reagent? What is its use?
17. What would have happened if human beings were to be able to digest the cellulose present in leafy salads?
18. The amount of lipids stored as an energy reserve far exceeds the energy stored as glycogen. Why?
19. What is an isoprene unit?
20. Write a note on plant and animal pigments.
21. Represent the influence of substrate concentration and pH on enzyme activity.
22. Derive the rate equation for an enzyme-catalysed reaction.
23. Draw the structure of the open chain form, cyclic form, and ring form of glucose.
24. Write a note on epimers.
25. Write a note on blood group antigens.

MULTIPLE-CHOICE QUESTIONS

1. Which of the following base pairing and hydrogen bonding is correct?
   (a) A=T
   (b) G≡C
   (c) A≡U
   (d) G=T
2. Which of the following is not present in RNA?
   (a) adenine
   (b) guanine
   (c) thymine
   (d) cytosine
3. Higher degree of hydration in DNA causes which of the following transformations?

(a) A to B form
(b) A to C form
(c) B to A form
(d) B to C form

4. Which of the following properties of the water stabilizes DNA structure by compromising the electrostatic repulsion among phosphate groups?
   (a) high solubility
   (b) high dielectric constant
   (c) high surface area
   (d) high specific heat
5. Eighteen carbon amino alcohol is the constituent of
   (a) phospholipid
   (b) glycolipid
   (c) sphingolipid
   (d) lipoproteins
6. Salt bridge is a
   (a) strong ionic bond
   (b) weak ionic bond
   (c) strong hydrogen bond
   (d) weak hydrogen bond
7. Alignment of the two strands of DNA double helix are
   (a) parallel
   (b) antiparallel
   (c) in the 3′ to 5′ direction
   (d) in the 5′ to 3′ direction
8. Cruciform is a type of
   (a) palindrome
   (b) stem-loop RNA
   (c) clover-leaf RNA
   (d) Z-form DNA
9. Catalase is an example of a
   (a) glycoprotein
   (b) metalloprotein
   (c) lipoprotein
   (d) nucleoprotein
10. Desmosine is a non-standard amino acid derived from
   (a) glycine
   (b) lysine
   (c) leucine
   (d) isoleucine
11. Cytochrome C is a protein with
   (a) sulphate
   (b) haeme
   (c) copper
   (d) magnesium
12. Protein in viral capsid exhibits
   (a) rotational symmetry
   (b) icosahedral symmetry
   (c) helical symmetry
   (d) tetrahedral symmetry
13. PRNP gene is located in
   (a) chromosome 20
   (b) chromosome 21
   (c) chromosome 22
   (d) chromosome 23

Answers to Multiple-choice Questions

1. (a)	2. (c)	3. (a)	4. (b)	5. (c)
6. (b)	7. (b)	8. (a)	9. (b)	10. (b)
11. (b)	12. (b)	13. (a)		

Jan Swammerdam was the first person to observe and describe red blood cells in 1668.

CHAPTER 5

Blood Cells and Haematopoiesis

After reading this chapter, students should be able to understand and appreciate the following:

- Process of haematopoiesis
- Haematopoietic growth factors
- Chemistry and functions of erythrocytes
- Role of nanorobots
- Chemistry and functions of leucocytes and platelets

Blood is a connective tissue made of cells and a liquid called plasma. There are three types of cells present in blood: white blood cells (lymphocytes and granulocytes), red blood cells, and platelets. White blood cells (WBCs) use blood as a transportation mode to access different parts of the body where they perform their function. Red blood cells (RBCs) function in the blood itself. All the three types of blood cells are derived from haematopoietic stem cells (HSCs) in the bone marrow. Plasma is a straw yellow fluid containing amino acids, proteins, sugars, clotting factors, mineral ions, and carbon dioxide. The components of blood can be separated by centrifuging the blood at around 3000 rpm for 10–15 minutes in a table-top centrifuge. Three layers are obtained in the process. Plasma forms the uppermost layer and red blood cells form the lowermost layer. White blood cells and platelets form the middle layer. In Chapter 3, we discussed that HSCs differentiate into two types of progenitors: lymphoid progenitor and myeloid progenitor (refer Fig. 3.1). Lymphoid progenitor produces lymphocytes, while myeloid progenitor produces granulocytes, red blood cells, and platelets. The role of lymphocytes and leucocytes in immunity has already been discussed in Chapter 3. This chapter will primarily focus on red blood cells and platelets.

5.1 HAEMATOPOIESIS

As defined earlier, haematopoiesis is the name given to the process of formation of blood cells from haematopoietic stem cells (HSCs) in the bone marrow. Haematopoietic stem cells are discussed in detail in Chapter 14. In essence, HSCs constitute a minor portion of the bone marrow cell population (1/1,000 cells) and yet are important in that such a tiny fraction can generate all the blood cell types. Administration of HSCs is capable of preventing the death of X-ray irradiated experimental animals whose bone marrow and haematopoietic system is damaged. This confirms that haematopoiesis is not possible without HSCs. This section focuses on the process of haematopoiesis. Haematopoietic growth factors used in the differentiation of HSCs into specific cell types are also discussed.

5.1.1 Process of Haematopoiesis

An HSC is pluripotent, it is capable of producing all blood cell types, but in a controlled manner. It initially produces a pluripotent progenitor which is committed to lymphoid progenitor or myeloid progenitor. The lymphoid progenitor is committed to lymphocyte production, and the myeloid progenitor is committed to granulocyte, erythrocyte, and thrombocyte (platelet) production. Commitment refers to the 'switching on' of the genes responsible for a particular differentiation with simultaneous 'switching off' of the genes required for other types of differentiation. For example, to differentiate into erythrocytes, the genes coding for GATA-1 are turned on while other differentiation modes are restricted. For the differentiation of B cells, PAX5 gene is switched on, simultaneously preventing other differentiation pathways. Although the stem cells are low in number, they undergo rapid division so that the population of the differentiated cell is high. Table 5.1 shows the amount of blood cells in human beings which is derived from the small fraction of stem cells.

TABLE 5.1 Blood cell profile in humans

Cell type	Subtype	Name of the cells	Normal values in humans (cells/L)
Red blood cells	–	Erythrocyte	5×10^{12}
White blood cells	Granulocytes	Neutrophil	5×10^{9}
		Eosinophil	2×10^{8}
		Basophil	4×10^{7}
	Monocytes	Monocytes	4×10^{8}
	Lymphocytes	B cells	2×10^{9}
		T cells	1×10^{9}
	Natural killer cells	Natural killer cells	1×10^{8}
Platelets	–	Thrombocytes	3×10^{11}

The myeloid progenitor can differentiate into the following five types of progenitors:

1. Erythroid progenitor
2. Megakaryocyte
3. Granulocyte–monocyte progenitor
4. Eosinophil progenitor
5. Basophil progenitor

Erythroid progenitor differentiates into erythrocytes, whereas megakaryocytes differentiate into platelets.

Formation of red blood cells or erythrocytes (erythropoiesis) from erythroid progenitor is driven by two colony stimulating factors (CSFs): interleukin-3 (IL-3) and a hormone called erythropoietin (EPO; refer Section 5.2 for discussion on CSF and EPO). Lack of oxygen or a reduction in red cell count in the blood is the signal for erythropoiesis. This signal stimulates the release of both IL-3 and EPO. IL-3 acts on and stimulates the proliferation of stem cells into larger colonies of 'early erythroid progenitor cells', in 7–10 days. These colonies are called erythrocytes burst forming cells (BFC-E) or erythrocyte burst forming units (BFU-E). BFU-E can proliferate to produce erythrocyte colonies only.

EPO from the kidney cells is released into the blood stream. Erythroid progenitor cell [also called erythrocyte colony-forming cell (CFC-E) or erythrocyte colony-forming unit (CFU-E)] is the target for EPO. When the targets are acted upon by EPO, they undergo six cycles of cell division and proliferation to produce immature erythrocyte colonies called erythroblasts. EPO is crucial as its absence induces apoptosis in vitro.

The erythroblasts produced by the CSF are immature. They cannot grow or divide further, but can mature. Erythroblasts posses a surface protein that binds to inhibitory receptors on the surface of the macrophages in the bone marrow. Macrophage-bound erythroblasts are not phagocytosed by the macrophages; instead, they extrude their nucleus to become a reticulocyte. The nucleus is digested by the macrophages. Reticulocytes leave the bone marrow and enter the blood stream where they extrude their mitochondria and ribosomes, and even endoplasmic reticulum, in a couple of days, thus, developing into mature erythrocytes. The integrity of the erythrocyte membrane is very crucial for maintaining the structure of erythrocytes. Membrane integrity is maintained and regulated by a transcription factor called GATA-1 (encoded by GATA-1 gene). GATA is a protein family named so because they bind to the consensus DNA sequence: (A/T) GATA (A/G). GATA-1 is responsible for the transcription of spectrin mRNA. Spectrin occupies the inner surface of the erythrocyte membrane and maintains its integrity. GATA-1 plays a role in erythropoiesis by converting foetal haemoglobin to adult haemoglobin. The process of erythropoiesis is summarized in Fig. 5.1.

Myeloid progenitor produces megakaryoblasts that develop into megakaryocytes, the actual precursor cells for platelets (thrombocytes). The production of platelets from megakaryocytes is called thrombopoiesis. The bone marrow is supplied with blood vessels called blood sinuses. Megakaryocytes, due to their huge size, cannot enter easily into the blood present in the sinuses, but they can extend some processes into the holes in the endothelial cells. These processes bud-off and pinch into the platelets and drop into the blood. Formation of platelets from a megakaryocyte is illustrated in Fig. 5.2. CSFs, such as cytokines (IL-3, IL-6, IL-11), leukaemia inhibitory factor, and thrombopoietin (TPO), are thought to stimulate thrombopoiesis. Few other signals that act as inhibitors of planet formation are platelet factor 4 (PF4), CXCL5 (a cytokine protein and chemokine ligand encoded by CXCL5 gene), CXCL7 (a cytokine protein and chemokine ligand 7 encoded by CXCL7 gene), and CCL5 (a cytokine protein and chemokine ligand encoded by CCL5 gene). Thus, it is clear that CSF regulates thrombopoiesis.

The very first generation of myeloid progenitor is called a myeloblast. Myeloblast is the general term usually prefixed with the name of the granulocytes they produce, for example, myeloblasts which generate eosinophils and basophils are called eosinophilic myeloblasts and basophilic myeloblasts, respectively. Similarly, myeloblasts that differentiate into neutrophils and monocytes are called granulocyte-monocytic myeloblast and granulocyte-macrophage progenitor cells (GMPC), respectively.

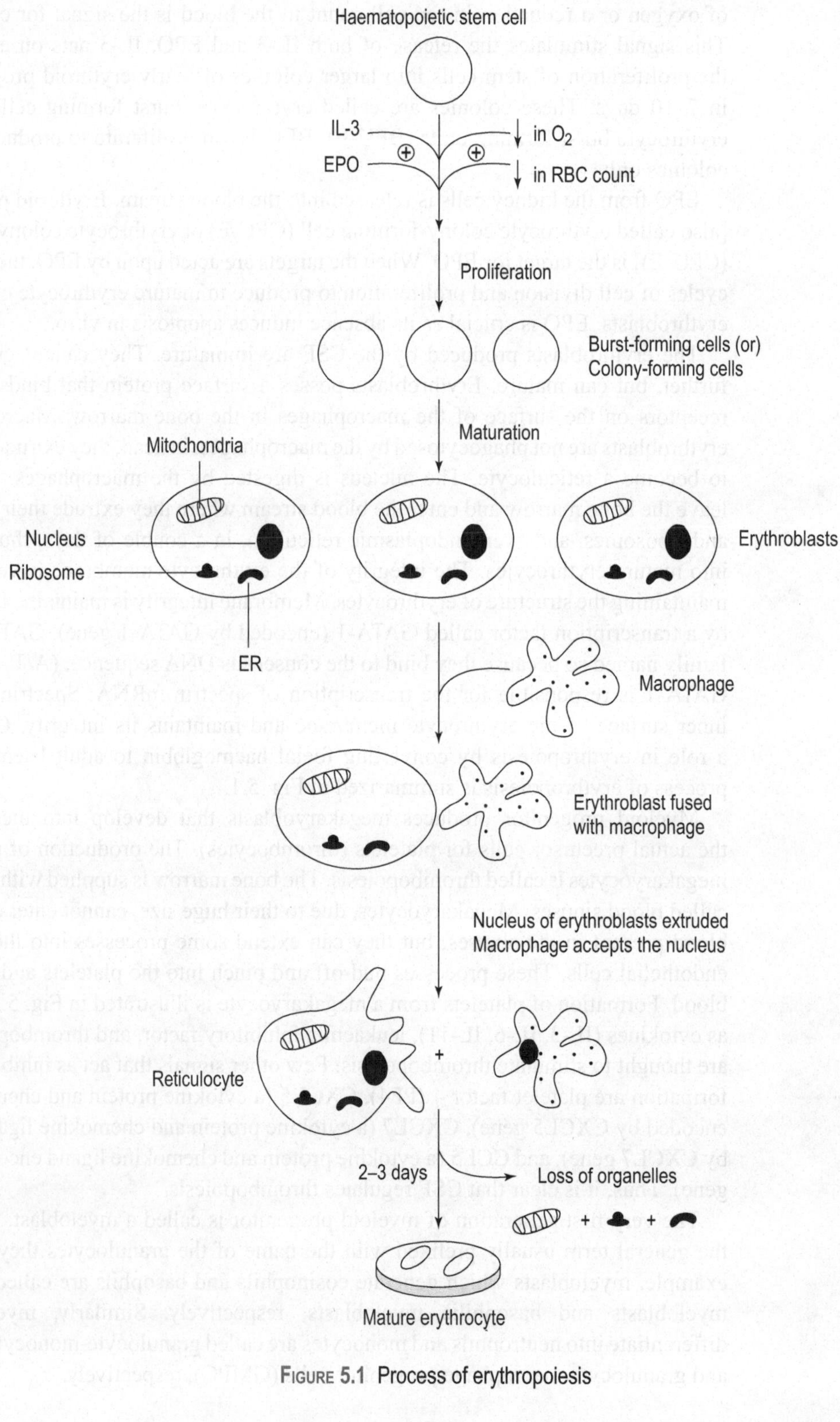

FIGURE 5.1 Process of erythropoiesis

FIGURE 5.2 Formation of platelets from megakaryocytes

Myeloblasts undergo a series of differentiation stages to form granulocytes (Fig. 5.3). The various stages are listed as follows:

Stage 1 Undifferentiated myeloblasts are around 10–20 μm in size with agranulated cytoplasm, prominent nucleolus, and dispersed chromatin.

Stage 2 As the myeloblasts develop, chromosomes in nucleus get condensed, and the nucleus undergoes segmentation. In this stage , the cytoplasm develops primary granules; these cells are called promyelocytes, measuring around 15–20 μm in diameter.

Stage 3 In this stage, the cells have a diameter of 12–18 μm with ellipsoidal or ovaidal nucleus without nucleolus. Secondary granules are synthesized. These cells are called myelocytes and are capable of mitotic division.

Stage 4 The daughter cells (diameter 10–5 μm) obtained from the mitosis of myelocytes are called metamyelocytes. They possess mature cytoplasm, bean-shaped eccentric nucleus, and more secondary granules than the primary granules.

Stage 5 Metamyelocytes develop into 'band form' (young granulocytes) having U-shaped nucleus with rounded ends. Band form develops into neutrophils, eosinophils, or basophils, depending on the CSF (Section 5.2).

GMPC differentiates into neutrophils and monocytes. The later accumulates in the tissues to form macrophages (refer Chapter 3). Neutrophils and macrophages should be synthesized to elicit immune response. Therefore, infection of body with a pathogen is the signal for the production of these cells. In response to infection, various cell types, including endothelial cells, fibroblasts, macrophages, and lymphocytes, synthesize many CSFs. CSFs in turn bind to their receptors (Section 5.2) even at a low concentration of 10–12 M and stimulate the production of neutrophils and monocytes in the bone marrow. The monocytic myeloblasts are converted into monoblasts by GM-CSF (see Chapter 3). Monoblasts are around 13–15 μm in diameter with basophilic cytoplasm, large nucleus, and one or more nuclei. The cytoplasm develops primary granules containing peroxidase, arylsulphatase, and acid phosphatase. Such cells are called promonocytes. Promonocytes develop into monocytes of 12–20 μm diameter each with indented central nucleus and vacuoles. Monocytes circulate in the blood. Development of monocytes into macrophages is discussed in Chapter 3 due to their role in phagocytosis. The formation of monocytes and macrophages from myeloblast is shown in Fig. 5.4.

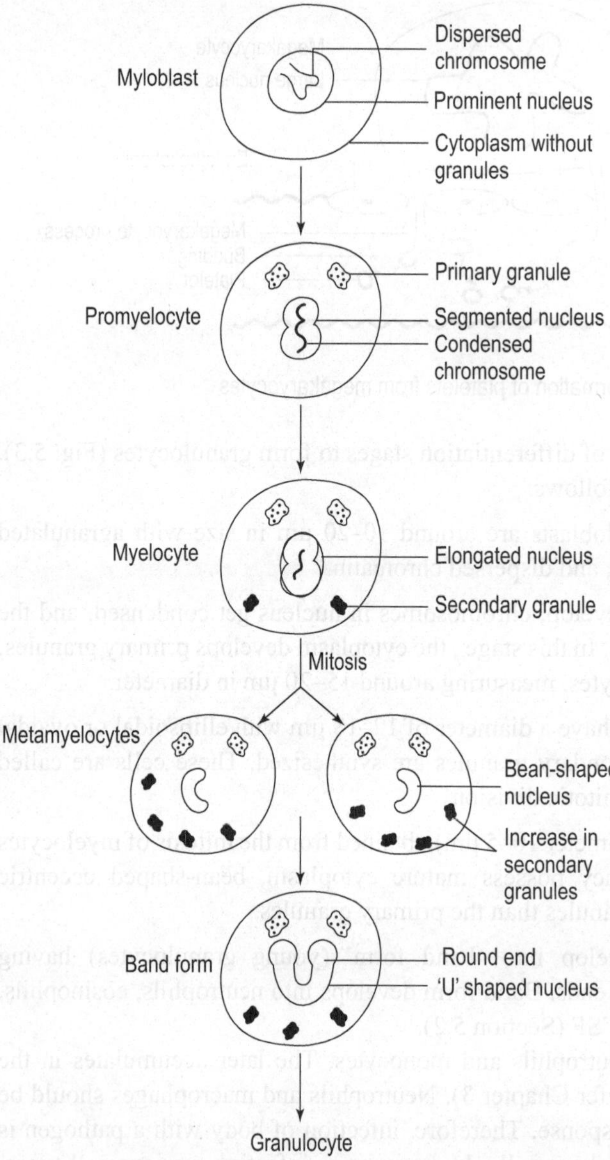

FIGURE 5.3 Myeloid lineage (stages in the differentiation of myeloblasts to granulocytes)

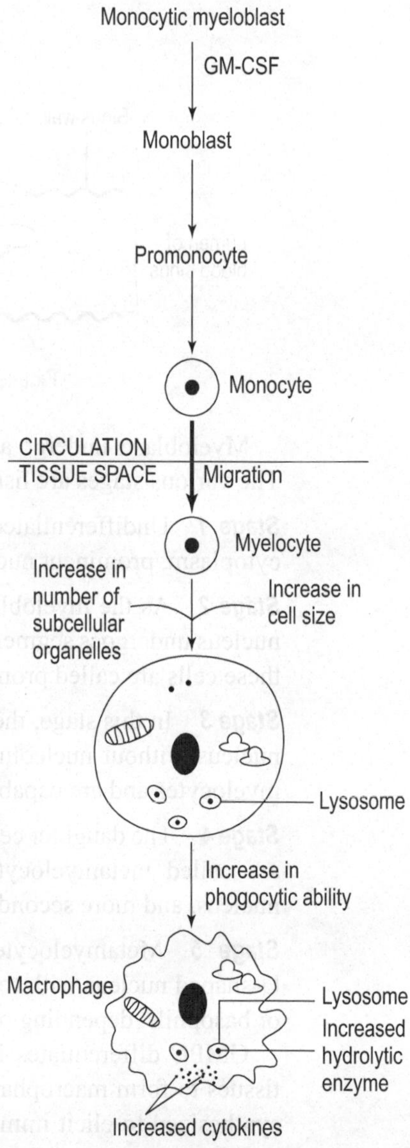

FIGURE 5.4 Development of monocytes

Eosinophilic and basophilic myeloblast develops into eosinophils and basophils, respectively. A cytokine called interleukin-5 (IL-5) or eosinophils differentiation factor is responsible for inducing the differentiation of eosinophils progenitor into eosinophils. IL-5 influences the production, survival, growth, and maturation of eosinophils. Basophil progenitors are called CD34+ progenitor cells. These CD34 receptors have a high binding affinity for IL-3, the differentiation factor for basophils.

Frequency of haematopoiesis varies with the bone tissues. In different stages of life, different bones are activated for haematopoiesis (Fig. 5.5). During the first five years of life, all bones in the human body undergo haematopoiesis. Till the age of 25 years, tibia and femur bones hold the function of haematopoiesis. The bones of the vertebrae, sternum, pelvis and ribs, and cranium carry out haematopoiesis through the remaining life time.

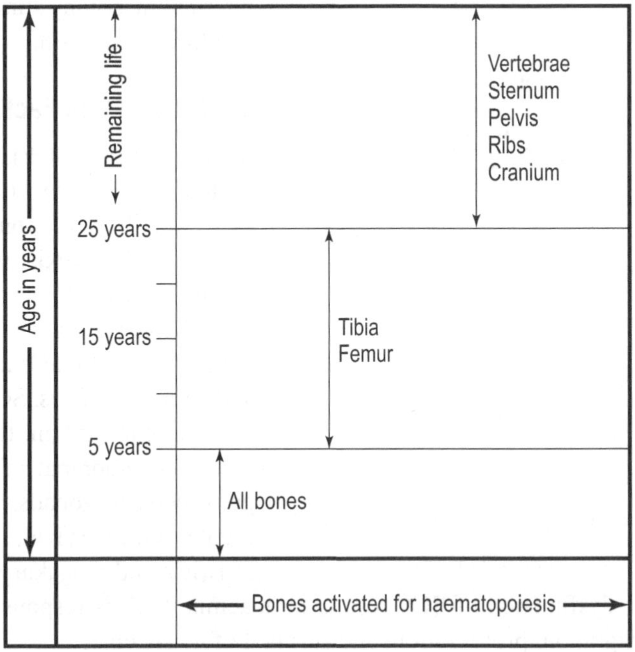

FIGURE 5.5 Profile of haematopoiesis in different bones

As the life time of blood cells is generally short, they need to be synthesized at regular intervals in order to replace the worn-out cells. At the same time, too many blood cells also result in diseases. Therefore, the process of haematopoiesis should be tightly regulated, primarily at three points. First, as the haematopoietic factors (HFs) are involved in haematopoiesis, they themselves can function as a key for the regulation too. Second, production of cytokines by the bone marrow stromal cells and other cell types, such as activated T cells and macrophages, can function as positive regulators, whereas induction of cell death by apoptosis can be a negative regulator. Receptors (present in the surface of the stem cells and progenitor cells) form the third regulatory factor. Activation or inactivation of receptors for haematopoietic cytokines can regulate haematopoiesis. Erythropoietin is well studied as a regulatory factor of haematopoiesis. When there is a decrease in the oxygen delivery, erythropoietin is released to stimulate erythropoiesis in the red bone marrow.

5.2 HAEMATOPOIETIC GROWTH FACTORS

Haematopoietic growth factors are signals required for haematopoiesis to occur. These signal molecules are usually provided by the stromal cells of the bone marrow. These factors are surface-bound or secreted extracellularly. In vitro haematopoiesis can be

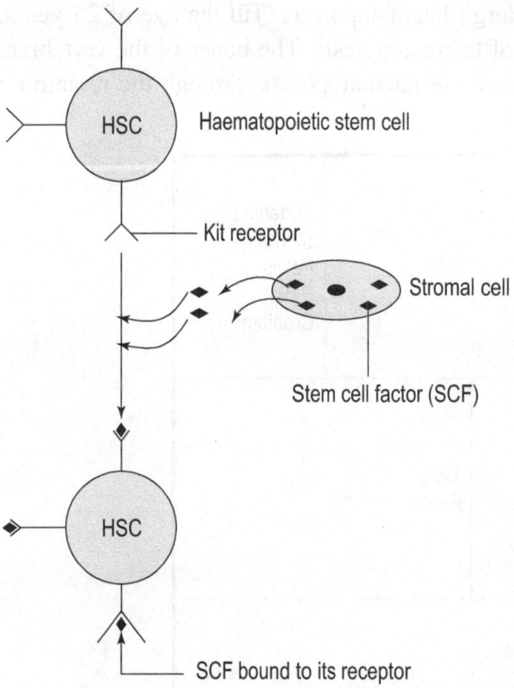

FIGURE 5.6 Stem cell factor

studied by providing growth factors to the cell culture or by allowing the cells to grow in a matrix of stromal cells which can supply the growth factors. In this section, we will discuss the stem cell factor, interleukin, colony stimulating factor, erythropoietin, and thrombopoietin.

5.2.1 Stem Cell Factor

Stem cell factor (SCF), also known as steel factor, is a protein produced by endothelial cells and fibroblasts, as well as by the stromal cells of the bone marrow (Fig. 5.6). They exist in two isoforms: soluble form and transmembrane form.

SCF itself is not a growth factor; instead, it enhances the response of HSCs to other factors, including CSF and cytokines. Thus, SCF is indirectly responsible for blood cell formation. It is also thought to play a role in the development of melanocytes and primordial germ cells in gonads. During haematopoiesis, SCF regulates the survival, self-renewal, and maintenance of HSCs and megakaryocytes. Therefore, deficiency or dysfunction of SCF results in anaemia. SCF is responsible for facilitating adhesion, migration, proliferation, and survival of mast cells.

SCF functions by binding to c-kit receptor or CD117 encoded by c-kit protoncogene, hence, they are also called c-kit ligand. These receptors are located on the surface of HSCs and are terminally differentiated mast cells. They are tyrosine kinase receptors present as transmembrane proteins. Each receptor is a dimer, which on activation by its ligand, cross links among each other and cross phosphorylates its tyrosine residues. This results in the receptor activation, thereby activating the following signalling cascades:

- RAS/ERK, (Rat sarcoma/Extracellular signal-regulated kinases) pathway
- PI3-kinase (Phosphatidyl inositol-3-kinase) pathway
- Src kinase pathway (Src is pronounced as 'sarc'. It refers to sarcoma)
- JAK/STAT pathways (JAK is janus kinase/just another kinase. STAT refers to Signal Transducers and Activators of Transcription protein)

5.2.2 Interleukin

While erythropoietin and thrombopoietin are the cytokines for the development of red blood cells and platelets, interleukin is the primary cytokine for the production of immune cells. The main function of interleukin (IL) is the differentiation of lymphoid progenitors to form immune cells. The types of interleukin and their function in the development of immune cells have already been discussed in Chapter 3. IL plays a significant role in the production of other blood cells too. Interleukin-3 (IL-3), produced by endothelium, binds to its receptor (CD123 or CD131), present on the HSCs and induces their differentiation into erythrocytes. IL-6, produced by endothelium, is responsible for haematopoiesis.

5.2.3 Colony Stimulating Factor

Colony stimulating factors (CSF) and their role in immunity have already been discussed in Chapter 3. Haematopoiesis requires haematopoietic inducing environment (HIM) and haematopoietic factors (HF). Stromal cell meshwork present in the bone marrow provides HIM by secreting certain HF, commonly called colony stimulating factor (CSF). The differentiation of HSCs is controlled by HIM and HF. Previous discussions on the process of haematopoiesis clearly suggest that CSF is crucial for the formation of blood cells, hence, this section focuses on CSF. The blood cells and CSF produced by them are shown in Table 5.2. However, the influence of CSFs/cytokines on the formation of the lymphoid lineages has already been discussed in Chapter 3 under Section 3.11.

CSFs may be a hormone (e.g., erythropoietin), local mediator, or membrane-bound ligand such as a stem cell factor (SCF).

TABLE 5.2 CSF in haematopoiesis

CSF	Blood cells produced
Erythropoietin	Erythrocytes
Interleukin-3	
Interleukin-3	Platelets
Interleukin-6	
Interleukin-11	
Leukaemia inhibitory factor	
Thrombopoietin (TPO)	
Granulocyte-macrophage CSF	Neutrophils
Monocytes/macrophages	
Neutrophil differentiation factor	Eosinophil
IL-5	
Basophil differentiation factor	Basophil
IL-3	

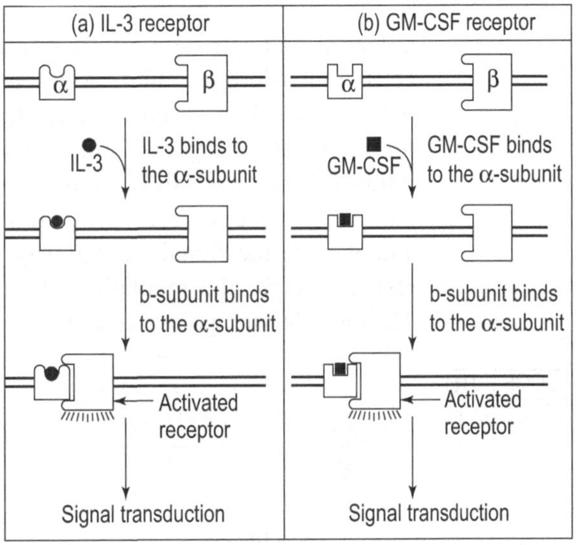

FIGURE 5.7 Receptors for colony stimulating factor

CSFs function by acting on their specific receptors located on the surface of their target cells. The CSF receptors are of two kinds: transmembrane tyrosine kinase and large cytokine receptor family. A CSF receptor is a heterodimer with two subunits: α-subunit and β-subunit (Fig. 5.7). The β-subunit is usually shared by different ligands (CSFs), whereas the α-subunit is specific for each ligand. (For example, the receptors for GM-CSF are called CD 116 clusters or simply CD116. Its β-subunit is analogous to that in IL-3 receptor.) In the absence of ligand, the two subunits are farther apart. The α-subunit has the ligand-binding domain or the CSF-binding domain, where CSF binds with low affinity. The binding triggers the β-subunit to move closer and pair with the α-subunit (dimerization). As a result, a heterodimer is formed which increases the affinity of the ligand. Finally, the β-subunit in the dimer transduces the signal to the intracellular region of the target cell to induce differentiation. Signal transduction starts with the phosphorylation of the tyrosine residues of the dimerized β-subunit by janus kinases family members and subsequent downstream signalling cascade, resulting in the differentiation of GMPCs into neutrophils and monocytes.

5.2.4 Erythropoietin

Erythropoietin (EPO) is a glycoprotein with 165 amino acid residues and four glycosyl moieties. The primary structure of EPO is shown in Fig. 5.8. The protein has two disulphide bridges formed between 29th and 33rd cysteine residues, and 6th and 161st cysteine residues. The S–S bridge between the 6th and 161st positions is involved in the orientation and binding of the EPO to its receptor. Three of the four glycosyl residues are N-linked to the asparagines residues in positions 24, 38, and 83. One of the glycosyl residues is O-linked to the serine residue in the 126th position. The glycosyl residues are attached with sialic acids, which determine the circulation of EPO. Sialic acid increases the circulatory time of the EPO, such that they are not cleared away by the liver, instead they act on their receptors.

EPO functions by endocrine signalling, that is, it is produced by one tissue (the kidney's) and acts on a different target (Chapter 11). It binds to erythroid progenitor cells. Burst-forming unit erythroid (BFU-E) and colony-forming unit erythroid (CFU-E) are the two types of erythroid progenitor cells possessing surface receptors for EPO. EPO receptors have extracellular domain, transmembrane domain, and the intracellular domain. EPO binds to the extracellular domain of the adjacent receptors so that the tyrosine residues on the intracellular domains cross phosphorylate each other and activate the signalling pathway leading to the expression of transcription factors responsible for the regulation of cell proliferation and survival (refer Chapter 11). The role of erythropoietin is summarized in Fig. 5.9. EPO initially binds to its receptors on

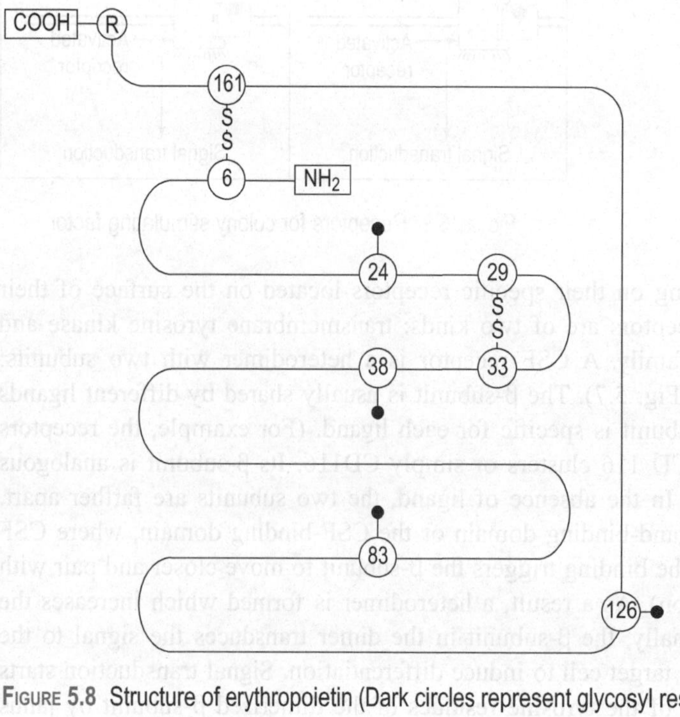

FIGURE 5.8 Structure of erythropoietin (Dark circles represent glycosyl residues)

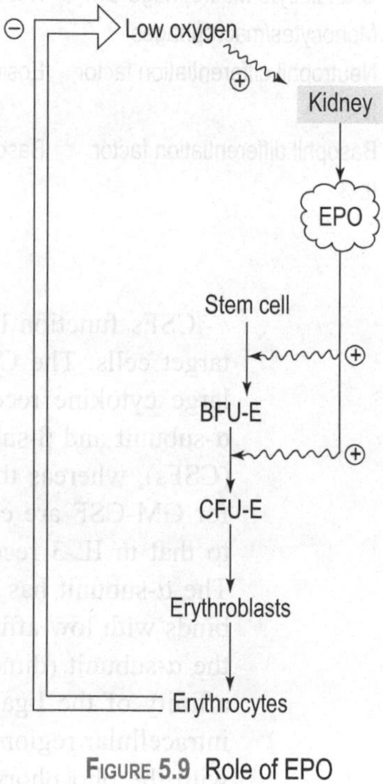

FIGURE 5.9 Role of EPO

BFU-E, and stimulates the intracellular signalling cascade to induce its proliferation into CFU-E (the proerythroblasts). Subsequently, EPO binds to proerythroblasts and induces its proliferation into erythroblasts, which further develop into reticulocytes. Reticulocytes migrate into the peripheral circulation where they mature into red blood cells.

5.2.5 Thrombopoietin

Thrombopoietin (TPO), also called leukaemia virus oncogene ligand, megakaryocyte growth and development factor, c-MpI ligand, Mpl ligand, and megapoietin, is a glycoprotein hormone. It is a cytokine and is produced by the following:

- Sinusoidal endothelial cells of the liver
- Parenchyma cells of the liver
- Proximal convoluted tubular cells of the kidney
- Stromal cells of the bone marrow

TPO is translated from its gene as a precursor with 353 amino acid residues, of which 21 residues are removed as signal peptides to form the active TPO with 332 residues, which is then glycosylated. In active TPO, the first 155 amino acid residues are homologous to that of the EPO. It has a molecular weight of 30 kD. Structurally, TPO has two domains: N-terminal domain (the receptor-binding domain) and C-terminal domain. The N-terminal domain has a sequence homology with EPO, and it binds to the c-Mpl receptor. The C-terminal domain enhances the bioavailability of the hormone. TPO is highly significant in that it has the longest half life of 30 h when compared to other haematopoietic factors.

The main function of TPO is to promote thrombopoiesis. In fact, TPO functions along with other cytokines to promote platelet formation. TPO binds to their receptors, called c-Mpl (encoded by c-Mpl gene), present on the megakaryocytes and platelets. Binding of TPO to the c-Mpl receptor activates the JAK–STAT pathway.

5.3 BLOOD-CELL LINEAGE

Blood-cell lineage includes various families of blood cells. HSCs are the origin of blood-cell lineage. In Chapter 3, the blood-cell lineage and the cytokines responsible for the production of the cells have already been discussed. To recall, HSC produces two major types of progenitors: lymphoid progenitor and myeloid progenitor. Lymphoid progenitor produces B-lymphocytes, T-lymphocytes, and natural killer cells. These cells have been discussed in Chapter 3. Myeloid progenitor produces granulocytes, monocytes, erythrocytes, and platelets. Granulocytes and monocytes have also been discussed in Chapter 3. The chemistry and functions of the other two cells—erythrocytes and platelets—and factors that produce this lineage are discussed in the following sections.

5.4 ERYTHROCYTES

Red blood cells (RBCs) or erythrocytes are the most common type of blood cells. Although they are produced at a rate of 2.4 million cells/sec, they have a short life span of just 120 days, after which they are destroyed in the spleen and liver. This section deals with the chemistry and the functions of RBCs.

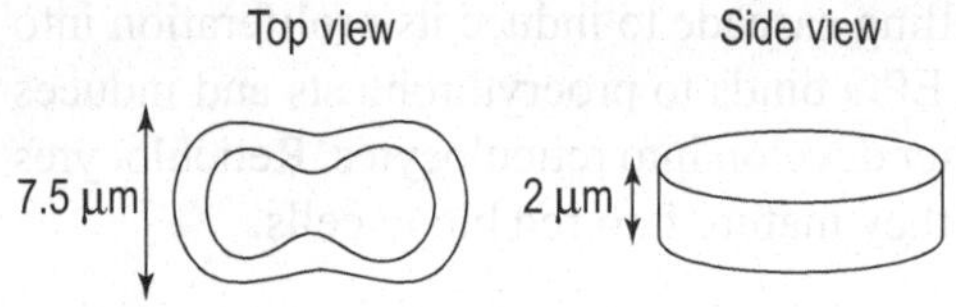

FIGURE 5.10 Morphology of erythrocyte

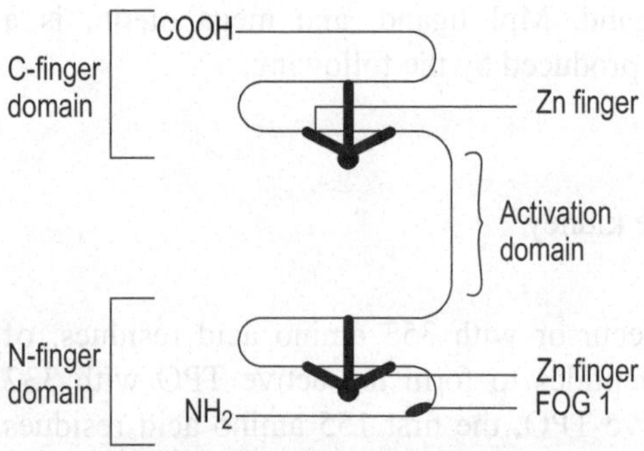

FIGURE 5.11 Structure of the GATA-1 transcription factor

5.4.1 Chemistry

Mammalian RBCs are circular with a side view resembling biconcave discs (Fig. 5.10). They lack nucleus and other organelles such as mitochondria. The biconcave disc morphology of the erythrocyte increases the surface-area-to-volume ratio of the cells, which is ideal for the diffusion of oxygen and carbon dioxide along the cell. A typical human RBC ranges between 6 and 8 μm in diameter. The RBC membrane is flexible, due to which they squeeze into smaller blood vessels. The outer surface of the RBC membrane has glycoproteins called blood group antigens, whereas the inner surface has two proteins called spectrin and actin, which are responsible for maintaining the integrity of the membrane. Blood group antigens are discussed in Chapter 4; spectrin has already been discussed in Chapter 2, and the role of actin is discussed in Chapter 9.

The formation of α-spectrin is regulated by GATA-1. GATA-1, also called erythroid transcription factor or GATA-binding factor 1, is a protein in human RBCs. Structurally, it has three domains (Fig. 5.11): C-finger domain, N-finger domain, and activation domain. The C-finger domain is made of a zinc finger domain and is found near the carboxy terminal of the protein. The N-finger domain is made of a zinc finger domain and FOG1 (friend of GATA) binding site. Both zinc finger domains are DNA-binding domains. The activation domain, found between the two fingers, is responsible for transcription of α-spectrin gene.

The membrane of an RBC is influenced by osmotic pressure. The effect of osmotic pressure on RBCs is shown in Fig. 5.12.

Hypertonic environment is one in which osmotic pressure is higher in the environment and lower in the RBCs. In this case, water will move out of the RBC membrane causing the cells to shrink. However, when the environment is hypotonic (more osmotic pressure in the cells), water will move through the membrane into the cell. Thus, the cell will swell and consequently burst. However, if the osmotic pressure inside and outside the cell are same, the membrane remains normal and the cell will maintain its biconcave disc shape. This condition is called isotonic, which usually prevails in the blood.

The extent to which the RBCs can withstand the hypotonicity is used to determine the integrity of the RBC membrane. This is used to predict a diseased state. For example, healthy RBCs when incubated in hypotonic solution, say hypotonic saline in vitro, can withstand the hypotonicity to a certain extent before they lyse. The RBCs from an individual subjected to oxidative stress (as in many diseases or due to exposure to toxic chemicals such as pesticides) can withstand the hypotonicity to a lesser extent. This reveals that the membrane integrity of the RBCs is deformed in diseased state, so that the cells are too fragile. In other words, the osmotic fragility is more for diseased cells than for normal cells. Figure 5.13 shows the osmotic fragility curve of RBCs from normal rat blood and pesticide-treated rat blood. The curve shows that the RBCs in pesticide-treated rats show 50% lysis as soon as the solution is made hypotonic, whereas the normal cells

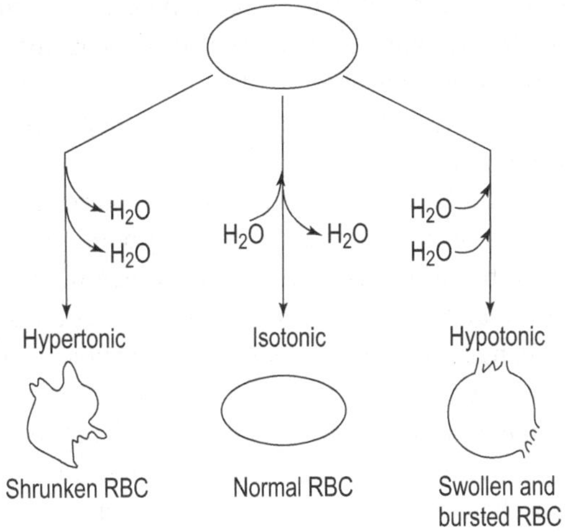

FIGURE 5.12 Influence of osmotic pressure on red blood cell architecture

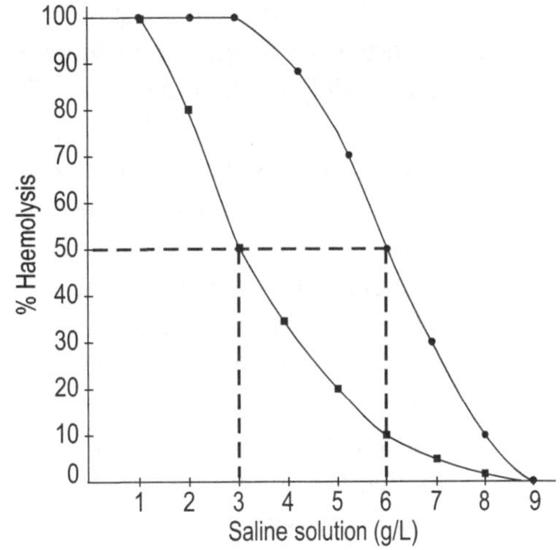

FIGURE 5.13 Osmotic fragility curve of normal red blood cells and pesticide-exposed red blood cells

withstand the hypotonicity to a larger extent. Thus, osmotic fragility is an indicator of RBC membrane damage.

5.4.2 Function

The primary function of RBCs is to transport oxygen. The functional component of RBCs is an iron-containing protein (metalloprotein) called haemoglobin (Hb). Haemoglobin is made of a protein component called globin and a prosthetic group called haeme. Globin is made of four subunits: two α and two β subunits (Fig. 5.14) represented as α2β2. The four subunits are held together by hydrophobic interactions, ionic bonds, and van der Waals interaction. Each subunit holds one haeme molecule almost at its centre. Each haeme molecule (Fig. 5.15) is made of an iron (Fe) atom at its centre. The Fe atom is capable of binding one oxygen atom. Thus, each haemoglobin molecule in RBCs can bind to four oxygen atoms.

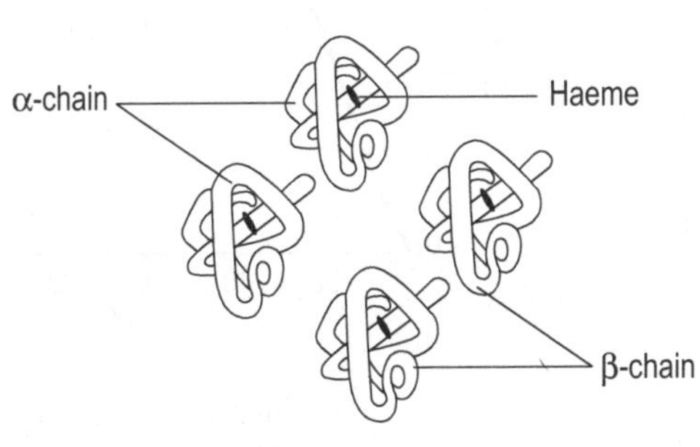

FIGURE 5.14 Structure of haemoglobin

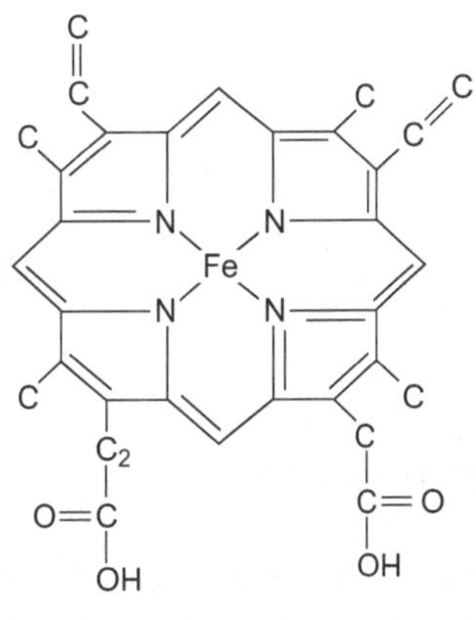

FIGURE 5.15 Structure of haeme

RBCs hold about 3.5 g of iron in the form of haemoglobin. Hb is responsible for the transport of oxygen from lungs or gills to the peripheral tissues due to its capacity of irreversibly binding to oxygen. They can also carry some amount of carbon dioxide back from the tissues. The affinity of Hb to its ligands, oxygen, or carbon dioxide depends on the partial pressure of oxygen (pO_2) in the environment.

As we breathe, oxygen moves from air sacs into the lungs. Alveolar capillaries have high pO_2, where the haemoglobin accepts oxygen, forming oxyhaemoglobin. In the presence of oxyhaemoglobin, the RBCs have bright red colour. As the oxygen binds to Hb, the domains of the globin undergo conformational changes and the haeme acquires a planar conformation. The conformational changes favour the binding of the subsequent oxygen atoms. This phenomenon is called positive cooperativity. Thus, oxyhaemoglobin has more affinity to oxygen atoms when compared to deoxyhaemoglobin. The net result is that the affinity for the fourth oxygen atom is 300 times greater than the affinity for the first oxygen atom. The conformational changes of haeme in oxyhaemoglobin and deoxyhaemoglobin are compared in Fig. 5.16.

As the blood flows, the haemoglobin gets access to the peripheral tissues, where pO_2 is low (pCO_2 is higher). Low pO_2 decreases the affinity of Hb for oxygen, and favours the detachment of oxygen from oxyhaemoglobin. Thus, in peripheral tissues, oxyhaemoglobin is converted to deoxyhaemoglobin and the haeme regains the non-planar conformation (Fig. 5.16). As the oxygen is released, the colour of the RBCs changes from bright red to purple.

The influence of pO_2 on the oxygen-binding ability of the Hb can be represented in the form of a saturation curve as shown in Fig. 5.17. This saturation curve is a plot of pO_2 versus percentage saturation of Hb by oxygen. The curve shows that the binding affinity increases with an increase in pO_2 as prevailing in the lung capillaries. On the other hand, the binding affinity decreases with a fall in pO_2 as prevailing in the peripheral tissues.

In the peripheral tissues, the RBCs eliminate the oxygen and bind to carbon dioxide to transport into the lungs. The carbon dioxide molecules bind to the amino acid residues of the globin. This form of haemoglobin is called carbamino haemoglobin. Carbamino

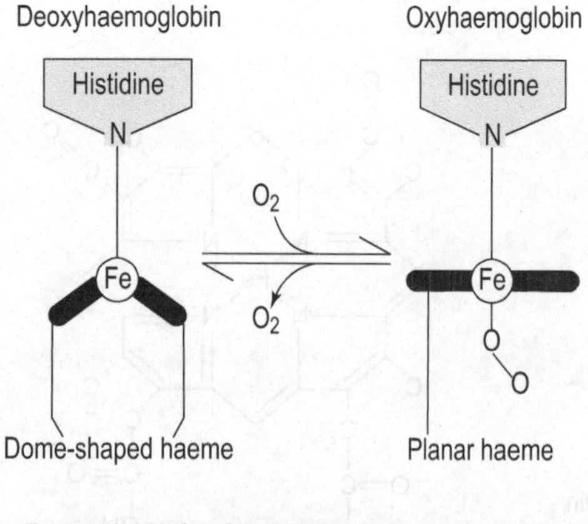

FIGURE 5.16 Conformational changes in the haeme moiety of oxyhaemoglobin and deoxyhaemoglobin (haeme moiety is shaded)

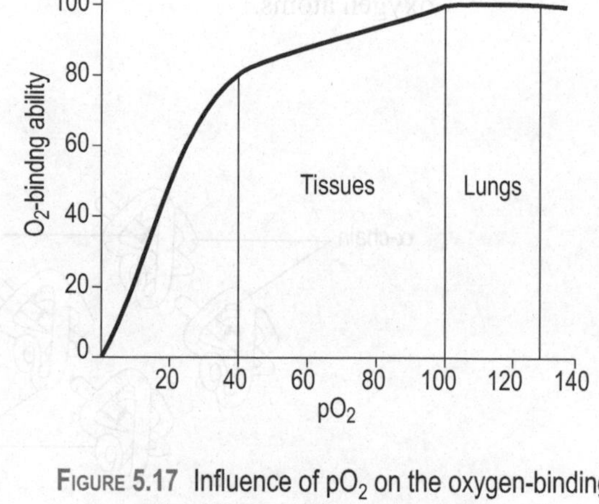

FIGURE 5.17 Influence of pO_2 on the oxygen-binding ability of haemoglobin (saturation curve)

haemoglobin is called the *reduced form of haemoglobin*, which means that they do not bind to oxygen. The carbamino haemoglobin travels in the reverse direction, that is, from the peripheral tissues to the lungs. In the lungs, where pCO_2 is low, the affinity of the Hb to CO_2 is reduced. Therefore, in the lungs, $HbCO_2$ is dissociated into Hb and CO_2. Around 23% of CO_2 is transported in the form of carbamino haemoglobin.

However, around 70% of the CO_2 is transported by an enzyme-dependent mechanism, catalysed by carbonic anhydrase. In this mechanism, the enzyme catalyses the coupling of CO_2 and water to produce carbonic acid. The carbonic acid dissociates into bicarbonate and proton. The proton is bound by haemoglobin to prevent acidic pH. The bicarbonate ion is transported through blood and it dissociates into CO_2 on reaching the lung capillaries.

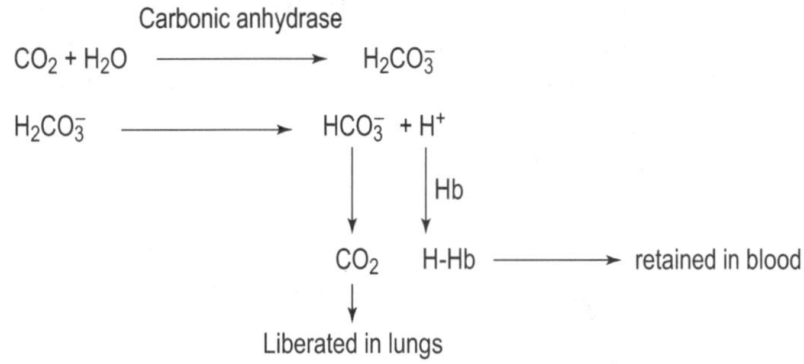

Thus, the CO_2 is brought to the lungs from the tissues. This cycle of oxygen transport to tissues and CO_2 transport to the lungs continues. The haemoglobin-mediated transport of oxygen from the lungs and carbon dioxide from the tissues is depicted in Fig. 5.18.

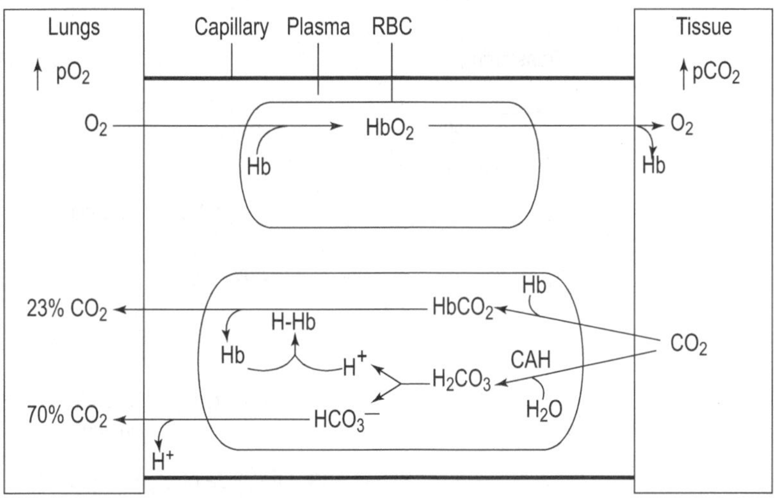

FIGURE 5.18 Function of haemoglobin in the transport of oxygen and carbon dioxide

5.4.3 Life Cycle

This section summarizes the life cycle (production and degradation) of erythrocytes based on the information discussed so far. The life cycle of erythrocytes is illustrated in Fig. 5.19. Erythropoiesis is stimulated by the following factors:

- haemorrhage

- damage to bone marrow
- exposure to high altitude
- exercise
- haemolytic disease
- low haemoglobin levels

These factors indicate that low oxygen (hypoxia) is the common stimulator of erythropoiesis. Hypoxia signals the juxtaglomerular cells of the kidney to release erythropoietin. Erythropoietin acts on the receptors of the myeloid progenitor cells to induce differentiation. To summarize, the myeloid progenitor cells are converted to erythroblasts, which grow into reticulocytes. Reticulocytes lose their nucleus, mitochondria, and ribosomes, and develop into mature anucleated erythrocytes. Each mature erythrocyte has a life span of around 120 days, after which they are destroyed in the spleen and liver. As erythrocytes are devoid of nucleus and DNA, they cannot divide; therefore, they just undergo ageing.

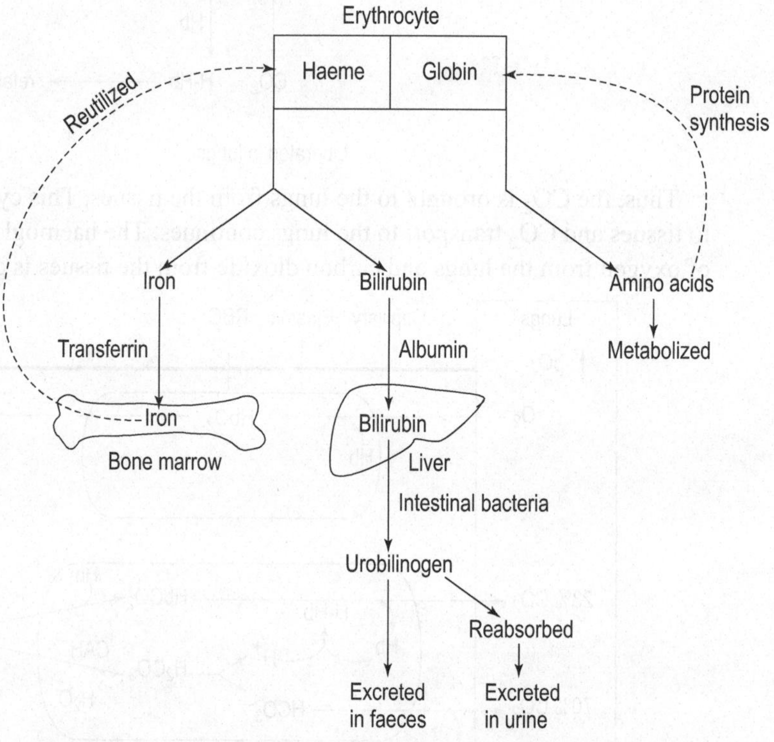

FIGURE 5.19 Life cycle of erythrocytes

Erythrocytes are damaged or degenerated due to ageing. Ninety per cent of the worn out erythrocytes are engulfed by macrophages in spleen, liver, and lymph nodes. The remaining 10% undergo lysis in the circulation. Macrophages undergo phagocytosis and break the content of the erythrocytes into globin and haeme. Globin is further degraded to amino acids, which are utilized by the macrophages for the synthesis of proteins. Haeme is converted into bilirubin and biliverdin with the release of iron (Fe). Bilirubin is released from the macrophages into the blood, where it binds to albumin and gets transported to the liver. In the liver, bilirubin is conjugated to glucuronic acid and released into the small intestine via bile. Bilirubin is converted into a brown-coloured pigment called urobilinogen by intestinal bacteria, some of which are excreted in the faeces and some of them are reabsorbed and excreted via urine. The iron produced in the macrophages may either be stored or released into the blood. Iron present in blood binds to transferrin (iron-binding protein) and gets transported to the bone marrow, where it is stored and reutilized for the synthesis of haemoglobin. Thus, erythrocytes are degraded and metabolized. After the cells are removed from the blood, new erythrocytes are produced in the bone marrow by erythropoiesis in about four days.

5.4.4 Abnormal Red Cells

This section discusses the diseases caused due to abnormal red blood cells. Some examples are listed as follows:

- Elliptocytosis
- Spherocytosis
- Sickle-cell anaemia
- Thalassaemia syndrome
- Erythroblastosis foetalis

Elliptocytosis is a disease in which the red blood cells have an abnormal shape, that is, they are either elliptical or oval rather than typical biconcave discs (Fig. 5.20a). This is due to a defect in the association of the cytoskeletal protein spectrin with its anchor ankyrin. Abnormal glycophorin also results in elliptocytosis. The susceptible proteins are affected due to mutation. In Chapter 2, membrane proteins of red blood cells have been discussed. Spectrin is a tetrameric protein with α and β subunits. The subunits are anchored to plasma membrane proteins, namely, band 3 and glycophorin C via ankyrin–protein 4.2 complex. The entire complex is associated with the cytoskeletal actin filaments. Mutation in the gene responsible for the formation of spectrin complex results in less stable membrane, affecting the integrity of the cells. As a result, the discoid cells can easily and permanently deform into elliptocytes. These deformed cells lack elasticity and are less resistant to shearing stress.

Spherocytosis is another disease characterized by abnormal red cells (Fig. 5.20b). The cells become spherical due to a defect in the spectrin or band 3 protein or both. Although spherical cells can function as normal cells, they are considered abnormal by spleen and destroyed. Thus, spherocytosis is accompanied by anaemia and jaundice. Removal of spleen is the only cure for ellitocytosis and spherocytosis.

Sickle-cell disease or sickle-cell anaemia is another disease which is caused due to the deformed shape of the red blood cells. RBCs acquire a crescent shape or a sickle shape (Fig. 5.20c). This disease is due to a point mutation in the gene coding for glutamic acid (6th residue) present in the β-globin chain of the haemoglobin. Mutation replaces glutamic acid with valine, resulting in abnormal haemoglobin called haemoglobin S (S for sickle). This results in a change in the shape of the red blood cells from biconcave disc to sickle. The sickle cells are so brittle that they cause excessive lysis of red cells leading to haemolytic anaemia. Normally, the biconcave red cells possess

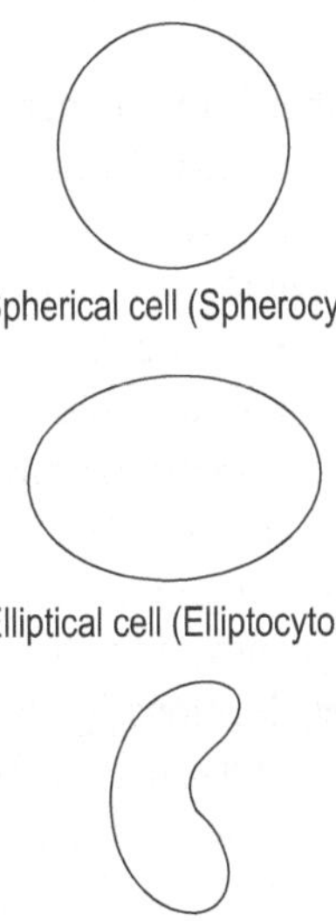

(a) Spherical cell (Spherocytosis)

(b) Elliptical cell (Elliptocytosis)

(c) Sickle cell

FIGURE 5.20 Abnormal erythrocytes

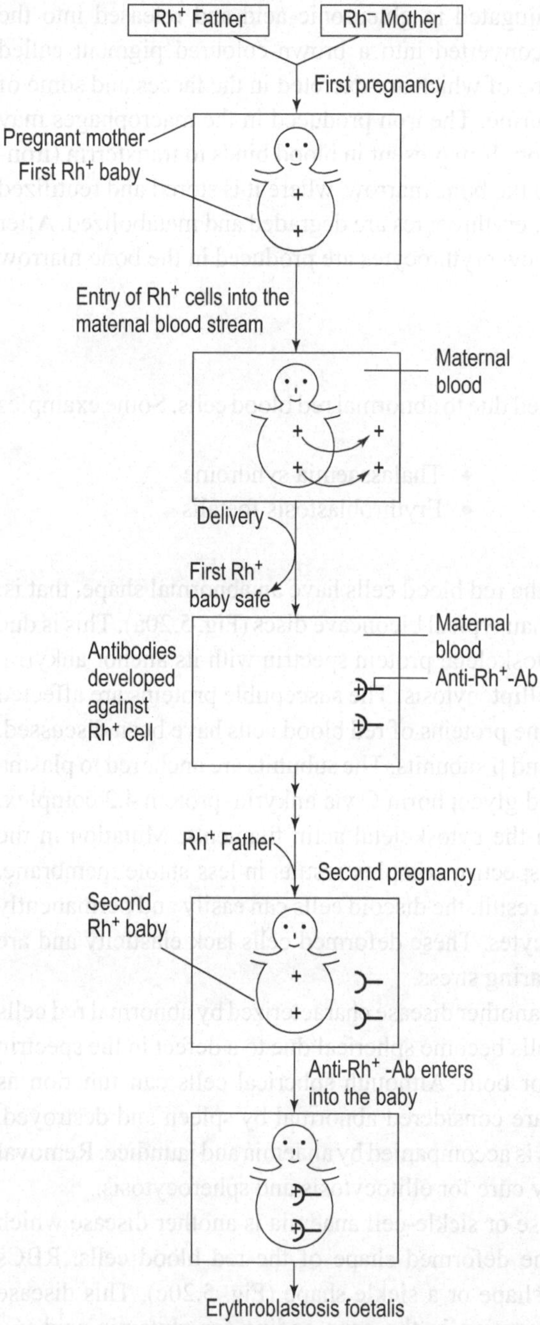

FIGURE 5.21 Erythroblastosis foetalis

haemoglobin A, responsible for carrying oxygen, as the red cells move along the blood vessels. Sickle cells, however, have haemoglobin S (HbS), which impairs the movement of the cells. The abnormal cells, thus, cannot pass through the smaller blood vessels, resulting in impairment in the blood flow and oxygen supply to major organs such as kidney, heart, brain, and bones. Thus, sickle cell anaemia can cause body pain, heart failure, and kidney failure.

Thalassaemia syndrome is yet another disease related to red blood cells. A defect in the synthesis of haemoglobin's α-chain or β-chain or both, probably due to mutation, results in thalassaemia syndrome. The cells lyse to cause anaemia. This syndrome is also called Cooley's anaemia. It is characterized by microcytic, hypochromic red blood cells, which implies that the cells are pale and smaller than normal.

Erythroblastosis foetalis (EBF) is a haemolytic disease occurring in newborn babies. This is due to antibody-induced haemolysis and anaemia. EBF is related to the blood group antigens present in the surface of the red blood cells (refer Chapter 4). When a foetus expresses the blood group antigen which is foreign to the mother, the mother's blood synthesizes antibodies against it. This concept is illustrated in Fig. 5.21 and can be explained as follows:

When an Rh-negative mother is impregnated by an Rh-positive father, and if the foetus in the mother's uterus is Rh-positive, then in such situations, during the time of delivery, the positive red cells cross the placenta and reach the mother's circulation. As positive cells are foreign to the mother's blood, she develops antibodies (Ig) against them. During the process of antibody formation, memory B cells are produced (refer Chapter 3). Although the first baby is safe, during the second pregnancy, if the foetus is again Rh positive, the mother develops Igs which cross the placenta and affect the red blood cells of the baby. This results in haemolytic anaemia and subsequent increase in blood bilirubin concentration. Because of anaemia, the bone marrow of the foetus is stimulated to release more immature red blood cells (erythroblasts) in the peripheral circulation (hence, the name erythroblastosis foetalis). EBF is characterized by the following symptoms in the foetus:

- jaundice
- hepatomegaly
- spleenomegaly
- oedema
- hydrops

5.4.5 Artificial Red Cells as Nanorobots

A nanorobot is a hypothetical microscopic device or molecular machine designed to perform certain tasks with precision at the nanometer scale. They are used in the field of medicine for treatment and therapy. The amount of nanorobots that needs to be injected into the body for the treatment of diseases is called therapeutic dose. The therapeutic dose depends on the severity of the disease. A single therapeutic dose has billions or trillions of nanorobots.

The following are the criteria for a nanorobot:

- It should be tiny and swift so that it can travel along the circulatory system.
- It must have the facility to load medicine.
- It should be equipped with miniature tools for its functioning.
- It should be excreted from the host after fulfilling its activity.
- It should perform its function at the nanoscale.

Tools or components of nanorobots

Carbon, in the form of diamond, diamondoid, and fullerene is the principal component of the nanorobot. Various tools or components found in nanorobots and their functions are summarized in Table 5.3.

Nanorobots get power from three sources to move in the blood:

1. from the patient's own body (internal power source)
2. source on board the robot itself (on board power source present designed in the robot itself)
3. from outside the patient's body (external power source)

A major advantage of nanorobots is their speed of action, precision, and durability. They can remain functional for years, decades, or centuries.

TABLE 5.3 Tools present in nanorobots

Tool	Function
Medicine cavity	To hold medicine and chemicals
Probes, knives, and chisels	To grab, break, and crush clots, plaques, and blockages in blood vessels.
Microwave emitters and ultra signal generator	To kill cancer cells
Electrodes	To generate electric current for killing cancer cells
Laser	To vaporize and burn away plaques and cancer cells

Components of artificial red cells

An artificial red cell, also called respirocyte or mechanical red cell or RBC nanorobot, is a classic example of a nanorobot. Respirocyte is a small device of around 4–6 μm in diameter, used to substitute the normal human red blood cell, and thereby function in the transport of respiratory gases, oxygen, and carbon dioxide. It is a hollow, spherical device (Fig. 5.22) constructed from 18 billion precisely arranged structural atoms. It has the following components:

- tanks/chambers for oxygen, carbon dioxide, or ballast water

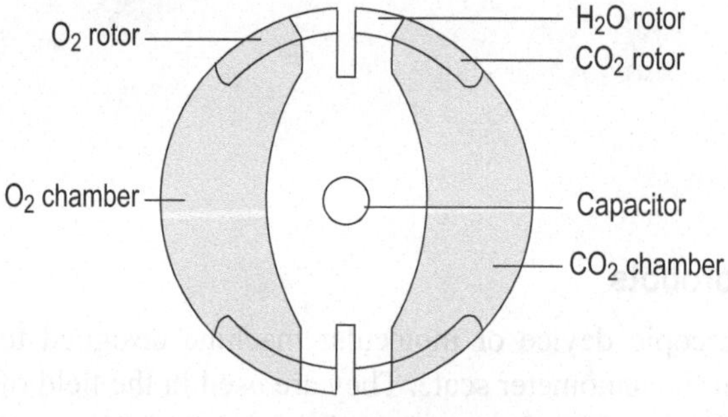

FIGURE 5.22 Components of artificial red cells (erythrocytes)

- rotors for collecting oxygen, carbon dioxide, and glucose, respectively
- an onboard chemo-mechanical turbine
- mechanical acoustic transducer embedded on the surface
- pressure and temperature sensors for regulating the function of the cells
- an onboard computer

Function Respirocytes are injected as a suspension, containing 5–6 trillion cells. It functions as an oxygen carrier and carries oxygen and delivers it to the peripheral tissues and takes up carbon dioxide. The oxygen-loading capacity of a respirocyte is much greater than natural red cells. The glucose tank collects glucose from the blood to power the robot. The onboard chemo-mechanical turbine converts the energy of glucose into useful work for powering the respirocyte. With this power, each cell can have an average pumping rate of 100 million molecules/s. Using the power, the oxygen tank loads oxygen (up to 1.5 billion molecules) with the help of an oxygen rotor to deliver to the tissues while circulating. The ballast water tank loads water (up to 6 billion molecules) using ballast rotor for counterbalance. The cells deliver oxygen to the peripheral tissues where the carbon dioxide tank loads carbon dioxide (up to 1.5 billion molecules) with the help of a carbon-dioxide rotor. Mechanical acoustic transducer functions as a pressure actuator. It converts pressure fluctuations into mechanical motion, depending on the commands broadcasted by the physician. Pressure sensors are used for monitoring the gas loading and unloading. A Ullage sensor is used for monitoring ballast and glucose chambers. The onboard computer (10,000 bit/s) controls the rotor field, gas loading, ballast-tank loading, throttling, and power distribution. It is also responsible for the interpretation of sensor data and external commands.

Potential applications

Artificial red cells can be used for the following purposes:

- as universal blood substitute
- in the preservation of living tissues
- in the treatment of anaemia
- in the prevention of choking
- in the management of lung and respiratory diseases and asphyxia
- in the management of carbon-monoxide poisoning
- to prevent Caisson disease, during deep sea diving
- by sportsmen to increase the oxygen supply to the tissues in conditions, such as exercise, running, swimming, football, and other endurance-oriented events.

5.5 LEUCOCYTES

Leucocytes are white blood cells (WBCs). They are formed from lymphoid progenitors of HSCs. They are of two types: granulocytes and agranulocytes. Granulocytes have primary and secondary granules on their cytoplasm, whereas agranulocytes have only primary granules.

Granulocytes release the contents of their granules (a process called degranulation). Compounds released from the granules play a role in eliciting immune response. The functions of granulocytes have already been dealt with in Chapter 3. As they are also a lineage of HSCs, some aspects of granulocytes are discussed here too.

5.5.1 Chemistry and Function

The cytoplasmic granules of granulocytes are membranous lysosomes that contain specific enzymes. Peroxidase is the marker for primary granules. The presence of alkaline phosphatase and absence of peroxidase are the markers for secondary granules. Contents of primary granules are used to kill and process the antigen, whereas those of secondary granules are used in extracellular processes such as regulation of the immune function. Primary granules are azurophilic. That is, the contents of primary granules, such as myeloperoxidase, phoshpholipase A2, lysozyme, elastase, defensins, acid hydrolases, neutral serine proteases, bactericidal/permeability increasing protein, cathepsin G, proteinase 3, and proteoglycans, readily stain with azure dye.

Neutrophil is the most predominant granulocyte and constitutes 50–60% of the total leucocytes, and hence, it is best studied and well understood. Markers for neutrophils are myeloperoxidase, lactoferin, and human neutrophils lipocalin. It was found that the release of the markers (responsible for immune function, refer Chapter 3) from the granules of neutrophils depend on the adhesion of the cells with ECM components, such as fibronectin and fibrinogen. That is, degranulation and immune response are determined by cell adhesion molecules.

Eosinophils are present in lesser amount compared to neutrophils, comprising around 1–5% of the total leucocytes. Eosinophilic granules are rich in major basic protein (MBP) and eosinophilic cationic protein (ECP). MBP kills the parasitic antigen by membrane disruption. ECP together with MBP helps in the fragmentation of parasitic antigens.

Basophils are the smallest and least common leucocytes, representing less than 1% of the total leucocytes. They have very few primary granules. The secondary granules are quite large. As mentioned earlier, the eosinohilic granules are rich in basic proteins (MBP), on the other hand, basophilic granules are rich in acidic proteins, which are usually sulphated or carboxylated, staining with toluidine blue. They play a role in allergic reactions, during complement-mediated degranulation.

Although granulocytes are produced in the bone marrow and released into the blood, they do not function in the blood, but travel via blood to the site of infection or injury. Thus, migration of leucocytes out of the blood vessels (extravasation) is crucial for the function of leucocytes. The types of leucocytes migrating from the blood vessels depend on the type of infection. For example, neutrophils extravasate after bacterial infection, whereas eosinophils extravasate after protozoal or parasitic infection. Blood vessel walls are lined with endothelial cells. Whenever there is an injury, inflammatory mediators are released, which signal the blood vessels to contract the endothelial cells (Fig. 5.23). The surface of the endothelial cells facilitates the rolling of leucocytes, whereby they escape via the space between the endothelial cells. The escape process is favoured by the chemokines produced by the damaged tissues. Thus, the cells are squeezed out of the blood vessel into the basal lamina (a layer lying superficially to the endothelium). Two types of adhesion molecules are involved in the migration process: selectin and integrins. The role of these adhesion molecules is discussed in detail in Chapter 10. As the migration process continues, the amount of leucocytes in the circulation is depleted. This is compensated by the production of more leucocytes in the bone marrow, probably via signalling molecules.

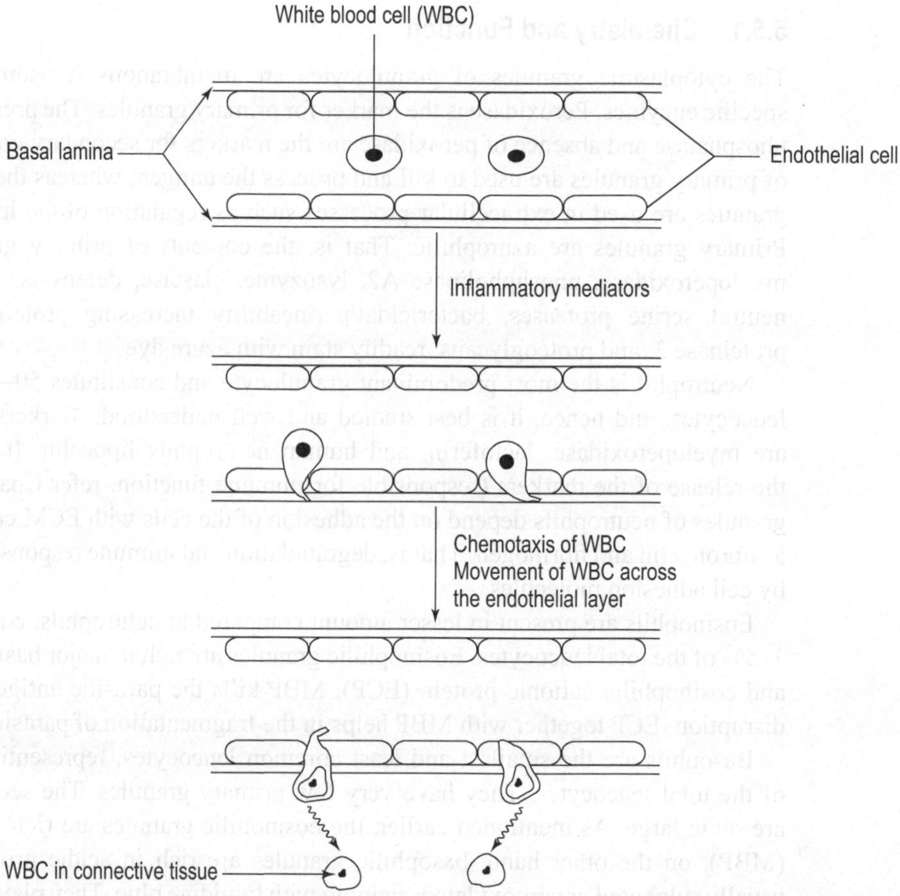

FIGURE 5.23 Migration of leucocytes across the blood vessels to reach their target site of action

Agranulocytes are characterized by absence of secondary granules. They include smaller lymphocytes, for example, B cells and T cells, and larger lymphocytes such as natural killer cells. The chemistry and the functions of these cells in immunity have already been discussed in Chapter 3.

5.6 PLATELETS

As mentioned in Section 5.1, platelets are blood cells (thrombocytes) derived from the myeloid progenitor cells. They are formed from megakaryoblasts. The process of platelet

TABLE 5.4 Components of platelet granules

Granule	Constituent
Dense granule/delta granule	ADP
	ATP
	Calcium
	Serotonin
Lambda granule	Hydrolytic enzymes
Alpha granule	Platelet factor 4
	Transforming growth factor-β1
	Platelet-derived growth factor
	Fibronectin, B-thromboglobulin
	von Willebrand factor
	Fibrinogen
	Clotting factors V
	Clotting factor XIII

formation (thrombopoiesis) is discussed in Section 5.1. The present section discusses the chemistry and function of platelets.

5.6.1 Chemistry

Platelets are small blood cells with a diameter of 2–4 μm. They resemble red blood cells in that they are discoid and anucleated. There are about 150 – 400 × 10^9 platelets per litre of blood, each having a life span of about 5 to 9 days. After 9 days, they are absorbed and degraded by spleen.

The structure of a platelet is shown in Fig. 5.24. It has a plasma membrane that invaginates into channels called open canalicular system. The outer surface of the membrane is made of Gp lb receptor and von Willebrand factor. The inner portion of the cell is broadly divided into two regions: the peripheral region called hyalomere and the central region called granulomere. Hyalomere is made of microfilaments and microtubules, which form the rim of the cell, whereas granulomere contains granules and other organelles, such as lysososme, mitochondria, endoplasmic reticulum, and Golgi apparatus. Granules are so significant that they mediate the primary function of platelets—the blood coagulation process. They have three types of granules: dense granules or delta granules, lambda granules, and alpha granules. The composition of these granules is shown in Table 5.4. Through their granular contents, platelets function in haemostasis (a process by which blood loss is prevented by clotting) and in inflammatory process.

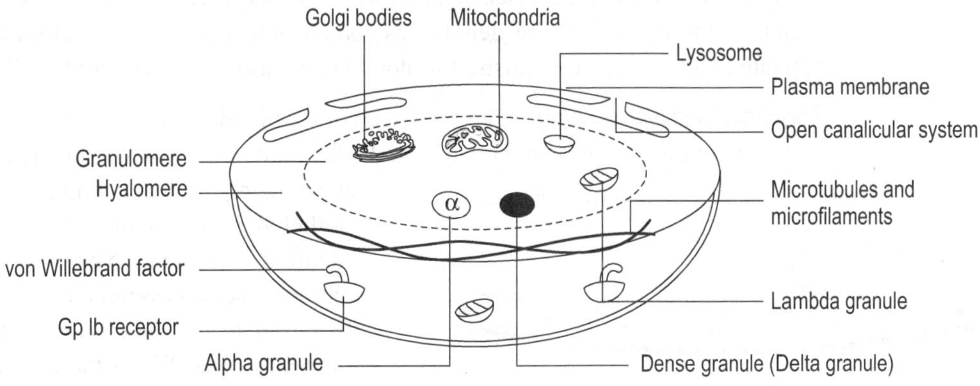

FIGURE 5.24 Structure of a platelet

The disease caused due to the abnormality in the number and function of platelets is called thrombocytopathy. Thrombocytopathy includes

- thrombocytosis (increase in the platelet count)
- thrombocytopenia (decrease in the platelet count)
- thrombasthenia (defect or decrease in the function of the platelets)
- grey-platelet syndrome (deficiency of alpha granules in platelets)

Thrombocytosis may be due to excess thrombopoietin or dysfunction of spleen, which fails to clear the blood platelets. Thrombocytopenia may be the result of overactivity of spleen (hyperspleenism) in which a large number of platelets are cleared away from the blood due to deficiency of thrombopoietin. This condition is common in Gaucher's disease and leukaemia; this can be managed by platelet transfusion.

5.6.2 Function

Platelets release the following growth factors and mediate tissue regeneration, tissue repair, and wound healing:

- Platelet-derived growth factor (PDGF)
- TGF beta
- Basic fibroblast growth factor
- Insulin-like growth factor 1
- Platelet-derived epidermal growth factor
- Vascular endothelial growth factor

However, the main function of platelet is to induce blood clotting (coagulation) and maintain haemostasis, whenever the blood vessels are damaged. Figure 5.25 shows the layers of the blood vessel. The following section deals with the role of a platelet in blood coagulation.

Blood coagulation

Blood coagulation is defined as the process of transformation of blood into a semisolid mass called clot to prevent blood loss during blood vessel damage. Blood coagulation is brought about by platelets in association with various proteins, collectively called clotting factors. Clotting factors are denoted as roman numerals, usually suffixed with 'a' for activated factor. The prerequisite for blood coagulation is a platelet plug (Fig. 5.26).

Formation of a platelet plug The inner wall of the blood vessels are lined with endothelial cells, containing a protein called von Willebrand factor (VWF) which functions in platelet aggregation at the injury site. Below the endothelial lining is the basement membrane containing collagen. Normal blood vessels (uninjured blood vessels) have their endothelial cells closer to the underlying collagen and the endothelial cells. When there is an injury, the following changes occur:

- Endothelial layer is damaged.
- VWF is released.
- Collagen in the wall of the blood vessel is exposed to the blood stream.

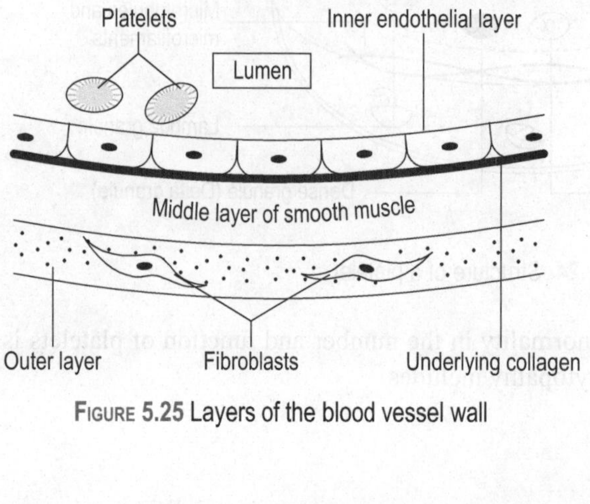

FIGURE 5.25 Layers of the blood vessel wall

- Tissue factor (also called thrombokinase or factor III or CD142) from the subendothelial layer is exposed to the blood stream.

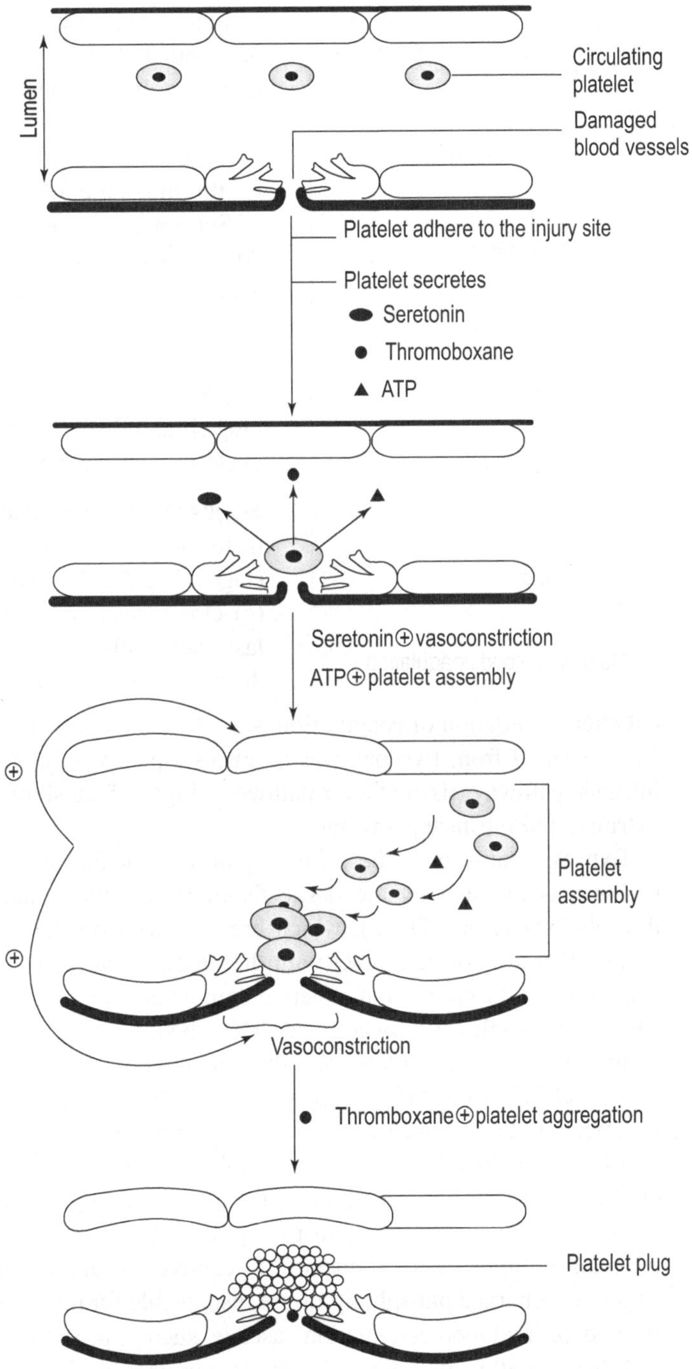

FIGURE 5.26 Formation of a platelet plug

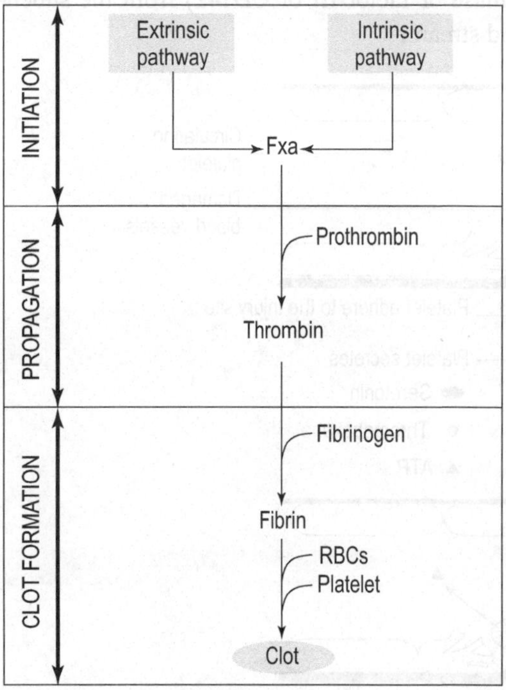

FIGURE 5.27 Stages in blood coagulation

These events together lead to the adherence of the platelets to the exposed collagen in the blood vessel wall via the glycoprotein Ia.

In adhered platelets, the arachidonic acid pathway is stimulated. The platelets undergo degranulation releasing the following:

- serotonin
- ADP
- thromboxane A2

Serotonin induces vasoconstriction, whereas ADP attracts other platelets to the site of injury. Thromboxane A2 promotes platelet aggregation via their surface integrin complex (receptors) called glycoprotein IIb/IIIa (GpIIb/IIIa) and fibrinogen. Aggregation leads to the formation of a platelet plug (Fig. 5.26). Platelet aggregation is inhibited by cyclo-oxygenase inhibitors, PGI2 and PGD2.

Stages in blood coagulation Once the platelet plug is formed, blood coagulation takes place in three stages (Fig. 5.27): (a) initiation, (b) propagation, and (c) clot formation. Platelet plug is involved in the last stage—clot formation. Each of these stages is discussed in detail here.

Initiation Initiation of coagulation is the first stage in which factor Xa is formed. Factor Xa is formed from two pathways: extrinsic pathway (contact activation pathway) and intrinsic pathway (tissue factor pathway). Figure 5.28 shows the formation of FXa from extrinsic and intrinsic pathways.

Extrinsic pathway is the primary pathway for the formation of factor Xa. Extrinsic pathway is initiated by the tissue factor (also called platelet tissue factor, factor III, thrombokinase, or CD142), which are produced on the external surface of the blood vessels that are not in contact with the blood stream. During injury, the tissue factor is expelled into the circulation, where it combines with the factor VII (FVII) forming active TF-FVIIa complex. FVIIa activates FX to FXa.

In the activated platelets, the negatively charged membrane phospholipids (phosphatidylserine and phosphatidylethanolamine) present in the inner surface are scrambled to the surface by the action of scramblase. Scramblase is an enzyme of flippase family, which functions as a transmembrane lipid transporter. In fact, the inner leaflet of the membrane has negatively charged phospholipids, whereas the outer leaflet has more of positively charged phospholipids. Scramblases can interchange the location of the phospholipids by the transport of negatively charged lipids to the outer surface. The negatively charged phospholipids are responsible for the activation of enzyme complexes formed during blood coagulation cascade, such as prothrombinase and tenase.

Intrinsic pathway is driven by the factors that are intrinsic components of the blood. High-molecular-weight kininogen (HMWK) is a blood protein, which functions as a cofactor for the activation of prekallikrein and FXII (Hageman factor) to kallikrein

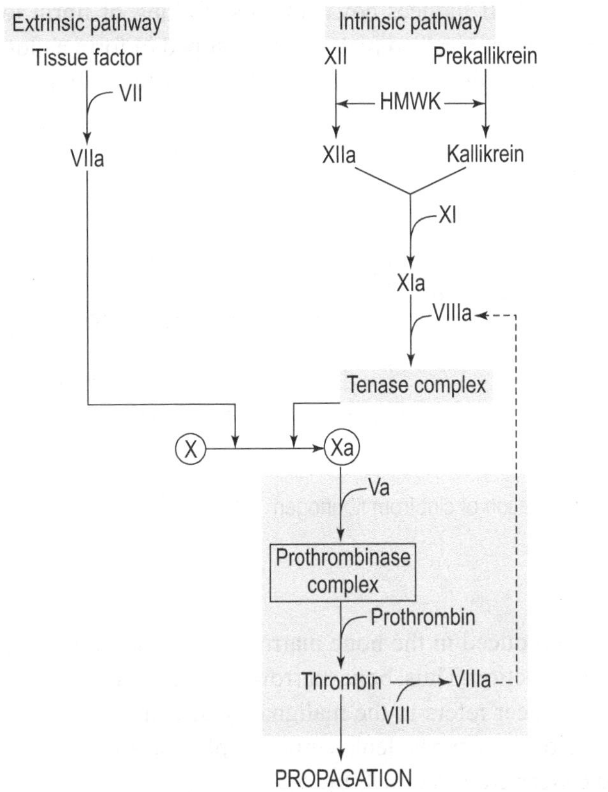

FIGURE 5.28 Formation of factor Xa from extrinsic and intrinsic pathways (tenase complex cycle and VIIIa cycle are shaded for clarity)

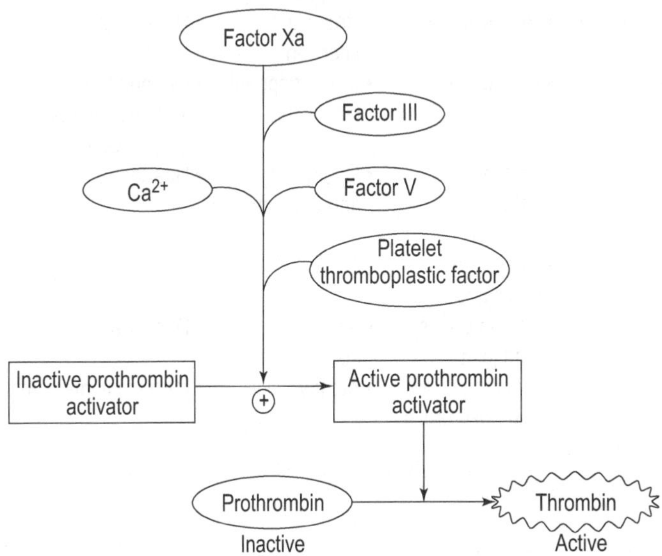

FIGURE 5.29 Propagation of blood clotting (formation of thrombin by FXa)

and FXIIa. FXIIa converts FXI to FXIa. FXIa activates FIX to FIXa, which in association with its cofactor FVIIIa forms the tenase complex. Tenase complex activates FX to FXa.

Tenase complex cycle (cycling of VIIIa) It is essential to know how the factor VIIIa required to drive the intrinsic pathway is cycled to form the tenase complex. FXa combines with its cofactor FVa to form a complex called prothrombinase. Prothrombinase cleaves inactiveprothrombin to active thrombin. Thrombin induces the release and activation of FVIII from its bound VWF. FVIIIa combines with FIXa, forming the tenase complex. Tenase complex in turn activates FX to FXa and the cycle continues. The cycling of FVIIIa is also shown in Fig. 5.28. Additional function of thrombin is discussed in the final stage (clot formation) of blood coagulation (Fig. 5.30).

Propagation Propagation is the second stage of blood coagulation during which the FXa forms thrombin (Fig. 5.29). FXa in combination with factor III, factor V, Ca^{2+}, and platelet thromboplastic factor (PF3) activates prothrombin activator (PA). PA in turn activates prothrombin to thrombin.

Clot formation Clot formation is the third and the final stage of the blood coagulation mechanism, where fibrin meshwork is formed to trap the red cells and platelets to form a clot. Thrombin acts on fibrinogen (Fig. 5.30). Fibrinogen, also called FI, is a plasma glycoprotein produced in the liver. Fibrinogen can cross link platelets by binding to their GpIIb/IIIa receptors. Thrombin cleaves fibrinogen by proteolysis of specific peptide bond to form active fibrin monomers, which form a meshwork around the platelet plug. FXIII induces covalent cross-linking of fibrin to form a dense aggregate in which the erythrocytes and platelets are trapped to form a clot. We have already discussed that thrombin is also used in the recycling of tenase complex, as discussed under initiation stage.

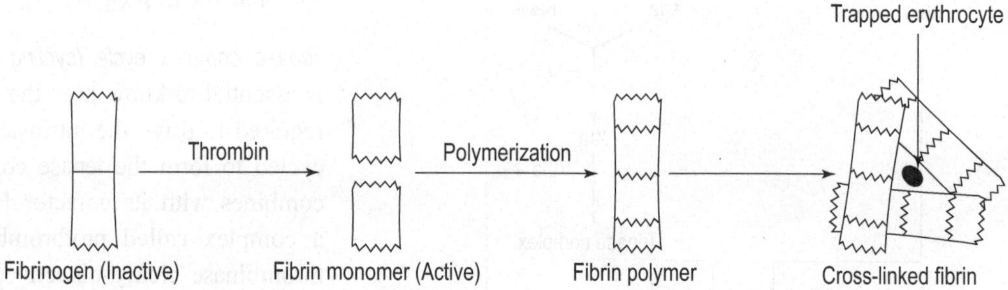

FIGURE 5.30 Formation of clot from fibrinogen

5.7 BLOOD CANCER

We have learned that blood cells are produced in the bone marrow and released into the blood and lymphatics to perform their function. Thus, bone marrow, blood, and lymphatics are the tissues of importance. Blood cancer refers to the malignancy affecting one or all of these three tissues. Blood cancer is of three types, leukaemia, lymphoma, and multiple myeloma. The symptoms of blood cancer are listed in Table 5.5.

TABLE 5.5 Symptoms of blood cancer

General symptom	• Fatigue (Bruise haematoma)	• Malaise • Recurrent fever	• Dyspnea • Recurrent infection	• Haemorrhage • Hyperhydrosis
	• Weight loss	• Nausea and vomiting	• Loss of appetite (anorexia)	• Lymphadenopathy (enlargement of lymph node)
	• Abdominal distension • Poor vision	• Abdominal pain • Oliguria	• Confusion and delirium	• Headache
Leukaemia	• Anaemia	• Recurrent infections		
Lymphoma	• Swelling in the organs	• Pain in bones and joints	• Abdominal distress	• Dyspnea
Multiple myeloma	• Bone pain	• Bleeding gums and nose bruise	• Kidney failure	• Recurrent infection
	• Weight loss			

5.7.1 Leukaemia

Leukaemia is the cancer of leucocytes, the white blood cells, caused by intense radiation exposure or due to heredity factors. In leukaemia, abnormally large numbers of leucocytes are produced in the bone marrow. Excess leucocytes accumulate in the bone marrow and also spread to liver, spleen, kidney, and to the fluids surrounding the brain. Leukaemia is of two types—acute leukaemia and chronic leukaemia. Acute leukaemia occurs suddenly and subsides for a short period. It is more common in children. Chronic leukaemia occurs suddenly and progresses very slowly. It is common in adults. If the elevated levels of leucocytes are lymphocytes, the cancer is called lymphocytic leukaemia.

5.7.2 Lymphoma

Lymphoma is the cancer of immune cells. It is the most common type of blood cancer. Abnormal multiplication of lymphocytes results in the accumulation of cells in the lymph nodes and lymphatic vessels. There are two types of lymphoma—Hodgkin's lymphoma (produced due to B-cell abnormality) and non-Hodgkin's lymphoma (produced due to abnormal B cells or T cells). The causes of lymphoma include ageing, viral infection (HIV, hepatitis, and EBV), exposure to toxic chemicals, and genetic factors. In lymphoma, lymph nodes and spleen swell causing abnormal pain.

5.7.3 Multiple myeloma

Multiple myeloma is the cancer of plasma cells (mature B cells capable of producing antibodies). The cancer cells accumulate in the bone marrow and affect the normal production of blood cells. This results in bone lesion and bone pain. They produce an unusual antibody called paraprotein, which affects the production of normal antibody, resulting in increased susceptibility to infection. Paraprotein is also known to cause kidney problems. Multiple myeloma is characterized by anaemia, hypercalcemia, and neurological symptoms.

SUMMARY

Blood is a specialized body fluid made up of cells and a liquid called plasma. White blood cells (lymphocytes and granulocytes), red blood cells, and platelets are the cells of the blood, all derived from a common cell called haematopoietic stem cell (HSC) by a process called haematopoiesis, which occurs in the bone marrow. HSCs constitute very minor portion of the bone marrow cell population (1/1000 cells), but they can still generate all types of blood cell. HSC initially produces a pluripotent progenitor, which is committed to lymphoid progenitor or myeloid progenitor. The lymphoid progenitor is committed to lymphocyte production, while the myeloid progenitor is committed to granulocyte, erythrocyte, and thrombocyte (platelet) production. The formation of red blood cells or erythrocytes is promoted by two colony-stimulating factors (CSFs): interleukin-3 (IL-3) and a hormone called erythropoietin (EPO). Myeloid progenitor produces megakaryoblasts that develop into megakaryocytes, which further differentiate into platelets or thrombocytes (by a process called thrombopoiesis).

The very first generation of myeloid progenitors is called a myeloblast. Myeloblasts that differentiate into neutrophils and monocytes are called granulocyte-monocytic-myeloblast or granulocyte-macrophage progenitor cells (GMPC). GMPCs differentiate into neutrophils and monocytes. Monocytes develop and accumulate in the tissues as macrophages. Eosinophilic and basophilic myeloblasts develop into eosinophils and basophils, respectively. A cytokine called, interleukin-5 (IL-5), or eosinophils differentiation factor, induces the differentiation of eosinophil progenitor into eosinophils. IL-5 influences the production, survival, growth, and maturation of eosinophils.

Haematopoietic growth factors produced by the stromal cells of the bone marrow induce haematopoiesis. Examples of haematopoietic growth factors include stem cell factor, interleukin, colony-stimulating factors, erythropoietin, and thrombopoietin. The primary function of RBCs is to transport oxygen by an iron-containing protein (metalloprotein) called haemoglobin (Hb). Erythrocytes are damaged or degenerated due to ageing after which they are engulfed by macrophages in spleen, liver, and lymph nodes.

Abnormal red cells result in diseases, such as elliptocytosis, spherocytosis, sickle cell anaemia, thalassaemia syndrome, and erythroblastosis foetalis. Sometimes, malignancies may occur in the bone marrow, blood, and lymphatics (the organs involved in the formation and function of blood cells). This condition is called blood cancer. Leukaemia, lymphoma, and multiple myeloma are the types of blood cancer. Artificial red cells, called respirocytes (nanorobots), are hypothetical devices which can be used as a substitute for erythrocytes. Platelets are small discoidal, anucleated blood cells with a diameter of 2–4 μm. The disease caused due to the abnormality in the number and function of platelets is called thrombocytosis. Thrombocytopathy include thrombocytosis, thrombocytopenia, thrombasthenia, and grey-platelet syndrome.

The main function of platelet is to induce blood clotting (coagulation) and maintain haemostasis. Platelets induce clotting by transforming the blood into a semisolid mass called clot. Clotting prevents blood loss during blood vessel damage. Blood coagulation is brought about by platelets in association with various proteins, collectively called clotting factors and a platelet plug.

TERMS TO REMEMBER

Alpha granules Granules containing platelet factor 4, transforming growth factor-β1, platelet-derived growth factor, fibronectin, B-thromboglobulin, von Willebrand factor, fibrinogen, and clotting factors V. They are present in platelets.

Band form The name given to young granulocytes.

Blood coagulation The process of transformation of blood into a semisolid mass called clot to prevent blood loss during blood vessel damage.

Blood A connective tissue made of cells and liquid called plasma.

Delta granules Granules of ADP, ATP, calcium, and serotonin, present in the platelets.

Elliptocytosis Disease in which the red blood cells have elliptical or oval shape rather than typical biconcave discs.

Erythroblastosis foetalis Haemolytic disease occurring due to antibody-induced haemolysis and anaemia in newborn babies.

Erythrocytes The most common blood cell with discoidal morphology that functions in oxygen transport.

Erythropoietin A glycoprotein produced by kidney and liver, responsible for erythropoiesis.

Granulocyte-monocytic myeloblast Cells that differentiate into neutrophils and monocytes.

Grey-platelet syndrome Disease caused by a deficiency of alpha granules in platelets.

Haematopoiesis The formation of blood cells from the haematopoietic stem cells in the bone marrow.

Haematopoietic growth factors Signals required for haematopoiesis to occur.

Interleukin The primary cytokine for the production of immune cells.

Lambda granules Granules of hydrolytic enzymes present in platelets.

Leukaemia Cancer of leucocytes characterized by the accumulation of leucocytes in the bone marrow and spreading to liver, spleen, kidney, and to the fluids surrounding the brain.

Lymphoid progenitor Cells produced from haematopoietic stem cells, responsible for differentiating into lymphocytes, namely B lymphocytes, T lymphocytes, and natural killer cells.

Lymphoma Cancer of immune cells where an excess accumulates in the lymph nodes and lymphatic vessels.

Megakaryocytes Giant cells which produce platelets.

Multiple myeloma Cancer of plasma cells that results in bone lesion and bone pain.

Myeloid progenitor Cells produced from haematopoietic stem cells, responsible for differentiating into erythrocytes, granulocytes, and platelets.

Osmotic fragility Test to detect the integrity of the red blood cell membrane.

Plasma A straw-yellow fluid containing amino acids, proteins, sugars, clotting factors, mineral ions, and carbon dioxide.

Reticulocytes Cells formed during erythropoiesis. They lose cellular organelles and mature into erythrocytes.

Sickle-cell disease Disease in which the red cells acquire a crescent shape or a sickle shape.

Spherocytosis Disease in which discoidal erythrocytes become spherical due to defect in the membrane proteins: spectrin or band 3 protein or both.

Stem cell factor Stem cell factor, also called steel factor, is a protein produced by endothelial cells and fibroblasts, as well as by the stromal cells of the bone marrow. It regulates the survival, self-renewal, and maintenance of HSCs and megakaryocytes.

Thrombasthenia Disease caused by a defect or decrease in the function of the platelets.

Thrombocytopenia Disease caused by a decrease in the platelet count.

Thrombocytosis Disease caused by an increase in the platelet count.

Thrombopoietin A glycoprotein that induces the formation of platelets.

REVIEW QUESTIONS

1. Erythrocytes cannot grow or divide, they can only be produced from stem cells. Discuss this statement.
2. How will you separate the components of blood?
3. Explain blood-cell lineage.
4. Tabulate the blood cell concentrations in humans.
5. What are the lineages produced from myeloid progenitor?
6. What is the role of GATA-1 in erythropoiesis?
7. Explain the stages leading to the differentiation of myeloblasts into granulocytes.
8. What is band form?
9. Frequency of haematopoiesis varies with bone tissues. Explain with the help of a diagram.
10. Define and classify haematopoietic growth factors.
11. What is a stem cell factor?
12. Write the role of interleukin in haematopoiesis.
13. Tabulate the functions of colony-stimulating factor.
14. Explain the structure and function of erythropoietin.
15. Describe the structure and function of thrombopoietin.
16. Write a note on the influence of hypotonic and hypertonic environment on red cell integrity. Comment on osmotic fragility.
17. Explain the structure of haemoglobin.
18. Haemoglobin is the functional component of erythrocytes. Explain.
19. Write the reactions to indicate the role of carbonic hydrase in gas transport.
20. List the factors that stimulate erythropoiesis.
21. How are erythrocytes degraded and metabolised?
22. Explain erythroblastosis foetalis with the help of a suitable diagram.
23. Write a note on the following:
    a. Elliptocytosis
    b. Spherocytosis
    c. Sickle-cell anaemia
    d. Thalassaemia syndrome
24. Describe the types of blood cancer.
25. Explain in detail the components of an artificial red blood cell and its applications.
26. Tabulate the factors produced by the granules of platelets.
27. Define the following:
    a. Thrombocytosis
    b. Thrombocytopenia
    c. Thrombasthenia
    d. Grey-platelet syndrome
28. Explain the structural alterations produced by blood vessel damage.
29. How is a platelet plug formed?
30. Explain the pathways for the formation of factor Xa.
31. Define tenase complex.
32. Explain the formation of clot during coagulation.

MULTIPLE-CHOICE QUESTIONS

1. Erythrocytes develop from
   (a) lymphoid progenitor
   (b) myeloid progenitor
   (c) megakaryocytes
   (d) memory cells

2. Which of the following are giant cells?
   (a) platelets
   (b) myeloblasts
   (c) megakaryoytes
   (d) lymphocytes
3. Clotting factors are
   (a) proteins
   (b) lipids
   (c) ions
   (d) polysaccharides
4. Deficiency of alpha granules in the platelet is called
   (a) alpha defect syndrome
   (b) agranular syndrome
   (c) grey-platelet syndrome
   (d) thrombocytic syndrome
5. Platelets store serotonin in
   (a) alpha granule
   (b) beta granule
   (c) gamma granule
   (d) delta granule
6. Reticulocytes are formed during
   (a) erythropoiesis
   (b) blood coagulation
   (c) thrombopoiesis
   (d) lymphopoieis
7. Young granulocytes are called
   (a) granuloblast
   (b) granulomyelocyte
   (c) band form
   (d) spherical form
8. Erythropoietin and thrombopoietin are
   (a) lipopolysaccharides
   (b) glycoprotein
   (c) asialoglycoprotein
   (d) metalloprotein
9. Signal for erythropoiesis is
   (a) reduced oxygen
   (b) reduced platelet count
   (c) reduced haemoglobin
   (d) reduced erythropoietin
10. Tenase complex converts
    (a) factor X to Xa
    (b) factor IX to IX a
    (c) thrombin to prothrombin
    (d) fibrinogen to fibrin
11. Erythropoietin has
    (a) four glycosyl residues
    (b) four disulphide bridges
    (c) two glycosyl residues
    (d) two lipid rafts
12. Which of the following is normal?
    (a) discoidal erythrocyte
    (b) spherical erythrocyte
    (c) elliptical erythrocyte
    (d) sickled erythrocyte
13. Carbamino haemoglobin transports around
    (a) 23% of carbon dioxide
    (b) 43% of carbon dioxide
    (c) 83% of carbon dioxide
    (d) 99.3% of carbon dioxide
14. Majority of oxygen is transported to peripheral tissues as
    (a) dissolved oxygen in the blood
    (b) oxyhaemoglobin
    (c) bicarbonate
    (d) carbonic acid

Answers to Multiple-choice Questions

1. (b)	2. (c)	3. (a)	4. (c)	5. (d)
6. (a)	7. (c)	8. (b)	9. (a)	10. (a)
11. (a)	12. (a)	13. (a)		

CHAPTER 6

Genetic Mechanisms in Cell

Friedrich Miescher—Swiss physician and biologist who discovered DNA.

After reading this chapter, students should be able to understand and appreciate the following:

- Prokaryotic and eukaryotic genomes
- Human genome project
- Process of replication
- Mechanism of transcription and post-transcriptional modification
- Steps and enzymes involved in translation
- Different types of post-translational modification

6.1 WHAT IS A GENOME?

From Chapter 4 it is clear that the cells of all living organisms contain DNA, the storehouse of genetic information. German biochemist Frederich Miescher first observed DNA in late 1800s. This amazing macromolecule encodes all the information needed to program a cell's activities, including reproduction, metabolism, and other specialized functions. As per the central dogma of life, DNA is converted to protein via mRNA. Each protein is encoded by a segment of DNA called gene. An organism's complete set of DNA, including its entire gene is known as its genome. Each genome contains all the information needed to build and maintain that organism. In human beings, a copy of the entire genome—more than 3 billion DNA base pairs—is contained in all cells that have a nucleus. DNA is condensed into 24 distinct molecules called chromosomes. Each chromosome is composed of about 50 million to 250 million base pairs. Within a genome, there are coding regions, which encode for proteins, and non-coding regions (also called 'junk' DNA), which do not encode proteins but do interact with other proteins and perform regulatory function. Genes comprise only about 2% of the human genome; the remainder consists

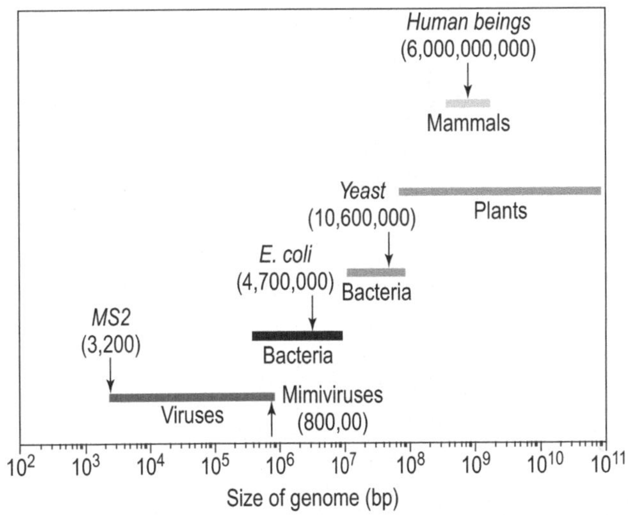

FIGURE 6.1 Comparison of genome size

of non-coding regions, whose functions may include providing chromosomal structural integrity and regulating where, when, and in what quantity proteins are made. The human genome is estimated to contain 20,000 to 25,000 genes. Each of the estimated genes in the human genome makes an average of three proteins. Comparison of genome size of different organisms is shown in Fig. 6.1.

6.2 PROKARYOTIC GENOME

Prokaryotic and eukaryotic genomes are distinguished by the amount of 'junk' DNA (non-coding DNA) contained in them. Prokaryotes tend to have very little junk DNA (typically less than 15% of the genome), whereas eukaryotes have substantial amounts of junk DNA. However, genome sequences of certain bacteria with very limited ecological niches (e.g., the human pathogen, *Mycobacterium leprae*, and the Aphid endosymbiont, *Buchnera*) indicate that, like in eukaryotes, a substantial amount of bacterial genomes can consist of junk DNA.

The human genome is composed of 23 pairs of linear chromosomes, and approximately 3,200,000,000 bases (3,200 Mb) of DNA (1 Mb = 1,000,000 b). Prokaryotic genomes are small and compact with closely packed genes and with minimal number of non-coding sequences. For example, the genome of the bacterium *Escherichia coli* consists of a single circular chromosome of 4,639,000 bases (4.6Mb) with 4,289 genes. The bacteria *Mycoplasma genitalium* possess the smallest genome of 580,070 bases (0.58Mb or 580 kbp) (1 kbp = 1,000 bp) with just 468 genes. The sizes of the various bacterial genomes sequenced so far are:

Organism	**Genome size**
Aquifex aeolicus	1.5 Mb
Archaeoglobus fulgidus	2.3 Mb
Aeropyrum pernix	0.66 Mb
Borrelia burgdorferi	1.3 Mb
Caulobacter crescentus	3.8 Mb
Chlorobium tepidum	2.1 Mb
Clostridium acetobutylicum	4.1 Mb
Dehalococcoides ethenogenes	2 Mb
Deinococcus radiodurans	3 Mb
Desulfovibrio vulgaris	1.7 Mb
Geobacter sulfurreducens	1 Mb
Halobacterium halobium plasmid	(Genome size 2 Mb, NRC100 plasmid size 190 kb)
Methanobacterium thermoautotrophicum	1.7 Mb
Methanococcus jannaschii	1.7 Mb
Methylococcus capsulatus	4.6 Mb
Mycoplasma genitalium	580 Mb
Nitrosomonas europaea	1.6 Mb
Pseudomonas putida	5 Mb
Pyrobaculum aerophilum	1.8 Mb

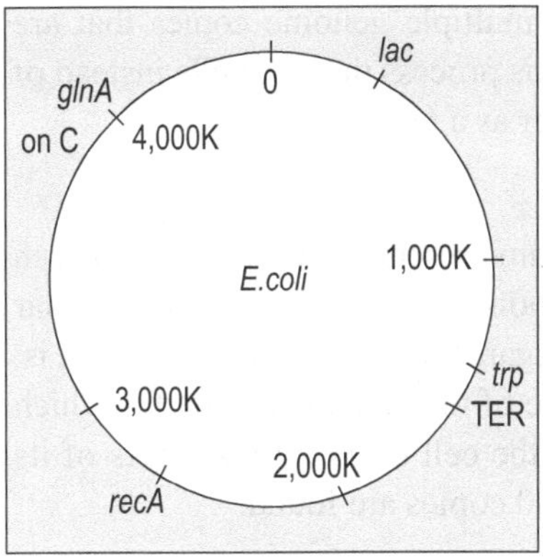

FIGURE 6.2 The genome of E.coli

Pyrococcus furiosus	2.1 Mb
Rhodopseudomonas palustris	4 to 5 Mb
Shewanella putrefaciens	4.5 Mb
Sphingomonas aromaticivorans	F199pNRC100 plasmid (genome size 4 Mb, pNL1 plasmid size 184 kb)
Thermotoga maritima	1.8 Mb
Thiobacillus ferrooxidans	2.9 Mb

6.2.1 E. coli Genome

The genome of E. coli is well characterized. In E. coli, the double-stranded DNA of 4,639,675 base pairs constituting 4,267 genes exist as a circular genome. The E. coli genome is a circular double-stranded molecule of DNA; this DNA molecule is 4,639,675 base pairs in length and encodes 4,267 genes. A schematic representation of the E. coli genome, showing the positions of replication origin (oriC), replication termination (TER), and several protein genes is given in Fig. 6.2.

E. coli O157:H7 is a major food-borne infectious pathogen that causes diarrhoea, haemorrhagic colitis, and haemolytic uremic syndrome. The genome sequence of an O157:H7 contains a chromosome, a large virulence plasmid (pO157) and a cryptic plasmid (pOSAK1). The general features of E. coli O157:H7 genome are listed in Table 6.1.

TABLE 6.1 Features of E. coli O157:H7 genome

Feature	Chromosome	pO157	pOSAK1
Length of sequence (bp)	5,498,450	92,721	3,306
G+C ratio (%)	50.5	47.6	43.4
Open reading frame (ORF)	5,361	83	3
Protein-coding region (%)	88.1	82.5	52.8
Average ORFs length (bp)	904	922	579
rRNA (16S-23S-5S)	7	0	0
tRNA and mRNA	103	0	0
Non-classical RNA	13	0	0

6.2.2 Prokaryotic Genomes of Interest

Deinococcus radiodurans

Deinococcus radiodurans, being an extremophile, can tolerate a radiation dose up to 1,500,000 rads. This radiation dose is 3,000 times more than the human fatal dose. It has a peculiar genome containing two chromosomes, (2,648,638 and 412,348 base pairs), a megaplasmid (extrachromosomal genetic material of size greater than 100 kB), and a small plasmid, with 45,704 base pairs. The factor that was thought to contribute to the radiation-resisting property of *D. radiodurans* is the multiple genome copies that are exchanged between two cells during cell division. In this process the two cells instead of separating physically, will remain attached to each other as a tetrad.

Azotobacter vinelandii

Azotobacter vinelandii is an anaerobic, obligate bacterium present in soil. It fixes nitrogen to metabolize carbohydrates, organic acids, and alcohols. It contains a single circular genome of 5,365,318 base pairs. The genome of this organism exhibits plasticity, that is, the number of genome copy varies during the life cycle of the cell, the reason for which is still not understood. During the exponential phase, the cell bears 2 to 4 copies of its chromosomes, whereas in the stationary phase 50 to 100 copies are found.

Agrobacterium tumefaciens

Agrobacterium tumefaciens is well known as a causative agent of crown gall tumour in plants. It is a Gram-negative, soil-dwelling bacterium, which transfers its own genome into plant cells. A circular chromosome, a linear chromosome, and two plasmids are the genetic components which form the genome of this organism. One of its plasmid, called the Ti plasmid, causes the crown gall tumour.

6.3 TYPES OF VIRAL GENOME

So far, we have discussed that genomes are made of DNA. However, in viruses, both DNA and RNA are packed into genomes. The viral genome ranges in size from 3,200 bp (as in hepadna virus) to nearly 1.2 million bp (as in mimivirus). The size of the largest viral genome of mimivirus is 1,181,404 bp. *Bacillus* phage G has a genome size of approximately 480,000 bp. Emiliania huxleyi virus 86 (EhV-86) is the type species of the genus *Coccolithovirus* within the family *Phycodnaviridae* which has a genome size of 407,339 bp. This is the largest *Phycodnaviridae* sequenced to date. *Paramecium bursaria Chlorella* virus NY-2A has a genome size of 369,000 b.

6.3.1 Viral Genome with Double-stranded DNA

Double-stranded viral genome is present in viruses infecting E. coli (phages), eubacteria (podoviridae), animals (pox virus, herpes virus, adenovirus, and papovaviridae), insects (baculovirus, iridovirus, and polyviridae), and algae (phycodnaviridae). The genome size ranges from 5 kb to over 1 million bases (Table 6.2.). Mimivirus is the largest and most complex virus known. This virus has a 1.2 Mbp genome, which contains 911 protein-coding genes.

6.3.2 Viral Genome with Single-stranded DNA

Single-stranded viral genome is possessed by bacteriophages (such as inoviridae and microviridae), mammalian viruses (such as circoviridae and parvoviridae), avian viruses (such as circovirus-like organisms), and plant viruses (such as geminiviridae and nanoviruses). These genomes may be linear or circular in nature. The size of various single-stranded viral genomes is shown in Table. 6.3.

TABLE 6.2 Diversity in viral genome size

Virus	Genome size (double stranded)
Papovaviridae	5–8 kb
Lambda phage and Adenoviridae	30–40 kb
Baculoviridae	200 kb
Herpes, Pox, Irido, and Phycodnaviridae	300 kb
Mimivirus	1 million bases

TABLE 6.3 Diversity in viral genome size

Virus	Genome size (single stranded)
Circoviridae	3 kb
Parvoviridae	4–5 kb
Geminiviridae	2.7–5.4 kb
Microviridae	4.5 kb
Inoviridae and nanoviruses	5–6 kb

6.3.3 Double-stranded RNA Virus

Double-stranded RNA is the genome of enveloped phages (cystoviridae, 13 kb); the animal-, plant-, and insect-infecting reoviridae (20–27 kb); the vertebrate- and invertebrate-infecting birnaviridae (6 kb); totiviridae (4–7 kb); in primitive eukaryotes (fungi and protozoa, and perhaps in insects), partitiviridae (4–9 kb) (which mainly infect fungi) and the cryptoviruses (family Partitiviridae), which occur in plants, transmissible only via seed or pollen.

The following are examples of single-stranded RNA viruses:

- viruses infecting eubacteria (leviviridae)
- viruses of insects (noda-, tetraviridae)
- viruses of fungi (barnaviridae)
- viruses of plants (bromo-, como-, poty-, sequi-, and tombusviridae, and 19 unassigned genera)
- viruses of vertebrates (astro-, calici-, corona-, flavi-, picorna-, and togaviridae, and genus Arterivirus)

The genome sizes of RNA virus lie in the following ranges:

- less than 5 kb (levi-, barna-, tombusviridae)
- 7 to 12 kb (astro-, como-, picorna-, calici-, bromo-, poty-, sequi-, toga-, flaviviridae, and tobraviruses)
- 13 kb (arteriviruses) to 15.5 kb (closteroviruses)
- 20 to 30 kb (coronaviridae)

Viruses with single-stranded RNA genome

Some viruses contain single-stranded RNA as their genetic material. Two types of RNAs are found—positive-sense RNA, which is similar to mRNA, and can be readily translated, and negative-sense RNA, which is complementary to mRNA that must be transformed into positive-sense RNA by the enzyme RNA polymerase for subsequent translation. Positive-sense RNA is present in picornaviruses (such as hepatitis A virus, hepatitis C virus, rhinovirus, poliovirus, and foot-and-mouth virus), SARS virus, yellow fever virus, rubella virus, and enteroviruses. Negative-sense single-stranded RNA genome is present in Ebola virus, Marburg virus, influenza virus, and in viruses causing measles, mumps, and rabies.

6.4 HUMAN GENOME PROJECT

The Human Genome Project (HGP, 1990–2003) refers to an international research effort to sequence the human genome and identify the genes that it contains.

6.4.1 Origin

The HGP formally began in 1990 and was completed in 2003, two years ahead of its original schedule. The yearwise accomplishment of HGP is summarized in Table.6.4. The first attempt to sequence the complete human genome (comprising 3 billion base pairs) was made in 1990 by the National Institutes of Health (NIH) and the Department of Energy (DOE). In 1990, the NIH and the DOE joined with international partners in a quest to sequence all 3 billion base pairs in the human genome, which is the complete set of DNA in the human body.

6.4.2 Goals of HGP

The main goals of the HGP were to:

- sequence (determine the exact order of the base pairs in a segment of DNA) the 3 billion DNA base pairs that constitute the human genome.
- map all of the estimated 20,000 to 25,000 human genes (gene mapping).
- sequence the genomes of several other organisms those are important to medical research such as mice and fruit flies.
- develop new tools to obtain and analyse data and to make this information widely available.
- explore the consequences of genomic research through its Ethical, Legal, and Social Implications (ELSI) program.

All these goals may reveal the causative factors for genetic diseases and help in the development of treatment method, prevention method, and diagnostic tools.

TABLE 6.4 Yearwise accomplishments of HGP

Year	Accomplishment of the HGP
1990	Human Genome Project launched in USA
1991	First genome centre established in USA
1992	Human genetic map developed
1993	New five-year plan for the HGP published in USA
1994	Human genetic mapping performed
1995	Physical mapping of human genome performed
1996	First human gene map established
1997	DOE formed joint genome institute
1998	30,000 genes incorporated into genome map
1999	Sequence of first human chromosome (chromosome 22) completed
2000	Human genome draft version completed with free access to the genome information
2001	Human genome draft version published
2002	Draft version of mouse genome sequence completed
2003	Final version human genome sequence completed

6.4.3 Protocol for Genome Mapping/sequencing

The protocol for genome mapping in HGP includes the following five steps:

(i) *Subcloning*

Chromosomes with 50 million to 250 million bases were subcloned by fragmentation into shorter bits.

(ii) *Template preparation and sequencing reaction*

Each fragment is a template to produce a different set of fragments, which vary in length from each other by a single base.

(iii) *Separation*
Gel electrophoresis and fluorescent dyes were used to separate the fragments in each set, in a single lane. The fragments in a set were separated by gel electrophoresis. New fluorescent dyes allow separation of all four fragments in a single lane on the gel.

(iv) *Base calling*
Base calling refers to the identification of the final base located at the end of each fragment. This process remakes the original sequence of As, Ts, Cs, and Gs for every fragment, produced in step (i).
After the bases were 'read', computers were used to assemble the short sequences (in blocks of about 500 bases each, called the read length) into long, continuous stretches that are analysed for errors, gene-coding regions, and other characteristics.

(v) *Data projection*
Finished sequences were submitted to major public sequence databases such as GenBank. HGP sequence data were, thus, made freely available to everyone around the world.

6.4.4 Accomplishments

After following the above protocol and sequencing the genome, the following accomplishments were made by HGP:

- HGP researchers published a working draft of the human genome in 2001.
- The project sequenced the entire human genome in 2003.
- Many human genes, their structure, and organization were identified.
- HGP designed tools to analyse the data freely available via the Internet.
- The genomes of several other organisms, including brewers' yeast, roundworm, and fruit fly were also mapped.
- A working draft of the mouse genome was completed.
- The similarities and differences between human genes and those of other organisms were identified and thereafter, the functions of particular genes were discovered. Genes critical for life were also discovered. By studying the similarities and differences between human genes and those of other organisms, researchers can discover the functions of particular genes and identify which genes are critical for life.
- HGP's ELSI program became world's largest bioethics program and a model for other ELSI programs worldwide.
- All data generated by HGP were made freely and rapidly available on the Internet, serving to accelerate the pace of medical discovery around the globe.

6.4.5 Results of HGP

The following results were obtained at the successful conclusion of HGP:

- The human genome contains 3,164.7 million bases and a total of 30,000 genes.
- An average gene consists of 3,000 bases, but sizes vary greatly, with the largest known human gene being dystrophin having 2.4 million bases.

- Almost all (99.9%) nucleotide bases are exactly the same (conserved) in all human beings.
- Over 50% of genes have unknown functions.
- About 2% of the genome is coding region (code for proteins) and about 50% of the human genome is non-coding region ('junk' DNA). They shed light on chromosome structure and dynamics. Over time, these repeats reshape the genome by rearranging it, creating entirely new genes, and modifying and reshuffling existing genes.
- Chromosome 1 has the most genes (2968), and the Y chromosome has the fewest (231).
- The genome has gene-rich 'urban centres' and gene-pool 'desert centres'. The former region is composed of G: C and the later of A: T.
- Unlike other organisms' genomes, human genome has large number of non-coding sequences and random distribution of genes.
- The ratio of germline (sperm or egg cell) mutations is 2:1 in males versus females.

The extension of HGP includes the following areas:

- Transcriptomics
- Proteomics
- Structural genomics
- Knockout studies
- Comparative genomics

6.4.6 Outcome of HGP

- The Human Genome project spurred a revolution in biotechnology innovation around the world and played a key role in making the US the global leader in the new biotechnology sector.
- As a result of HGP, researchers can today find a gene suspected of causing an inherited disease in a matter of days, rather than the years it took before the genome sequence was at hand.
- Many biotechnology-based products resulting from HGP are currently being subjected to clinical trials.
- HGP's ELSI program has become a model for other research efforts seeking to address ethical issues in a proactive manner.

6.5 COMPARATIVE GENOMICS

Comparative genomics refers to the comparison of genome sequences of different species, including yeast, mice, rat, chimpanzees, and human beings. Scientists depend on computer programs to align multiple genomes and to identify the genomic similarities; BLAST (from National Centre for Biotechnology Information) is a widely used sequence-similarity tool that is accessible via Internet. VISTA is yet another tool for comparative genome analysis; mVISTA is used to compare genomes from multiple species, while rVISTA recognizes regulatory sequence and analyses transcription-factor-binding sites.

Using comparative genomics it is possible to:

- identify genetic similarities and differences among different species

- identify the structure and function of human genes
- develop new strategies to combat human diseases after understanding the structure and function of human genes
- identify species-specific conserved genes
- identify species-specific genes responsible for conferring the unique characteristics to that species

To cite an example, comparative genomics of fruit flies and human beings suggest that both the species share 60% of genes. Moreover, researchers have discovered that human genes responsible for Parkinson's disease are conserved when inserted into fruit fly and hence the later developed the symptoms of the disease as human beings do. This discovery has led to the recognition of fruit fly as a model for studying and investigating Parkinson's disease.

6.6 EPIGENOMICS

In Greek, epigenome means 'above' the genome. The term epigenome refers to chemical compounds that tag, label, modify, or mark the genome in a way that it determines the timely tunable function of the DNA in specific type of tissues. In other words, epigenome turn the genes 'on' or 'off', playing a vital role in disease and heredity. Epigenomes, though not a part of the DNA itself, can be passed on from cell to cell during cell division, and thus, from one generation to the next. An epigenome map can be correlated with the genomic structure and genomic modification, consequently leading to the development of a better tool for diagnosis and treatment of diseases. Scientists have already revealed the relation between human epigenetic changes and cancer. Hypermethylation and consequent gene silencing are features of cancer cells. Here, zebularine, the inhibitor of DNA methylation, which complexes with the DNA methyl transferase can be used to prevent hypermethylation. It was also reported that epigenetic changes are turned on due to environmental changes; this might lead us to understand environment-related health issues.

Epigenomics is the study of alterations in the regulation of activity and expression of genes. This is usually independent of the gene sequence. Methylation of cytosine residues and methylation/acetylation/phosphorylation/sumoylation of histones are two types of modifications in DNA. Cellular differentiation is the best example for the epigenetic changes in eukaryotic cells. Many diseases, including cancer, are caused due to aberrant epigenetic changes.

6.6.1 Human Epigenome Project

The Human Epigenome Project (HEP) is a public/private collaboration run by the members of the Human Epigenome Consortium. In 2003, scientists from the United Kingdom's Wellcome Trust Sanger Institute in Cambridge and the biotechnology company Epigenomics formed the Human Epigenome Project. The primary aim of HEP is to identify and interpret the methylation pattern in the genome of all major tissues. The natural targets for methylation are cytosine bases at CpG sequences. Methylated bases regulate the gene expression. However, there are also unusual targets, referred to as methylation variable positions (MVPs), which lead to a diseased state. Therefore, MVPs are the best epigenetic markers which may enable better understanding of human diseases.

The objective of HEP involves treatment of the genomic DNA with sodium bisulphite to transform unmethylated cytosines into uracil without affecting methylated cytosines. The bisulphate-treated DNA is amplified in a subsequent PCR reaction by using bisulphate-specific primers. The PCR products are sequenced. The degree of methylation can be found after amplification and sequencing of selected amplicons from the disulphate-treated DNA. As CpG is the predominant target for DNA methylation, determination of methylation is done by comparing the corresponding signal ratios at CpG dinucleotides.

Major histocompatibility complex (MHC) is a region of chromosome 6. As this region is associated with majority of the diseases compared to other regions, methylation pattern within the MHC was studied by HEP. The preliminary results reveal the location of MVPs in the vicinity of the promoter regions of loci 150. About 135,000 CpG methylation values were obtained from the analysis of 235 amplicons across the 4 Mb MHCs on chromosome 6 in 32 samples (derived from 7 different tissues). Analysis of nearly 2,524 amplicons across chromosomes 6, 20, and 22 revealed around 1.9 million CpG methylation values.

6.7 REPLICATION OF DNA

Replication is a process by which each strand of a double-helical DNA is duplicated into a daughter strand in the presence of various enzymes. Thus, two new strands are formed after replication. The parent strand is called template. A simplified representation of replication is given in Fig. 6.3. Each parent strand acting as template is shown in solid line and the newly synthesized strand is shown in dotted line. Arrow represents 5′ to 3′ direction of replication.

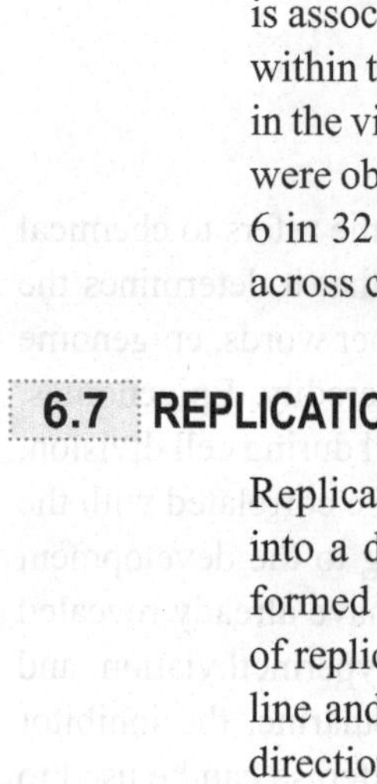

FIGURE 6.3 Schematic representation of replication of DNA double helix

There are three possible modes of replication: conservative, dispersive, and semiconservative (Fig. 6.4). In conservative replication, the original parental DNA strands (represented by solid line) remain intact and the helix of the next generation is made from entirely new strands (shown as dotted lines). In dispersive replication, each DNA strand is made of both old and new sequences interspersed or dispersed along the strand. Semiconservative replication produces a DNA helix composed of one old strand and one new strand. The Meselson–Stahl experiment proved that DNA replication in living cells follows the semiconservative scheme.

In Meselson–Stahl experiment, *Escherichia coli* were cultured for several generations in a medium containing heavy nitrogen (15N) in the form of ammonium chloride ($15NH_4Cl$). After many generations, the heavy nitrogen gets incorporated into the nitrogenous bases of the bacterial DNA making them denser than the normal DNA. The growth medium is then replaced by another medium containing light nitrogen ($_{14}N$) and the organism is allowed to grow. The cells are harvested for subsequent generations and the density is measured using caesium chloride density-gradient centrifugation. It is observed that in the first generation, the entire DNA has intermediate density; in the second generation, bacteria

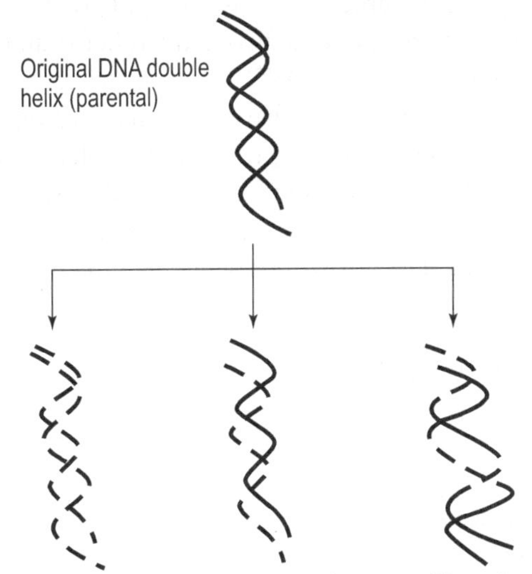

FIGURE 6.4 Three modes of replication (original strand is shown as solid line and the new strand is shown as dotted line)

possess DNA, of which, 50% have intermediate density and the remaining 50% have light density; and in the third generation, the cells have 25% DNA with intermediate density and 75% with light density. This experiment proves that living organisms follow the semiconservative model of DNA replication shown in Fig. 6.4.

6.7.1 Mechanism of Replication

The mechanism of replication is substantially best understood in prokaryotes. The location or sequence of DNA where the replication starts is called 'origin of replication', commonly called oriC. Replication always starts at oriC and does not occur at random. Prokaryotic genome has a single origin, whereas eukaryotic genome has multiple origins. For example, the human genome has 20,000 origins and yeast genome has 300 origins.

Requirements

The basic requirements for replication are listed as follows:

1. Substrate: The four deoxynucleoside triphosphates (dNTPs)—dATP, dGTP, dCTP, and dTTP.
2. Enzymes: Helicase, primase, topoisomerase/gyrase, DNA polymerase, and ligase.
3. Template strands: The two parent strands function as templates.
4. Primer: A short strand of RNA that provides free 3′-OH end to prime the replication in 5′–3′ direction is called primer.

All the enzymes and factors involved in replication processes which are listed here exist in the form of a large macromolecular complex called replisome.

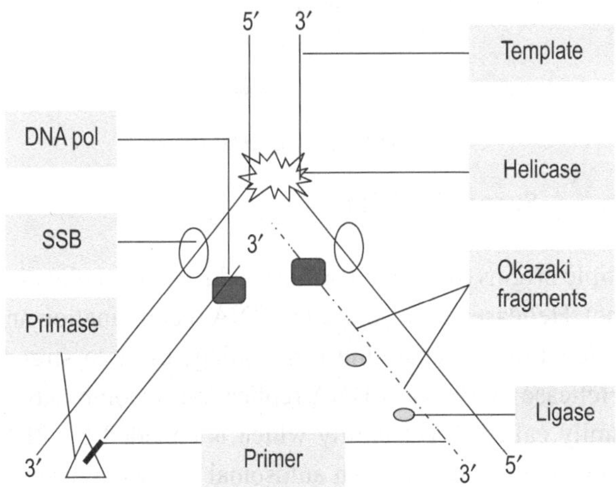

FIGURE 6.5 Mechanism of replication

The mechanism of replication along with various components involved is illustrated in Fig. 6.5, whereas the events involved in replication mechanism are summarized in Fig. 6.6. These events are grouped under three steps: initiation, elongation, and termination.

Initiation

Initiation is the first step in replication; it involves the following events:

Formation of replication fork by helicase Template strands (the original parental DNA strand) unwind to form a fork-like structure called replication fork.

The enzyme helicase is responsible for catalysing the process of unwinding. Unwinding occurs due to breaking of hydrogen bonds between bases of the two antiparallel strands. The breaking happens in places where the chains are rich in A-T. This is because there are only two bonds between adenine and thymine (whereas there are three hydrogen bonds between cytosine and guanine). Helicase unwinds DNA by translocating along one strand in the 5′–3′ direction while simultaneously separating the second strand as shown in Fig. 6.7. In the figure, helicase (represented as an explosion) translocates along one strand simultaneously separating the other strand. The arrow represents the direction of translocation of helicase.

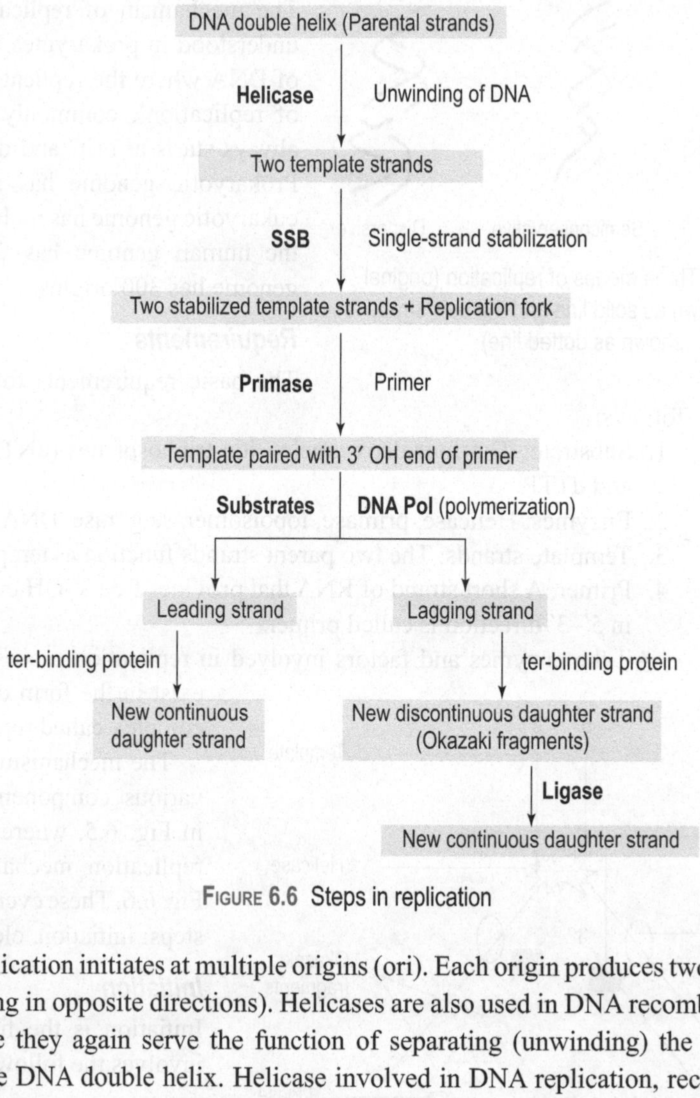

FIGURE 6.6 Steps in replication

DNA replication initiates at multiple origins (ori). Each origin produces two replication forks (moving in opposite directions). Helicases are also used in DNA recombination and repair where they again serve the function of separating (unwinding) the two strands that form the DNA double helix. Helicase involved in DNA replication, recombination, and repair belongs to a protein family called RecQ family which is encoded by BLM gene (BLM is the causative gene for bloom syndrome, an autosomal recessive disorder characterized by short stature and facial rashes). There are five known human RecQ helicases represented as RECQL, WRN, BLM, RECQL4, and RECQL5. Helicase can catalyse any one of the following two reactions to drive unwinding:

(i) $ATP + H_2O \rightarrow ADP + Pi$
(ii) $NTP + H_2O \rightarrow NDP + Pi$

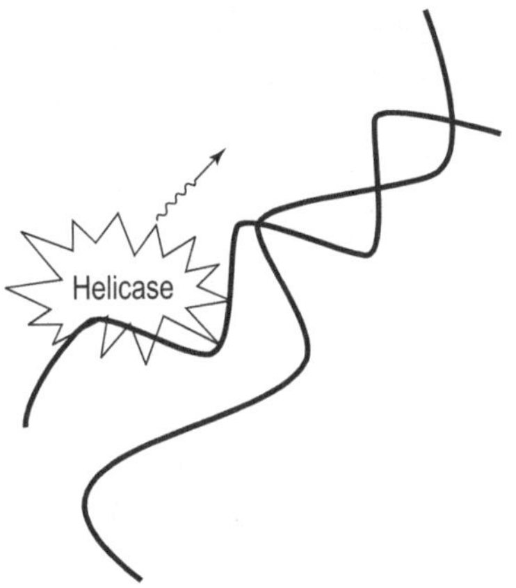

FIGURE 6.7 Unwinding of DNA helix by helicase

Stabilization of template strand by single-strand-binding protein Single-stranded templates generated by helicase at the replication fork are stabilized by a protein called single-strand-stabilizing protein (SSB). Unpaired polynucleotide templates generated by helicase activity are sticky and are capable of reunion. They are also susceptible to degradation by nuclease enzymes. Thus, each of the single strands of the helix should be stabilized to prevent reunion and nuclease attack. SSB binds to each side of the strand forming a channel-like structure and stabilizes the strand (Fig. 6.8). SSB binds to either side of the strand forming a channel in which the strand is trapped.

Eukaryotic SSB, called RPA (re-replication protein A), is a tetramer whose beta-pleated structures are involved in the formation of channels to hold the strand.

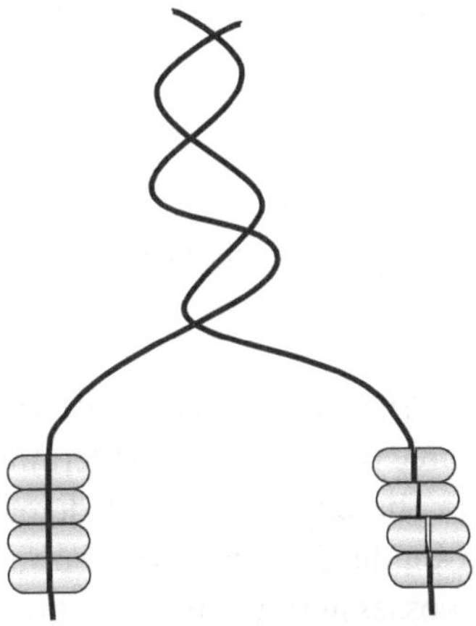

FIGURE 6.8 Single-strand-binding protein (shaded) stabilising the unwound template of the DNA helix

Topoisomerase Topoisomerase enzyme is responsible for catalysing the interconversion of DNA strand topologies forming a covalently closed strand of different topology without a change in the chemical structure. The enzyme has a tyrosine residue at its active site by means of which it binds to one of the DNA strands to induce a transient cut or nick. After changing the topology, the nick is religated (resealed). Studies on bacterial cells reveal that topoisomerase catalyses three types of topological changes:

(i) Supercoiling ↔ Relaxation
(ii) Knotting ↔ Unknotting
(iii) Catenation ↔ Decatenation

Three types of DNA topologies are shown in Fig. 6.9. Topoisomerases are of two types: Type I enzyme that induces single-strand break, and type II that induces double-strand break at the expense of ATP. Type I enzyme has two subfamilies: Type IA and Type IB. Type IA covalently binds to the 5′ end of the DNA break by phosphotyrosine bond, whereas type IB forms covalent linkages with the 3′ end of the DNA break. Types of topoisomerases (TIA, type IA; TIB, type IB; and TII, type II) are shown in Fig. 6.10. TIA binds to single strand at the 3′ end of the break, TIB binds to single strand at the 5′ end of the break, and TII binds to the double strands. DNA topoisomerase I (E. coli and eukaryotic) and DNA topoisomerase III (E. coli and eukaryotic) are members of type I family, whereas topoisomerase IV and DNA gyrase are members of type II family.

Assembly of primer by primase Unlike RNA polymerase, no DNA polymerase can initiate DNA synthesis de novo. Therefore, a short stretch of RNA called the primer which can provide 3′ OH is required to prime the polymerization reaction. Primase is the enzyme

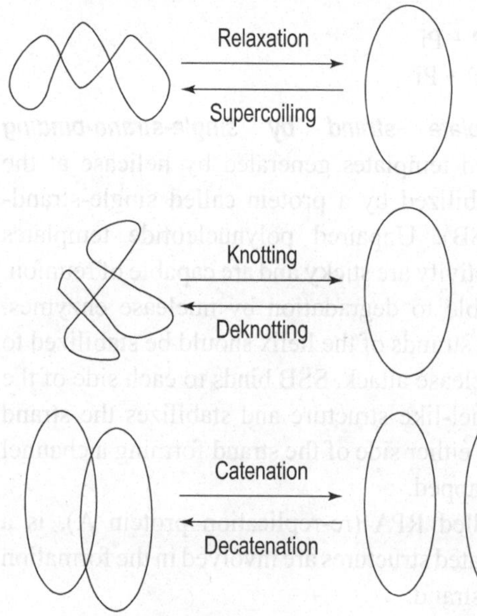

FIGURE 6.9 Types of DNA topological changes catalysed by topoisomerase

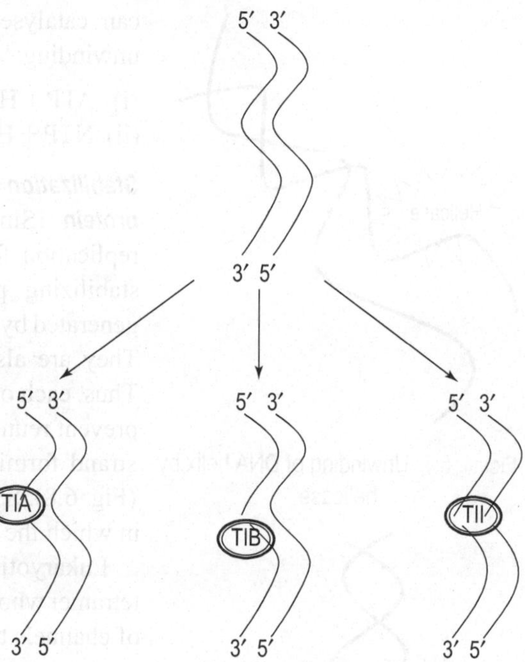

FIGURE 6.10 Types of topoisomerases

responsible for the synthesis of primer in vivo (Fig. 6.11). Primase is actually part of an aggregate of proteins called the primesome and contains three domains—amino terminal domain, carboxyl terminal domain, and a core domain. The amino terminal domain contains 110 amino acid residues and a zinc finger DNA-binding domain. The carboxy terminal domain is formed by 151 amino acid residues, possessing the binding site for DnaB helicase. The core domain is made of 322 residues. It is the catalytic domain. RNA primase can attract RNA nucleotides which bind to the DNA nucleotides of the 3′–5′ strand due to the hydrogen bonds between the bases. Primase is strongly connected with the DnaB helicase so that it is placed to make RNA primers on the ssDNA of the lagging strand (one of the daughter strands, which is discontinuous). It is not clear exactly how primase is oriented with respect to DnaB.

Elongation

Elongation involves the following steps: (i) actual polymerization reflected by the elongation of the primer in 5′ to 3′ direction, (ii) removal of primers, and (iii) sealing of nicks.

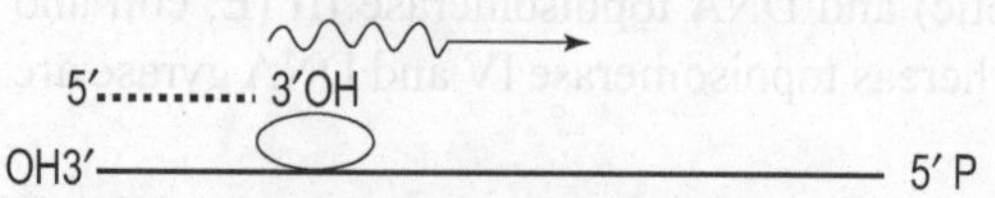

FIGURE 6.11 Priming of replication of DNA (solid line) by primase (oval) using primer (dotted line) (arrow represents the 5′–3′ direction of replication)

Polymerization of primer by DNA polymerase Complementary substrates are added to the primers and polymerization takes place by the enzyme DNA polymerase (DNA Pol) via phosphodiester bond formation. The replication of DNA is guided by the base pairing principle that A pairs with T and G with C. One of the primer parallel to the template strand which is in 5′ to 3′ direction is polymerized continuously forming leading strand. The other primer, parallel to the template strand which is in 3′ to 5′ direction, is polymerized discontinuously to produce many short bits or fragments called Okazaki fragments (named after their discoverer, Reiji Okazaki). Okazaki fragments occur as a result of the fact that the direction of replication in the lagging strand is opposite to the direction in which the replication fork is moving. Therefore, in the lagging strand, DNA replication can only occur in small sections of nucleotides at a time. These fragments must be ligated by DNA before the DNA can be functional.

Removal of RNA primers in lagging strand by DNA Pol I—exonuclease activity Each fragment in the lagging strand remains linked with the RNA primer. The primer is removed by DNA Pol I—exonuclease activity. The gaps produced by the primer removal are closed by the DNA polymerase activity.

Ligation of Okazaki fragments by ligase DNA ligase ligates the Okazaki fragments together by bonding their 3′-hydroxyl and 5′-phosphate groups via phosphodiester bond. Two types of ligases are known. One is ATP-dependent, and the other is NAD+ dependent. ATP-dependent ligases are present in eukaryotes, viruses and archae, whereas the NAD-dependent ligases are predominant in prokaryotes. Ligase has two domains: the larger N-terminal domain and the smaller C-terminal domain. N-terminal domain is the site of binding for ATP hence the name adenylation domain. N-terminal domain is responsible for recognizing double-stranded DNA, and hence, the name recognition domain. Larger ligases, such as DNA ligase II in human cells, have an additional subunit called zinc finger motif that has been found to form a complex with nicked DNA or to provide a structural role by supporting other subdomains.

Mechanism of action of ligase is shown Fig. 6.12. Ligating 3′-OH and 5′P ends of Okazaki fragments involves the following steps:

1. ATP and enzyme (E) reacts forming an enzyme-AMP complex (E-AMP) and pyrophosphate (PPi) .
2. AMP molecule is next transferred to the free 5′ phosphate group of the nicked DNA (5′-P –DNA).
3. E catalyses the attack by the 3′ hydroxyl group (DNA-3′–OH) on the 5′ phosphate group, losing the AMP in the process.

The same mechanism works in the case of NAD-dependent ligase also.

ATP-dependent ligase

E + ATP → E-AMP + PPi

E-AMP + 5′P–DNA → AMP–P5′–DNA + E

DNA–3′–OH + AMP–5′P–DNA → DNA–P–DNA + AMP

NAD-dependent ligase

E + NAD^+ → E-AMP + NMN^+

E–AMP + 5′P–DNA → AMP–P5′–DNA + E

DNA–3′–OH + AMP–5′P–DNA → DNA–P–DNA + AMP

FIGURE 6.12 Mechanism of ligase action

By the end of elongation, a new double helix is formed. This consists of one old strand (template) and one new strand. This process has already been discussed in this chapter; it is called semiconservative replication.

Termination

Termination is the last step of DNA replication. Polymerization (replication) continues until termination sequences of template are encountered. At this sequence, a specific protein called ter-binding protein binds and prevents helicase from further unwinding DNA. This results in termination of replication.

In the last part of the lagging strand, when the RNA primer is removed, it is not possible for the DNA polymerase to seal the gap (because there is no primer). So, the end of the parental strand where the last primer binds is not replicated. This would reduce the length of the lagging strand and the chromosome would get shorter at each replication. This problem can be circumvented by telomers. Telomers are repeats of a specific sequence of six nucleotides and are normally present at the ends of chromosomal DNA. Telomerase is a eukaryotic enzyme (not found in prokaryotes) which has an RNA component and a protein component. The RNA strand of the telomerase binds to the template end of telomeric DNA and the overhang protein subunit adds DNA nucleotides to the template, polymerizing it beyond its normal length. This leads to sufficient extension of the template end allowing DNA polymerase to complete synthesis of a normal-length lagging strand.

The final part of termination is proofreading and repair. The 3′ to 5′ exonuclease activity (proofreading activity) of DNA pol removes the wrong nucleotides (mismatched bases) and the DNA polymerase fills the gaps with the correct nucleotides.

6.7.2 Comparison of Prokaryotic and Eukaryotic Replication

Although the mechanism of replication in eukaryotes is fundamentally similar to that of eukaryotes, there are a few differences, as summarized in Table 6.5. Prokaryotic DNA has a single origin, whereas eukaryotic DNA has multiple origins of replication. Replication in eukaryotic DNA proceeds bidirectionally from each origin until it meets the replication fork from the adjacent origin. It requires five polymerases, termed alpha, beta, gamma, delta, and epsilon. Alpha polymerase is responsible for lagging-strand synthesis, and delta polymerase catalyses leading-strand synthesis, both in nucleus. The alpha and delta polymerases function in proofreading and correction as well. The beta and epsilon polymerases perform nuclear DNA repair. Gamma polymerase catalyses the replication of mitochondrial DNA. It lacks proofreading and error-correction activities leading to higher mutation rate in mitochondrial DNA than the nuclear DNA. Eukaryotes have two ligases, one involved in replication and the other involved in DNA repair. Table 6.6 summarizes the function of various enzymes and proteins involved in replication.

6.7.3 Replication Rate

The speed of DNA replication varies between prokaryotes and eukaryotes. In prokaryotes, the replication rate is 1000 nucleotides/s, which contributes to higher error rate. In eukaryotes, the replication rate is slower with the speed of just 50 nucleotides/s/replication fork.

TABLE 6.5 Comparison of prokaryotic and eukaryotic replication

Prokaryotes	Eukaryotes
Possess single origin of replication	Possess multiple origins
Replication requires DNA Pol I, II, and III	Replication requires DNA pol alpha, beta, gamma, delta, and epsilon
DNA pol has only exonuclease activity	DNA pol has both endonuclease and exonuclease activities
Have only one type of ligase, which is involved in replication	Have two types of ligases: one involved in replication and the other involved in DNA repair
Require NAD for ligase activity	Require ATP for ligase activity

TABLE 6.6 Enzymes involved in replication

Name of the enzyme	Function
Topoisomerase	Relaxes the supercoils in the helical structure
Helicase	Unwinds parent DNA helix to form replication fork
SSB	Stabilizes the unwound DNA strand
Primase	Brings the primer to the site of replication to prime DNA synthesis
DNA pol	Polymerizes the substrates against the template strand via phosphodiester bond formation to form new strand
Ter-binding protein	Binds and prevents helicase from further unwinding DNA resulting in termination of replication
Ligase	Seals the gaps found in lagging strands (between Okazaki fragments)

6.7.4 Proofreading of DNA Replication

DNA replication cannot occur with 100% accuracy. Error in DNA replication can occur due to alteration in base-pairing specificity. Base-pairing specificity has an error rate of 1 in 10,000 base pairs. Whenever a new base comes into the strand, DNA polymerase proofreads the DNA for mismatches. If mismatches are found, the enzyme immediately inhibits the process of further addition of bases, removes the mismatched bases, and incorporates the correct base finally allowing the polymerization to resume. This is called $3' \rightarrow 5'$ exonuclease activity or proofreading activity of the DNA polymerases as this activity enables the enzyme to remove nucleotides from the $3'$ end of the strand that it has just synthesized. In addition, DNA strands are also exposed to environmental hazards, such as UV rays, X-rays, carcinogens, and mutagens. In such cases, the damaged or altered bases are removed by nucleases, the gaps are filled by polymerases, and the nick is sealed by ligases. Thus, polymerase, nuclease, and ligase participate in DNA repair activity also.

6.8 TRANSCRIPTION

Transcription refers to the enzyme-catalysed synthesis of RNA from DNA template in $5' \rightarrow 3'$ direction. The new RNA formed from DNA is called nascent RNA or primary transcript.

6.8.1 Requirements

The following requirements must be fulfilled for DNA transcription to occur:

1. Substrates—The four ribonucleoside triphosphates (rNTPs)—ATP, GTP, CTP, and UTP.
2. Template—Single DNA strand. One of the strands of DNA is used as a template; it is called coding strand, while the opposite strand is called non-coding strand or non-template strand. Thus, the new RNA formed in the transcription will have base pairs complementary to the coding strand and similar to the non-coding strand.

   Example for coding strand (non-template), template strand, and RNA transcript are shown in Fig. 6.13. Template strand is responsible for RNA synthesis. Template and RNA are antiparallel.

(5′) CGCTATAGCGTT(3′) DNA non-template (coding) strand
(3′) GCGATATCGCAA (5′) DNA template strand
(5′) CGCUAUAGCGUU (3′) RNA transcript

FIGURE 6.13 The two strands of DNA, namely coding strand (non-template) and template strand

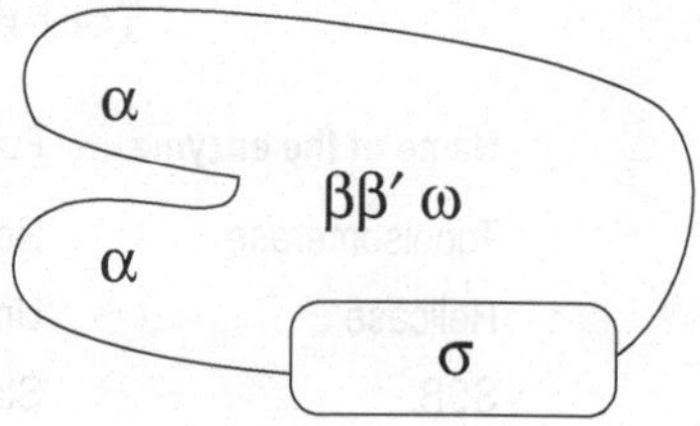

FIGURE 6.14 Structural subunits (α, β, σ, and ω) of E. coli RNA polymerase

3. Enzyme: RNA polymerase (RNA pol, also called DNA-dependent RNA pol). RNA pol of E. coli is well characterized. It is a protein of molecular weight 448 kDa with five subunits. The subunits are designated as α, β, β′, σ, and ω (Fig. 6.14). Functions of these subunits in transcription process are listed in Table 6.7. Subunits β, β', α2 form the core enzyme, whereas the subunit σ forms the holoenzyme. Eukaryotic RNA polymerases have three classes: I, II, and III, and they are compared in Table 6.8. Class I is found in nucleolus, whereas classes II and III are found in nucleoplasm. Class I enzyme catalyses the translation of rRNA, and the enzymes of classes II and III produce mRNA and tRNA.

TABLE 6.7 Subunits of RNA polymerase involved in transcription process

Enzyme subunit	Number	Gene encoding the subunit	Function
α	2 (a2)	Rpo A	Binding to the template DNA at the promoter sequence
β	1	Rpo B	Binding to the incoming substrate ribonucleotide
β′	1	Rpo C	Binding to the template DNA
σ	1	Rpo D	Initiation of transcription
ω	1	Rpo Z	Regulation of transcription

TABLE 6.8 Three classes of eukaryotic RNA polymerase involved in transcription process

Class	Nuclear location	Transcription product	Target template DNA
I	Nucleolus	rRNA	Nicked or gapped
II	Nucleoplasm	premRNA, snRNA	Nicked or gapped
III	Nucleoplasm	tRNA, 5S RNA	Duplex

4. Promoter sequence: Specific conserved sequence that promotes accurate initiation of transcription.
   There are three types of promoter sequences:
   (i) TATAAT box (Pribnow box)
   (ii) TATA box
   (iii) CAAT box
5. Transcription factors: These are low-molecular weight proteins involved in transcription process. These are of two types:
   (i) Initiation factor—σ factor is an initiation factor which directs RNA pol to recognize promoter sequence and initiate transcription. TFIID, TFIIE, TFIIB, and TFIIA are initiation factors in eukaryotes.
   (ii) Termination factor—Rho factor is a termination factor which helps in terminating the transcription process.

6.8.2 Mechanism of Transcription

The mechanism of transcription is illustrated in Fig. 6.15.

- DNA helix unwinds near the promoter sequence and two strands are separated.
- RNA pol binds to the promoter region of the DNA wrapping both the strands. The enzyme moves forward, and as it reaches the coding region it reads the single strand for transcription.
- Substrates complementary to the template bases are added and polymerized in 5′ to 3′ direction to form new RNA. The newly synthesized RNA strand breaks free from the DNA template and the DNA folds back into double helix. This reaction involves liberation of PPi. No primer is required for RNA synthesis.
- New RNA is called nascent RNA or primary transcript. The new RNA formed in the transcription will have base pairs complementary to coding strand and similar to non-coding strand. DNA template and the new RNA strand are antiparallel.
- Newly synthesized RNA extends towards unwinding point.
- The complex formed by DNA, RNA pol, and new RNA is called transcription bubble.
- As transcription continues, the transcription bubble moves along the DNA at a rate of 20–50 nucleotides/s.
- Transcription is terminated when the termination sequence is encountered by the bubble. At this termination site, the mRNA folds on itself. This prevents further continuation of transcription. RNA pol and the folded mRNA fall off the DNA resulting in termination.

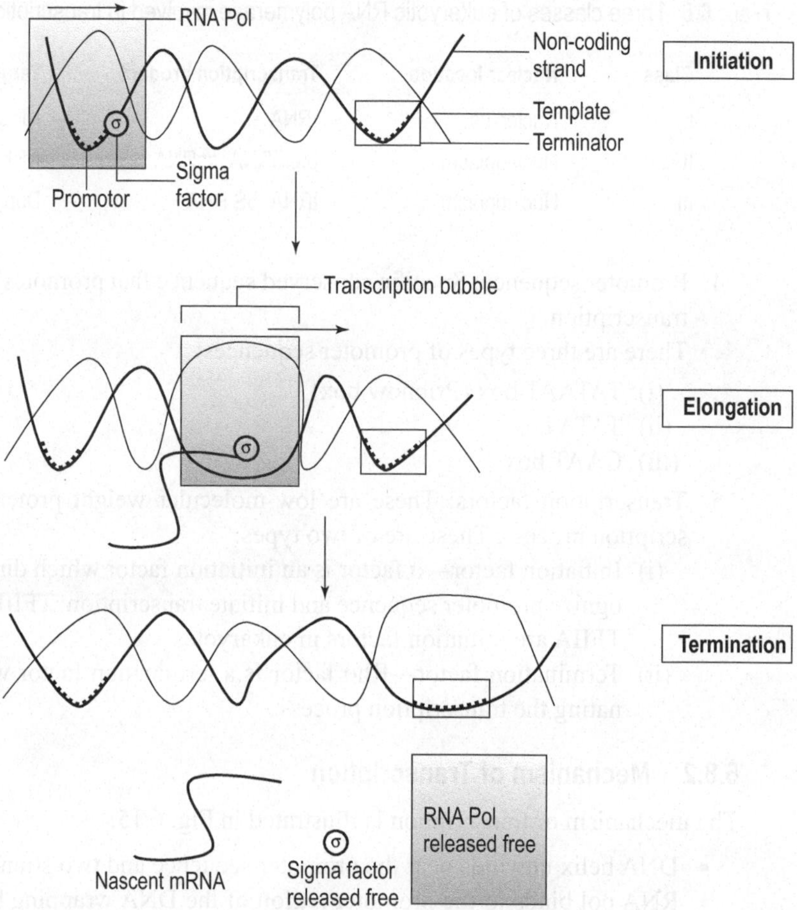

FIGURE 6.15 Mechanism of transcription involving three steps: initiation, elongation, and termination (arrow indicates movement of RNA polymerase)

RNAs produced by transcription fall into three categories (Fig. 6.16) which have their own function (as mentioned in Table 6.9):

- Messenger RNA (mRNA)
- Ribosomal RNA (rRNA)
- Transfer RNA (tRNA)

TABLE 6.9 Functions of different types of RNA

RNA type	Function
mRNA	Carrier of genetic information to be translated into proteins, functions as template for translation
rRNA	Forms a part of ribosomes to construct protein-synthesizing machinery
tRNA	Carries amino acids to the protein-synthesizing machinery

6.8.3 Post-transcriptional Modification of RNA

Post-transcriptional modification refers to several structural changes that RNA undergoes to become functional. It is also called post-transcriptional processing or RNA editing. Processing may occur in nucleoside or nucleotide of coding or non-coding sequences of an RNA. The modification can simply be represented as follows:

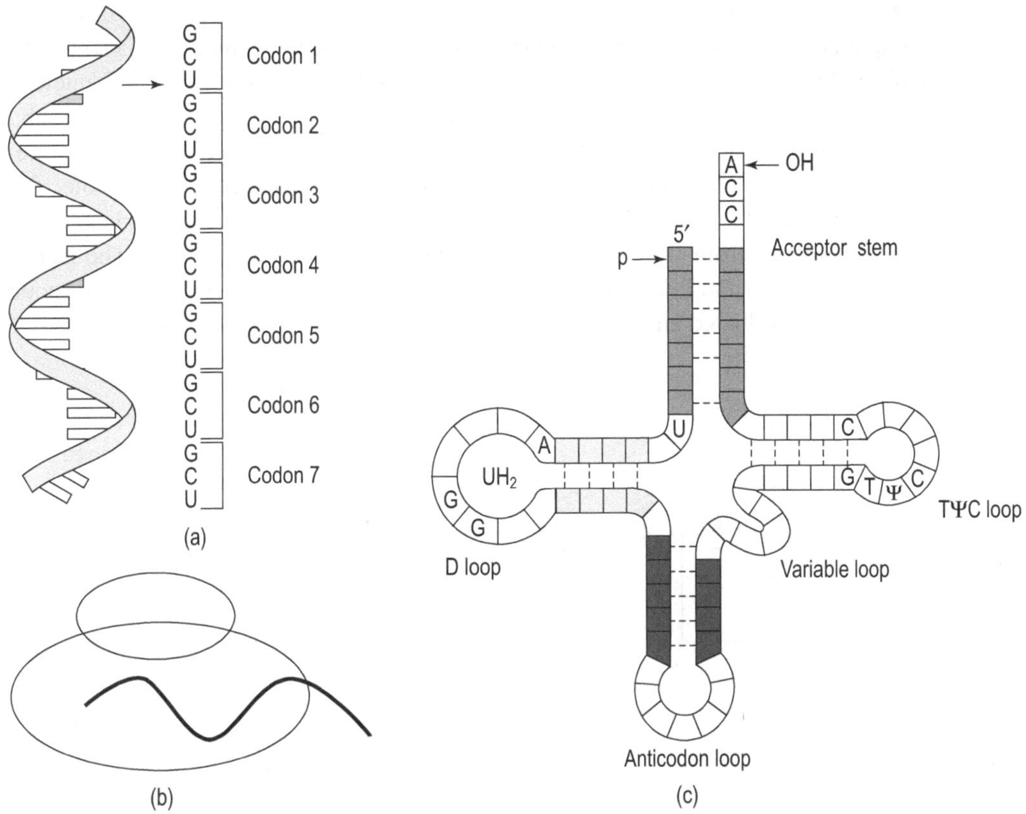

FIGURE 6.16 Structure of different types of RNAs: (a) mRNA, (b) rRNA, (c) t RNA

$$\text{Nascent RNA} \xrightarrow[\text{Post-transcriptional modification}]{\text{Structural changes}} \text{Functional RNA}$$

TABLE 6.10 Types of post-transcriptional modification

Prokaryotic processes	Eukaryotic processes
Trimming of tRNA	5′ capping
Removal of spacers	3′ polyadenylation
Addition of CCA	Splicing

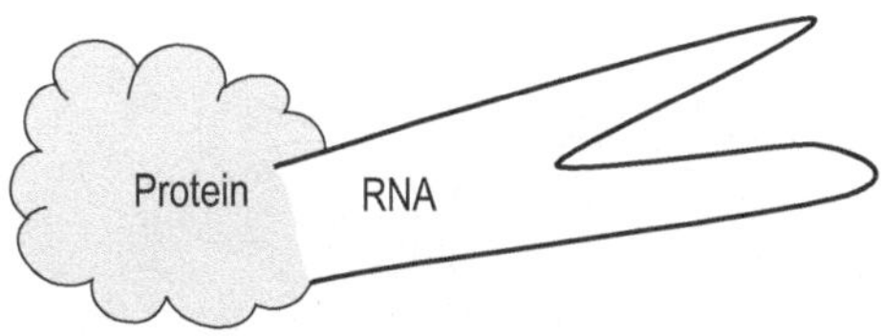

FIGURE 6.17 Schematic of the protein subunit and RNA component of the RNAse P involved in trimming of 5′ end of tRNA during post-transcriptional modification

Methods for post-transcriptional modification differ between prokaryotic RNAs and eukaryotic RNAs. Removal of spacers, trimming of tRNA, and addition of CCA sequence are common types of post-transcriptional modification in prokaryotes. On the other hand, 5′ capping, 3′ polyadenylation, and splicing are three types of post-transcriptional modification in eukaryotes. Different types of post-transcriptional modification are summarized in Table 6.10.

Trimming

Newly synthesized tRNA has a non-functional sequence at its 5′ end which should be trimmed. Ribonuclease P (RNAse P) is involved in trimming the terminal sequence in order to form an active functional tRNA. RNAse P, discovered by Sidney Altman, is a ribozyme made of a protein component and an RNA component (Fig. 6.17). The protein component is a polypeptide with molecular weight of 14 kDa and the RNA component is made of 377 nucleotides. RNAse P has one protein subunit, whereas the eukaryotic enzyme has eight subunits. Evidences suggest that the protein subunit is positioned at the site of the cleavage near the 5′ end of the tRNA. The RNA interacts with the TΨC loop of the tRNA molecule. The RNA component possesses the catalytic activity and is capable of snipping or trimming the head region from the inactive precursor tRNA, rendering it active. The process of trimming is shown in Fig. 6.18. Target sequence of the enzyme (non-functional sequence) at the 5′ end is shaded and the cleavage site is shown by a dotted arrow.

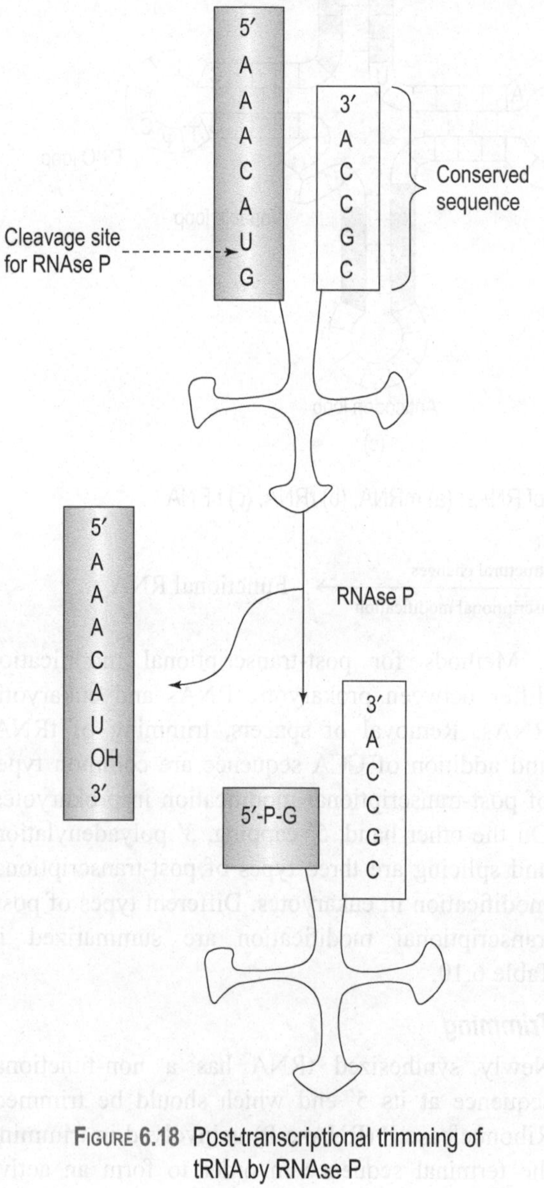

FIGURE 6.18 Post-transcriptional trimming of tRNA by RNAse P

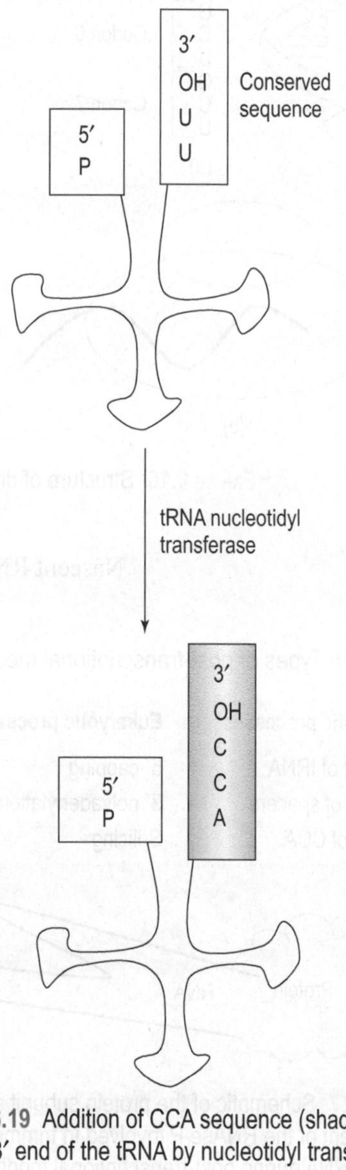

FIGURE 6.19 Addition of CCA sequence (shaded box) to the 3′ end of the tRNA by nucleotidyl transferase

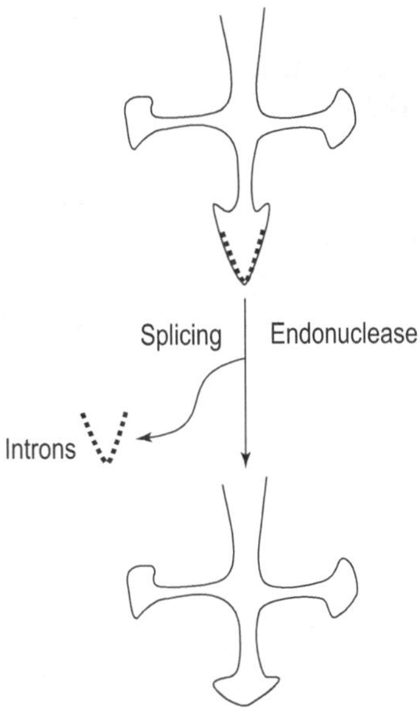

FIGURE 6.20 Removal of introns (shown as dark squares) from the anticodon loop of tRNA by endonuclease

Addition of CCA to tRNA

CCA sequence (cytosine, cytosine, and adenine) is a terminal sequence found in the 3′ end of the tRNA. This sequence binds to the RNA component of the larger ribosomal subunit and functions as a peptide acceptor at the time of peptide-bond synthesis. Some of the tRNAs do not possess the CCA sequence at their 3′ end which is common to all tRNAs. Addition of CCA sequence to the 3′ end is another mode of post-transcriptional modification (Fig. 6.19). The tRNA nucleotidyl transferase catalyses the addition of CCA making tRNA functional.

Splicing of tRNA

In the nascent tRNA, anticodon is located on the anticodon loop in association with introns. Introns are removed by endonuclease and the gap is linked by ligase (as shown in Fig. 6.20). As a result, the anticodon is positioned in a loop-like structure forming the anticodon loop.

Removal of spacers

Spacer sequence indicates non-functional sequences. rRNA is processed by the removal of spacer sequences. RNAse P and ribonuclease III (RNAse III) are involved in post-transcriptional modification of RNA in prokaryotes in order to render them functional. RNAse P is an endonuclease. It has a protein component and an RNA component. The RNA component is active even in the absence of protein component, hence, RNAse P is a ribozyme. RNAse III is an endonuclease for processing of the precursor RNA. These two enzymes remove the intervening spacer sequence from the nascent prokaryotic ribosomal RNA-30S precursor rRNA to make it functional. This process of spacer removal is shown in Fig. 6.21. The 30S precursor rRNA is composed of the following components, in order from 5′ to 3′ direction.

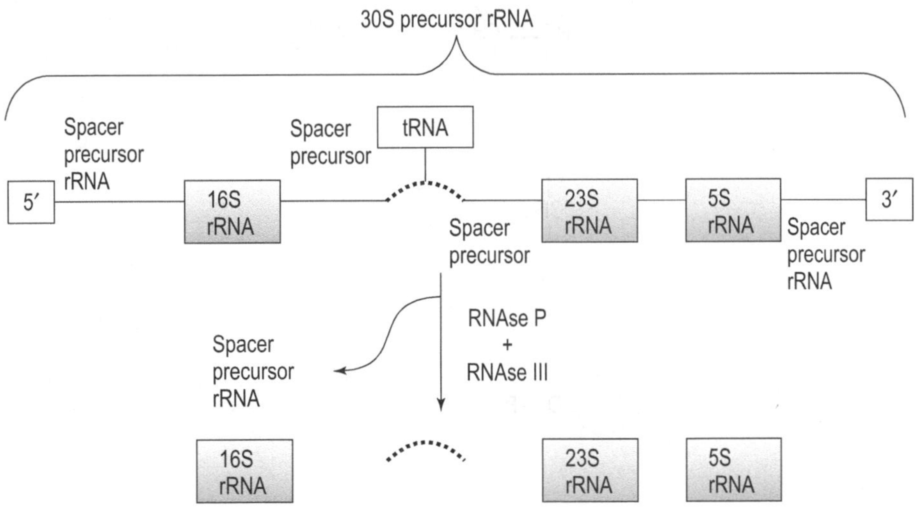

FIGURE 6.21 Removal of spacers in RNA by RNAse P and RNAse II

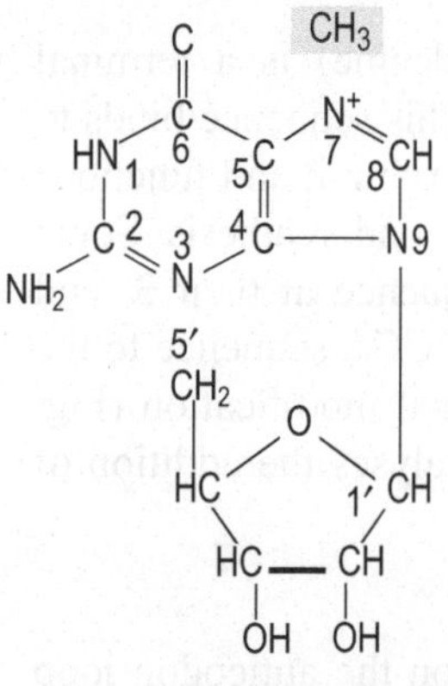

FIGURE 6.22 Structure of the 7-methylguanylate cap

- 16S rRNA
- Spacer sequence
- tRNA
- Spacer sequence
- 23S rRNA
- Spacer sequence
- 5S rRNA

The spacer sequences flanking both ends of tRNA and those lying between 23S and 5S rRNA are cleaved by RNAse P and RNAse III. This process of spacer removal results in the release of 16S RNA, tRNA, 23SRNA, and 5SRNA. Similarly, the 45S precursor rRNA of eukaryotes is composed of following sequences in order from 5′ to 3′ direction:

- 18S RNA
- Spacer sequence
- 5.8S RNA
- Spacer sequence
- 28S RNA

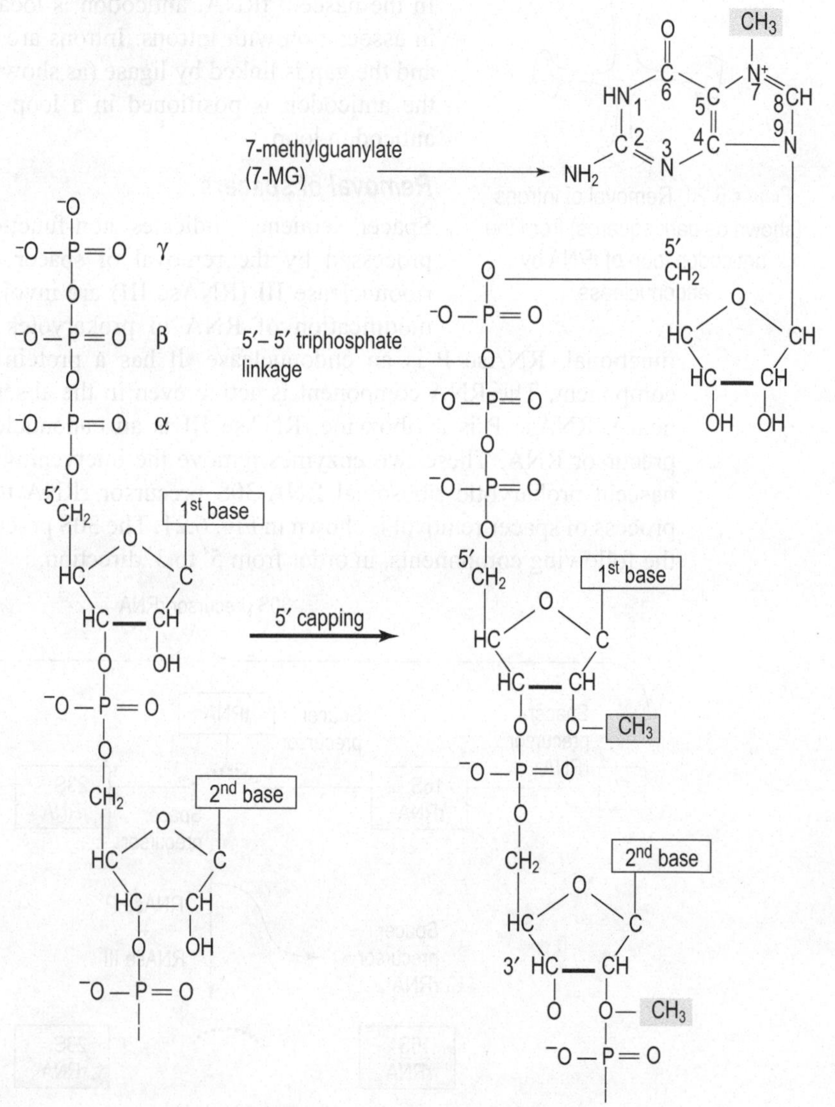

FIGURE 6.23 Capping of 5′ end of a RNA nucleotide by 7-methylguanylate

The spacer sequences are removed by endonucleolytic cleavage by endonucleases.

5′ capping

5′ capping is the common method of post-transcriptional modification in eukaryotes. Capping occurs soon after the beginning of the transcription process. The cap is represented chemically as 7-methylguanylate (7-MG) (Fig. 6.22) which is added to the 5′ end of mRNAs via 5′ to 5′ triphosphate linkage (Fig. 6.23). 5′ capping stabilizes mRNAs and results in more efficient translation.

3′ polyadenylation

Poly-A tail (polyadenylate tail) is a polymer of 200 to 300 adenylate residues (–AAAAAA–). It is linked via phosphodiester bond to the 3′ end of the RNA by the enzyme polyadenylyl transferase. This process results in stabilization of mRNA. There are two major signals for locating the poly-A adenylation site. They are AAUAAA sequence and GU- or U-rich sequence. The site is located in between the two regions: 10–35 nucleotides upstream from AAUAAA region and 50 nucleotides downstream from GU- or U-rich region. The process of polyadenylation is depicted in Fig. 6.24. Polyadenylation site located between two signals is shown as a dark rectangle. The site is cleaved and subsequently the tail is added.

Splicing

An mRNA has two types of sequences: coding sequences (also called exons), which are expressed or translated into proteins, and non-coding sequences (also called introns or intervening sequence), which are not expressed or converted to proteins. The process of removing the non-coding sequences after transcription and before translation is called splicing. Introns are characterized by the presence of guanine–uracil (GU) sequence on the 5′ border and adenine–guanine (AG) sequence on the 3′ border of the intron:exon junction. Splicing is catalysed by a large ribonucleoprotein complex called spliceosome. Spliceosome recognizes the introns by GU and AG borders and removes them, resulting in the formation of a completely expressible mRNA for consequent translation. The process of splicing is shown in Fig. 6.25.

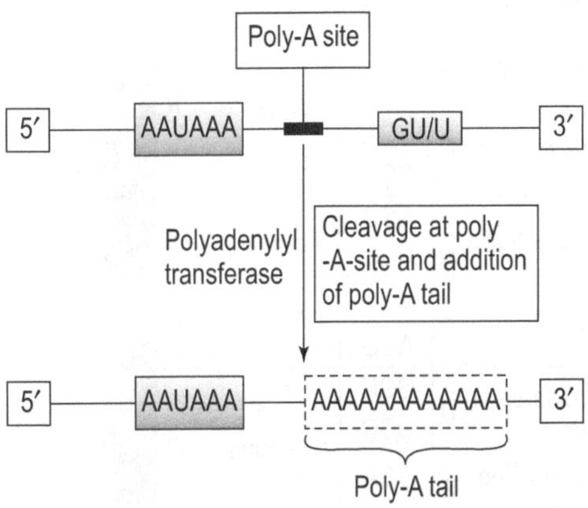

FIGURE 6.24 Process of polyadenylation

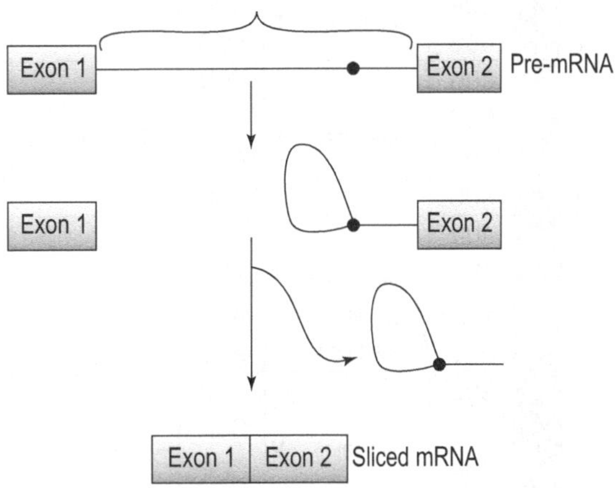

FIGURE 6.25 Process of splicing

6.9 TRANSLATION

Translation is a process by which the information encoded in the mRNA is converted into proteins by enzymes, ribosomes, and tRNA. Before understanding the process of translation, it is necessary to know about codons and anticodons (Fig. 6.26) which are specific nucleotide sequences involved in translation.

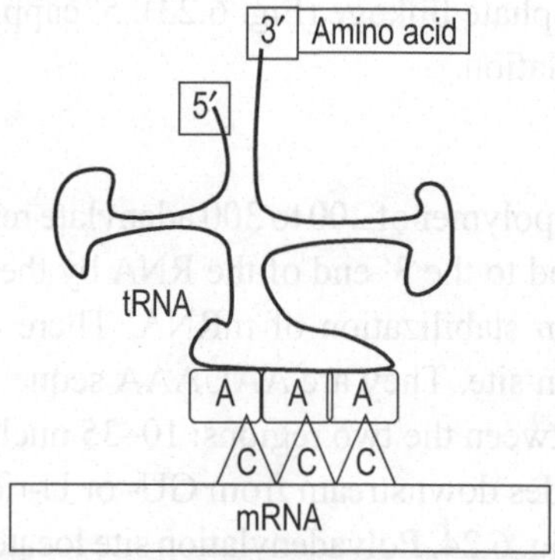

FIGURE 6.26 Codon (triangle) of mRNA and the anticodon (rectangle) of tRNA

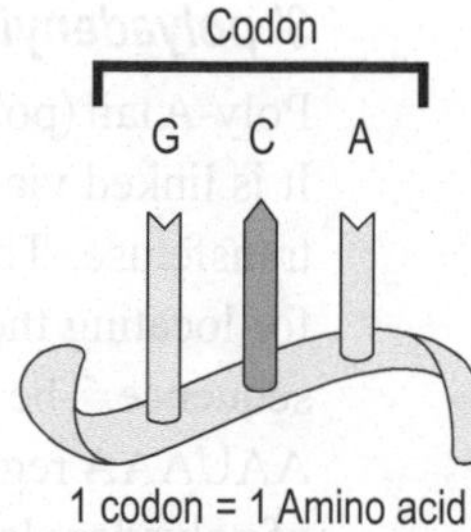

FIGURE 6.27 Picture of a codon

6.9.1 Codons and Anticodons

A set of three mRNA nucleotides are collectively called a codon (Fig. 6.27). Each codon is specific for a particular amino acid. For example, codon UCU corresponds to serine, codon CUU corresponds to praline, and codon GUU corresponds to valine. Codons corresponding to all the 20 amino acids are shown in Fig. 6.28.

First base	Second base: U		C		A		G		Third base
U	UUU	Phenyl alanine F	UCU	Serine S	UAU	Tyrosine Y	UGU	Cysteine F	U
	UUC		UCC		UAC		UGC		C
	UUA	Leucine L	UCA		UAA	Stop codon	UGA	Stop codon	A
	UUG		UCG		UAG	Stop codon	UGG	Tryptophan W	G
C	CUU	Leucine L	CCU	Proline P	CAU	Histidine H	CGU	Arginine R	U
	CUC		CCC		CAC		CGC		C
	CUA		CCA		CAA	Glutamine Q	CGA		A
	CUG		CCG		CAG		CGG		G
A	AUU	Isoleucine I	ACU	Threonine T	AAU	Asparagine N	AGU	Serine S	U
	AUC		ACC		AAC		AGC		C
	AUA		ACA		AAA	Lysine K	AGA	Arginine R	A
	AUG	M Methionine start codon	ACG		AAG		AGG		G
G	GUU	Valine V	GCU	Alanine A	GAU	Aspartic acid D	GGU	Glycine G	U
	GUC		GCC		GAC		GGC		C
	GUA		GCA		GAA	Glutamine acid E	GGA		A
	GUG		GCG		GAG		GGG		G

FIGURE 6.28 Genetic code showing the codons that represent specific amino acids

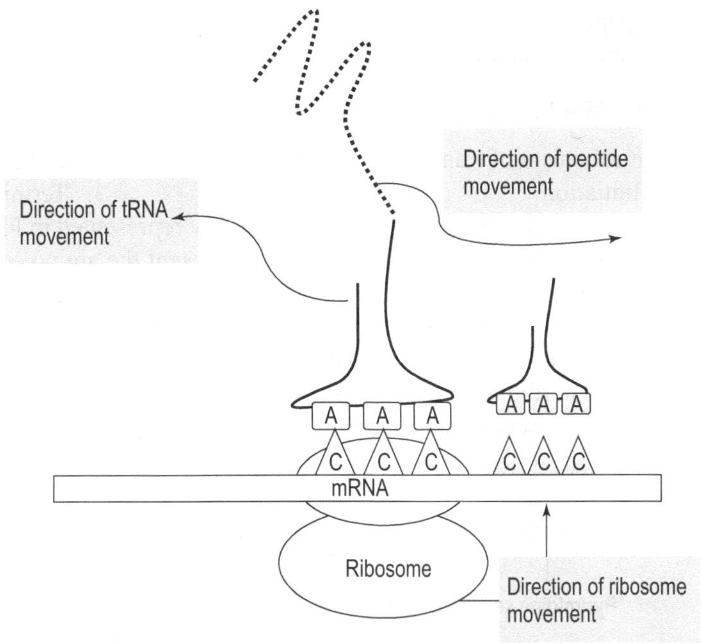

FIGURE 6.29 Schematic representation of translation

An anticodon refers to a set of three adjacent nucleotides in a tRNA; it pairs with the complementary codon (code) of mRNA. An anticodon determines the amino acid specificity. For example, the anticodon for glycine is CCC that binds to the codon (GGG) of mRNA.

A schematic representation of the different components involved in translation process is illustrated in Fig. 6.29. C represents the codon in mRNA, A represents the anticodon in tRNA, the dark square chain represents the new polypeptide chain, and the arrow represents the movement of the components after every addition of amino acid to the peptide chain. tRNA carries a specific amino acid and binds to the codon of the ribosome-bound mRNA via its anticodon. When the second tRNA carries the second amino acid, the ribosome moves three codons right, so that the anticodon of the second tRNA binds to the codon of the mRNA. The amino acid(s) in the first tRNA moves to the amino acid of the second tRNA forming a peptide. Simultaneously, the first tRNA moves away. Sequentially, the third amino acid moves to the second amino acid.

6.9.2 Components Needed for Translation

The various components needed for translation process are listed as follows:

- 30S ribosomal subunit containing 16S rRNA (for prokaryotes), 40S ribosomal subunit containing 18srRNA (for eukaryotes)
- mRNA to be translated
- formyl methionine tRNA (Fmet tRNAf)
- aminoacyl tRNA synthetase
- ATP
- transformylase
- GTP

- peptidyl transferase
- release factors

6.9.3 Mechanism of Translation

The mechanism of translation involves the following three stages:

1. Initiation 2. Elongation 3. Termination

The translation process is diagrammatically represented in Fig. 6.30 and schematically described in Fig. 6.31, where aa1 to aa5 represent the amino acids serially present in the peptide; and aa6 and aa7 are carried by the next incoming tRNA, which is ready to bind to the codons c5 and c6, respectively. Amino acids that have already delivered the first five amino acids move away from the translation machinery. Every time an amino acid is added, the ribosome moves right spanning one codon distance. Both the figures can be compared for better understanding.

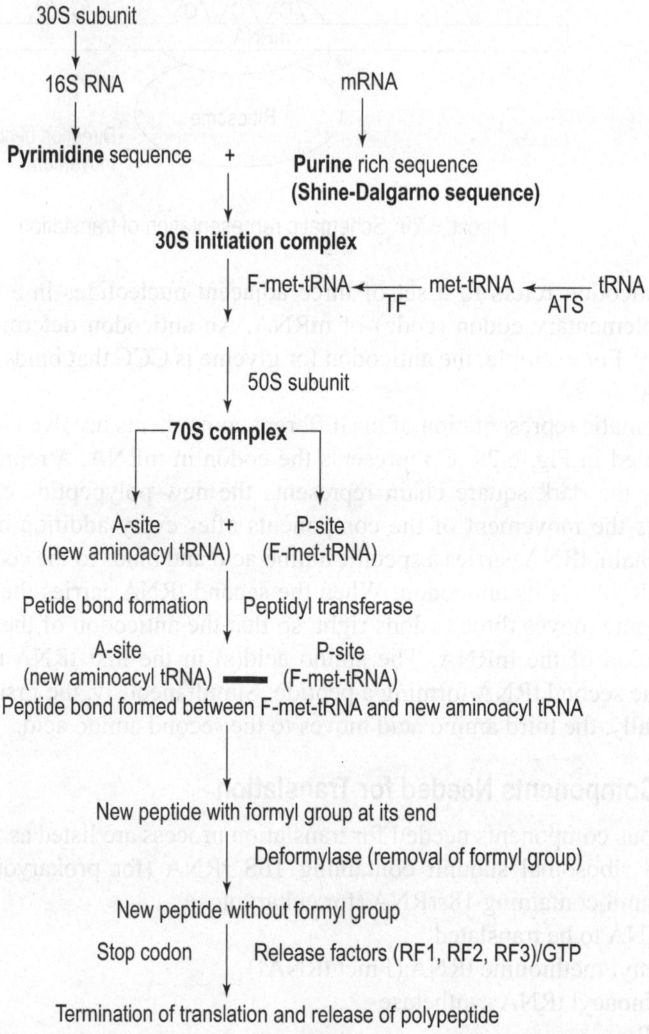

FIGURE 6.30 Schematic representation of translation process (TF = Transformylase, ATS = aminoacyl tRNA synthetase)

Initiation

Initiation of translation occurs in the ribosomes of cytoplasm. Ribosomes have two dissociable subunits, one larger and one smaller. Each subunit is formed by the supramolecular assembly of proteins and RNA, bearing a total mass equal to or greater than 106 Da.

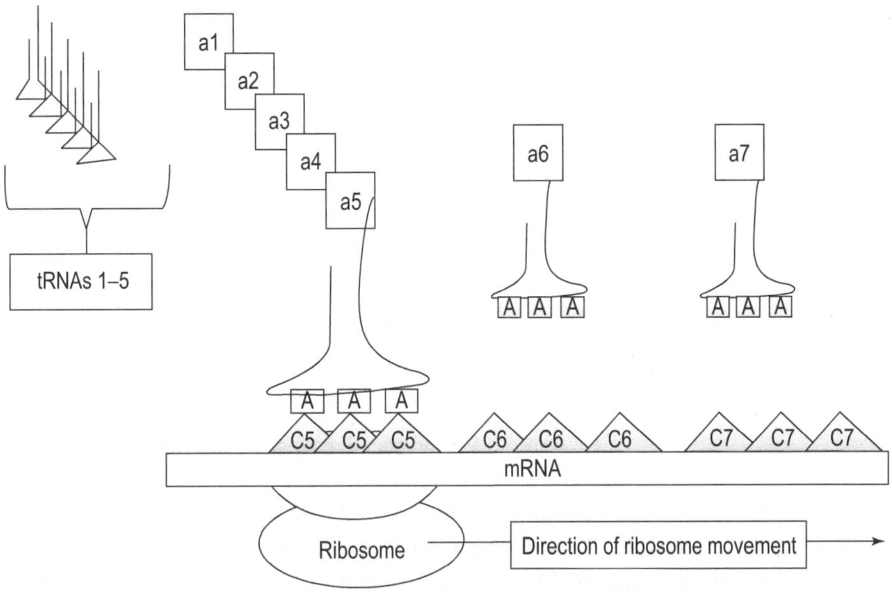

FIGURE 6.31 Diagrammatic representation of translation process

TABLE 6.11 Comparison of prokaryotic and eukaryotic ribosomes

Organism	Prokaryote				Eukaryote		
Ribosome	70				80		
Ribosomal subunit	30S	50S			40	60	
RNA component of ribosomal subunit	16S	23S	5S	18S		28S+5.85S	5S
Number of protein components in the ribosomal subunit	21	31	33			49	

30S subunit has 16S RNA of 1542 nucleotides and 21 proteins. 50S subunit possesses 23S RNA, 5S RNA, and 31 proteins. Eukaryotic ribosome (80S) is dissociable into 40S (with 33 proteins) and 60S subunits (49 proteins). 40S unit has 18S RNA, whereas the 60S unit has two RNA assemblies: 28S + 5.85S and 5S RNAs (Table 6.11 and 6.32).

30S initiation complex is needed for initiation of translation. This is formed by binding of pyrimidine sequence of 16S rRNA component of 30S subunit with the purine-rich sequence of mRNA which is called Shine–Dalgarno sequence.

tRNA is converted to methionyl-tRNA (met-tRNA) by aminoacyl tRNA synthetase at the expense of ATP. Aminoacyl tRNA synthetase has three sites (Fig. 6.33): Groups

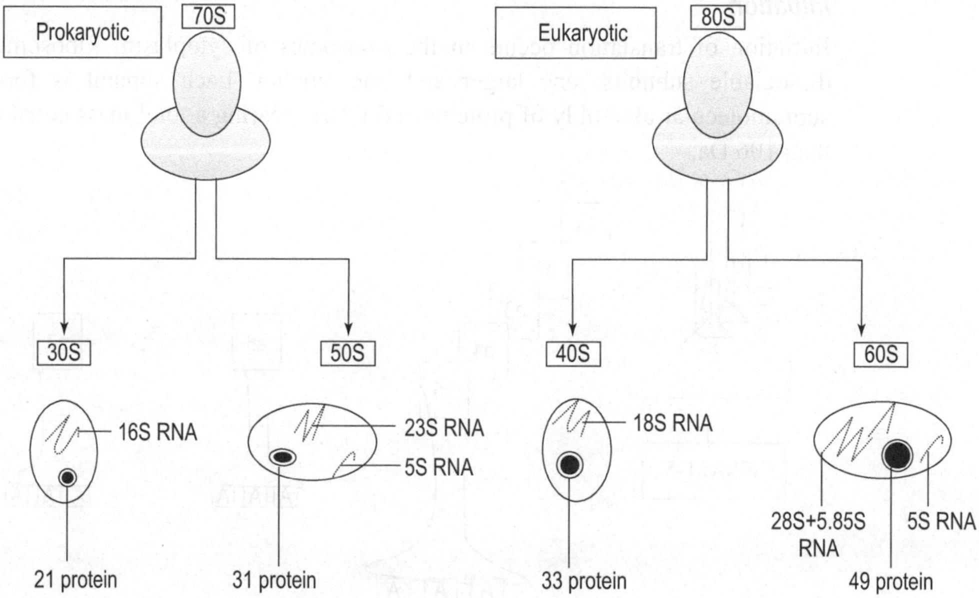

FIGURE 6.32 Prokaryotic (E. coli) and eukaryotic (rat) ribosomes

involved in the formation of high-energy ester bond between the tRNA and the amino acid are shaded. Bond formation occurs in the active site of the enzyme, the amino-acid binding site, the tRNA-binding site, and the active site. Bonding of tRNA to the amino acid occurs at the active site via ester bond at the expense of ATP. The reaction can be represented as follows:

Reaction 1: Amino acid + ATP → aminoacyl-AMP + PPi

Reaction 2: Aminoacyl-AMP + tRNA → aminoacyl-tRNA + AMP

Sum of 1 and 2: Amino acid + tRNA + ATP → aminoacyl-tRNA + AMP + PPi

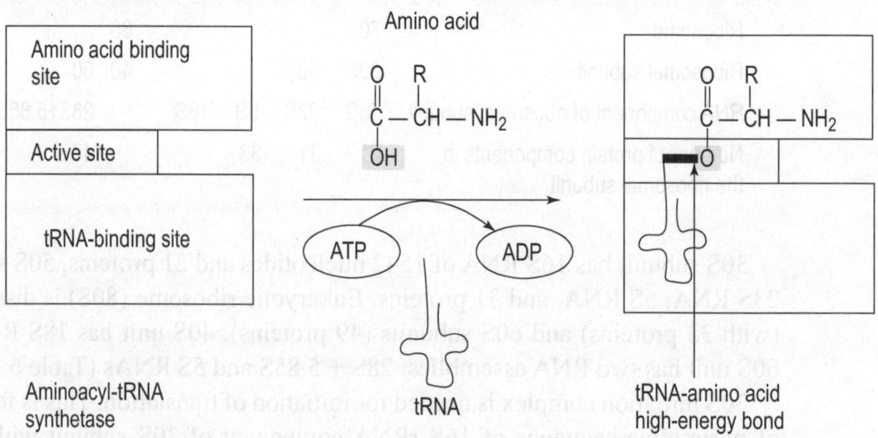

FIGURE 6.33 Role of aminoacyl tRNA synthetase in binding of amino acid to tRNA

met-tRNA is converted to formyl-methionyl-tRNA (F-met-tRNA or initiator t-RNA) by transformylase (Fig. 6.34). This reaction can be represented as follows:

10-Formyltetrahydrofolate + L-Methionyl-tRNAfMet + H_2O ⟶ Tetrahydrofolate + N-Formylmethionyl-tRNAfMet

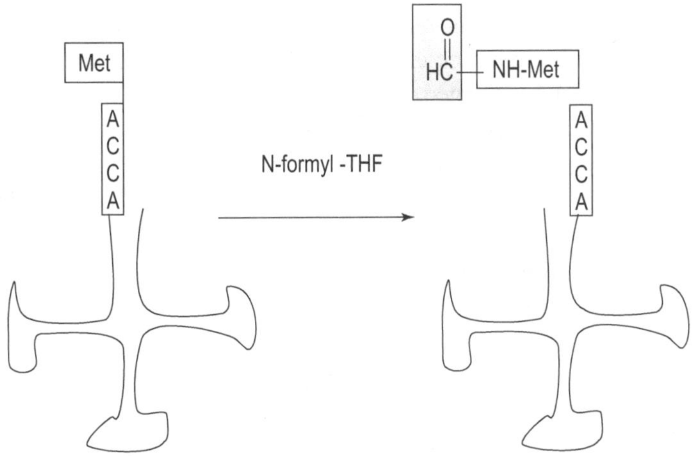

FIGURE 6.34 Conversion of methionyl tRNA to Formyl methionyl tRNA by transformylase

Finally, the initiator tRNA binds to the 30S complex. The 30S complex then binds to 50S subunit of the ribosome forming 70S complex. 70S complex has two sites: A site or the amino site and P site or the peptide site. Initiator tRNA carries methionine and has an anticodon (UAC) that is complementary to the start codon AUG of mRNA at the P site. Thus, initiator tRNA carrying methionine is attached to the P site.

Elongation

Let us consider for convenience that the codons lying right to the start codon (AUG) are the following (Fig. 6.28):

- CAU (specific for histidine)
- GGG (specific for glycine)
- CAG (specific for glutamine)

With these examples, the translation process involving P and A sites can be discussed as follows with reference to Fig. 6.35.

- The first incoming aminoacylated tRNA (the tRNA with His) that is complementary to the codon adjacent to the initiator codon (CAU) binds to the A site (the amino site) of the 70S ribosomal subunit.
- Methionine from initiator tRNA in the P site is transferred to histidine at A site. Peptide bond is formed between the two amino acids in the presence of peptidyl transferase thus starting elongation. This enzyme has an RNA component which is actually responsible for the catalytic action. Therefore, peptidyl transferase is a ribozyme (catalytic RNA or enzymatic RNA).
- Simultaneously, the uncharged initiator tRNA relieves out of the ribosome vacating the P site.
- Then the ribosome moves along three nucleotides in 5′–3′ direction along the mRNA. (i.e., moves one codon distance to right).
- As a result, methionine–histidine dipeptide will be at the P site and the A site will be vacant.

- Now, the next amino acid glycine comes with its tRNA to the A site.
- Then the dipeptide is transferred to glycine and attached via peptide bond as before. Uncharged rRNA is released.
- Ribosome moves one codon distance further, leaving a free A site.
- tRNA with glutamine enters A site. Tripeptide at P site moves to glutamine and a peptide bond is formed.
- Elongation of peptide chain continues.

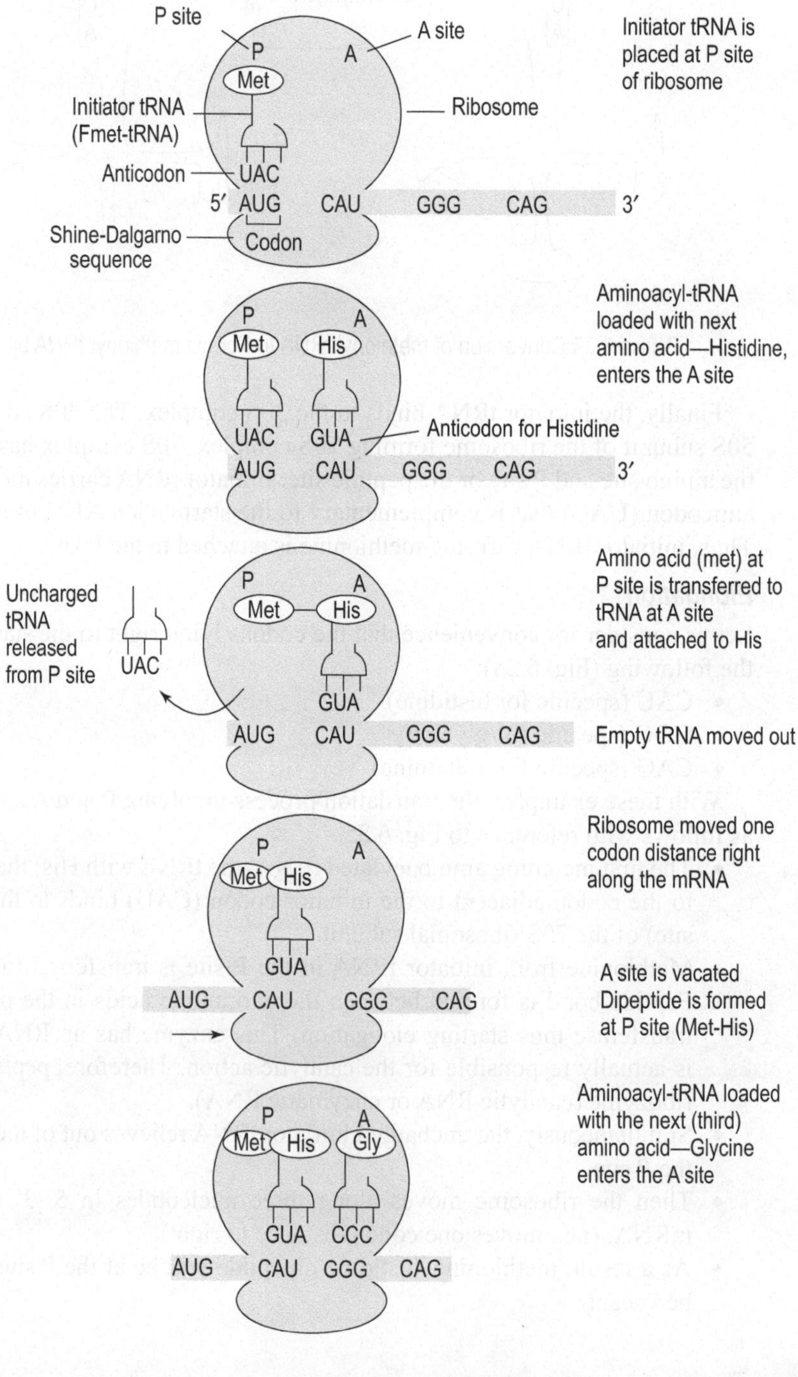

(*Contd*)

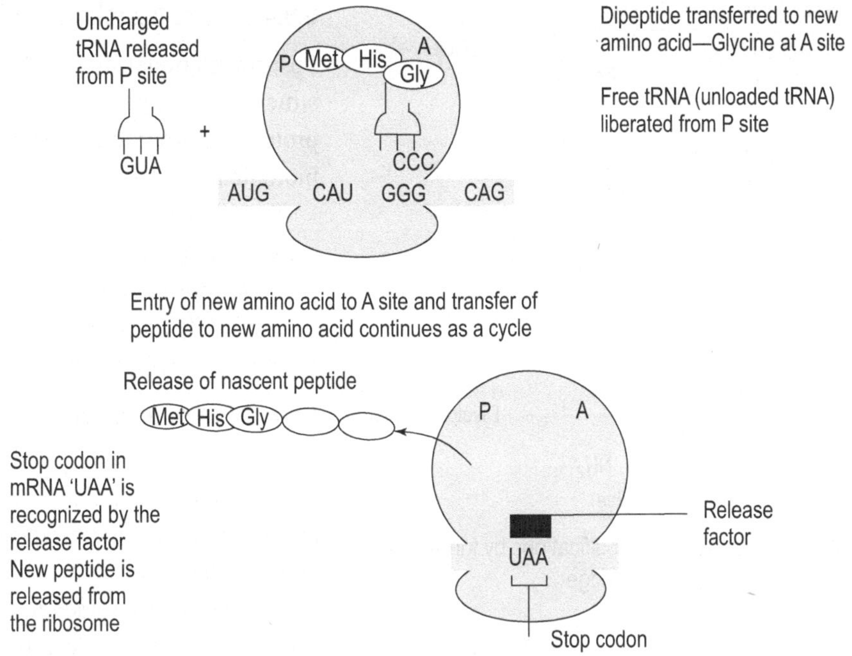

FIGURE 6.35 Ribosomes showing the role of A site and P site in translation process

Termination

Elongation continues until the termination sequence or stop codon is encountered. Three stop codons are found in RNA: UAG, UAA, and UGA. Release factors, such as RF1, RF2, and RF3 are involved in the termination process. RF1 and RF2 recognize stop codons, such as UAA, UAG, and UGA.

RF3 directs peptidyl transferase to release the polypeptide from the tRNA in the P site by hydrolysis in the presence of GTP. Deformylase removes the formyl group from the growing polypeptide chain in the presence of water:

$$\text{Formyl-L-methionyl peptide} + H_2O \longrightarrow \text{Formate} + \text{methionyl peptide}$$

Then the ribosomal subunit is also dissociated at the expense of GTP hydrolysis. The functions of enzymes catalysing translation are summarized in Table 6.12.

TABLE 6.12 Enzymes catalysing translation

Enzyme	Function
Aminoacyl tRNA synthetase	Catalyses the esterification of a specific amino acid to form an aminoacyl tRNA at the expense of ATP
Transformylase	Catalyses the transfer of formyl group to the methionine which is carried by the tRNA
Peptidyl transferase	Catalyses the formation of peptide bond between adjacent amino acids in the ribosome machinery
Deformylase	Removes the formyl group from the growing polypeptide chain

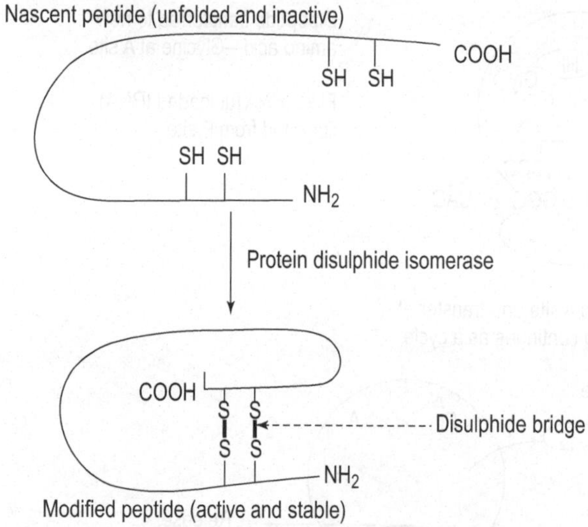

FIGURE 6.36 Post-translational modifications by formation of disulphide bridge

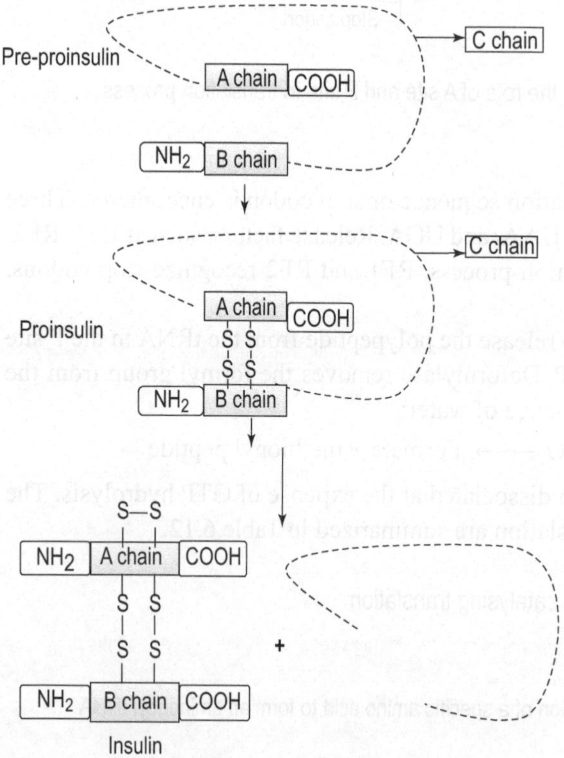

FIGURE 6.37 Post-translational modification of insulin chain (Pre-proinsulin has three chains designated as A, B, and C. Three disulphide bridges are formed to generate proinsulin. Two are interchains—formed between A and B chains, whereas one is an intrachain formed in the A chain. C chain is cleaved away to generate functional insulin.)

6.9.4 Post-translational Modification

Post-translational modifications refer to those structural changes that take place in native proteins in order to render them functional and biologically active.

Three types of post-translational modification are common in proteins:

(i) Formation of disulphide bridges
(ii) Glycosylation
(iii) Proteolytic cleavage

Formation of disulphide bridge

The protein disulphide isomerase catalyses the formation of disulphide bridges in the peptide chain where cysteine residues are present. This results in the conversion of protein or peptide chain into properly folded form which forms the tertiary and quaternary conformation (Fig. 6.36). Pre-proinsulin is transcribed as a 110 amino acid chain. Removal of the signal peptide produces proinsulin. Formation of disulphide bonds between the A- and B-chain components, and removal of the intervening C chain, produces a biologically active insulin molecule containing 51 amino acids, less than half of the original translation product (Fig. 6.37).

Glycosylation

Glycosylation means addition of carbohydrates in the form of oligosaccharides to the proteins. This process is catalysed by an enzyme called glycosyl transferase. This enzyme recognizes asparagine residues in vicinity to serine or threonine in the protein and attaches the oligosaccharides (Fig. 6.38).

Proteolytic cleavage Proteolytic cleavage refers to the hydrolytic breakage of peptide bonds in the proteins, in the presence of protease enzyme. Proteolytic cleavage is a well-known method of post-translational modification in hormones and enzymes. Some enzymes are synthesized as biologically inactive proteins called zymogens or proenzymes. Example, chymotrypsin (protein digesting enzyme) is derived from chymotrypsinogen (inactive form of chymotrypsin), and carboxypeptidase is derived from its inactive precursor procarboxypeptidase. Proelastase is the zymogen for elastase. Pepsinogen is the inactive precursor for pepsin. Similarly, inactive hormones are called pre-prohormones or prohormones. Insulin, the active hormone, is derived from the proteolytic cleavage of pre-proinsulin and proinsulin in pancreatic cells. Examples of proteolytic cleavage are shown in Fig. 6.39.

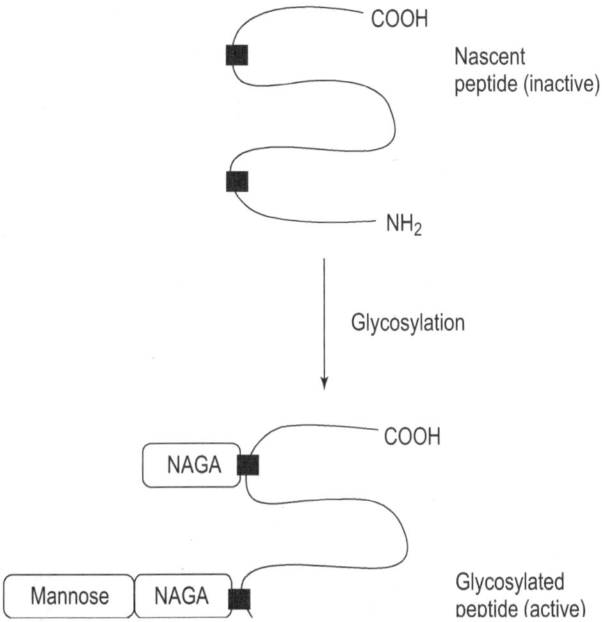

FIGURE 6.38 Post-translational modifications by glycosylation

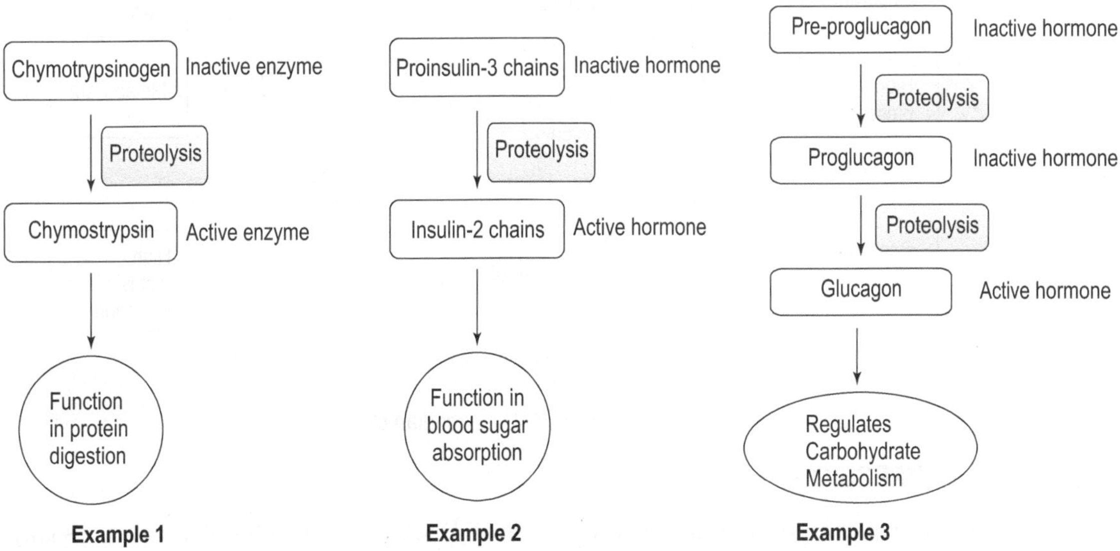

FIGURE 6.39 Examples for proteolytic cleavage

6.10 CIRCADIAN CLOCK

In living organisms, the biochemical, physiological, or behavioural processes take place as a 24-hour cycle. This is known as the circadian clock or circadian rhythm (Fig. 6.40). The processes constituting the circadian clock are broadly divided into two types: behavioural processes and autonomous processes. Locomotor activity, feeding, excretion, sensory processing, and learning capability are the behavioural processes that repeat in the circadian clock. Body temperature, cardiovascular function, melatonin secretion, cortisol secretion, metabolism, and sleep come under autonomous process. The supra chiasmatic nucleus (SCN) located above the optic chiasma of the brain controls the circadian clock. SCN has a core region and the shell. The core region uses gamma amino butyric acid (GABA) and vasoactive intestinal polypeptide (VIP) as neurotransmitters. The shell region uses GABA and arginine vasopressin as neurotransmitters. Genetic factors have a major role in determining the circadian clock. The genes responsible for the function of circadian clock have been revealed by mutation studies. Table 6.13 represents the genes influencing the circadian clock.

TABLE 6.13 Genes controlling the circadian clock

Organism	Name of the gene
Drosophila melanoglaster	Per gene
Mould	Frq gene
Golden hamster	Tau gene
Domestic mice	Clock gene

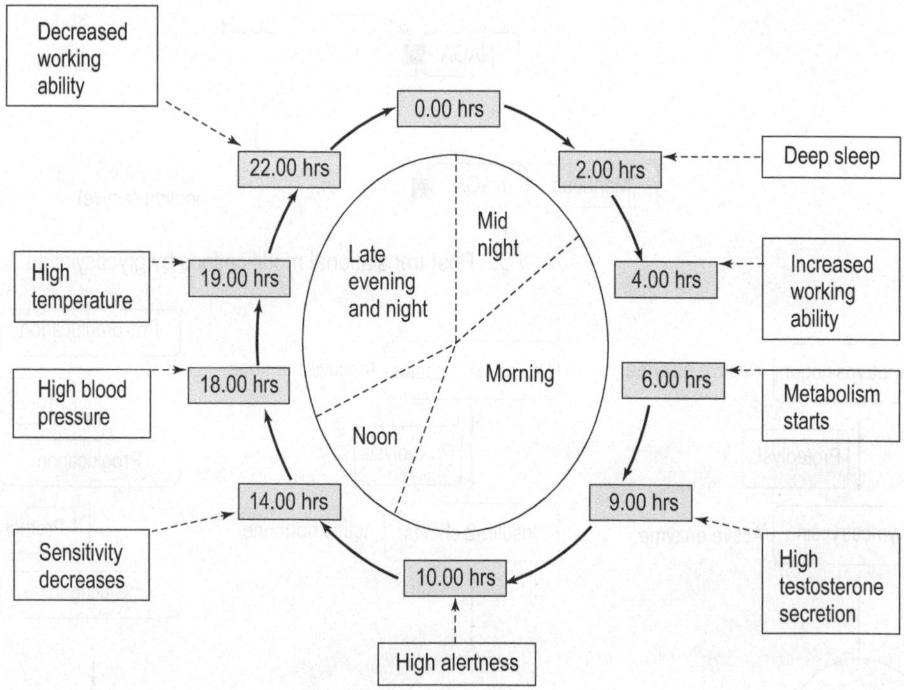

FIGURE 6.40 The circadian clock

6.11 GENE EXPRESSION

Gene expression refers to the expression of DNA into a functional product, usually protein, mRNA, or tRNA. In short, the three major genetic mechanisms—replication, transcription, and translation—together form the gene expression process. This section deals with the gene expression in prokaryotes and eukaryotes.

6.11.1 Gene Expression in Prokaryotes

Prokaryotic genome has two types of genes: constitutive gene and inducible gene. Constitutive genes encode the products that are constantly required by the cells, for example, TCA cycle enzymes. Inducible genes encode the products that are required under certain environmental conditions. These products are formed by the induction of the specific gene by signalling molecules (refer Chapter 11). Expression of a particular gene is regulated either by induction via inducer or positive regulator or by suppression via negative regulator or repressor. Gene expression is well studied in E. coli, by the operon concept. Operon is a cluster of functional and regulatory genes all controlled by a single regulatory signal.

Structural gene encodes a protein or RNA molecule; regulatory gene is one that encodes a product (regulatory protein) that is capable of regulating the structural gene. The regulatory protein binds to the structural gene and influence the expression of the structural gene.

The role of operon in E. coli gene expression is best understood from the Lac operon. The components of a Lac operon are shown in Fig. 6.41. It is made of regulatory genes designated as i, CAP, P, and O, and structural genes designated as Z, Y, and A (Table 6.14). Regulatory genes 'i' and 'CAP' codes for repressor protein and catabolite activator protein (CAP), respectively. The structural gene 'Z' codes for β-galactosidase. Z gene is induced or suppressed by the presence or absence of lactose in the nutrient medium. LacY expresses β-galactoside permease, a membrane-bound transport protein used for pumping lactose into the cell. LacA codes for β-galactoside transacetylase, an enzyme that transfers an acetyl group from acetyl-CoA to β-galactosides. It was found that the lacZ and the lacY genes are necessary for lactose catabolism.

In the presence of lactose medium, the cells have to metabolize the lactose by galactosidase as follows:

$$\text{Lactose} \xrightarrow{\beta\text{-galactosidase}} \text{Galactose} + \text{glucose}$$

For this reaction to take place, the LacZ gene should be induced. Induction of Z gene is stimulated by the induction of CAP gene. CAP gene produces CAP, which in turn enhances the binding of RAN polymerase to the promoter gene. This leads to the induction of the adjacent Z gene. As a result, β-galactosidase is synthesized for the further metabolism of lactose.

In the absence of lactose, the 'i' gene is activated to synthesize repressor protein. Repressor protein binds to the operator and suppresses the Z gene.

TABLE 6.14 Genes of Lac operon

Nature of the gene	Name of the gene	Product of the gene
Regulatory genes	'i' gene	Repressor protein
	'CAP' gene	Catabolite activator protein
	Promoter gene	–
	Operator gene	–
Structural genes	'Z' gene	β- galactosidase enzyme
	'Y' gene	β-galactoside permease
	'A' gene	β-galactoside transacetylase

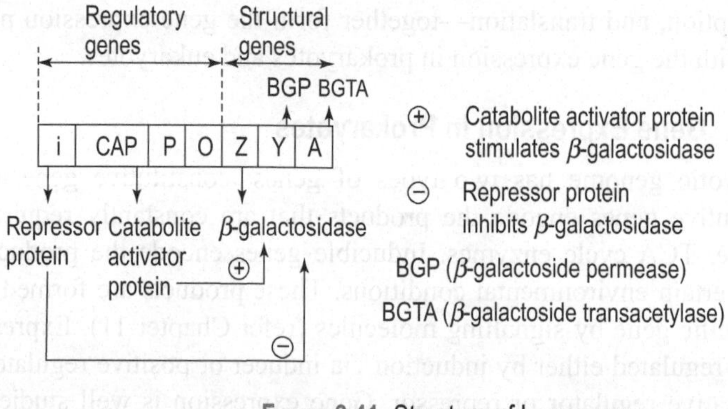

FIGURE 6.41 Structure of Lac operon

Gene expression was also studied in tryptophan operon in prokaryotes. Tryptophan (Trp) is an aromatic amino acid constituent of many of the biological proteins. Trp gene is induced or suppressed based on the availability of the Trp in the cell. The Trp operon has five structural genes (Trp A, Trp B, Trp C, Trp D, Trp E) and five regulatory genes (primary promoter or Trp P, operator or Trp O, attenuator or Trp A, secondary internal promoter or Trp P2, and terminator or Trp T). The structural genes code for the following enzymes:

- Anthranilate synthase I
- Anthranilate synthase II
- Phosphoribosyl anthranilate isomerase
- Tryptophan synthetase β
- Tryptophan synthetase α

These enzymes catalyse the stepwise synthesis of tryptophan from the substrate chorismate.

When sufficient amount of Trp is present, further synthesis of Trp is prevented by tryptophan repressor in association with tryptophan itself. Trp and Trp repressor bind to the Trp O gene and suppress the transcription of the genes coding for the enzymes that catalyse the synthesis of the tryptophan. In the presence of Trp, the attenuator gene is also suppressed and the transcription of synthetic enzymes is inhibited, but in the absence of trp, attenuator gene has no role in transcription.

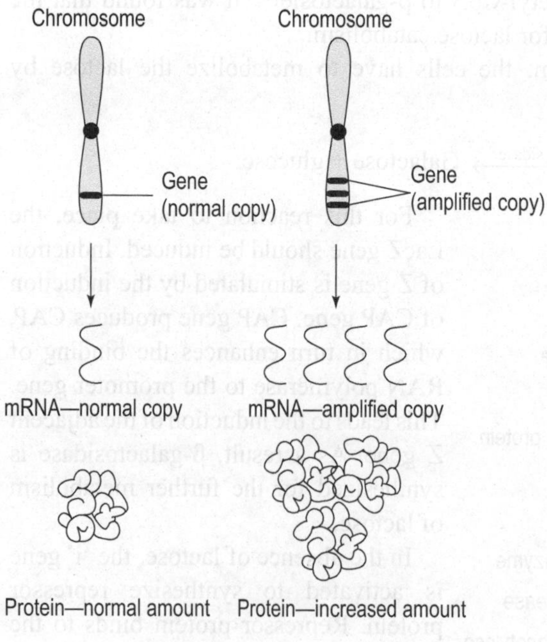

FIGURE 6.42 Gene amplification

Eukaryotic gene expression is determined by the following factors:

Gene amplification

Gene amplification refers to the production of multiple copies of a particular gene, leading to an increase in the amount of the corresponding mRNA and the protein product (Fig. 6.42). For example, genes encoding histone proteins or ribosomal RNA are always in amplified state. The metallothionein gene on the other hand is selectively amplified, that is, the genes are amplified and expressed when the cells are exposed to heavy metal toxicity. Other examples for selective gene amplification include multidrug resistance gene and dihydrofolate resistance gene.

Antibody formation

Antibodies are formed and glycosylated after translation. They are involved in eliciting immunity against foreign antigens. The genes encoding the antibodies are amplified only at the time of infection. Moreover, there are diverse classes of antibodies specific for a group of antigen. This diversification is carried out by genetic recombination in the gene locus.

Transcriptional factor

Transcription of a gene is regulated by chromatin. Chromatin is the region of a gene that is complexed with proteins called histones. The interaction between the DNA and histones is very high due to positive and negative charges on histones and DNA. Chromatin is responsible for silencing the transcription. The chromatin–DNA interaction is determined by transcription factors, such as activators and repressors.

DNA methylation is another way of regulating gene expression. De novo methyltransferases methylate cytosine residues. Maintenance methyltransferases methylate one strand of DNA when the one strand is already methylated. Methylation results in the silencing of the promoter gene.

SUMMARY

Genome of an organism is its complete set of DNA, bearing all the information required to build and maintain that organism. The human genome was studied by the Human Genome Project. The genome undergoes several molecular processes such as replication, transcription, and translation, collectively called genetic mechanisms.

The DNA of the genome is replicated by the process of replication, catalyzed by various enzymes and proteins, including helicase, single strand stabilizing protein, primase, ligase, and DNA polymerase. The semiconservative mode of replication was proved by Meselson Stahl experiment. The rate of DNA replication varies between prokaryotes and eukaryotes. In prokaryotes, the replication rate is 1000 nucleotides per second, which contributes to a higher error rate. In eukaryotes, the replication rate is slower with the speed of just 50 nucleotides per second per replication fork.

DNA is converted into RNA by a process called transcription. Three types of RNAs are produced by transcription: messenger RNA (mRNA), ribosomal RNA (rRNA), and transfer RNA (tRNA). Transcription is catalyzed by RNA polymerase and transcription factors.

The nascent RNA undergoes post-transcriptional modification to become biologically active and functional. Removal of spacers, trimming of tRNA and addition of CCA sequence are common types of post transcriptional modification in prokaryotes. 5' capping, 3' polyadenylation, and splicing are three types of post- transcriptional modification in eukaryotes. The information encoded in the m-RNA is converted into proteins by enzymes, ribosomes, and tRNA in a process called translation. The proteins undergo post-translational modification, such as of disulphide bridges, glycosylation, and cleavage. The expression into a functional product,, usually protein, mRNA or tRNA is referred to as gene expression.

TERMS TO REMEMBER

3′ flanking region A portion of DNA that is not transcribed into the mature mRNA, but which is present adjacent to 3′ end of the gene.

Circular chromosome DNA arranged in a closed circle, which is negatively supercoiled allowing for the compact nature of many bacterial genomes.

Coding strand One of the strands of the DNA double helix which is transcribed into RNA.

Codon A set of three nucleotides which is specific for a single amino acid to be incorporated into peptide.

Comparative genomics Analysis of DNA sequence patterns of human beings and well-studied model organisms side-by-side for identifying human genes and interpreting their function.

Genome The total DNA present in each cell of an organism.

Introns Specific DNA sequence in a genome which are transcribed but spliced out later.

Kb/Mb A kilobase (Kb) is 1000 bases of DNA, while a megabase (Mb) is 1,000,000 bases.

Knockout studies Inactivation of genes in living organisms and monitoring any changes that could reveal their functions.

Ligase Enzyme that ligates two pieces of DNA.

Linear chromosome A non-closed chromosome, which has inverted repeats at the ends, similar to teleomeres in eukaryotic chromosomes.

Megaplasmid A very large plasmid ranging in size from 100 to 1700 Kb.

Plasmid Extrachromosomal DNA which replicates independently of the chromosome and regulates its own replication.

Poly-A tail A stretch of adenine polymer with around 100 residues, which is added to the mRNA after transcription.

Primary transcript Immature RNA that is formed immediately after transcription and exists without any post-transcriptional modification.

Primer A short RNA segment of 5 to 60 nucleotide length used to prime the synthesis of DNA.

Promoter A segment of approximately 100 nucleotides located upstream to a gene that controls the transcription of that gene.

Proteomics Studying protein expression and function to aid in drug design.

Restriction endonuclease Enzyme usually obtained from bacteria, for cutting the DNA at specific sequence.

Ribosomes A supramolecular assembly of proteins and RNA, present in the cytoplasm to initiate protein synthesis.

Structural genomics Study and investigation to generate the 3-D structures of one or more proteins from each protein family, thus, offering clues to function and biological targets for drug design.

Transcription factor Proteins that bind to DNA for controlling the transcription process.

Transcriptomics Large-scale analysis of messenger RNAs transcribed from active genes to follow when, where, and under what conditions genes are expressed.

Zinc finger Structural motif of DNA-binding proteins, with cysteine-binding and zinc-binding **fingers.**

REVIEW QUESTIONS

1. Which is the smallest genome known?
2. When a cell divides, what is the source of the extra DNA?
3. Tabulate the functions of enzymes involved in replication.
4. During replication, unwinding of DNA helix by helicase occurs in A:T rich areas. Why?
5. Which factor is responsible for preventing the rewinding of DNA during replication process?
6. How can DNA direct the synthesis of RNA?
7. Of the three types of RNA formed by transcription, which one is responsible for protein synthesis?
8. Using the sequence of the DNA strand shown below, work through all of the steps to get to a polypeptide:

   3′ TCCACGTCAGACAATCCGAAGGTAGGC5′
   5′ AGGTGCAGTCTGTTAGGCTTCCATCCG 3′
9. From the given DNA strand identify the template that could drive RNA synthesis:

   3′ AATTATGGCCGCTAT 5′
   5′ TTAATACCGGCGATA 3′
10. Differentiate codons from anticodons.
11. Write codons and anticodons for the following amino acids: Phe, Arg, Leu.
12. Name the triplet bases of mRNA and tRNA.
13. What is the initial amino acid attached to the initiator tRNA?
14. Explain the role of topoisomerase in solving the topological problems in DNA.
15. What is semiconservative replication?

16. Compare the ribosomes of prokaryotes and eukaryotes.
17. Name the enzymes involved in producing and hydrolyzing the RNA primers in replication process.
18. Explain the types of post-transcriptional modifications in prokaryotes.
19. How is eukaryotic mRNA modified after transcription?
20. Cite a few examples for proteins that are post-translationally modified by proteolytic cleavage.
21. Write the functions of enzymes involved in translation process.
22. What is an extremophile?

MULTIPLE-CHOICE QUESTIONS

1. The combination of a ribose and a base is called
   (a) ribonucleotide
   (b) ribonucleoside
   (c) deoxyribonucleotide
   (d) deoxyribonucleoside
2. Double helical DNA is also called
   (a) Watson and Crick model
   (b) clover leaf model
   (c) zinc finger model
   (d) all the above
3. Enzyme that joins the Okazaki fragment is
   (a) endonuclease
   (b) exonuclease
   (c) helicase
   (d) ligase
4. D loop is present in
   (a) mRNA
   (b) tRNA
   (c) rRNA
   (d) snRNA
5. Which of the following is not a post-transcriptional modification?
   (a) proteolysis
   (b) CCA addition
   (c) disulphide bridge formation
   (d) all the above
6. Crown gall tumor in plants is caused by
   (a) agrobacterium tumifaceans
   (b) deinococcus radiodurans
   (c) aspergillus niger
   (d) staphylococcus aureus
7. Which of the following is the enzyme used in translation?
   (a) transformylase
   (b) deformylase
   (c) helicase
   (d) topoisomerase
8. A short stretch of RNA used to initiate replication is termed
   (a) promoter
   (b) primer
   (c) primase
   (d) replisome
9. Messelson Stahl's experiment proved
   (a) semiconservative replication
   (b) transcription
   (c) post-translational modification
   (d) gene expression
10. Lac operon is best known for
   (a) gene expression in prokaryotes
   (b) gene expression in eukaryotes
   (c) post-translational modification in prokaryotes
   (d) post-transcriptional modification ineukaryotes
11. LacZ gene encodes
   (a) beta galactosidase
   (b) permease
   (c) transacetylase
   (d) anthranilate synthetase

Answers to Multiple-choice Questions

1. (b) 2. (a) 3. (d) 4. (b) 5. (d)
6. (a) 7. (a) 8. (b) 9. (a) 10. (a)
11. (a)

Sir Hans Adolf Krebs, German scientist, awarded Nobel Prize for the discovery of TCA cycle (Krebs cycle) in 1953.

CHAPTER 7

Bioenergetics, Metabolic Pathways, and Energy Production in Cells

After reading this chapter, students should be able to understand and appreciate the following:

- Principles of bioenergetics
- Nucleic acid metabolism
- Metabolism of proteins and amino acids
- Metabolism of carbohydrates
- Metabolism of fatty acids and lipids
- Energy metabolism
- Pyruvate dehydrogenase pathway
- TCA cycle
- Electron transport chain
- Oxidative phosphorylation and ATP production
- Integration of metabolism
- Diseases due to abnormal metabolism

7.1 INTRODUCTION

Biochemical pathways in the cells of the living organisms are organized into different pathways called *metabolic pathways*. Through these pathways food is converted to energy, and necessary macromolecules are also synthesized. The two aspects of metabolism are (a) catabolism and (b) anabolism. *Catabolism* refers to the degradation of larger molecules into smaller ones, whereas *anabolism* refers to the biosynthesis of complex molecules from their precursors. This chapter deals with catabolism and anabolism of nucleic acids, proteins, carbohydrates, lipids, and finally discusses the integration of these pathways and production of energy.

In catabolism, complex macromolecules (proteins, carbohydrates, and lipids) are converted into simple molecules (with little energy trapped). Simple molecules, in turn, are catabolized into a few molecules, which are then oxidized to carbon dioxide (CO_2) and water (H_2O) along the following final common pathway: (1) Krebs cycle, (2) electron transport chain, and (3) oxidative phosphorylation. During these pathways, acetyl CoA is oxidized to CO_2 and water, and the energy is trapped as ATP.

During anabolism, building blocks are combined into complex molecules with the help of energy. Two forms of energy are used here: (1) ATP, obtained from the catabolic pathway, and (2) high-energy hydrogen, in the form of reduced NADPH (nicotinamide adenine dinucleotide phosphate). The combined effect of all the metabolic pathways that interconvert precursors, metabolites, and low-molecular-weight compounds is known as intermediary metabolism. Both catabolism and anabolism are multistep pathways, each step being catalysed by specific enzymes and cofactors, as discussed in subsequent topics. Metabolisms occurring in the cytosol, such as glycolysis, fatty acid synthesis, and pentose phosphate pathway, are catalyzed and regulated by cytosolic enzymes.

Mitochondrial matrix contains enzymes catalysing TCA cycle, oxidative phosphorylation, fatty acid oxidation, and ketone body formation. Because of the compartmentalization of enzymes, metabolic processes that proceed in opposite directions are separated. For example, biosynthesis of fatty acid from the acetyl CoA is confined to cytosol, whereas the oxidation of fatty acid is confined to mitochondrial matrix.

Before discussing the metabolic pathways of biomolecules, an introduction to bioenergetics is essential, dealing with the basic concepts of metabolic pathways. The following section deals with bioenergetics.

7.2 BIOENERGETICS

Bioenergetics is the study of the processes of energy production and transduction which occur as a result of catabolic pathways. Sometimes, the term *thermodynamics* is used while referring to bioenergetics.

Two laws, known as the *laws of thermodynamics*, were proposed to explain the basic concept of energy.

First law of thermodynamics

Energy can change its form, but the amount of energy is conserved. This is known as the law of conservation of energy.

Second law of thermodynamics

The universe always tends towards more and more disorder. That is, in all natural processes, the entropy of the universe increases.

7.2.1 Systems and Surroundings

To apply the second law of thermodynamics to biological systems, it is essential to understand about the 'system' and the 'surrounding'. In biology, the term *system* refers to an organism, a cell or any two biochemical components of a metabolic pathway, and the term *surrounding* refers to the environment of the system. The reacting system and its surroundings together constitute the universe. When the reaction takes place without exchanging material or energy with the surroundings, it is called a *closed system*. On the other hand, the process that is accompanied by the exchange of energy with the surroundings is called an *open system*.

7.2.2 The Three Thermodynamic Quantities—Enthalpy, Entropy, and Free Energy

There are three thermodynamic quantities that describe the energy changes occurring in a system. They are enthalpy (H), entropy (S), and Gibbs free energy (G). Quantitative approach to metabolism can be achieved by these three fundamental quantities. In a biochemical system, usually, the changes in these quantities are considered. The changes are represented as ΔH, ΔS, and ΔG (Table 7.1).

In the standard state, these three components are written as $\Delta H^{\circ\prime}$, $\Delta S^{\circ\prime}$, and $\Delta G^{\circ\prime}$. [The standard state of a biochemical system is defined as the one with pH 7, temperature 25°C (288 K), solute concentration of 1 M, and pressure of all gases at 1 atm.]

TABLE 7.1 Units of the three thermodynamic quantities

Quantity	Unit
ΔG	joules/mole (or) cal/mole
ΔH	joules/mole (or) cal/mole
ΔS	joules/mole.degree Kelvin

Enthalpy is defined as the heat content of a system. It reflects the number and kinds of chemical bonds in the reactants and products of the reaction system. It is represented by the following equation:

$$H = E + PV$$

where

E = internal energy

PV = product of pressure and volume

The reacting system that takes up heat from its surroundings is called an endothermic system and has positive values of ΔH.

Entropy is defined as the degree of randomness of a system. It is a measure of the energy that is not accessible to perform useful work. It is denoted by 'S'. Entropy and enthalpy can together be used to define the third quantitiy 'free energy'. This predicts the direction of spontaneous reactions. Entropy alone does not predict if the reaction is favourable.

Gibbs free energy is defined as the amount of *energy capable of doing work* during a reaction at constant temperature, volume, and pressure. When a reaction proceeds with the release of free energy (i.e., when the system changes, so as to posses less free energy), the free energy change ΔG has a negative sign and the reaction is said to be exergonic. When a reaction proceeds with the gain of free energy (i.e., when a system changes, so as to possess more free energy), the free energy change ΔG has a positive sign and the reaction is said to be endergonic .

ΔG of a chemical reaction is represented as follows:

$$\Delta G_{\text{reaction}} = \Delta G_{\text{(product)}} \Delta G_{\text{(reactant)}}$$

Positive values of ΔG (i.e., $\Delta G_{\text{(product)}} > \Delta G_{\text{(reactant)}}$) indicate endergonic reaction. Negative values of ΔG (i.e., $\Delta G_{\text{(product)}} < \Delta G_{\text{(reactant)}}$) indicate exergonic reaction. In other words, an exergonic reaction proceeds with the release of energy, whereas an endergonic reaction needs energy input for proceeding. For example, ATP synthesis (anabolism) is an endergonic reaction, whereas ATP hydrolysis (catabolism) is an exergonic reaction. These reactions are as follows:

$$\text{ADP} + \text{Pi} + 7.3 \text{ kcal} \rightarrow \text{ATP (Endergonic)}$$

$$\text{ATP} \rightarrow \text{ADP} + \text{Pi} - 7.3 \text{ kcal (Exergonic)}$$

ΔG and $\Delta G^{o\prime}$ can be related in the following manner:

Consider the following reaction

$$A + B \leftrightarrow C + D$$

For this reaction,

$$\Delta G = \Delta G^{o\prime} + RT\ln\frac{[C]\,[D]}{[A]\,[B]} \qquad (7.1)$$

where

R = gas constant (1.987×10^{-3} kcal × deg^{-1} × mol^{-1})

T = temperature in degree Kelvin (degree Celsius + 273)

TABLE 7.2 Relation between standard free energy change ($\Delta G^{o\prime}$) and equilibrium constant

K_{eq}	$\Delta G^{o\prime}$ (kJ mol^{-1})
0.001	17.1
0.01	11.4
0.1	5.7
1.0	0.0
10	−5.7
100	−11.4
1000	−17.1

ln = natural logarithm, which can be converted to log 10 by multiplying with 2.303

The change in free energy during the synthesis of 1 mol of a compound (x) from its precursor under standard condition is called *standard free energy of formation* of that compound and it is denoted as $\Delta G_F^{o\prime}$. Here, F = formation of the compound (anabolism).

At equilibrium, when the rate of forward and reverse reactions is the same, there is no change in free energy, that is

$$\Delta G = 0$$

Also, $$\frac{[C]\,[D]}{[A]\,[B]} = K_{eq}$$

where K_{eq} is a physical constant, characteristic of a reaction. When $K_{eq} > 1$, $\Delta G^{o\prime}$ is negative, indicating that the reaction proceeds in the forward direction. When $K_{eq} < 1$, $\Delta G^{o\prime}$ is positive, indicating that the reaction proceeds in the reverse direction. If $K_{eq} = 1$, the reaction is said to be in equilibrium. The relation between K_{eq} and $\Delta G^{o\prime}$ is given in Table 7.2. Thus, by substituting the values of ΔG and K_{eq} in Eq. (7.1), we obtain

$$0 = \Delta G^{o\prime} + 1.987 \times 10^{-3}\,(37 + 273) \times 2.303 \log_{10} K_{eq}$$

$$\Delta G^{o\prime} = 1.418 \log_{10} K_{eq}$$

If $\Delta G^{o\prime}$ is known, the actual free energy change ΔG can be calculated by substituting the actual temperature and the concentration of the reactants and products in Eq. (7.1).

The changes in the three thermodynamic quantities are related to each other by the following equation:

$$\Delta G = \Delta H - T\Delta S$$

where T = absolute temperature.

By convention, ΔS is positive when entropy increases and ΔH is negative when heat is released by the system to the surroundings. However, in favourable processes, ΔG or $\Delta G^{o\prime}$ is negative. In fact, ΔG or $\Delta G^{o\prime}$ of a spontaneously reacting system is always *negative.*

The coupling of a thermodynamically unfavourable reaction (endergonic reaction with positive $\Delta G^{o\prime}$ value) with a thermodynamically favourable reaction (exergonic reaction with negative $\Delta G^{o\prime}$ value) is called a *coupled reaction.* As an example, consider the following two reactions related to the phosphorylation of glucose and hydrolysis of ATP.

$$\text{Glucose} + \text{Pi} \rightarrow \text{Glucose-6-P} + H_2O\text{: } (\Delta G^{o\prime} = +3.3 \text{ kcal/mol})$$

$$\text{ATP} \rightarrow \text{ADP} + \text{Pi: } (\Delta G^{o\prime} = -7.3 \text{ kcal/mol})$$

The formation of glucose-6-phsophate is a thermodynamically unfavourable reaction. However, when coupled with ATP hydrolysis—a favourable reaction—the phosphorylation of glucose can occur with a net $\Delta G^{o\prime}$ of –4.0 kcal/mol. The coupled reaction can be represented as follows:

$$\text{Glucose} + \text{ATP} \rightarrow \text{Glucose-6-P} + \text{ADP: } (\Delta G^{o\prime} = -4.0 \text{ kcal/mol})$$

7.2.3 ATP and Its High-energy Phosphate Bonds

As shown in the preceding reaction, ATP has a highly negative $\Delta G^{\circ\prime}$ value for the hydrolysis of phosphoanhydride bond to terminal gamma phosphate (γP) group, which leads to the formation of ADP and Pi. That is, it has a high group transfer potential for phosphate. Such compounds are called high-energy phosphate compounds. Phosphate compounds found in living organisms can be divided into two groups: high-energy compounds (possessing $\Delta G^{\circ\prime}$ value of hydrolysis that is more negative than –25 kJ/mol) and low-energy compounds with lower $\Delta G^{\circ\prime}$ value. ATP, which has $\Delta G^{\circ\prime}$ value of –30.5 kJ/mol (i.e., 7.3 kcal/mol) is a high-energy compound. On the other hand, glucose-6-phosphate, which has $\Delta G^{\circ\prime}$ value of –13.8 kJ/mol (i.e., –3.3 kcal/mol) is a low-energy compound. Examples of high- and low-energy phosphate compounds are given in Table 7.3. The free energy released by hydrolysis of phosphate compounds does not come from the specific bond that is broken, but it results from the product of the reaction, which has smaller free energy content than the reactants.

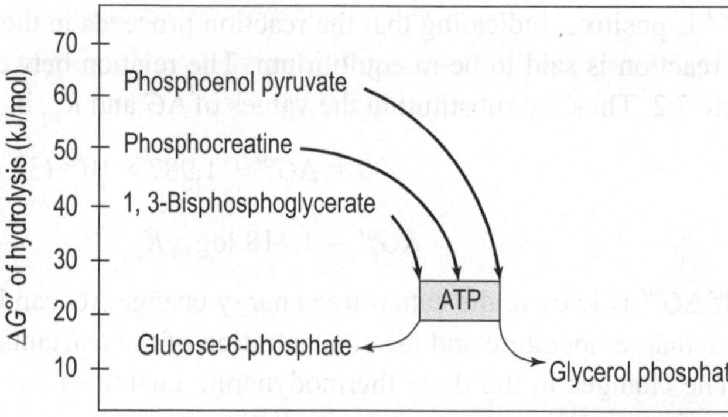

FIGURE 7.1 Flow of phosphate group from compounds with high $\Delta G^{\circ\prime}$ value to compounds with low $\Delta G^{\circ\prime}$ value (arrow indicates the flow of phosphate)

Phosphoenolpyruvate (PEP) has the highest phosphate group transfer potential, whereas ATP has high potential and glucose-6-phosphate has low potential (Fig. 7.1). In other words, synthesis of any phosphorylated compound always occurs due to breakdown of another phosphorylated compound which has more negative free energy of hydrolysis. Phosphate group flows from high-energy phosphate donors, such as PEP, via ATP to low-energy acceptor molecules, such as glycerol and glucose-6-P. This flow is catalysed by kinases.

The excellent phosphate group transfer potential of ATP makes it the energy currency of the cell. This is attributed to the following factors, which are schematized in Fig. 7.2:

- ATP has four negative charges which contribute to electrostatic repulsion. After hydrolysis of the terminal γP, however, the repulsive force is relieved by the charge separation. As a result, Pi and ADP^{2-} are formed.
- Pi is stabilized by the formation of a resonance hybrid (which is not possible in ATP). The hybrid has four P–O bonds with the same degree of double bond character, and the hydrogen ion is not permanently associated with any one of the oxygen.

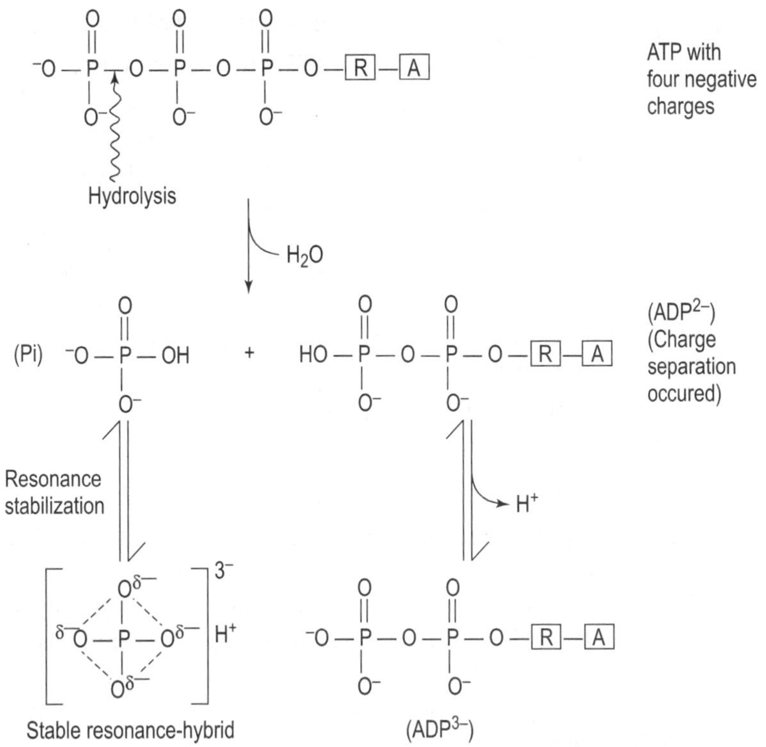

FIGURE 7.2 Factors attributing to larger free energy change of ATP hydrolysis (R = Ribose; A = Adenine)

- ADP^{2-} immediately ionizes, releasing a proton into a medium of very low hydrogen ion concentration at pH 7.
- The greater degree of solvation (hydration) of the products Pi and ADP relative to ATP further stabilizes the products relative to the reactants.

ATP donates some of the energy to the following endergonic reactions:

(a) Assembly of informational macromolecules
(b) Bioluminescent reaction in fireflies
(c) Active transport across membranes
(d) Muscle contraction

Each of these reactions is briefly described as follows:

Assembly of informational macromolecules

ATP is required for the polymerization of monomeric units into high-molecular-weight polymers and for the assembly of orderly sequence. For example, nucleoside triphosphates are the precursors for DNA and RNA synthesis by polymerization. In this process, the anhydride bond between α and β phosphates is hydrolyzed with the release of PPi and nucleoside monophosphate (NMP). NMP moieties are added to the growing polymers. Similarly, during the synthesis of proteins, activation of amino acids occurs by the donation of adenylate groups from ATP. Formation of peptide bond in the ribosome is also accompanied by GTP hydrolysis.

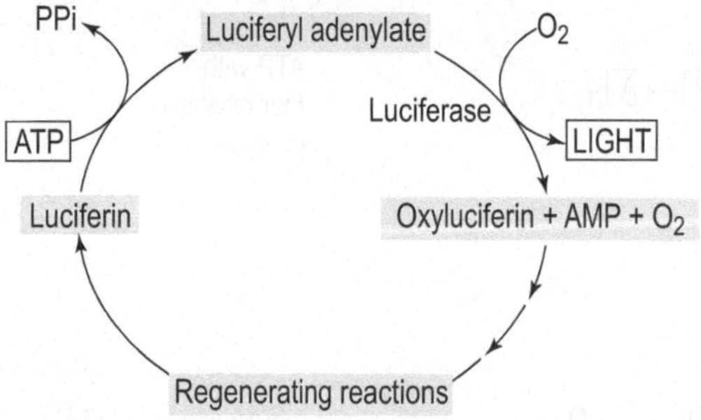

FIGURE 7.3 Role of ATP as the energy source in the conversion of chemical energy into light energy

Bioluminescent reaction in fireflies

ATP is the energy source for the conversion of chemical energy into light energy in fire flies (Fig. 7.3). In this reaction, a complex carboxylic acid, luciferin, is activated by the enzyme luciferase in the presence of ATP to form PPi and luciferyl adenylate, the later being acted upon by the molecular oxygen and luciferase to bring about oxidative decarboxylation of luciferin to produce oxyluciferin. This reaction is accompanied by light emission, the colour of which depends on the species of the firefly and the enzyme structure. The intensity of the light flash is proportional to determine ATP quantity. Therefore, in the laboratory, luciferin and luciferase are used to quantify ATP with high sensitivity.

Active transport across membranes

ATP can supply energy for transporting ions or other molecules across a membrane into another aqueous compartment where the concentration of the transported molecules is higher. In tissues, such as human kidney and brain, as much as two-thirds of the energy consumed at rest is used to pump sodium and potassium ions across plasma membranes via the sodium–potassium ATPase enzyme. Sodium and postassium transport is driven by cyclic phosphorylation and dephosphorylation of the transporter protein with ATP as the phosphate donor.

Sodium-dependent phosphorylation of the sodium and potassium ATPase forces a change in the protein's conformation and the potassium-dependent dephosphorylation favours return to the original conformation. Each cycle in the transport process results in the conversion of ATP to ADP and Pi, and it is the free energy change of ATP hydrolysis that drives the pumping of sodium and potassium. In animal cells, the net hydrolysis of one ATP is accompanied by the outward transport of three sodium ions and the uptake of two potassium ions. The free energy change for the transport of a non-ionic solute from one compartment to another is represented as follows:

$$\Delta G_t = RT\ln\left(\frac{C2}{C1}\right)$$

where

$C1$ = molar concentration of solute in the compartment from which the ion or the molecule moves

$C2$ = molar concentration in the compartment into which the ion or the molecule moves

When a charged species move across a membrane without a counter ion, the separation of electrical charge requires extra electrical work beyond the osmotic work against a concentration gradient. The extra electrical work is denoted as $ZF\Delta\Psi$, where

Z = unitless electrical charge of transported

F = Faraday constant = 97.48 kJ/Vmol

$\Delta\Psi$ = transmembrane electrical potential volts

Therefore, the total energy cost of moving a charged species against an electrochemical gradient is

$$\Delta G_t = RT\ln\left(\frac{C2}{C1}\right) + ZF\Delta\Psi$$

Muscle contraction

ATP hydrolysis drives the cross bridge cycle during the sliding motion of actin–myosin complexes in the muscle. In skeletal muscle cells, myosin and actin are involved in the transduction of chemical energy derived from ATP into mechanical energy (i.e., muscle movement). This mechanism is described in Chapter 11.

TABLE 7.3 Standard free energy of hydrolysis of phosphate-containing compounds

Compound	$\Delta G^{\circ\prime}$ (kcal/mol)
Phosphoenolpyruvate	–14.8
Carbamoyl phosphate	–12.3
Creatinine phosphate	–10.3
ATP	–7.3
Glucose-1-phosphate	–5.0
Glucose-6-phosphate	–3.3

The standard free energy of hydrolysis of some physiologically relevant phosphate-containing compounds is shown in Table 7.3.

The value of free energy of hydrolysis ($\Delta G^{\circ\prime}$) for ATP is intermediate to the value for a compound with high phosphate transfer potential—phosphoenolpyruvate (PEP)—and that for a compound with low phosphate transfer potential—glucose-6-P. So, ATP can function as an *energy carrier* between the catabolic pathway (where it is formed) and the anabolic pathway (where it is used). Thus, ATP acts as the *universal currency* for energy in cells.

In PEP, the hydrolysis of phosphate–ester bond produces pyruvate. Initially, the pyruvate exists in enol form which then undergoes tautomerization (see Terms to Remember) to highly stable keto form. This results in $\Delta G^{\circ\prime}$ value of –61.9 kJ/mol.

High-energy phosphate compounds that store energy for use during heavy work over limited periods of time are called *phosphogens*. In humans, creatine phosphate fills this role. During rest, creatine phosphate is regenerated at the expense of ATP.

$$\text{Creatine} + \text{ATP} \rightarrow \text{Creatine phosphate} + \text{ADP}$$

Apart from phosphate compounds, thioesters also have large free energy of hydrolysis. Thioesters are compounds that do not release Pi on hydrolysis but nevertheless have large, negative standard free energy of hydrolysis. For example, acetyl CoA is hydrolyzed to acetic acid which in turn is ionized by the removal of proton into acetate that is resonance stabilized. $\Delta G^{\circ\prime}$ value is –32.2 kJ/mol. Thus, it can be summarized that compounds with large negative standard free energy of hydrolysis ($\Delta G^{\circ\prime}$) give products that are more stable than the reactants.

7.2.4 Biological Oxidation–Reduction Reaction

The transfer of phosphate groups and electron transfer reactions are pivotal features of metabolism. Electron transfer reactions are called oxidation–reduction (redox) reactions. Flow of electrons in an oxidation–reduction reaction is responsible for all the work done by the living organisms. Oxidation and reduction reactions must occur together. The electron-donating molecule in a redox reaction is called the *reducing agent* or reductant and it is said to be oxidized. On the other hand, the electron-accepting molecule in a redox reaction is called the *oxidizing agent or oxidant,* and it is reduced. Both the agents can function as a conjugate redox pair, for example, Fe^{2+}/Fe^{3+} and Cu^{+}/Cu^{2+}. The general equation for redox reactions is written as follows:

Electron donor ↔ electron + electron acceptor

Electrons flow in the cells in the following four forms:

1. as direct electrons (e.g., $Fe^{2+} + Cu^{2+} \rightarrow Fe^{3+} + Cu^{+}$)

2. as hydrogen atoms (e.g., $AH_2 \rightarrow 2H^+ + 2e$)
3. as hydride ions (equivalent of one proton and two electrons, represented as ':H '), for example, NAD-linked dehydrogenase
4. direct combination with oxygen (e.g., oxidation of hydrocarbon to alcohol)

The term *reducing equivalent* is used to describe the electron equivalent which may be any one of the listed forms.

Electrons flow from the intermediates of metabolic pathway to the electron carriers via enzyme catalysis. In turn, the electrons flow from the electron carriers to electron acceptors with the release of energy. Four classes of molecules function as universal electron carriers: (i) nucleotides—NAD^+, $NADP^+$, Fad, and FMN; (ii) lipid-soluble quinines—ubiquinone and plastoquinone; (iii) iron–sulphur proteins; and (iv) cytochromes.

When a substrate undergoes oxidation (dehydrogenation), $2H^+ + 2e^-$ are produced. (As mentioned earlier, $1H^+ + 2e^-$ form a hydride ion.) $NAD^+/NADP^+$ accepts a hydride ion (:H and gets transformed into NADH/NADPH). The second H^+ is released to the aqueous solvent. Thus, NAD^+ and $NADP^+$ utilize hydride as electron flows. This reaction is represented as follows:

$$NAD^+ + 2e^- + 2H^+ \rightarrow NADH + H^+$$

$$NADP^+ + 2e^- + 2H^+ \rightarrow NADPH + H^+$$

NAD^+ is the cofactor for catabolic reactions, whereas $NADP^+$ is the usual cofactor for anabolic reactions. The roles of all the reducing equivalents are discussed under electron transport chain and oxidative phosphorylation in Sections 7.11 and 7.12, respectively.

Cells contain a variety of molecular energy transducers which convert the energy of electron flow into useful mechanical or biological work. Spontaneous electron flow between two chemical species is driven by a force proportional to the difference in electron affinity between those species. Such driving force is called electromotive force (EMF). EMF can accomplish work if an appropriate energy transducer, such as a motor protein or mitochondria, is present in the circuit.

In a biological circuit, the source of electron is a relatively reduced compound (with low affinity for electron), such as glucose. Glucose is oxidized to electron which flows spontaneously through a series of electron carriers (intermediates) to another chemical with high affinity for electrons, such as oxygen. Such a flow is spontaneous and exergonic because oxygen has a higher affinity for electrons than the intermediate carriers. The resulting EMF provides energy to molecular transducers that do biological work. Some examples are discussed in the following subsection.

7.2.5 Conversion of EMF to Chemical Work via Protomotive Force

In mitochondria, membrane-bound transducers couple electron flow to the production of a transmembrane pH difference, accomplishing osmotic-electrical work. This generates a proton gradient which has potential energy called *protomotive force* (PMF).

Another molecular transducer in mitochondrial membrane utilizes the PMF to do chemical work: ATP is synthesized from ADP and Pi as protons flow spontaneously across the membrane. In *E. coli*, membrane-bound transducers convert EMF into PMF which is used to power flagellar motion.

So far, we have discussed the bioenergetics with reference to ATP hydrolysis and redox reaction. We now move on to the metabolic pathways of individual biomolecules where ATP is either used or synthesized in one or more reactions of the pathway. As long as there is a balance in the pathways, the cells will be normal. Whenever there is an abnormality in the metabolic pathways due to enzyme deficiency or genetic defects, it leads to specific disorders or diseases. These diseases or disorders are also discussed under the respective pathways. Finally, we discuss how the electrons obtained from the metabolic pathways are used up in the resynthesis of ATP.

7.3 METABOLISM OF NUCLEIC ACIDS

Nucleotides required for the synthesis of nucleic acids can primarily be salvaged from in vivo sources. They can also be synthesized from low-molecular weight precursors, de novo leading to a lesser need for the supply of nucleotides from dietary sources. During the synthesis of purine and pyrimidine nucleotides (in salvage as well as in de novo pathways) nucleoside-5-phosphates are formed by the utilization of an activated sugar called 5-phosphoribosyl-1-pyrophosphate (PRPP) and an enzyme called phosphoribosyltransferases. PRPP itself is produced from ribose-5-P by an enzyme PRPP synthetase at the expense of ATP hydrolysis. The formation of PRPP may be depicted in the following manner:

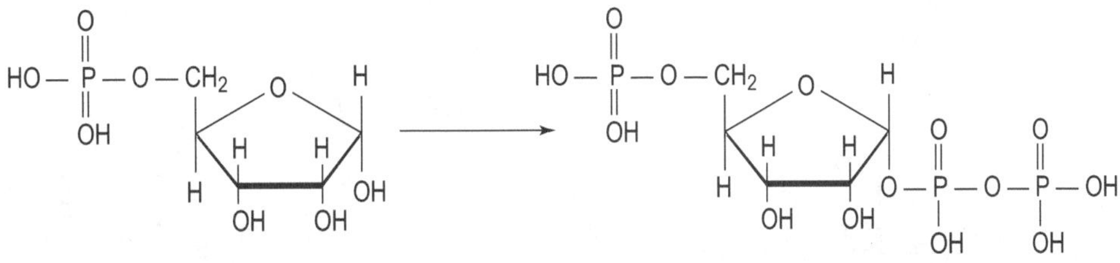

The concerted action of three enzymes catalyses the catabolism of nucleotides—endonucleases, phosphodiesterases, and nucleoside phosphorylases. Endonucleases cleave DNA and RNA at the bonds located in the internal regions, yielding oligonucleotides. Phosphodiesterases hydrolyze the oligonucleotides from the free ends, releasing free nucleosides. The free nucleosides are further acted upon by phosphorylases to generate free bases and sugars. The bases can either be reutilized or further catabolized. Purine bases are catabolized to uric acid and pyrimidine bases to β-aminoiosobutyrate, NH_3, and CO_2.

7.3.1 De Novo Synthesis of Purine Nucleotides

Purine nucleotides are predominantly synthesized in the liver. As mentioned earlier, the synthesis process starts with PRPP. PRPP is converted to inosine-5′-monophosphate (IMP) via ten steps, in the presence of 5 mol of ATP, 2 mol of glutamine, 1 mol of glycine, 1 mol of CO_2, 1 mol of aspartate, and 2 mol of formate. The atoms of purine ring contributed by these compounds are shown by arrows in Fig. 7.4. The purine base of IMP is hypoxanthine. The pathway is shown in Fig. 7.5. Step 1, catalysed by glutamine PRPP amidotransferase, is the rate-limiting step in IMP formation. IMP is the common intermediate in the purine nucleotide and it can be converted into AMP or GMP, depending on the cellular demand.

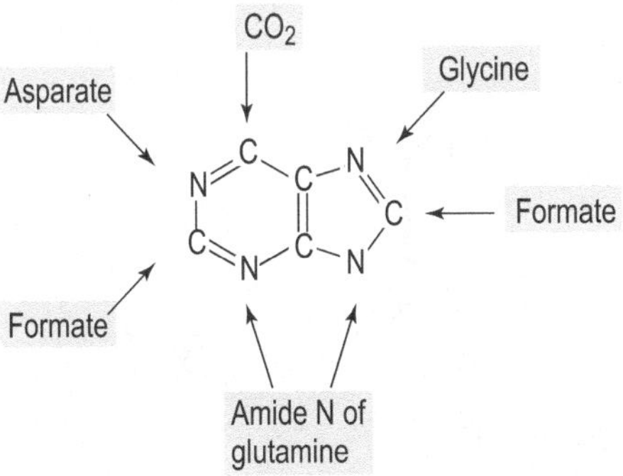

FIGURE 7.4 Compounds contributing the purine atoms

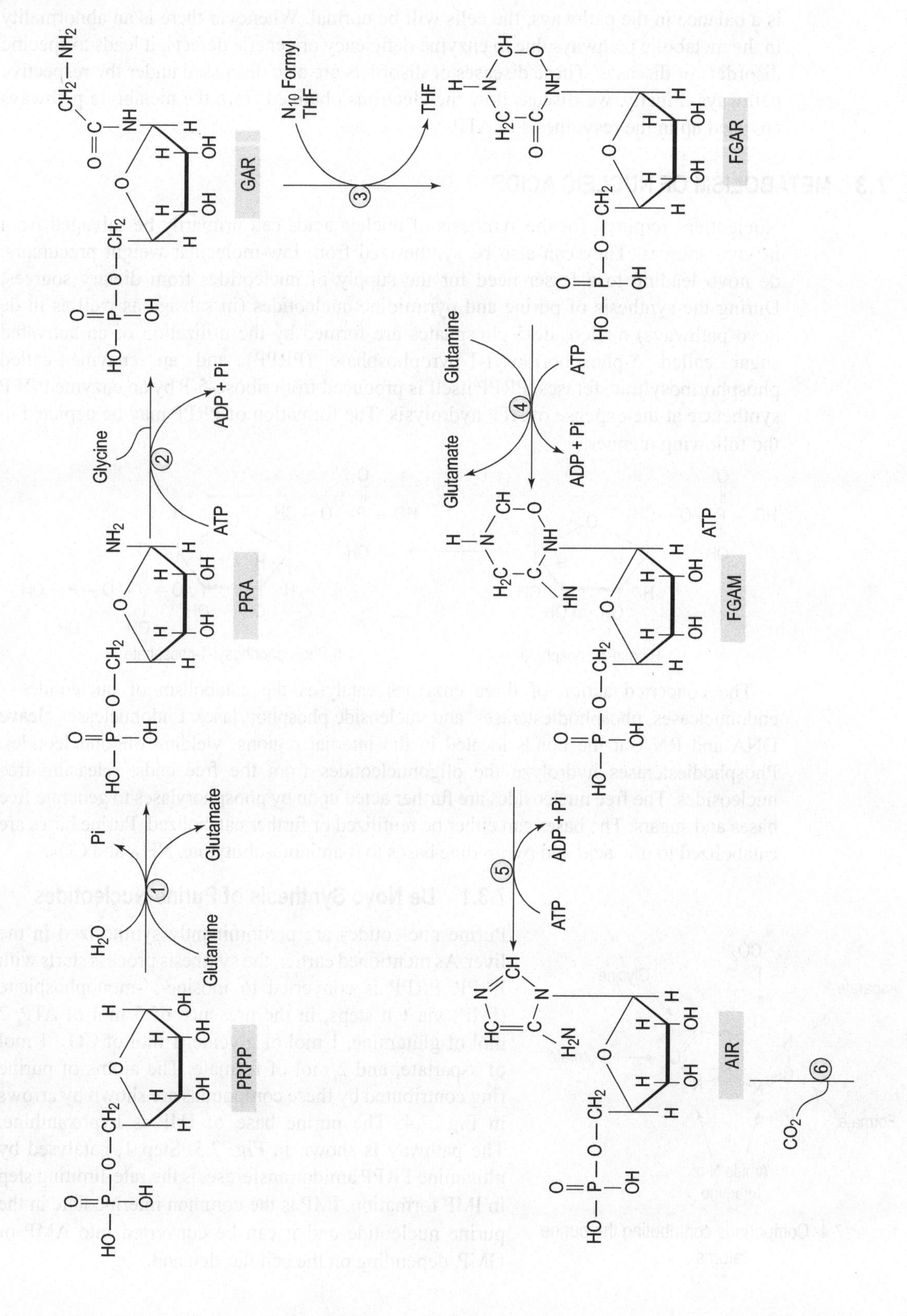
PRPP
H₂O
Glutamine
Glutamate
Pi
PRA
Glycine
ATP
ADP + Pi
GAR
N₁₀ Formyl THF
THF
FGAR
Glutamine
Glutamate
ATP
ADP + Pi
FGAM
ATP
ADP + Pi
AIR
CO₂

(Contd)

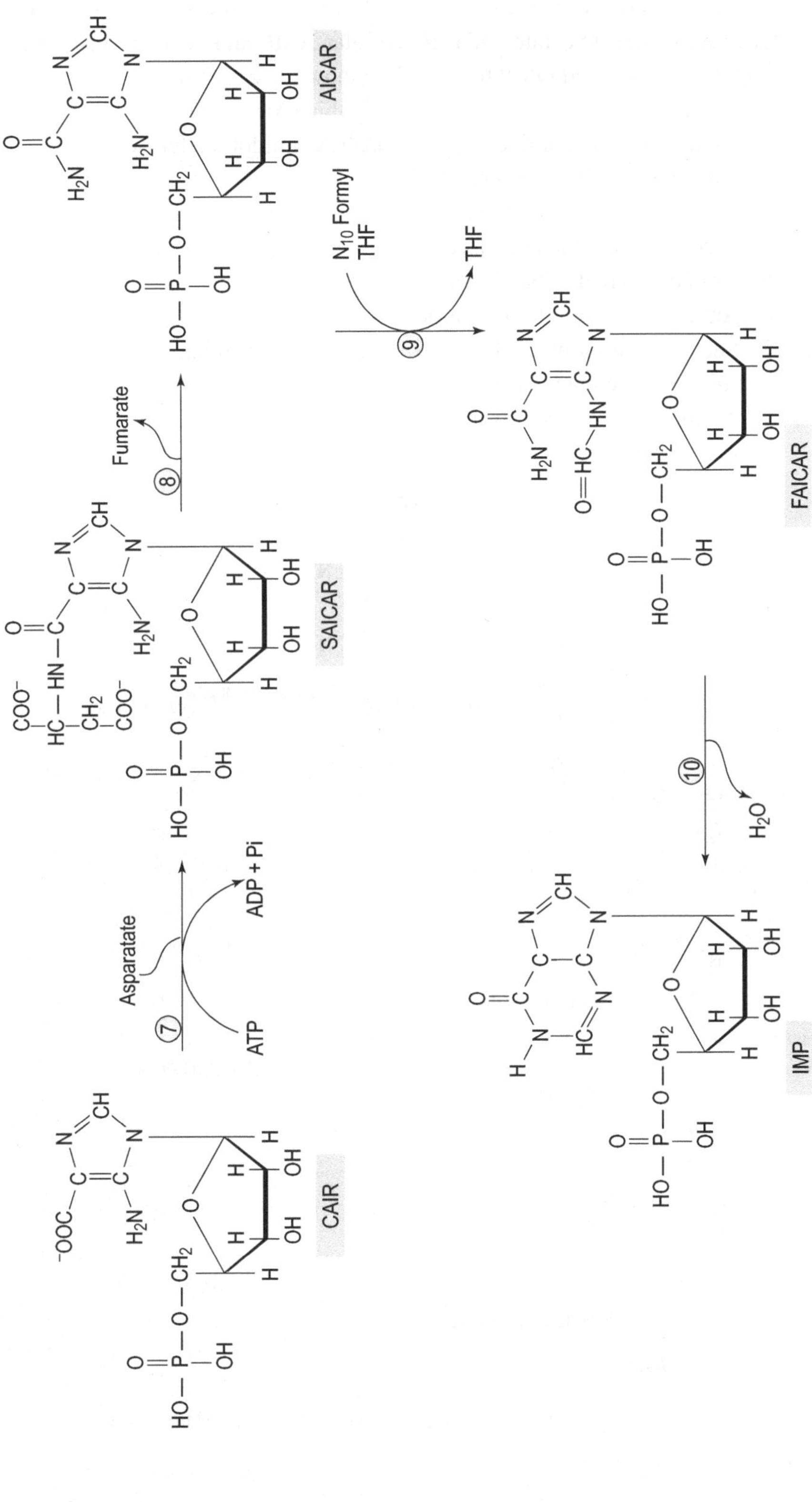

FIGURE 7.5 Formation of IMP from PRPP

In the presence of GTP, IMP forms AMP. In the presence of ATP, IMP forms GMP. Because of these two fates of IMP, cellular AMP and GMP levels are controlled. The synthesis of AMP and GMP from IMP is shown in Fig. 7.6.

The following enzymes are involved in this process:

1. Glutamine phosphoribosylpyrophosphate amidotransferase
2. Glycinamide ribotide synthase
3. Glycinamide ribotide transformylase
4. Formylglycinamide synthase
5. Aminoimidazole ribotide synthase
6. Aminoimidazole ribotide carboxylase
7. Succinylaminoimidazolecarboxamide ribotide synthase
8. Adenylosuccinate lyase
9. Aminoimidazole carboxamide ribotide transformylase
10. IMP cyclohydrolase

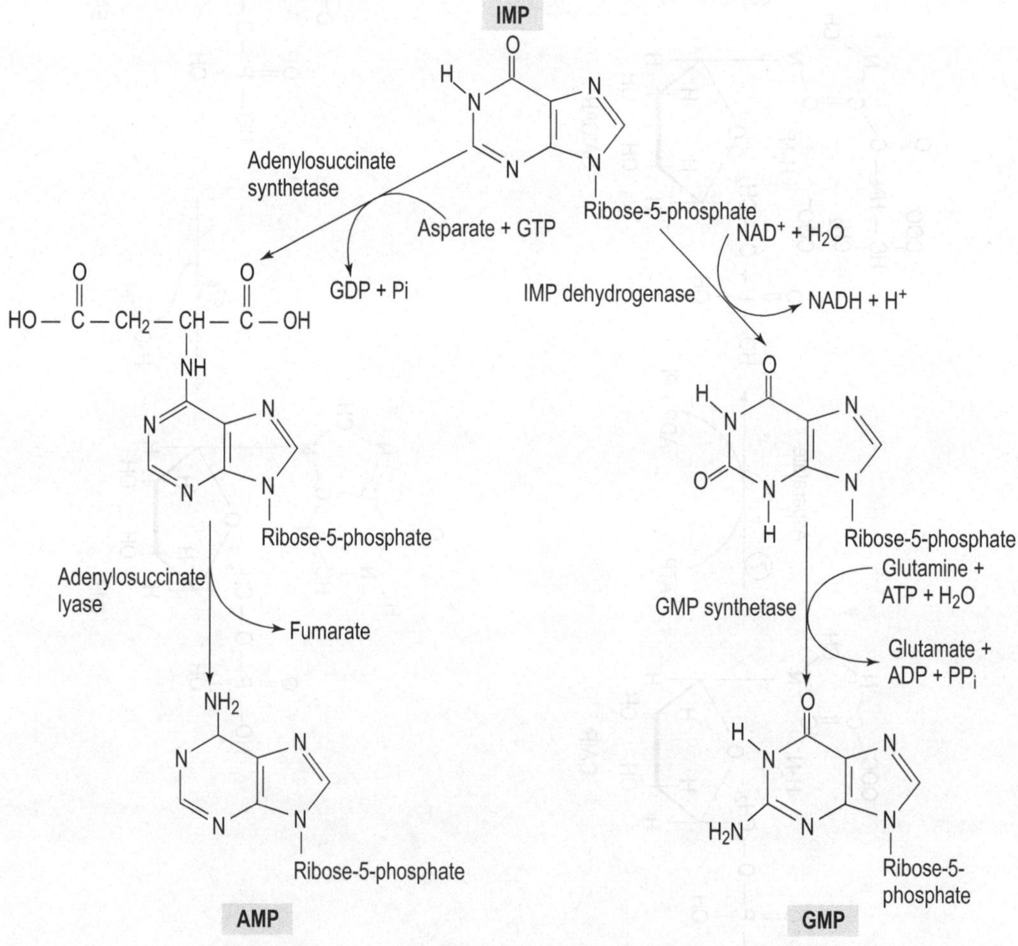

FIGURE 7.6 Formation of AMP and GMP from IMP

7.3.2 Salvage of Purine Nucleotides

In salvage pathway, purine nucleotides are salvaged from purine bases and purine nucleosides. Salvage from adenine is catalysed by adenosine phosphoribosyltransferase (APRT). Salvage from guanine and hypoxanthine is catalysed by hypoxanthine–guanine phosphoribosyltransferase (HGPRT) in a process called phosphoribosylation. The reactions are listed as follows:

Adenine + PRPP ⟶ AMP + PPi

Hypoxanthine + PRPP ⟶ IMP + PPi

Guanine + PRPP ⟶ GMP + PPi

In another process, called *deamination*, adenosine deaminase (ADA) converts adenosine to inosine, which in turn can be converted into hypoxanthine. The outline of salvage pathway for the synthesis of purine nucleotides is shown in Fig. 7.7.

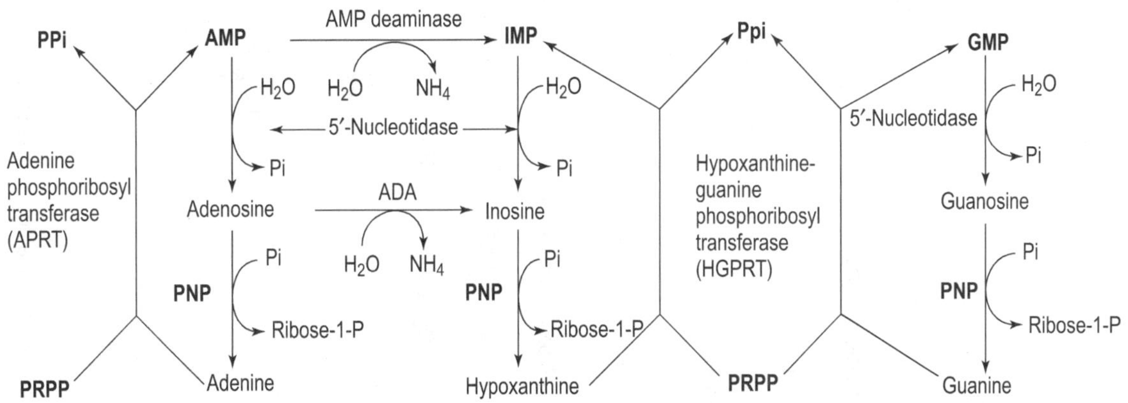

FIGURE 7.7 Outline of salvage pathway for the synthesis of purine nucleotides

7.3.3 Catabolism of Purine Nucleotides

Purine nucleotides are converted to uric acid which is excreted in the form of sodium salts. During this degradation process, the respective nucleosides and xanthine are formed as intermediates. Catabolic pathway of purine nucleotides is shown in Fig. 7.8.

7.3.4 Regulation of Purine Synthesis

Purine synthesis is regulated at three levels. The first regulatory point is the enzyme PRPP synthetase. AMP and GMP exert feedback inhibition over PRPP synthetase. An appropriate ratio of AMP and GMP results in maximal inhibition of the enzyme. PRPP amidotransferase is the second check point. This enzyme has a guanine nucleotide binding site and an adenine nucleotide binding site. GTP, GDP, GMP, ATP, ADP, and AMP can bind to their respective sites and inhibit the enzyme allosterically. The third regulatory point is the branching point, where IMP branches to produce AMP and GMP. Excess ATP can stimulate GMP synthesis, whereas excess GTP accelerates the synthesis of AMP.

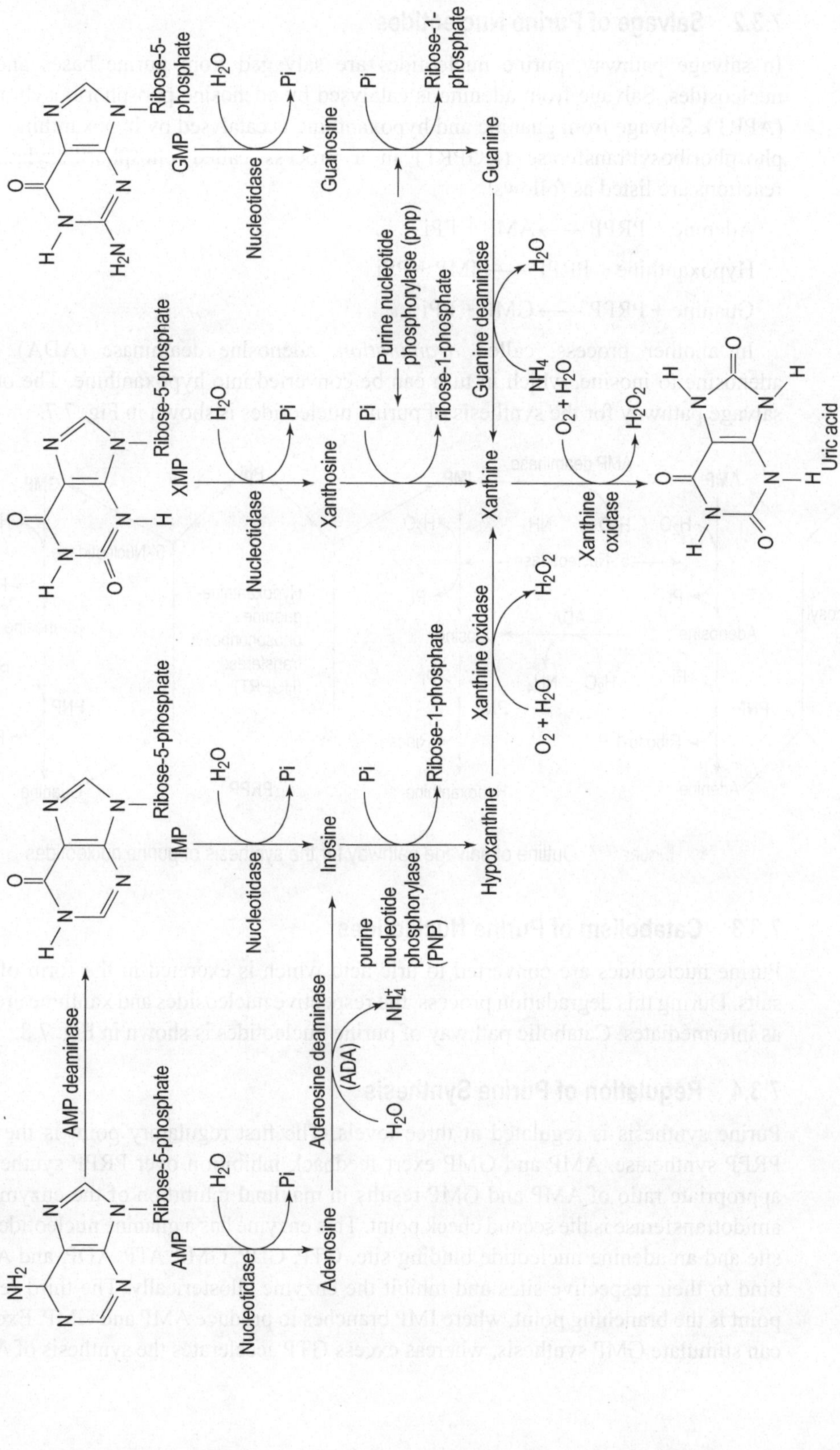

FIGURE.7.8 Catabolism of purine nucleotides to uric acid

7.3.5 Disorders of Purine Metabolism

Abnormal metabolism of purine results in clinical and genetic disorders. A defect in the purine catabolism may result in various disorders. Excessive catabolism of purine nucleotides and deficiency in the salvage enzyme HGPRT results in excessive accumulation of monosodium urate in the synovial fluids of the joints. Lesch–Nyhan syndrome is another disorder resulting from abnormal purine metabolism characterized by gout and malfunction of central nervous system. In this disorder, the gene encoding HGPRT located in the chromosome X is lost.

Severe combined immunodeficiency syndrome (SCID) is another consequence of abnormal purine metabolism, the syndrome being characterized by total loss of immunity due to defective production and function of T cells and B cells. This is mainly because of the deficiency in the enzyme ADA.

The von Gierke disease is another cause of hyperuricemia. It is a glycogen storage disease in which the deficiency of glucose-6-phosphatase raises the level of its substrate glucose-6-phosphate. This leads to an increase in influx of this substance into the pentose phosphate pathway, resulting in elevation of ribose-5-phosphate. Increase in ribose-5-phosphate causes a concomitant increase in the level of PRPP, which in turn increases the amount of purine.

7.3.6 De Novo Synthesis of Pyrimidine Nucleotides

Biosynthesis of pyrimidine nucleotides is simpler than that of purine nucleotides. This section discusses the biosynthesis of uracil and cytosine nucleotides, followed by thymine nucleotide.

Uracil and cytosine nucleotides are formed in three stages. Glutamine is the starting precursor, which is converted to carbamoyl phosphate in the presence of ATP and carbon dioxide. Carbamoyl phosphate is converted to uridine monophosphate (UMP) in the presence of aspartate, NAD, and PRPP. This is a five-step pathway. UMP is then phosphorylated to UTP. UTP forms CTP in the presence of glutamine and ATP. The steps and enzymes involved in the formation of UTP and CTP are given in Fig. 7.9.

The following enzymes are involved in this process:

1. Aspartate transcarbamoylase, ATCase
2. Carbamoyl aspartate dehydratase
3. Dihydroorotate dehydrogenase
4. Orotate phosphoribosyltransferase
5. Orotidine-5′-phosphate carboxylase

First, dUMP (deoxy uridine monophosphate) is formed from dUDP (deoxy uridine diphosphate) or dCDP (deoxy cytidine diphosphate). The dUMP is then acted upon by thymidylate synthase and converted into dTMP in the presence of N5, N10-methylene tetrahydrofolate (THF) (Fig. 7.10). The methyl group needed for the formation of thymine is contributed by N5, N10-methylene THF, which is then converted into dihydrofolate (DHF). DHF in turn, can be reduced by dihydrofolate reductase (DHFR) and THF is regenerated. In the next step, N5, N10-THF is formed from THF by serine hydroxymethyl transferase.

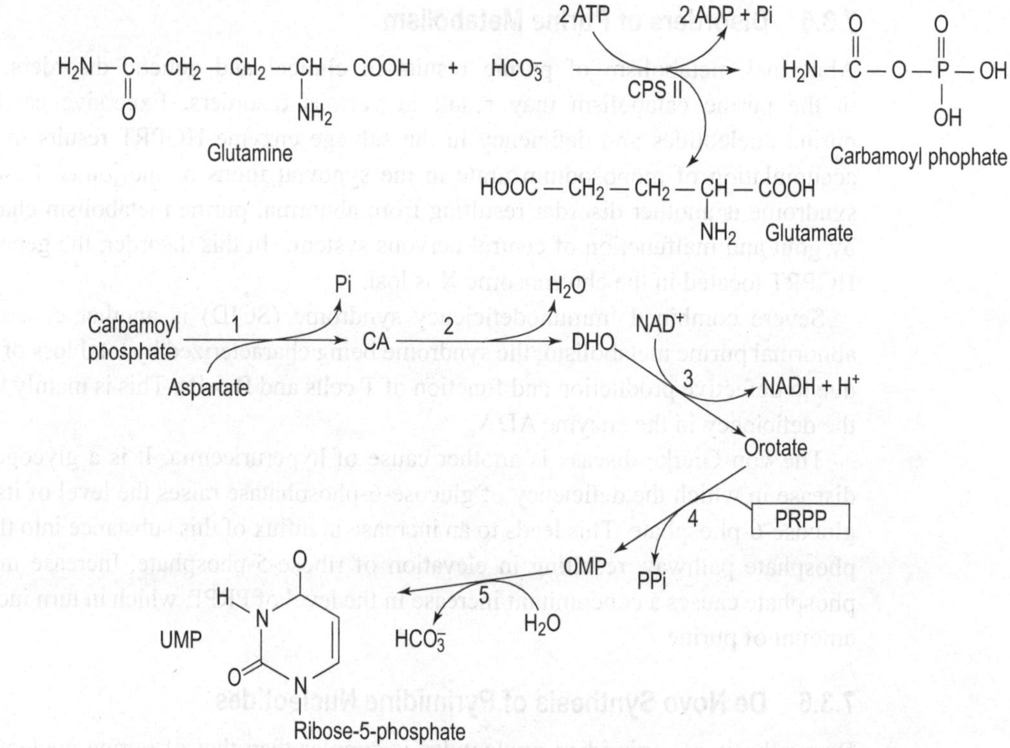

FIGURE 7.9 Steps involved in the biosynthesis of pyrimidine nucleotides (UTP and CTP)

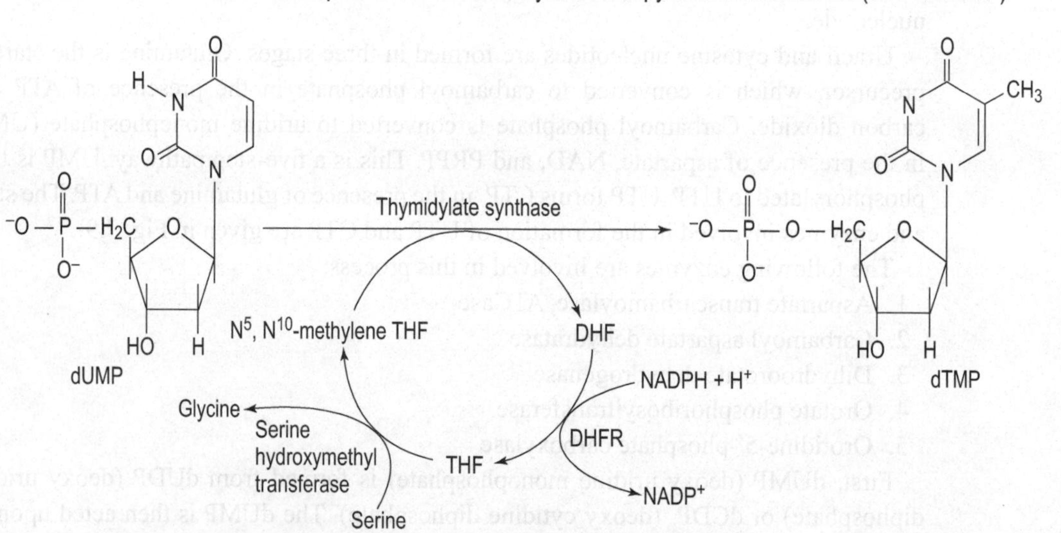

FIGURE 7.10 Formation of dTMP

7.3.7 Salvage of Thymine Nucleotide

The dTTP (deoxy thymidine triphosphate) can be salvaged from either thymidine or deoxyuridine by the action of thymidine kinase. The rate of this enzymatic synthesis is very high during the S phase of the cell cycle. The reaction is as follows:

$$\text{Thymidine} + \text{ATP} \rightarrow \text{TMP} + \text{ADP}$$
$$\text{Deoxyuridine} + \text{ATP} \rightarrow \text{dUMP} + \text{ADP}$$

7.3.8 Regulation of Pyrimidine Biosynthesis

Pyrimidine biosynthesis is regulated at three levels. Aspartate transcarbamoylase (ATCase) and CTP synthetase are the regulatory enzymes of the pyrimidine biosynthesis pathway. ATCase is a multisubunit allosteric enzyme, which performs two important activities. It catalyses the formation of carbamoyl phosphate from glutamine and ATP. This is called the carbamoyl phosphate synthetase II (CPS II) activity of ATCase. Glycine competes with the substrate glutamine for the glutamine-binding site of CPS II, and thus plays a regulatory role. ATCase also catalyses the conversion of carbamoyl phosphate to carbamoyl aspartate. ATP is the allosteric activator, whereas CTP is the allosteric inhibitor of the ATCase. Pyrimidine synthesis is also regulated at the level of PRPP formation. Excessive PRPP results in excessive synthesis of pyrimidine. CTP synthetase, which is responsible for converting UTP to CTP, is feedback inhibited by CTP and activated by GTP.

7.3.9 Catabolism of Pyrimidine Nucleotides

Pyrimidine nucleotides from nucleic acids are acted upon by nucleotidases and pyrimidine nucleoside phosphorylase to release free pyrimidine bases—cytosine, uracil, and thymine. The bases undergo deamination and ring cleavage to form β amino acids, ammonia, and carbon dioxide (Fig. 7.11). Cytosine is degraded to uracil, with the release of ammonia. 5-Methyl cytosine is deaminated to thymine. The ring in uracil and thymine are reduced in the presence of NADH and then cleaved with the release of ammonia and carbon dioxide to form β-alanine and β-aminoisobutyrate, respectively. β-Alanine is either excreted or incorporated into carnosine or anserine, whereas the β-aminoisobutyrate is excreted. The amino group produced during the ring cleavage takes part in the transamination reaction.

7.4 METABOLISM OF AMINO ACIDS AND PROTEINS

Body protein is contributed by two sources—the exogenous source (or the dietary source) and the endogenous source. Both of these are sources of the body's amino acid pool. As amino acids are the building blocks of proteins, this section starts with the formation and utilization of the amino acid pool.

7.4.1 Formation of the Amino Acid Pool

Dietary proteins are degraded in the digestive system by proteolysis to form amino acids and absorbed into the intracellular compartment of the intestinal cells. In this process, the proteins are first acted upon by the enzyme pepsin at the acidic pH in the gastric juice of the stomach. This results in the formation of larger peptides, which stimulate the production of cholecystokinin–pancreozymin and secretin hormone system from the intestinal endocrine cells. Cholecystokinin induces pancreatic acinar cells to secrete trypsinogen, which is activated to trypsin by the enteropeptidase produced by the intestinal mucosal epithelial cells. Trypsin, in turn, activates the inactive endopeptidases (chymotrypsinogen and

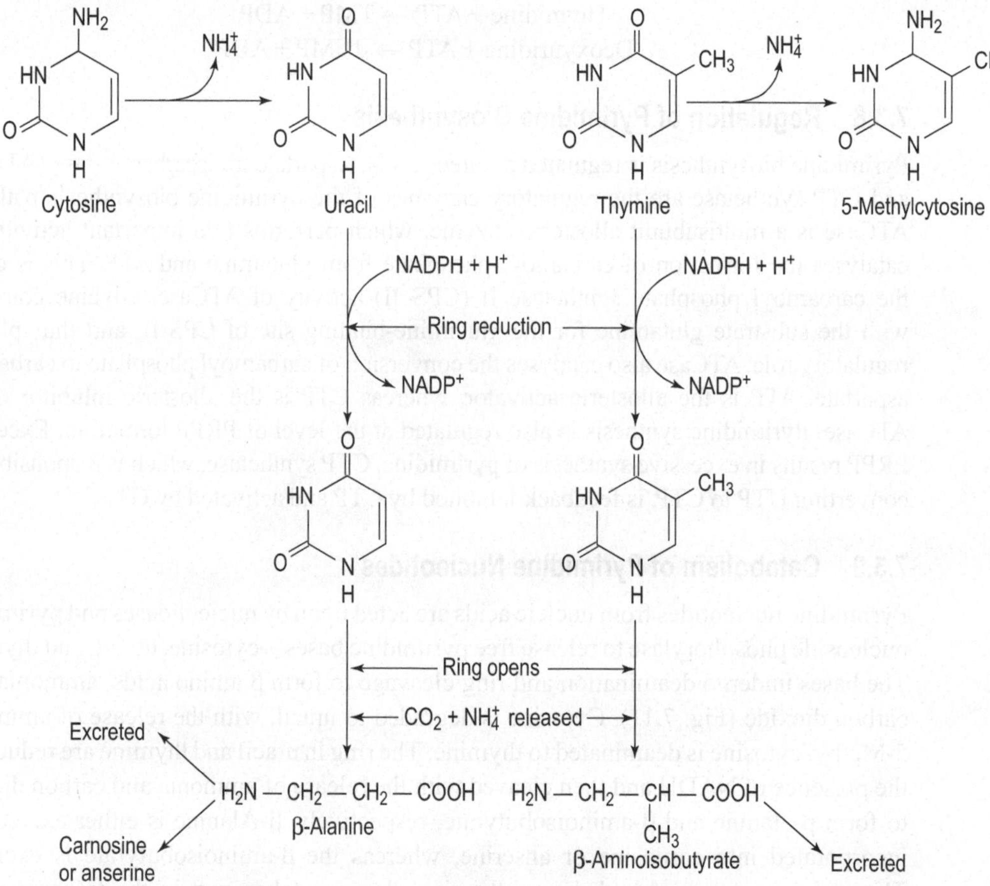

FIGURE 7.11 Catabolism of pyrimidine nucleotides

proelastase) and the inactive exopeptidase (carboxypeptidase). Finally, endopeptidases, exopeptidases along with intestinal aminopeptidases and dipeptidases cleave the larger peptides, liberating dipeptides and free amino acids, both of which are absorbed from the intestinal lumen into the intestinal epithelial cells through amino acid transport system. There are five different amino acid transport systems for the uptake of the following: (a) small, neutral amino acids (b) large, neutral, and aromatic amino acids (c) acidic amino acids; (d) basic amino acids and cystine and (e) iminoacids and glycine.

Any defect in this transport system results in specific disorders. These disorders are discussed as follows:

Hartnup's disease

Hartnup's disease is caused due to the defect in the transport of large, neutral aromatic amino acids into the intestinal cell. This disease was first observed in the family of 'Hartnup'. It is, therefore, given the name Hartnup's disease. Because of this defect, the amino acids are retained within the gut and later excreted in the urine. During their accumulation in the gut, tryptophan is acted upon by the intestinal bacteria to form indole derivatives, such as indoxyl-sulphate, which are also excreted in urine in higher levels.

Cystinuria

Cystine is the dimer of cysteine, formed by disulphide linkage. A defect in the transport of basic amino acids results in the accumulation of cystine. This is probably due to the defect in the absorption of the amino acid in the intestine and renal tubules. Cystine is insoluble in urine. Therefore, it is deposited in the urinary tract, and leads to the formation of stones and urinary tract infection.

Glycinuria

Glycinuria results from block in the transport of glycine and iminoacids (proline and hydroxy proline) in the intestine and renal tubules. These amino acids are, therefore, excreted in the urine without any clinical abnormalities.

Endogeneous amino acid pool is formed by the degradation of body protein, the rate of which varies depending on the nature of the protein. For example, the half lives of hepatic enzymes are just few hours, whereas that of haemoglobin is around 120 days, and that of collagen is too long to determine. In addition to the nature of the protein, other factors are also responsible for the degradation of protein into its constituent amino acids. They are protein denaturation, lysosomal activation, and glucocorticoid-induced proteolysis (in muscle cells).

7.4.2 Utilization of Amino Acid Pool

The amino acid pool is utilized within the body as follows:

- As precursors for protein synthesis, that is, the anabolism of protein (This process is called recycling of amino acids, because amino acids formed from digestion and degradation of proteins are reused for the synthesis of proteins).
- Catabolized further to form urea, in a pathway called the urea cycle. In this pathway, the amino skeleton of the amino acid is degraded to urea and CO_2.

The nitrogen from the body is lost by amino acid degradation. Ideally, for an individual, the total nitrogen loss from the body should be equal to the total daily nitrogen intake. This is called *nitrogen balance*, which indicates that the particular individual is healthy. Higher nitrogen intake is called positive nitrogen balance, which is common in growing children. Higher nitrogen loss is called negative nitrogen balance, which occurs during diseased state, tissue wasting, and starvation. Negative nitrogen balance is fatal if it continues for a long time, at a value of one third of the total body protein. The metabolic pathway used in the degradation of amino acids is discussed separately in the following section.

7.4.3 Catabolism of Amino Acids

This section discusses the degradation of amino group and the carbon skeleton of amino acids.

Degradation of amino group of amino acids

The amino group of amino acids degrades via two stages—*transamination* (Fig. 7.12) and *urea cycle* (Fig. 7.13). In the transamination reaction, the amino acid (say AA_1) combines with a ketoacid (KA_1) and form a new ketoacid (KA_2) and a new amino acid (AA_1). This reaction is catalysed by the enzyme transaminase in the presence of pyridoxal phosphate as the cofactor. The new amino acid (AA_1) then regenerates the first ketoacid and liberates ammonia, the oxidative deamination reaction being catalysed by the enzyme

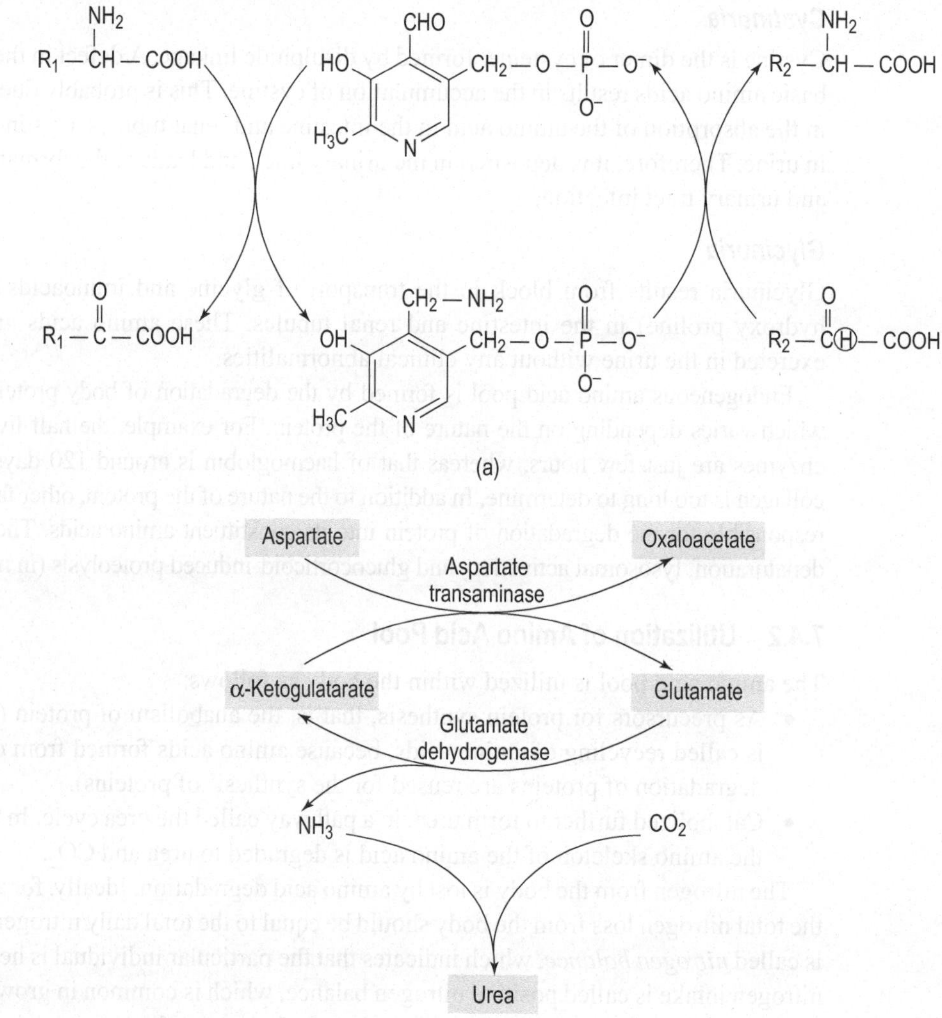

FIGURE 7.12 Degradation of amino acids: (a) Degradation of amino acid by transamination reaction (b) Degradation of aspartate, an example for transamination reaction

dehydrogenase in the presence of NAD $(P)^+$. For example, the amino acid aspartate and α-ketoglutarate undergo transamination to form oxaloacetate and glutamate, in the presence of aspartate transaminase. Glutamate in turn deaminates, regenerating α-ketoglutarate and liberating ammonia. The enzyme catalysing the oxidative deamination reaction is glutamate dehydrogenase in the mitochondrial matrix. The transamination reaction is regulated by the enzyme amino acid dehydrogenase in the presence of allosteric regulators. ADP and GDP (guanosine diphosphate) are positive modulators, whereas ATP and GTP (guanosine triphosphate) are negative modulators of glutamate dehydrogenase.

Utilization of ammonia—the urea cycle The ammonia liberated by the transamination reaction is converted into urea in the liver, in the presence of ATP and aspartate. Urea

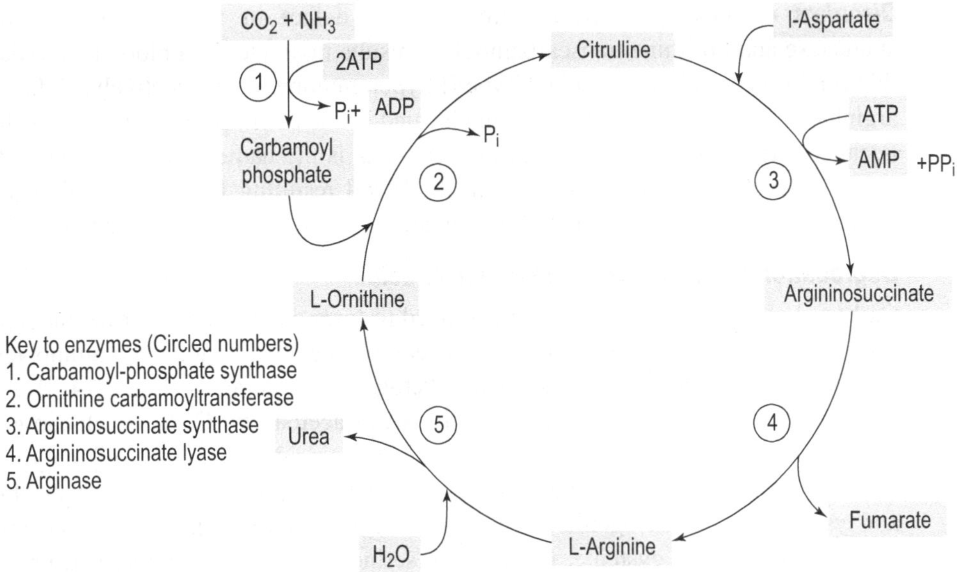

FIGURE 7.13 Urea cycle

contains two atoms of nitrogen, one derived from ammonia, and the other from aspartate. The overall reaction that produces urea may be written as follows:

$$CO_2 + NH_3 + 3ATP + Aspartate + 2H_2O \rightarrow Urea + 2ADP + 2Pi + AMP + PPi + Fumarate$$

The pathway leading to the formation of urea is outlined as follows:

- Ammonia is first incorporated into carbon dioxide to form carbamoyl phosphate in the presence of carbamoyl phosphate synthetase and two molecules of ATP.
- Carbamoyl phosphate combines with L-ornithine and forms L-citrulline in a reaction catalysed by ornithine transcarbamoylase.
- L-Citrulline reacts with L-aspartate that is generated from transamination reaction to form argininosuccinate, and this reaction is catalysed by synthetase in the presence of ATP.
- Argininosuccinate is decomposed by argininosuccinate lyase to form fumarate and L-arginine. The fumarate being a Krebs cycle intermediate (Section 7.10) links the urea cycle with Krebs cycle.
- Finally, arginine is cleaved into urea and L-ornithine. L-ornithine further reacts with the next molecule of carbamoyl phosphate and continues the cycle. Urea is a highly soluble compound and it enters the blood and gets excreted in the urine.

Five enzymes are involved in catalysing the urea cycle. Among them, the first two enzymes—carbamoyl phosphate synthetase and ornithine transcarbamoylase—are compartmented in the liver mitochondria. The other three enzymes (argininosuccinate synthetase, argininosuccinate lyase, and arginase) are distributed in the cytosol of the liver cells.

Disorders of urea cycle metabolism Genetic defect in the carbamoyl phosphate synthetase and ornithine transcarbamoylase results in an elevated blood ammonia level; this condition is called Type I and Type II hyperammonemia, respectively. Deficiency in argininosuccinate synthetase results in elevated levels of citrulline in urine; this disorder is called citrullinuria. Argininosuccinic academia is the name given to the condition of elevated levels of argininosuccinate in the blood resulting from genetic defect in the enzyme argininosuccinate lyase. Defect in arginase causes hyperargininemia.

Degradation of carbon skeleton of amino acids

Carbon skeleton of amino acids are degraded to form either glucose or ketone bodies or both. Depending on the end product of the carbon skeleton, degradation amino acids are classified as glucogenic amino acids and ketogenic amino acids (Fig. 7.14). Ketogenic amino acids are degraded to either acetyl CoA or acetoacetyl CoA, which can generate ketone bodies (refer Section 7.8). Glucogenic amino acids are catabolized to pyruvate or Krebs cycle intermediates, which can form glucose via phosphoenolpyruvate. Leucine and lysine are purely ketogenic. Glycine, alanine, valine, asparagine, glutamine, arginine, aspartate, glutamate, cysteine, methionine, serine, threonine, histidine, and proline are purely glucogenic. On the other hand, isoleucine, phenylalanine, tyrosine, and tryptophan are both ketogenic and glucogenic.

We shall now discuss the degradation of the carbon skeleton of different kinds of amino acids.

Conversion of amino acids to pyruvate The simplest amino acid, glycine; three-carbon amino acids serine, alanine, and cysteine; and the four-carbon amino acid threonine are

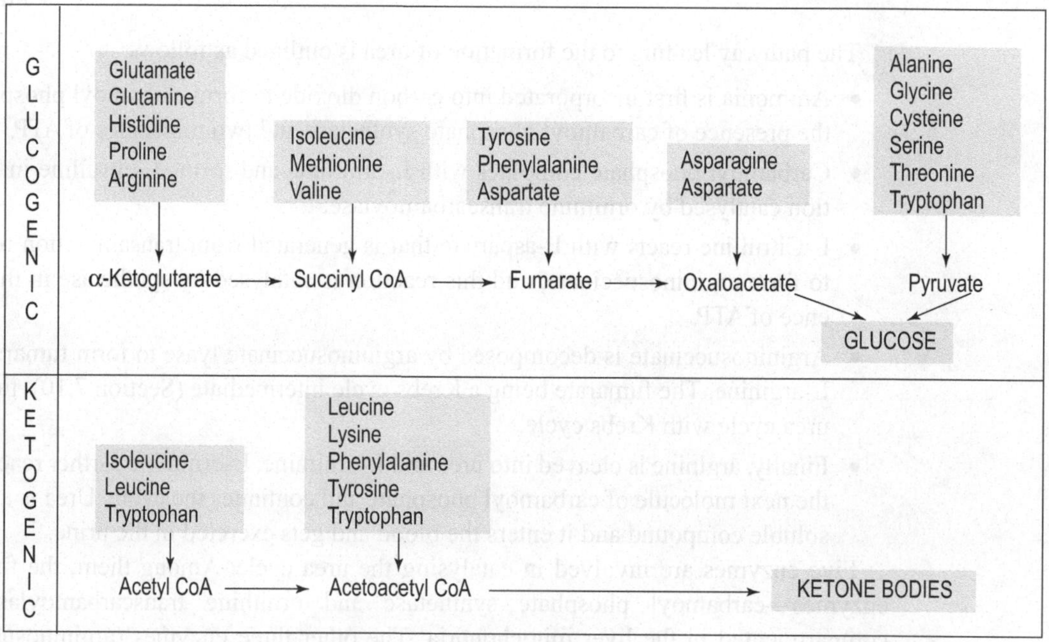

FIGURE 7.14 Glucogenic and ketogenic amino acids and the products of their carbon skeleton degradation

catabolized to form pyruvate (Fig. 7.15). Glycine forms serine in the presence of serine hydroxymethyltransferase and N5, N10-methylene THF. Serine is then converted to pyruvate by non-oxidative deamination. Alanine undergoes transamination with α-ketoglutarate, serine undergoes deamination by serine dehydratase, whereas threonine is converted into aminoacetone intermediate; all finally forming pyruvate. Cysteine is degraded in different ways, forming pyruvate or CoA or tauraine or 5-phosphoadenosine-5-phosphosulphate (PAPS), the active sulphate. These pathways are shown in Fig. 7.16.

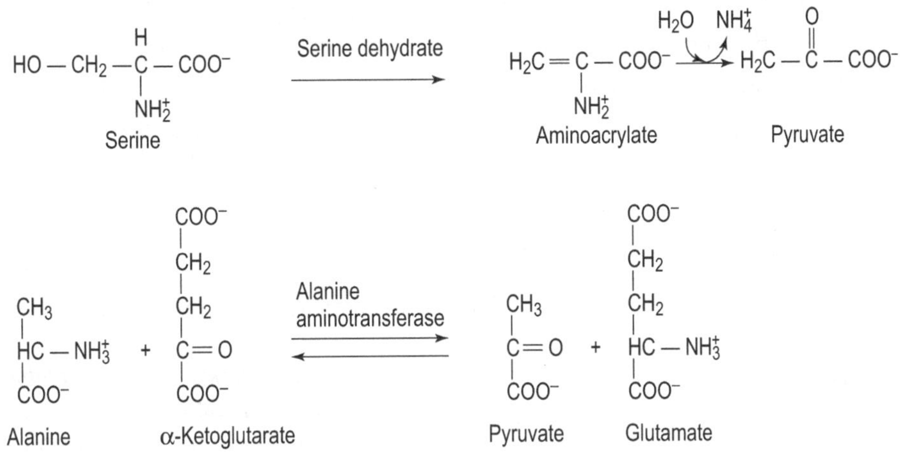

FIGURE 7.15 Catabolism of amino acids to pyruvate

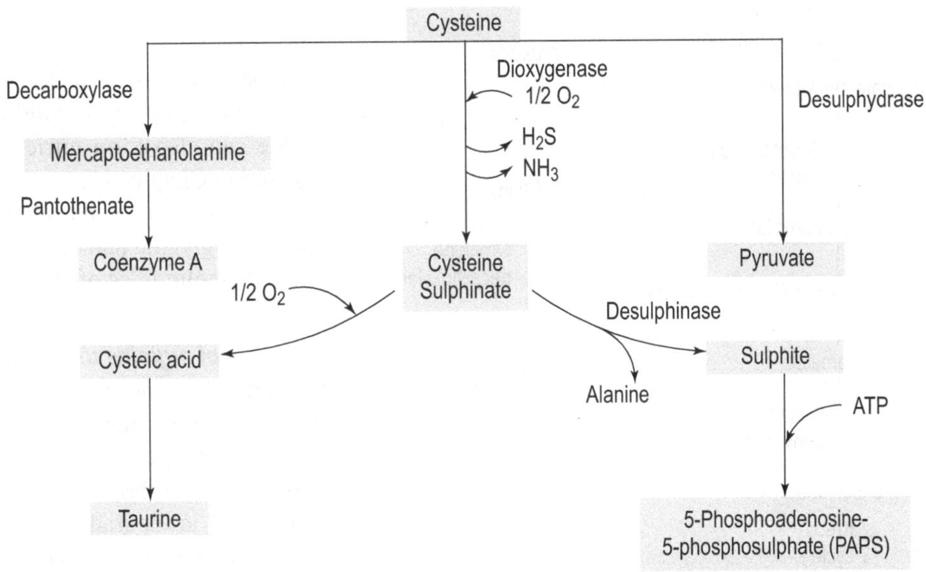

FIGURE 7.16 Degradative pathways of cysteine

Conversion of amino acids to oxaloacetate Aspartate and asparagine are converted to oxaloacetate, which is a Krebs cycle intermediate. Asparagine is acted upon by asparaginase to aspartate. Aspartate reacts with α-ketoglutarate and gets transaminated into oxaloacetate. Degradation of asparagines and aspartate is shown in Fig. 7.17.

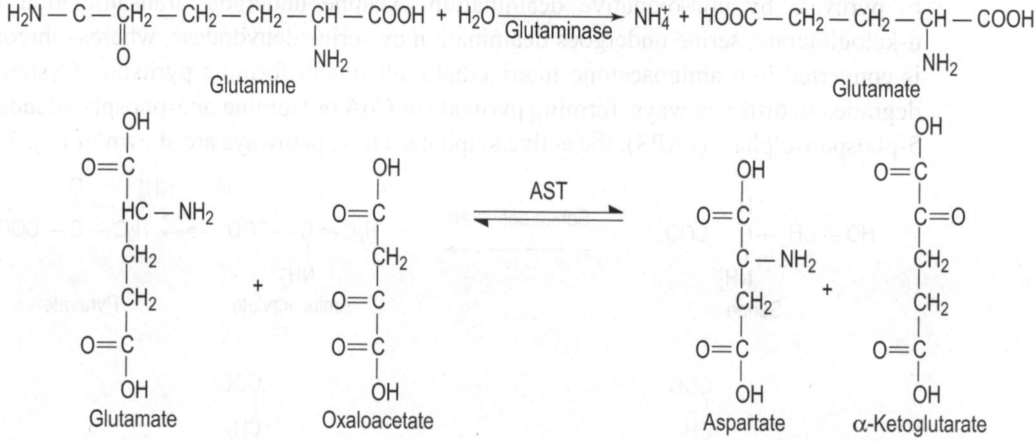

FIGURE 7.17 Conversion of asparagine and aspartate to oxaloacetate

Conversion of amino acids to α-ketoglutarate Some of the five-carbon amino acids are catabolized to glutamate, which is then oxidatively deaminated to α-ketoglutarate. Arginine, glutamine, histidine, and proline come under this category.

Arginine forms ornithine and urea (urea cycle). Ornithine is converted to glutamate. Ornithine can also form putrescine which, in turn, is converted to spermidine. In addition, arginine can also form nitric oxide (functions discussed in Chapter 11) or creatine phosphate. Formation of creatine phosphate occurs in the presence of glycine and SAM.

The catabolic fates of arginine are shown in Fig. 7.18.

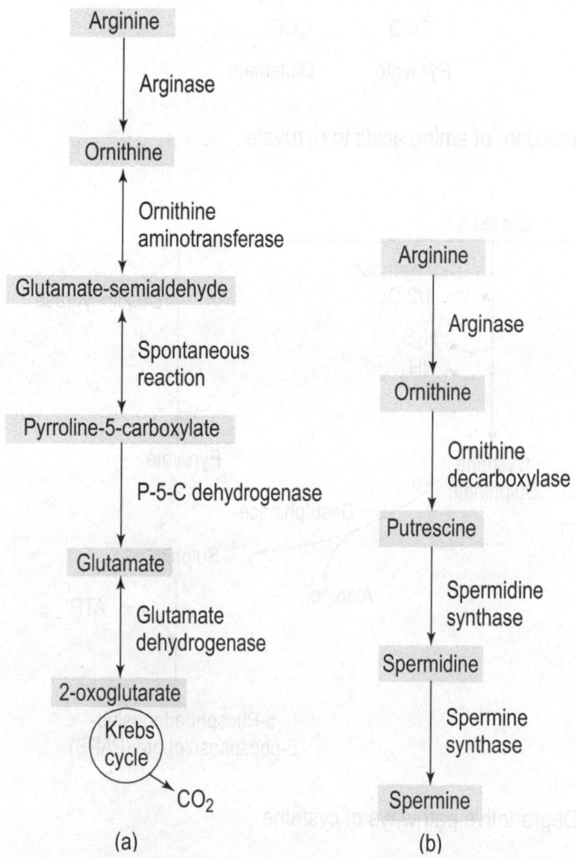

FIGURE 7.18 Catabolism of arginine (a) Catabolism of arginine to glutamate (b) Catabolism of arginine to spermidine

Catabolism of arginine to spermidine

Glutamine is converted to glutamate by glutaminase. Glutamate is then converted to α-ketoglutarate by oxaloacetate and transaminase. The reactions are shown in Fig. 7.19.

Histidine is degraded by histidase in a non-oxidative deamination reaction, forming urocanic acid. Urocanase converts urocanic acid to 4-imidazole-5-propionate which is further hydrolyzed to N-formiminoglutamate (FIGLU). THF accepts the formimino moiety from FIGLU and releases the glutamate. THF

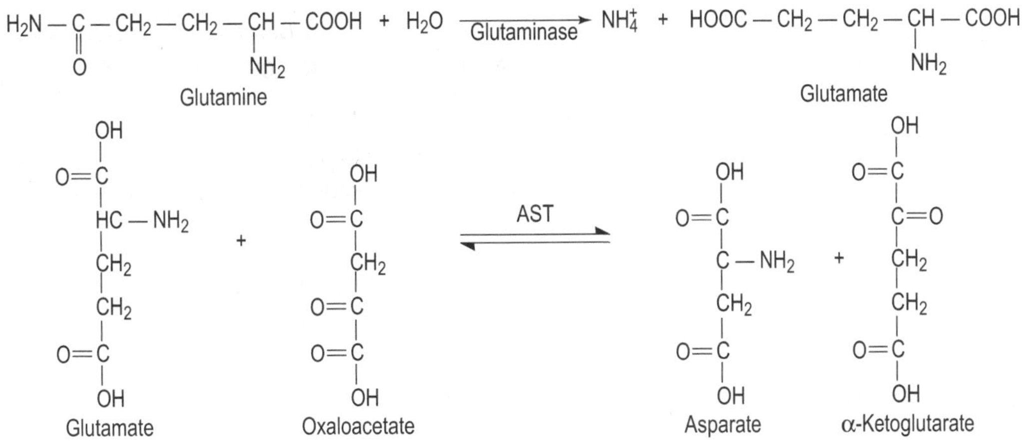

FIGURE 7.19 Catabolism of glutamine (AST, aspartate transaminase)

is converted to N5–formimino THF. Metabolic degradation of histidine is shown in Fig. 7.20. Deficiency of histidase leads to histidinemia, the accumulation of histidine in the blood. As a result, urinary excretion of histidine is elevated. Mental retardation and defective speech are the features of histidinemia.

Proline oxidase converts proline to proline-5-carboxylate, which forms glutamate via glutamate semialdehyde intermediate. Formation of glutamate in the last step is catalysed by the oxidation of glutamate semialdehyde in the presence of dehydrogenase and NAD^+ (Fig. 7.21). The deficiency of proline oxidase results in hyperprolinemia Type I, whereas that of glutamate-semialdehyde dehydrogenase results in hyperprolinemia Type II. Both of these are autosomal recessive disorders.

Conversion of branched-chain amino acids to acetoacetyl CoA and succinyl CoA Branched-chain amino acids, such as valine (glucogenic), leucine (ketogenic), and isoleucine (both glucogenic and ketogenic), are degraded by transaminases in the presence of α-ketoglutarate to their corresponding α-ketoacids—α-ketoisovaleric acid, α-ketoisocaproate, and α-ketomethylvalerate, respectively. A defect in branched-chain ketoacid dehydrogenase results in maple syrup urine disease, in which the branched-chain ketoacids are excreted in the urine. The ketoacids imparts a characteristic odour to the urine, hence the name. The three α-ketoacids are acted upon by a common enzyme, branched-chain ketoacid dehydrogenase, NAD^+, TPP, and lipoate to catalyse their oxidative decarboxylation, followed by further oxidation by FAD. The products of these reactions are methacrylyl CoA, β-methyl crotonyl CoA, and methyl acetoacetyl CoA for valine, leucine, and isoleucine, respectively. These reactions are shown in Fig. 7.22. Valine also yields L-methyl malonyl CoA. L-Methyl malonyl CoA undergoes intramolecular

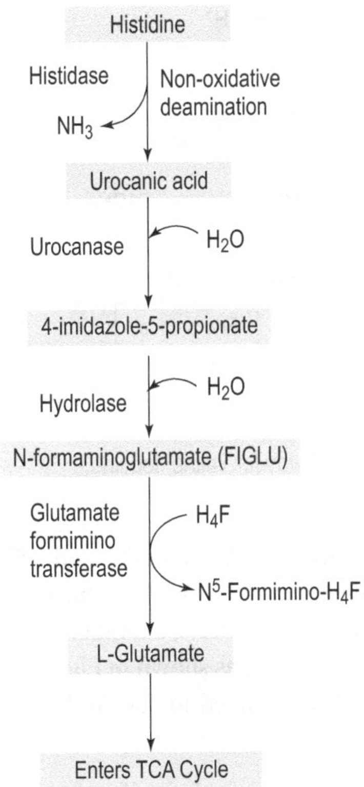

FIGURE 7.20 Catabolism of histidine (H_4F = Tetrahydro folate)

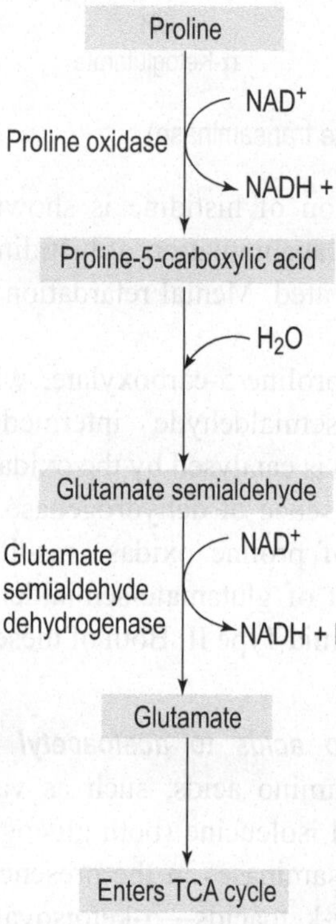

FIGURE 7.21 Catabolism of proline

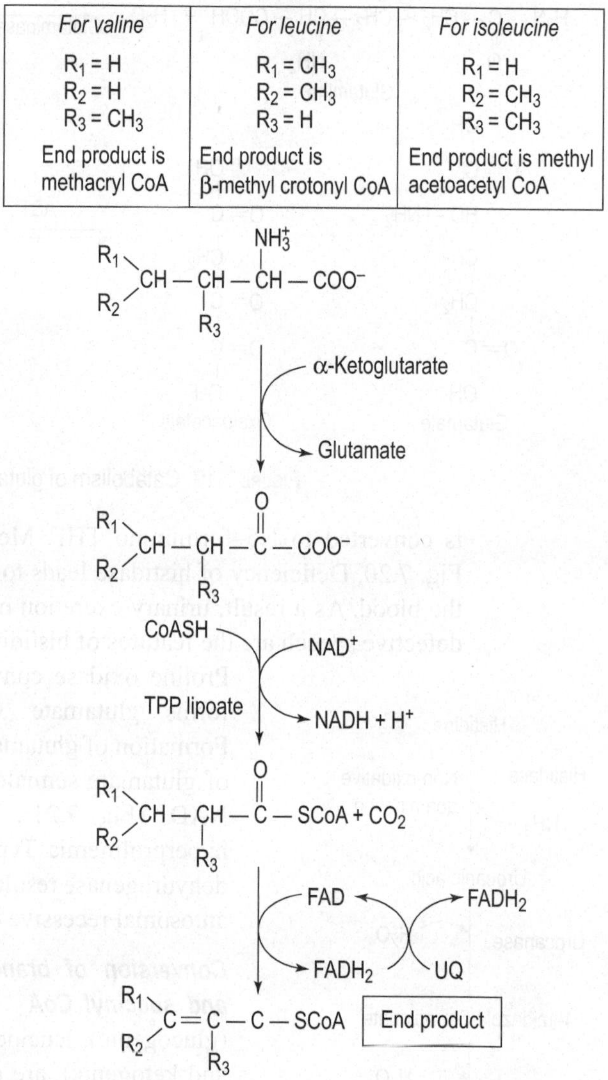

FIGURE 7.22 Common metabolic pathway for valine, leucine, and isoleucine

rearrangement reaction, catalysed by methyl malonyl CoA mutase and 5′-deoxyadenosylcobalamin in the presence of ATP to form the succinyl CoA. Methylmalonic aciduria is the disorder of the deficiency in the enzyme methyl malonyl CoA mutase.

The CoA products formed from the branched-chain amino acids, as shown in Fig. 7.22, are further individually degraded. Valine and isoleucine are degraded to succinyl CoA, whereas leucine is metabolized to acetyl CoA (Fig. 7.23).

Conversion of methionine to succinyl CoA Methionine has biological significance as a methyl group donor in methylation reactions after it is activated to *S*-adenosyl methionine

(SAM). Compounds that depend on SAM for methyl group and the product formed from them are shown in Table 7.4.

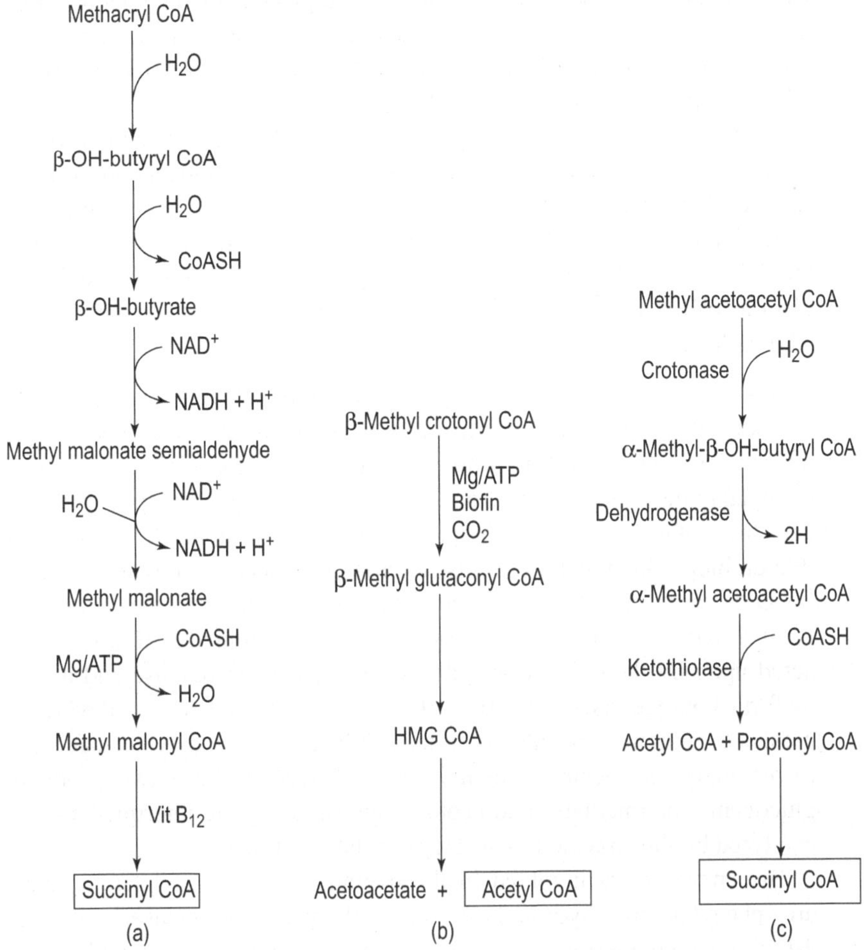

FIGURE 7.23 Individual metabolic pathway of the products formed from (a) Valine, (b) Leucine, and (c) Isoleucine

TABLE 7.4 Compounds methylated by activated methionine

Methyl-group-accepting compound	Product formed by methylation
DNA	Methylated DNA
RNA	Methylated RNA
Protein	Protein methylated in histidine and lysine residues
Norepinephrine	Epinephrine
Niacinamide	N-Methyl niacinamide
Phosphatidyl ethanolamine	Phosphatidyl choline

The first step in the metabolism of methionine is its activation to SAM by SAM synthetase at the expense of ATP. SAM liberates its methyl group for methylation reactions and gets itself converted to *S*-adenosylhomocysteine (SAH) in the presence *O*-methyl transferase. SAH is hydrolyzed to adenosine and homocysteine. Homocysteine has two fates—it can react with serine to form cystathionine under the catalysis of cystathionine synthase and pyridoxal phosphate. (SAH can also regenerate methionine by accepting a methyl group from methylcobalamin. By this reaction, methylcobalamin forms cobalamin, which regains its methyl group from N_5-methyl THF and regenerates methylcobalamin.) Cysteine is degraded into cysteine and homoserine, and the latter is deaminated to α-ketoglutarate. α-ketoglutarate is decarboxylated by CoASH to form propionyl CoA, which is then converted to succinyl CoA.

A defect in the synthesis of cystathionine synthase elevates the level of homocysteine in the blood and ruine. This condition is called homocystinuria Type-1, which has features such as mental retardation, hepatomegaly, skeletal deformities, and ectopia lentis. Homocystinuria Type-2 is another disorder with mental retardation caused due to the deficiency of N_5-methyl THF-homocysteine methyl transferase. There is yet another disorder called homocystinuria Type-3 in which the deficient enzyme is N_5, N_{10}-methylene THF reductase.

Conversion of amino acids to glucogenic and ketogenic intermediates Tyrosine is catabolized to fumarate, a glucogenic intermediate, and acetoacetate, a ketogenic intermediate (Fig. 7.24). Phenylalanine is hydroxylated to tyrosine (hydroxyl phenylalanine), and the reaction catalysed by the enzyme phenylalanine hydroxylase and the cofactor tetrahydrobiopterin. In this reaction, the cofactor is oxidized to dihydrobiopterin, which in turn is essentially reduced by dihydrobiopterin reductase and NADPH to regenerate the cofactor. Tyrosine is transaminated with α-ketoglutarate to form 4-hydroxyphenylpyruvate by tyrosine transaminase. Hydroxylphenylpyruvate is acted upon by the enzyme 4-hydroxyphenylpyruvate oxidase and ascorbate coenzyme to form homogentistate. Homogentisate is converted to 4-maleylacetoacetate, in the presence of homogentisate oxidase. Further, this 4-maleylacetoacetate is isomerized to 4-fumarylacetoacetate which is then cleaved to form two products—fumarate (a glucogenic intermediate) and acetoacetate (a ketogenic intermiediate). The cleavage is catalysed by the enzyme fumarylacetoacetate hydrolase.

Six enzymes are involved in the degradation of phenylalanine and tyrosine. They are phenylalanine hydroxylase, dihydrobiopterin reductase, tyrosine transaminase, 4-hydroxyphenylpyruvate oxidase, homogentisate oxidase, and fumarylacetoacetate hydrolase. Defects in one or more of these enzymes cause genetic disorders or diseases.

- Defect in phenylalanine hydroxylase leads to accumulation of phenylalanine derivatives, such as phenylpyruvate, which are toxic. This disease is called classic phenylketonuria (PKU), which causes mental retardation.
- Defect in the enzyme dihydrobiopterin reductase, which is essential for the regeneration of the cofactor tetrahydrobiopterin, causes atypical PKU, which causes neurologic problems.
- Defect in tyrosine transaminase results in the accumulation of tyrosine, causing tyrosinemia Type II.
- Defect in 4-hydroxyphenylpyruvate oxidase causes neonatal tyrosinemia.
- Defect in the enzyme homogentisate oxidase causes alkaptonuria, which is characterized by excessive excretion of homogentisate in the urine. Excessive urinary homogentisate darkens on exposure to air. This disease is not life threatening.
- Defect in the enzyme fumarylacetoacetate hydrolase causes tyrosinemia Type I or tyrosinosis.

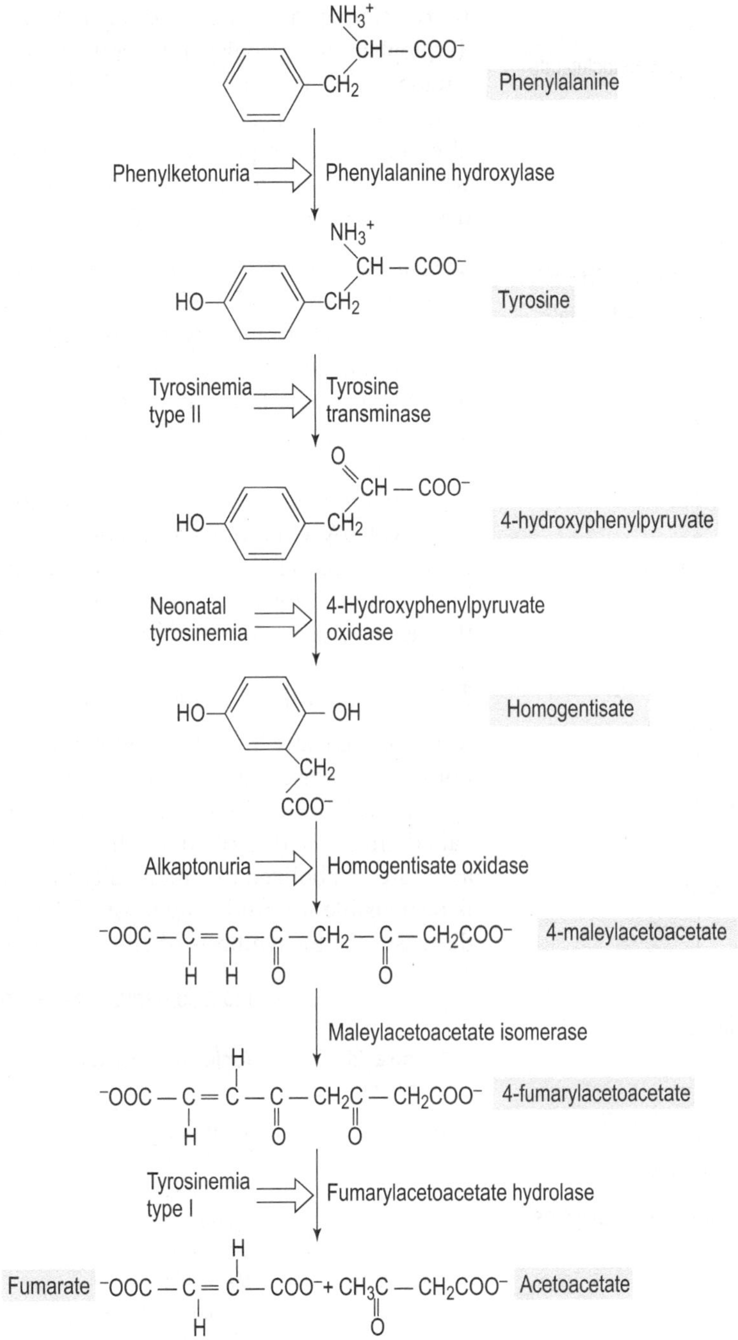

FIGURE 7.24 Degradation of phenylalanine and tyrosine

Tyrosine is the precursor for many amino acid derivatives. That is, tyrosine is catabolized to form biologically important compounds, which are discussed in Section 7.4.4.

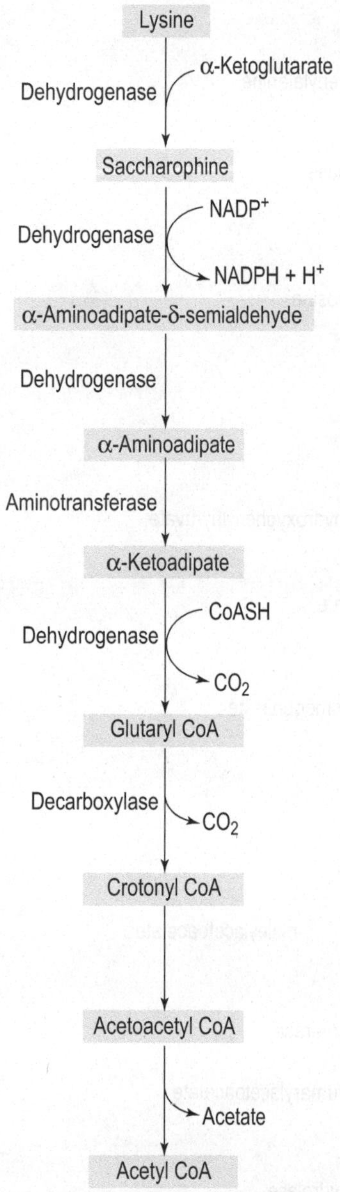

FIGURE 7.25 Degradation of lysine

Conversion of tryptophan to ketogenic and glucogenic intermediates Tryptophan is catabolized to crotonyl CoA, formate and alanine. Crotonyl CoA is an intermediate in the fatty acid oxidation pathway. Thus, tryptophan is ketogenic. As a producer of alanine, tryptophan is also glucogenic. Tryptophan is also a precursor for several derivatives, which include serotonin, melatonin, and NAD^+. Formations of these derivatives are discussed in the following section.

Conversion of lysine to ketogenic intermediate Catabolism of lysine is shown in Fig. 7.25. Lysine reacts with α-ketoglutarate to form saccharopine, which is further converted to glutamate and α-aminoadipate. Both the reactions are catalysed by NADP-dependent dehydrogenase. α-Ketoadipate is formed from α-aminoadipate by the catalysis of aminotransferase. CoASH and dehydrogenase act upon α-ketoadipate to form glutaryl CoA with the liberation of carbon dioxide. Glutaryl CoA decarboxylates to crotonyl CoA. Crotonyl CoA is converted to acetoacetyl CoA which then forms acetyl CoA. In the intestine, lysine undergoes bacterial putrefaction to form cadaverine, which is subsequently excreted in the faeces. During putrefaction, carbon dioxide is liberated.

7.4.4 Biosynthesis of Amino Acid Derivatives

Many biologically important compounds, such as neurotransmitters, hormones, vasodilator, vasoconstrictor, and pigments, are synthesized from amino acids (Table 7.5). These compounds are called amino acid derivatives. In other words, some of the amino acids are catabolized to biologically important derivatives. Tyrosine is responsible for producing most of these compounds. This section discusses the biosynthesis of amino acid derivatives.

TABLE 7.5 Amino acid derivatives of biological significance

Amino acid precursor	Amino acid derivative	Primary function of the amino acid derivative
Tyrosine	Thyroxine	
	Dopamine	Neurotransmitter
	Norepinephrine	Neurotransmitter
	Epinephrine	Hormone
	Melanine	Pigment
Tryptophan	Melatonin	Hormone
	Serotonin	Vasoconstrictor
Glutamate	Gamma amino butyric acid (GABA)	Neurotransmitter
Histidine	Histamine	Vasodilator
Lysine	Carnitine	Transport of fatty acids into mitochondria in the form of carnitine shuttle.

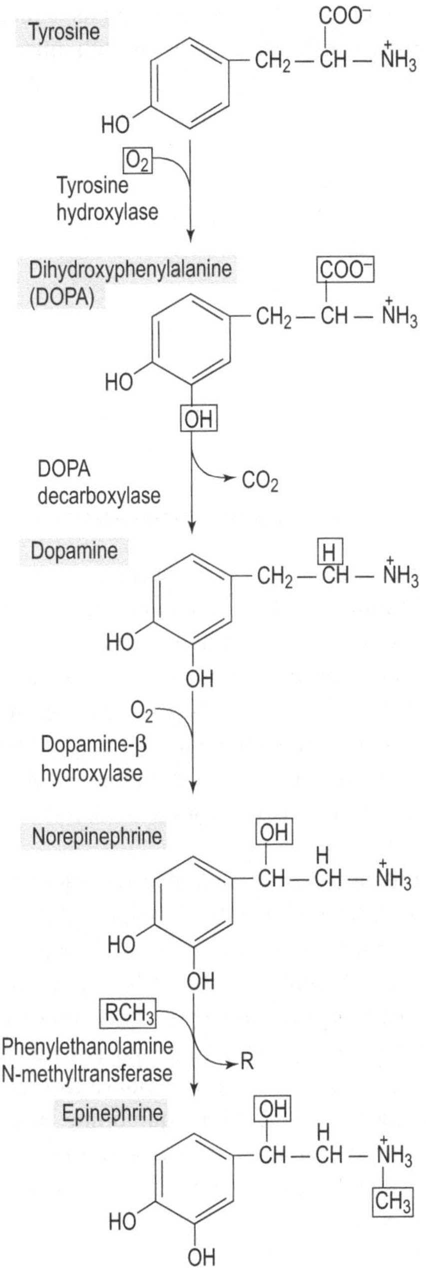

FIGURE 7.26 Synthesis of tyrosine derivatives (catecholamines)

Tyrosine derivatives

Tyrosine is an excellent precursor for several biologically important derivatives. In fact, the three catecholamines—dopamine, norepinephrine, and epinephrine—are derived from tyrosine (Fig. 7.26). Tyrosine (hydroxylphenylalanine) is hydroxylated to dopa (dihydroxyphenylalanine) by the enzyme tyrosine hydroxylase, in the presence of tetrahydrobiopterin as a cofactor. Dopa is decarboxylated by aromatic amino acid decarboxylase in the presence of pyridoxal phosphate cofactor to form dopamine. Dopamine is the neurotransmitter in the basal ganglia. The side chain of dopamine is hydroxylated to norepinephrine by the copper-containing enzyme dopamine-β-hydroxylase, essentially in the presence of ascorbate and molecular oxygen. This enzyme is a mixed function oxygenase, such as tyrosine hydroxylase. The side chain carbon that is directly attached to the benzene ring is involved in the hydroxylation. Norepinephrine is converted to epinephrine by the methylation of its amino group by phenylethanolamine-N-methyl transferase. S-Adenosylmethionine donates methyl group for this reaction. This reaction primarily takes place in the adrenal medulla, where the epinephrine molecules are stored as granules.

The rate-limiting enzyme that regulates the synthesis of tyrosine derivatives is tyrosine hydroxylase. It is an allosteric enzyme feedback inhibited by dopamine, norepinephrine, and epinephrine.

Melanin, being a pigment, is synthesized in the pigment-producing cells called melanocytes. Tyrosine is converted to dopa by the enzyme tyrosinase. Dopa is converted to dopoquinone by the same enzyme. Tyrosinase functions in the presence of copper as a cofactor. Dopaquinone is converted to melanin. Defect in tyrosinase results in lack of melanin pigment leading to a disorder called albinism. Derivation of melanin from tyrosine is shown in Fig. 7.27.

Thyroid hormones (Fig. 7.28) are derived from tyrosine. The follicle cells of thyroid gland have a protein called thyroglobulin. The tyrosine residues of this protein are iodinated to form 3, 3′, 5, 5′-tetraiodothyronine (thyroxine or T_4) and triodothyronine (T_3).

Tryptophan derivatives

Serotonin, melatonin, and NAD$^+$ are derivatives of tryptophan. Serotonin (Fig. 7.29) is synthesized by the pineal gland. Tryptophan is hydroxylated to 5-hydroxytryptophan by tryptophan hydroxylase in the presence of tetrahydrobiopterin. Further, 5-hydroxytryptophan is decarboxylated by aromatic amino acid decarboxylase and

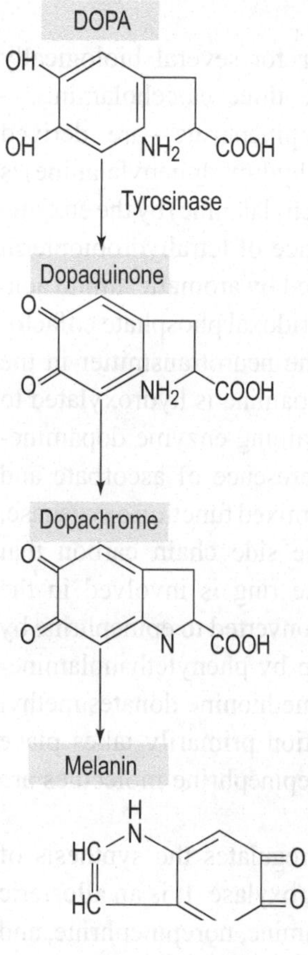

FIGURE 7.27 Formation of melanin pigment from tyrosine

pyridoxal phosphate to from 5-hydroxytryptamine, which is called serotonin. Serotonin is a vasoconstrictor that stimulates smooth muscle contraction. Serotonin is further catabolized by monoamine oxidase (MAO) to 5-hydroxyindoleacetic acid which is excreted in urine.

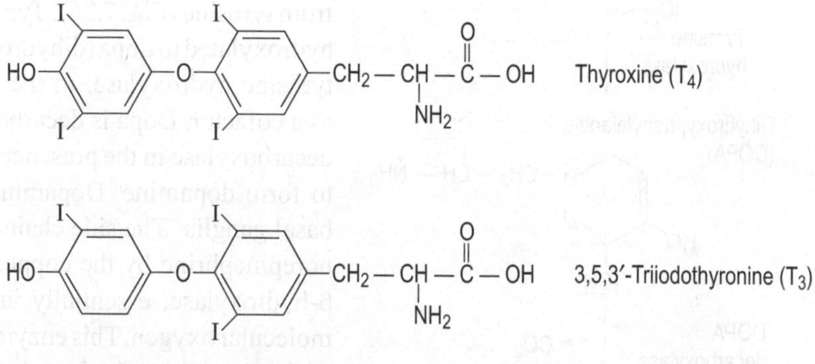

FIGURE 7.28 Thyroid hormone produced from iodination of tyrosine residues

Melatonin (N-acetyl-5-methoxytryptamine) is a hormone which functions as an antioxidant, immune system activator, and as a synchronizer of the biological clock. It is actually formed from serotonin which itself is a tryptophan derivative. Serotonin, which is 5-hydroxytryptamine, is converted to 5-hydroxy–N-acetyltryptamine by the enzyme acetyltransferase (serotonin N-acetylase). The product is then methylated by O-methyltransferase to form melatonin. SAM donates the methyl group for this reaction. Formation of melatonin from serotonin is shown in Fig. 7.30.

Tryptophan is the precursor for nicotinamide adenine dinucleotide (NAD^+), the primary electron acceptor in the oxidative phosphorylation. Tryptophan is converted to N-formylkynurenine by tryptophan pyrrolase in the presence of oxygen and ferrous ion. Kynurenine formylase converts N-formylkynurenine into kynurenine. Kynurenine is hydroxylated to 3-hydroxykynurenine by hydroxylase and NADPH. Kyrurenine is cleaved by kynureninase and pyridoxal phosphate to form alanine and anthranilate; the latter is acted upon by oxidase to form 2-acroleyl-3-amino fumarate. This compound is then dehydrated to quinolinic acid. Quinolinic acid reacts with PRPP; a molecule of carbon dioxide and pyrophosphate is removed to form niacin mononucleotide. The mononucleotide is converted to NAD^+. Figure 7.31 shows the metabolism of tryptophan to NAD^+.

Synthesis of glutamate derivative

Gamma amino butyric acid (GABA) is a neurotransmitter and regulator of muscular tone. It is synthesized from glutamate by decarboxylation reaction catalysed by glutamate decarboxylase and pyridoxal phosphate (Fig. 7.32).

Synthesis of histidine derivative

Histidine is the precursor for histamine, a potent vasodilator and a neurotransmitter, released in response to allergic reactions. Histidine is decarboxylated by histidine decarboxylase to form histamine (Fig. 7.33).

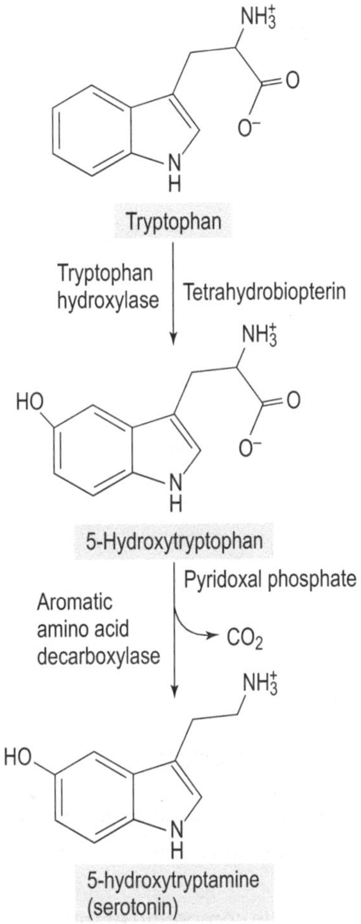

FIGURE 7.29 Formation of serotonin (5-hydroxytryptamine)

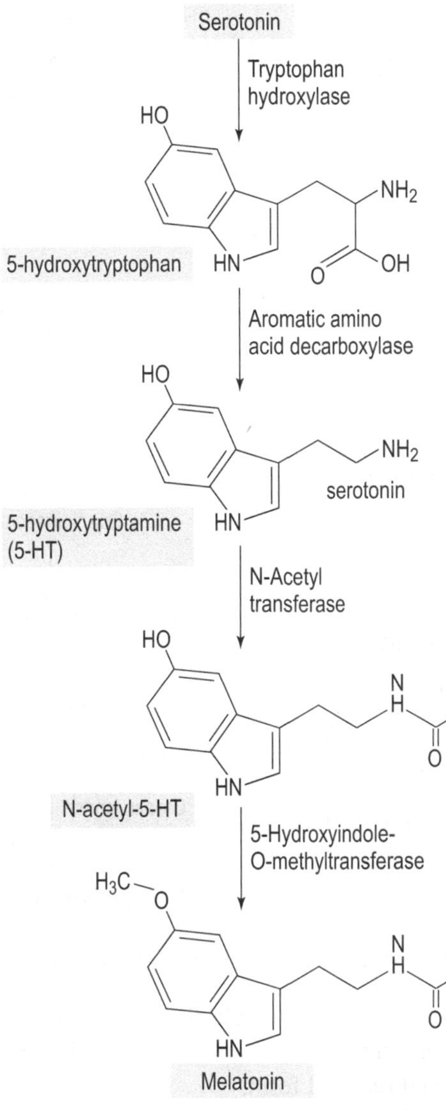

FIGURE 7.30 Formation of melatonin from serotonin

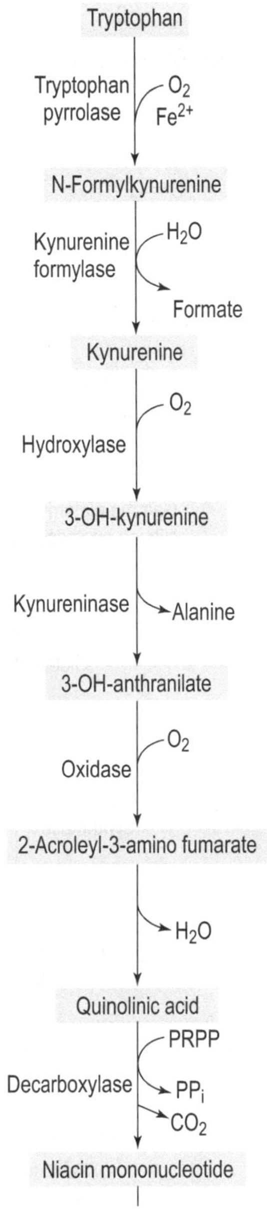

FIGURE 7.31 Conversion of tryptophan to NAD^+

Formation of glutathione and creatine phosphate

Glutathione is a tripeptide synthesized by peptide linkage formed between glutamate, cysteine, and glycine (Fig. 7.34). It is peculiar that the peptide bond of glutamate is contributed by the gamma carboxyl group and not by the carboxyl group attached to the alpha carbon atom. Glutathione plays three major roles in cells. It is the cofactor for glutathione peroxidase which is involved in detoxification reaction. Thus, it functions

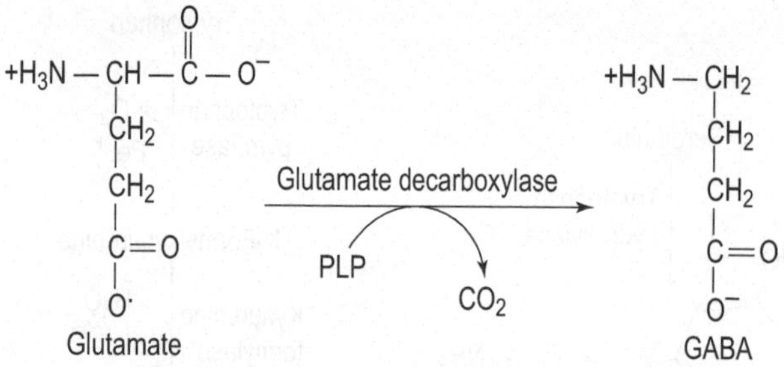

FIGURE 7.32 Metabolism of glutamate to GABA

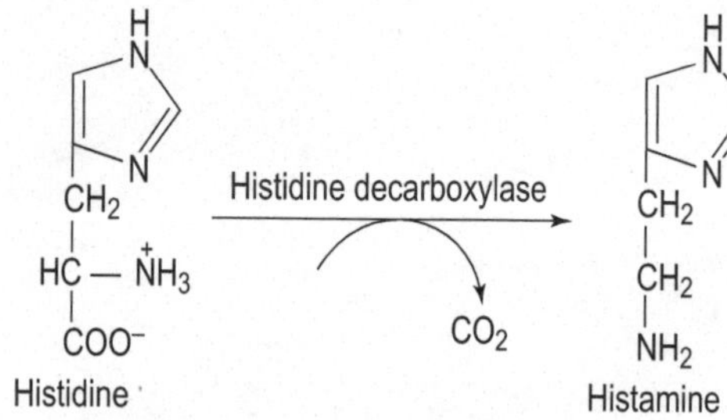

FIGURE 7.33 Metabolism of histidine to histamine

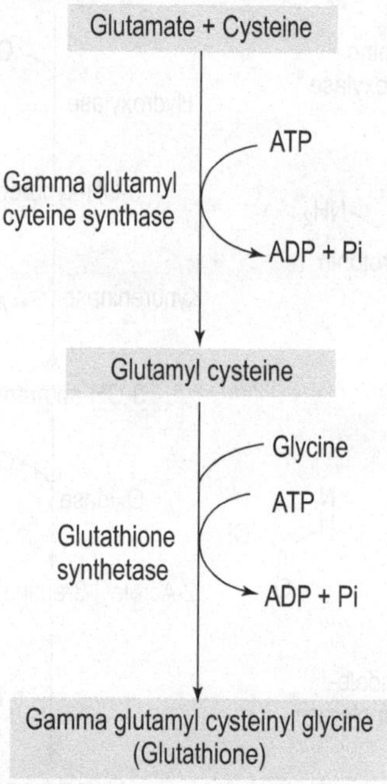

FIGURE 7.34 Formation of glutathione

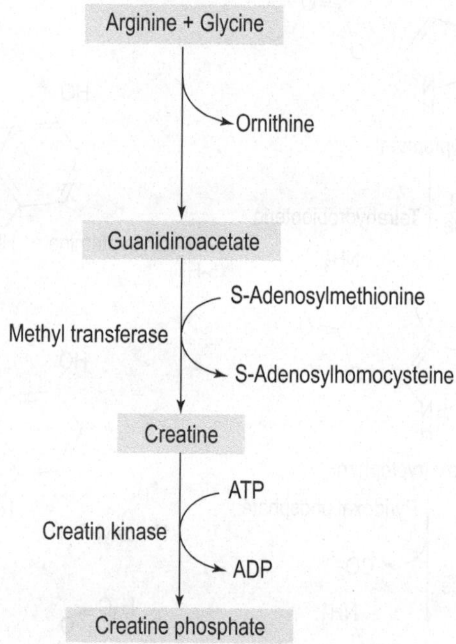

FIGURE 7.35 Formation of creatine phosphate

as an antioxidant. It is responsible for the transport of amino acids across the plasma membrane in certain cells. It functions as a sulphydryl buffer due to its ability to exist in oxidized form and reduced form.

Creatine phosphate (phosphocreatine), a high-energy phosphate compound of muscle is formed from three amino acids—glycine, arginine, and S-adenosylmethionine (Fig. 7.35). The guanidium ring of arginine is incorporated into glycine to form guanidinoacetate. The guanidinoacetate is methylated at the N-group by S-adenosylmethionine to form creatine, which is then phosphorylated by ATP. Creatine phosphate is the reserved source of high-energy phosphate bond in myocytes and neurons. It has a high phosphate group transfer potential.

Synthesis of polyamines

Polyamines are a group of organic compounds possessing two or more primary amine groups (hence the name). They are known to regulate cell proliferation, differentiation, and

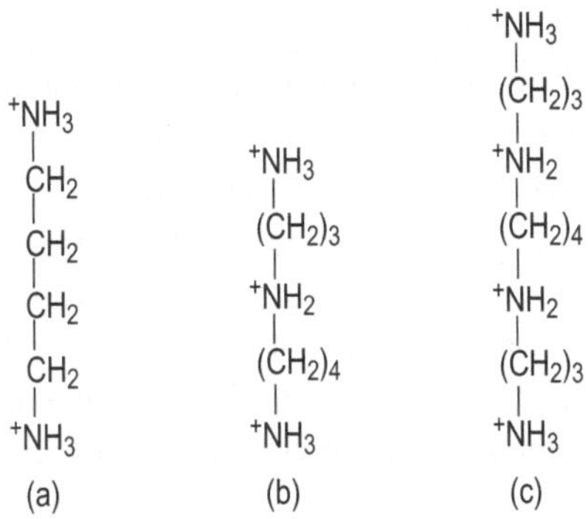

FIGURE 7.36 Polyamines produced from ornithine (a) Putrescine, (b) Spermidine, (c) Spermine

apoptosis. Putrescine, spermidine, and spermine are important polyamines (Fig. 7.36). Ornithine, a non-essential amino acid is the precursor for polyamines. Ornithine is decarboxylated by ornithine decarboxylase to form putrescine, a foul smelling amine. S-Adenosylmethionine propylamine (obtained from decarboxylation of S-adenosylmethionine) donates aminopropyl residue to the putrescine forming spermidine. Spermidine condenses with yet another residue of aminopropyl to form spermine (Fig. 7.37).

Synthesis of carnosine and anserine from histidine and alanine

Histidine condenses with β-alanine to form carnosine. Carnosine (β-alanyl-L-histidine) is a dipeptide composed of beta-alanine and histidine. It is predominantly distributed in muscle and brain. It can scavenge reactive oxygen species and function as an antioxidant. Carnosine accepts a methyl group from S-adenosylmethionine to form another dipeptide, anserine (beta-alanyl-N-methylhistidine). Anserine is also an antioxidant.

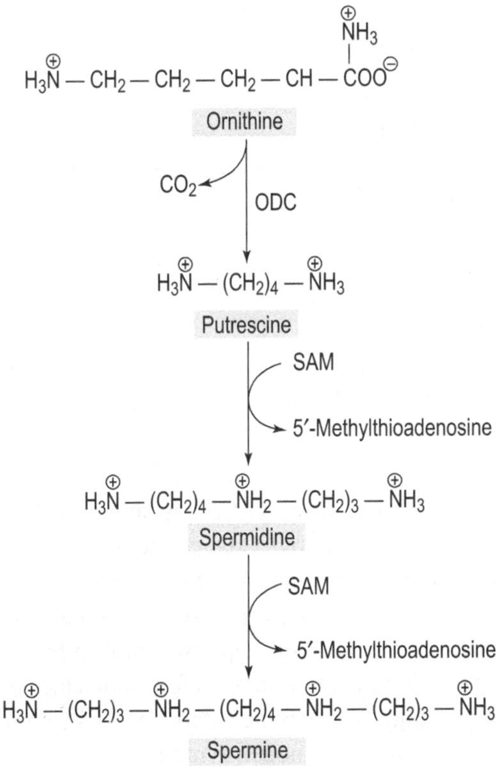

FIGURE 7.37 Formation of polyamines from ornithine

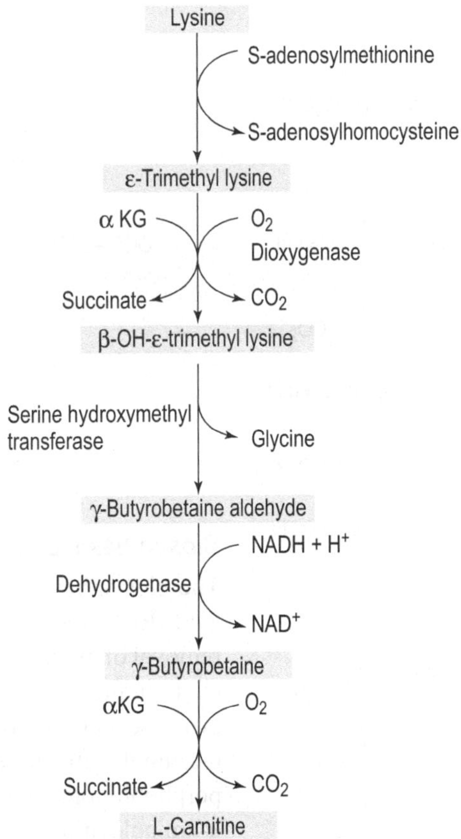

FIGURE 7.38 Carnitine formation from lysine

Carnitine synthesis from lysine

Lysine is catabolized to carnitine in liver and kidney. Lysine residues are first methylated by SAM to form ε-trimethyl lysine. Trimethyl lysine is converted to β-hydroxy-ε-N-trimethyl lysine. This reaction is catalysed by dioxygenase, which is accompanied by an uptake of α-ketoglutarate and release of succinate and carbon dioxide. β-hydroxy-ε-N-trimethyl lysine is acted upon by serine hydroxymethyl transferase to form glycine and γ-Butyrobetaine aldehyde, the later is converted into γ-butyrobetaine by dehydrogenase. γ-butyrobetaine is converted to L-carnitine in the presence of α-ketoglutarate with the liberation of succinate and carbon dioxide. Formation of carnitine is shown in Fig. 7.38.

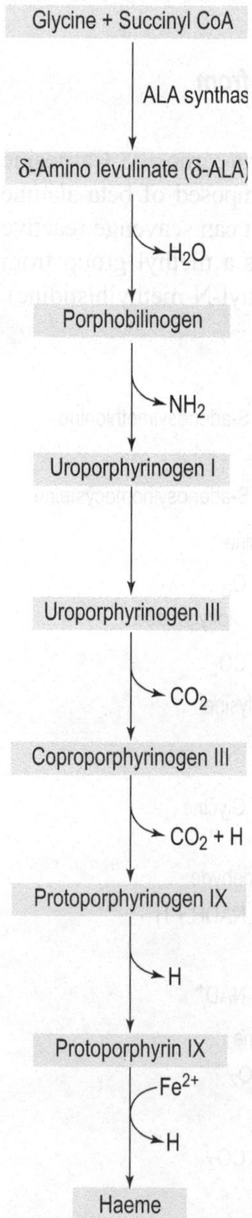

FIGURE 7.40 Conversion of glycine to haeme

Haeme metabolism

Haeme is the prosthetic group of several proteins, including haemoglobin, cytochrome c, catalase, and certain peroxidases. It is composed of a porphyrin ring formed from four pyrole rings (Fig. 7.39). Glycine is one of the precursor for the biosynthesis of haeme, the other precursor being succinyl CoA.

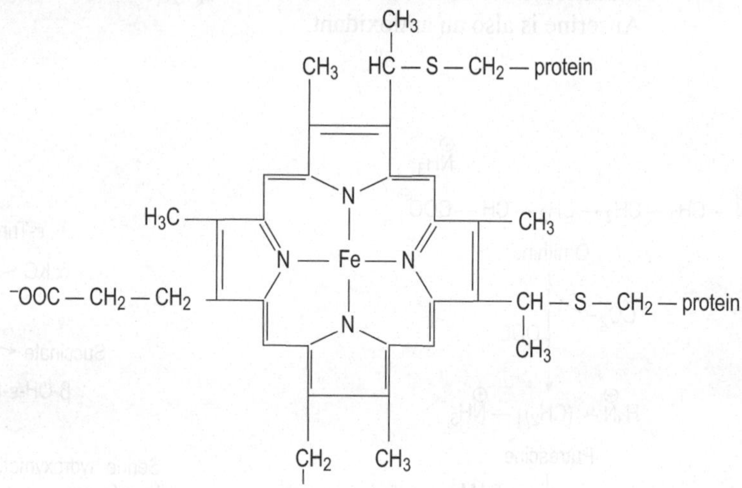

FIGURE 7.39 Structure of haeme

Biosynthesis of haeme

The biosynthesis of haeme from glycine and succinyl CoA is shown in Fig. 7.40. Eight molecules of each glycine and succinyl CoA condense with the removal of eight molecules each of CoA and CO_2. This results in the formation of α-aminolevulinate. The catalysts for this step are α-aminolevulinate synthase and pyridoxal phosphate. In the second step, two molecules of α-aminolevulinate react with the liberation of water, to form four molecules of porphobilinogen that contain pyrrole ring. The enzyme catalysing this step is α-aminolevulinate synthase, which is a porphobilinogen synthase.

Four molecules of porphobilinogen condense to a tetrapyrole called uroporhyrinogen I, which is then isomerized to uroporphyrinogen III.

Uroporphyrinogen synthase and uroporphyrinogen III cosynthase catalyse the condensation reaction and isomerization reaction, respectively. Deficiency of uroporphyrinogen synthase results in excessive accumulation of the α-aminolevulinate and porphobilinogen, a condition called acute intermittent porphyria, characterized by severe abdominal pain and neurologic abnormalities.

Deficiency of uroporphyrinogen III cosynthase causes congenital erythropoietic porphyria, that is, increased accumulation and excretion of uroporphyrinogen I, symptoms of which include photosensitivity, skin lesions, and haemolytic anaemia.

In the fifth step, the four side chains of uroporphyrinogen III are decarboxylated to coproporphyrinogen III by uroporphyrinogen decarboxylase.

In the sixth step, the two side chains of coproporphyrinogen III are decarboxylated to protoporphyrinogen IX by coproporphyrinogen oxidase. Deficiency of uroporphyrinogen decarboxylase and uroporphyrinogen decarboxylase cause porphyria cutanea tarda and coproporphyria, respectively, both showing symptoms of abdominal pain, neural disorder, and photosensitivity. In both the conditions, the respective substrates are accumulated and excreted.

Protoporphyrin IX is formed by the dehydrogenation of protoporphyrinogen IX. Potoporphyrinogen IX catalyses the removal of six hydrogen atoms to carry out the reaction.

In the last step, iron ($Fe2^+$) is inserted into protoporphyrin IX to form haeme. Ferrochelatase catalyses this reaction. Deficiency of ferrochelatase causes protoporphyria and dermatological problems.

The first enzyme of the haeme synthesis, α-aminolevulinate synthase, is the rate-limiting enzyme of the pathway. The enzyme is inhibited by the product haeme itself. In fact, haeme represses the gene encoding α-aminolevulinate synthase.

Catabolism of haeme In spleen, and to a lesser extent in the liver, haeme is degraded between the pyrrole rings by haeme oxygenase to form biliverdin, with the liberation of carbon monoxide. The iron atom is then dissociated to produce free tetrapyrrole called biliverdin that is green in colour. Biliverdin is reduced to a reddish brown pigment called bilirubin by NADPH. Bilirubin, due to its poor solubility, is transported by serum albumin to liver, where it is conjugated to glucuronic acid to form bilirubin diglucuronide. Uridinie-glucuronyl transferase catalyses this conjugation reaction. Conjugated bilirubin is excreted into the bile, which, on reaching the bowel is hydrolyzed to free bilirubin and converted to urobilinogens and stercobilinogens that are excreted in the urine and faeces. The iron liberated from the haemoglobin is transported by a carrier protein transferrin and stored in the cells in conjugation with a protein called ferritin. High amount of iron is stored in combination with protein and polysaccharides, as granules called hemosiderin.

Deficiency of uridine glucuronyltransferase leads to Crigler–Najar syndrome and neonatal jaundice, both of which are characterized by very high levels of the serum bilirubin. Jaundice is a condition in which excessive bilirubin gets deposited in the skin and mucous membrane giving yellowish colour to the eyes. There are three types of jaundice. Prehepatic jaundice, which is caused by diseases or intoxicants which induce abnormal red blood cell destruction. Hepatic jaundice, which is due to diseases in the liver, such as hepatitis and cirrhosis, in which there is impairment in the conjugation and hepatic uptake of bilirubin. Posthepatic jaundice, which is due to physical obstructions that prevent the bile from reaching the digestive tract, as in gall stone.

7.4.5 Biosynthesis of Amino Acids

Out of twenty amino acids, nine are essential ones, and should be supplied to the body through diet. The other eleven amino acids are non-essential and are synthesized in vivo. The eleven non-essential amino acids and their precursors are shown in Fig. 7.41.

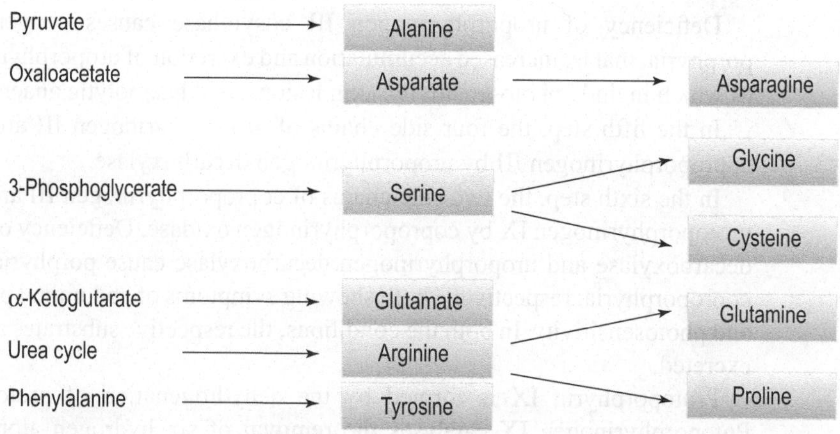

FIGURE 7.41 Non-essential amino acids (shaded) and their precursors

Alanine is synthesized from the pyruvate obtained from glycolysis. This is a transamination reaction catalysed by alanine aminotransfersase as shown in Fig. 7.42.

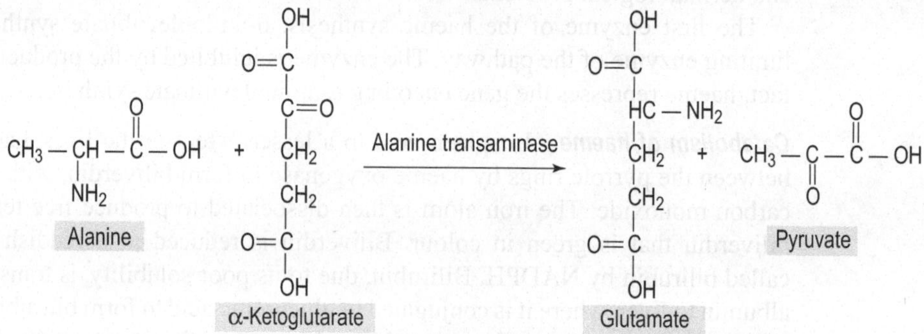

FIGURE 7.42 Formation of alanine from pyruvate by transamination reaction

Aspartate is synthesized from the ketoacid oxaloacetate by transamination reaction catalysed by aspartate aminotransferase and pyridoxal phosphate. This reaction is involved in the interconversion of glutamate and aspartate, that is, glutamate and oxaloacetate react to form aspartate and α-ketoglutarate. Aspartate donates its atoms for other biosynthesis processes. For example, it donates its amino group for urea synthesis and its nitrogen and carbon atoms for the synthesis of purines and pyrimidines. Aspartate is used in shuttling the reducing equivalents from the cytosol to mitochondria. Aspartate links urea cycle with Krebs cycle.

- Asparagine is synthesized from aspartate by the catalytic action of asparagines synthetase.
- Glutamate is synthesized from α-ketoglutarate by transamination reaction catalysed by glutamate dehydrogenase.

- Glutamate generates glutamine, predominantly in the liver by amidation reaction, catalysed by glutamine synthetase.

Biosynthesis of aspartate, asparagines, glutamate, and glutamine are shown in Fig. 7.43.

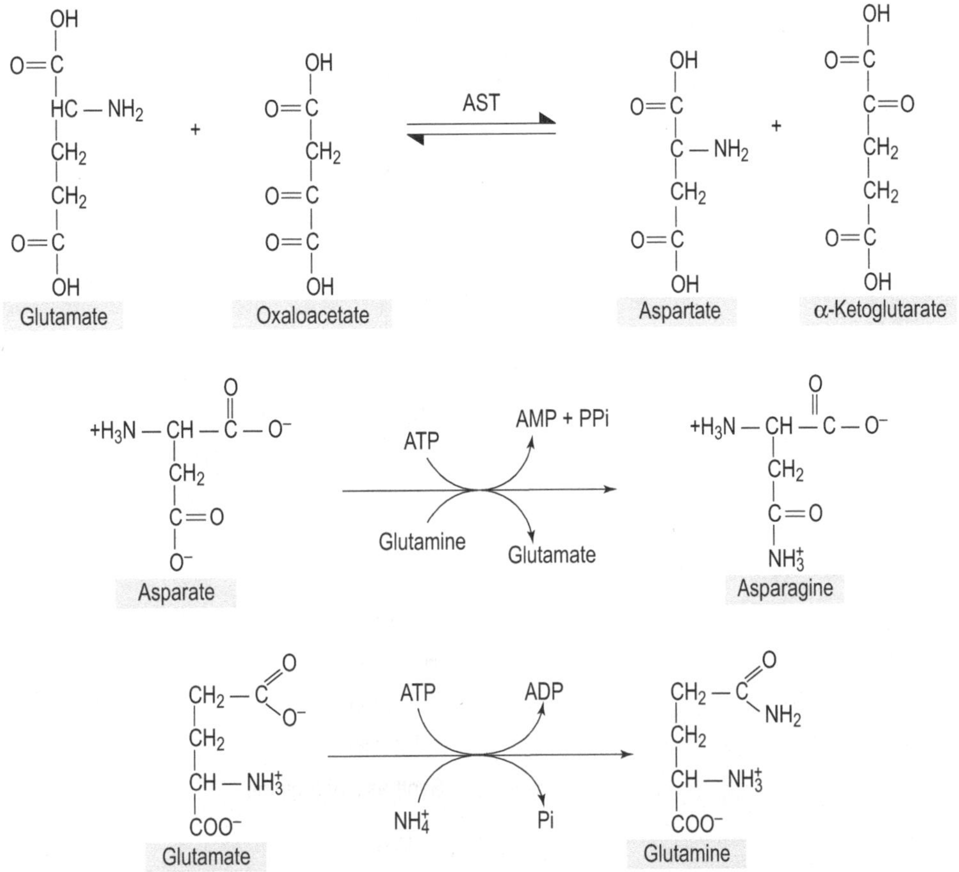

FIGURE 7.43 Formation of aspartate, glutamate, asparagine, and glutamine (AST, aspartate transaminase; AS, Asparagine synthetase; GS, glutamine synthetase)

Arginine is synthesized in the liver during urea cycle. Urea cycle has five steps (Fig. 7.13). In the fourth step, argininosuccinate is cleaved by the argininosuccinate lyase to form arginine and fumarate. This reaction is represented in Fig. 7.44.

Cysteine is synthesized from two pathways—one is from cystine and the other from methionine as shown in Fig. 7.45. Cystine is a dimer of two cysteine residues linked by a disulphide bridge. One molecule of cystine is reduced to form two molecules of cysteine, by the action of NAD^+-dependent dehydrogenase. In the second pathway, methionine is converted to homocysteine by THF. Homocysteine reacts with serine and forms cystathionine under the catalysis of cystathionine-β-synthase. Cystathionine is degraded and deaminated by γ-cystathionine to form cysteine and α-ketobutyrate.

Glycine, the simplest amino acid is very essential for the synthesis of haeme, purines, creatin, bile salt, hippuric acid, and glutathione. Glycine is synthesised from four sources—glyoxylate, threonine, serine, and ammonia (Fig. 7.46).

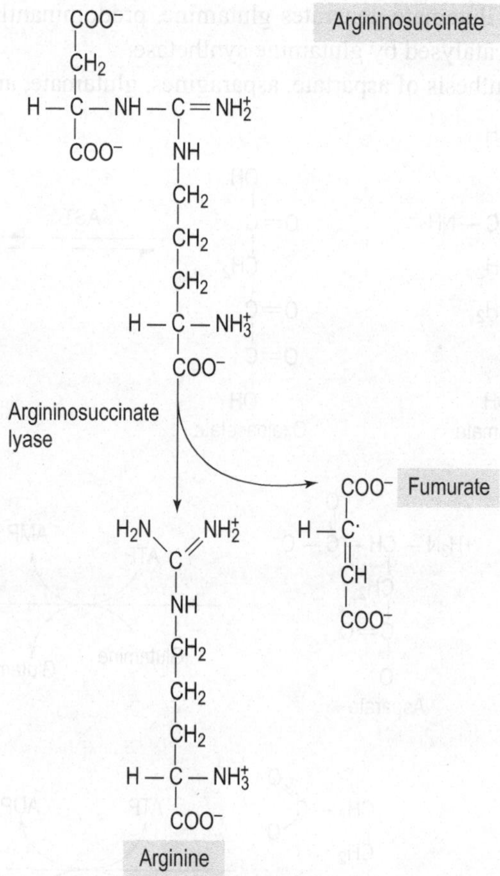

FIGURE 7.44 Synthesis of arginine

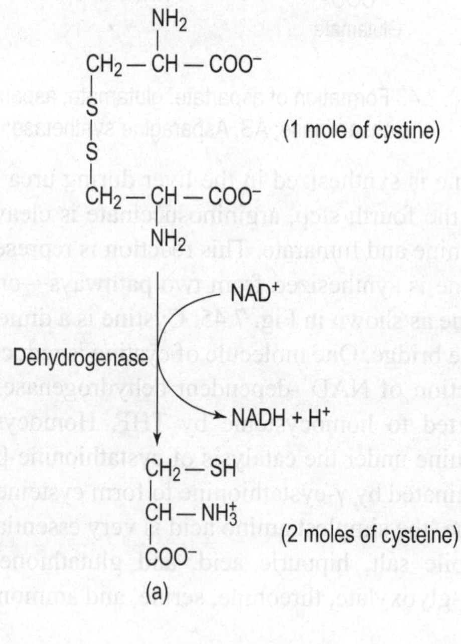

(a)

(*Contd*)

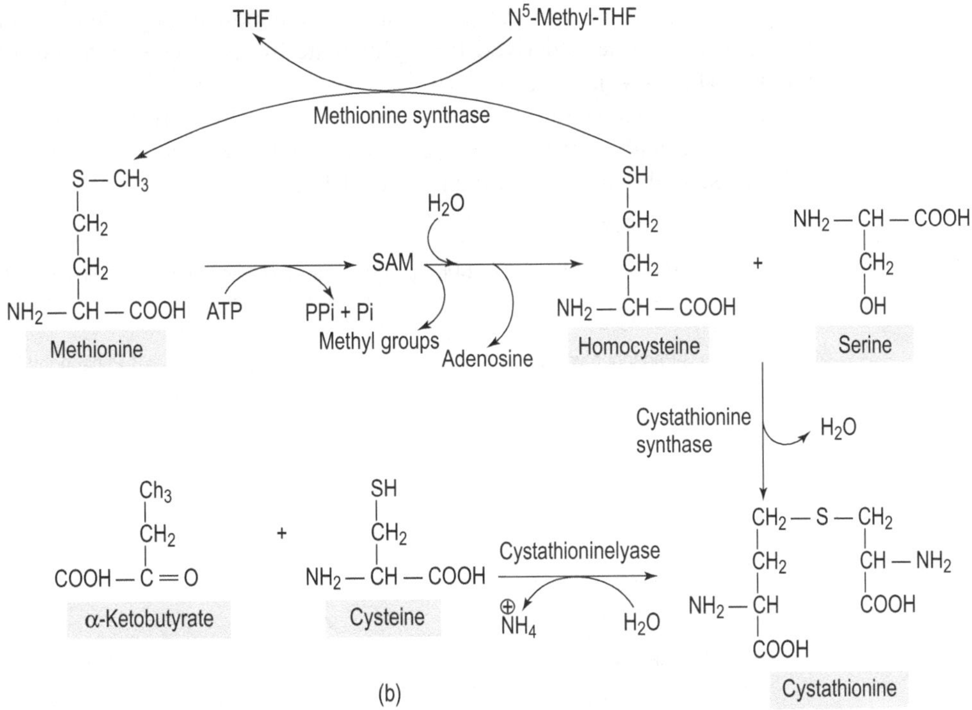

FIGURE 7.45 Synthesis of cysteine from (a) cystine and (b) methionine

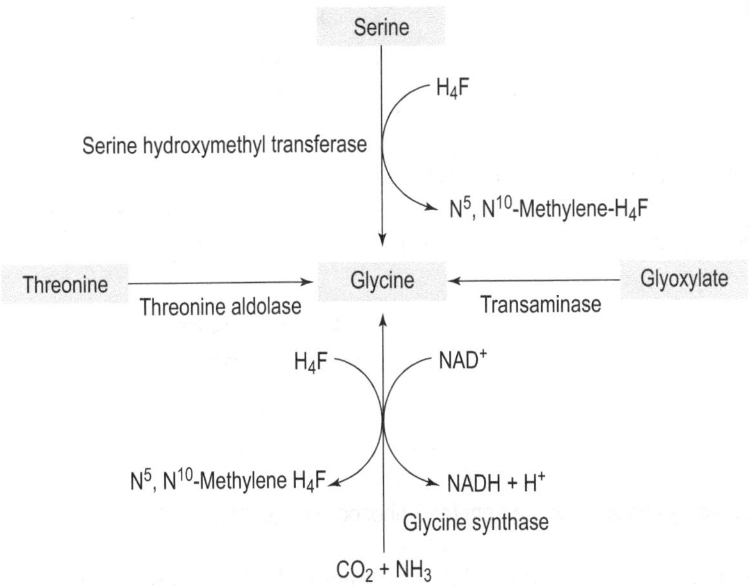

FIGURE 7.46 Biosynthesis of glycine

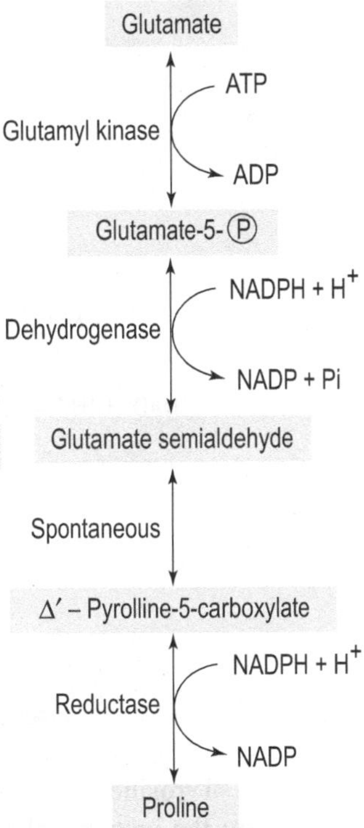

FIGURE 7.47 Synthesis of proline

To understand the synthesis of proline, the catabolism of proline to glutamate should be recalled. Proline is derived from glutamate by the reverse reaction of the catabolic pathway (Fig. 7.47).

Serine is also synthesized in a two-step reaction from 3-phosphoglycerate, an intermediate in glycolysis. Serine can also be synthesized from glycine by hydroxymethyl transferase in the presence of methylene THF (Fig. 7.48).

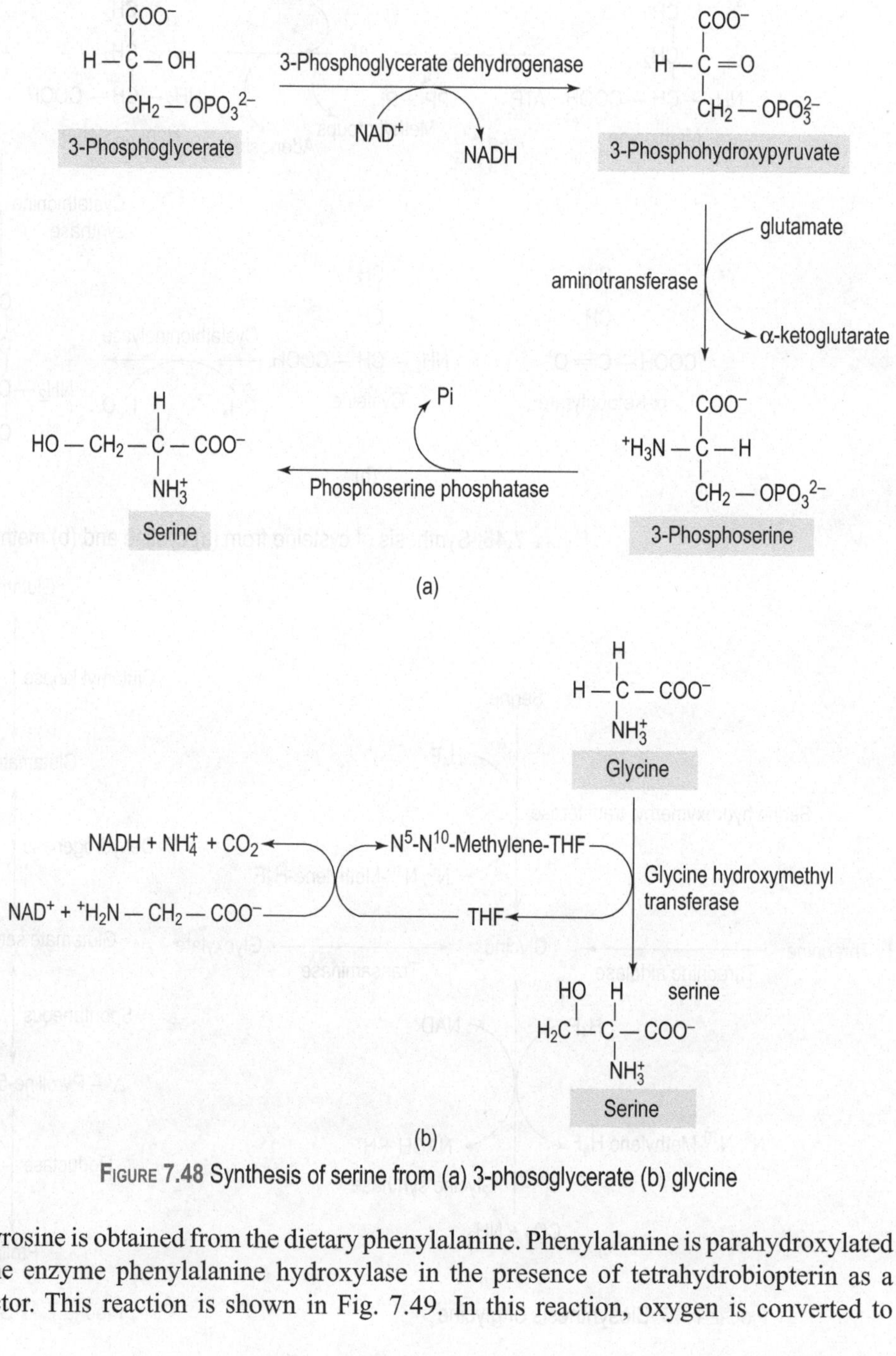

FIGURE 7.48 Synthesis of serine from (a) 3-phosoglycerate (b) glycine

Tyrosine is obtained from the dietary phenylalanine. Phenylalanine is parahydroxylated by the enzyme phenylalanine hydroxylase in the presence of tetrahydrobiopterin as a cofactor. This reaction is shown in Fig. 7.49. In this reaction, oxygen is converted to water, and tetrahydrobiopterin is converted to dihydrobiopterin. Tetrahydrobiopterin is regenerated by the reduction of dihydrobiopterin by NADPH.

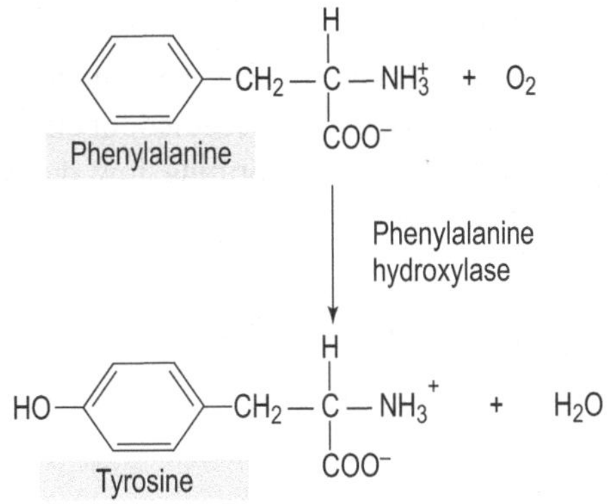

FIGURE 7.49 Hydroxylation of phenylalanine to tyrosine

7.5 METABOLISM OF FATTY ACIDS AND LIPIDS

Triacylglycerol or triglycerides (TG), phospholipids, and cholesterol are the lipids present in diet, which contribute to the exogenous lipids. When a dietary lipid enters the digestive tract, digestion is initiated in the small intestine. Basically, the enzymatic digestion of all these three lipids produces free fatty acids (FFA). This section, therefore, deals with the digestion of lipids in common, and the metabolism of FFA, TG, phospholipids, and cholesterol as individual topics.

7.5.1 Digestion and Absorption of Lipids

When the lipids enter the small intestine, the cells in the mucosa of the lower duodenum secrete a peptide hormone called cholecystokinin, the function of which is to contract the gall bladder for releasing the bile into the intestine and to stimulate the exocrine cells of the pancreas to release the digestive enzymes in the intestinal tract. Bile salts, sodium glycocholate, and sodium taurocholate increase the surface area for the subsequent enzyme action.

Pancreatic lipase hydrolyzes the C1 or C3 acyl chains of TG to form FFA and diacylglycerol or monoacylglycerol. Hydrolysis of TG is shown in Fig. 7.50. Cleavage of the acyl chain of either C1 or C3 produces diacylglycerol (DG). In the next step, the other terminal acyl chain (either C3 or C1) is hydrolyzed to form monoacylglycerol (MG). MG can be further hydrolyzed into FFA and glycerol. The TG products are then absorbed into the intestinal mucosal cells. FFA present in the diet is also absorbed into the intestinal cells. Within the mucosal cells, TG is resynthesized and conjugated with a specific protein, the apoprotein called 'apo-b-48', to form a complex called chylomicrons. The endogenous TG, synthesized by the liver, associates with other apoproteins, such as apo C, forming a complex called very low density lipoprotein (VLDL).

Dietary phospholipids entering the intestinal tract are acted upon by the pancreatic phospholipases (PL), which catalyse the cleavage of fatty acids from the phospholipids.

Depending on the bond that is cleaved, PL is named as PLA1, PLA2, PLC, and PLD. The sites of cleavage by these enzymes are depicted in Fig. 7.51. Cleavage by PLA1 or PLA2 liberates lysophospholipid and FFA. PLC produces DG and a phosphorylated alcohol, whereas PLD forms a phosphotidate and an alcohol. The products formed by the hydrolysis of phospholipids are absorbed into the intestinal mucosal cells. Phospholipids entered into the intestinal cells are readied for circulation by conjugation with apoproteins, such as apo-B-100, or apo-C, forming low density lipoprotein (LDL) or high density lipoprotein (HDL). FFA is transported in conjugation with plasma albumin.

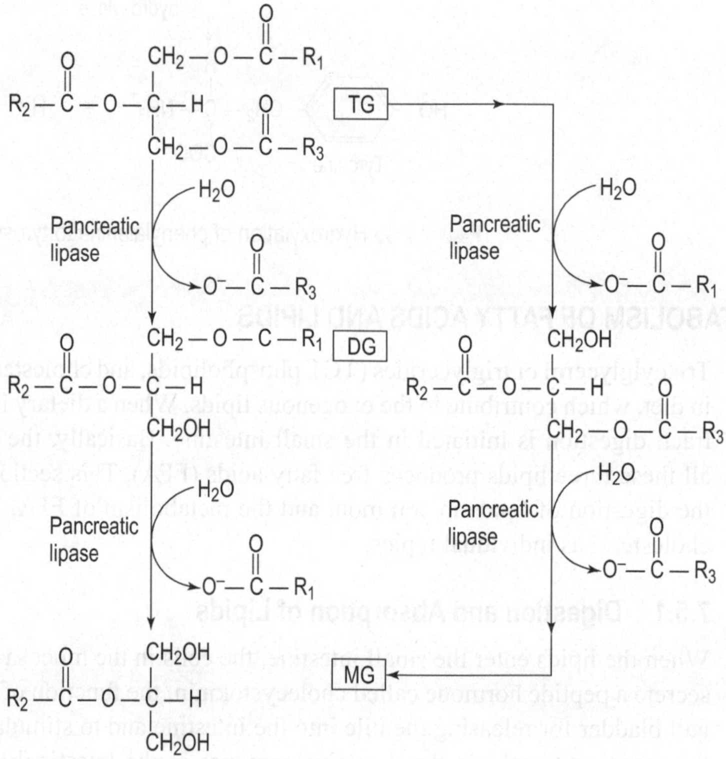

FIGURE 7.50 Intestinal hydrolysis of Triglycerides (TG = Triglyceride; DG = Diglyceride; MG = Monoglyceride)

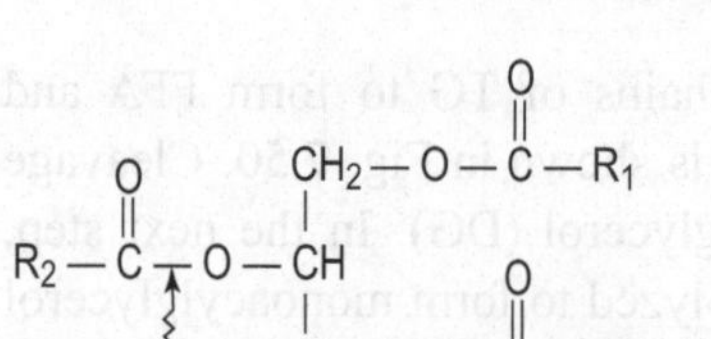

FIGURE 7.51 Site of cleavage of phospholipids by phospholipases (PL) (X = nitrogen base)

Cholesterol ester is converted to free cholesterol and FFA, both absorbed into the intestinal cells, where the cholesterol is re-esterfied. Like phospholipids, cholesterol is transported as LDL or HDL. The density of various lipoprotein forms of circulatory lipids are shown in Table 7.6, which indicates that the difference in the density is due to the density of the apoprotein. It should also be noted that though each lipoprotein is specific for each lipid, the other lipids are also present in minor amounts, as shown in the second column.

Abnormal levels of lipoproteins result in several diseases. Increased level of one or more of the lipoproteins in the blood is called hyperlipidemia, which results in atherosclerosis that blocks blood vessels causing heart attack and stroke. HDL deficiency (marked decrease in HDL level in blood) is called Tangier disease. It is a rare familial disorder characterized by recurrent polyneurapthy, lymphadenopathy, tonsillar hyperplasia, and hepatosplenomegaly.

TABLE 7.6 Constituents of plasma lipoproteins

Name of the lipoprotein	Percentage of lipid				Percentage of apoprotein	Density of apoprotien
	TG	PL	Ester cholesterol	Free cholesterol		
Chylomicron	90	3	4	1	2	<0.96
VLDL	60	15	15	5	5	0.96–1.006
LDL	8	22	40	10	20	1.019–1.063
HDL	5	30	20	5	40	1.063–1.210

7.5.2 Metabolism of Fatty Acid

From the previous section, it is clear that FFAs are released by the hydrolysis of dietary lipids. Apart from the source, FFAs are also produced within the body from the basic raw material, that is, FFA is synthesized de novo, from acetate and acetyl CoA. De novo synthesis of FFA occurs primarily in the liver and the adipose tissue, and to a lesser extent, in the kidney. In pregnant and lactating women, FFA are synthesized in the mammary gland, for providing fat to the milk.

De novo *synthesis of fatty acids*

Fatty acid synthesis occurs in the cytosol, whereas the raw material acetyl CoA is produced in the mitochondria. Therefore, acetyl CoA has to travel to the cytosol. Acetyl CoA cannot cross the mitochondrial membrane; however, citrate formed by the condensation of acetyl CoA and oxaloacetate in the mitochondria can easily be shuttled to the cytosol. In the cytosol, the citrate can be lysed by citrate lyase to regenerate the acetyl CoA and oxaloacetate in the presence of ATP. Acetyl CoA shuttling is shown in Fig. 7.52. Oxaloacetate formed in the cytosol is converted to malate and then pyruvate, which enters the mitochondria and reforms oxaloacetate by the action of pyruvate carboxylase and ATP.

Figure 7.52 represents the conversion of acetyl CoA to fatty acids. Synthesis of fatty acids from acetyl CoA has eight steps:

Step 1 Acetyl CoA is carboxylated to malonyl CoA. Bicarbonate is the carbon dioxide donor. Acetyl CoA carboxylase and biotin are the catalysts, whereas ATP drives the reaction. Acetyl CoA carboxylase is the key enzyme for the regulation of fatty acid synthesis. The enzyme is inhibited by palmitoyl CoA or glucagon and activated by citrate or insulin.

Step 2 From step 2 to step 7 all the reactions are catalysed by the fatty acid synthase enzyme complex. It is a large protein with three domains—I, II, and III. In total, the protein has seven enzyme activities distributed in three domains (Table 7.7).

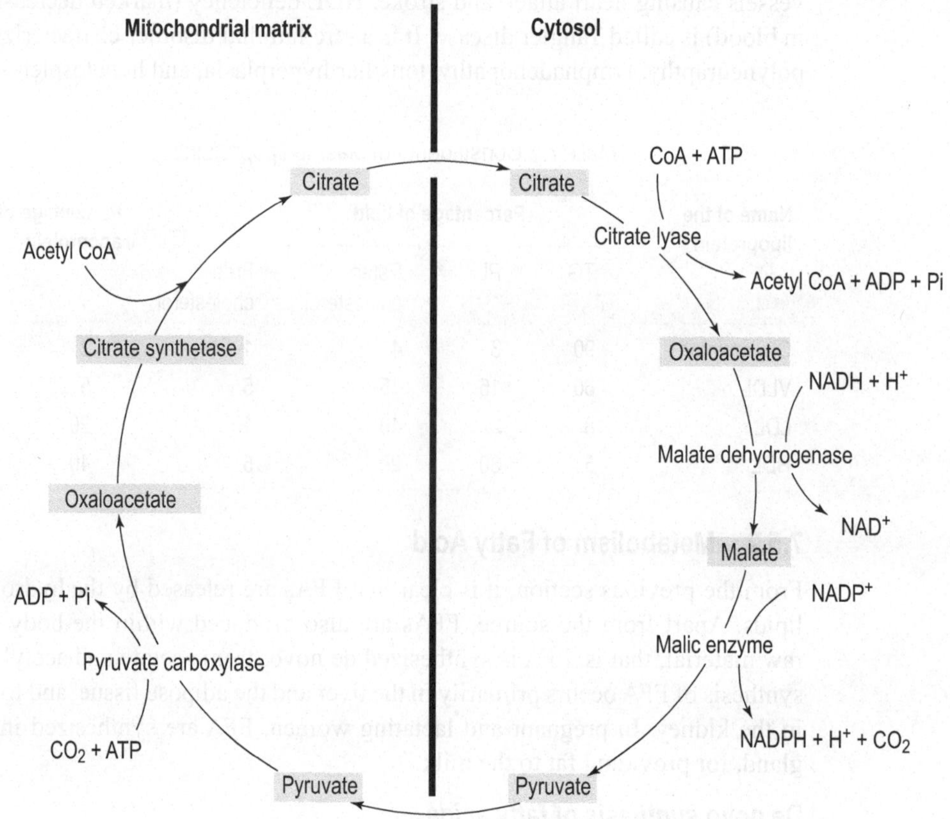

FIGURE 7.52 Shuttling of acetyl CoA to cytosol for fatty acid synthesis

TABLE 7.7 Activities of fatty acid synthase complex

Domain I	Domain II	Domain III
Acetyl transacylase (ATA)	3-Ketoacyl reductase (KAR)	Thioesterase (TE)
Malonyl transacylase (MTA)		
3-Ketoacyl synthase (KAS)	Enoyl-ACP-reductase (EAR)	
	*Acyl carrier protein (ACP)	
	3-Hydroxylbutyryl-ACP dehydratase (HBADH)	

* ACP has phosphopantetheine prosthetic group, derived from the vitamin pantothenic acid and this group is identical to that in the CoA.

Another molecule of acetyl CoA (apart from one that is involved in step 1) reacts with sulphydryl group of ACP via thiester linkage to form acetyl-ACP and CoA. This step is catalysed by the ATA.

Step 3 The malonyl CoA, formed in step 1 reacts with ACP to form malonyl-ACP under the catalysis of MTA.

Step 4 Malonyl-ACP and acetyl-ACP are condensed to form acetoacetyl-ACP. This reaction is catalysed by KAS.

Step 5 Acetoacetyl-ACP is reduced by NADPH-dependent enzyme KAR, to form 3-hydroxybutyryl-ACP.

Step 6 HBADH catalyses the conversion of 3-hydroxybutyryl-ACP to crotonyl-ACP.

Step 7 Crotonyl-ACP is converted to butyryl-ACP by NADPH-dependent EAR.

Step 8 Butyryl-ACP (a four-carbon unit formed in step 7) reacts with the malonyl-ACP (a two carbon unit formed in step 3) to form a six-carbon unit. This reaction occurs seven times with a two-carbon unit incorporated at each time. This results in the formation of palmitic-ACP. Palmitic acid is released from the ACP and from the fatty acid synthase enzyme complex by TE (palmityl deacylase).

The overall reaction of all eight steps for the formation of palmitic acid can be represented as follows:

$$\text{8 acetyl CoA} + \text{7 ATP} + \text{14 NADPH} + \text{14 H}^+ + \text{H}_2\text{0}$$

$$\text{Acetyl CoA Carboxylase} \ \Big\downarrow \ \text{Fatty acid synthase complex}$$

$$\text{Palmitate} + \text{7 ADP} + \text{7 Pi} + \text{14 NADP}^+ + \text{8 CoASH}$$

Two fatty acid synthase complexes are arranged as a dimer, enabling the synthesis of two fatty acids at a time. NADPH required for the reaction is supplied by the pentose phosphate pathway (discussed in Section 7.8).

Normally, the fatty acid synthesis pathway produces the 16-carbon fatty acid, palmitic acid. To derive the longer chain fatty acids, palmitic acid is elongated further by elongase system. Endoplasmic reticulum elongase system extends the palmitic acid by adding malonyl CoA to form the 18-carbon fatty acid—stearic acid. Mitochondrial elongase system elongates the fatty acids by adding acetyl CoA rather than malonyl CoA.

The discussion under this section reveals the synthesis of saturated fatty acids. We will now discuss how unsaturated fatty acids are formed the saturated fatty acids. The two most common unsaturated fatty acids in mammals are palmitoleic acid and oleic acid (Fig. 7.53). These two acids can be formed by the introduction of a double bond, that is, by the desaturation of the saturated fatty acid. Fatty acid oxygenase (mixed function oxygenase) of endoplasmic reticulum catalyses the formation of double bond between carbons 9 and 10. This reaction requires molecular oxygen and NADPH.

$$H_3C-(CH_2)_5-CH=CH-(CH_2)_7-COOH$$

(a)

$$H_3C-(CH_2)_7-CH=CH-(CH_2)_7-COOH$$

(b)

FIGURE 7.53 Two common unsaturated fatty acids in mammals: (a) palmitoleic acid and (b) oleic acid

7.5.3 Biosynthesis of Triacylglycerol or TG

TG is triglycerol acylated with fatty acids via ester linkage. In order to synthesize TG, the fatty acid must first be activated

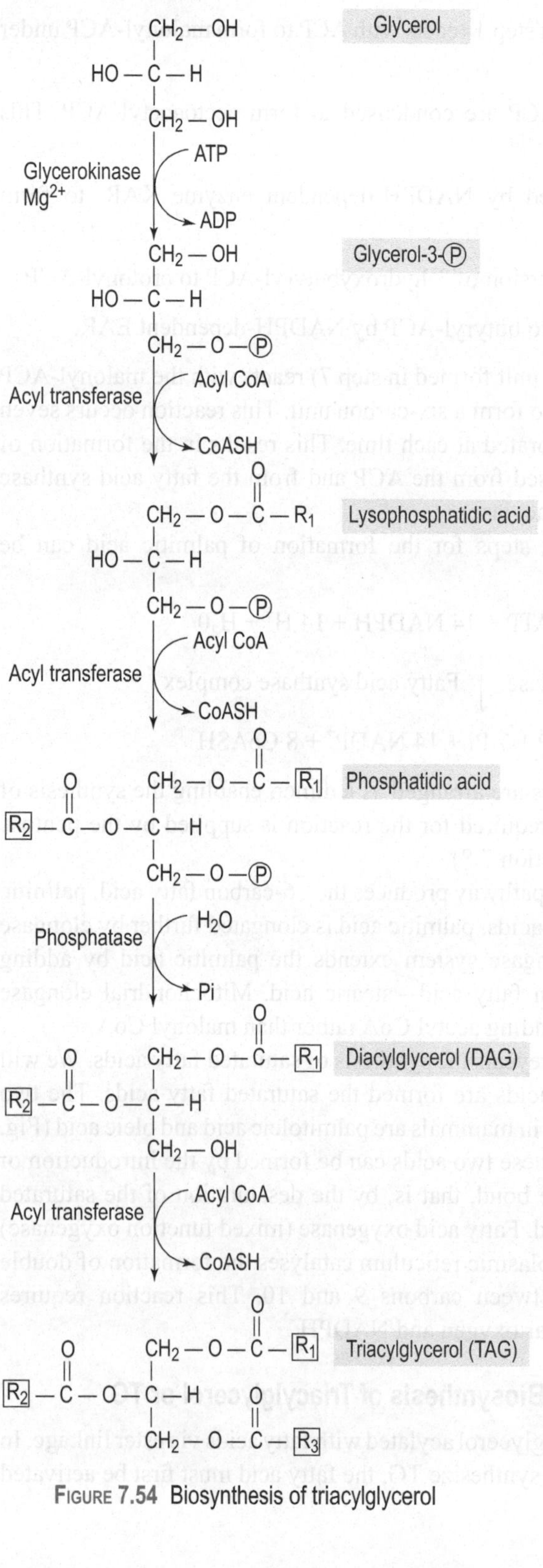

FIGURE 7.54 Biosynthesis of triacylglycerol

to its CoA form, and the glycerol should be phosphorylated. Fatty acid is converted to fatty acyl CoA by acyl CoA synthetase. The glycerol is phosphorylated to glycerol-3-P by glycerokinase. Glycerol-3-P is also formed by the reduction of dihydroxyacetone phosphate (DHAP), an intermediate in the glycolytic pathway. Glycerol-3-P is acylated by acyl transferase with the liberation of CoA to form lysophosphatidic acid. DHAP is yet another source of lysophosphatidic acid. DHAP is acylated to acyl DHAP by acyl transferase. Acyl DHAP is then reduced by NADPH-dependent reductase to form lysophosphatidic acid. Overall, lysophosphatidic acid is formed from glycerol, DHAP, and acyl DHAP. The formation of lysophosphatidic acid and subsequent steps in the synthesis of TG are shown in Fig. 7.54.

Saturated fatty acids are incorporated into lysophosphatidic acid at the second carbon. This reaction is catalysed by acyltransferase with the liberation of CoA to form phosphatidic acid. Phosphatidic acid is then cleaved by phosphatase to form diacylglycerol. The third fatty acyl moiety is then added to the third carbon to form TG.

There are two methods by which the DHAP is acylated, both producing lysophosphatidic acid as the product. In one method, DHAP reacts with acyl CoA in the presence of acyl transferase liberating CoASH and acyl DHAP. Acyl DHAP is then reduced by NADPH to form lysophosphatidic acid. In the second method, DHAP is first reduced to glycerol-3-P (glycerol-3-P is also obtained from the phosphorylation of glycerol by glycerokinase mediated by magnesium ions and ATP). Glycerol-3-P is then acylated to form lysophosphatidic acid. Lysophosphatidic acid is the glycerol with its first carbon, acylated, and the third carbon phosphorylated.

In the next step, lysophosphatidic acid is acylated at the second carbon hydroxyl to form phosphatidic acid. In the third step, the phosphatidic acid is cleaved by phosphatise to liberate Pi and form diacylglycerol. Diacylglycerol is then acylated at the third carbon hydroxyl to form TG. TG is stored in the adipose tissue as a fat depot.

7.5.4 Degradation of TG (Lipolysis of TG)

In the last two sections, we discussed how the fatty acids are synthesized and subsequently incorporated into glycerol to form TG. This section describes how TG is degraded to its constituent fatty acid and glycerol.

Whenever there is a need for energy, hydrolysis of TG is initiated by the hormone epinephrine or glucagon. Epinephrine or glucagon activates adenylate cyclase. Active adenylate cyclase hydrolyzes ATP to cyclic AMP (cAMP), which in turn activates protein kinase. Protein kinase catalyses the phosphorylation of TG lipase by ATP to form active TG lipase in the presence of magnesium ion as a cofactor. TG lipase cleaves TG to form DG and FFA. DG is acted upon by lipase to liberate MG and FFA. Finally, MG is hydrolyzed to FFA and glycerol. Catabolism of TG is shown in Fig. 7.55.

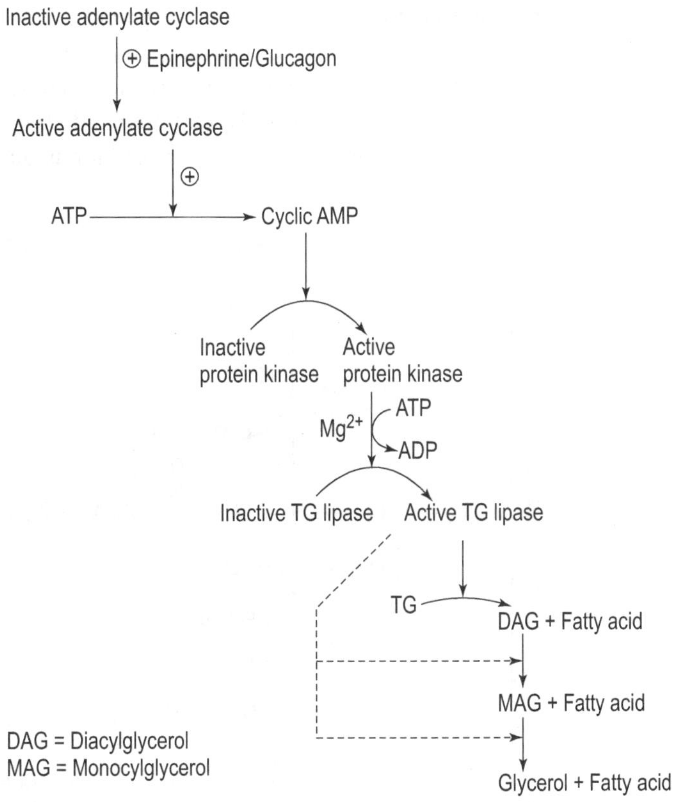

FIGURE 7.55 Catabolism of triacylglycerol (TG)

7.5.5 Oxidation of Free Fatty Acids

FFA liberated from the hydrolysis of TG (as discussed in Section 7.5.4) is transported in plasma by albumin to enter into the energy deficient tissues via sodium-dependent transport system (refer Chapter 2) for oxidation. (Glycerol, the other product of TG degradation is utilized by liver for glucose synthesis.) Oxidation of FFA is significant because FFA is an important source of energy. It can form high-energy compounds, such as NADH and FADH2, and release acetyl CoA. Acetyl CoA enters the Krebs cycle (Section 7.10). FFAs are oxidized by three methods—β-oxidation, α-oxidation, and ω-oxidation. Each of these methods are discussed in this section.

β-Oxidation

β-Oxidation is the principal pathway for the degradation of fatty acids containing even number of carbon atoms. In this pathway, a fatty acid is degraded by the removal of two carbon atoms at a time from the β position, that is, β-carbon is oxidized to form a β-ketoacid. However, for β-oxidation to occur the fatty acid is first activated to acyl CoA by the enzyme acyl CoA synthetase. This reaction is driven by ATP as the energy source and magnesium ion as the cofactor. Cells have two activation systems for activating long-chain, medium-chain, and short-chain fatty acids. The endoplasmic reticulum acyl CoA synthetase (thiokinase) activates the long-chain fatty acids, that is, fatty acids with greater than 12 carbons, whereas the inner mitochondrial acyl CoA synthetase activates the medium-chain fatty acids (with 4–10 carbons) and short chain fatty acids (with two and three carbons). Activation and β-oxidation reaction is outlined in Fig. 7.56.

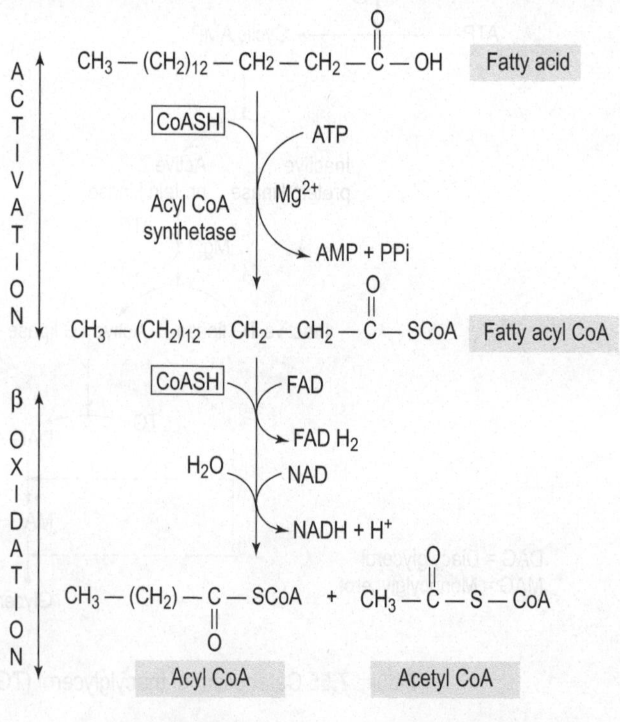

FIGURE 7.56 Activation and β-oxidation of a fatty acid

Carnitine shuttle Although the fatty acyl CoA can readily be oxidized, the length of the fatty acids determines the initiation of oxidation. For oxidation to start, the fatty acyl CoA should be transported from the cytosol to the mitochondria. Medium- and short-chain fatty acyl CoA cross the mitochondrial membrane freely, whereas long-chain fatty acids cannot do so. Still, cells have a shuttling system to transport the long-chain fatty acyl

CoA into mitochondrial matrix. This is known as the carnitine shuttle system. Carnitine (β-hydoxy-γ-trimethylammonium butyrate) is an amino acid derivative obtained from the catabolism of lysine. Carnitine shuttle is shown in Fig. 7.57. Two enzymes located in the mitochondrial membranes catalyse the carnitine-mediated shuttling. The outer mitochondrial membrane has carnitine–palmitoyl transferase I (CPT I), and the inner membrane has carnitine–palmitoyl transferase II (CPT II). During the catalysis of CPT I, carnitine accepts the acyl group from the fatty acyl CoA and gets itself converted to acyl carnitine. CPT II catalyses the transfer of the acyl carnitine to a CoA molecule drawn from the mitochondrial matrix. Thus, free acyl CoA will be shuttled to mitochondrial matrix and the free carnitine will start the shuttling of next fatty acyl CoA molecule.

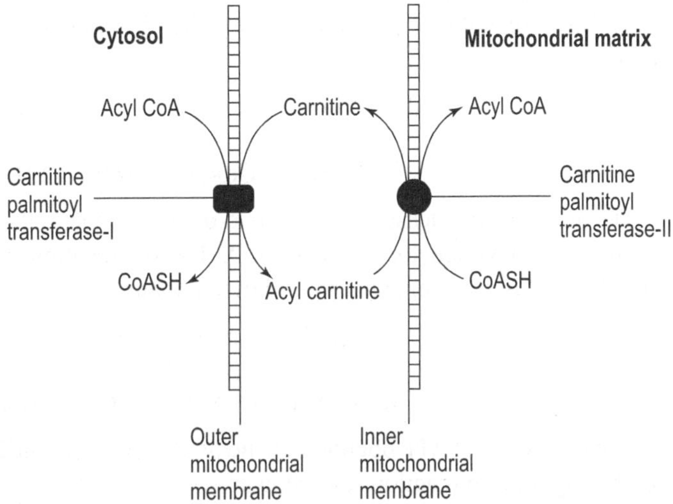

FIGURE 7.57 Carnitine shuttle

β-oxidation pathway The fatty acyl CoA undergoes a repeated four-step reaction, which removes two-carbon units from the chain every time. The four-step reaction that removes two-carbon units from the palmitoyl CoA is shown in Fig. 7.58.

Step 1 Palmitoyl CoA (Acyl CoA) is converted to enoyl CoA by acyl CoA dehydrogenase and FAD. FAD is reduced to FADH2 in this dehydrogenation reaction.

Step 2 Enoyl CoA hydratase hydrates the enoyl CoA to 3-hydroxyacyl CoA.

Step 3 3-Hydroxyacyl CoA dehydrogenase converts 3-hydroxyacyl CoA to 3-ketoacyl CoA. NAD+ is reduced to NADH during this dehydrogenation reaction.

Step 4 3-Ketoacyl CoA is cleaved to acetyl CoA and tetradecanoyl CoA by thiolase by the uptake of a CoASH molecule.

In this step, a two-carbon unit is removed as acetyl CoA. The remaining product, tetradecanoyl CoA (myristyl CoA), has an acyl moiety with just 12 carbon atoms, which is less than the initial compound palmityl CoA by two carbon atoms. The myristyl CoA is then subjected to the preceding four reactions, again forming a product with 10 carbon atoms and an acetyl CoA. For its complete oxidation, palmitoyl CoA should undergo the four-step reaction seven times. Thus, seven molecules of acetyl CoAs will be liberated during the oxidation of palmitoyl CoA.

The overall reaction of palmitate oxidation may be written as follows:

$$\text{Palmitoyl CoA} + 7\ \text{CoA} + 7\ \text{FAD} + 7\ H_2O \downarrow 8\ \text{Acetyl CoA} + 7\ \text{FADH}_2 + 7\ \text{NADH} + 7\ H^+$$

Energy yield from β-oxidation pathway The reaction given in the preceding discussion shows that 7 moles of $FADH_2$ and 7 moles of NADH are produced from the oxidation of palmitic acid. Each $FADH_2$ can yield 2 moles of ATP, whereas each NADH can yield 3 moles of ATP by electron transport chain and oxidative phosphorylation (Section 7.11). Thus, oxidation of palmitate can yield 35 moles of ATP. The energy yield can be represented as follows:

$$\text{Palmitoyl CoA} + 7\ \text{CoA} + 7\ O_2 + 35\ \text{ADP} + 35\ \text{Pi} \downarrow 8\ \text{Acetyl CoA} + 35\ \text{ATP} + 42\ H_2O$$

The acetyl CoA released by fatty acid oxidation enters the Krebs cycle (Section 7.10) where it is converted to CO_2 and H_2O. The degradation of acetyl CoA derived from palmitic acid is represented in the following manner:

$$8\ \text{acetyl CoA} + 16\ O_2 + 96\ \text{ADP} + 96\ \text{Pi} \downarrow 8\ \text{CoASH} + 96\ \text{ATP} + 16\ CO_2 + 104\ H_2O$$

The total yield of ATP obtained by the oxidation of palmitic acid will be determined by combining the two reactions as follows:

$$\text{Palmitoyl CoA} + 23O_2 + 131\ \text{ADP} + 131\ \text{Pi} \downarrow 8\ \text{CoASH} + 131\ \text{ATP} + 16\ CO_2 + 146\ H_2O$$

The activation of palmitate, however, requires 2 ATPs. Therefore, the net energy yield of β-oxidation of palmitate is 131 – 2 = 129 mol.

Respiratory quotient is defined as the moles of carbon dioxide produced, divided by the moles of oxygen consumed during the complete oxidation of a metabolic fuel to CO_2 and H_2O. The respiratory quotient for palmitate is 0.7. Respiratory quotient for palmitate can be obtained by using the above equation as follows:

$$\frac{16 \text{ moles of } CO_2}{23 \text{ moles of } O_2 \text{ consumed}} = 0.7$$

Fatty acids containing odd number of carbon atoms are not common, they occur only in minor amount. They are oxidized as their acyl CoA until a three-carbon unit, propionyl CoA, is formed. Propionyl CoA is converted to D-methylmalonyl CoA in a carboxylation reaction catalysed by propionyl CoA carboxylase and biotin. The D-form of methylmalonyl CoA is converted to L-form by methylmalonyl CoA racemase. The L-form of methylmalonyl CoA is isomerized by vitamin B12-dependent isomersase to form succinyl CoA, which enters the Krebs cycle. The oxidation of fatty acids containing odd number of carbon atoms is shown in Fig. 7.59.

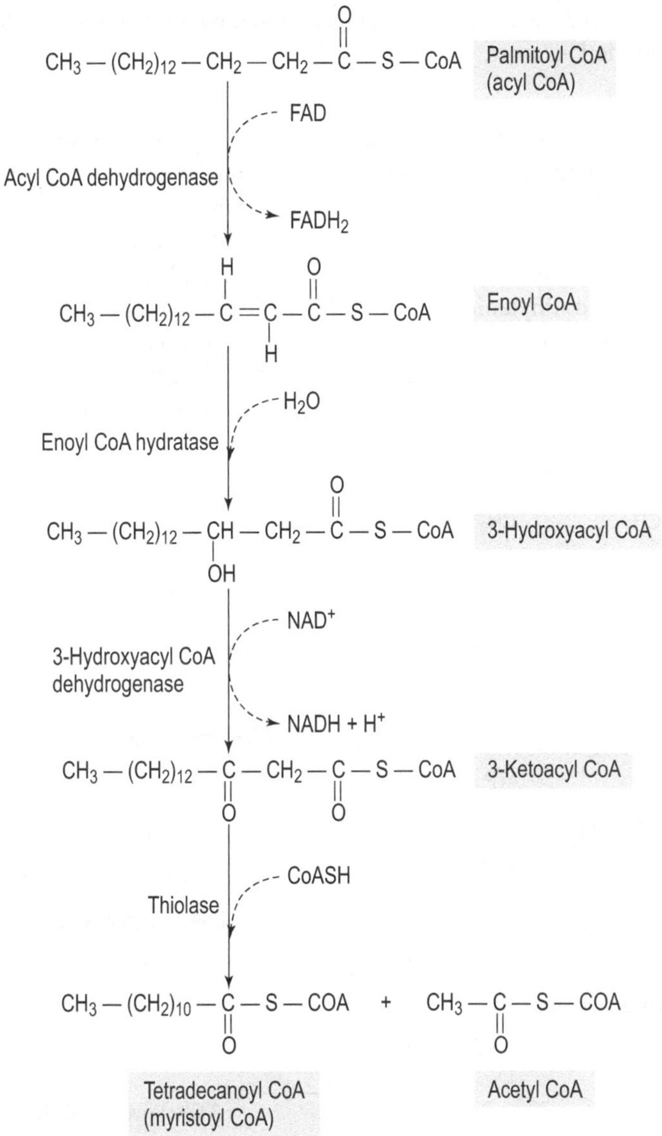

FIGURE 7.58 Four steps in the β-oxidation pathway

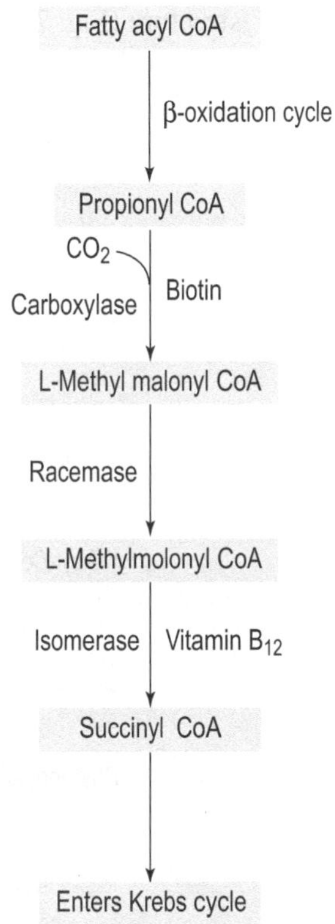

FIGURE 7.59 Oxidation of fatty acids containing odd number of carbon atoms

Oxidation of unsaturated fatty acids Though the basic steps and enzymes involved in the β-oxidation of fatty acids are similar for both saturated and unsaturated forms, the latter requires two additional enzymes for complete oxidation. Nearly 50% of the fatty acids in the human body are unsaturated. Naturally occurring unsaturated fatty acids have their double bonds in the cis configuration. However, the enzyme enoyl CoA hydratase can act upon the fatty acid with trans configuration. Therefore, the cis form of the fatty acid is isomerized to trans form by the cis-trans-enoyl CoA isomerase. Moreover, oxidation of polyunsaturated fatty acid requires another enzyme called epimerase. Actually, hydration of double bond yields the D-isomer of β-hydroxylacyl CoA, which is not the substrate for

L-β-hydroxylacyl CoA dehydrogenase. At this stage, epimerase converts the D-isomer of β-hydroxylacyl CoA to the L-isomer.

α-Oxidation

α-Oxidation is the process in which a fatty acid is degraded by the removal of a single carbon, known as the α-carbon. In this process, the α-carbon is hydroxylated and cleaved off. α-oxidation occurs in the peroxisomes. Hence, the pathway is also called peroxisomal oxidation. α-Oxidation usually takes place in the fatty acids that cannot undergo the conventional β-oxidation. In human beings, a well-known example for α-oxidation is the degradation of phytanic acid. Phytanic acid (3, 7, 11, 15-tetramethyl hexadecanoic acid) is the major constituent of ruminant fat and dairy products. As it has methyl group in its structure (Fig. 7.60), it cannot be degraded by acyl CoA dehydrogenase, the first enzyme of β-oxidation. Phytanic acid is, therefore, diverted to α-oxidation. Hence, α-oxidation can be learned in terms of phytanic acid oxidation. The pathway for the α-oxidation of phytanic acid is shown in Fig. 7.61.

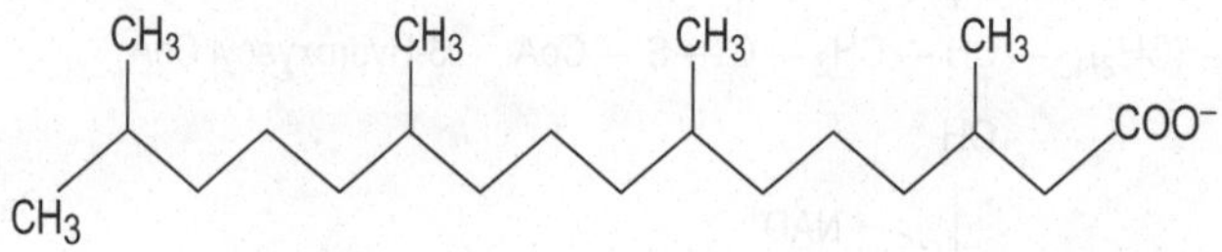

FIGURE 7.60 Structure of phytanic acid

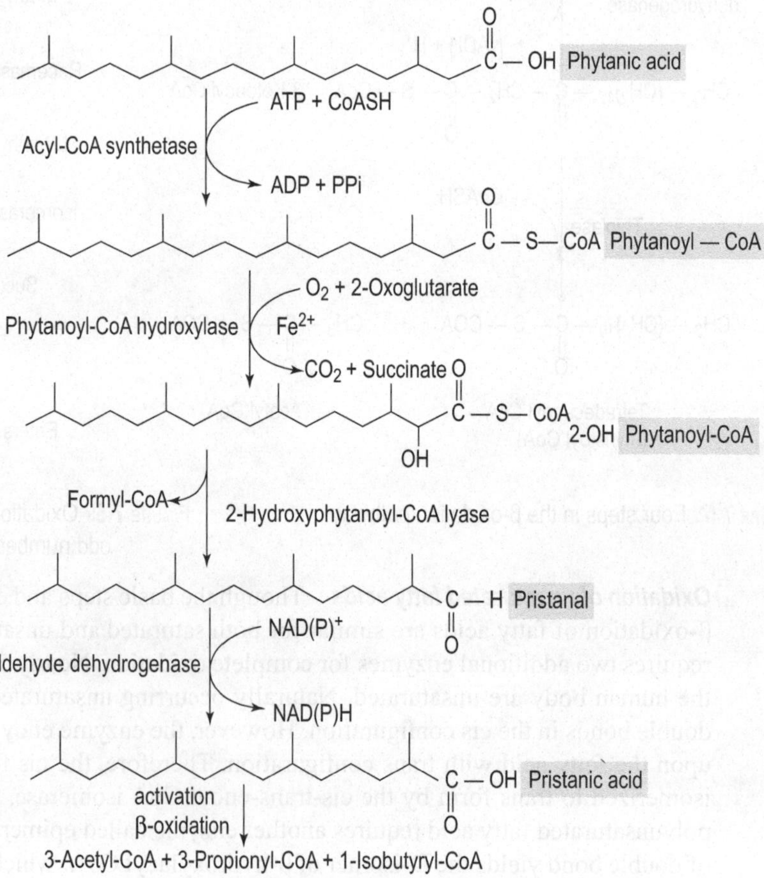

FIGURE 7.61 α-Oxidation pathway of phytanic acid

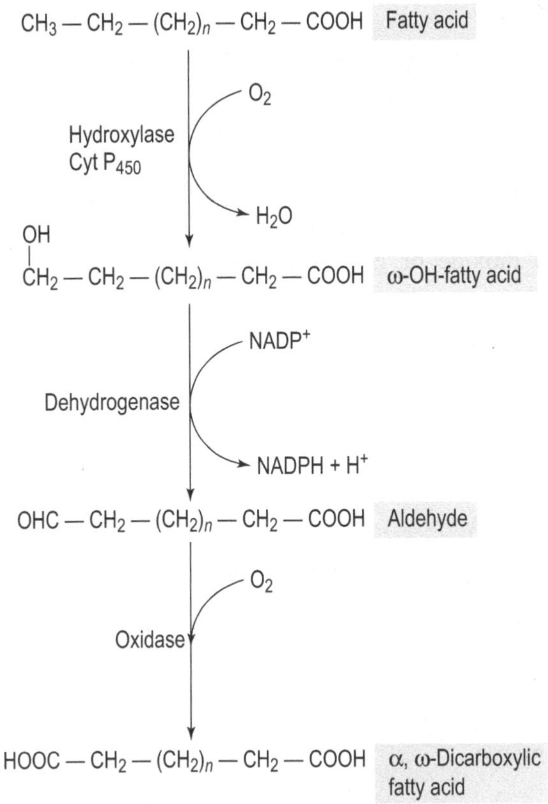

FIGURE 7.62 ω-Oxidation of fatty acids

Phytanic acid is activated to phytanoyl CoA by CoASH. ATP drives this reaction under the catalysis of acyl CoA synthetase.

Phytanoyl CoA is oxidized to 2-hydroxy-phytanoyl CoA. Phytanoyl-CoA hydroxylase (phytanoyl-CoA dioxygenase) catalyses this reaction in the presence of molecular oxygen, ferrous ion, and 2-oxoglutarate.

Hydroxy phytanoyl CoA is cleaved into formyl CoA and pristanal. TTP-dependent lyase catalyses the cleavage of the alpha carbon into formyl CoA. Subsequently, the formyl CoA is converted into formate and then into carbon dioxide. The other product pristanal is acted upon by aldehyde dehydrogenase and converted to pristanic acid, which is capable of undergoing β-oxidation. Activation and β-oxidation of pristanic acid produces 3 mol each of acetyl CoA and propionyl CoA and 1 mol of isobutyryl CoA.

Deficiency of phytanoyl CoA dioxygenase results in accumulation of phytanoyl CoA, which causes severe brain damage. This condition is called Refsum's disease.

ω-Oxidation

ω-Oxidation occurs in the endoplasmic reticulum of liver and kidney cells. It involves oxidation of the ω-carbon. This pathway occurs in medium fatty acids containing 10 to 12 carbon atoms. It has the following three steps (Fig. 7.62):

Step 1 ω-Carbon of the fatty acid is hydroxylated by molecular oxygen and NADPH. This reaction is catalysed by cyt p450 dependent mixed function oxidase. Hydroxylation converts fatty acid into fatty-ω-hydroxy acid.

Step 2 The hydroxyl group of the fatty-ω-hydroxy acid is oxidized to aldehyde group by NAD^+-dependent alcohol dehydrogenase. A fatty-ω-aldoacid is formed.

Step 3 The aldehyde group of fatty-ω-aldoacid is oxidized to carboxyl group by NAD^+-dependent alcohol dehydrogenase. A dicarboxylic acid is formed. This acid can undergo β-oxidation to form succinic acid and adipic acid. Succinic acid enters the Krebs cycle.

7.6 METABOLISM OF KETONE BODIES

Acetoacetate, acetone, and β-hydroxy butyrate are collectively called ketone bodies. They are synthesized in the liver and are used as energy sources in brain, heart, kidney, and muscles. They are synthesized from acetyl CoA in the mitochondria.

7.6.1 Biosynthesis of Ketone Bodies

The synthesis of ketone bodies is known as ketogenesis. Figure 7.63 shows the steps involved in ketogenesis. Two molecules of acetyl CoA condense in a reaction catalysed by thiolase; this reaction involves the cleavage of a high-energy thioester bond to form acetoacetyl CoA. Acetoacetyl CoA reacts with acetyl CoA forming β-hydroxy-β-methylglutaryl CoA (HMG CoA) under the catalysis of HMG CoA synthase. HMG CoA is lysed by HMG CoA lyase to form acetoacetate. Acetoacetate can either decompose spontaneously to acetone and carbon dioxide or get reduced by NADH-dependent reductase to form β-hydroxy butyrate.

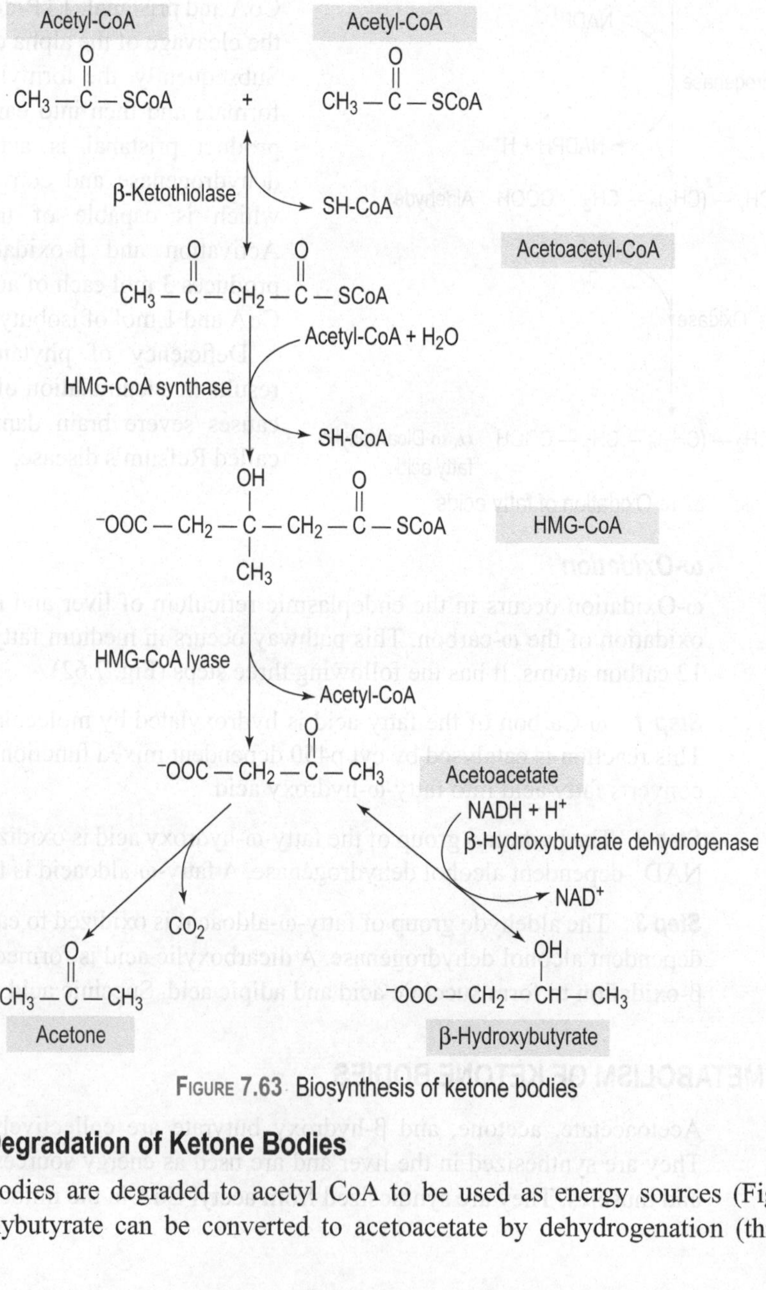

FIGURE 7.63 Biosynthesis of ketone bodies

7.6.2 Degradation of Ketone Bodies

Ketone bodies are degraded to acetyl CoA to be used as energy sources (Fig. 7.64). β-Hydroxybutyrate can be converted to acetoacetate by dehydrogenation (this is the reverse of the reduction reaction shown in Fig. 7.63). Acetoacetate can then be oxidized to acetyl CoA in just two steps—first, acetoacetate is activated to acetoacetyl CoA by acetoacetyl CoA synthetase or 3-ketoacid CoA transferase, at the expense of ATP. Second, the acetoacetyl CoA is lysed by thiolase to liberate two molecules of acetyl CoA, which enter the Krebs cycle (Section 7.10).

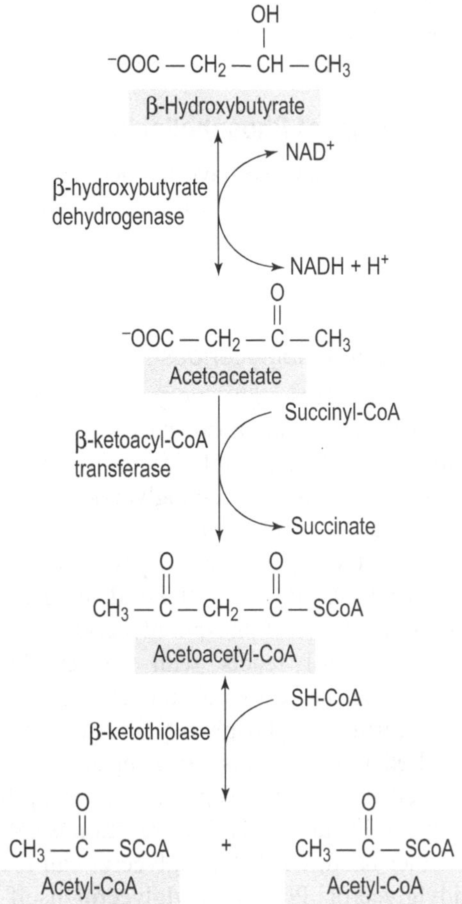

FIGURE 7.64 Degradation of ketone bodies

When the rate of synthesis of ketone bodies is greater than their degradation, excess ketone bodies accumulate in the blood, leading to a condition called ketonemia. As a result of this accumulation, ketone bodies are excreted in the urine in large amounts, a condition called ketonuria. Excess acetoacetate lower the blood pH, causing ketosis and metabolic acidosis. Ketonemia and ketosis are more common in starvation and diabetes mellitus.

7.6.3 Energy Yield by Ketone Bodies

Conversion of β-hydroxy butyrate to acetoacetate generates NADH molecule. NADH enters the electron transport chain and produces three ATPs. In the next step, one molecule is ATP is utilized for the activation of acetoacetate to acetoacetyl CoA. Finally, acetoacetyl

CoA is cleaved to two acetyl CoA, each of which in turn produces 12 ATPs/molecules via electron transport and oxidative phosphorylation.

Thus, we can write

ATP formed by the oxidation of β-hydroxy butyrate to acetoacetate = 3

ATP formed by the oxidation of acetoacetyl CoA = 24

Total ATP = 27

ATP consumed for activation of acetoacetate to acetoacetyl CoA = 1

Net ATP produced = 27 – 1 = 26 mol

7.6.4 Metabolism of Phospholipids

In our discussion so far, we have learned about the metabolism of fatty acids and TG. We will now turn our attention to the metabolism of the next major group of lipids, the phospholipids. The structure, types, and functions of phospholipids have already been discussed in Chapter 4. Phospholipids are synthesized in the smooth endoplasmic reticulum from where they are carried to the golgi apparatus and then to the membranes.

Metabolism of phosphoglycerides

This section discusses the biosynthetic pathway of the six phosphoglycerides (phosphatidylethanolamine, phosphatidylserine, phosphatidylcholine, phosphatidyl-inositol, phosphatidylglycerol, and cardiolipin) from a common precursor, lysophosphatidate. Figure 7.65 shows the synthesis of phosphoglycerides from lysophosphatidate.

Lysophosphatidate reacts with an acyl CoA to form phosphatidate. Phosphatidate has two fates—it is either hydrolyzed by phosphatise to form 1, 2-diacylaglycerol and Pi, or it is acted upon by CTP to form CDP-diacylglycerol and PPi. Both these products are diverted to the synthesis of phosphoglycerides.

1, 2-Diacylaglycerol reacts with the cytidine diphosphate-ethanolamine (CDP-ethanolamine) via its C3-hydroxyl group, in the presence of CDP-ethanolamine transferase, forming phosphatidylethanolamine. The ethanolamine moiety of phosphatidyl-ethanolamine is exchanged for serine to form phosphatidylserine. Phosphatidylethanolamine accepts methyl group from three molecules of SAM to form phosphatidylcholine and SAH. This reaction is catalysed by N-methyl transferase.

On the other hand, CDP-diacylglycerol can react with glycerol-3-phosphate to form phosphatidylglycerol. Phosphatidylglycerol itself can react with another molecule of CDP-glycerol and produce cardiolipin (diphosphatidylglycerol). CDP-diacylglycerol reacts with inositol under the catalysis of CDP-diacylglycerol-inositol transferase and liberates CMP along with phosphatidylinositol.

Metabolism of sphingolipids

Sphingosine, the precursor for all the sphingolipids (ceramide, sphingomyelin, cerebrosides, sulphatid , and ganglioside) is obtained from palmitoyl CoA and serine in a two-step reaction (Fig. 7.66). Palmitoyl CoA and serine react in the presence of NADH to form sphingosine, which in turn reacts with acyl CoA and FAD to form ceramide. Ceramide generates sphingomyelin and cerebroside by two different pathways. Ceramide reacts with CDP-choline to form sphingomyelin. Excess sphingomyelin is normally hydrolyzed to ceramide and phosphatidyl choline by the lysosomal sphingomyelinase. Genetic defect in this enzyme causes a disease called Niemann–Pick disease in which there is excessive accumulation of sphingomyelin.

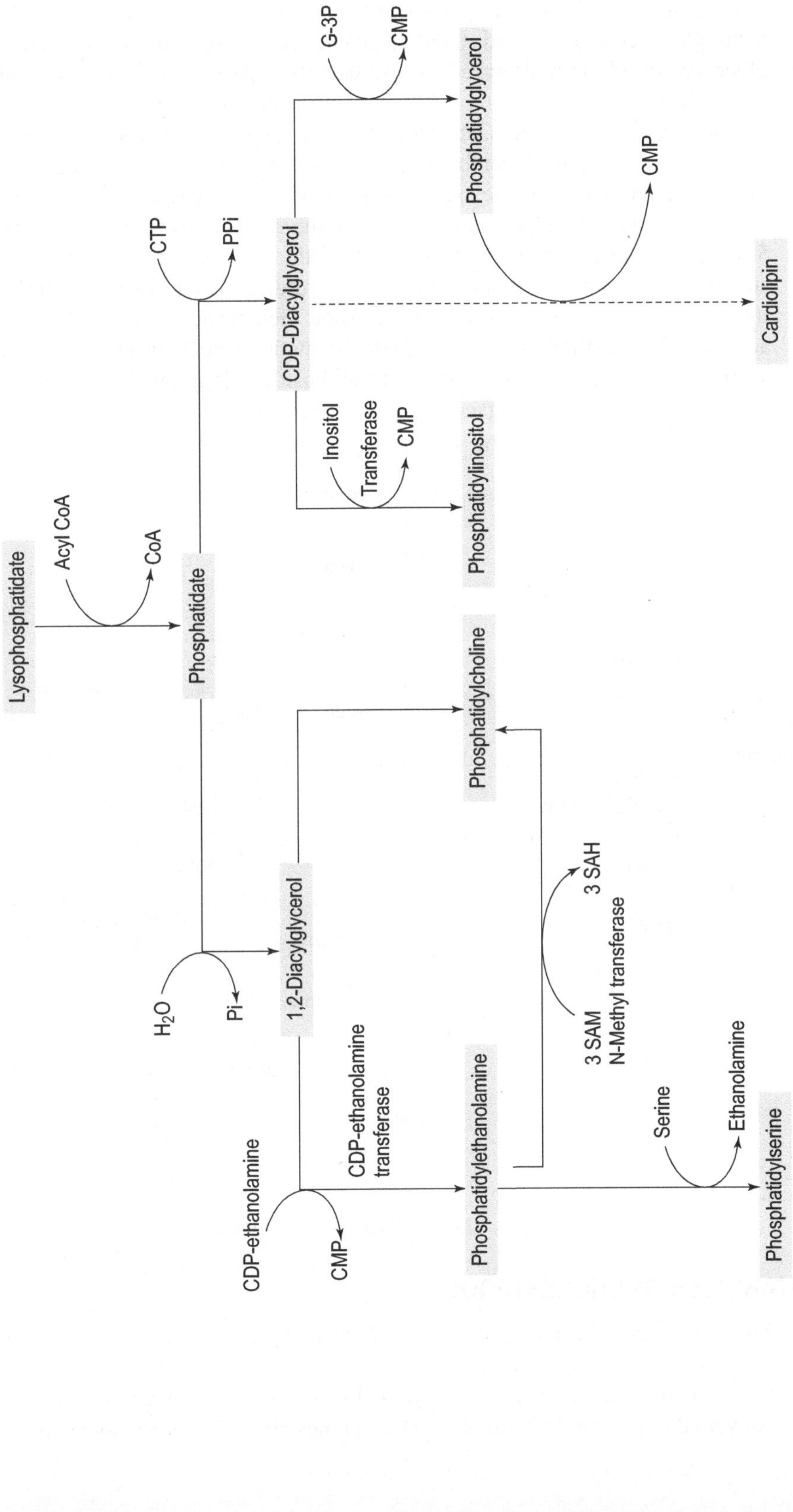

FIGURE 7.65 Synthesis of phosphoglycedrides from lysophosphatidate

Ceramide can also react with UDP-sugar to form cerebroside. UDP glucose can form glucocerebroside and UDP-galactose can form galactocerebroside. Excessive glucocerebroside is hydrolyzed by glucocerebrosidase to form glucose and regenerate ceramide. Defect in glucocerebrosidase results in pathologic accumulation of the substrate glucocerebroside causing Gaucher's disease. Similarly, excessive galactocerebroside is hydrolysed by galactocerebrosidse. Genetic defect in this enzyme causes Krabbe's disease, characterized by the excessive accumulation of galactocerebroside.

Cerebroside gets sulphated by accepting the sulphate from the 3′-phosphoadenosine-5′-phosphosulphate (PAPS), to form sulphatid. Microsomal transferase catalyzes the sulphation of cerebroside. Sulphatid can be hydrolysed by lysosoma hydrolase to liberate its sulphate moiety and reform the cerebroside. Genetic deficiency in this enzyme leads to accumulation of sulphatid, causing metachromatic leukodystrophy. Stepwise addition of sugar residues to cerebrosides can form gangliosides. Gangliosides can also be hydrolyzed with the liberation of its sugar moieties to regenerate the cerebroside.

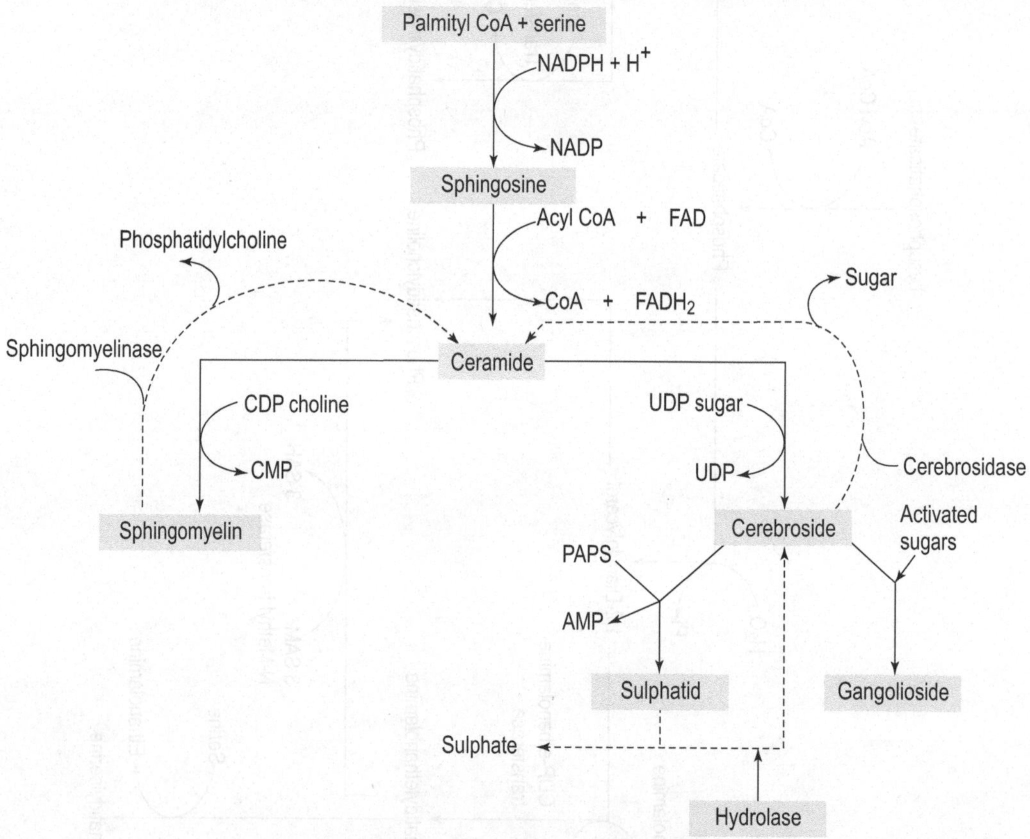

FIGURE 7.66 Metabolism of sphingolipids

7.7 METABOLISM OF CHOLESTEROL

The next major dietary lipid that needs to be discussed is cholesterol. Cholesterol is synthesized primarily in the liver and then esterified for storage. Cholesterol is degraded to bile acids in the liver or it is degraded to steroid hormones in kidney and gonads, or converted to vitamin D in the skin. Thus, cholesterol is the precursor for bile acids, major

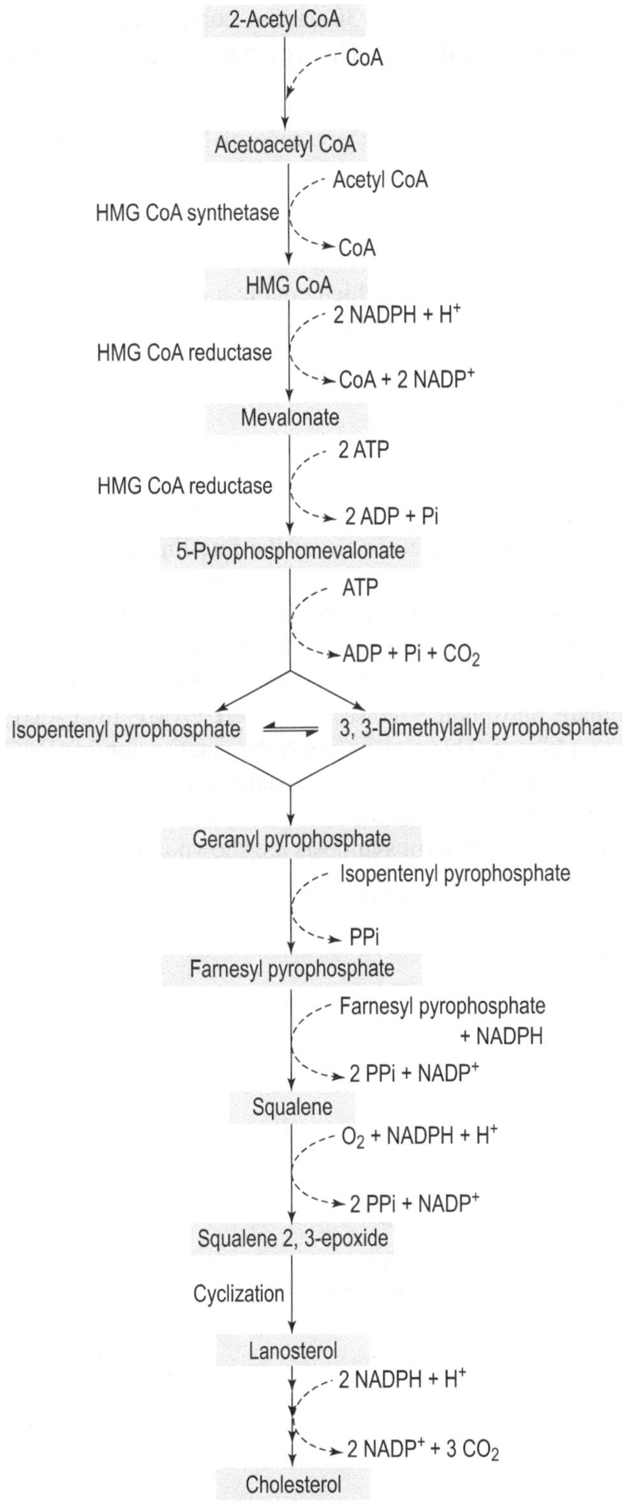

FIGURE 7.67 Biosynthesis of cholesterol

steroid hormones, including sex hormones, and vitamin D. This section discusses the biosynthesis and catabolism of cholesterol. Catabolism of cholesterol is nothing but the synthesis of bile acids and steroid hormones.

7.7.1 Biosynthesis of Cholesterol

Acetyl CoA is the precursor for cholesterol synthesis. Acetyl CoA contributes to all the 27 carbon atoms of the cholesterol. The synthesis pathway of cholesterol is shown in Fig. 7.67.

The pathway has eleven steps that are listed as follows:

Step 1 Two molecules of acetyl CoA condense into acetoacetyl CoA in the presence of another CoA and thiolase enzyme.

Step 2 Acetoacetyl CoA reacts with acetyl CoA forming HMG CoA. HMG CoA synthetase catalyses this reaction.

Step 3 HMG CoA is reduced by NADPH to form mevalonate and liberate CoA. This reaction is catalysed by HMG CoA reductase, the rate-limiting enzyme of cholesterol synthesis.

Step 4 Mevalonate accepts a phosphate group from ATP and gets activated to 5-phosphomevalonate, in the presence of mevalonate kinase.

Step5 5-Phosphomevalonateisdecarboxylated at the expense of ATP, to form a five-carbon isoprene unit, isopentenyl pyrophosphate (IPP) and 3, 3′-dimethyl allyl pyrophosphate (DPP). In addition, IPP is isomerized to DPP by isomerase; equilibrium between IPP and DPP is maintained.

Step 6 IPP and DPP are condensed to form a 10-carbon compound, geranyl pyrophosphate (GPP). This reaction is catalysed by geranyl pyrophosphate synthetase.

Step 7 GPP condenses with another molecule of IPP forming a 15-carbon unit, farnesyl pyrophosphate (FPP). Farnesyl pyrophosphate synthetase catalyses this step.

Step 8 Two molecules of FPP are conjugated to a 30-carbon compound, squalene, in the presence of NADPH and squalene synthetase. This reaction is called reductive synthesis.

Step 9 Squalene forms squalene-2-, 3-epoxide in the presence of NADP+-dependent oxygenase.

Step 10 Squalene-2, 3-epoxide is cyclized to form lanosterol under the catalysis of cyclase.

Step 11 Lanosterol undergo a multistep reaction, in which double bonds are rearranged and three methyl groups are removed reducing the total carbons to 27. Thus, cholesterol is formed in this step.

Regulation of cholesterol synthesis

Cholesterol is primarily regulated by the inhibition of the rate-limiting enzyme, HMG CoA reductase. Excess cholesterol inhibits the enzyme HMG CoA reductase by feedback inhibition to prevent further synthesis. Hormonal regulation is also possible, where glucagon is the inhibitor, whereas insulin is the activator of the rate-limiting enzyme. Glucogan activates adenylate cyclase to form cAMP from ATP. cAMP activates protein kinase A (PKA). PKA inactivates HMG CoA reductase by phosphorylation. This is a reversible reaction. Therefore, HMG CoA reductase can also be reactivated by the removal of the phosphate group by phosphatase, the reaction stimulated by insulin hormone. Insulin can also stimulate phosphodiesterase to convert cAMP to 5′AMP, thus preventing activation of protein kinase. This also prevents the inactivation of HMG CoA reductase. Antihypercholesterolemic drugs, such as lovastatin and mevastatin, are reversible competitive inhibitors of HMG CoA reductase.

Feedback and hormonal regulation of cholesterol synthesis are shown in Fig. 7.68.

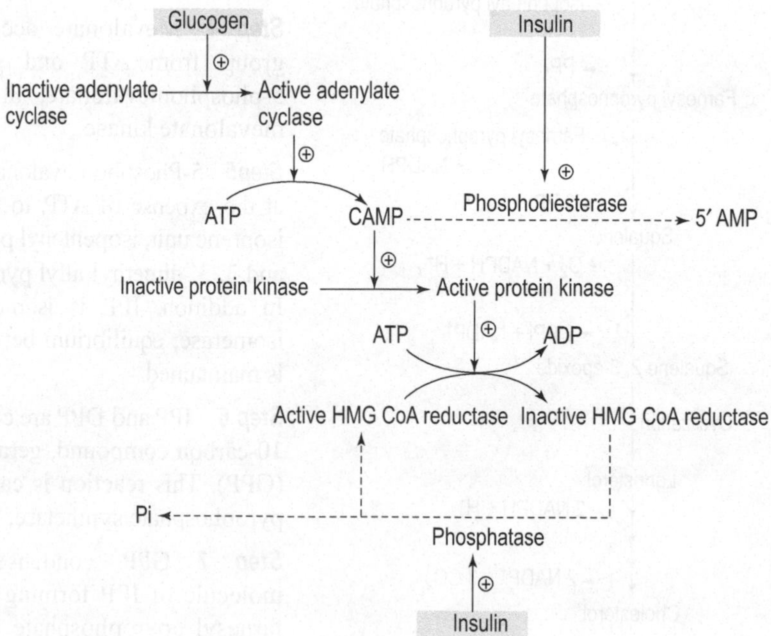

FIGURE 7.68 Regulation of cholesterol synthesis

The cholesterol synthesized as per the steps listed in the preceding discussion are esterified at the 3-β-hydroxyl group. Long-chain fatty acids are used in the esterification after getting activated to long-chain fatty acyl CoA. Free cholesterol is taken up by HDL and esterified by lecithin in a reaction catalysed by lecithin-cholesterol acyl transferase (LCAT).

7.7.2 Catabolism of Cholesterol

Catabolism of cholesterol results in the formation of the following: bile acids, steroid hormones (glucocorticoids, mineralocorticoids, sex hormones), and vitamin D.

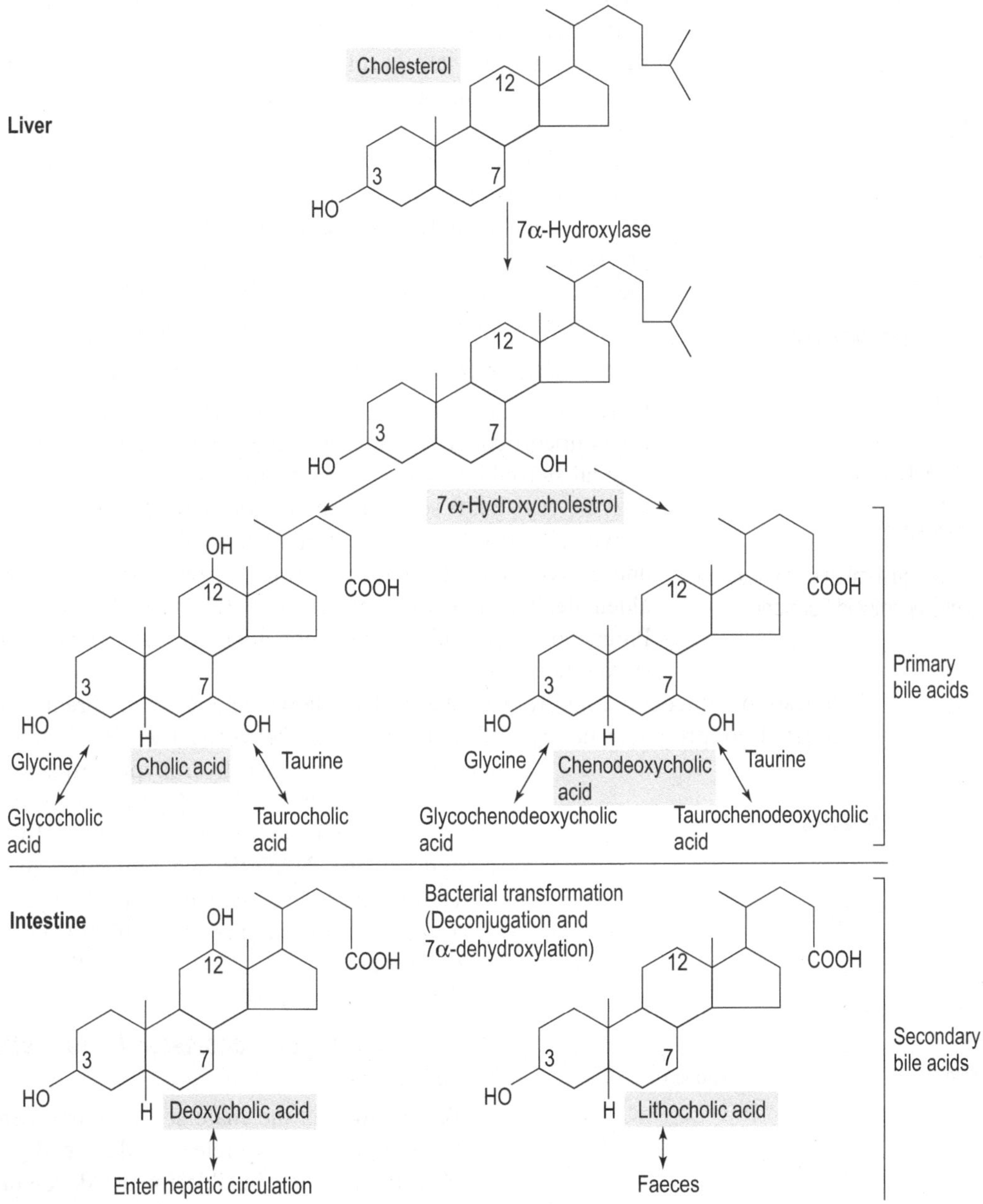

FIGURE 7.69 Bile acids

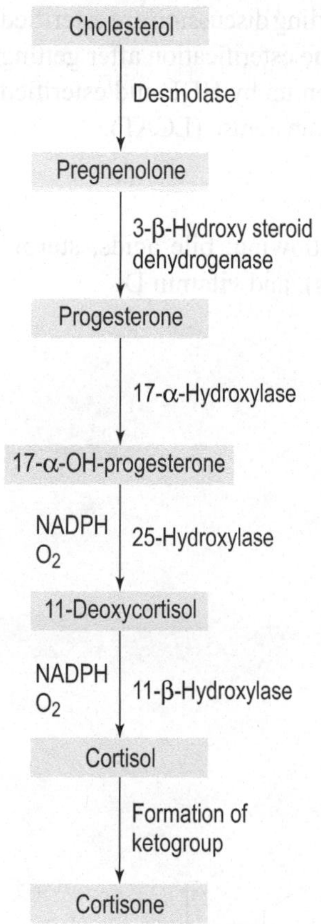

FIGURE 7.70 Conversion of cholesterol to glucocorticoid hormone

Conversion of cholesterol to bile acids

Cholesterol produced in the liver undergoes two types of transport: forward transport and reverse transport. In forward transport, it is carried in the plasma as LDL to reach the peripheral tissues. About one-fifth of the cholesterol that is not utilized by the peripheral tissues comes back to the liver in the form of HDL. This is called reverse transport. HDL cholesterol (C-27 steroid) is converted into bile acids (C-24 steroid) in the liver. The side chain of the cholesterol is cleaved to form a C-24 carboxyl group with the loss of three carbon atoms. The double bond in the fifth position is then saturated and the steroid ring is hydroxylated. This results in the formation of cholic acid and chenodeoxycholic acid (Fig. 7.69), which are called primary bile acids. They give rise to deoxycholic acid and lithocholic acid, which are secondary bile acids.

Primary bile acids are secreted from liver into the bile to reach the intestine. Secondary bile acids are formed in the intestine from primary bile acids, by bacterial action. Bile acids are excreted in the faeces. Thus, catabolism of cholesterol to bile acids presents the excessive cholesterol accumulation in blood and tissues.

Conversion of cholesterol to adrenal hormones

Glucocorticoids (cortisol, cortisone, and corticosterone) and mineralocorticoids (aldosterone and corticosterone) are major adrenal steroid hormones. Cytochrome-P450 side-chain cleavage enzyme (cyt-P450 sce), also called desmulase, a mitochondrial enzyme, hydroxylates the cholesterol at the C20 and C22 position, and cleaves the side chain to form pregnenolone and isocaproic aldehyde. This reaction requires NADPH and molecular oxygen. Pregnenolone is the common precursor for the synthesis of all steroid hormones.

Synthesis of glucocorticoid from cholesterol is shown in Fig. 7.70. Pregnenolone is converted to progesterone by 3-β-hydroxysteroid dehydrogense. Progesterole is converted to 17-α-hydroxyprogesterone by 17-α-hydroxylase. 17-α-Hydroxyprogesterone is converted to 11-deoxycortisol by 25-hydroxylase, NADPH and molecular oxygen. 11-β-Hydroxylase converts 11-deoxycortisol to cortisol in the presence of NADPH and molecular oxygen. Cortisone is the 11-keto derivative of cortisol.

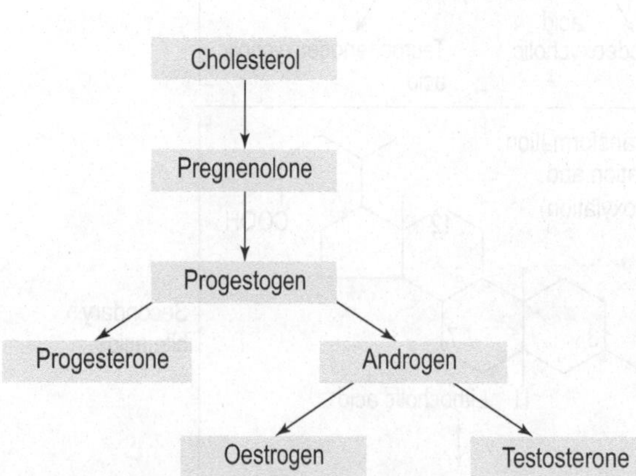

FIGURE 7.71 Degradation of cholesterol to sex hormones

Conversion of cholesterol to steroid hormones

Testosterone is the male sex hormone derived from cholesterol, secreted by the Leydig cells of testis. It is responsible for the development of reproductive system, and secondary sexual characters in males. Estradiol is the female sex hormone secreted by the thecal cells of the graffian follicle in the ovary. It is involved in the development of the reproductive system and secondary characters in females. Progesterone is formed by the placenta during pregnancy. Hence, it is also called as pregnancy hormone. Formation of sex hormones from cholesterol is shown in Fig. 7.71.

Conversion of cholesterol to vitamin D

7-Dehydrocholesterol is an intermediate formed during the conversion of lanosterol to cholesterol. It accumulates in the skin and gets converted into vitamin D_3 on exposure to ultraviolet light. Structures of compounds involved in vitamin D_3 formation are shown in Fig. 7.72.

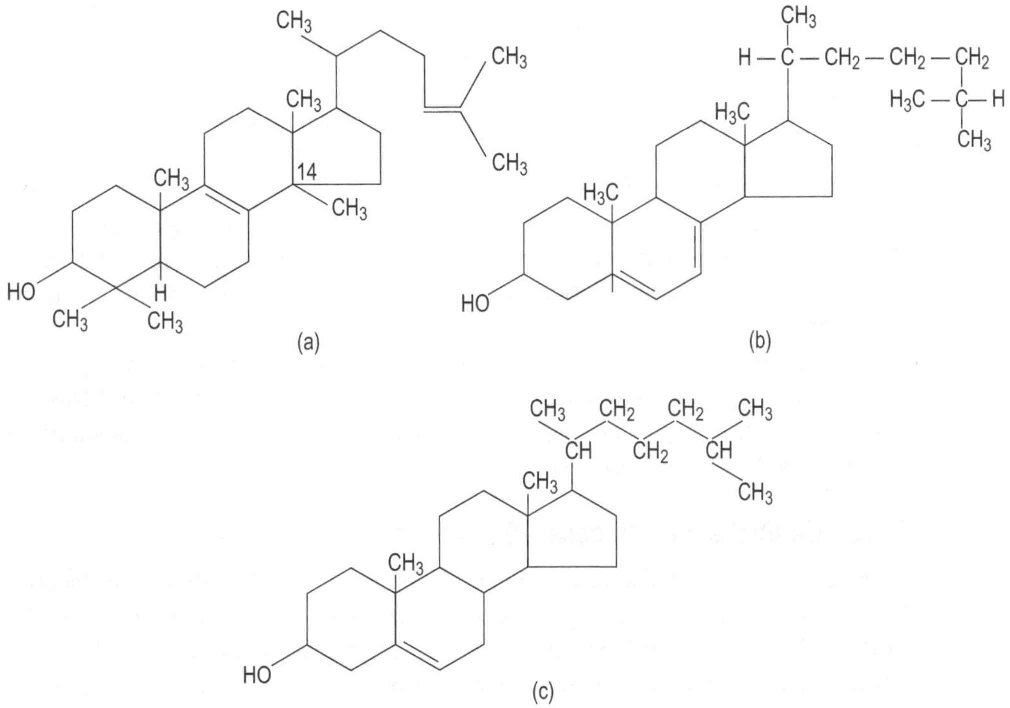

FIGURE 7.72 Structures of compounds involved in the formation of vitamin D_3 (a) Lanosterol, (b) 7-Dehydrocholestrol, (c) Cholesterol

7.8 METABOLISM OF CARBOHYDRATES

There are different pathways by which carbohydrates are synthesized and catabolised. Major pathways and alternative pathways of carbohydrate metabolism are dealt with in this section.

7.8.1 Digestion of Carbohydrates

Dietary carbohydrates occur in different forms: starch (plant foods from cereals and millets), lactose (milk), fructose (fruits) and sucrose (table sugar and cane sugar), dextrin and maltose (from malt), and glycogen (animal-based foods).

Saliva and gastric juice contain amylase. However, the digestion of carbohydrates cannot occur due to less incubation in mouth and acidic pH in the stomach. Therefore, digestion of carbohydrates actually starts in the small intestine by the action of pancreatic amylase. The alkaline pH and presence of calcium ions as activators are highly suitable for the action of pancreatic amylase. Starch is converted to dextrin and then to maltose. Maltose and other dietary disaccharides, such as lactose and sucrose, are further hydrolyzed by disaccharidases located on the outer membrane of the intestinal epithelial cells. Disaccharides are hydrolysed as follows:

$$\text{Maltose} \xrightarrow{\text{Maltase}} \text{D-glucose} + \text{D-glucose}$$

$$\text{Lactose} \xrightarrow{\text{Lactase}} \text{D-glucose} + \text{D-galactose}$$

$$\text{Sucrose} \xrightarrow{\text{Sucrase}} \text{D-glucose} + \text{D-fructose}$$

Maltase converts maltose to two glucose molecules. Lactase converts lactose to glucose and galactose. Deficiency of lactase in the intestinal cell membrane causes 'lactose intolerance' where the individual develops abdominal cramps and diarrhoea and cannot tolerate ingested milk. Sucrase hydrolyzes sucrose to glucose and fructose. Thus, the dietary carbohydrates are digested finally to D-glucose, D-galactose, and D-fructose. D-galactose and D-glucose are absorbed from the lumen into the intestinal epithelial cells by ATP-dependent active transport. Fructose is absorbed by facilitated diffusion (refer to Chapter 2 for transport mechanism). We will now discuss the metabolism of glucose, galactose, and fructose individually.

7.8.2 Catabolism of Glucose—Glycolysis

A three-member scientist team, Gustav Embden, Otto Meyerhof, and Jacob Parnas discovered the catabolism of glucose in the cytosol. This catabolic pathway is called Embden–Meyerhof–Parnas pathway or glycolytic pathway or simply glycolysis. Glycolysis may be aerobic or anaerobic. Aerobic glycolysis occurs in the liver, forming pyruvate, with the generation of 6 moles of ATP/glucose. On the other hand, anaerobic glycolysis takes place in the muscle cells, forming lactate with the production of 2 moles of ATP/glucose.

Glycolytic pathway has ten steps, as shown in Fig. 7.73. These steps are explained as follows:

Step 1 Glucose is converted to glucose-6-phosphate (G6P) by ATP under the action of hexokinase. Phosphorylated glucose is retained in the cells, as they are less permeable to the membrane. This reaction is essentially irreversible. Hexokinase is allosterically inhibited by G6P. In liver, the enzyme glucokinase carries out this reaction.

Step 2 G6P is isomerized to fructose-6-phosphate (F6P) by phosphoglucose isomerase. This is a reversible reaction.

Step 3 F6P is phosphorylated at its first carbon to form fructose-1, 6-bisphosphate (F1, 6BP) in a irreversible reaction catalysed by phosphofructokinase (PFK). PFK is

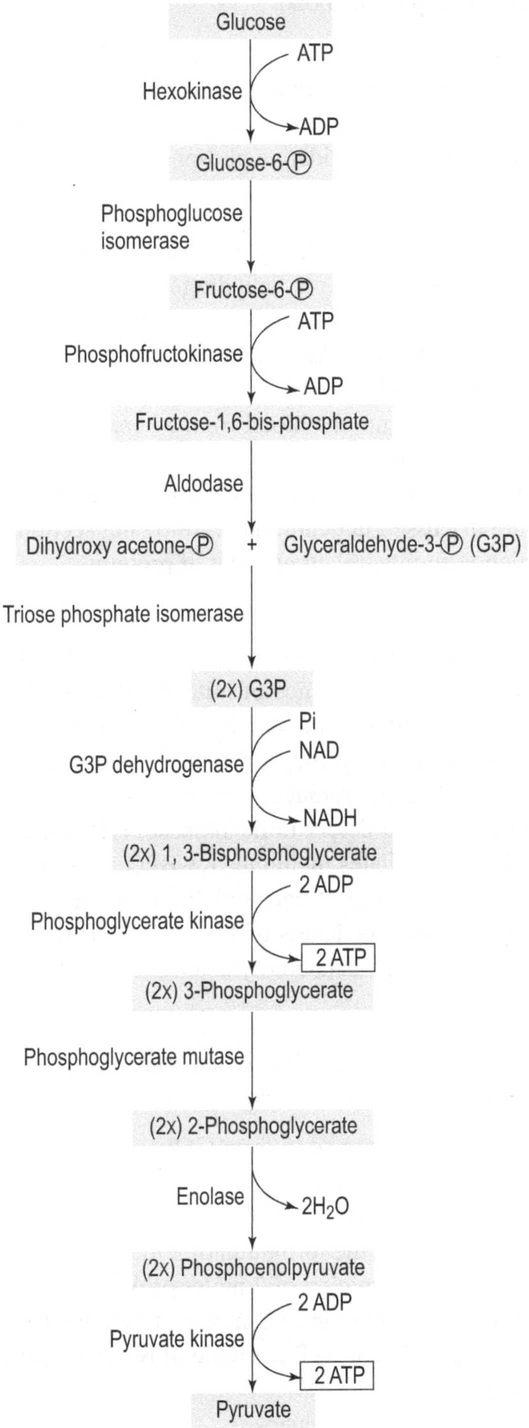

FIGURE 7.73 Glycolytic pathway

significant since it is the rate-limiting enzyme and major regulatory enzyme of glycolysis. AMP and F2, 6BP are activators of PFK. ATP and citrate inhibit PFK and prevent glycolysis.

Step 4 In this step, F1, 6BP, a 6-carbon compound is converted into dihydroxy acetone phosphate (DHAP) and glyceraldehyde-3-phosphate (G3P), both 3-carbon compounds. This reaction is catalysed by aldolase.

Step 5 Reversible interconversion of DHAP and G3P is catalysed by triose phosphate isomerase. G3P is involved in the subsequent step of the glycolysis, therefore, the equilibrium is shifted towards formation of G3P from DHAP. Thus, two molecules of G3P are formed. From subsequent steps, two moles of products are formed (represented as 2x).

Step 6 This reaction is called substrate-level oxidative phosphorylation. G3P accepts a phosphate in group in the form of inorganic phosphate (Pi), in the presence of NAD^{+}. The enzyme is G3P dehydrogenase and the product is 1, 3-bisphosphoglycerate (1, 3-BPG). In this step, a high-energy phosphate bond is formed, that is, 1, 3-BPG is a high-energy compound with high phosphate group transfer potential. The NADH formed in this reaction should be reoxidized to NAD^{+} for the continuation of glycolysis.

Step 7 This step is very important because this is the first reaction that generates ATP in the glycolytic pathway. 1,3-BPG donates one of its phosphate atoms to ADP forming ATP and 3-phosphoglycerate (3-PG). This is a second example for substrate level oxidative phosphorylation, and the catalyst is phosphoglycerate kinase. As two moles of 1, 3-BPG are present, two ATPs are formed in this step. However, recall that up to step 5, that is, up to triose phosphate formation, two moles of ATP are consumed. Therefore, at this step, utilization and production of ATP are balanced.

Step 8 3-PG is converted to 2-phosphoglycerate (2-PG) in a reversible reaction catalysed by phosphoglycerate mutase.

Step 9 2-PG is hydrolyzed by enolase to form phosphoenolpyruvate (PEP), a high-energy phosphate compound possessing high-energy phosphoenol bond.

Step 10 This step involves an irreversible reaction, in which, PEP reacts with ADP to form pyruvate and ATP. The enzyme used here is pyruvate kinase. Thus, two moles each of pyruvate and ATPs are formed per mole of glucose entering the glycolytic pathway. This reaction is yet another example for substrate-level oxidative phosphorylation.

Fate of pyruvate

The pyruvate formed in the glycolysis has three fates—entry into gluconeogenesis, entry into TCA cycle, and formation of alanine. Pyruvate is dehydrogenated to lactate in anaerobic glycolysis by lactate dehydrogenase. In this reaction, NADH is oxidized to NAD. Thus, anaerobic glycolysis is a way of oxidizing NADH generated in step 7. This reoxidation is very essential for the continuation of glycolysis. The lactate formed in the muscle cells, under anaerobic condition, is released into the blood from where it reaches the liver and gets converted to glucose by gluconeogenesis.

Pyruvate can also be converted to acetyl CoA by pyruvate dehydrogenase. Acetyl CoA enters TCA cycle. These pathways are discussed in detail in the following sections. Alanine converts pyruvate to alanine, which is an integral component of proteins.

Red cell glycolysis

Defect in pyruvate kinase results in haemolytic anaemia. This is because, in red cells, glycolysis is the only source of energy because of the absence of mitochondria. Defect in pyruvate kinase prevents ATP production and reduces the activity of ATP-dependent ion pump, which is essential for the maintenance of the biconcave shape of red cells. As a consequence, the cells swell and lyse leading to premature and excessive red cell destruction. Haemolytic anaemia is also caused by a defect in hexokinase. Hexokinase deficiency leads to low levels of glycolysis intermediates including 1,3-BPG. Normally, 1,3-BPG binds to haemoglobin for lowering the affinity of haemoglobin to oxygen, for the subsequent delivery of oxygen to tissues. Thus, low levels of 1,3-BPG leads to low levels of oxygen supply to the tissues. This also results in haemolytic anaemia.

Energetics of glycolysis

The net ATP yield for aerobic glycolysis is 6 moles, whereas that for anaerobic glycolysis is 2 moles. This can be calculated for both methods as shown below.

Aerobic glycolysis In step 1, one mole of ATP is consumed to form G6P.

In step 3, one mole of ATP is consumed to form F1,6-BP.

Total consumption = 2 moles of ATP per mole of glucose

In step 7, phosphoglycerate kinase reaction leads to the formation of 2 moles of ATP.

In step 10, pyruvate kinase reaction leads to the formation of 2 moles of ATP.

In step 6, G3P dehydrogenase reaction causes formation of 2 moles of NADH. G3P shuttle (Fig. 7.74) results in the production of 2 moles of ATP per mole of NADH. Thus, 4 moles of ATP are formed per mole of glucose.

Total production = 8 moles of ATP per mole of glucose.

Net ATP yield = 8 – 2 = 6 moles/glucose.

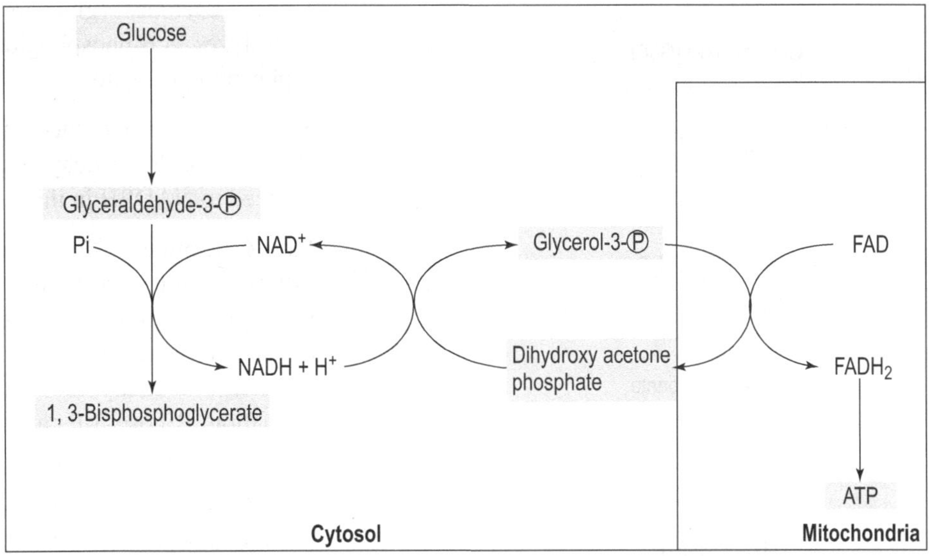

FIGURE 7.74 Glycerol-3-phosphate (G3P) shuttle and formation of ATP from NADH

Anaerobic glycolysis In step 1, one mole of ATP is consumed to form G6P.

In step 3, one mole of ATP is consumed to form F1, 6-BP

Total consumption = 2 moles of ATP per mole of glucose.

In step 7, phosphoglycerate kinase reaction leads to the formation of 2 moles of ATP.

In step 10, pyruvate kinase reaction leads to the formation of 2 moles of ATP.

Total production = 4 moles of ATP per mole of glucose.

Net ATP yield = 4 – 2 = 2 moles/glucose.

In addition to glycolysis, glucose can also be catabolized by an alternative pathway, the pentose phosphate pathway (hexose monophosphate shunt or HMP shunt).

7.8.3 HMP Shunt

HMP shunt takes place in tissues that require NADPH, such as liver, lactating mammary glands, eye lens, thyroid gland, erythrocytes, adipose tissue, and adrenal cortex. This shunt is the source of NADPH required for the endergonic reductive biosyntheses (e.g., synthesis of cholesterol and fatty acids) and ribose-5-phosphate required for the biosynthesis of nucleic acids. It is also responsible for the utilization of pentoses to convert them into F6P and G3P. In erythrocytes, the shunt provides NADPH or glutathione cycle that is required for detoxification process and for maintaining the integrity of the red cells. HMP shunt is a seven-step pathway, shown in Fig. 7.75.

Step 1 G6P is converted to 6-phosphogluconolactone by G6P dehydrogenase. This enzyme is $NADP^+$ dependent. Therefore, NADP is reduced to NADPH in this step. Magnesium ions functions as a cofactor for this reaction. The deficiency of G6P causes haemolytic anaemia. This is the rate-limiting step, and G6P is the regulatory enzyme of HMP shunt.

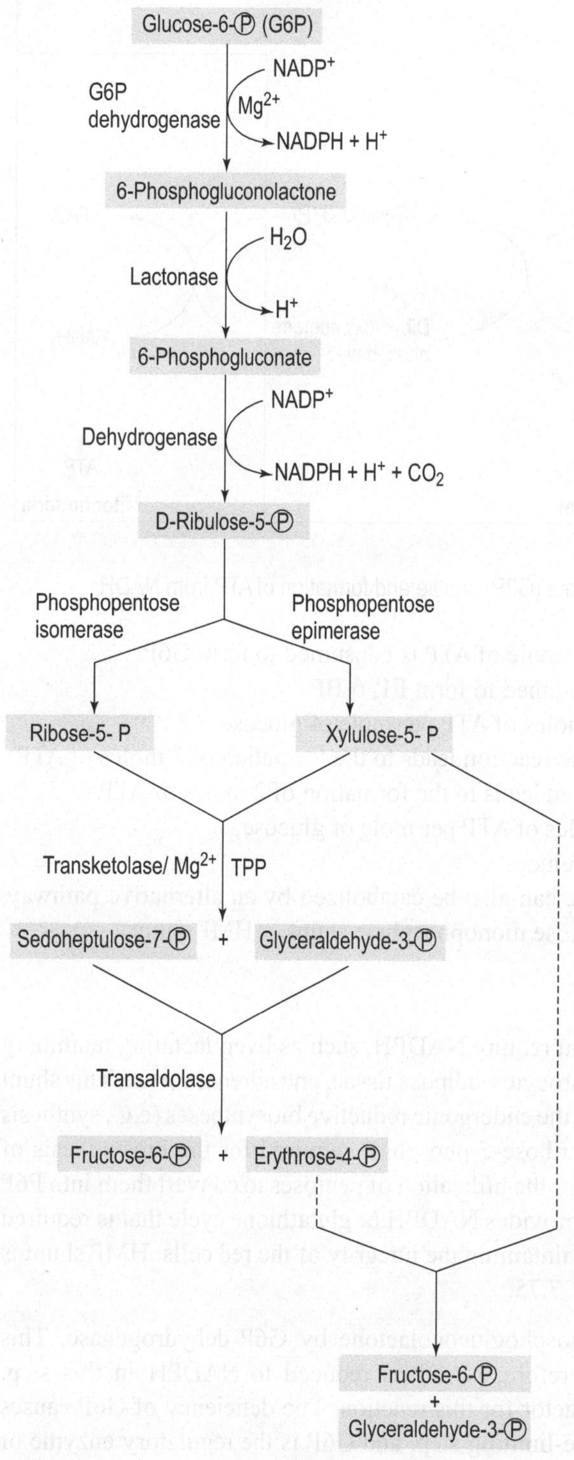

Figure 7.75 HMP shunt (pentose phosphate pathway)

Step 2 Lactonase (gluconolactone hydrolase) hydrolyzes 6-phosphogluconolactone to 6-phosphogluconate.

Step 3 6-Phosphogluconate is oxidized to D-ribulose-5-P by dehydrogenase. NADP is reduced to NADPH in this reaction.

Step 4 Ribulose-5-P is converted to ribose-5-P and xylulose-5-P, both by isomerization reaction. Phosphopentose isomerase catalyses ribose-5-P formation. Phosphopentose epimerase catalyses the xylulose-5-P formation. Ribose-5-P can be utilized for the synthesis of nucleic acids.

Step 5 Ribose-5-P and xylulose-5-P react to form sedoheptulose-7-P and glyceraldehyde-3-P (Gly3P) under the catalysis of transketolase, TPP (thiamine pyrophosphate), and magnesium ions. In this reaction, a two-carbon unit is transferred from a ketose to an aldose to form a seven- and a three-carbon sugar. That is, two five-carbon sugars react to form one seven-carbon sugar and one three-carbon sugar.

Since this step depends on TPP, chronic thiamine deficiency impairs the activity of transketolase, which leads to Wrnicke-Korsakoff syndrome with symptoms of weakness or paralysis and impaired mental function.

Step 6 Sedoheptulose-7-P and Gly3P are acted upon by transaldolase to form F6P and erythrose-4-P. Here, a six-carbon sugar and a four-carbon sugar are formed from a seven-carbon sugar and three-carbon sugar.

Step 7 Erythrose-4-P formed in step 6 and xylulose-5-P formed in step 4 combine to form F6P and Gly3P. TTP-dependent transketolase catalyses this reaction.

Steps 1 to 7 can be represented as an overall reaction of HMP shunt pathway as follows:

$$3\text{ Gly 3P} + 6\text{ NADP}^+ \rightarrow 2\text{ F6P} + 3\text{ CO}_2 + \text{G3P} + 6\text{ NADPH} + 6\text{ H}^+$$

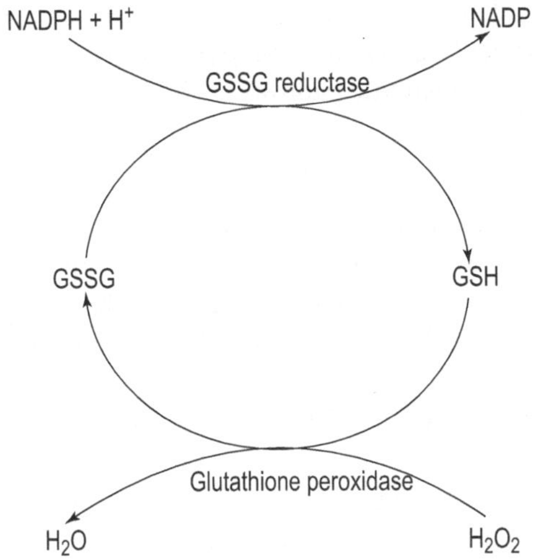

FIGURE 7.76 NADPH-dependent glutathione cycle in red cells

Thus, HMP shunt is the source of NADPH. NADPH is very essential for the maintenance of glutathione cycle in the red blood cells. Glutathioine cycle is shown in Fig. 7.76. Glutathione cycle detoxifies hydrogen peroxide that is formed in the red cells. Glutathione peroxidase, a selenium-dependent detoxifying enzyme, converts hydrogen peroxide to water. In this reaction, reduced glutathione (GSH) is converted to oxidised glutathione (GSSG). At this stage, NADH donates its hydrogen atoms to GSSG and regenerates GSH to continue the cycle. Thus, the red cell membrane is prevented from deleterious effects of hydrogen peroxide, and the membrane integrity is maintained. Also, oxidation of ferrous ion to ferric ion in haemoglobin is also prevented.

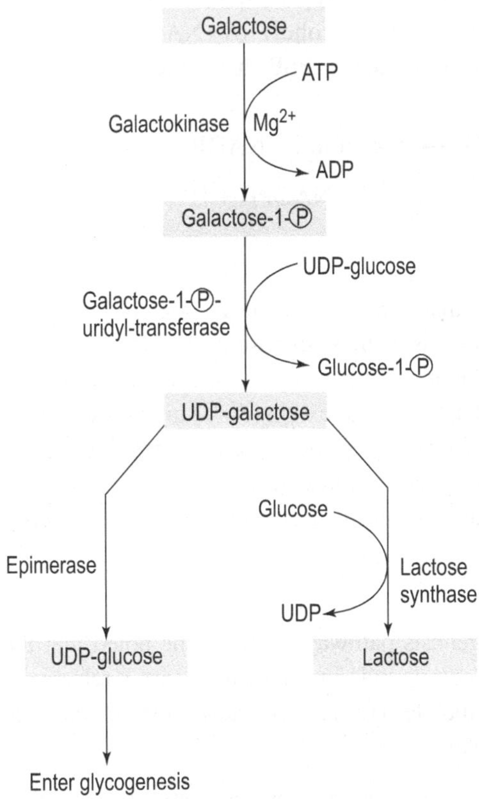

FIGURE 7.77 Metabolism of galactose

7.8.4 Metabolism of Galactose

Galactose is phosphorylated to galactose-1-P. Phosphorylation is catalysed by galactokinase in the presence of magnesium ions. Galactose-1-P reacts with UDP-glucose. Galactose-1-P accepts the UDP and donates its phosphate to glucose. Thus, UDP-galactose and glucose-1-P are formed. This reaction is catalysed by galactose1-P-uridyl transferase. UDP-galactose is epimerized by epimerase to UDP-glucose, which in turn can enter the glycogenesis pathway (Section 7.8.7). In lactating women, UDP-galactose is acted upon by lactose synthase to form lactose with the liberation of free UDP. Galactose metabolism is shown in Fig. 7.77.

Deficiency of galactokinase (the first enzyme of galactose metabolism) leads to accumulation of galactose in blood and tissues. In the eye lens, galactose is reduced to dulcitol by aldose reductase. Dulcitol is retained in the lens and forms cataract. The second enzyme of galactose metabolism is galactose-1-P-uridyl transferase. Deficiency of this enzyme results in the accumulation of both galactose and galactose-1-P in tissues. This condition is called classic galactosemia.

7.8.5 Metabolism of Fructose

As mentioned earlier, fructose is supplied to cells from fruits as well as from dietary sucrose by the action of sucrase. Apart from this, fructose is also obtained from glucose by isomerization. Fructose metabolism is different in muscle cells and liver cells. This is because hexokinase enzyme is present in muscle cells but not in liver.

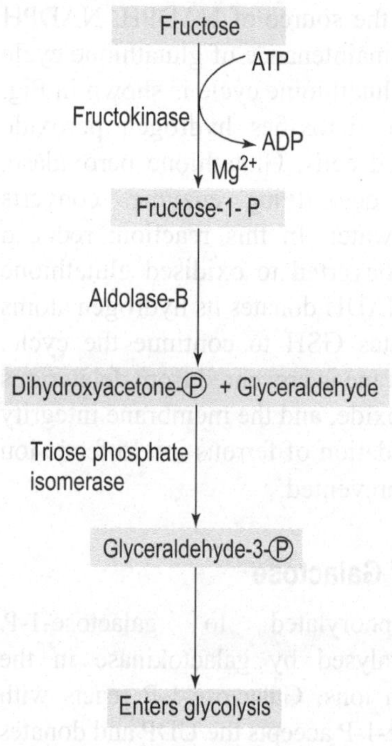

FIGURE 7.78 Metabolism of fructose

In muscle cells, hexokinase converts fructose to F6P at the expense of ATP in the presence of magnesium ions as the *cofactor*. F6P then enters the muscle glycolytic pathway.

$$\text{Fructose} + \text{ATP} \xrightarrow{Mg^{2+}} \text{Fructose-6-P} + \text{ADP}$$

In liver cells, fructose is converted to fructose-1-P (F1P) by fructokinase (Fig. 7.78). However, F1P cannot be converted to F1,6 BP by hexokinase or PFK (phospho fructokinase). Therefore, another enzyme called aldolase-B (phosphofructoldolase) continues the metabolism. It converts F1P to DHAP and glyceraldehydes. Triosephosphate isomerase converts DHAP to Gly3P. Gly3P enters glycolysis.

Deficiency of fructokinase in liver results in excessive accumulation of fructose which is excreted in the urine. This condition is called fructosuria. Defect in aldolase causes accumulation of F1P leading to liver and kidney damage. This condition is called fructose intolerance and is characterized by hypoglycemia and vomiting.

Fructose is the major energy source for sperm cells. In fact, fructose is synthesized to form glucose by aldose reductase in the seminal vesicle to supply energy to sperm cells. D-glucose is reduced to sorbitol (sugar alcohol) by NADPH. Sorbitol dehydrogenase converts sorbitol to fructose as cofactor. F6P then enters the muscle glycolytic pathway.

$$\text{D-Glucose} + \text{NADPH} + H^+ \rightarrow \text{D-sorbitol} + \text{NADP}^+$$

$$\text{D-sorbitol} + \text{NAD}^+ \rightarrow \text{D-fructose} + \text{NADPH} + H^+$$

7.8.6 Gluconeogenesis

The term 'gluco' means glucose, 'neo' means new, and 'genesis' means synthesis; therefore, glucogneogenesis refers to the synthesis of new glucose molecules from non-carbohydrate precursors, such as pyruvate, lactate, glycerol, and gluconeogenic amino acids. It takes place primarily in the liver (95%), to a lesser extent (50%) in kidney (during starvation), and very lesser extent (5%) in the intestinal epithelial cells. Cytosol is the site for gluconeogenesis. Gluconeogenesis pathway contributes glucose whenever there is low amount of storage glucose (glycogen), such as during starvation, strenuous exercise, and metabolic acidosis.

Substrate for gluconeogenesis

Figure 7.79 shows the reactions of gluconeogenesis pathway. It should be noted that the starting precursor is pyruvate and two of the compounds in the pathway are Krebs cycle intermediates, which include oxaloacetate and malate. This is to emphasize that Krebs cycle intermediates can enter gluconeogenesis pathway.

Pyruvate Pyruvate is the principal substrate for gluconeogenesis, whereas the other substrates join the pathway as intermediates. Table 7.8 summarizes the substrates for gluconeogenesis and their entry point in the pathway.

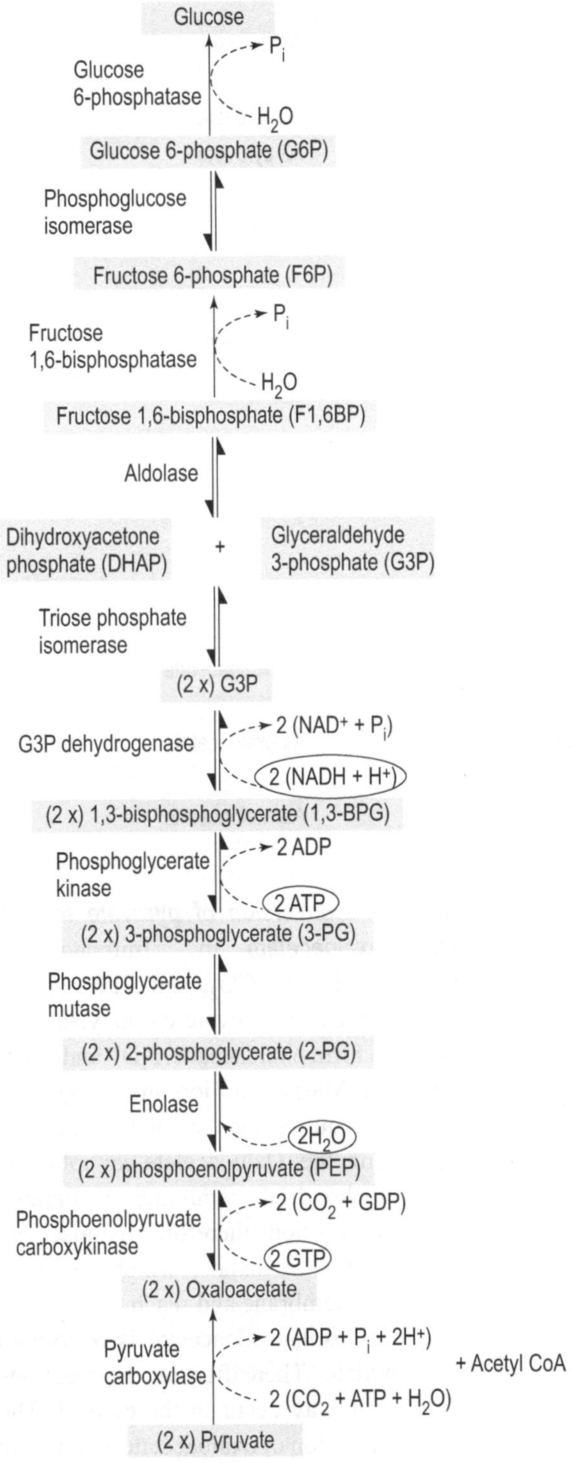

FIGURE 7.79 Gluconeogenesis pathway

Lactate Lactate formed by glycolysis, during muscle contraction, enters the blood and reaches the liver. Liver uses lactate for gluconeogenesis to synthesize glucose. This glucose then enters the blood and a part of it reaches the muscle, where it is degraded to glucose. The cycle, thus, continues in a clockwise direction. This is called Cori cycle (Fig. 7.80).

Gluconeogenic amino acids Serine, glycine, cysteine, and tryptophan are metabolized to pyruvate itself. Some other amino acids are converted to Krebs cycle intermediates, which in turn are metabolized to oxaloacetate (an intermediate of gluconeogenic pathway). Glutamate, glutamine, proline, arginine, and histidine are converted to α-ketoglutarate. Valine, threonine, methionine, and isoleucine are metabolized to succinyl CoA. Phenylalanine and tyrosine form fumarate. Asparagione and aspartate are converted to oxaloacetate.

Alanine Pyruvate formed during glycolysis is not only diverted to form lactate, but also to form alanine by transamination reaction. Alanine is released by the muscle into the blood, and then taken up by the liver. In the liver, alanine is converted to pyruvate by reverse transamination.

Glycerol In adipose tissues, hydrolysis of TG forms glycerol. Glycerol reaches the liver via circulation. In the liver, it is phosphorylated to 3-PG, an intermediate in the gluconeogenic pathway.

Gluconeogenesis pathway

Gluconeogenesis is essentially the reverse of glycolyis. For convenience, the pathway can be divided into three stages:

Stage 1: Conversion of pyruvate to PEP via oxaloacetate

Stage 2: Conversion of PEP to F1, 6BP via six steps

Stage 3: Conversion of F1, 6BP to glucose via isomerization and dephosphorylation

We will now discuss these stages in detail.

TABLE 7.8 Substrates for gluconeogenesis

Substrate	Intermediate pathway	Entry into gluconeogenic
Pyruvate	-	As pyruvate
Lactate	Pyruvate	As pyruvate
Alanine		
Serine		
Glycine		
Cysteine		
Tryptophan	-	As pyruvate
Glutamate		
Glutamine		
Proline		
Arginine		
Histidine	α-ketoglutarate	As oxaloacetate
Valine		
Threonine		
Methionine		
Isoleucine	Succinyl CoA	As oxaloacetate
Phenylalanine		
Tyrosine	Fumarate	As oxaloacetate
Asparagine		
Aspartate	Oxaloacetate	As oxaloacetate

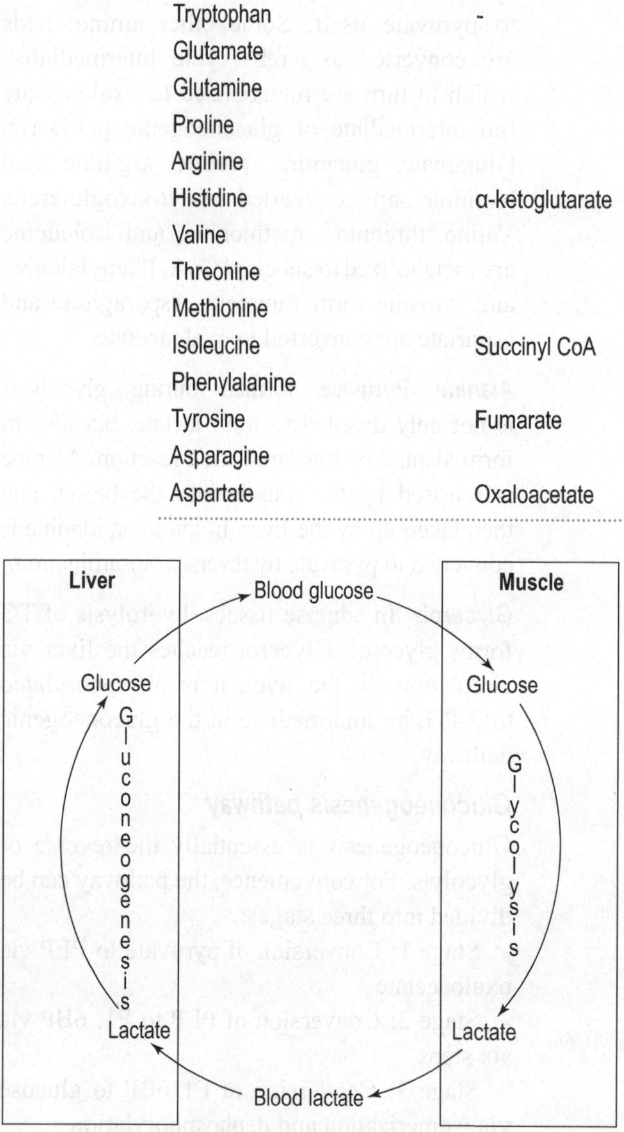

FIGURE 7.80 Cori cycle

Stage 1: Conversion of pyruvate to PEP via oxaloacetate In mitochondrial matrix, pyruvate, CO_2, and ATP react in the presence of pyruvate carboxylase and biotin to form oxaloacetate. ADP and Pi are liberated. Magnesium ion and manganese ions are required for the catalytic activity of the enzyme. Oxaloacetate cannot cross the mitochondrial membrane to continue the next reaction, therefore, oxaloacetate is converted to malate, which can then cross the membrane and reach the cytosol. In the cytosol, oxaloacetate is re-formed from malate. Thereafter, all the reactions of the pathway occur in the cytosol. The interconversion of oxaloacetate and malate is catalyzed by malate dehydrogenase.

$$\text{Oxaloacetate} + \text{NADH} + \text{H}^+ \xrightarrow{\text{Malate dehydrogenase}} \text{Malate} + \text{NAD}^+$$

Oxaloacetate is converted to PEP by GTP (Guanosine triphosphate). CO_2 and GDP are formed as byproducts. PEP carboxykinase (PEPCK) catalyses this reaction in the presence of manganese ions. Reaction 1 and 2 convert pyruvate to PEP, which is the reverse of pyruvate-kinase-catalysed reaction of glycolysis.

Stage 2: Conversion of PEP to F1,6BP via six steps PEP is converted to F1,6-BP by six enzymes. All these reactions are simply the reverse of glycolysis, and the enzymes are also same.

F1,6BP is converted to F6P by F1,6B phosphatase (F1,6BPase). This reaction is the reverse of the reaction catalysed by PFK during glycolysis. F1,6BPase is the major regulatory enzyme of gluconeogenesis. It is an allosteric enzyme. Citrate is the allosteric activator, whereas AMP and F2,6BP are allosteric inhibitors. The effects of these allosteric modulators are exactly opposite in glycolysis. This phenomenon is called reciprocal control of opposing metabolic pathways.

Stage 3: Conversion of F1,6BP to glucose via isomerization and dephosphorylation F6P is isomerized to G6P by phosphoexose isomerase. G6P is dephosphorylated to glucose. This reaction is the reverse of the first reaction of the glycolytic pathway, but the enzyme differs. Here the enzyme is glucose-6-phosphatase (G6Pase) that is specifically distributed in gluconeogenic tissues.

Genetic defect in G6Pase results in accumulation of its substrate G6P, which is converted to excess glycogen and stored in liver. This condition is called von Gierke's disease (glycogen storage disease type). This disease is characterized by enlarged liver, hypoglycaemia, lacticacidosis, hyperuricemia, and hyperlipidemia.

Gluconeogenesis is subjected to hormonal and non-hormonal regulation. Acetyl coA regulates the activity of pyruvate carboxylase. The level of substrates—lactate, pyruvate, and alanine—also determines the rate of gluconeogenesis. Glucagon, epinephrine, and glucocorticoids are stimulators of PECK, F1,6Bpase, and G6Pase, respectively. Insulin inhibits these three enzymes, and stimulates the key glycolytic enzymes (hexokinase, PFK, and pyruvate kinase).

Energetics of gluconeogenesis

Gluconeogenesis is purely an energy-dependent pathway, that is, energy is consumed, but not formed, in the pathway. Two moles of pyruvate are converted to one mole of glucose at the expense of 4 moles of ATP and 2 moles of GTP (an equivalent of ATP). ATP/GTP consumption is as follows:

Conversion of pyruvate to oxaloacetate = 2 ATPs
Conversion of oxaloacetate to PEP = 2 GTPs
Conversion of 3PG to 1,3BPG = 2 ATPs
Total consumption = 6 ATPs

Thus, gluconeogenesis requires 6 moles of ATP for the formation of one mole of glucose.

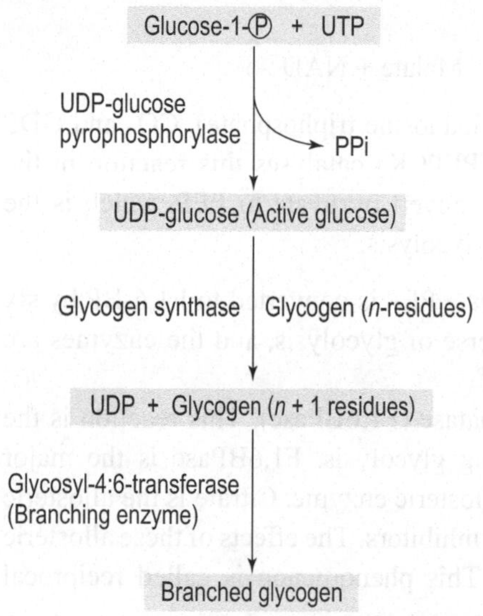

FIGURE 7.81 Glycogenesis

7.8.7 Glycogenesis

Glycogenesis is the synthesis of glycogen, the polymeric storage form of glucose. Glycogen is synthesized and stored in liver and muscle. The main difference between the liver glycogen and the muscle glycogen is that liver glycogen can supply glucose units to rest of the body, whereas muscle glycogen can use the glucose units for its own need.

The three enzymes involved in glycogenesis are the following:

1. UDP-glucose pyrophosphorylase
2. Glycogen synthase
3. Branching enzyme

The outline of glycogenesis depicting the role of these three enzymes is shown in Figure 7.81.

Glycogenesis is a three-step pathway. In step 1, uridine diphosphate glucose (UDP-glucose or active glucose), the precursor for glycogen, is formed by the esterification reaction between the C1 of glucose-1-P and the diphosphate moiety of UTP under the catalysis of UDP-glucose pyrophosphorylase (Fig. 7.82). This reaction results in the formation of UDP-glucose and PPi. PPi is immediately hydrolyzed by cellular pyrophosphatases, making this reaction essentially irreversible.

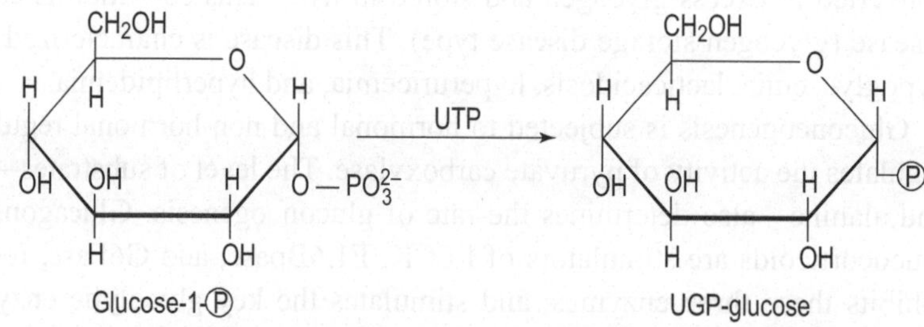

FIGURE 7.82 Formation of UDP-glucose for glycogenesis

Glycogenesis can occur only if there is a pre-existing glycogen chain, with *n* residues of glucose. This pre-existing glycogen is called the primer glycogen. In step 2, UDP-glucose donates its glucose unit to the primer glycogen in the presence of glycogen synthase. As a result, UDP and glycogen with $n + 1$ glucose residues are formed.

In step 3, branching of glycogen takes place. In the presence of branching enzyme (glycosyl-4:6 transferase), glucose chain of seven residues is transferred to C-6 hydroxyl group of a glucose residue that is four residues away from an existing branch. Branching of glycogen is shown in Fig. 7.83. A genetic defect in the branching enzyme results in Anderson's disease (Type IV glycogen storage disease) in infants. This disease is fatal and is characterized by cirrhosis and hepatosplenomegaly.

The overall reaction of glycogenesis is as follows:

Glycogen (n residues) + UDP-glucose → Glycogen (n + 1 residues) + UDP

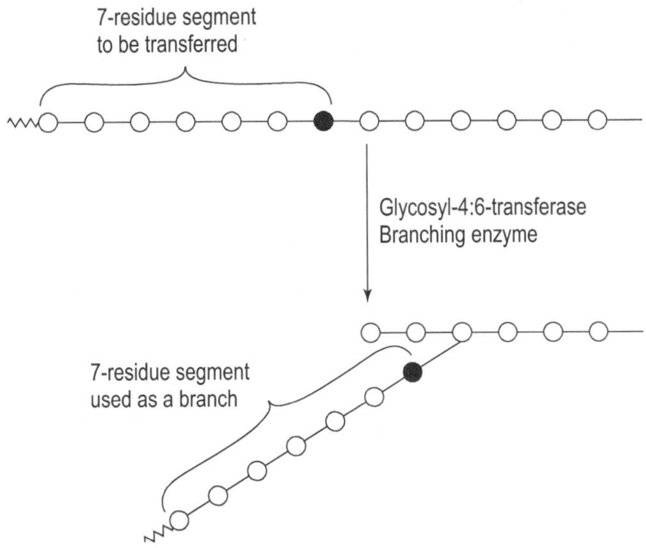

FIGURE 7.83 Branching of glycogen chain

Cascade regulation of glycogenesis

Glycogen synthase is the rate-limiting enzyme of glycogenesis. It is regulated by a phosphorylation cascade which involves phosphorylation of different enzymes and compounds. Hormones are responsible for initiating the regulation cascade. Epinephrine or glucagon activates adenylate cyclase, which in turn converts ATP to cAMP. cAMP activates protein kinase. Active protein kinase phosphorylates glycogen synthase. Phosphorylated glycogen synthase is inactive, therefore, glycogen is not synthesized in response to glucagon.

Insulin, on the other hand, activates the phosphoprotein phosphatase which dephosphorylates the glycogen synthase and activates it. Thus, glycogen proceeds in response to insulin signal.

The hormonal regulation of glycogenesis by switching on/off of glycogen synthase is depicted in Fig. 7.84.

7.8.8 Glycogenolysis

Glycogenolysis is the hydrolysis of the α-1,4 and 1-6-glycosidic bonds in glycogen in order to liberate free glucose residues. This occurs during fasting or hypoglycaemic state. The two enzymes involved in glycogenolysis are the following:

1. Glycogen phosphorylase
2. Debranching enzyme

Phosphorylase mainly cleaves most of the 1-4 linked residues, whereas debranching enzyme cleaves the left out 1-4 linked residues in addition to the branching 1-6 linked residues. Activities of these two enzymes are shown in Fig. 7.85.

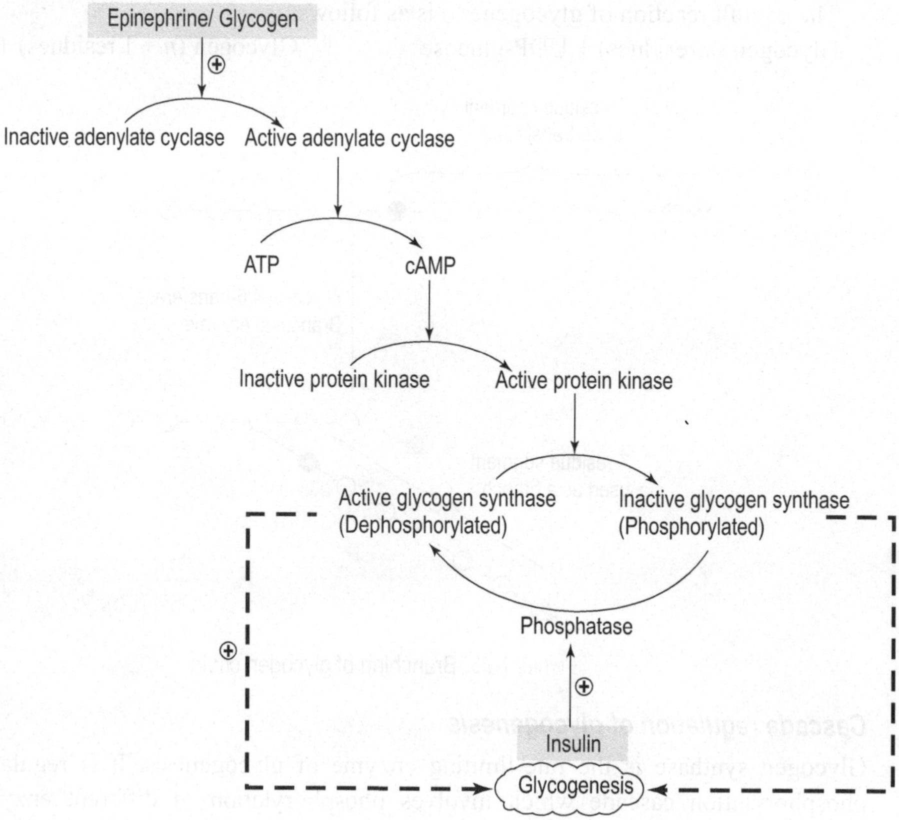

FIGURE 7.84 Regulation of glycogenesis

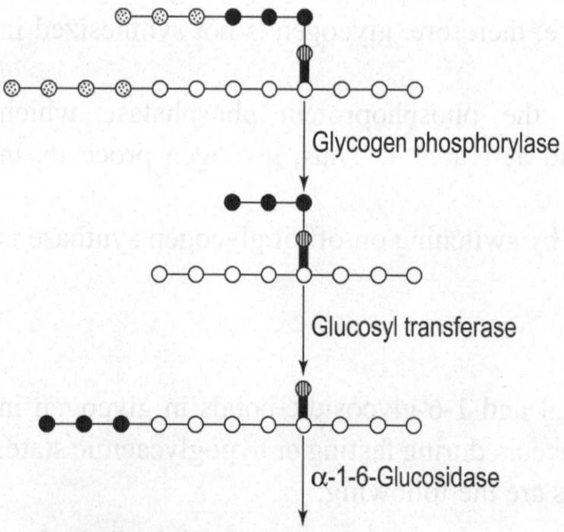

FIGURE 7.85 Activities of glycogen phosphorylase and debranching enzyme during glycogenolysis

For better understanding, let us assume that two long chains of glycogen (chain 1 and chain 2), both with 1-4 linked glucose residues, are linked by 1-6-linked glucose (the branching glucose). Glycogen phosphorylase cleaves the 1-4-glycosidic bonds and phosphorylates the excised glucose, thereby liberating glucose-1-P from chain 1. As a result, the chain with n residues is converted to glycogen with $n-1$ residues. The reaction of glycogen phosphorylase may be written in the following manner:

Glycogen (n residues) + Pi
Glycogen ($n-1$ residues) + Glucose-1-P

This reaction continues till three residues remain in chain 1. Next, the debranching enzyme comes into action.

The debranching enzyme has two activities—1,4-glucosyl transferase activity and 1,6-glucosidase activity. Transferase transfers the remaining 3 glucose residues from chain 1 to the terminus of the neighbouring chain 2.

This reaction exposes the branching residue which links chain 1 and chain 2. The exposed residue is called 1-6-linked residue. Glucosidase cleaves the exposed residue to yield a free glucose.

Regulation of glycogenolysis

Glycogen phosphorylase is the regulatory enzyme for glycogenolysis. Regulation cascade of glycogenolysis is similar to glycogenesis, but the effect is opposite. This is because glycogen synthase is active in the dephosphorylated form, whereas glycogen phosphorylase is active in phosphorylated form. Epinephrine or glucagon activates adenylate cyclase, which in turn converts ATP to cAMP. cAMP activates protein kinase. Active protein kinase phosphorylates glycogen phosphorylase. Phosphorylated glycogen phosphorylase is active. Therefore, glycogenolysis is stimulated in response to glucagon. So, we say that epinephrine and glucagon inhibit glycogenesis, but stimulate glycogenolysis. In both pathways, insulin has the effects opposite to that of epinephrine and glucagon, that is, it stimulates glycogenesis and inhibits glycogenolysis. Cascade regulation of glycogenolysis is shown in Fig. 7.86.

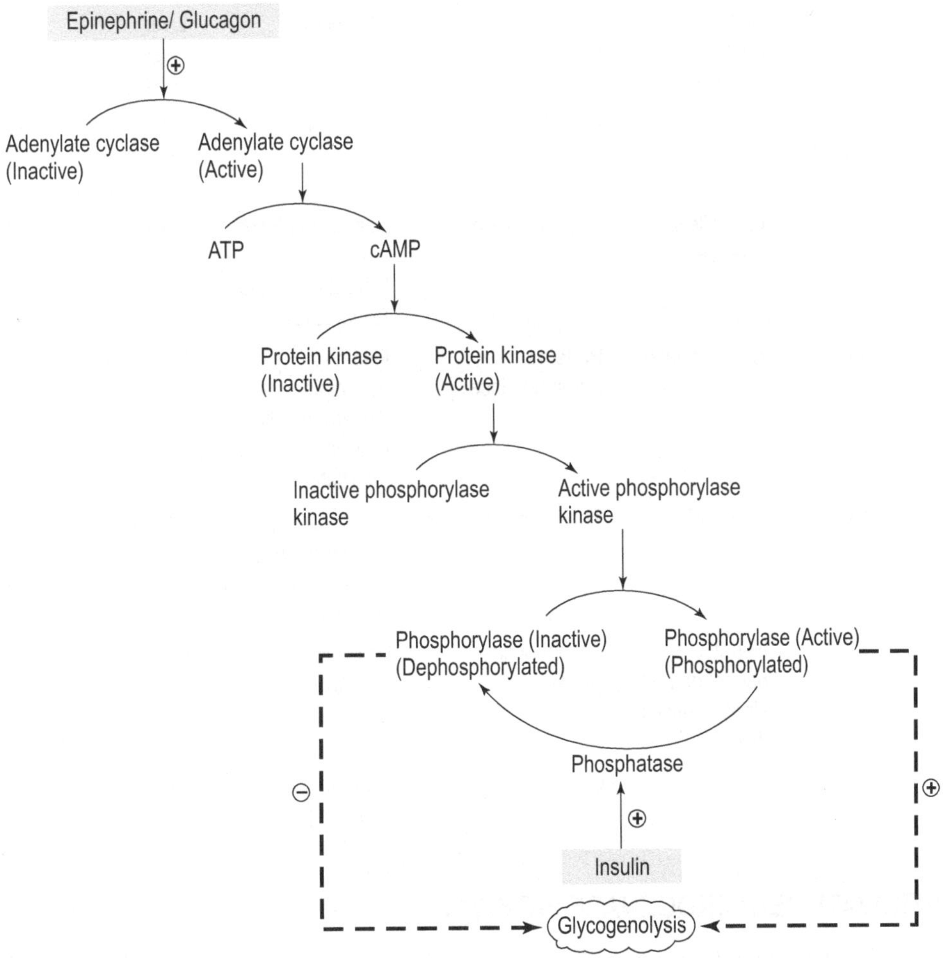

FIGURE 7.86 Regulation of glycogenolysis

Glycogen storage disease

Glycogen storage disease (GSD) is the disease caused by abnormal glycogen metabolism. This is due to defect in the one or more enzymes involved in glycogen metabolism and related pathways. Different types of GSD are assigned roman numerals. Types of GSD and the defective enzymes are listed in the Table 7.9.

TABLE 7.9 Glycogen storage disease

GSD Type	Eponym of the GSD	Defective enzyme	Symptoms	Principal tissue affected	Glycogen level
I	von Gierke's disease	Glucose-6 -phosphatase	Lactic acidosis Hyperuricemia	Liver Kidney	High
II	Pompe's disease	α-1,4-Glucosidase	Muscle weakness Heart failure	All organs	Very high
III	Cori's, Forbe's disease	Debranching enzyme	Myopathy	Liver Heart Muscle	High
IV	Anderson's disease	Branching disease	Cirrhosis	Liver Spleen Muscle	Normal
V	Mc Ardle's disease	Phosphorylase	Exercise induced cramps Rhabdomyolysis Renal failure	Muscle	High
VI	Her's disease	Protein kinase A Phosphorylase	Mild hyperlipidemia Hepatomegaly Growth retardation	Liver	High
VII	Tauri's disease	Phosphofructokinase	Muscle cramps and weakness Growth retardation Haemolytic anaemia	Muscle	High
VIII	Hepatic phosphorylase kinase deficiency disease	Phosphorylase kinase	Growth retardation	Liver	High
0	-	Glycogen synthase	Rare muscle cramping	Liver	Low

7.9 PYRUVATE DEHYDROGENASE PATHWAY

Pyruvate dehydrogenase (PDH) is a mitochondrial enzyme located in the matrix. Pyruvate formed from glycolysis enters mitochondria to be acted upon by PDH for forming

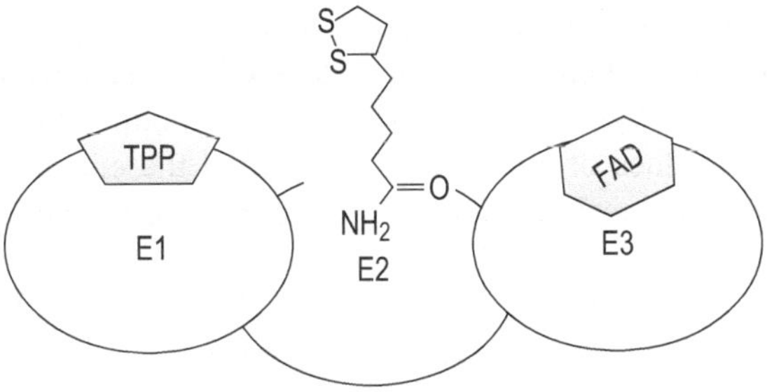

FIGURE 7.87 Subunits of pyruvate dehydrogenase enzyme complex

acetyl CoA. Acetyl CoA then enters the TCA cycle. Thus, PDH pathway is a link between the glycolysis pathway and the TCA cycle. Pyruvate dehydrogenase is a multienzyme complex with multiple copies of three subunits, E1, E2, and E3 (Fig. 7.87). The E2 subunit forms the core of the enzyme and functions as dihydrolipoyl transacetylase. El and E2 subunits form the periphery of the PDH, and they possess pyruvate dehydrogenase and dihydrolipoyl dehydrogenase, respectively. Thiamine pyrophosphate, lipoamide, and FAD are cofactors for E1, E2, and E3, respectively.

PDH, in the presence of NAD^+, catalyses the oxidative decarboxylation of pyruvate into acetyl CoA simultaneously forming NADH and H+. The reaction is written as follows:

$$2\ CH_3-CO-COO^- + NAD^+ + COASH \longrightarrow 2\ CH_3-CO-S-COA + NADH + H^+ + CO_2$$

Pyruvate → **Acetyl CoA**

Acetyl CoA → TCA cycle → $CO_2 + H_2O$

$NADH + H^+$ → Electron transport chain → ATP

This reaction takes place via five steps as shown in Fig. 7.88.

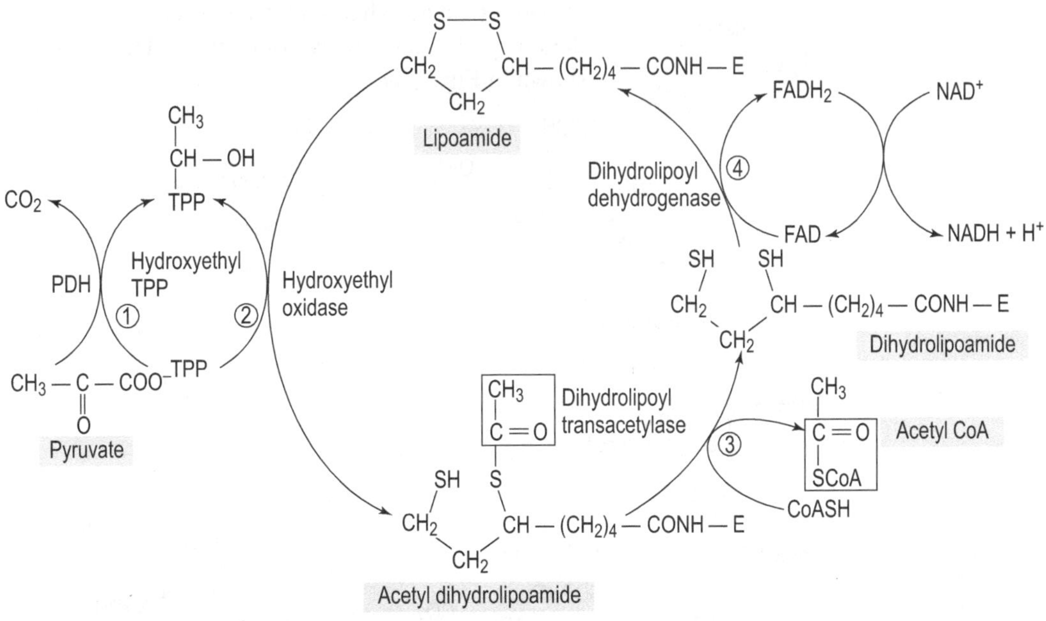

FIGURE 7.88 Steps leading to the formation of acetyl CoA by pyruvate dehydrogenase reaction

Acetyl CoA is the source of a two-carbon unit—acetate. It has a high-energy thiosester bond that enables the acetyl CoA to function as an excellent acetate donor for the anabolism of fatty acids, ketone bodies, and cholesterol. Acetyl CoA enters into the TCA cycle, whereas NADH enters into the electron transport chain (explained in Section 7.11). PDH is regulated by phosphorylation/dephosphorylation process as a covalent modification. PDH kinases phosphorylate the serine residues of E1 subunit and inhibit the enzyme. PDH phosphatases dephosphorylate the serine residues and enable the enzyme to regain its activity.

The availability of the substrates NAD and CoA or the products acetyl CoA and NADH also determines the rate of the PDH-catalysed reaction. In the adipose tissues, insulin activates PDH.

7.10 TRICARBOXYLIC ACID CYCLE OR THE TCA CYCLE

TABLE 7.10 Source of acetyl CoA entering the TCA cycle

Primary biomolecule	Acetyl CoA source derived from the biomolecule
Carbohydrate (Glucose)	Pyruvate
Lipid	Fatty acids
Proteins (Amino acids)	Leucine
	Isoleucine
	Tryptophan
	Lysine
	Phenylalanine
	Tyrosine

TCA cycle, also called Krebs cycle or respiratory cycle, is a central crossroad pathway of eight steps, where the final oxidation of all the metabolic fuel takes place in aerobic condition. TCA, in close association with the electron transport chain, generates ATP, the energy currency of the cell. The substrate for TCA cycle is acetyl CoA, which is derived from various sources as listed in Table 7.10.

TCA cycle occurs in the mitochondrial matrix. It is called amphibolic pathway as it is involved in both catabolism and anabolism. As a catabolic pathway, TCA cycle degrades two carbon acetyl residues. As an anabolic pathway, TCA cycle is responsible for providing intermediates that are used as precursors in the biosynthesis of many compounds. The TCA cycle is shown in Fig. 7.89.

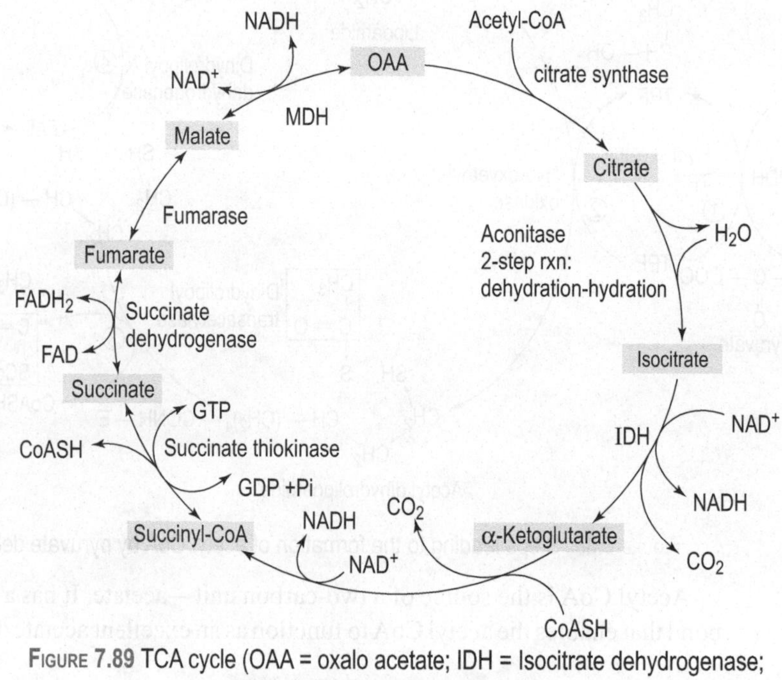

FIGURE 7.89 TCA cycle (OAA = oxalo acetate; IDH = Isocitrate dehydrogenase; MDH = Malate dehydrogenase)

7.10.1 Steps in TCA Cycle

In step 1, the acetyl CoA (derived from carbohydrates, fats, and amino acids) reacts with oxaloacetate, starting the cycle and forming citrate and CoA. This is a condensation reaction catalysed by citrate synthase.

In step 2, citrate is isomerized to isocitrate by aconitase.

In step 3, isocitrate is oxidatively decarboxylated to α-ketoglutarate by isocitrate dehydrogenase in the presence of NAD^+. NADH, H^+, and carbon dioxide are liberated in this step.

In step 4, α-ketoglutarate forms succinyl CoA in the presence of α-ketoglutarate dehydrogenase, CoA, and NAD. During this step, NADH, H^+, and carbon dioxide are liberated. Succinyl CoA is an energy-rich thioester, similar to acetyl CoA.

In step 5, succinate thiokinase converts succinyl CoA to succinate. During this reaction, GDP is converted to GTP in the presence of magnesium ions and CoA is formed. This reaction is called substrate level phosphorylation. Because of this reaction, energy is conserved in the form of GTP, which can be converted into ATP.

In step 6, succinate is dehydrogenated by succinate dehydrogenase and FAD to form fumarate and $FADH_2$. Succinate dehydrogenase is tightly integrated with the inner mitochondrial membrane. The $FADH_2$ formed in this reaction can directly transfer its electrons to the electron transport chain of the inner mitochondrial membrane and get itself oxidized.

In step 7, hydration of fumarate to malate is catalysed by fumarase.

In step 8, malate is dehydrogenated by malate dehydrogenase and NAD to form oxaloacetate and NADH + H^+. Oxaloacetate can react with the next molecule of acetyl CoA, thus, continuing the TCA cycle. The other fate of oxaloacetate is its transfer into the cytoplasm, where it enters the gluconeogenesis pathway.

Overall, the TCA cycle completely oxidizes the acetate group of acetyl CoA to carbon dioxide and water. During this process, it generates NADH and $FADH_2$. NADH and $FADH_2$ are continuously supplied to the electron transport chain and the electrons are transported to molecular oxygen with the liberation of NAD and FAD. NAD and FAD liberated are given back to mitochondrial matrix to start another TCA cycle. It makes the mitochondria the bulk consumer of respiratory oxygen. Thus, TCA cycle cannot occur in the absence of oxygen.

Among the eight TCA cycle enzymes, four are dehydrogenases. Two of them catalyse oxidative decarboxylation and the other two catalyse dehydrogenation. All these four dehydrogenases and the succinate thiokinase deserve significance because the reaction catalysed by these five enzymes ends up in the synthesis of ATP. For every NADH produced in the TCA cycle, three ATPs are formed. For every $FADH_2$, two ATPs are formed, and for every GTP liberated, one ATP is formed. Thus, for one complete TCA cycle, 12 ATPs are synthesized as outlined in Table 7.11.

Some of the reactions of TCA cycle shown in Fig. 7.89 increase the concentration of intermediate compounds, allowing an increase in the rate of oxidation of the acetyl CoA. These reactions are called anaplerotic reactions and the process is called anaplerosis. As more intermediates become available, more moles of acetyl CoA can be processed. Intermediates formed by anaplerosis are listed in Table 7.12.

TABLE 7.11 ATP generation by TCA cycle per acetyl CoA molecule

Step	Name of the reaction	Reactant	Product	Number of ATPs formed
3	Oxidative decarboxylation	Isocitrate + NAD+	α-Ketoglutarate + CO_2 + NADH + H^+	3
4	Oxidative decarboxylation	α-Ketoglutarate + CoA+ NAD+	Succinyl CoA + CO_2 + NADH + H+	3
5	Substrate level phosphorylation	Succinyl CoA + GDP + Pi	Succinate + CoA + GTP	1
6	Dehydrogenation	Succinate + FAD	Fumarate + FADH2	2
8	Dehydrogenation	Malate + NAD+	Oxaloacetate + NADH + H+	3
	Total ATP formed in TCA cycle		12	

TABLE 7.12 Intermediates formed by anaplerotic reactions during TCA cycle

Intermediate	Source for the intermediate
Pyruvate	Alanine
	Glycine
	Serine
	Cysteine
	Threonine
	Tryptophan
Acetyl CoA	Leucine
	Isoleucine
	Tryptophan
	Threonine
Oxaloacetate	Pyruvate carboxylase
	Transamination of aspartate
α-ketoglutarate	Transamination of glutamate
	Glutamine
	Histidine
	Arginine
	Proline
Succinyl CoA	Oxidation of odd-chain fatty acids
	Isoleucine
	Valine
	Methionine
	Threonine
Fumarate	Phenylalanine
	Threonine

7.10.2 Utilization of TCA Cycle

This section describes the utilization of TCA cycle intermediates and how these intermediates are diverted to other metabolic pathways in the cell. Malate enters the cytosol and initiates gluconeogenesis.

Fatty acid biosynthesis and cholesterol biosynthesis are cytosolic processes requiring acetyl CoA as precursor. However, acetyl CoA cannot enter the cytosol from mitochondria; so, citrate—the intermediate of TCA cycle—crosses the mitochondrial membrane to reach the cytosol where it is split into oxaloacetate and acetyl CoA by ATP-citrate lyase (citrate cleavage enzyme). α-Ketoglutarate, the intermediate formed in the third step of TCA cycle is a precursor for glutamate. Glutamate is formed by the following reactions:

α-Ketoglutarate + NH_4 + NADH + H^+ → NAD^+ + Glutamate (catalysed by dehydrogenase)

α-Ketoglutarate + Alanine → Glutamate (catalysed by transaminase and pyridoxal phosphate)

Succinyl CoA reacts with glycine to form ALA from which haeme is synthesized.

7.10.3 Other Sources of TCA Cycle Intermediates

The TCA cycle intermediates shown in Fig. 7.89 are also produced by other metabolic fuels. These fuels are produced by metabolism of various biomolecules. These intermediates also enter the TCA cycle. Figure 7.90 shows the entry point for TCA cycle intermediate diverted from other sources. This figure reveals that the major biomolecules—carbohydrates (as glucose), lipids (as fatty acids), and proteins (as amino acids)—finally enter the TCA cycle. This is why TCA cycle is called as the central *crossroad pathway.*

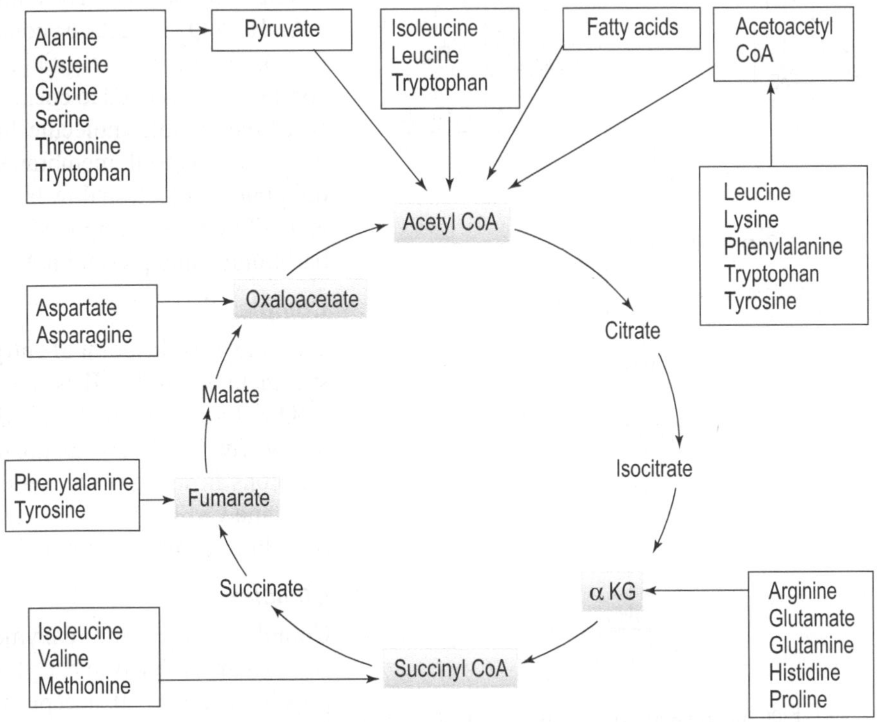

FIGURE 7.90 Entry of TCA cycle intermediates from other metabolic fuels (intermediates entering the cycle are shaded)

Depending on the metabolic needs of the cell, TCA cycle is regulated by its own enzymes. Principal regulatory enzymes are citrate synthase, isocitrate dehydrogenase, and α-ketoglutarate dehydrogenase.

7.11 ELECTRON TRANSPORT CHAIN

The electron transport chain (ETC) is the common pathway in the aerobic cells by which electrons derived from various pathways are transferred to molecular oxygen. It is a series of highly organized redox enzymes. A variety of substrates utilize the electron transport chain as a common pathway because many of them are oxidized by enzymes that use NAD^+ or FAD^+ as electron acceptor cofactor. In aerobic organisms, the electrons derived from fuel molecules are first transferred to specialized electron carriers, NAD^+ and FAD^+, from which the electron reaches the molecular oxygen via mitochondrial electron transfer system. The active portion of NAD^+ is the nicotinamide ring which accepts a proton and two electrons from the substrate. The active portion of FAD molecules is the alloxazine ring which accepts two protons and two electrons from the substrate.

The ETC is composed of complex I, complex II, coenzyme Q, complex III, cytochrome C, and complex IV. All these complexes are enzymes. The electron flow through ETC is shown in Fig. 7.91. The enzymes of ETC are tightly embedded in the inner membrane in association with the enzymes of oxidative phosphorylation.

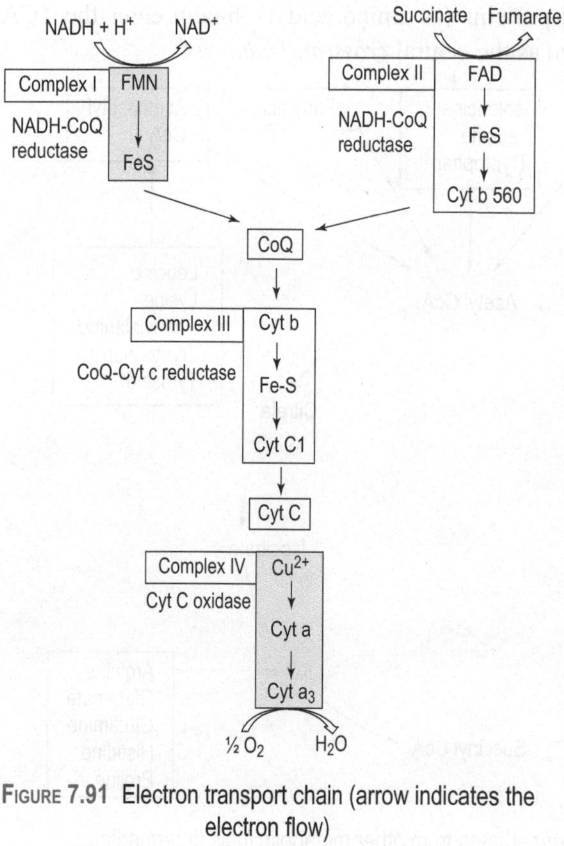

FIGURE 7.91 Electron transport chain (arrow indicates the electron flow)

Complex I

Complex I is the point of entry for electrons from NADH into the ETC. The complex is the enzyme NADH-CoQ reductase containing FMN and FeS as the prosthetic group. The electron from complex I is donated to coenzyme Q (CoQ). CoQ is a lipid-soluble molecule firmly embedded in the mitochondrial membrane. It is also called ubiquinone. Electron flow from complex I to CoQ is inhibited by compounds, such as rotenone, barbiturate, and piericidin A.

Complex II

Complex II is the point of entry of electrons from succinate. Complex II is the enzyme succinate-CoQ reductase containing FAD, FeS centre, and haeme (in Cyt b650) as prosthetic groups. The electrons flow from succinate to CoQ via FAD, FeS, and Cyt b650. CoQ, thus, accepts electrons from both complex I and II.

Complex III

Complex III is the enzyme CoQ reductase with haeme (Cyt b and Cytc1) and FeS as the prosthetic group. It accepts the electrons from CoQ and passes it to cytochrome C via Cytb, FeS, and CytC1.

Cytochrome C

Cytochrome C is a soluble protein found in association with the inner membrane to perform electron transfer role. It mediates the transfer of electrons received from complex III to complex IV.

Complex IV

Complex IV is cytochrome c oxidase containing copper and haeme of Cyt a and Cyt a3 as the prosthetic groups. It donates the electrons to the molecular oxygen through copper ion, Cyt a, and Cyt a3. The electrons react with molecular oxygen forming water. Carbon monoxide, hydrogen sulphide, azide, and cyanide are competitive inhibitors of complex IV.

Oxygen toxicity

Oxygen toxicity occurs due to its own partial reduction into superoxide anion (O^{2-}), a free radical. It is formed in mitochondria due to the reduction of oxygen by $FADH_2$ and reduced CoQ. The superoxide anion is a highly reactive unstable molecule that can react with and damage DNA, proteins, and membranes. This is involved in aging. Still, detoxification mechanism exists in the body to nullify the toxicity. Superoxide dismutase, a metalloenzyme, converts O^{2-} to hydrogen peroxide (H_2O_2), which in turn is detoxified to water by the enzyme peroxidase. Vitamin C and vitamin E are also reported to act as detoxifiers.

Respiratory burst

Respiratory burst is an immunologic phenomenon that occurs after microbial infection. It usually occurs during phagocytosis of the antigen. This process results in increased oxygen consumption, increased formation of O^{2-} in order to destroy the antigen, and activation of pentose phosphate pathway.

Before discussing the synthesis of ATP, an understanding of biological phosphorylation is essential. There are three types of biological phosphorylation that generate phosphate compounds, including ATP, the energy currency of the cell: (1) direct phosphorylation, (2) substrate-level phosphorylation, and (3) oxidative phosphorylation. In all these types, the inorganic phosphate moiety required for phosphorylation is donated by yet another molecule of ATP.

Direct phosphorylation is the phosphorylation of a compound to form a biologically significant phosphorylated derivative. Enzymes are regulated by direct phosphorylation of their amino acid residues. Examples of pathways involving phosphorylation reactions are given in Fig. 7.92.

Substrate-level phosphorylation is defined as the phosphorylation of one substrate by the other substrate to form high-energy compounds, such as ATP or GTP. One of the substrates is dinucleotide, which is converted to the high-energy trinucleotide form by the other substrate. Four substrate-level phosphorylation reactions are found in the metabolic pathways—two reactions in glucose metabolism, one in TCA cycle, and one in muscle metabolism. Substrate-level phosphorylation reactions are shown in Fig. 7.93.

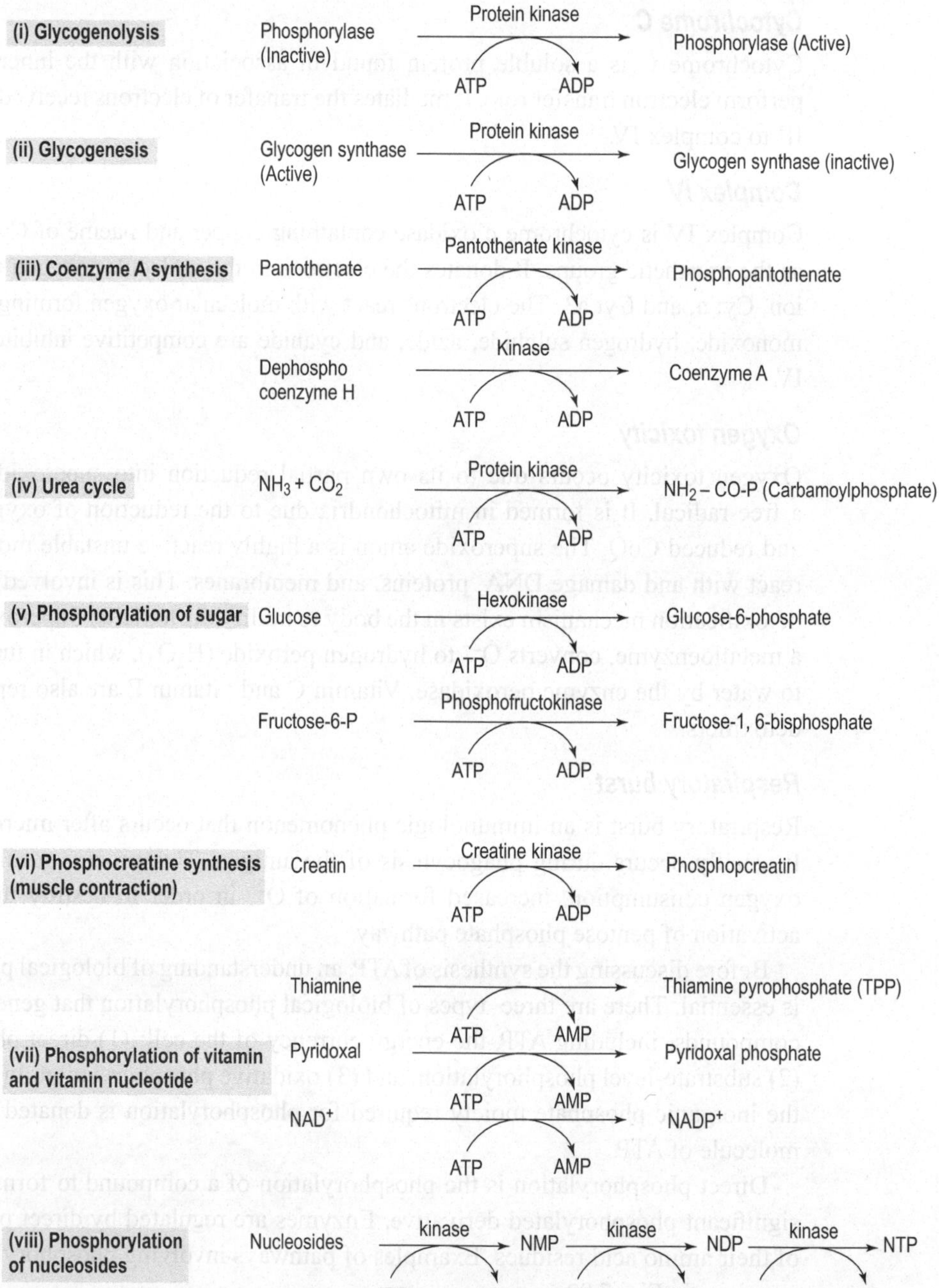

FIGURE 7.92 Direct phosphorylation reactions in metabolic pathway

Oxidative phosphorylation implies the coupling of ATP synthesis with the oxidation of NADH or $FADH_2$. This process is shown in Fig. 7.94. In this process, the free energy that is released during ETC is used for coupling the oxidation process to the phosphorylation process (i.e., the formation of ATP). Thus, oxidative phosphorylation is the main source of energy in cells.

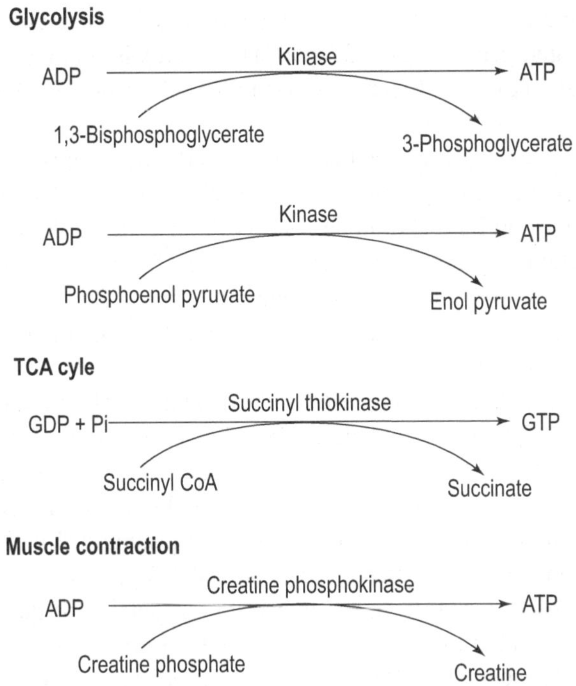

FIGURE 7.93 Substrate-level phosphorylation reactions of metabolism

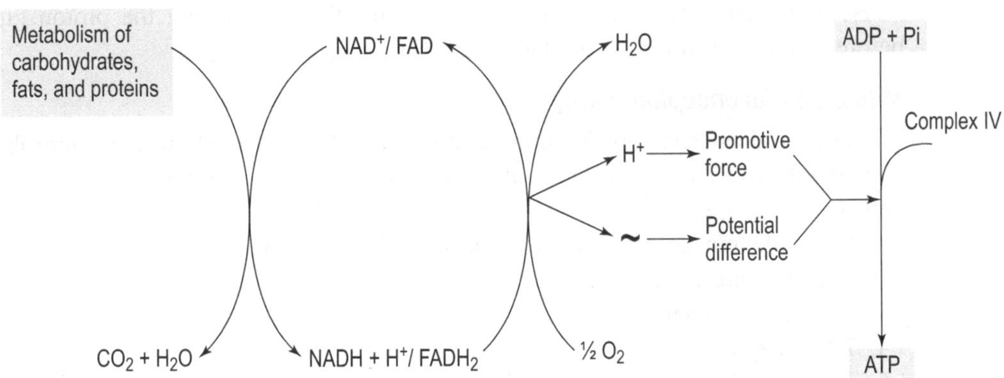

FIGURE 7.94 Oxidative phosphorylation process (~ = Energy)

7.12 OXIDATIVE PHOSPHORYLATION AND ATP PRODUCTION

The redox potential of electron carriers in ETC is in the following order, down the chain:

$$-0.32\text{ V} \rightarrow +0.04\text{V} \rightarrow +0.25\text{V} \rightarrow +0.82\text{ V}$$

This indicates that the potential of electron carriers in ETC becomes more positive on going from oxidation of NADH to the reduction of molecular oxygen. From the difference in potential between the first couple (NAD^+/NADH) and the last couple

(O_2/H_2O) of the ETC, that is, 1.14V, it is reported that each electron pair passing through the ETC can generate energy of 52 kcal. This energy is coupled to the synthesis of ATP, by phosphorylation of ADP. Electron entering the ETC from NADH and $FADH_2$ generates 3 and 2 molecules of ATP, respectively.

Coupling of energy produced in ETC to ATP formation is explained by chemiosmotic coupling hypothesis, proposed by Peter Mitchell in 1961. This hypothesis states that the energy produced in ETC should be conserved for endergonic reactions. This is facilitated by translocation of protons (vectorial pumping of protons by electron carriers) from the mitochondrial matrix into the intermembrane space. This translocation creates a proton gradient called protomotive force (PMF). This PMF is a mode of coupling the energy with the formation of ATP.

Formation of ATP is catalysed by the enzyme ATP synthase, which is also called H+ ATPase or F_0F_1 ATPase. This enzyme has two subunits—F_0 subunit which forms a channel by spanning the membrane, and F_1 subunit, which is tightly bound to F_0, facing the matrix side of the mitochondrial membrane. F_0 subunit functions as a passage for the proton ions to cross the membrane, whereas F_1 subunit functions as the catalytic site. ADP/ATP transporter catalyses the transport of ADP into the matrix for phosphorylation, and the transport of ATP from the matrix to the outside of the mitochondria for use by the cells. The number of moles of ATP formed per gram of oxygen atoms consumed for a given substrate is called the P:O ratio. Substrates that donate electrons to NAD^+ and FAD^+ have P:O values of 3:1 and 2:1, respectively. In other words, the P:O value is 3:1 and 2:1 for NADH oxidation and $FADH_2$ oxidation, respectively.

ATP synthesis is inhibited by a family of compounds called uncouplers (Table 7.13). These compounds allow the leakage of protons, thus, collapsing the proton gradient before it can be used for the production of ATP.

Mitochondrial encephalopathy

Mitochondrial encephalopathy is a group of disorders characterized by morphologic abnormalities of muscle mitochondria. This is due to one or more of the following defects:

1. Substrate transport from the cytoplasm to the mitochondria
2. Substrate utilization
3. TCA cycle enzymes
4. ETC enzymes
5. ETC and phosphorylation coupling

TABLE 7.13 Uncouplers that inhibit ATP synthesis by oxidative phosphorylation

Name of the compound	Function of the compound
2, 4-Dinitrophenol	Weight loss drug
Dicoumarol	Anticoagulant
Chlorocarbanoyl cyanide phenyl hydrazone (CCCP)	Leaks out proton
Bilirubin	Natural metabolite of haeme degradation

7.13 INTEGRATION OF METABOLISM

Through metabolism, the major biomolecules—amino acids, carbohydrates, fatty acids, and ketone bodies—are processed to metabolic fuels. Metabolic fuels are substances that are oxidized to release energy, which is used to support anabolic processes and other cellular functions. The energy of the metabolic fuel is its calorific value, expressed as kcal/g. Calorific values of different sources of metabolic fuels are listed in Table 7.14.

TABLE 7.14 Calorific values of metabolic fuels

Metabolic fuel	Calorific value (kcal/g)
Carbohydrates	4
Amino acids	4
Ketone bodies (hydroxybutyrate, acetoacetate)	4
Fatty acids	9

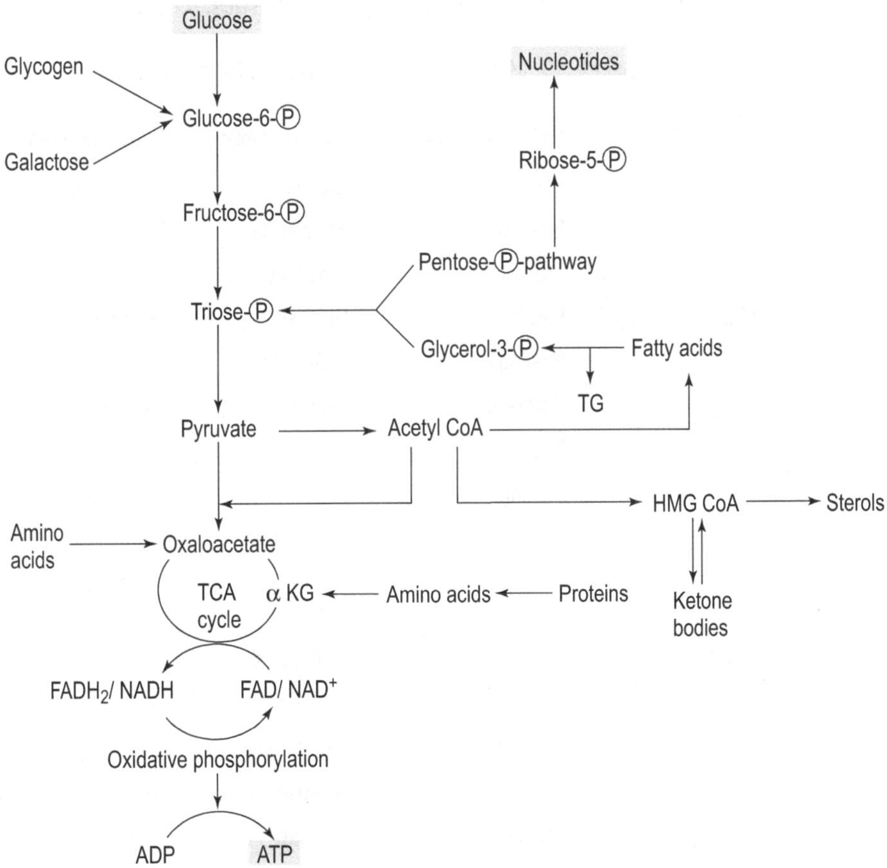

FIGURE 7.95 Integration of metabolic pathways

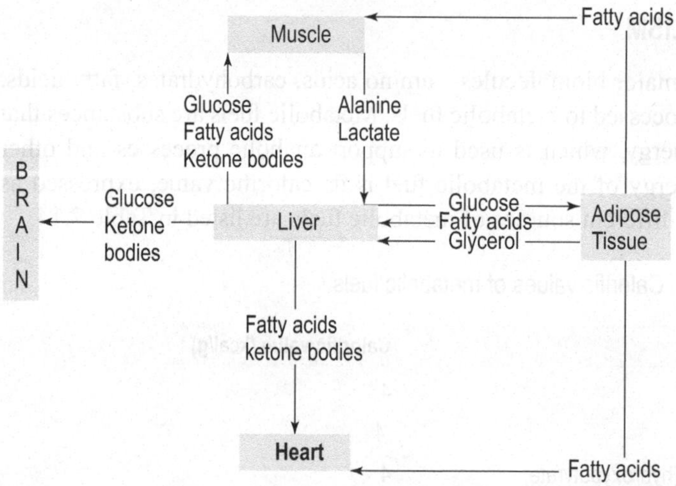

FIGURE 7.96 Integration of metabolic pathways of different organs

During the process of metabolism and energy production, certain metabolites of one pathway are linked to other pathways, thus, functioning as intermediates or intermediate metabolites. This forms the point for the integration of different metabolic pathways. Figure 7.95 represents the integration of metabolism.

Distribution of metabolic pathways differs among the major tissues, that is, all pathways do not occur in all the cells. The specific metabolic pathways of major tissues are shown in Table 7.15. Some metabolites flow from one organ to the other through the blood, resulting in integration of metabolic pathways among major organs or tissues, as depicted in Fig. 7.96.

TABLE 7.15 Metabolic pathways specific for major organs

Organ	Metabolic state	Fuel imported	Fuel exported	Fuel stored
Liver	Fed	Glucose	Fatty acids	Glycogen
	Fasting	Fatty acids	Glucose	–
	Starvation	Amino acids	Ketone bodies	–
Muscle	Fed	Glucose	Lactate	Glycogen
	Fasting	Fatty acids	–	–
	Starvation	Fatty acids	Amino acids	–
Adipose tissue	Fed	–	–	Triaceylglycerol
	Fasting	–	Fatty acids	–
	Starvation	Fatty acids	Glycerol Fatty acids Glycerol	–
Heart	Fed	Fatty acids	–	–
	Fasting	Fatty acids	–	–
	Starvation	Ketone bodies	–	–
Brain	Fed	Glucose	–	
	Fasting	Glucose	–	
	Starvation	Ketone bodies	–	

SUMMARY

Biochemical pathways in the cells of the living organisms are organized into different pathways called metabolic pathways.

In catabolism, complex macromolecules (proteins, carbohydrates, lipids) are converted into simple molecules (with little energy trapped). Simple molecules, in turn, are catabolized into a few molecules which are oxidized to carbon dioxide (CO_2) and water along the following final common pathway: (1) Krebs cycle, (2) electron transport chain, and (3) oxidative phosphorylation. During these pathways, acetyl CoA is oxidized to CO_2 and water and the energy is trapped as ATP.

During anabolism, the building blocks are combined into complex molecules with the help of energy. Two forms of energy are used here: (1) ATP obtained from the catabolic pathway and (2) high-energy hydrogen, in the form of reduced NADPH (reduced nicotinamide adenine dinucleotide phosphate).

Nucleotides required for the synthesis of nucleic acids can primarily be salvaged from in vivo sources. They can also be synthesized from low molecular weight precursors, de novo leading to a less need for the supply of nucleotides from the dietary source. Dietary proteins are degraded in the digestive system by proteolysis to form amino acids and absorbed into the intracellular compartment of the intestinal cells by various transport systems. The pool is reused for the synthesis of body proteins or catabolized further via urea cycle and carbon skeleton degradation to form TCA cycle intermediates, peptide, and hormone derivatives and haemoglobin. Lipids are degraded to fatty acids and TGs. Fatty acids are oxidized to form acetyl CoA which enters TCA cycle. TGs are diverted to ketogenesis. Ketone bodies are further converted to acetyl CoA and energy in the form of ATP. Cholesterol is synthesized primarily in the liver from acetyl CoA and then esterified for storage. Cholesterol is degraded to bile acids in the liver or it is degraded to steroid hormones in kidney and gonads, or converted to vitamin D in the skin. Dietary carbohydrates are digested finally to D-glucose, D-galactose, and D-fructose. D-galactose and D-glucose are absorbed from the lumen into the intestinal epithelial cells by ATP-dependent active transport for individual catabolism.

Glucose is degraded by glycolysis to form pyruvate or lactate and ATP. In addition to glycolysis, glucose can also be catabolized by HMP shunt, producing NADPH and ribose-5-phosphate required for the biosynthesis of nucleic acids. Galactose is converted to UDP-glucose, which enters the glycogenesis pathway. Fructose is converted to fructose-6-P, which enters glycolysis. Glucose is stored as glycogen in the liver. Glucose is polymerized to glycogen by glycogenolysis. Glucose is also formed from non-carbohydrate precursors by a process called gluconeogenesis. During starvation or fasting, glucose is liberated from the glycogen store by glycogenolysis. Acetyl CoA, derived from the degradation of glucose, lipids, and amino acids, functions as the precursor for TCA cycle and form-reducing equivalents, which in turn are passed through electron transport chain and used up for the synthesis of ATP by oxidative phosphorylation, in the presence of ATP synthetase.

TERMS TO REMEMBER

5-Phosphoribosyl-1-pyrophosphate (PRPP) An activated sugar that is the precursor for the synthesis of purine and pyrimidine nucleotides.

Anabolism Biosynthesis of complex molecules from their precursors.

Anserine A dipeptide obtained from methylation of carnosine, it is also called methyl carnosine.

Bioenergetics Study of the energy production and transduction which occur as a result of catabolic pathways.

Carnosine A dipetpide composed of beta-alanine and histidine. Hence, it is also called beta-alanyl-L-histidine.

Catabolism Degradation or breaking of larger molecules into smaller ones.

Cystinuria Urinary excretion of excessive cysteine. This is due to the defect in the transport of basic amino acids resulting in the accumulation of cysteine, which is excreted in the urine. Excess cysteine is also deposited in the urinary tract, forming stones and urinary tract infection.

De novo synthesis Synthesis of a compound from low-molecular-weight precursors.

Entropy The degree of randomness of a system. It is a measure of the energy that is not accessible to perform useful work.

First law of thermodynamics Energy can change its form, but the amount of energy is conserved.

Gamma amino butyric acid (GABA) GABA is a derivative of glutamate that functions as a neurotransmitter and regulator of muscular tone.

Gibbs' free energy The amount of energy capable of doing work during a reaction at constant temperature, volume, and pressure.

Glutathione Glutathione is a tripeptide synthesized by peptide linkage formed between glutamate, cysteine, and glycine.

Glycinuria Glycinuria refers to a loss of excessive glycine urine due to the block in the transport of glycine and imino acids in the intestine and renal tubules.

Gout Accumulation of uric acid crystals in the joints and synovial fluids. It is caused due to excessive catabolism of purine nucleotides and deficiency of HGPRT.

Hartnup's disease Disease caused due to the defect in the transport of large neutral aromatic amino acids into the intestinal cell. Indole derivatives, such as indoxyl-sulphate, which are also excreted in urine in higher levels.

High-energy phosphate compound Compound that has a high group-transfer potential for phosphate, for example, ATP.

Histamine Histamine is a potent vasodilator and a neurotransmitter obtained by the decarboxylation of histidine. It is released in response to allergic reactions.

Hydride ions Hydride ion is an equivalent of one proton and two electrons, represented as ':H⁻'.

IMP A compound derived from PRPP, which functions as the common intermediate in the purine nucleotide synthesis. It can be converted into AMP or GMP, depending on the cellular demand.

Lesch–Nyhan syndrome A disorder of abnormal purine metabolism characterized by gout and malfunction of central nervous system.

Melatonin (N-acetyl-5-methoxytryptamine) It is a hormone derived from tryptophan. It functions as an antioxidant, immune system activator, and as a synchronizer of biological clock.

Oxidant/oxidising agent Electron-accepting molecules in a redox reaction.

Polyamines Group of organic compounds possessing two or more primary amine groups.

Protomotive force In mitochondria, membrane-bound transducers couple electron flow to the production of a transmembrane pH difference, accomplishing osmotic electrical work. This generates a proton gradient which has potential energy called protomotive force (PMF).

Reducing equivalent A common term used to refer the electron equivalent which may be NADH, FADH2, electrons, or hydride ions.

Reductant/reducing agent Electron-donating molecules in a redox reaction.

Salvage pathway Synthesis of a compound from its constituent building blocks, usually salvaged from the body.

Second law of thermodynamics Universe always tends towards more and more disorder. That is, in all natural processes, the entropy of the universe increases.

Severe combined immunodeficiency syndrome (SCID) SCID is a consequence of abnormal purine metabolism, the syndrome being characterized by total loss of immunity due to defective production and function of T cells and B cells.

REVIEW QUESTIONS

1. Discuss the enzymes involved in the formation of body's amino acid pool.
2. How is the amino acid pool utilized in the body?
3. Define nitrogen balance and its types.
4. Write about the disorders associated with the defective transport of amino acids from the gut lumen.
5. Define anaplerotic reactions.
6. Give reason for calling TCA cycle as the central crossroad pathway.
7. What is substrate-level phosphorylation? How does it occur in TCA cycle?
8. How many ATPs are generated in TCA cycle per molecule of acetyl CoA?
9. Discuss the role of citrate cleavage enzyme.
10. Give the order of redox potential of electron carriers in the ETC.
11. What is the amount of energy generated per electron pair flowing through the ETC?
12. Define system and surroundings with reference to cell biology.

13. State the first and second law of thermodynamics.
14. Tabulate the order of phosphate group transfer potential of phosphorylated compounds in the cell.
15. The group transfer potential of PEP is higher than that of ATP. Still, ATP is termed the energy currency of the cell. Give reason.
16. Define the three thermodynamic quantities with their units.
17. Write an equation to show the relation between the three thermodynamic quantities.
18. Write an equation to reveal the relation between ΔG and $\Delta G^{\circ\prime}$.
19. Describe the factors that contribute to the highly negative free energy change of ATP hydrolysis.
20. Phosphate compounds are not the only source of energy. Justify this statement.
21. Explain the major endergonic reactions driven by ATP hydrolysis.
22. Write the equation to represent the free energy change for the ATP-mediated transport of a non-ionic solute between different cellular compartments.
23. Define jaundice and its types.
24. Write about the fate of iron liberated from haeme degradation.
25. Comment on the rate-limiting step of haeme biosynthesis.
26. Tabulate the enzyme deficiency disorders of haeme biosynthesis.
27. Why is glucose converted to glucose-6-phosphate in the first step of glycolysis?
28. Write the energetics of glycolysis and arrive at the net ATP production in aerobic and anaerobic glycolysis.
29. Phytanic acid cannot be degraded by alpha oxidation. Give reason.

MULTIPLE-CHOICE QUESTIONS

1. The $\Delta G^{\circ\prime}$ value for ATP is
   (a) –7.3 kcal/mol
   (b) +7.3 kcal/mol
   (c) –7.3 kJ/mol
   (d) +7.3 kJ/mol
2. ATP phosphorylates glucose in an aqueous medium to form glucose-6-phosphate. Which of the following statements is valid?
   (a) ATP is the surrounding
   (b) ATP and glucose are the system
   (c) aqueous medium is the system
   (d) aqueous medium and glucose are the surrounding
3. Entropy is the
   (a) degree of randomness, denoted as S
   (b) capacity to do work, denoted as S
   (c) heat liberated during work, denoted as H
   (d) degree of randomness, denoted as E
4. Gibbs' free energy is the energy
   (a) needed to do work
   (b) liberated during a reaction
   (c) needed to induce randomness
   (d) liberated as heat
5. Which of the following compounds has the highest phosphate-group-transfer potential?
   (a) ATP
   (b) PEP
   (c) Glucose-6-P
   (d) GTP
6. The term hydride ion is used to refer to
   (a) one proton and two electrons
   (b) two protons and one electron
   (c) two protons
   (d) two electrons
7. Electron transport chain is in the
   (a) mitochondrial membrane
   (b) mitochondrial matrix
   (c) plasma membrane
   (d) cytosol
8. The precursor for the synthesis of purine and pyrimidine nucleotides is
   (a) PRPP
   (b) ATP
   (c) GTP
   (d) MNP
9. Deficiency of HGPRT leads to
   (a) gout
   (b) SCID
   (c) Hartnup's disease
   (d) cystinuria

10. Carnosine is the precursor of
   (a) anserine
   (b) PRPP
   (c) uric acid
   (d) purine
11. The energy required for the bioluminescent reaction is donated by
   (a) PEP
   (b) ATP
   (c) GTP
   (d) CTP
12. Which of the following amino acids is excreted in Hartnup's disease?
   (a) basic
   (b) acidic
   (c) neutral
   (d) polar
13. The hormone responsible for initiating the digestion of protein in the intestine is
   (a) trypsin
   (b) cholecystokinin
   (c) insulin
   (d) carboxypeptidase
14. Indole derivative is excreted in urine during
   (a) Hartnup's disease
   (b) cystinuria
   (c) cysteinuria
   (d) indoluria
15. Which of the following is fatal?
   (a) positive nitrogen balance
   (b) cystinuria
   (c) negative nitrogen balance
   (d) glycinura

Answers to Multiple-choice Questions

1. (a)	2. (b)	3. (a)	4. (a)	5. (b)
6. (a)	7. (a)	8. (a)	9. (a)	10. (a)
11. (b)	12. (c)	13. (a)	14. (a)	15. (c)

CHAPTER 8

Germ Cells

British physiologist Dr Robert G. Edwards received Nobel Prize (2010) for producing the first baby via in vitro *fertilization.*

After reading this chapter, students should be able to understand and appreciate the following:

- Germ-cell lineage
- Germ-cell cycle and sexual reproductive cycle
- Haploid and diploid cells
- Meiosis and gamete production
- Sperm and ovum
- In vivo fertilization and artificial fertilization
- Germ-cell tumour

8.1 GERM-CELL LINEAGE

Germ-cell lineage represents all the cells involved in the formation of gametes. The precursor cells develop into mature germ cells via different processes, including proliferation, differentiation, and migration. This process is called *germ-cell specification.* The germ cells colonize the gonads, either ovary or testes, before sexual differentiation (a process called germ-cell colonization). This section discusses germ-cell specification and germ-cell colonization, which are crucial for germ-cell lineage.

8.1.1 Germ-cell Specification

Germ-cell specification occurs in the epiblast. Primordial germ cells (PGCs) are the precursors for the germ cells and are formed in the epiblast, which in turn was formed between the blastulation and the gastrulation stages of embryogenesis (embryonic development). Figure 8.1 shows the differentiation and developmental pattern of embryo, where the epiblast can also be seen. A five-day old embryo (denoted E.5) is called blastocyst. A blastocyst contains trophoblast and inner cell mass (ICM). The trophoblast develops into cytotrophoblast, whereas the ICM delaminates to form a bilayer of cuboidal cells (called epiblast) and a single layer of flat cells called hypoblast (Fig. 8.2). The formation of epiblast occurs just before gastrulation and lies closer to the trophoblast, whereas the hypoblast lies below the epiblast. Epiblast gives rise to the embryonic ectoderm, mesoderm, and endoderm (i.e., the embryo proper), and the hypoblast forms the yolk sac. The organs developed from these three layers are listed in Table 8.1. The gonads (the testes and the ovary, where the germ cells are produced) are derived from the mesoderm. Once the gonads form, the PGCs migrate to the gonads for further differentiation.

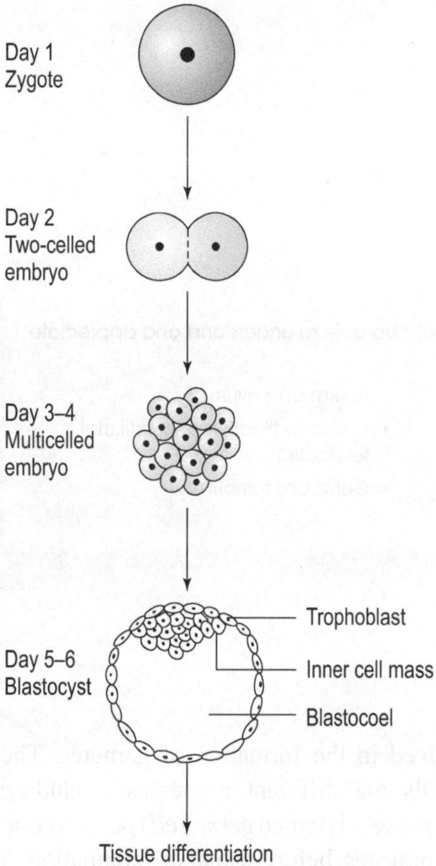

FIGURE 8.1 Stages of human embryogenesis

TABLE 8.1 Developmental fate of the embryonic germ layer

Layer	Organs differentiated from the layer
Ectoderm	Nervous system Skin Cornea and eye lens Epithelial lining of canals and cavities Epithelium of pituitary gland, pineal gland, and adrenal medulla Neural crest
Mesoderm	Reproductive organs (testes and ovary) Epithelium of the reproductive ducts Muscles Kidney Bone marrow Connective tissues of spleen, blood, and blood vessels Bone Cartilage Connective tissues lining the body cavity
Endoderm	Gut epithelium Epithelium of ureter and bladder Epithelium of respiratory tract (including larynx, trachea, bronchi, lungs) Liver Pancreas Thyroid Parathyroid Thymus

Once the epiblast is formed, in a 7.2-day old embryo, the PGCs appear as a cluster of about 20 cells, (E 7.2). The cells divide rapidly reaching a total of 26,000 on the 13th day. PGCs have a characteristic feature of over-expressing oct-4 (an octamer-binding transcription factor that binds to 'ATTTGCAT' sequence with high affinity). PGCs migrate to the genital ridge, a tissue that develops into gonads (Fig. 8.3). The migration, survival, and proliferation of PGCs are promoted by the signalling molecules of the surrounding somatic cells.

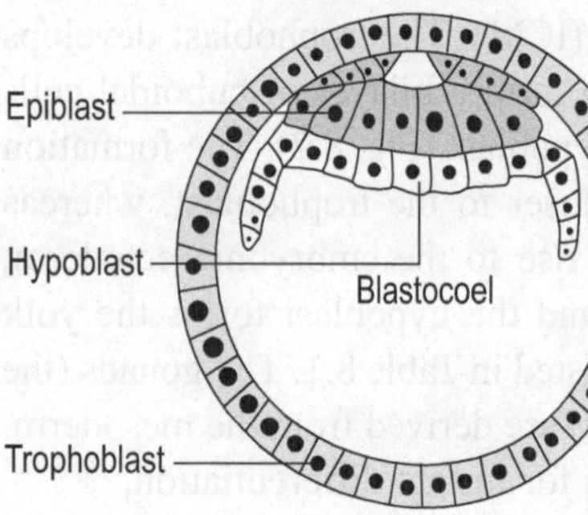

FIGURE 8.2 Epiblast

8.1.2 Germ-cell Colonization

The PGCs migrating into the genital ridge colonize the male or female germ cells. In males, the genital ridge develops into testes where the PGCs produce colonies of spermatogonia developing into a sperm. In females, the genital ridge develops into ovary where the PGCs produce colonies of oogonia, which form the ovum or the matured egg. Formation of matured sperm and ovum from the PGCs occurs through a series of steps as described in Section 8.4. Differentiation of PGCs into germ cells requires leukaemia inhibitory factor (LIF). LIF is a glycoprotein that functions as a pleiotrophic cytokine (a cytokine that influences multiple cell types). It is required for the growth and survival of PGCs. It facilitates growth and survival by suppressing programmed cell death. Development of genital ridge into gonad is determined by the Y chromosome in the somatic cell of the genital ridge. The Y chromosome contains sry gene (sex determining region of Y), which is responsible for the development of testes. Actually, the sry gene induces the differentiation of somatic cells in the genital ridge into sertoli cells. Sertoli cells have the following functions:

- They produce signals that stimulate the PGCs to produce the sperm.
- They secrete anti-mullerian hormone, which suppresses the development of the female reproductive system.
- They induce the differentiation of adjacent somatic cells into Leydig cells, which secrete the testosterone (male sex hormone) that is responsible for the male characteristics.

In the absence of Y chromosome, however, the genital ridge develops into ovary, where the somatic cells differentiate into follicle cells. Follicle cells signals the PGCs to commit to the development of ovum or egg. The entire germ-cell lineage is shown in Fig. 8.4.

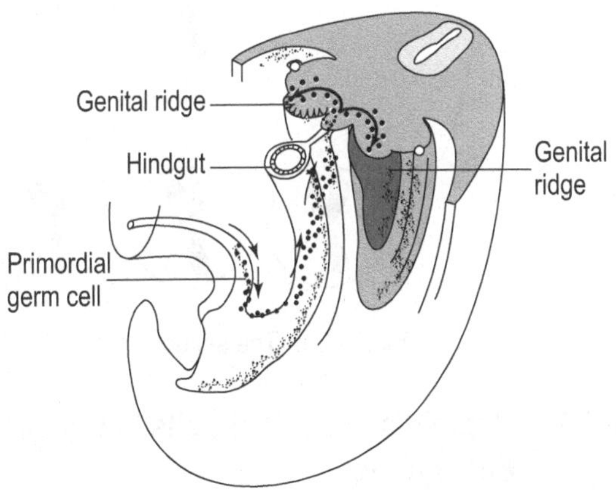

FIGURE 8.3 Migration of primordial germ cells to the genital ridge (Arrow indicates the migration path)

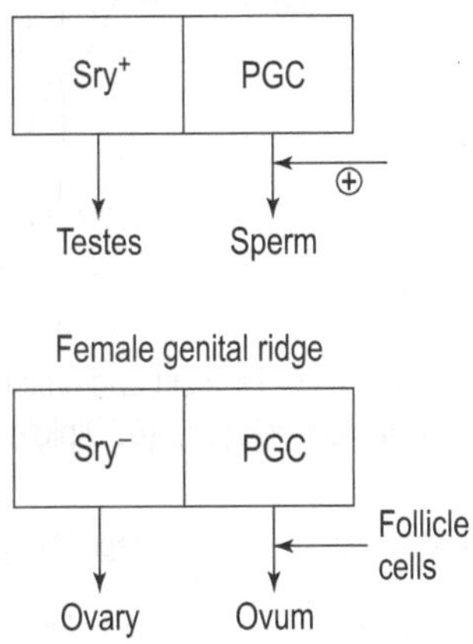

FIGURE 8.4 Germ-cell lineage

8.2 SEXUAL REPRODUCTIVE CYCLE

The sexual reproductive cycle of an organism refers to the formation of haploid cells (with a single set of chromosomes, denoted as $1n$) and diploid cells (with double set of chromosomes, denoted as $2n$) alternatively. Haploid and diploid cells are shown in Fig. 8.5. The diploid cells from a diploid organism undergo meiosis (refer Chapter 1) to produce haploid cells. Two such haploid cells (from male and female) fuse (fertilize) together forming a diploid cell (the zygote). The diploid cells undergo genetic recombination to form a genome with genes of both the haploid cells involved in the zygote formation. The diploid cells then divide by mitosis producing a large number of diploid cells which form the diploid organism. The diploid organism again produces haploid cells by mitosis, and the cycle continues. The sexual reproductive cycle is shown in Fig. 8.6. In general, all the somatic cells divide by mitosis. The cells which divide by meiosis are called germ cells or gametes. Gametes are haploid, and they fuse to form diploid cells called zygotes, which eventually develop into a diploid adult.

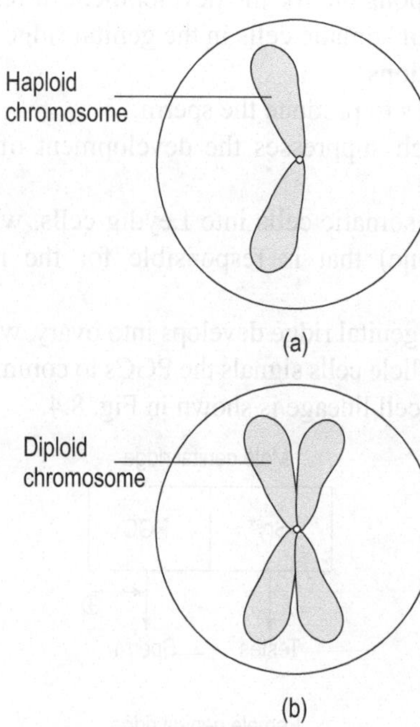

FIGURE 8.5 Haploid and diploid cells
(a) Haploid cell (1*n*), (b) Diploid cell (2*n*)

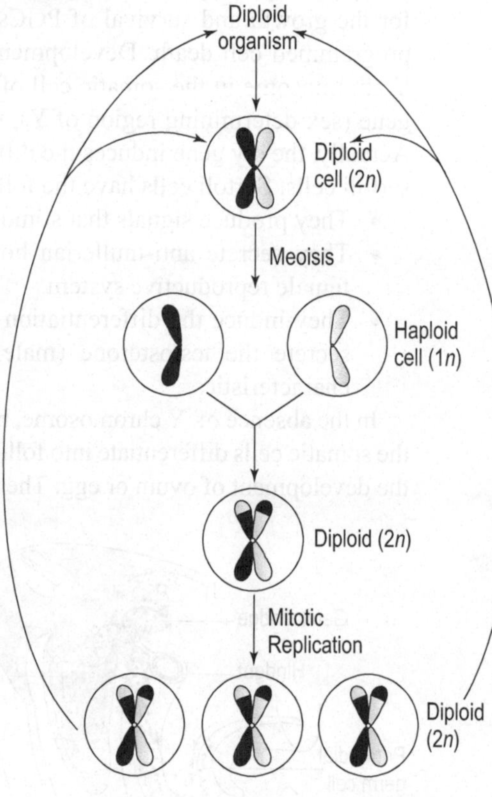

FIGURE 8.6 The sexual reproductive cycle

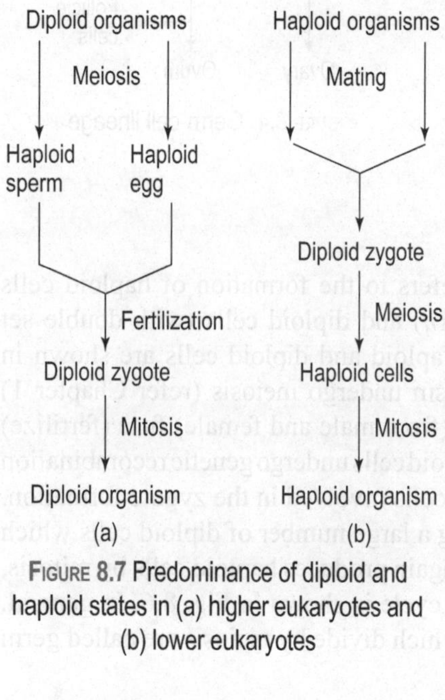

FIGURE 8.7 Predominance of diploid and haploid states in (a) higher eukaryotes and (b) lower eukaryotes

8.2.1 Haploid and Diploid Cells in the Life Cycle of Eukaryotes

The reproductive cycle clearly indicates that haploid and diploid cells are found in the life cycle of eukaryotes. Haploid cells are usually germ cells or gametes. In adult human beings, for example, the germ cells in males and females are called sperm and ovum, respectively, both of which are haploid. They are produced by meiotic division of the cells present in the sex organs, the testes and the ovary. These haploid germ cells fuse to form zygote, a diploid cell. The diploid cell undergoes mitotic division to develop into an adult. Details of sperm and ovum are discussed in Section 8.4. In an adult organism, the cells which form the germ cells are called germ-line cells and those that differentiate into other cells of the body are called somatic cells. The predominance of diploid and haploid cells, however, varies between the life cycles of different organisms. This difference that exists between a higher and lower eukaryote is illustrated in Fig. 8.7.

In most of the multicellular organisms, the major part of the life exists in diploid state. The haploid state is found only during reproductive stage, for producing the next generation. In higher eukaryotic organisms, such as mice and other mammals, the organisms proliferate in the diploid state via mitosis. The gametes produced by them are haploid, which are converted into diploid state by fusion with other gametes from opposite sex. However, in lower eukaryotes, such yeast and chlamydomonas, most of the life cycle is in haploid state, the only diploid state being the zygote state.

8.3 MEIOSIS AND GAMETE PRODUCTION

Meiosis is a type of cell division in which the chromosome is exactly halved. This is in contrast to mitosis in which the chromosome is exactly duplicated (Fig. 8.8). Germ cells are produced by two meiotic divisions, meiosis I and meiosis II, undergone by the PGCs in the testes and the ovary (refer Section 8.1). The process of meiosis has already been discussed in Chapter 1. During germ-cell formation, the diploid cells undergo one round of DNA replication and two rounds of cell division, meiosis I and meiosis II. As a result, four haploid cells are formed (Fig. 8.9). The two meiotic divisions can be explained stepwise as follows:

- The diploid cell has maternal DNA (shown in white) and paternal DNA (shown in black). They both undergo replication and duplicate.
- The DNA aligns along the equator of the cell, and the chromosomes are paired.
- The chromosomes undergo genetic recombination, and are pulled towards the periphery (pole) of the cell.
- The cell has equal number of maternal and paternal chromosome on both the poles. The cell divides, so that the chromosome is equally shared by the two daughter cells. This finishes the first meiosis.
- The chromosomes of the each daughter cell are again realigned along the equator and pulled towards the pole, and the cells divide again by meiosis (so called meiosis II). Thus, four daughter cells are formed with equal sharing of the chromosomes. These daughter cells are called germ cells or gametes, and they are haploid.

Details of the steps involved in the formation of ovum and sperm are discussed in Section 8.4.

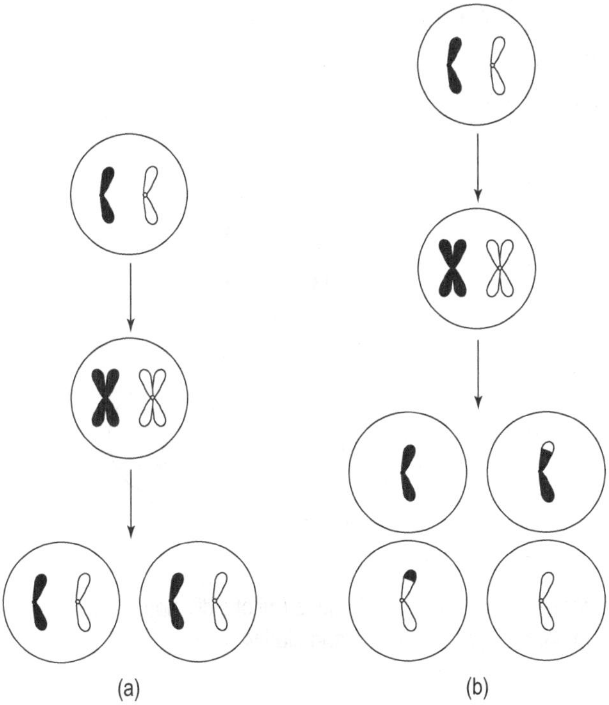

FIGURE 8.8 The outline of (a) mitosis and (b) meiosis with reference to the chromosome

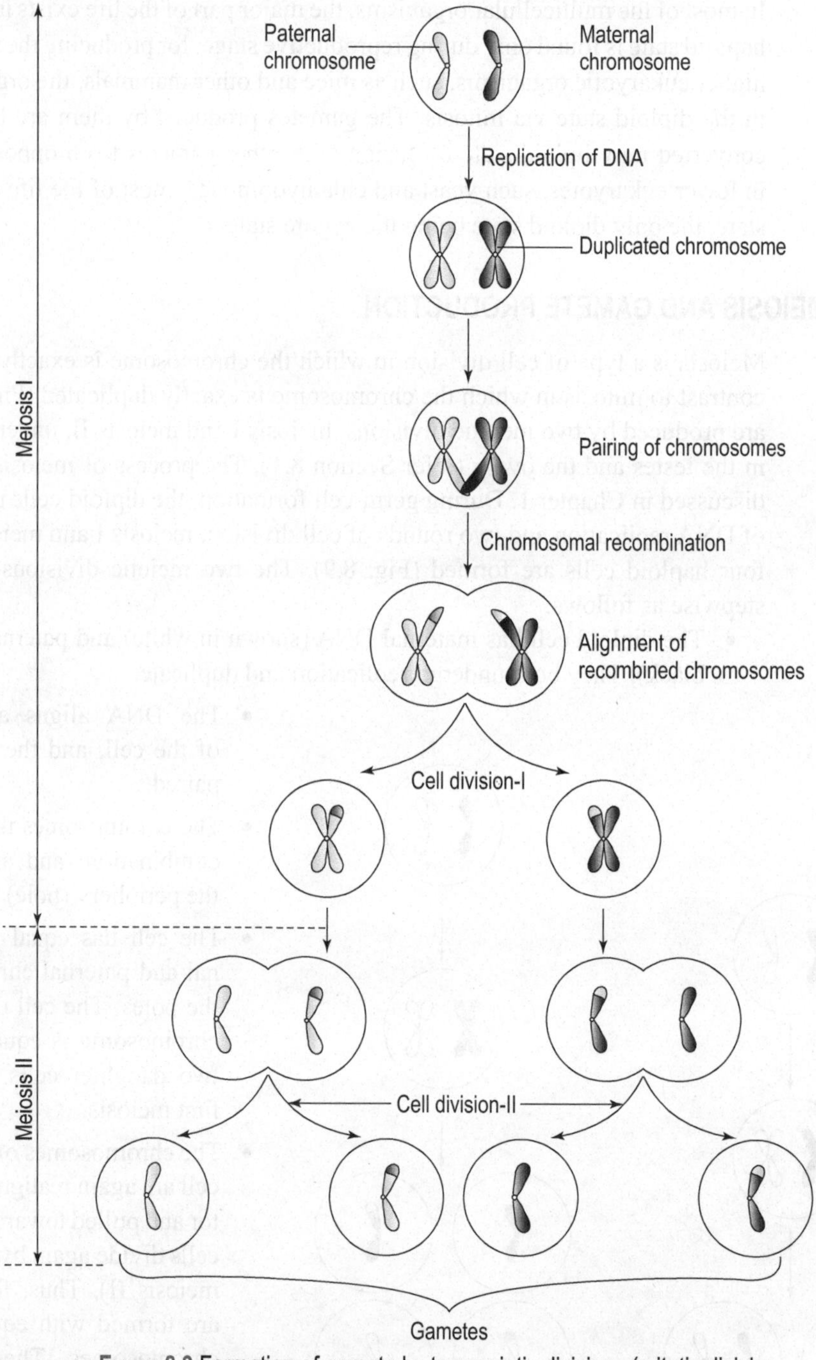

FIGURE 8.9 Formation of gamete by two meiotic divisions (mitotic division is also shown in parallel for comparison and clarity)

8.4 GAMETES

In Section 8.1, we discussed that the germ cells—sperm and ovum—are produced by the PGCs. The details of these cells are discussed in this section.

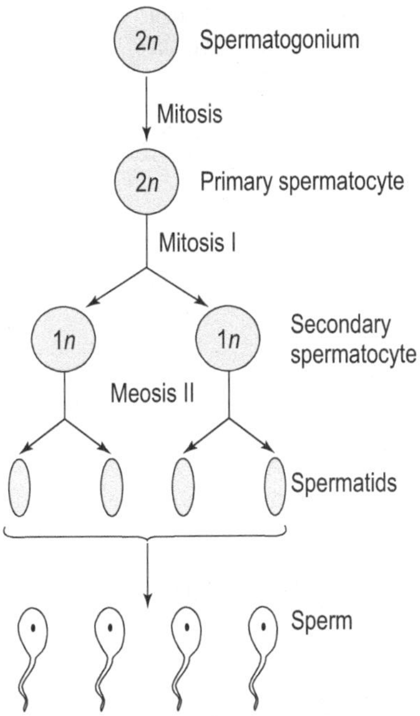

FIGURE 8.10 Spermatogenesis

8.4.1 Sperm

Sperm is the motile male germ cell that penetrates into the ovum for fertilization. Sperm is obtained from the PGCs in the testes by a process called spermatogenesis (Fig. 8.10). The testes have small tubular structures called seminiferous tubules, where the PGCs differentiate into diploid cells called spermatogonia (singular spermatogonium). Some of the spermatogonia renew themselves to form spermatogonial stem cells, whereas the others develop into sperm by a process called spermatogenesis. The spermatogenesis occurs by following steps:

- The spermatogonium contains 46 single chromosomes which are converted to 46 sister chromatids by mitosis during the interphase. The cells formed by mitosis grow into diploid primary spermatocytes (2*n*).
- The primary spermatocyte undergoes meiosis I forming two haploid cells, each with 23 sister chromatids. The haploid cell is called secondary spermatocyte (1*n*).
- The secondary spermatocyte undergoes meiosis II to form spermatids (1*n*).
- Some part of the spermatid differentiates into sperm (1*n*), which are detached from the remaining portion called residual bodies.
- Thus, each spermatogonium gives rise to four mature sperms (Fig. 8.10).

During the entire process of spermatogenesis, the cells remain interconnected via cytoplasmic bridges (Fig. 8.11). This connection remains until the free sperm is formed and detached. During spermatogenesis, the developing gametes are found in association with the sertoli cells (Fig. 8.12), which project from the basal lamina of the seminiferous tubules into the lumen. Sertoli cells are analogous to follicle cells of the ovary. Sertoli cells, in association with another somatic cell population called Leydig cells, are responsible for the growth and survival of the germ cells. Leydig cells secrete the male sex hormone—testosterone—to promote spermatogenesis.

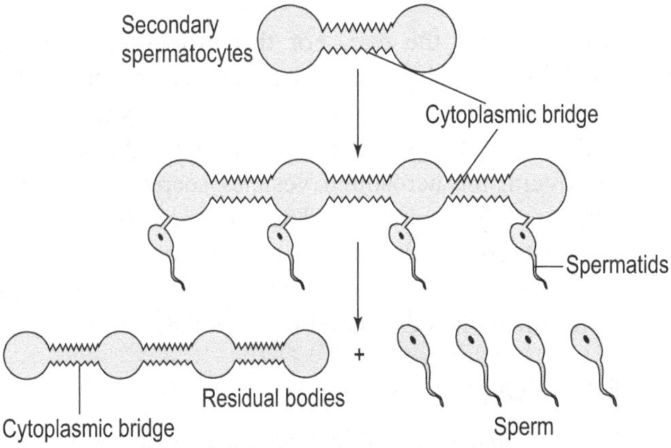

FIGURE 8.11 Presence of cytoplasmic bridge in the sperm forming cells during spermatogenesis

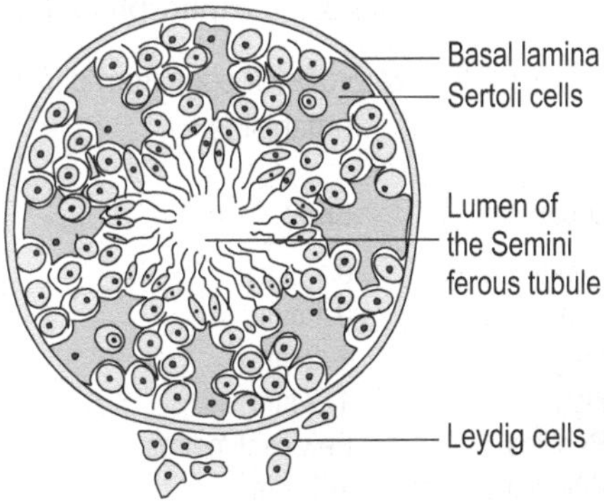

FIGURE 8.12 Sertoli cells and Leydig cells involved in spermatogenesis

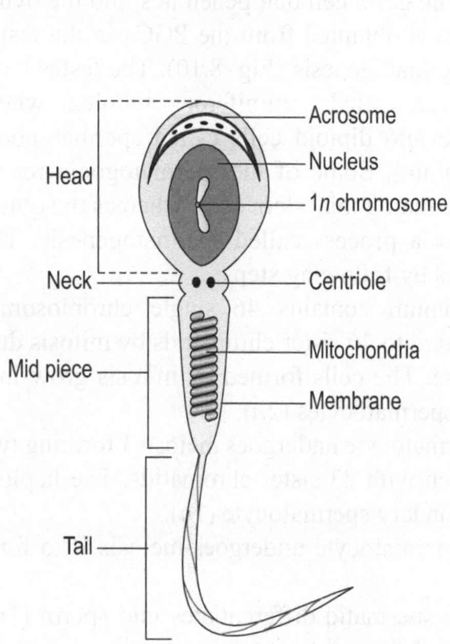

FIGURE 8.13 Structure of a human sperm

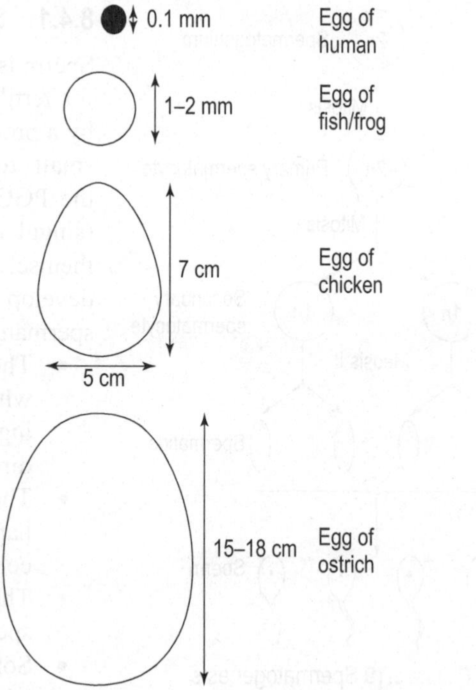

FIGURE 8.14 Size profile of eggs from different organisms

Matured sperm is a motile structure stored in epididymis. It consists of a head and a tail, linked by a mid piece (Fig. 8.13). The head connects to the mid piece by a small neck. The entire structure is enclosed in a single plasma membrane. The head is made of nucleus at the centre and acrosomal vesicle at the anterior end. The nucleus is made of condensed $1n$ chromosomes, which are tightly packed with positively charged proteins called protamines. The acrosomal vesicle contains proteins and enzymes. The mid piece contains mitochondria. The tail is like a flagellum and is made of axoneme and dynein. The axoneme is made of microtubules, and the dynein is a motor protein (refer Chapter 9).

The tail helps to propel the sperm towards the egg. For this propelling motion, adenosine triphosphate (ATP) is needed. The mitochondria of the mid piece produce ATP which promotes the sliding motion of the dynein along with microtubules to promote motion. On reaching the ovum, the acrosomal vesicles secrete hydrolytic enzymes via exocytosis. The enzymes dissolve the egg coat (zona pellucida), thereby promoting the entry of sperm. This results in fertilization. The mechanism of fertilization is discussed in Section 8.5.

TABLE 8.2 Size profile of eggs from different organisms

Organism	Size *
Human being	0.1 mm
Sea urchin	0.1 mm
Frog	1–2 mm
Fish	1–2 mm
Reptiles	Few centimetres
Whooping crane	102 mm

* The size of the egg is usually larger than that of the somatic cell (10–20 μm).

8.4.2 Ovum

Ovum is the non-motile matured egg or the female gamete, spheroid or ovoid in shape. The size of the egg ranges from 0.1 mm to few centimetres (Fig. 8.14

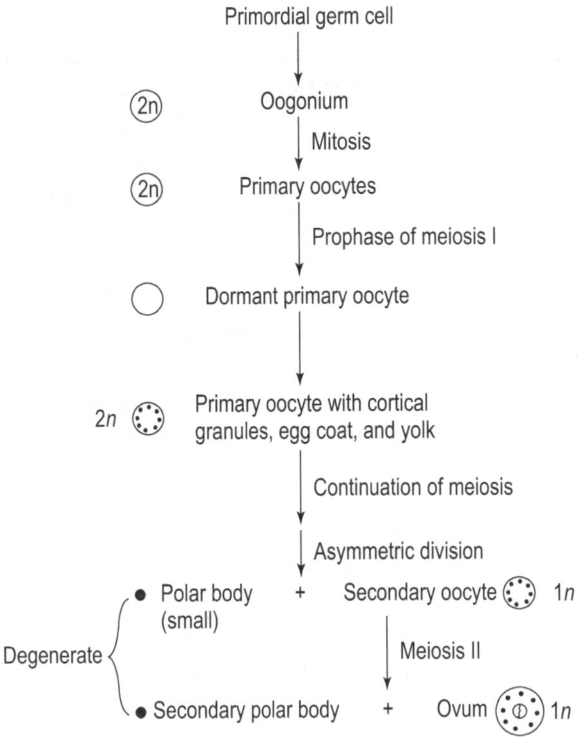

FIGURE 8.15 Steps in oogenesis

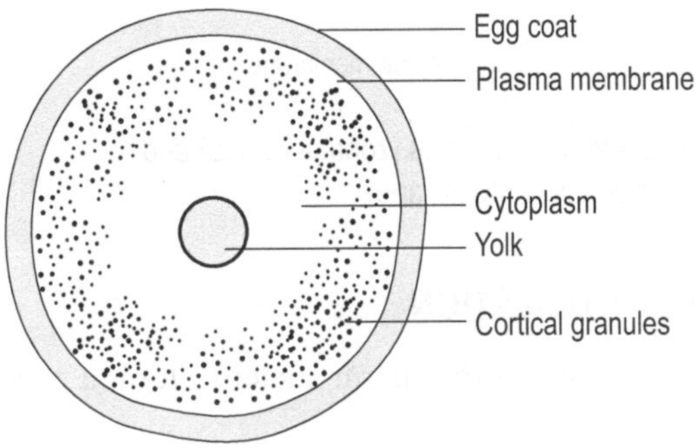

FIGURE 8.16 Structure of an ovum

and Table 8.2), but generally larger than the size of somatic cell.

Ovum is obtained from the PGCs by a process called oogenesis (Fig. 8.15) which occurs via the following steps:

- The PGCs in the gonad develops into oogonium, which is a diploid cell (2*n*).
- Oogonium proliferate by several mitotic divisions within the ovary forming primary oocytes (2*n*).
- The primary oocytes undergo meiosis I, and replicate their DNA. During prophase I of meiosis, the cells remain dormant for a few days to many years, depending on the species.
- During the dormant stage, the primary oocytes acquire the cortical granules and the egg coat. The cells also develop the yolk and other macromolecules.
- During sexual maturation, the hormones signal the primary oocytes to resume meiosis I. By the end of meiosis I the primary oocytes divide asymmetrically generating two cells of different sizes, both haploid. The smaller cell is called polar body, and the larger one is called secondary oocyte (1*n*).
- The secondary oocytes divide by meiosis II to form the mature egg or the ovum (Fig. 8.16). Simultaneously, a smaller cell called the secondary polar body is also formed. The polar body and the secondary oocytes have haploid chromosomes. The polar bodies eventually degenerate.

During maturation, the oocytes are nourished in different ways. The yolk present in the ovum is usually synthesized in the liver and then imported. Sometimes, the accessory cells produced in the ovary are also responsible for the nourishment. For example, nurse cells pass their cytoplasmic contents (ribosomes, mRNA, proteins, and lipids) to the oocytes via cytoplasmic bridges. Another population of somatic cell called the follicle cell forms an epithelial layer surrounding the oocyte, and communicates with the oocyte via small gap junction (refer Chapter 10 for gap junction). Through the gap junction, small precursor molecules of follicle cells are transported into the oocyte for the subsequent formation of macromolecules. Follicle cells also produce some macromolecules which help in the formation of egg coat as well as in the control of patterning and axial asymmetries of the egg.

The outer part of the egg has a coat which functions as a mechanical barrier and also a species-specific barrier to the sperm cells. This barrier is called the egg coat, which surrounds the plasma membrane. It is a specialized extracellular matrix rich in glycoprotein. It is produced by the surrounding cells or by the egg cell itself. The egg coat of mammals and non-mammals are called the zona pellucida and vitelline layer, respectively (Fig. 8.17). Below the egg coat is the plasma membrane, which surrounds the cytoplasm. The outer region of the cytoplasm is the cortex, which contains secretory granules. When the sperm activates the egg, the cortical granules come out of the egg by exocytosis and alter the egg coat in order to prevent the further entry of sperm. The cytoplasm has a nutrition rich region called yolk. The yolk contains lipids, proteins, and polysaccharides. These molecules are in the form of membrane-bound granules or unbound granules, called yolk granules. The yolk is responsible for nourishing the embryo. In some organisms, such as chicken, the yolk forms the major part of the cell, whereas in some others, such as mammals, the yolk is very little or sometimes absent. In mammals, therefore, the nourishment is given by the placenta.

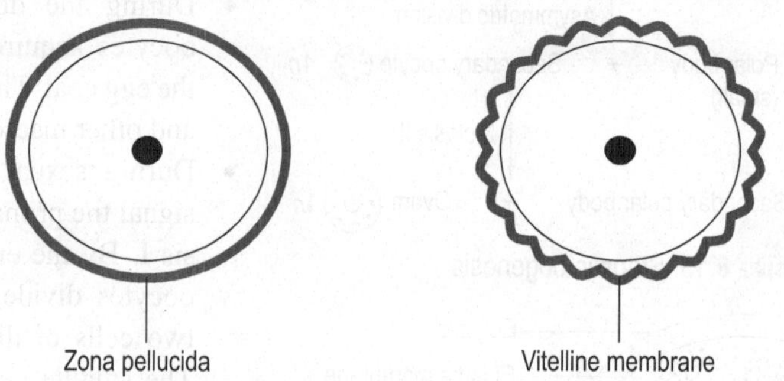

FIGURE 8.17 Comparison of egg coat in (a) mammals (zona pellucida) and (b) non-mammals (vitelline membrane)

8.5 IN VIVO FERTILIZATION AND ARTIFICIAL FERTILIZATION

Fertilization is the process of fusion of the egg and the sperm nuclei to form a zygote, which develops into a new organism.

8.5.1 In Vivo Fertilization

In vivo fertilization takes place via various steps, listed as follows:

- Capacitation of sperm
- Sperm–egg binding
- Sperm acrosomal reaction
- Egg cortical reaction
- Fusion of pronuclei and formation of zygote

Capacitation of sperm

Capacitation is the activation of sperm cells that occur in the cervical tract of the female, in order to make the cells competent to fuse with the egg. Capacitation occurs just after the ejaculation of the sperm into the female reproductive tract, and it takes about 5–6 hours. Capacitation is initiated by bicarbonate ion (HCO_3^-) in the vagina. The HCO_3^- activates the adenylate cyclase of the sperm (Fig. 8.18) and thereby increasing the concentration of cyclic AMP (cAMP). As a result, the following changes occur in the sperm cell:

- Loss of membrane cholesterol, which destabilizes the acrosomal sperm head, allowing greater binding between sperm and oocyte.
- Loss of membrane glycoprotein, which increases the fluidity of the membrane resulting in an increase in calcium influx.
- Increase in mobility to reach the egg. Motility is increased by the calcium-mediated raise in cAMP levels.
- Increase in sperm metabolism, to penetrate the egg.
- Decrease in membrane polarity, to facilitate the subsequent acrosomal reaction.

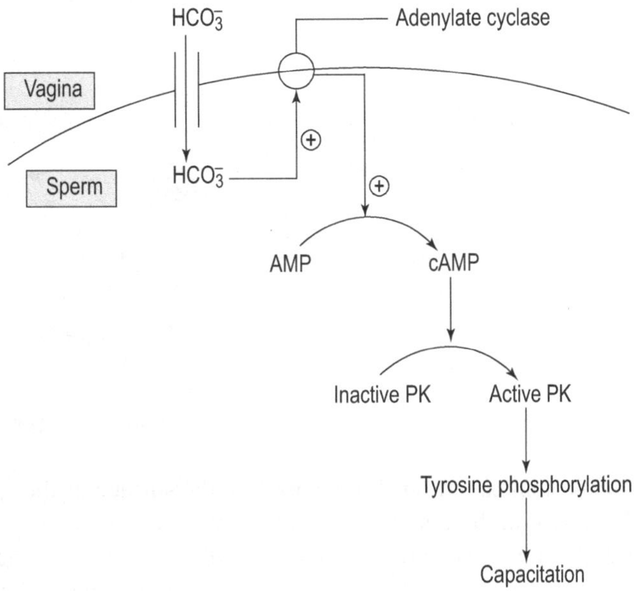

FIGURE 8.18 Capacitation of sperm

Sperm–egg binding

The sperm is attracted towards the egg cell via chemoattractants, nature of which is still not known. It is, however, known that the chemical signals are produced by the follicle cells surrounding the ovum. The zona pellucida (ZP) or the egg coat has three layers: ZP1, ZP2, and ZP3. ZP2 and ZP3 are composed of long fibrils that are cross linked by ZP1 forming a three-dimensional network. ZP3 has linked oligosaccharides that attach with the membrane protein receptors on the sperm. Thus, ZP3 mediates the attachment of the sperm to the egg. ZP3 is so crucial that its absence results in infertility due to impairment in sperm–egg binding.

Sperm–acrosomal reaction

Once the sperm is bound to the egg's ZP, acrosomal reaction starts. Acrosomal reaction is the release of the acrosomal contents (enzymes and proteins) from the sperm via exocytosis

(Fig. 8.19). ZP3 triggers the reaction by increasing the influx of sperm's cytosolic calcium ions. Consequent to the acrosomal reaction, the hydrolytic enzymes allow the sperm to tunnel through the egg coat, and to fuse the plasma membranes of the egg and the sperm. The sperm head binds tangentially with the plasma membrane of the egg resulting in fusion. The sperm crosses the ZP layers and fuses with the plasma membrane in about 6 minutes. After fusion, the sperm releases its nucleus into the egg's cytoplasm.

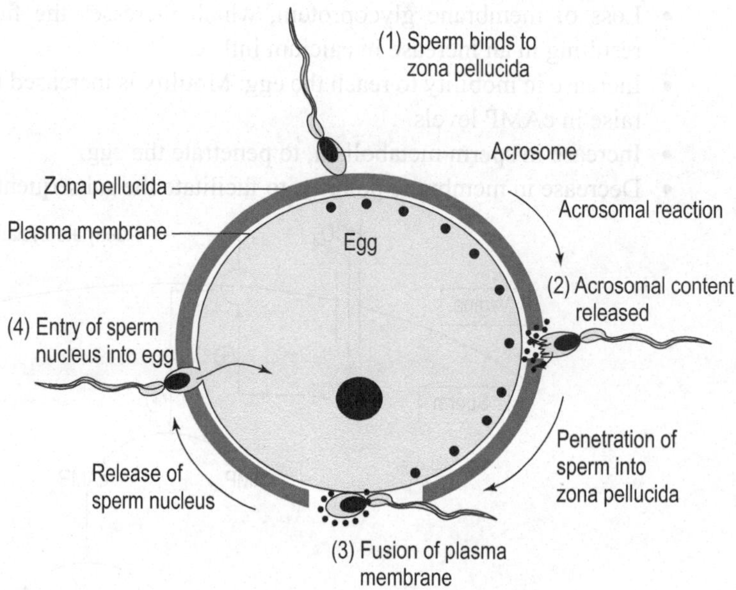

FIGURE 8.19 Acrosomal reaction

A protein called fertilin exposed on the surface of the sperm (Fig. 8.20) is thought to play a role in the fusion. Fertilin is a transmembrane glycoprotein with two subunits—α and β—non-covalently associated with each other. The N-terminal ends form the extracellular domain and the C-terminal end forms the intracellular domain. The extracellular domain has integrin-binding domain and the fusogenic region. Through the integrin-binding domain, the sperm binds to the integrin present on the surface of the egg. The four-pass transmembrane protein associated with the integrin and the fusogenic region of the fertilin helps in the efficient fusion of the sperm and the egg. Mutant mice deficient in the four-pass membrane protein and fertilin show infertility, proving the role of these proteins in fertilization.

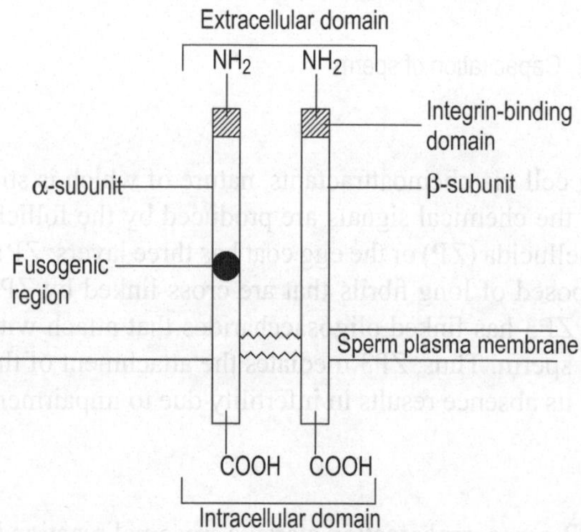

FIGURE 8.20 Structure of fertilin

Egg cortical reaction

On fusion, the sperm injects a protein into the egg cell (the nature of the protein is still unknown).

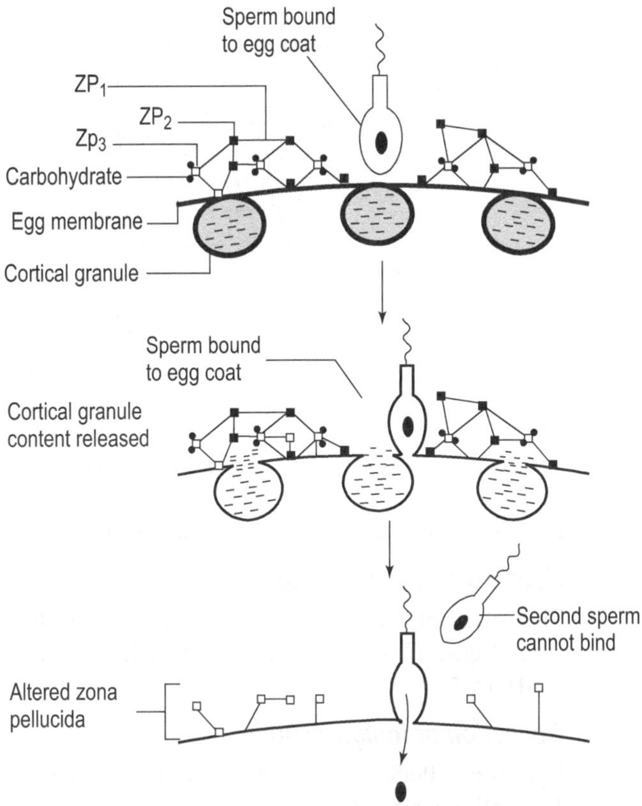

FIGURE 8.21 Prevention of polyspermy by egg cortical reaction

This protein induces an increase in the cytosolic calcium ion concentration, which spreads along the cell like a wave. Elevated calcium ions lead to two events:

1. Activation and further development of egg
2. Induction of egg cortical reaction

Egg cortical reaction is the degranulation of the cortical granules by exocytosis in order to prevent polyspermy (Fig. 8.21). The enzymes of the cortical granules lead to proteolytic cleavage of ZP2 layer and hydrolysis of oligosaccharide chains in ZP3. The proteins of the cortical granules alter the egg coat making it harder, preventing further entry and binding of other sperms to the egg, thus, preventing polyspermy. Therefore, egg cortical reaction ensures that the egg fuses with only one sperm.

Fusion of pronuclei and formation of zygote

Pronucleus is the haploid nucleus of the germ cell. On reaching the egg's cytoplasm, the sperm releases its centriole and the nucleus. The following events then occur, which altogether promote the fusion of pronuclei.

- Centrosome is formed around the centriole
- The nuclear membrane interdigitates.
- The centrosomes replicate and mitotic spindle is formed from the centriole (refer Chapter 1)
- The haploid pronuclei fuse, and the chromosomes are aligned along the mitotic spindle.
- The diploid zygote is formed.

Formation of diploid zygote from haploid pronuclei is illustrated in Fig. 8.22.

8.5.2 Artificial Fertilization

Artificial fertilization (AF) is the process of inducing the fusion of egg and sperm in vitro. This technique is also called in vitro fertilization. The fusion results in zygote formation, which is then introduced into the female reproductive system. AF is used to compensate the defect in the female reproductive system or the defective sperm quality in the male, both of which prevent fertilization. Thus, AF is the way of treating infertility in males and females. The woman should be less than 35 years old, in order to undergo a successful AF. Before discussing about the protocol for AF, it is necessary to know the structure of female reproductive system, which is shown in Fig. 8.23.

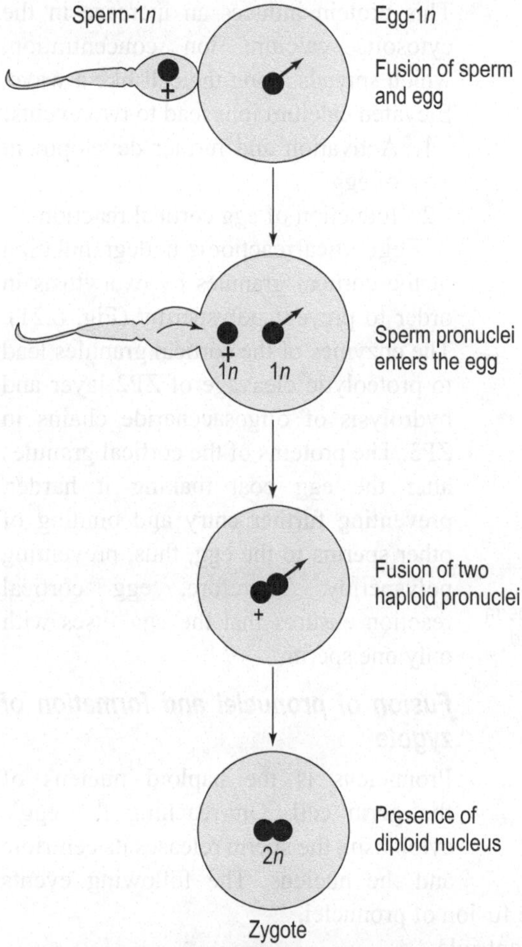

FIGURE 8.22 Formation of diploid zygote from haploid pronuclei

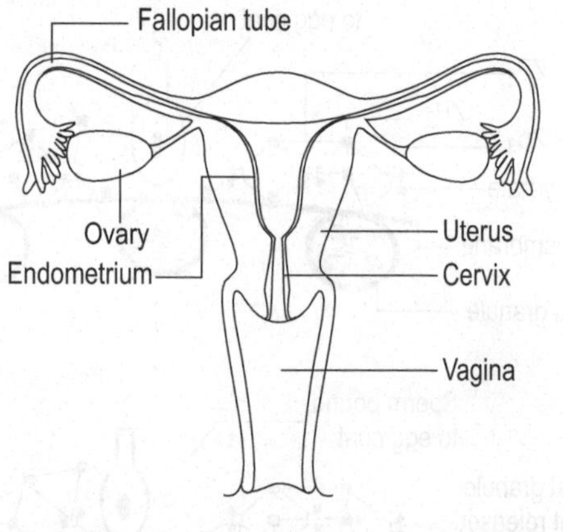

FIGURE 8.23 Reproductive organs of a female

Following are the steps in the AF protocol:

Hyperstimulation of ovary

The ovary of the woman is stimulated to produce lot of eggs. This can be done using gonadotropins (FSH analogues).

Collection of follicular fluid

Eggs are collected by aspirating the follicular fluid. The follicular fluid is collected from the ovary by transvaginal egg retrieval technique. An ultrasound guided needle is inserted via vagina to reach the ovary, and aspirate the follicular fluid. This procedure is done under anaesthesia or sedation. The fluid is tested for the presence of eggs.

Collection of sperm

The semen is collected from the male and the inactive cells and the seminal fluid are removed by a technique called sperm washing. If the sperm is collected from a sperm bank, it should be thawed before use.

Fusion of sperm and egg

The sperm is either incubated with the egg under stringent laboratory conditions in a culture media for about 18–20 hours. Alternatively, the sperm may be introduced into the cytoplasm of the egg by intracytoplasmic injection (Fig. 8.24). In this protocol, a micropipette of 60 micron width is used to hold the egg in position and a still smaller micropipette of just 6–7 micron width is used to aspirate and inject the sperm directly into the cytoplasm of the egg cell. After 18 hours, the culture dish may be tested for diploid zygote formation, which is the conformation for fertilization. The zygote is observed for 6- to 8-celled embryo stage, for the subsequent transfer into the uterus of the woman. Within the uterus, the embryo develops as usual, and germ cell cycle will continue (refer Section 8.6).

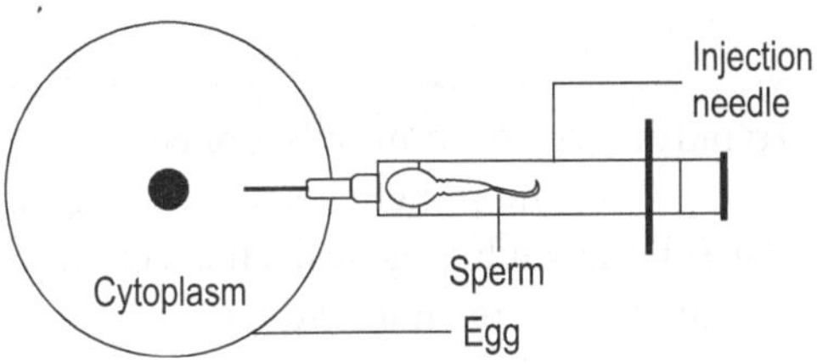

FIGURE 8.24 Intracytoplasmic injection for artificial fertilization

Intravaginal transfer of embryo

The embryo is transferred into the uterus of the woman using a plastic catheter. The catheter is inserted into the cervix and then into the vagina. After the introduction, the woman is advised to lie still for at least an hour. If the embryo is implanted, it will develop into a baby. Figure 8.25 summarizes the artificial fertilization protocol.

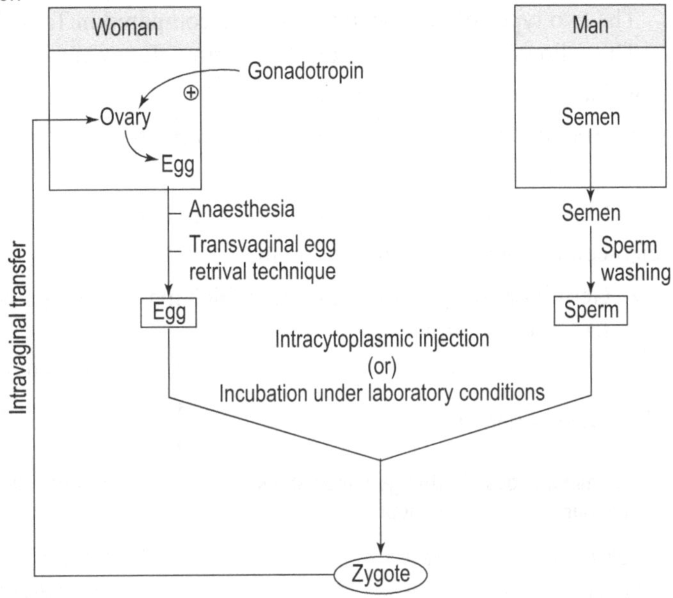

FIGURE 8.25 Steps in artificial fertilization

8.6 GERM-CELL CYCLE

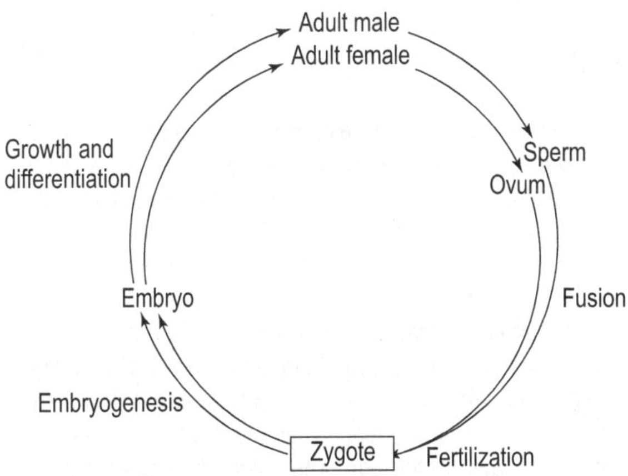

FIGURE 8.26 The germ-cell cycle

From the discussions in the previous sections, it is clear that the fusion of the male and the female germ cells produce a zygote, the zygote develops into a foetus, and then into an adult, who can produce the germ cell (the egg or the sperm) again. The cells can fuse after the fertilization to form zygote and the cycle continues. This is called germ-cell cycle. The product of the germ-cell cycle is the egg or the sperm. The germ-cell cycle of a mammal is shown in Fig. 8.26.

8.7 GERM-CELL TUMOUR

Germ-cell tumour is caused by the abnormal and rapid division of germ cells, a phenomenon termed as neoplasia. It is usually found outside the gonad, and sometimes in other organs too. It is caused due to abnormal migration of PGCs during embryogenesis. It is very rare, occurring in 1 per million people. Germ-cell tumour is classified into two types:

1. Germinomatous or seminomatous tumour
2. Non-germinomatous or non-seminomatous tumour

The two types of germ-cell tumour are compared in Table 8.3.

The following are the symptoms of germ-cell tumour:

- Painless mass in the scrotum
- Abdominal pain and constipation in women
- Asphyxia and wheezing in later stage
- Mass surrounding the anus in new-born babies

Examples for germ-cell tumour include the following:

- Germinoma (a germ-cell tumour which may be benign or malignant and have undifferentiated morphology)
- Dysgerminoma (tumour in ovary)
- Seminoma (tumour in testes)
- Embryonal carcinoma (uncommon germ-cell tumour occurring in sperm and ovary)
- Choriocarcinoma (rapidly growing tumour of uterus)
- Teratoma (germ-cell tumour composed of multiple cell types derived from one or more germinal layers)
- Polyembryoma (aggressive form of tumour, usually found in ovaries).

Table 8.3 Types of germ-cell tumour

Parameter	Germinomatous tumour	Non-germinomatous tumour
Growth rate	Slow	Fast
Survival rate	High	Low
Sensitivity to radiation	High	Low
Response to chemotherapy	More	Less

SUMMARY

Primordial germ cells (PGCs) are the precursors for germ cells. Primordial germ cells are formed in the epiblast. Epiblast is formed during embryogenesis, between the blastulation and the gastrulation of the development. Epiblast gives rise to the embryonic ectoderm, mesoderm, and the endoderm. The gonadal ridge develops from the mesoderm. The PGCs migrate to the gonadal ridge for further differentiation into germ cells.

The germ cells are haploid with unpaired or single set chromosome represented as 1*n*. The male and the female germ cells fuse, and fertilize producing a diploid cell called zygote represented as 2*n*. The zygote develops into an organism. During the development, the organism produces haploid germ cells, which can be used to produce diploid organism, and the cycle continues. This is called the sexual reproductive cycle. The germ cells are produced from the primordial germ cells via one meiotic process and two mitotic processes.

During fertilization, the sperm is capacitated by the bicarbonate ions present in the female reproductive system. The capacitated sperm is active with high mobility, and it undergoes acrosomal reaction and tunnel through the zona pellucida layer of the egg. The sperm then fuses with the plasma membrane of the egg, which is thought to be facilitated by a transmembrane protein called fertilin. The sperm then enters the cytoplasm of the egg cell and releases its centriole and the nucleus. Consequently, the chromosomes of the sperm and the egg pairs, and aligns in the mitotic spindle to form a diploid zygote. In case of infertility, the sperm and the ovum can be collected from the donors and fertilized in vitro and then introduced into the uterus. This process is called in vitro fertilization or artificial fertilization.

TERMS TO REMEMBER

Acrosomal reaction The release of the acrosomal contents (enzymes and proteins) from the sperm via exocytosis, in order to tunnel through the egg coat.

Artificial fertilization The process of inducing the fusion of egg and sperm in vitro, for subsequent development in vivo.

Capacitation Capacitation is the bicarbonate ion-mediated activation of sperm cells in the cervical and vaginal tract in order to initiate fertilization.

Differentiation The process by which a less-specialized cell becomes a more-specialized cell type.

Diploid cell Cell with paired chromosomes, for example, somatic cells (i.e., all body cells except the germ cells).

Egg cortical reaction Egg cortical reaction is a calcium-ion-mediated event in the egg cell, where the granules of the egg cortex release their contents.

Epiblast A structure formed between the blastulation and the gastrulation stage of the embryogenesis. It develops into the embryo proper.

Fertilin Transmembrane protein of sperm involved in the fusion of sperm with the egg.

Fertilization The process of fusion of the egg and the sperm nuclei to form a zygote, which develops into a new organism.

Follicle cells A type of somatic cell that signals the primordial germ cells to differentiate into egg cells.

Genital ridge Undeveloped gonad formed of mesenchymal cells.

Germ-cell lineage All the cells involved in the formation of gametes or the reproductive cells.

Germ-cell tumour Abnormal and rapid division of germ cells.

Haploid cell Cells with single or unpaired chromosomes, for example, germ cells.

Oogonium Precursor cell which develops into egg in the ovary.

Ovum Ovum is the non-motile matured egg or the female gamete.

Primordial germ cells The precursor cells for the sperm and the egg. It is formed in the epiblast.

Sertoli cells A type of somatic cell in the seminiferous tubule that signals the primordial germ cells to differentiate into sperm cells.

Sperm The motile male germ cell that penetrates into the ovum for fertilization.

Spermatogonium Precursor cell which develops into sperm in the testes.

The sexual reproductive cycle The formation of haploid cells (one with a single set of chromosome, denoted as 1n) and diploid cells (one with double set of chromosome, denoted as 2n) alternatively.

Vitelline layer A protective layer, surrounding the egg cell of non-mammals.

Zona pellucida A protective layer of protein and oligosaccharides, surrounding the egg cell of mammals.

REVIEW QUESTIONS

1. What is the role of epiblast in germ-cell lineage?
2. What is a gonadal ridge?
3. How are germ cells specified?
4. Diagrammatically explain the sexual reproductive cycle.
5. Explain the process of spermatogenesis with a diagram.

6. How is a PGC converted into an egg cell in the ovary?
7. Write about the capacitation of sperm.
8. Write about the structure and function of fertilin.
9. What is the role of fusogenic region in fertilization?
10. Write about the fusion of pronuclei.
11. What do you know about artificial fertilization?
12. Describe germ-cell tumour.

MULTIPLE-CHOICE QUESTIONS

1. Germ-cell specification initiates in
   (a) ICM
   (b) epiblast
   (c) hypoblast
   (d) genital ridge
2. Which of the following is the precursor for germ cells?
   (a) primordial germ cell
   (b) blastula
   (c) gastrula
   (d) zygote
3. Gonads are derived from
   (a) ectoderm
   (b) endoderm
   (c) mesoderm
   (d) ICM
5. Anti-mellerian hormone
   (a) suppresses the male reproductive system
   (b) suppresses the female reproductive system
   (c) stimulates the epidydymis
   (d) stimulates fertilization
6. Cytokines that influence multiple cell types are called
   (a) polycytokines
   (b) pleiocytokines
   (c) chemokines
   (d) interleukin
7. Epidydymis stores
   (a) ovum
   (b) sperm
   (c) zygote
   (d) acrosomes
8. Protamines are
   (a) positively charged proteins
   (b) negatively charged proteins
   (c) egg granules
   (d) sperm granules
9. Capacitation takes place in
   (a) cervical tract
   (b) testes
   (c) ovary
   (d) uterus
10. Acrosomes contain
   (a) proteins
   (b) lipids
   (c) carbohydrates
   (d) enzymes
11. Follicle cells are
   (a) germ cells
   (b) somatic cells
   (c) primordial germ cells
   (d) mesodermal cells
12. Vitelline layer is present in
   (a) mammalian egg
   (b) non-mammalian egg
   (c) cervical tract
   (d) gonads
13. Zona pellucida is rich in
   (a) proteins and oligosaccharides
   (b) lipids
   (c) enzymes
   (d) granules
13. Egg cortical reaction is mediated by
   (a) calcium
   (b) fertilin
   (c) protamine
   (d) cytokine

Answers to Multiple-choice Questions

1. (b)	2. (a)	3. (c)	4. (a)	5. (b)
6. (b)	7. (a)	8. (a)	9. (d)	10. (b)
11. (b)	12. (a)	13. (a)		

Albert Szent-Gyorgyi von Nagyrápolt discovered proteins, actin and myosin and their complex.

CHAPTER 9

Cytoskeleton

After reading this chapter, students should be able to understand and appreciate the following:

- Types and functions of cytoskeleton
- Phases involved in polymerization of cytoskeletal filaments
- Treadmilling and dynamic instability
- Regulation of cytoskeletal structure and function by accessory proteins
- Role of protein cross bridges in cytoskeleton
- Molecular motors and their function
- Drugs targeting the cytoskeleton

9.1 NATURE AND TYPES OF CYTOSKELETON

Cytoskeleton refers to a network of structurally diverse protein filaments of nano size present in the cytoplasm of cells. In other words, they are filamentous polymers. Each filament is called a protofilament, and it is an assembly of smaller and easily diffusible protein subunits. The protofilaments associate laterally with one another forming an intricate network because of weak non-covalent bonds and hydrophobic interactions. This network is responsible for shaping and moving the cell, and rearranging the cellular components during the process of cell division and growth. Cytoskeleton extends from one end of the cell to the other covering a few micrometers. The overall function of the cytoskeleton is regulated by yet another group of proteins called cytoskeleton-associated accessory proteins (CSAAP). The parameters regulated by CSAAP include spatial distribution, dynamic behaviour, site of assembly and disassembly of the constituent protein of the filaments, filament–filament interaction, and energy harnessing to generate force. The regulatory action of CSAAP helps cells to maintain a highly organized and flexible internal structure. Basically, the filaments forming the cytoskeletal network are of three types—microtubules, microfilaments (or actin filaments), and intermediate filaments. Although all the filaments are proteins, they differ in their mechanical properties and dynamics. These three types of filaments are discussed in the following sections.

9.2 MICROTUBULES

The building block of a microtubule is tubulin. It is a heterodimer made of α tubulin and β tubulin, which interact with each other via non-covalent interactions. Each of the tubulin molecules is bound to a GTP molecule. In α tubulin, GTP is tightly bound and trapped between the α–β interface forming an integral part of the tubulin, whereas in β

tubulin, GTP is loosely bound and susceptible to hydrolysis into GDP and responsible of tubule dynamics. Figure 9.1 illustrates the structure of α tubulin and β tubulin with GTP (GTP molecule is shown as a dark cloud. GTP in a tubulin is trapped in the a–b interface forming an integral part of the tubulin.)

9.2.1 Protofilament

The tubulin heterodimer assembles one above the other to form a protofilament. Structurally, the protofilament has alternate α and β tubulins (Fig. 9.2a). Thirteen protofilaments align in a parallel fashion to form a rigid, elongated, hollow cylindrical structure called microtubule (Fig. 9.2b). The cross section of the microtubule appears as a ring of 13 tubulins with a diameter of approximately 25 nm (Fig. 9.2c). The protofilaments within the microtubule are stabilized by two types of protein–protein interactions: (i) longitudinal interaction between α and β tubulins and (ii) lateral contact between adjacent protofilaments via α–α and β–β interactions. Within the cell, one end of the microtubule is attached to a centrosome—the microtubule-organizing centre (MTOC) (Fig. 9.3).

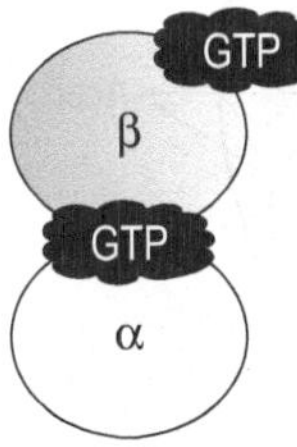

FIGURE 9.1 Structure of tubulin heterodimer with α tubulin and β tubulin

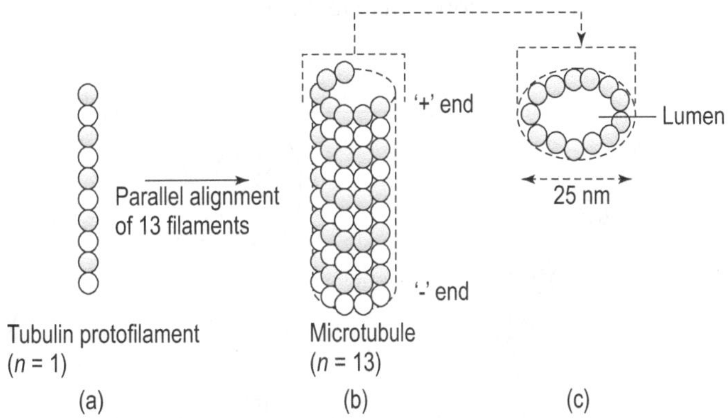

FIGURE 9.2 (a) Structure of protofilament of a microtubule (b) Microtubule formed by the parallel alignment of 13 protofilaments (c) Cross-sectional view of a microtubule showing a ring made of 13 tubulin units with 25 nm diameter

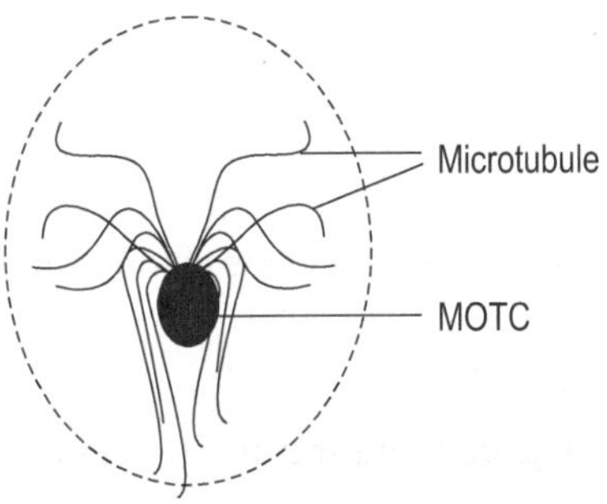

FIGURE 9.3 Organization of microtubule (wavy lines) with microtubule-organizing centre (MTOC, dark circle) in the cell

Polarity of the microtubular end

The presence of different structural molecules at the two ends leads to a difference in polarity in the microtubule. α tubulin is bared at one end and β tubulin is exposed at the other end of the microtubule leading to a different structural polarity. The end that is very dynamic, showing a faster rate of dissociation (shrinkage) and addition (growth), is called the 'plus end', whereas the other end is called the 'minus end'. In a microtubule, the β end is the plus end, whereas the α end is the minus end (Fig. 9.2b).

9.2.2 Functions of Microtubules

Microtubules are mainly responsible for the movement of cell and subcellular organelles. Special protein elements called molecular motors drive the motion at the expense of ATP. During the process of cell division, both mitosis and meiosis, microtubules align into spindle fibres, which is a track for the migration of chromosome. Kinesin and dynein are molecular motors that drive the chromosomal migration. In locomotive organelles, such as cilia and flagella, microtubules provide driving force in the presence of molecular motors and ATP to elicit 'beating' and 'wavy' motions.

9.3 ACTIN FILAMENT OR MICROFILAMENT

Actin is a globular monomer with a central groove or cleft for binding to ATP molecule (Fig. 9.4b). Actin molecules are arranged in a head-to-tail fashion to form a chain-like protofilament. The protofilament is usually made of two chains intertwined to form a helix of 5 to 9 nm diameter. The ATP-binding cleft forms the minus end and the other end forms the plus end (Fig. 9.4a). Within the cell, actin filaments are organized into linear bundles, linear networks, and three-dimensional gels. They are predominant in the cortical region of the cytoplasm (Fig. 9.5).

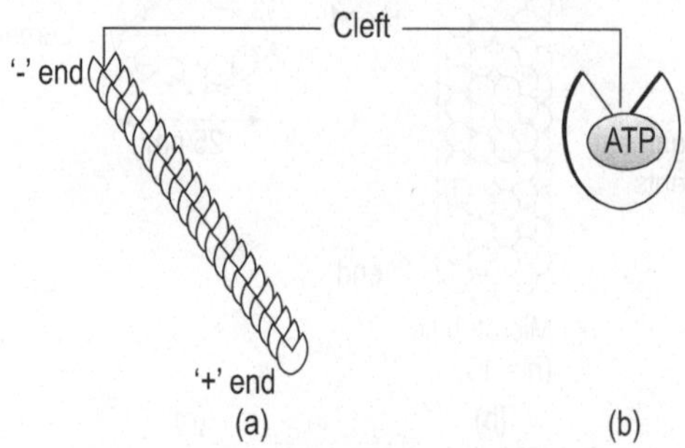

FIGURE 9.4 (a) Actin protofilament formed from polymerization of actin monomers (the end of the filament exposing the cleft forms the minus end) (b) Actin monomer with a cleft is projected (ATP molecule bound to the cleft of the monomer is shown as a shaded cloud)

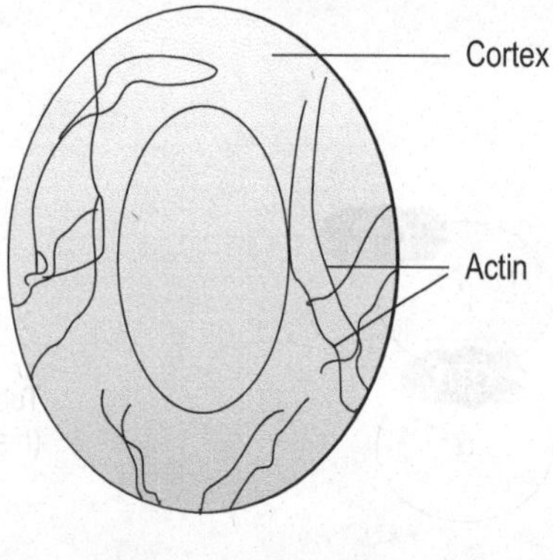

FIGURE 9.5 Distribution of actin filaments within the cell—the network is predominant in the cortical region (shaded region) of the cytoplasm

9.3.1 Functions of Actin Filaments

Actin filaments:

- serve as bridging elements between the transmembrane proteins and the cytoplasmic proteins.
- are anchors for centrosomes during the process of mitosis.
- function as partitions to segregate the two dividing cells during the process of cytokinesis.
- help in mechanical movement in leucocytes.
- promote cytoplasmic streaming and gliding.

9.4 INTERMEDIATE FILAMENTS

Intermediate filaments are present in the cells of metazoans, that is, vertebrates, nematodes, and molluscs. They are present in a few cell types and not in others. The individual unit of intermediate filament is a polypeptide, a monomer which forms lateral contact with another monomer forming a dimer. Two dimers subsequently associate to form a tetramer. Eight tetramers are associated to form a protofilament. Several protofilaments associate in a head-to-tail fashion forming a rope-like structure. Formation of intermediate filament from a single polypeptide is schematized in Fig. 9.6.

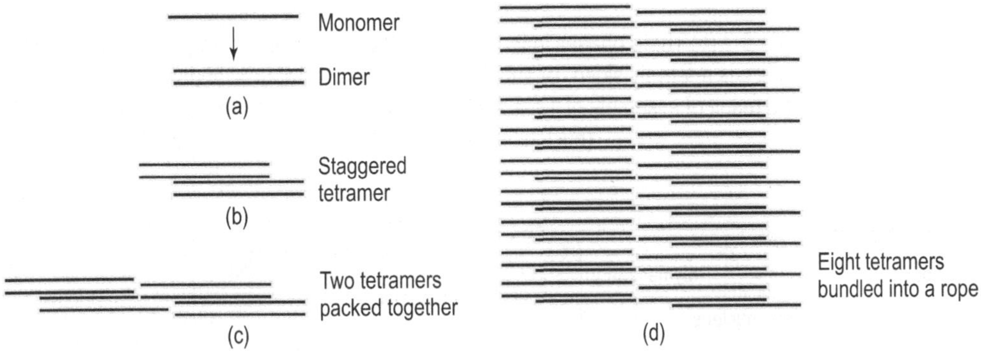

FIGURE 9.6 Formation of intermediate filament from a single polypeptide helix

Intermediate filaments are found to be easily deformed and withstand excess stress and strain without rupture. Thus, they are responsible for maintaining the cellular integrity. Intermediates filaments are distributed in different types of cells. They are shown in Table 9.1. Intermediate filaments act as a supporting framework and provide tensile strength to the cells.

9.5 NUCLEATION AND POLYMERIZATION

Formation of protofilaments for subsequent building of cytoskeletal network occurs by the process of polymerization, that is, by assembly of protein monomers into filaments. In vitro studies indicate that the polymerization takes place in three different stages or phases: (i) Lag phase or stationary phase, (ii) growth phase or elongation phase, and (iii) steady-state or equilibrium phase as shown in Fig. 9.7. This process is called triphasic polymerization.

TABLE 9.1 Types of intermediate filaments

Name of the filament	Cells expressing the filament
Lamin	Nucleus
Vimentin	Mesenchymal cells
Desmin	Myocytes
Glial fibrillary acidic protein	Glial cells
Peripherin	Neurons
Keratin	Epethilial cells, hair, and nails
Neurofilaments	Neurons

Lag phase represents the time course for a process called *nucleation*. Nucleation is a process in which two or more subunits bind to form an oligomer, which remains stable without dissociation and acts as a nucleus for further polymerization. For example, for the polymerization of tubulin protofilament, at least

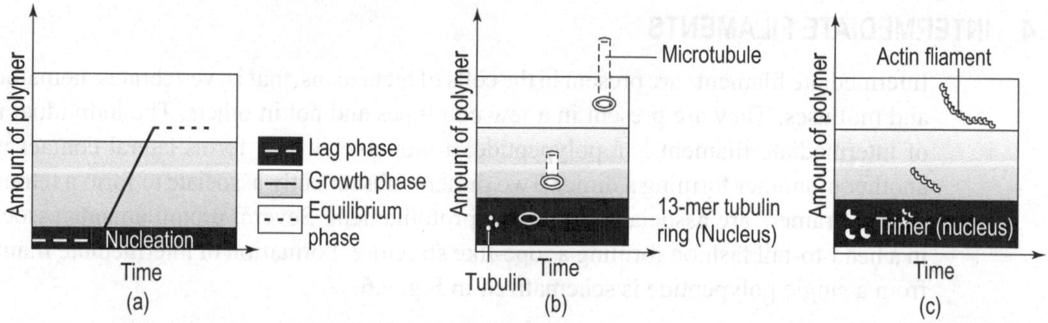

FIGURE 9.7 (a) Three phases in the polymerization of protofilament (nucleation occurs in the lag phase, which is relatively slow) (b) Tubulin profile during the course of polymerization (nucleus is formed as a result of assembly of 13 subunits as a ring) (c) Actin profile during the course of polymerization (nucleus is formed as a result of assembly of a trimer)

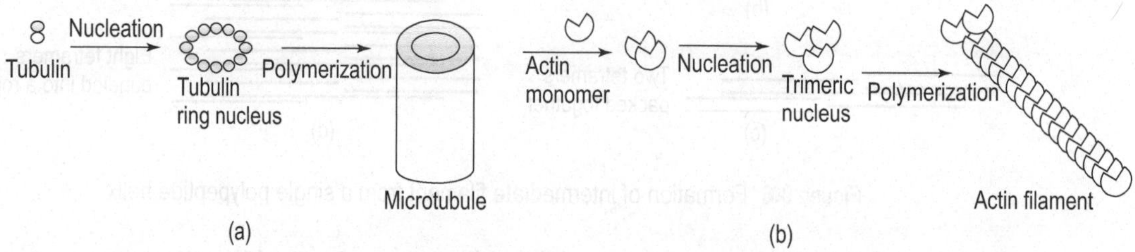

FIGURE 9.8 Nucleation of (a) tubulin and (b) actin (an oligomer of 13 tubulin subunits is the nucleus for the formation of microtubule; a trimer of actin functions as a nucleus for the formation of microfilament. In either case, the nucleus remains stable enough to continue polymerization, but the process of nucleation is relatively slow.)

13 monomers should assemble as a ring to function as a nucleus (Fig. 9.8a). In the case of actin protofilament, a trimer can be stable enough to drive polymerization (Fig. 9.8b). If the number of monomer is reduced to less than 13 or 3 for tubulin and actin, respectively, the structure will not be stable, leading to disassembly and inhibition of polymerization. Nucleation is a relatively slow process. The lag phase representing the nucleation process can be decreased or skipped if preformed nuclei exist. Cells make use of proteins to catalyse the nucleation at a specific site to form the cytoskeleton at that site. Thus, it is clear that nucleation is the rate-limiting step in the formation of cytoskeleton and hence the cell depends on it to regulate its own shape and movement.

Elongation phase represents the formation of a polymer from the stable nucleus. In this phase, the protein subunits add rapidly to elongate the filaments. At the end of the process, a balance is achieved between the addition of monomer and the dissociation of monomer at the filament ends. This phase is called the equilibrium phase or steady-state phase, where the filament remains constant or the concentration of the subunit remains unchanged. At the equilibrium phase, the concentration of the free subunits left in the solution is called as the critical concentration, represented as C_c. C_c is equal to the ratio of the rate constants for subunit dissociation and subunit addition, that is, $C_c = k_{\text{off}}/k_{\text{on}}$. The addition and dissociation of subunit in a protofilament and the corresponding rate constants are shown in Fig. 9.9. The rate of addition is called 'on rate' and the rate of dissociation is called 'off rate'.

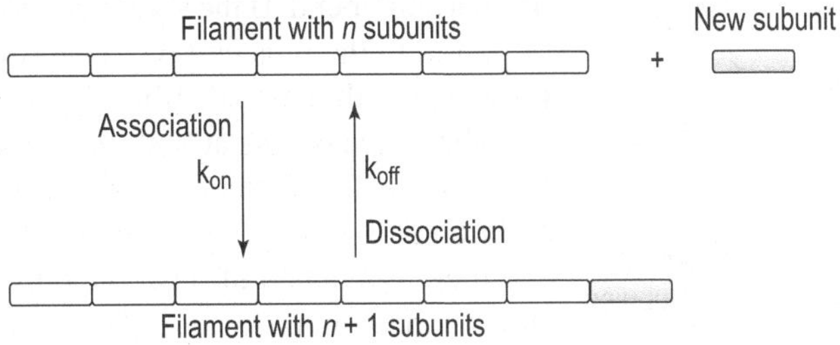

FIGURE 9.9 Addition and dissociation of subunits at the end of protofilament (k_{on} is the rate constant for addition; k_{off} is the rate constant for dissociation. The subunit that is added or dissociated is shaded for clarity.)

9.6 TREADMILLING

Treadmilling is a process by which the cytoskeletal polymer filament is maintained at constant length by the addition of subunits at one end and severing of subunits at the other end. This process is shown in Fig. 9.10. This process seems to be like a filament migrating across the stratum or the cytosol. The treadmilling process predominates in actin filaments.

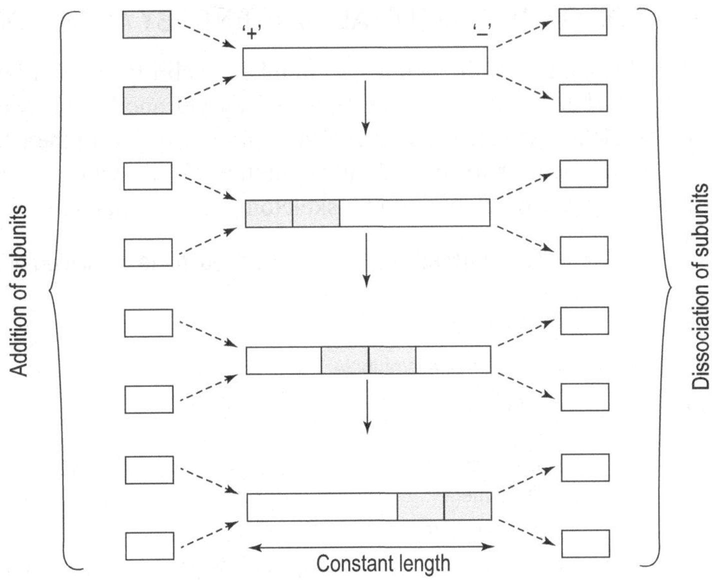

FIGURE 9.10 Process of treadmilling [Addition of subunits (dotted arrows) occurs at the plus end and dissociation of subunits (dotted arrows) occurs at the minus end. The subunits that are added initially are shaded to enable understanding the seeming motion of the subunits along the length of the filament. Bold arrow indicates that the net result of treadmillng is the constant filament length maintained due to equal rate of addition and dissociation.]

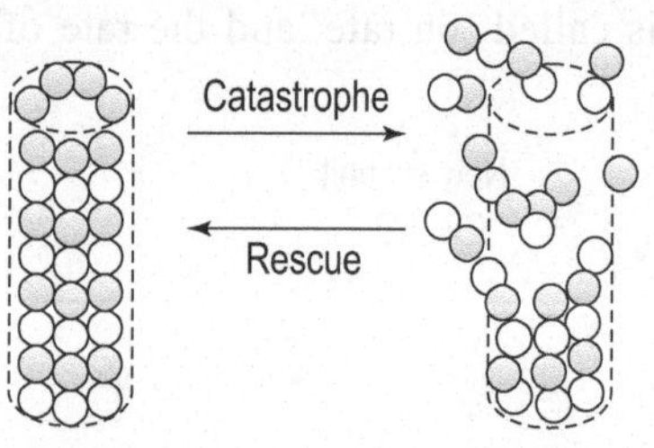

FIGURE 9.11 Process of dynamic instability predominant in microtubules

The cytosolic concentration (Cyc) of the free subunits of the filament determines the status (the shrinking status or growth status) of a particular end. If the Cyc is greater than the critical concentration (C_c) at both the plus end (C_{c+}) and the minus end (C_{c-}), it results in growth at both the ends. If the C_{yc} value is lesser than both Cc+ and C_{c-}, shrinkage occurs at both the ends. However, in the general case, C_{c+} will be lesser than C_{c-} because of the probability of subunit addition at the plus end. When the Cyc value falls between C_{c+} and C_{c-}, there is a net addition at the plus end and net disassembly at the minus end, at an equal rate. This leads to a constant filament length and, thus, treadmilling predominates.

9.7 DYNAMIC INSTABILITY

Dynamic instability is the process of abrupt transformation of a filament from growth stage to shrinkage stage or vice versa (Fig. 9.11). When the cytosolic concentration of a free monomer subunit is intermediate between C_c for the plus end and C_c for the minus end, any plus end will grow and any minus end will shrink. Sometimes, a single filament can grow for a while and then suddenly begin to shrink rapidly even when the free subunit concentration is constant. This sudden interconversion is called dynamic instability. Transformation of a filament to rapid shrinkage state is called a catastrophe, whereas the transformation to rapid growth is called a rescue. Dynamic instability predominates in microtubules.

9.8 REGULATION OF CYTOSKELETAL FILAMENTS BY ACCESSORY PROTEINS

The location, length, geometry, number, stability, nucleation, and the dynamics of the cytoskeletal filaments are regulated by yet another class of diverse proteins called cytoskeletal associated proteins. These proteins perform their function by binding either to the filament or to the individual subunit. This section reveals the role of accessory proteins in the regulation of cytoskeleton as summarized in Table 9.2.

TABLE 9.2 Function of accessory proteins involved in the regulation of cytoskeleton

Accessory protein	Function	Binding site
Gamma tubulin ring complex	Nucleation of microtubule	End of the filament
Actin regulatory protein complex	Nucleation of actin filament	Sides of the filament
Thymosin	Locking the actin monomer and slowing down the growth of actin filament	Sides of the filament
Profilin	Regulation of thymosin	Sides of the filament
Stathmin	Binding to tubulin and slowing down the polymerization of microtubule	Sides of the filament
Microtubule-associated proteins (MAPs)	Stabilization of microtubules Inhibition of disassembly and prevention of the dynamic instability Promote the interaction of microtubule with other cellular components	Ends of the filament
Tropomyosin	Regulates the binding of myosin head to actin filament Regulates muscle contraction	Sides of the filament
Cofilin	Severes and depolymerizes actin filament	Sides of the filament

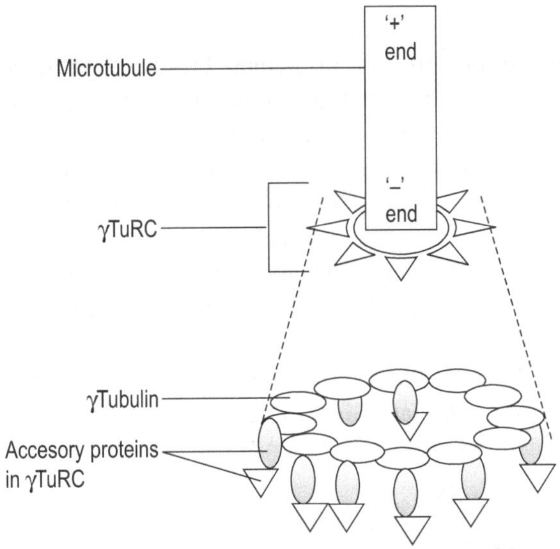

FIGURE 9.12 γTuRC located at the minus end of the microtubule (upright rectangle) is enlarged to show the presence of γ tubulin ring and other accessory proteins

9.8.1 Gamma Tubulin Ring Complex

Gamma tubulin ring complex (γTuRC) is a ring made of gamma tubulin to which several other proteins are anchored. γTuRC is concentrated at the MTOC where they nucleate tubulin heterodimers at the negative end. Subsequently, the filament grows along the plus end direction. γTuRC is found in yeast, human beings, insects, and vertebrate cells where they serve as a template to initiate the nucleation and polymerization of microtubules. The ring complex attached to the microtubule is shown in Fig. 9.12.

9.8.2 Actin Regulatory Protein Complex

The actin regulatory protein (ARP) complex is made of ARP2, ARP3, and an assembler protein (Fig. 9.13a). ARP2 and ARP3, which have 45% structural identity to actin monomer itself, are held or assembled by the assembler protein in the cortical region of the cytoplasm, almost beneath the plasma membrane. At this region, the complex initiates the nucleation of the minus end leading to the elongation of the plus end of the filament (Fig. 9.13b). The complex remains more active if it is bound to the side of the pre-existing actin filament. This results in the formation of actin filaments at 70° angle to the original filament appearing in a web-like or tree-like fashion (Fig. 9.13c).

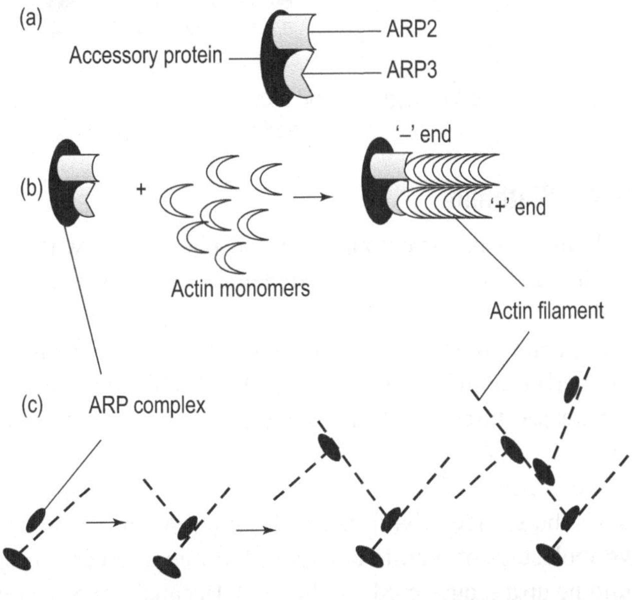

FIGURE 9.13 (a) Structure of actin regulatory protein complex (ARP2 and ARP3 are shaded) (b) Nucleation of actin at the minus end and elongation at the plus end of the filament (c) Formation of actin filament web, like a tree (This occurs when the ARP is attached to a pre-existing filament; the actin monomers in the filament are simplified as dark bands and the ARP complex are shown as dark ovals.)

9.8.3 Thymosin and Profilin

Thymosin is a small polypeptide hormone, predominant in thymus. Profilin is an actin-binding protein with specificity for proline-rich sites. Both thymosin and profilin regulate the actin filament assembly by binding to actin monomers. Inside the cell, 50% of actin is in filamentous form, whereas the remaining 50% is in free form. Free actin is locked by thymosin. Thymosin-bound actin occurs as a sequestered pool in the cytoplasm, and it cannot bind to either ends of the pre-existing actin filament. This makes polymerization less favourable; thus, mimicking the action of latrunculin. Thymosin itself is regulated by profilin, whose role is schematized in Fig. 9.14. When cells get signal to build actin filament, profilin binds to the actin monomer opposite to the ATP-binding cleft. Profilin-bound actin can readily bind to the free plus end of the actin filament. On binding to the filament, a conformational change takes place in actin such that it loses its affinity for profilin, and profilin detaches from actin, leaving the actin filament one monomer unit longer.

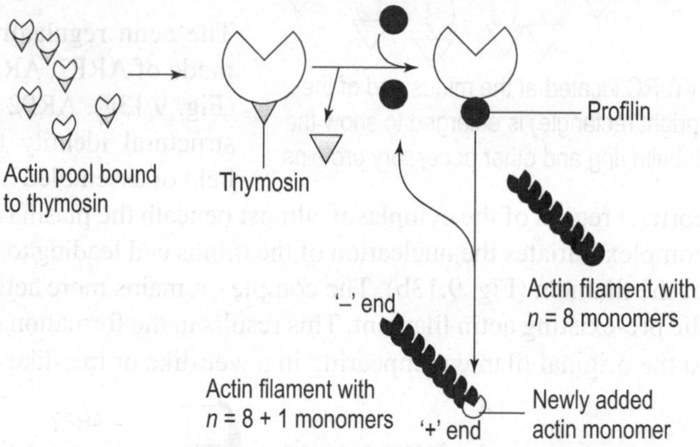

FIGURE 9.14 Role of thymosin and profilin in regulating the filamentation of actin (Free actin is not shaded and filamentous actin is shaded dark)

9.8.4 Stathmin

Stathmin is a major cytoplasmic phosphoprotein, with a molecular weight of 17 kDa, which regulates microtubule dynamics and remodelling, depending on the cellular needs.

The action of stathmin is analogous to that of colchicine. That is, stathmin makes the polymerization of microtubule less favourable and creates dynamic instability resulting in increased turnover of microtubules. A model for the function of stathmin in creating dynamic instability is shown in Fig. 9.15.

The concentration of free tubulins in cell is responsible for the polymerization of microtubule. Free tubulin maintains microtubules in a growing state. Stathmin binds to two molecules of tubulin at a time, forming a ternary complex represented as T2s, thus, building up a sequestered tubulin pool. Because of sequestration, concentration of the free tubulin shrinks. Consequently, the elongation of microtubule slows down (the process is called catastrophe). Consequently, the GTP cap at the plus end will be hydrolyzed at a rapid rate, shifting the microtubule from the growing state to shrinking state. Thus, we can say *as thymosin prevents the polymerization of actin, stathmin prevents the polymerization of microtubule.*

Stathmin is regulated by the phosphorylation. Phosphorylation of the serine residues (Ser16, Ser25, Ser38 and Ser63) detatches the protein from the tubulin, thereby increasing the free tubulin concentration leading to polymerization.

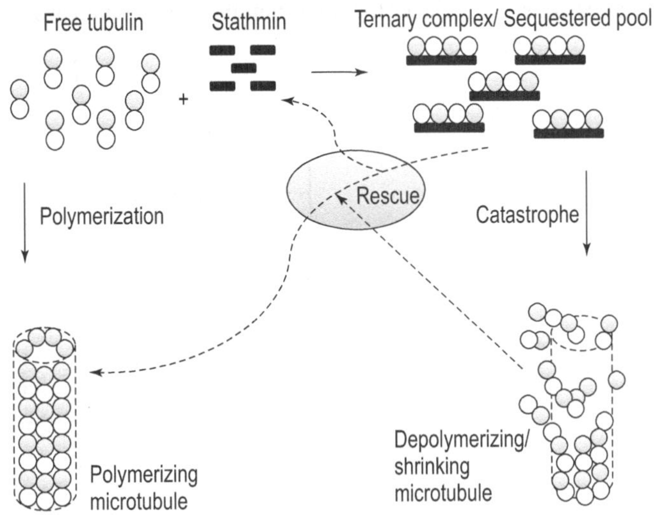

FIGURE 9.15 A model for the function of stathmin in the regulation of microtubule dynamics (Stathmin is shown as a dark rectangle. Binding of stathmin to free tubulin results in catastrophy, leading to depolymerization of the microtubule. This causes dynamic instability. The rescue process, which results in detachment of tubulin from stathmin resulting in continuation of polymerization, is shown as dotted arrows)

9.8.5 Microtubule-associated Proteins

Microtubule-associated proteins (MAPs) are a group of tissue-specific regulatory proteins. MAPs interact with the sides of the microtubules. MAPs have two domains: microtubule-binding domain (MBD), which is the C-terminal domain, and a projecting domain. MBD is composed of multiple tubule-binding motifs (Fig. 9.16a). Functionally, MAP is similar to taxol, which leads to stabilization of microtubules and inhibition of disassembly, thereby favouring the growing state and preventing the dynamic instability. MAPs also perform other functions which include the following:

- Association of different microtubules with each other (cross linking)
- Linking of microtubules with other proteins or cellular components
- Steering the microtubules towards a definite cellular location

In vitro studies show that MAP accelerates the nucleation of tubulins, which is critical for polymerization of microtubules. MAP itself is regulated by protein kinase enzymes via phosphorylation. Phosphorylated MAPs cannot accelerate nucleation and polymerization leading to dynamic instability, which is observed during the mitosis phase.

Examples for MAPs are MAP2 and tau protein, both present in vertebrates, and XMAP 215, which is a *Xenopus*-associated protein with molecular weight 215, present in all eukaryotes. MAP2 has a long projection domain capable of binding to other microtubules present nearby (Fig. 9.16b). Because of the length of the projection domain, the microtubules in a bundle linked by MAP2 are not spaced closer (Fig. 9.16c). In contrast to MAP2, tau protein has a shorter projection domain. Both the MBD and the projection domain of tau protein are attached to the same tubule filament forming a loop. Therefore, the microtubular bundles are spaced closer together than they exist when associated with MAP2.

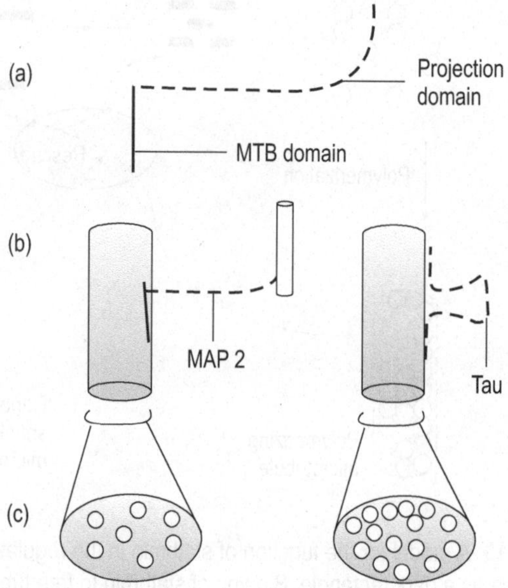

FIGURE 9.16 (a) Domains of MAPs (projection domain is shown as dotted line and MTB domain as solid line) (b) Association of MAP2 and tau protein with the microtubule showing the difference in the domain length (c) Hypothetical representation of the cross section of microtubule bundles showing the spatial distribution [tubule associated with MAP2, spaced apart (shown on left), and tau protein, spaced closer together (shown on right)]

9.8.6 Tropomyosin

Tropomyosin is a class of accessory protein that binds to the side of actin filaments, spanning approximately seven actin subunits at a time. Tropomyosin regulates the binding of myosin to actin in association with troponin, thereby regulating muscle contraction.

9.8.7 Cofilin

Cofilin is an actin-binding protein of molecular weight 14 kDa, smaller than its target protein actin. Cofilin is known to depolymerize the minus end of the actin filament and to prevent further reassembly. Both the filamentous and free forms of actin are the targets for cofilin. Cofilin binds preferentially to the ADP-containing actin at 1:1 stoichiometry, making the twist of the filament tighter, thereby inducing a mechanical stress. Mechanical stress leads to (i) decrease in the distance spanned by the filament, that is, decrease in length of the filament and (ii) dissociation of ADP–actin from the minus end of the filament, followed by severing of actin filament and depolymerization. This results in a net increase in the rate of treadmilling. Influence of cofilin on length of the actin filament

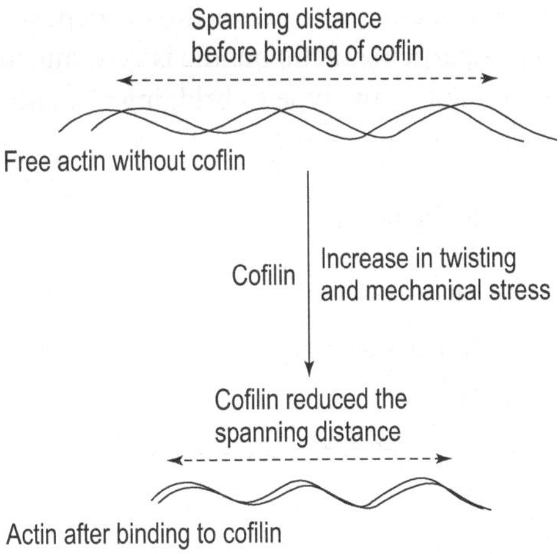

FIGURE 9.17 Influence of cofilin on actin filament (Actin is shown as a double helix. Spanning length of actin is indicated by dotted arrow. Cofilin decreases the spanning length by increasing the twist and mechanical stress leading to severing and treadmilling.)

is schematized in Fig. 9.17. Because of its function, cofilin is also called actin-depolymerizing factor. ATP-containing actin are resistant to cofilin-induced severing.

9.8.8 Plus-end Capping Protein—CapZ

Proteins that bind to the end of the filaments can regulate the cytoskeleton even at lower concentrations, for example, CapZ, a protein predominant in the Z band of the muscle cells. CapZ has an α subunit and a β subunit. Both the subunits cap the filament end, but the β subunit actually binds to the barbed plus end of the actin filament, making the end inactive and slowing down the growth and depolymerization processes that usually take place at the end. In the absence of the capping proteins, the actin filament undergoes addition (polymerization) and severing (depolymerization) at the plus and minus ends leading to rapid elongation or shrinkage depending on the amount of free monomers available. However, in the presence of a plus-end cap, such as CapZ, only the minus end will be active for growth and shrinkage. Therefore, the rate of both growth and shrinkage will be slow at all monomer concentrations. The rate of filament growth and shrinkage in the presence and absence of a plus-end cap is represented in Fig. 9.18.

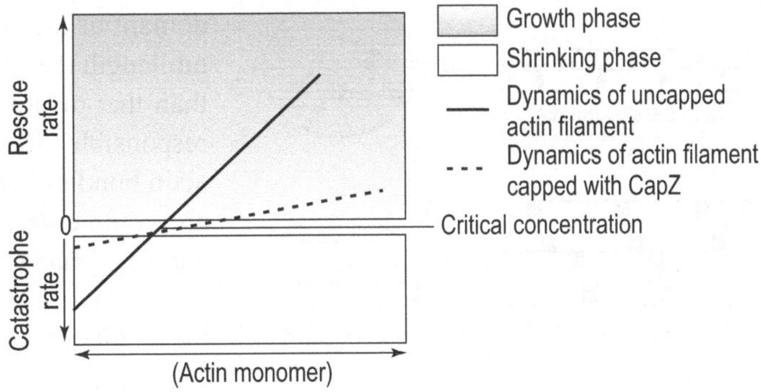

FIGURE 9.18 Effect of a cap protein on filament dynamics

9.8.9 Catastrophin

Catastrophin is a capping protein capable of binding to the ends of microtubules. It lowers the normal activation energy barrier which blocks dismantling of protofilaments. Consequently, it induces catastrophes, resulting in shrinking of the tubule.

9.9 PROTEIN CROSS BRIDGE

Several proteins cross link cytoskeletal filaments into parallel arrays or bundles and form cross bridges, which add to the tensile strength and mechanical properties of the cell. Protein cross bridges are found in actin filaments and intermediate filaments.

Actin filaments are cross linked by a range of proteins. The way of cross linking depends on the subcellular location of actin. The function of a particular actin bundle is determined by the type of cross linking, which in turn is determined by the type of bridging proteins. Proteins responsible for cross linking actin filaments are shown in Table 9.3.

TABLE 9.3 Proteins used in cross linking of actin filaments

Protein	Type of cross link	Cellular location
Fimbrin	Tight parallel bundle	Plasma membrane projection, e.g., filopodium
Villin	Tight parallel bundle	Intestinal villi
α Actinin	Loose contractile bundle	Contractile fibres, e.g., stress fibres
Filamin	Gel-like network	Cortex
Spectrin	Web-like network	Cortex, e.g., erythrocytes

Fimbrin is a monomer with a tiny actin-binding domain of 14 nm and 27 kD. It is found in intestinal epithelial cells, leucocytes, and mesenchymal cells of solid tissues. It cross links actin filaments into tight parallel bundles with plus ends aligned at one end (Fig. 9.19 a). This type of cross link is found in projections of plasma membrane, such as filopodia, which help cell to explore its surroundings. Villin is a protein similar to fimbrin in its function. Both fimbrin and villin cross link actin filaments in microvilli in the intestinal epithelial cells, which is responsible for increasing the absorptive surface area. α Actinin is an anti-parallel dimer with actin-binding domain at the N-terminal ends. It spans 30 nm length (Fig. 9.19b), comparatively larger than that of fimbrin (14 nm). This length is responsible for making loose contractile actin bundles as found in stress fibres, which are contractile and exert stress. The looser packing made by actinin freely allows myosin II molecules to accommodate into the bundle which gives the ability to contract. Filamin is a giant cytoplasmic protein with molecular weight of 250 kD. It has two actin-binding sites interconnected by a 'v' shaped domain, spanning approximately 25 nm. The structural feature of filamin enables cross linking of actin filaments, almost at right angles to each other resulting in a highly viscous gel-like or web-like network, usually an orthogonal network predominant in the cortex of the cytoplasm. The network enables the cell membrane to project into a structure called lamellipodia with which the cells can crawl. Spectrin is another bundling

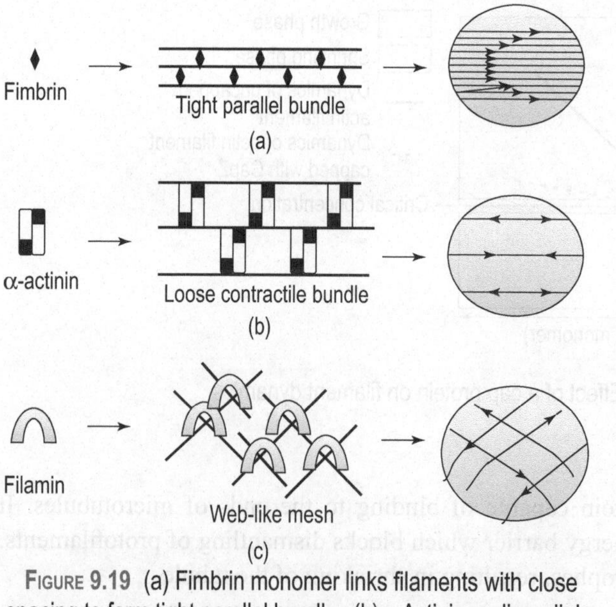

FIGURE 9.19 (a) Fimbrin monomer links filaments with close spacing to form tight parallel bundles (b) a Actinin, a dimer links filaments with wide spacing to form loose contractile bundles (c) Filamin dimer cross links filaments at almost right angles to each other forming a gel-like network (orientation of filament is shown in shaded circle)

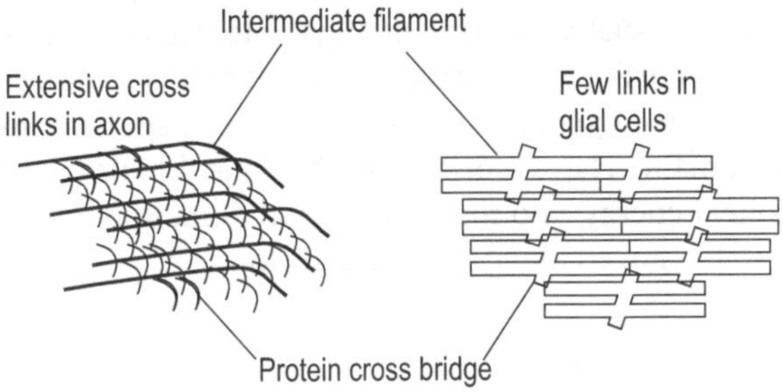

FIGURE 9.20 Intermediate filaments with different degrees of protein cross linking

protein, analogous to filamin in that it forms a gel-like actin network in red blood cells. The structure of this protein is discussed in Chapter 7. Spectrin–actin network helps to maintain the red cell integrity during circulation.

The extent of cross linking among intermediate filaments depends on the function of the cells. For example, the axon of neural cells shows high degree of cross linking among intermediate filaments resulting in high tensile strength. Whereas in glial cells, few cross linkings are found (Fig. 9.20).

Plectin is a huge protein with a molecular weight of 500 kD and is composed of a central alpha coil linked to two globular domains, the latter of which is responsible for cross linking. Plectin is peculiar since it cross links intermediate filaments to other intermediate filaments, microtubules, actin arrays, and myosin thick filaments. It also links cytoskeleton to junctions existing in the plasma membrane. Plectin is responsible for providing cells with necessary tensile strength needed to withstand mechanical stress. Another protein capable of forming cross bridge is flaggrin. Flaggrin is a predominant protein in keratohyalin granules. This protein is rich in basic amino acids, such as arginine, lysine, and histidine. Flaggrin bundles keratin filaments, the structural unit of epidermis, to give toughness to the outermost layer of the skin.

9.10 SEVERING PROTEINS

Severing proteins break longer filaments into smaller fragments. Katanin severes microtubules, and gelsolin family proteins severe actin filaments. Katanin (meaning 'sword' in Japanese) is a dimer with one large (80 kD) and one small (60 kD) subunit. The larger subunit directs the protein to the MTOC, whereas the smaller subunit severes the longitudinal bonds in the 13 protofilaments in the microtubule at the expense of ATP. The smaller subunit is also called ATPase domain. Severing results in the formation of many small filaments which have their own plus and minus ends and therefore can nucleate into many other larger filaments. Sometimes, severing also causes a rapid rise in depolymerization, leading to a change in physical and mechanical properties of the cytoplasm.

Members of gelsolin super family severe actin in the presence of high calcium ion as the signal, and without the input of energy by ATP. The protein has two domains. One domain binds to the external surface of the actin filament and remains there for encountering a thermal fluctuation which generates a small fissure between the neighbouring subunits of the protofilament. The protein insinuates its other domain into the fissure, severing the filament. Local rise in PIP2 concentration detaches the severing protein from the filament.

9.11 MOLECULAR MOTORS

Molecular motors are cellular protein molecules that harness energy from ATP to perform mechanical work. Basically two types of mechanical work are carried out: (i) movement of cellular contents, such as lipids, synaptic vesicles, proteins, and cell organelles (e.g., mitochondria) and (ii) sliding of filaments relative to one another. Molecular motors function along the cytoskeleton track. Cells make use of three types of molecular motors for moving the contents—myosin, kinesin, and dynein. All these motors consist of a head domain (or motor domain) and a tail domain. The motor domain binds and hydrolyzes ATP to convert its energy into mechanical energy into motion. The tail domain binds to the cellular component (called cargo) to be moved along the cytoskeleton.

9.11.1 Myosin

Myosin is the molecular motor for actin filament which results in muscle contraction. Myosin has three structural domains: dead domain, neck domain, and tail domain. The head domain has a binding site for ATP and is known as the moor domain. It is responsible for binding to the actin filament and causes hydrolysis of ATP through its ATPase activity. This results in the conversion of the energy derived from ATP to mechanical force leading to the movement of myosin head over the actin filament.

Myosin in muscle contraction

A signal is passed from the motor nerve to the muscle cell in the form of action potential. This results in the depolarization of the T tubule in the muscle cell membrane, which in turn opens the calcium channel present in the membrane of the sarcoplasmic reticulum. Consequently, calcium ions effluxes from the sarcoplasmic reticulum into the cytosol. Calcium mediates the movement of tropomyosin present on the surface of the actin filament, which otherwise blocks the myosin-binding site presents in the actin filament. Myosin head slides past the actin filament, both the components not undergoing any change in length (Fig. 9.21). This sliding motion is the muscle contraction, which occurs in the presence of ATP. Muscle contraction involves the following steps (Fig. 9.22):

- One molecule of ATP binds to the cleft at the back side of the myosin head, that is, behind the actin-binding site.
- ATP binding induces a conformational change in the actin-binding site, thereby reducing the affinity between actin and the myosin head.
- Using the energy derived by ATP hydrolysis, the head moves along the filament by a distance of approximately 5 nm, thus, taking a new position in the filament. The products of hydrolysis, ADP and Pi, still remain with the head.
- Weak binding of head to the new position results in the release of Pi, concomitantly leading to a tight binding.
- Release of Pi triggers a 'power stroke' which help myosin to regain its original conformation with the simultaneous release of ADP.
- Then the next sliding cycle starts with binding of another ATP molecule as before and the cycle continues.

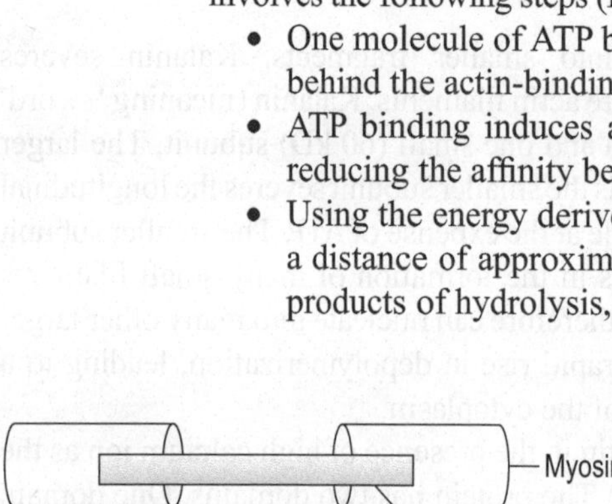

FIGURE 9.21 Sliding motion of myosin over actin filament—The muscle contraction

9.11.2 Kinesin

Kinesin is made of two motor domains with bound ATP and two cargo-binding domains. These two domains are linked by a long stalk (Fig. 9.23). The entire structure falls between 70 to 80 nm in length. Kinesin is responsible for the movement of cellular material or cargo along the microtubules, away from the nucleus, that is, from the centre of the cell to the periphery. This type of movement is called anterograde transport. Kinesin moves towards the plus end of the microtubule (Fig. 9.24).

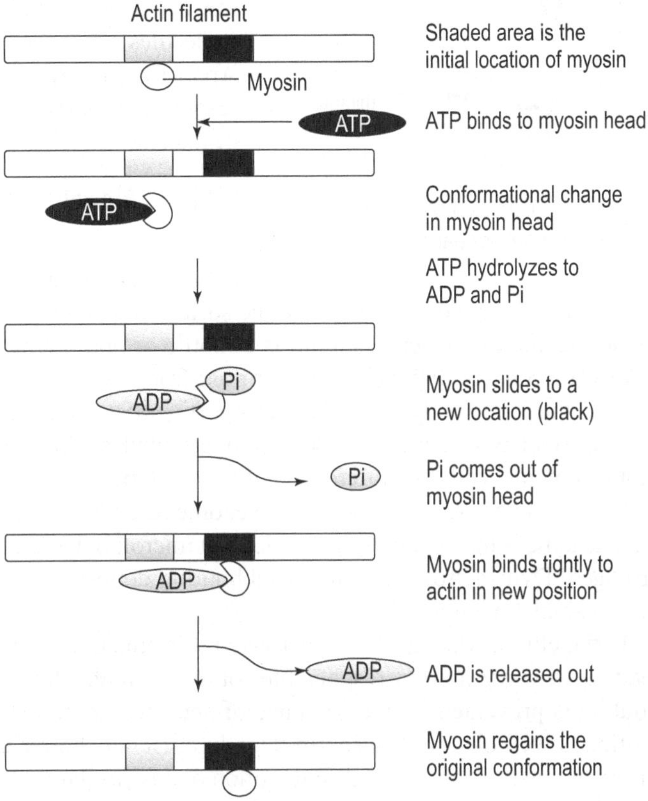

FIGURE 9.22 Steps involved in sliding movement of myosin over actin at the expense of ATP to cause muscle contraction

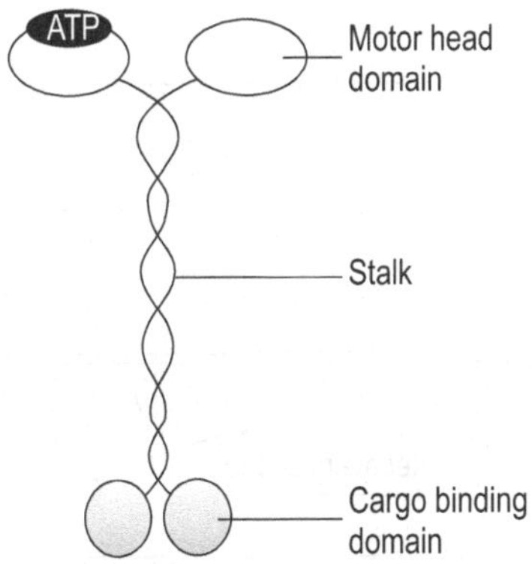

FIGURE 9.23 Structure of kinesin

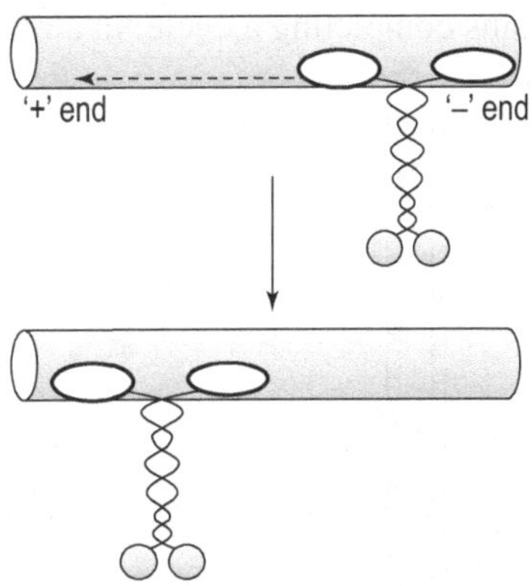

FIGURE 9.24 Association of kinesin with microtubule (dotted arrow indicates movement of kinesin towards the plus end of the tubule)

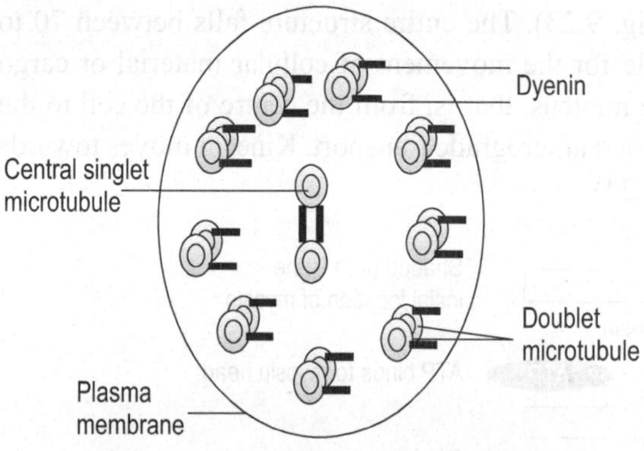

FIGURE 9.25 Cross section of an axenome in flagellum and cilium (cartwheel appearance)

9.11.3 Dynein

Dynein functions as a molecular motor and moves towards the minus end of the microtubules, which is usually directed towards the centre (nucleus) of the cell. Thus, dynein transports the cellular content towards the nucleus. Because of its function, dynein molecules are called minus-end-directed motors. Apart from transporting cargo in the cell, it is also responsible for the whipping and propelling motion of flagella and cilia.

9.11.4 Flagella and Cilia

Flagella and cilia are motility organelles constructed from microtubules. Flagellum is a tail-like appendage protruding from the surface of certain prokaryotic and eukaryotic cells to aid locomotion, whereas cilium is a motile or non-motile hair-like projection found in eukaryotic cells. Both flagella and cilia arise from a structure in the cell surface called 'basal body'. Thus, basal body forms the lower portion of the flagellum and cilia. Structurally, both these organelles are similar, made from microtubules which can be moved mechanically by motor proteins, with the difference only in the type of motion. The core of the structure is called axenome. Cross section of an axenome resembles a cartwheel as shown in Fig. 9.25. An axenome is made of nine sets of doublet microtubules (two fused microtubules) radially arranged. Dynein cross links the neighbouring doublets. The central portion is made of a pair of singlet microtubule.

In flagellum, when dynein is activated it results in bending of microtubule (Fig. 9.26), leading to wavy motion responsible for swimming. Sliding motion of the microtubule doublet is prevented by the presence of accessory proteins, such as nexin. This type of motion is well observed in sperm flagellum. Being shorter in comparison, the motion of the axenome in cilium is a beating motion that is similar to the breast stroke in swimming. Dynein generates a force perpendicular to the axis of the axenome which leads to a fast power stroke during which the fluid is driven over the surface of the cell, followed by a slow recovery stroke during which the cilium returns to the original position (Fig. 9.27), thus completing a cycle, in almost 0.1 to 0.2 s.

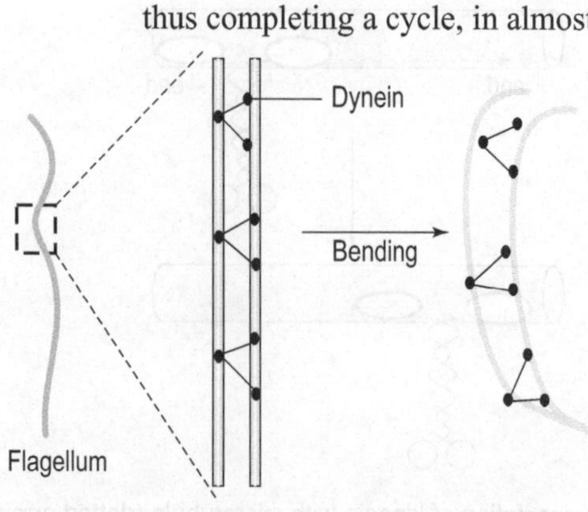

FIGURE 9.26 Role of dynein in bending of flagellum to create wavy motion

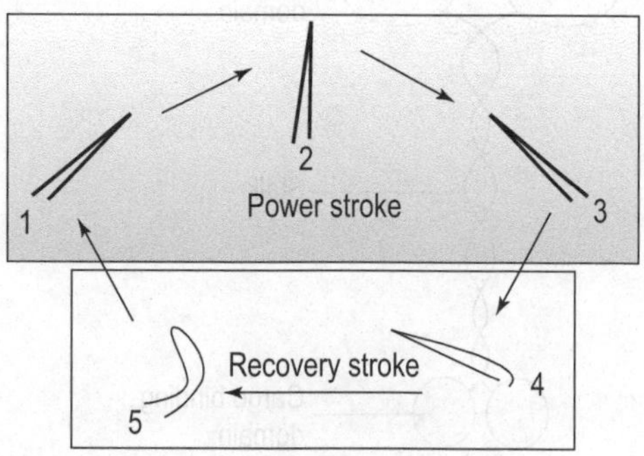

FIGURE 9.27 Bending of cilium by dynein to create beating motion

9.12 DRUGS INFLUENCING CYTOSKELETON

Cytoskeletal filaments are targets for various drugs and toxins (Table 9.4). Two groups of drugs influence the polymerization process of cytoskeletal filaments. One group of drugs prevents polymerization and the other group leads to enhancement in polymerization. A balance between the polymerization–depolymerization processes is essential for the proper organization of the cytoskeletal filaments and proper functioning of the cells. Therefore, drugs and toxins which induce assembly and disassembly of filaments by polymerization and depolymerization process, respectively, impair the structure and function of the cells. Table 9.4 represents the various drugs influencing the filament assembly and disassembly.

TABLE 9.4 Drugs influencing the cytoskeleton

Drugs	Target	Mode of action
Lantrunculin	Actin monomers	Depolymerization of actin filaments
Phalloidin	Actin filaments	Polymerization of actin filaments
Colchicine	Tubulin heterodimer	Depolymerization of microtubules
Taxol	Microtubules	Increase in tubulin polymerization

9.12.1 Latrunculin

Latrunculin (Fig. 9.28) is a natural toxin produced from the sea sponge *Latrunculia magnifica*. It is capable of rapidly binding to the ATP-binding cleft of the actin monomer and prevents further polymerization of the subunit into filaments.

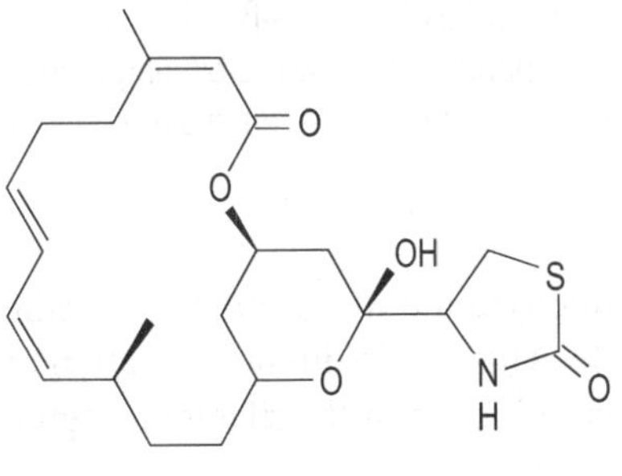

FIGURE 9.28 Structure of latrunculin

9.12.2 Phalloidin

Phalloidin is a bicyclic, heptapeptide toxin produced by the fungus *Amanita phalloides*. The cyclic structure of phalloidin is shown in Fig. 9.29. Phalloidin binds to actin filament with much high affinity than free actin monomer. The interface between the monomeric unit on the filament is the target for the toxin. On binding to the interface, the toxin decreases the rate constant for the dissociation of actin monomers from the filament. This results in the stabilization of the actin filament. Affinity of phalloidin to actin filament is used to image the actin within live and fixed cells. Phalloidin is tagged with a fluorescent tag, allowed to react with the cell. When diaminobenzindine is added, it is oxidized by eosin to form a red-coloured product which can be made electron dense and viewed in the cytoplasm by electron microscopy.

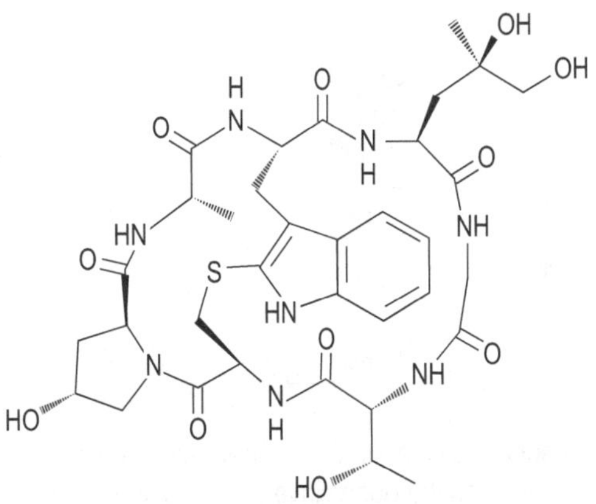

FIGURE 9.29 Structure of phalloidin

9.12.3 Colchicine

Colchicine is a secondary metabolite (Fig. 9.30) produced by the plants belonging to the genus Colchicum. It binds to the tubulin protein at the α–β interface, forming a tubulin–colchicine complex (Fig. 9.31). Consequently, the toxin copolymerizes into the microtubule lattice, suppressing microtubule dynamics, and prevents the polymerization of tubulin into microtubular filament. Also, higher concentration of colchicine causes depolymerization of microtubules. The net result is the inhibition of microtubule formation. As microtubule is essential for the spindle formation during mitosis and cell division, colchicine causes inhibition of mitosis. Hence, colchicine has the potential to function as an anti-cancer agent, but its toxicity matters.

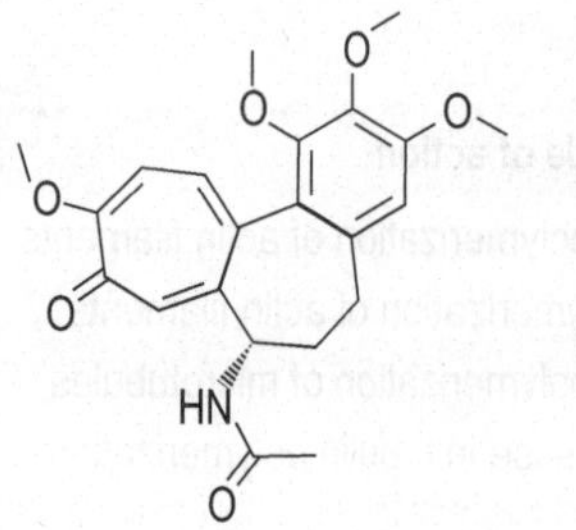

FIGURE 9.30 Structure of colchicine

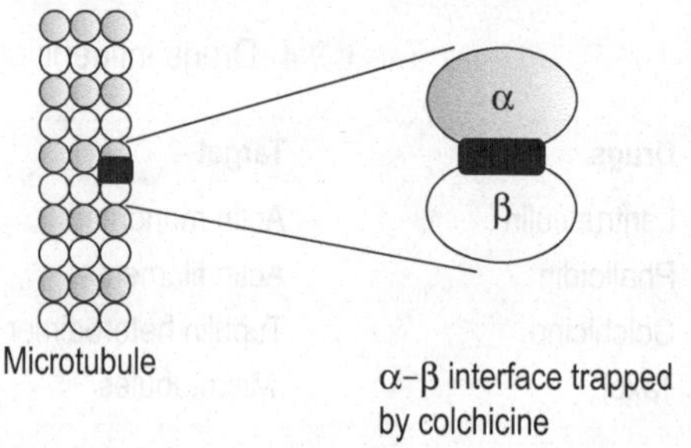

FIGURE 9.31 Interaction of colchicine with microtubule [the α–β het erodimer targeted by colchicine (black rectangle) is expanded to show the binding of the drug to the interface between α (shaded circle) and the β (open circle) subunit]

9.12.4 Vinblastine

Vinblastine (Fig. 9.32) binds to the plus end of the microtubule in the mitotic spindle and kinetochore present in the M phase of the cell cycle. This results in the formation of vinblastine–tubulin paracrystal, resulting in the depolymerization of the filament. Figure 9.33 shows the vinblastine bound to the tubulin at the plus end.

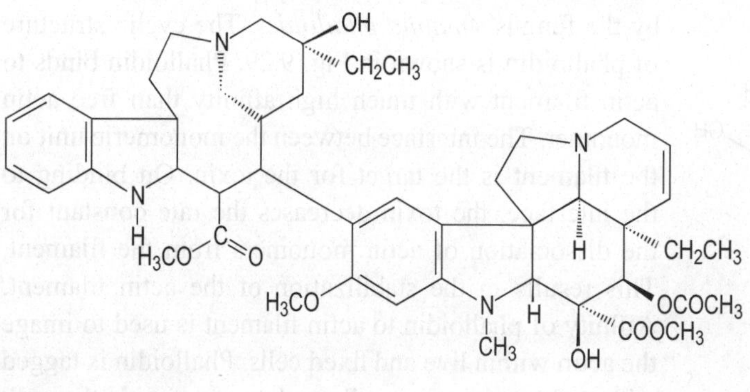

FIGURE 9.32 Structure of vinblastine

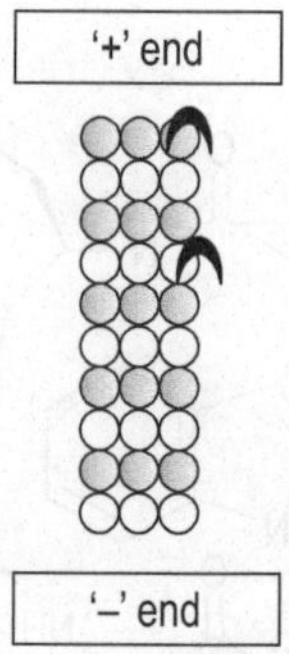

FIGURE 9.33 β subunit of the tubulin is shown as shaded circle and the α subunit as the open circle (vinblastine (shown as dark arc) bound to the microtubule at the plus end)

9.12.5 Taxol

Taxol is a plant alkaloid (Fig. 9.34) which targets and binds to the pocket in the β tubulin on the inner surface of the microtubule filament. Drug binding counteracts the effects of GTP hydrolysis occurring on the other side of the monomer. This results in the stabilization of microtubule against depolymerization associated with an inhibition in the tubule dynamics (growth and shrinkage). As a result, the drug blocks the cell cycle by interfering with the G1 or M phase. Hence, this is used as an anti-cancer agent or anti-mitotic agent.

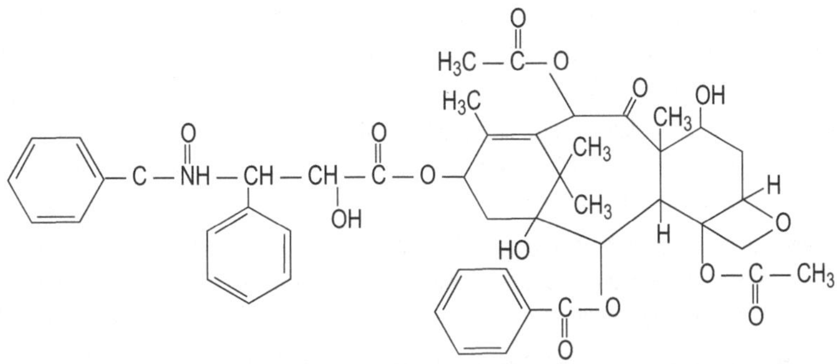

FIGURE 9.34 Structure of taxol

SUMMARY

Cytoskeleton is an intricate network of interconnected protein filaments responsible for regulating the size, movement, and mechanical function of the cells. There are three types of cytoskeleton in the cell, namely microtubules, actin filaments, and intermediate filaments. Microtubule is made by the assembly of tubulin heterodimers, and actin filaments are formed by the polymerization of actin monomers. Intermediate filaments are formed by the association of several helical polypeptides. Microtubules are predominantly distributed on the centrosome, actin filaments on the cytosol, and intermediate filaments are found across the cytoplasm or surrounding the nucleus as nuclear lamina. The filaments differ in polarity at their ends. They have plus and minus ends.

Polymerization of filaments to form cytoskeleton is a triphasic process which involves lag phase, exponential phase or elongation phase, and equilibrium phase. Nucleation is the crucial process involved in polymerization. It is a slow process that occurs at the lag phase. Polymerization and the assembly of monomers occur only after nucleation. For proper functioning, the cells have to maintain the filament length by regulating the rate of polymerization and depolymerization. Cytoskeletal filaments are targets for various drugs and toxins. Influence of drugs on filaments is used to study about abnormal cellular diseases, such as cancer, cell behaviour, and structure. These drugs either enhance or suppress polymerization or disassembly of monomomeric units; examples include latrunculin, phalloidin, vinblastine, taxol, and colchicine. A group of regulatory proteins, such as gamma tubulin ring complex, ARPC, thymosin, profilin, stathmin, MAP, tropomyosin, and cofilin regulate the cytoskeletal formation and function, by binding to the monomer or to the filament.

Capping protein and catastrophin are yet another group of proteins that binds to end of the filaments and regulate them. Within the cell, actin and intermediate filaments are held together by protein bridges which form different types of bundles that determine the function of the filament. Fimbrin and villin cross link actin to form tight parallel bundles that form projections of membrane. Alpha actinin bundles actin into loose contractile bundles responsible for contractile function. Filamin connects actin filaments at almost right angles to each other, forming web-like or gel-like network. Plectrin is involved in linking intermediate filaments. Severing proteins, such as katanin and gelsolin, are involved the depolymerization of microtubules and actin, respectively. Cytoskeletal filaments function along with proteins called molecular motors to aid mechanical function and transport of cargo within the cell. For example, myosin is a molecular motor which functions along with actin to aid muscle contraction. Dynein is yet another molecular motor responsible for the motion of cilia and flagella.

TERMS TO REMEMBER

α Actinin An antiparallel dimer that cross links actin to form loose contractile bundles as found in stress fibres which are contractile and exert stress.

Actin regulatory protein complex A protein complex that initiates the nucleation of the minus end leading to the elongation of the plus end of the filament.

CapZ A protein that binds to the barbed plus ends of the actin filament, making the end inactive, and slowing down the growth and depolymerization processes.

Catastrophin A capping protein which induces catastrophes, resulting in shrinking of the tubule.

Cofilin An actin-binding protein which depolymerizes the minus end of the actin filament and prevents further reassembly.

Colchicine A secondary metabolite produced by the plants belonging to the genus Colchicum. It binds to the tubulin protein at the α–β interface, suppressing microtubule dynamic, and preventing the polymerization of tubulin.

Cytoskeleton A network of structurally diverse protein filaments of nano size.

Dynein A molecular motor responsible for moving flagellum and cilium.

Dynamic instability The process of abrupt transformation of a filament from growth stage to shrinkage stage or vice versa

Filamin A giant cytoplasmic protein that promotes cross linking of actin filaments, almost at right angles to each other resulting in a highly viscous gel-like or web-like network, usually orthogonal network predominant in the cortex of the cytoplasm.

Fimbrin A monomer which cross links actin filaments into tight parallel bundles with plus ends aligned one end, as in filopodia.

Gamma tubulin ring complex A ring made of gamma tubulin and several other accessory proteins. It is concentrated at the MTOC where they nucleate tubulin heterodimers at the negative end.

Gelsolin A family of severing protein used to depolymerize actin filaments.

Katanin A protein used to severe microtubules.

Latrunculin A natural toxin produced from the sea sponge Latrunculia magnifica. It is capable of rapidly binding to the ATP-binding cleft of the actin monomer, and prevents the further polymerization of the subunit into filaments.

Microtubule A type of cytoskeleton made up of 13 protofilaments constructed from tubulin heterodimers.

Microtubule-associated proteins (MAPs) A group of regulatory proteins which interact with the sides of the microtubules. They stabilize microtubules, inhibit disassembly and favours the growing state.

Molecular motors Cellular protein molecules that harness energy from ATP to perform mechanical work.

Myosin The molecular motor for actin filament which results in muscle contraction.

Nucleation A slow process that occurs in the initial stage (lag phase) of polymerization. It results in the formation of a stable oligomer that can elongate further.

Phalloidin A bicyclic, heptapeptide toxin which binds to and stabilizes actin filaments.

Plectin A huge protein that cross links intermediate filaments to other intermediate filaments, microtubules, actin arrays, and myosin thick filaments. It also links cytoskeleton to junctions existing in the plasma membrane.

Stathmin A major cytoplasmic phosphoprotein with molecular weight of 17 kDa which regulates microtubule dynamics by creating dynamic instability resulting in increased turnover of microtubules.

Taxol A plant alkaloid that stabilizes microtubule against depolymerization.

Thymosin A small polypeptide hormone, which locks free actin monomer and prevents polymerization.

Treadmilling A process by which the cytoskeletal polymer filament is maintained at constant length by the addition of subunits at one and severing of subunits at the other end.

Vinblastine A drug that binds to the plus end of the microtubule resulting in the depolymerization of the filament.

REVIEW QUESTIONS

1. Give reasons for the association of nucleotides at the end of actin filaments and microtubules.
2. Discuss the difference in polarity at the end of cytoskeletal filaments.
3. Define treadmilling and retrieve the ratio of critical concentration for plus and minus ends during treadmilling.
4. Draw the phases contributing to the filament polymerization and comment on nucleation process.
5. Tabulate the name and mode of action of the drugs influencing the activity of cytoskeleton.
6. Describe the task of regulatory proteins that bind to the end of the filaments.
7. Name the proteins that bind to the sides of filaments to influence the activity of the cytoskeleton.
8. Construct a flow chart to reveal the function of protein cross bridges.
9. Draw the mechanism of muscle contraction and highlight the role of myosin.
10. Explain the 'cartwheel' structure formed by microtubules in flagella and cilia.
11. In intestinal epithelial cells, cytoskeletal bridging by protein plays a crucial role – explain.
12. Theorize the process of treadmilling and dynamic instability.

MULTIPLE-CHOICE QUESTIONS

1. Cytoskeleton is located in
   (a) plasma membrane
   (b) cytoplasm
   (c) nucleus
   (d) mitochondrial matrix
2. Filaments of cytoskeleton are held together by
   (a) non-covalent bonds
   (b) covalent bonds
   (c) salt bridges
   (d) disulphide bridges
3. Each tubulin is bound to a molecule of
   (a) ATP
   (b) GTP
   (c) actin
   (d) myosin
4. The cross section of microtubule is made of
   (a) 13 tubulins
   (b) 12 tubulins
   (c) 11 tubulins
   (d) 10 tubulins
5. Alpha end of the microtubule is called the
   (a) plus end
   (b) minus end
   (c) hydrophobic end
   (d) hydrophilic end
6. The intermediate filament found in the nucleus is called
   (a) laminin
   (b) fimbrin
   (c) myosin
   (d) kinesin
7. Desmin is found in
   (a) myocytes
   (b) hepatocytes
   (c) neurons
   (d) epithelial cells
8. Cytoskeletal network is built by a process called
   (a) triphasic polymerization
   (b) biphasic polymerization
   (c) treadmilling
   (d) dynamic instability
9. Treadmilling is a process by which the cytoskeletal filaments are
   (a) maintained with constant length

(b) degraded continuously
(c) grow continuously
(d) severed into fragments

10. The process of transformation of filament from a growth stage to shrinkage or vice versa is called
(a) treadmilling
(b) dynamic instability
(c) nucleation
(d) polymerization

11. GTRC influences the
(a) nucleation of microtubule
(b) polymerization microtubule
(c) nucleation of microfilament
(d) polymerization of microfilament

12. Catastrophin is a
(a) capping protein for microtubule
(b) capping protein for intermediate filament
(c) severing protein for microtubule
(d) severing protein for intermediate filament

13. The actin filament found in the intestinal cells is called
(a) laminin
(b) fimbrin
(c) villin
(d) kinesin

14. The severing protein for microtubule is
(a) katanin
(b) gelsolin
(c) actin
(d) tubulin

15. The severing protein for microfilament is
(a) katanin
(b) gelsolin
(c) actin
(d) tubulin

16. Myosin is the molecular motor for
(a) actin
(b) villin
(c) GTP
(d) ATP

17. Kinesin is the molecular motor for
(a) microtubule
(b) microfilament
(c) intermediate filament
(d) ATP

Answers to Multiple-choice Questions

1. (b)	2. (a)	3. (b)	4. (a)	5. (b)
6. (a)	7. (a)	8. (a)	9. (a)	10. (b)
11. (a)	12. (a)	13. (c)	14. (a)	15. (b)
16. (a)	17. (a)			

CHAPTER 10

Cell Junction, Cell Adhesion, and Extracellular Matrix

Dr Gerald M. Edelman received Nobel Prize for Physiology/Medicine in 1972 for the discovery of cell-adhesion molecules (CAMs).

After reading this chapter, students should be able to understand and appreciate the following:

- Types and functions of cell junctions
- Structure and function of plasmodesmata
- Role of cadherin, atenin, selectin, and N-CAM in cell adhesion
- Components and functions of extracellular matrix (ECM)
- Basal laminae and their function
- Regulation of ECM turnover
- Role of integrin in cell adhesion
- Structure and function of plant cell wall

Cells have different shapes, sizes, and functions. They are aggregated into tissues and organs of specific function and special activities. They also allow for bidirectional communication and signal transfer, leading to cellular processes such as survival, proliferation, differentiation, migration, and apoptosis. All these processes occur due to cell–cell adhesion and cell–extracellular matrix (ECM) interaction.

10.1 CELL JUNCTIONS

There are specialized cell–cell and cell–matrix junctions, collectively known as 'cell junctions', which usually appear like discrete patches. They are usually formed of strands of proteins. This section discusses cell junctions and their types.

10.1.1 Types of Cell Junctions Based on Function

Cell junctions influence the function of a specific cell type. Functionally, cell junctions are classified into three categories—occluding junctions, anchoring junctions, and communicating junctions. These three types of cell junctions are basically proteins that are anchored to their ligands via anchor proteins (Table 10.1).

Occluding junctions

Occluding junctions are found in epithelial cells, such as intestine and kidney tubules. They seal the two cells tightly. Tight sealing functions as diffusion barrier or permeability barrier for specific molecules and ions, depending on the function of the cell. For example, let us consider the intestinal epithelial cell (Fig. 10.1). It has two surfaces: apical surface, facing the gut lumen, and the basolateral surface, facing the extracellular fluid. The function of the cell is to retain the gut contents for a certain period and allow some of the

TABLE 10.1 Types of cell junctions

Type	Subtype	Function protein	Structural protein	Anchor
Occluding junction	Tight junction (Vertebrate) Septate junction (Invertebrates)	Confinement of transport protein and establishment of diffusion barrier in intestinal cells	Claudin Occludin	Zonula occludens (ZO) proteins
Anchoring junction	Adherens junctions cell	Links actin with the neighbouring	E-cadherin	Catenin Actinin Vinculin
	Focal adhesions	Links actin with the extracellular matrix (ECM)	Integrin heterodimer	Talin Vinculin Actinin Filamin
	Desmosomes	Links intermediate filaments with the neighbouring cell	Desmoglein Desmocollin Plakoglobin	Desmoplakin
	Hemidesmosomes	Links intermediate filaments with the ECM	Integrin	Plectin
Communicating junction	Gap junction	Direct intercellular communication to transmit metabolites and signals	Connexin	–
	Plasmodesmata (in plants alone)	Transport molecules and virus	Desmotubule	–

molecules to enter into the cell from the gut and from the cell to the extracellular fluid. In addition, the cells should also prevent the reverse movement of molecules on either side. The region between the plasma membrane of the epithelial cells is made of occluding junctions, which are strands of proteins. These proteins are transmembrane adhesion proteins. They bind the two membranes so tightly that a diffusion barrier is established. Because of this reason, this junction is called *tight junction*. Tight junctions are made of two types of proteins—claudin and occludin. Claudin is essential for the formation and function of tight junction, whereas the function of occludin is unknown. Both claudin and occludin are anchored to the actin cytoskeleton via zonula occludens proteins (ZO proteins). In intestine, tight junctions confine the transport proteins to their corresponding surfaces or membrane domains, thereby establishing the diffusion barrier. For example, tight junction model is explained by using transcellular transport of glucose in intestine.

Tight junction model in transcellular transport Figure 10.2 illustrates the tight junction and transcellular transport of glucose. Sodium-dependent glucose symport is needed for the transport of glucose from the lumen into the cell. The symport protein is confined to the apical surface. Subsequently, the glucose has to come out of the cell to reach the extracellular fluid via facilitated diffusion mediated by a glucose-carrier protein. This

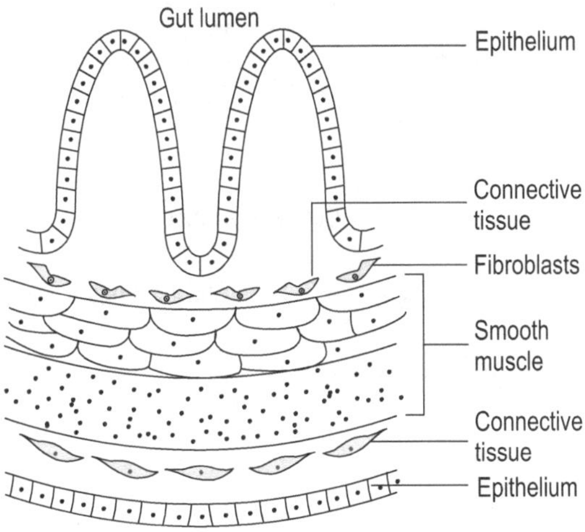

FIGURE 10.1 Intestinal epithelium

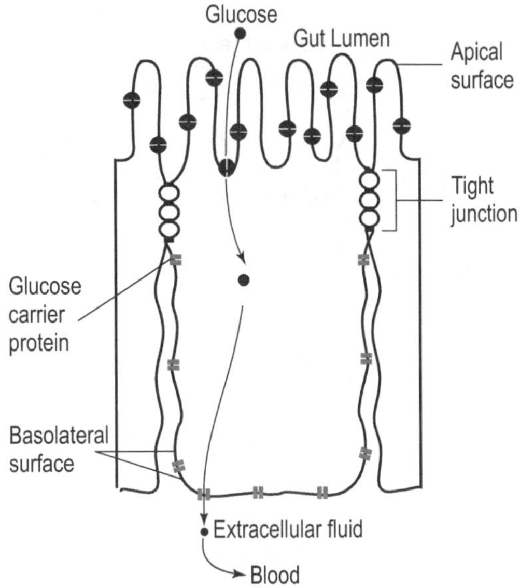

FIGURE 10.2 Tight junction and transcellular transport of glucose (Arrow indicates transport)

protein is confined to the basolateral region in the membrane. The tight junctions are between the two surfaces so that they not only confine the transport proteins, but also prevent the backflow of molecules.

In invertebrates, the occluding junctions are called septate junctions. Septate junctions are made up of parallel aligned protein strands that form band-like structure. They are formed and made functional by a protein called 'disc-large'.

Anchoring junctions

Anchoring junctions are transmembrane proteins which connect cytoskeletal filaments of a cell to the ECM or to the cytoskeleton of the neighbouring cells. In doing so, anchoring junctions transduce forces from the cytoskeleton to the ECM or to the other cell, which is not possible by the flimsy lipid bilayer of the membrane. Therefore, anchoring junction is predominant in tissues, such as muscles and heart, which are in mechanical stress. A model of anchoring junction is shown in Fig. 10.3.

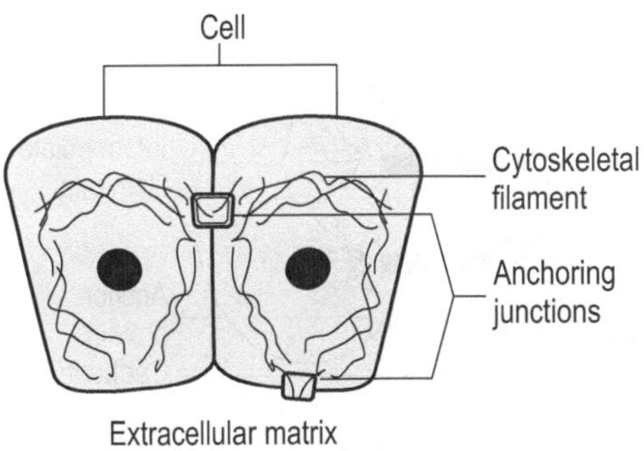

FIGURE 10.3 Anchoring junction

Proteins in anchoring junction Anchoring junctions are constructed from two types of proteins—intracellular anchor proteins and transmembrane adhesion proteins (Fig. 10.4). Anchor proteins are distributed in contact with the cytoskeleton, actin, or intermediate filaments. Transmembrane adhesion protein spans the membrane and projects on the either side of the membrane. The intracellular domain of the adhesion protein is associated with the anchor protein, whereas the extracellular domain is associated with either the extracellular domain of adhesion protein from another cell or with the ECM

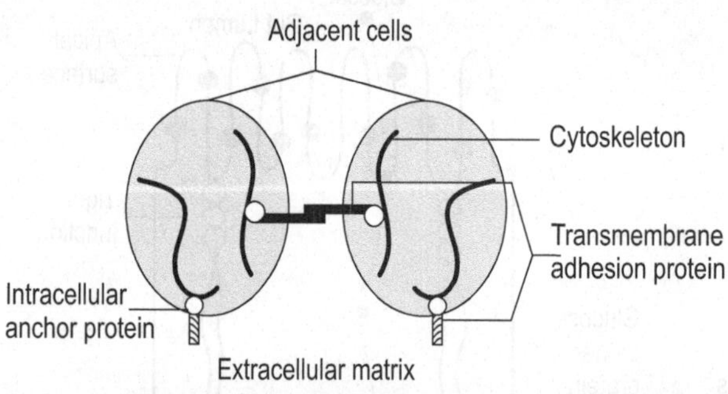

FIGURE 10.4 Structural proteins of an anchoring junction

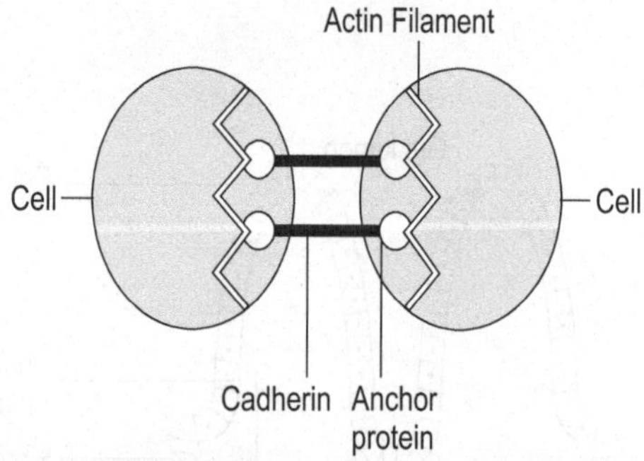

FIGURE 10.5 Adherens junctions

There are four types of anchoring junctions (as listed in Table 10.1):

- Adherens junctions
- Focal adhesion
- Desmosomes
- Hemidesmosomes

Adherens junctions Adherens junctions promote cell–cell communication by interlinking their actin filaments. The transmembrane adhesion protein in adherens junctions are from cadherin family. The cadherin proteins are linked to actin bundles via anchor proteins, such as α-catenin, β-catenin, γ-catenin, α-actinin, and vinculin. Adherens junctions lie below the tight junction (Fig. 10.5), encircling the interaction cells, forming adhesion belt called *zonula adherens*. Formation of adhesion belt is the prerequisite for the formation of tight junction. Adhesion belt promotes looping and rolling of epithelial sheets during morphogenesis to form structures, such as neural tubes.

Focal adhesions Focal adhesions connect the actin filaments of the cytoskeleton with the ECM. Focal adhesions are composed of transmembrane adhesion proteins of integrin family. The extracellular domains of the integrin subunits are attached to the protein component of ECM. The intracellular domain is attached to the actin filaments via anchor protein complex formed from α-actinin, talin or filamin, and vinculin (Fig. 10.6). Focal adhesions are found in muscle cells where they are attached to tendons.

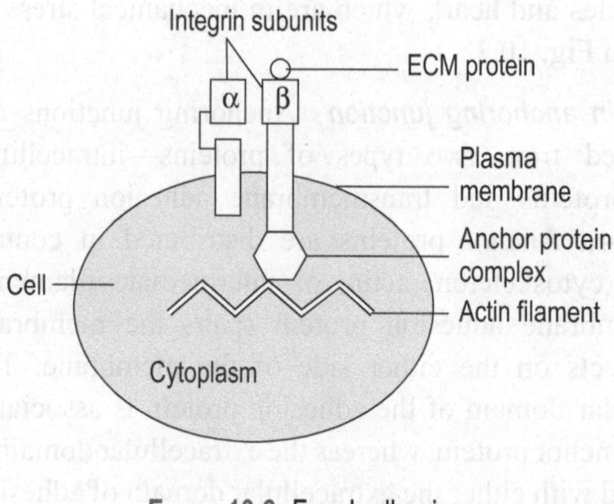

FIGURE 10.6 Focal adhesion

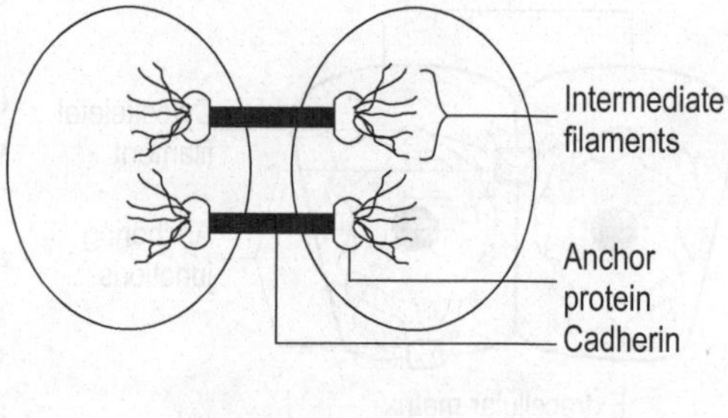

FIGURE 10.7 Desmosome

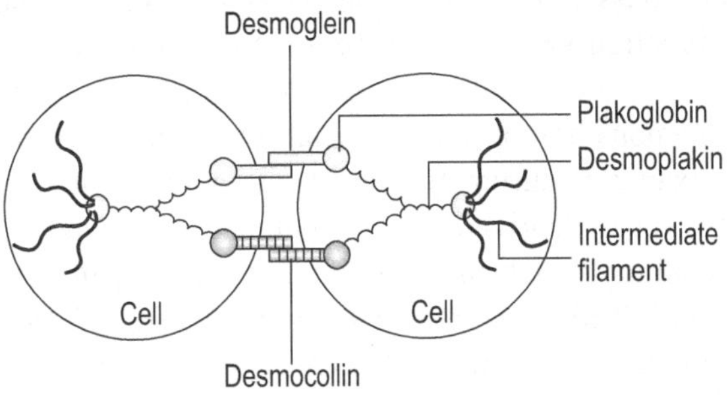

FIGURE 10.8 Structural component of desmosome

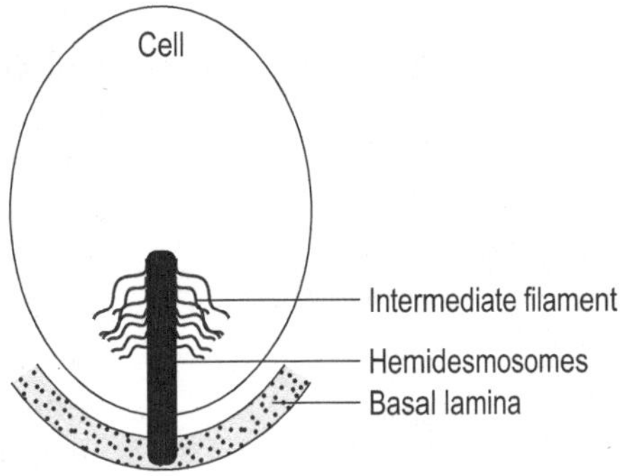

FIGURE 10.9 Structural component of hemidesmosome

Desmosomes Desmosomes, like adherens junctions, are made of proteins from cadherin family, but they are connected to the intermediate filaments via anchor proteins (Fig. 10.7). They are responsible for maintaining the integrity of the skin epithelial cells. Desmosomes are button-like junctions made of two types of cadherin family proteins: desmoglein and desmocollin (Fig. 10.8). The cytoplasmic domains of these two proteins are attached to a plaque of anchor protein made of γ-catenin (plakoglobin) and desmoplakin connected to each other. Desmoplakin makes contact with intermediate filaments. The type of the intermediate filaments connected to the desmoplakin depends on the function of the cells, for example, keratin in epithelial cells and desmin in cardiocytes. The extracellular domain of desmoglein and desmocollin holds the two cells together.

Hemidesmosomes Hemidesmosomes are constructed from the transmembrane proteins of integrin family. They connect the intermediate filaments of the cells with the ECM. Hemidesmosomes are morphologically similar to desmosomes, but they connect basal surface of the epithelial cell to the underlying basal lamina (Fig. 10.9). The integrin protein of the desmosome has its extracellular domain connected to laminin protein of the basal lamina. The intracellular domain is connected to keratin filaments through an anchor protein called plectin.

Communication junction

Gap junction is the communicating junction in animal cells. In plant cell, the communicating junction is called plasmodesmata. Structure and functions of gap junction and plasmodesmata are discussed here.

Gap junctions—Structure, function, and regulation Gap junctions are small pores of 2 to 4 nm, interconnecting two cells, in order to pass smaller molecules or signals between the cells. Gap junction occurs as a patch between the cells. Like other cell junctions, gap junctions are made of proteins called *connexin*. Connexin is a four-pass transmembrane protein. Six molecules of connexin are aligned in a parallel fashion to form a cylinder-like structure. Two such cylinders from the adjacent cells are interconnected to form a channel called connexion, which serves as a junction between the cells (Fig. 10.10). In invertebrates (*Drosophila*), the gap junction protein is called shaking-B, which belongs to the innexin family. The size of the pore in gap junction is, however, so small that only small molecules, such as amino acids or inorganic ions, are admissible, whereas

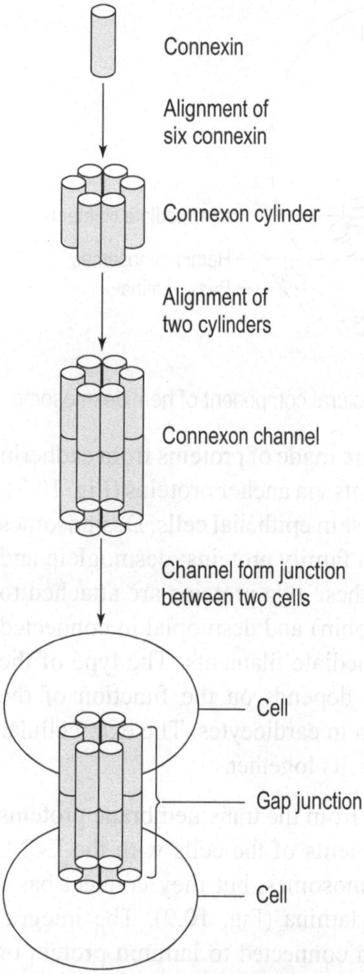

FIGURE 10.10 Formation of a gap junction

macromolecules, such as proteins, are not admissible. In fact, molecules with a mass of less than 1,000 Da pass through the gap junction. Gap junction serves as a medium for communicating action potentials in electrically excitable cells, such as nerve cells. It is also responsible for passing metabolic signals between the cells. For example, during fasting or starvation, when the blood glucose level is deprived, the sympathetic nerve endings secrete noradrenaline, which in turn stimulates the hepatocytes to breakdown the glycogen store. This stimulus is passed from one cell to the other by gap junctions to enhance glycogenolysis. Gap junction is also responsible for the synchronization of heart muscle contraction and peristaltic movement in intestinal smooth muscles. Gap junction is thought to play a role in fertility. In ovary, the oocyte is surrounded by a thick layer of ECM called zona pellucida (refer Chapter 8). This zona pellucida is surrounded by granulosa cells. Granulosa cells are interconnected by gap. Moreover, the granulosa cells are projected through the zona pellucida into the occyte via gap junctions. Although the exact function is still unknown, mutation study reveals that absence of gap junction protein results in infertility.

Gap junctions are highly dynamic structures that can undergo conformational changes to close or open in response to extracellular and intracellular signals. For example, low intracellular pH and high calcium ion concentration reduce the pore size of the gap junction. This is true especially when the cells are damaged and the membrane becomes leaky. The neurotransmitter dopamine is the extracellular signalling molecule for the gap junctions in retina. In response to bright light stimulus, the dopamine restricts the pore size of the gap junction in the neurons of the retina; this results in the stimulation of cone photoreceptors which are responsible for the detection of bright light.

Plasmodesmata—Gap junction in plant cells Plasmodesmata is the only intercellular junction found in plants. It is the gap junction establishing direct contact between two cells. Plasmodesmata is, however, structurally and morphologically different from the gap junctions in animal cells (Fig. 10.11). Plasmodesmata exists as a channel of 20 to 40 nm, piercing the cell wall, thus, connecting the cytoplasm of the two adjacent cells. A narrow cylindrical protein structure called desmotubule runs along the channel and interconnects the smooth endoplasmic reticulum of the two cells. The function of plasmodesmata is similar to the gap junction in animal cells. The molecular weight cut-off for plasmodesmata is more or less similar to gap junction. Plasmodesmata serves as a mean for communicating signals between cells, including the passage of gene regulatory proteins between cells. Sometimes, plant virus and viral RNAs also spread among cells via plasmodesmata. Plant virus secretes proteins which, in turn, increase the pore size of the junction transiently to give way for the virus. The mechanism of transport function of plasmodesmata is unknown.

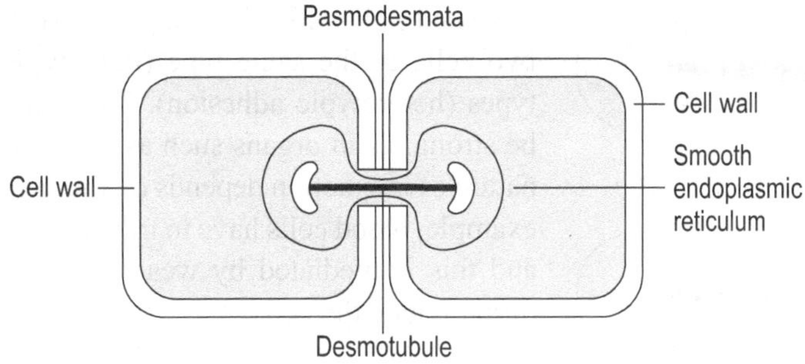

FIGURE 10.11 Plasmodesmata

10.2 CELL ADHESION

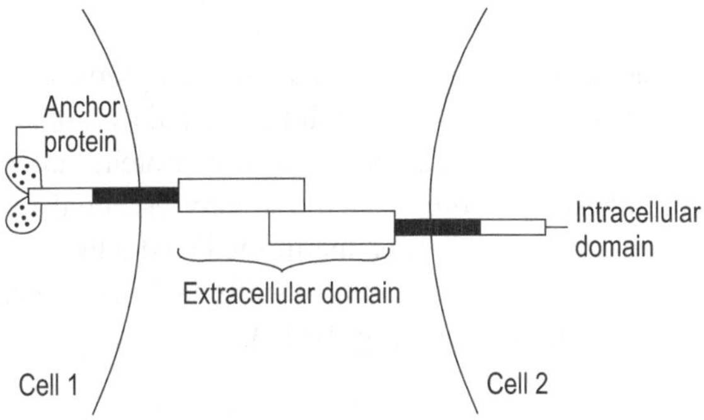

FIGURE 10.12 Cell-adhesion molecule

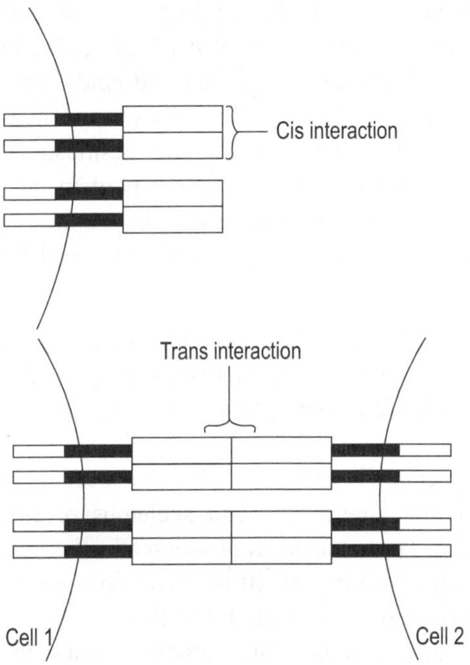

FIGURE 10.13 Cis-trans interaction between cell-adhesion molecules

Cell adhesion is mediated by cell-adhesion molecules (CAMs). CAMs can mediate cell–cell adhesion or cell–matrix adhesion. There are many CAMs, but they fall into two families: cathedrin and immunoglobin (Ig) superfamilies. CAMs are distributed throughout the plasma membrane as integral proteins. As shown in Fig. 10.12, they have a cytosolic domain and an extracellular domain. The cytosolic domain makes contacts with several adaptor proteins (refer Chapter 11) and communicates with signalling pathways to enable the contact with cytoskeleton. The extracellular domain is physically attached to other CAMs of the same cell or with other cells of same or different type. CAMs interact with each other in two modes: cis interaction and trans interaction (Fig. 10.13). CAMs within the plasma membrane of the same cell associate by lateral contact, called cis interaction. This results in dimer and larger oligomers of CAMs. CAM oligomers of adjacent cells join via their distal domains to facilitate trans-interaction. Due to cis-trans interaction, a Velcro-like adhesion is established between two cells. Thus, the cis interaction is intracellular and the trans interaction is intercellular. Adhesion between two cells may be due to CAMs of same type (homophilic adhesion) or different types (heterophilic adhesion). Adhesion mediated by cadherins and Ig superfamilies are homophilic type, whereas those mediated by integrins and selectins are of heterophilic type. Sometimes, the CAMs of two cells are interconnected via multivalent linker molecules; this is called linker-

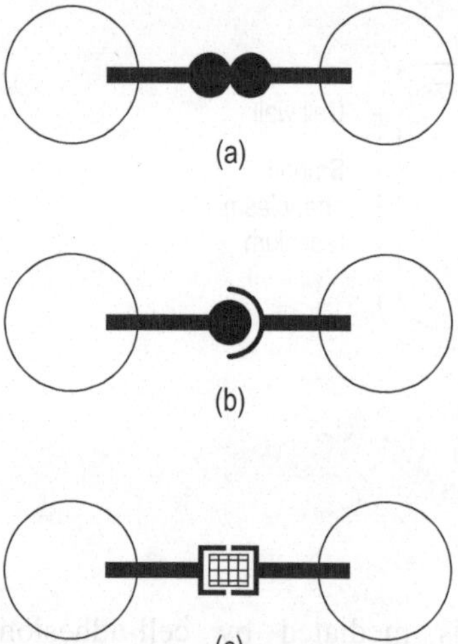

FIGURE 10.14 Types of cell–cell adhesion (a) homophilic (b) heterophilic (c) linker mediated

mediated cell–cell adhesion (Fig. 10.14). CAMs can interconnect two cells of the same type (homotypic adhesion) or different cell types (heterotypic adhesion). The nature of cell–cell adhesion may be strong, as in organs such as liver, or weak as in blood cells. The nature of interaction depends on the specific function of the cells. For example, blood cells have to travel to the area of damage or infection, and this is mediated by weak cell–cell adhesion which can easily disaggregate the cells.

10.2.1 Role of Cadherin and Catenin in Cell–Cell Adhesion

Cadherin refers to a family of glycoproteins, often called cadherin superfamily (Table 10.2) which is crucial for cell adhesion in embryogenesis. All members of cadherin superfamily are generally single-pass transmembrane proteins with 700 to 750 amino acid residues. Cadherin has an intracellular domain or the cytoplasmic tail and an extracellular domain. The intracellular domain is linked to the actin cytoskeleton via an intracellular anchor protein called catenin, probably in adheren junction. Catenin is very essential for cadherin action. Absence of catenin impairs the strength of cadherin-mediated adhesion. The extracellular domain is found on the cell surface as a dimer, with each monomer possessing five foldings or repeats called cadherin repeats (Fig. 10.15).

Members of cadherin superfamily

There are two types of cadherin—classical and non-classical cadherins. Classical cadherins are calcium-dependent CAMs, and they are named after the tissues of their predominance, for example, E-cadherin (predominant in epithelial cell), N- cadherin (predominant in nerve cell), and P-cadherin (predominant in placenta and epidermis).

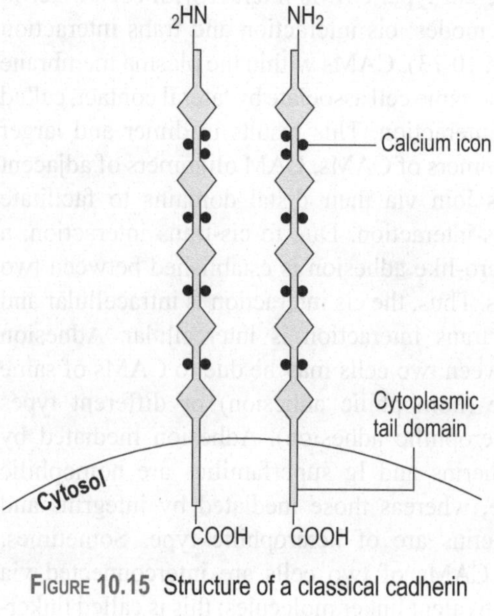

FIGURE 10.15 Structure of a classical cadherin

Non-classical cadherins are either adhesive or non-adhesive. Adhesive cadherins include desmosomal cadherins, such as desmocollin and desmoglein, and procadherins expressed in brain tissues. Non-adhesive cadherins include T-cadherin, protocadherins, and fat cadherin.

Functions of cadherin The primary function of cadherin is cell adhesion. Apart from this, it is also involved in cell signalling. These two functions are discussed here.

Cadherin-mediated cell–cell adhesion occurs in adheren junction, where the actin skeletons of two cells are linked. The mechanism of action of a classical cadherin is illustrated in Fig. 10.16. The two subunits of the cadherin dimer are paired and flexible. In the presence of calcium ions, which usually occupy the spaces between the cadherin repeats, the dimers acquire

TABLE 10.2 Cadherin superfamily members

Type	Member	Cell predominance	Location
Classical cadherin	E-cadherin	Epithelial cells	Adheren junction
	N-cadherin	Neurons, muscle cells, lens, fibroblasts	Chemical synapses and adheren junction
	P-cadherin	Placenta, epidermis	Adheren junction
	VE cadherin	Endothelial cells	Adheren junction
Non-classical cadherin	Desmocollin	Skin	Desmosomes
	Desmoglein	Skin	Desmosomes
	T-cadherin	Nerve cells and muscle cells	Adheren junction
	Fat protein	Epithelia and central nervous system	Tight junction synapses
	Protocadherins	Nerve cells	Chemical synapses

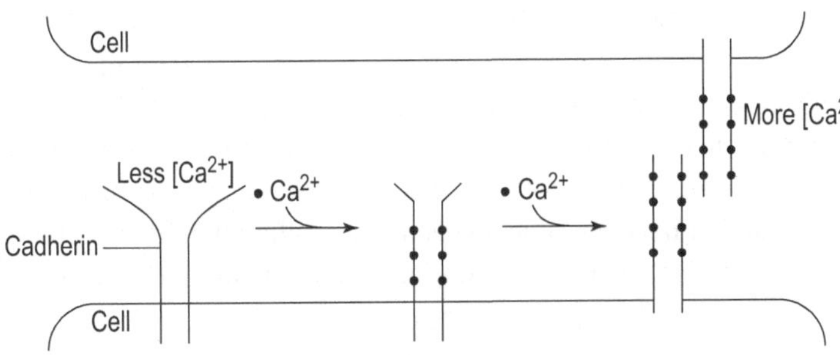

FIGURE 10.16 Mechanism of action of classical cadherin

a rigid conformation, lock with each other, and mediate cell–cell adhesion, which is of homophilic type. Cadherins of the same type aggregate strongly. For example, E-cadherin adheres to E-cadherin strongly than it does to N-cadherin. This type of adhesion leads to organization of cells into specific tissues.

Cadherin-mediated cell–cell adhesion is crucial in both embryogenesis and early stages of vertebrate development. In both stages, the expression pattern of cadherin varies. In embryogenesis, expression of cadherin results in compaction, a process in which the blastomers adhere to each other and get packed tightly by intercellular junctions. Ectoderm expresses E-cadherin. However, when it is differentiated into neural tube, E-cadherin is lost, but N-cadherin is expressed. Again, when neural crest cells move from the neural tube, cadherin 7 is expressed. During the formation of ganglion, N-cadherin is expressed.

A group of cadherins, called vasculoendothelial cadherin (VE cadherin), functions in cell signalling and promotes cell survival. In this task, the VE cadherin functions as a co-receptor, the conventional receptor being RTK (Chapter 11). Vasculoendothelial growth factor (VEGF) is the ligand for RTK. RTK along with VE cadherin activates intracellular signalling cascade leading to cell survival.

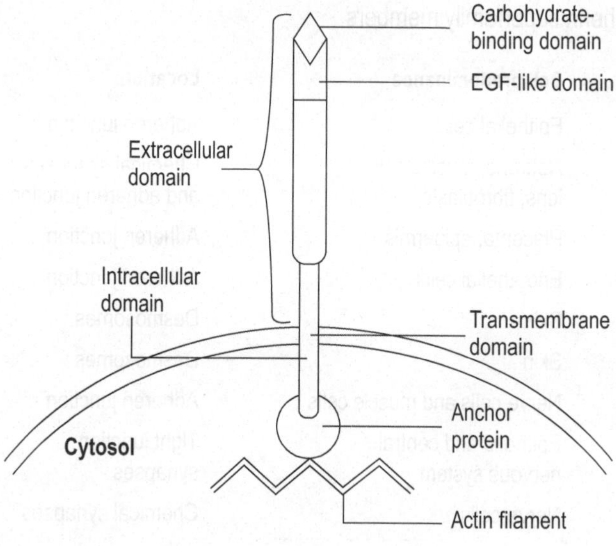

FIGURE 10.17 Structure of selectin

10.2.2 Role of Selectin in Cell–Cell Adhesion

Selectin is a transmembrane protein with specificity for oligosaccharides. Each selectin molecule has three domains—extracellular domain, transmembrane domain, and intracellular domain. The extracellular domain has a carbohydrate-binding site and an EGF-binding site. The transmembrane domain spans the membrane, and the intracellular domain indirectly attaches to actin filaments via anchor proteins (Fig. 10.17). Selectin is distributed in different tissues, for example, leucocytes (L-selectin), blood platelets (P-selectin), and activated endothelial cells (E-selectin). The function of selectin is well understood in propelling white blood cells along the surface of the endothelial lining of the blood vessels to move them into the tissues (to extravasate). Migration and extravasation is crucial whenever there is infection or inflammation or both. In this process, integrin is also involved. Therefore, cell–cell interaction between leucocytes and endothelial cells is mediated by selectin and integrin in a stepwise fashion, resulting in migration and extravasation (Fig. 10.18).

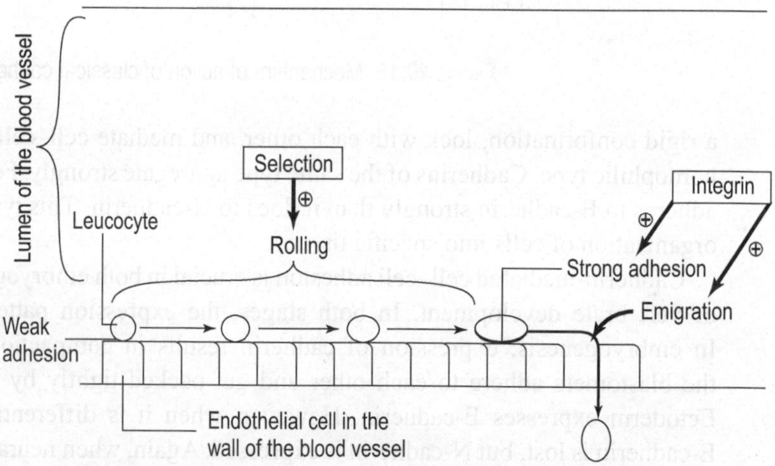

FIGURE 10.18 Role of selectin and integrin in cell–cell adhesion for mediating white blood cell migration and extravasation

Steps involved in the cell–cell interaction between leucocytes and endothelial cells

In the resting stage, that is, in the absence of infection of inflammation, leucocytes and endothelial cells are in minimal or no contact. Endothelial cells store P-selectin in a vesicle. Leucocytes express a specific carbohydrate with affinity for P-selectin. This carbohydrate is called *selectin ligand.*

1. When there is infection or inflammation, inflammatory signals are produced which activate the endothelial cells. Activation results in the:
   - movement of the P-selectin to the endothelial cell surface, facing the blood
   - synthesis and expression of platelet-activating factor (PAF) in the cell surface
   - synthesis and expression of ICAM-1 in the cell surface
2. P-selectin interacts weakly and transiently with its ligand present in the leucocyte surface. This weak interaction, along with the blood flow, literally 'roll' the leucocytes along the surface of the endothelial lining.
3. PAF alters the shape of the leucocyte and activates the integrins present in the leucocyte surface.
4. Active integrin mediates tight binding of leucocyte with the intercellular adhesion molecules (ICAMs) on the endothelial cells. This results in firm adhesion and subsequent extravasation into the underlying tissue.

Cell types, such as neutrophils, monocytes, and T and B lymphocytes undergo extravasation as described in the preceding steps.

10.2.3 Neural Cell-adhesion Molecule

Yet another protein superfamily called immunoglobulin superfamily (Ig superfamily) encompasses proteins containing one or more Ig-like structural domains. Members of this superfamily can function as calcium-independent homophilic CAMs or heterophilic CAMs. Neural cell-adhesion molecule (N-CAM) and fasciclin III (FAS3) are homophilic type, whereas ICAMs are heterophilic CAMs. N-CAM is the best-studied member of the Ig superfamily.

N-CAM has an extracellular part with five Ig-like domains, interlinked by a sisulphide bridge (similar to Ig, Fig. 10.19). The N-terminal end of the protein projects outward. The C-terminal end has one or two 'fibronectin-type III' domains. The remaining part of the protein either spans the membrane and reaches the intracellular region or anchors to the membrane via GPI anchor.

N-CAM is thought to function in just fine tuning of cell adhesion and subsequent cell sorting, both of which are required for tissue organization. For example, during the formation of pancreatic islets of Langerhans, cells aggregate and thereafter they are sorted by N-CAM for organization. Thus, N-CAM is primarily responsible for cell sorting and organization and not for actual cell aggregation.

N-CAM also functions as a relay protein (Chapter 11) in cell signalling. N-CAM, along with *src family* tyrosine kinases, phosporylates the tyrosine residues in signalling proteins and relay the signals onward.

FAS3 is responsible for adhering motor neuronal junctions to the muscle cells, thus making a synaptic connection.

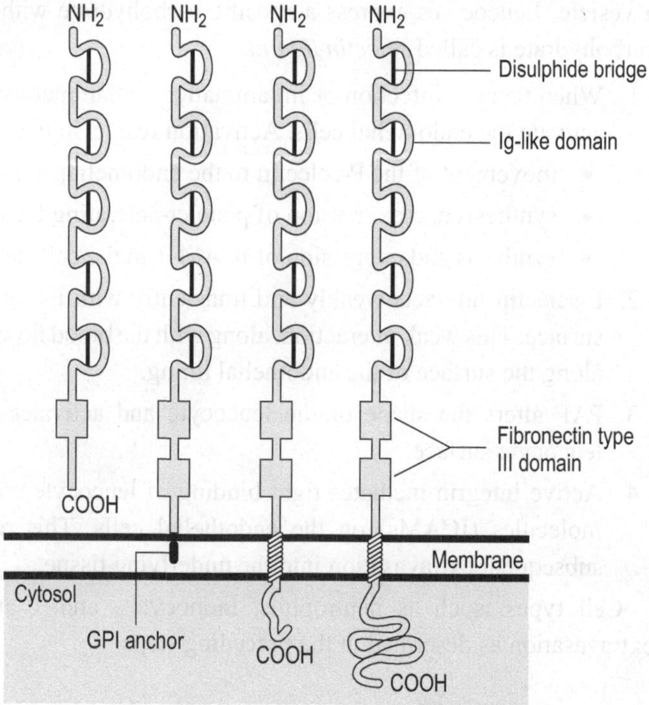

FIGURE 10.19 Structure of N-CAM

10.2.4 Cell Adhesion Mediated by Cell-surface Molecules

Cell surface-adhesion receptors (CSARs) are proteins that mediate contact between the cell and the ECM, thus establishing the *cell–matrix adhesion*, for example, integrin and immunoglobulin superfamily CAMs. ECM is a meshwork of proteins, carbohydrates, and glycoproteins, which holds the tissues together and regulates the architecture and function of the tissues. Cell–matrix adhesion is very crucial because it determines various functions of ECM and its major components (Table 10.3). Thus, it is logical to say that ECM facilitates many cellular functions via cell–matrix adhesion. ECM components and their adhesion with the cell contribute to high tensile strength and rigidity in tendons, teeth, and bones. ECM being a storehouse of signalling molecules is responsible for cell survival and differentiation. ECM can also function as a lattice over which cells move during organogenesis. ECM also serves as a medium for the movement, migration, and rearrangement of cells during morphogenesis. It is also responsible for skeletal development. Hence, deficiency in ECM-specific components or a defect in the expression of the genes encoding the ECM proteins results in malfunction or deformation of a particular organ. For instance, deficiency in collagen and the proteoglycan perlecan results in deformation and disfiguring of skeletal system, characterized by dwarfism. All these functions of ECM critically depend on cell–matrix adhesion via CSARs. Moreover, CSARs are responsible for the cell-signalling pathways. CSARs interact with intracellular signalling pathways via adaptor proteins and signalling enzymes, such as kinases (refer Chapter 11), thereby influencing cell survival, cell proliferation, cell migration, cell branching (as in blood vessels, mammary glands, and salivary glands), and cytoskeletal organization. For example, fibronectin is an ECM component protein, which communicates with the cell via the CSAR, integrin, and stimulates branching in glandular cells (Fig. 10.20). It is proved that deficiency or dysfunction of fibronectin and integrin blocks branching thereby affecting the specific function of the cells.

TABLE 10.3 Major components of ECM

Component type	Name	Function
Proteoglycan	Perlecan	Development of skeletal system
		Interlinking of type IV collagen and laminin Organization of basal lamina
Collagen	Type I, II, and III	Formation of fibres
	Type IV	Formation of sheath structures
Multi-adhesive matrix proteins	Fibronectin	Cell signalling Cell branching
	Laminin	Structural component of basal lamina Associates with integrin for cell signalling
	Nidogen/entactin	Cross links basal laminal proteins

10.2.5 Structure of Integrin

Integrin is a transmembrane glycoprotein. It is a heterodimer with two subunits, α and β, both linked by non-covalent interaction. So far, 24 types of α subunits and 9 types of β subunits were found in humans. The α subunit has four binding sites of divalent cations, such as calcium and magnesium. The β subunit has a single site for binding calcium or magnesium ions. The α subunit has two segments linked by a disulphide bridge, and the β subunit has cysteine-rich moiety. The extracellular domain of integrin binds to the ligands of ECM. Majority of the principal protein components of ECM including collagen, fibronectin, and laminin are ligands for integrin. The intracellular domain of the integrin binds indirectly to actin filaments via anchor proteins, that is, α-actinin, filamin, talin, and vinculin. The binding of the intracellular domain to actin makes the integrin more active and provides a conformation to the integrin, enabling tighter binding with its ligand.

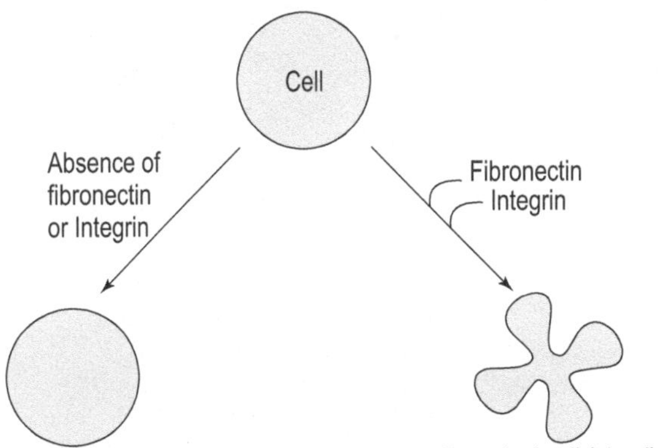

FIGURE 10.20 ECM component (fibronectin) and CSAR (integrin) in cell branching in glandular tissues

Types and distribution of integrin

Integrin is divided into eight subfamilies, designated as β1 to β8. Different types of integrin and their tissue distribution are given in Table 10.4.

TABLE 10.4 Type of integrins and their tissue distribution

Type	Tissue distribution
$\alpha_1\beta_1$	Various tissues
$\alpha_2\beta_1$	Various tissues
$\alpha_4\beta_1$	Haematopoietic cells
$\alpha_5\beta_1$	Various tissues
$\alpha_6\beta_1$	Various tissues
$\alpha_L\beta_2$	T lymphocytes
$\alpha_M\beta_2$	Neutrophils
	Monocytes
$\alpha_V\beta_3$	Thrombocytes
$\alpha_B\beta_5$	Fibroblasts
	Epithelial cells
$\alpha_V\beta_6$	Proliferating lung epithelial cells
$\alpha_6\beta_4$	Epithelial cells

Binding function of integrin

Integrin plays a role in the final step of leucocyte migration. During the migration of leucocytes along the endothelial cells, integrin along with ICAM mediates a firm adhesion of leucocyte to the endothelial cells for subsequent extravasation into the underlying tissue. This process is explained in Section 10.2.4 and Fig. 10.20.

Regulation of integrin activity

Regulation of integrin activity is very essential for the control of cell adhesion, aggregation, and movement. Integrin activity is regulated by conformational changes (Fig. 10.21). Integrin exists in two conformations: inactive form and active form. Inactive integrin is flexible and bent, like a loop. It has its cytoplasmic C-terminal domain very closely bound. Inactive form has a low affinity for its ligand. Hence, it is also called 'low affinity form'. Active integrin is rigid and straight. It has its cytoplasmic C-terminal domain further apart. Active form has a high affinity for its ligand. Hence, it is also called 'high affinity form'. The ability of integrins to exist in the two forms permits transduction signal on either side of the cell, that is, outside-in and inside-out signalling.

Outside-in signalling Extracellular ligands in ECM initiate outside-in signalling. Binding of ECM ligands activates the integrin by separating the cytoplasmic tails. This separation is recognized by intracellular anchor proteins or adaptor proteins which undergo changes, that is, either bind to or detach from the tails. This change alters the cytoskeleton and consequently changes (inhibit or stimulate) the signalling pathway.

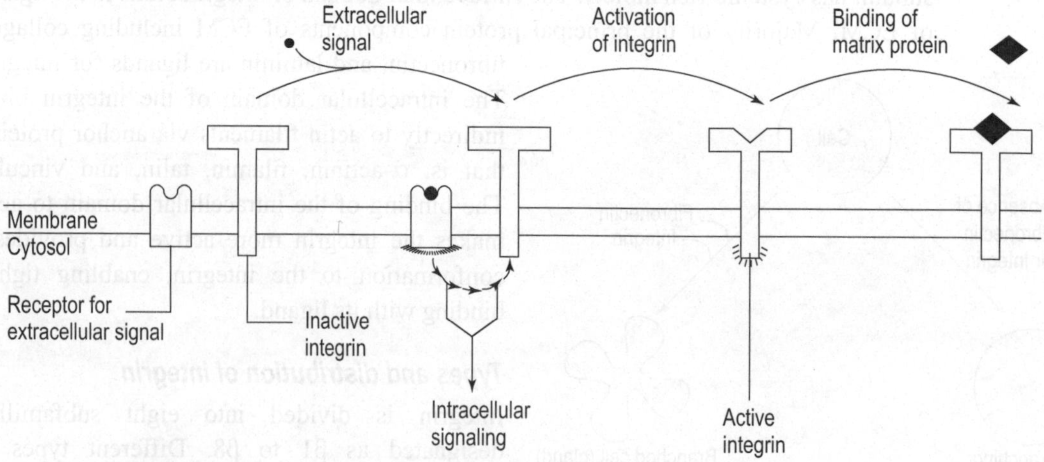

FIGURE 10.21 Regulation of integrin activity

Inside-out signalling Metabolic state of the cell determines the inside-out signalling. Depending on the metabolic state of the cell, the intracellular proteins either bind to or detach from the cytosolic tail of the integrin. This results in separation or adhesion of the two cytoplasmic tails. Consequently, the integrin may be activated or inactivated.

Role of integrin in intracellular signalling pathway

In addition to cell–matrix adhesion, integrins also mediate intracellular signalling pathway. The previous section has already described that the ability of integrin to undergo conformational changes is responsible for relaying signals. This section focuses on the role of integrin in relaying signal via focal adhesion kinase (FAK). FAKs are cytoplasmic tyrosine kinases. Integrins form cluster at the cell–matrix adhesion point. As a result, the adaptor proteins associated with the integrins recruit FAK to the focal adhesions. Talin, which is bound to β subunit, and paxillin, which is bound to the α subunit of the integrin, recruit FAKs and form FAK clusters. FAKs cross phosphorylate each other at their tyrosine residues. The phosphotyrosines of FAKs further recruit src family tyrosine kinases, which actually dock to FAKs, and phosphorylate some more tyrosines of FAK, as well as other signalling proteins. This results in the relaying of signal onward. Integrin signals are transduced via other molecules, such as integrin-linked kinases (IKK) and particularly interesting new cysteine-rich protein (PINCH). Integrin-mediated signalling pathway results in cell survival, cell growth, cell division, cell differentiation, or apoptosis. Integrin can function as a co-receptor. It can coordinate with conventional receptors in mediating signalling pathway. Integrin associates with RTK and mediates signal pathways, which lead to tumour invasion and metastasis.

10.3 EXTRACELLULAR MATRIX

Extracellular matrix (ECM) is an intricate network of macromolecules, such as proteins and polysaccharides, occupying the extracellular space. ECM is more predominant in connective tissues than in epithelial tissues. The amount of ECM varies depending on the nature of the tissues. For example, ECM is plentiful in cartilage and bone and very less in brain and spinal cord. ECM is secreted by the cells called fibroblasts located in the extracellular space. For example, in cartilage, chondrocytes are the fibroblasts which secrete ECM, whereas in bone, ECM is secreted by osteoblasts.

The polysaccharides of the ECM are in the form of glycosaminoglycan (GAG) and proteoglycans. The structure of GAG and proteoglycan are discussed in Chapter 4. Protein component of ECM include fibrous proteins, such as collagen, elastin, fibronectin, and laminin.

10.3.1 Functions of GAGs and Proteoglycans

GAGs form a jelly network which is porous. The pores act as sieve and enable the movement of molecules across the cells, depending on size, charge, or both. Proteoglycans play a major role in cell signalling by binding to extracellular ligand and mediating the appropriate response within the cell (Refer Chapter 11). For example, fibroblast growth factor (FGF) ligand binds to proteoglycan and is consequently oligomerized by cross linking

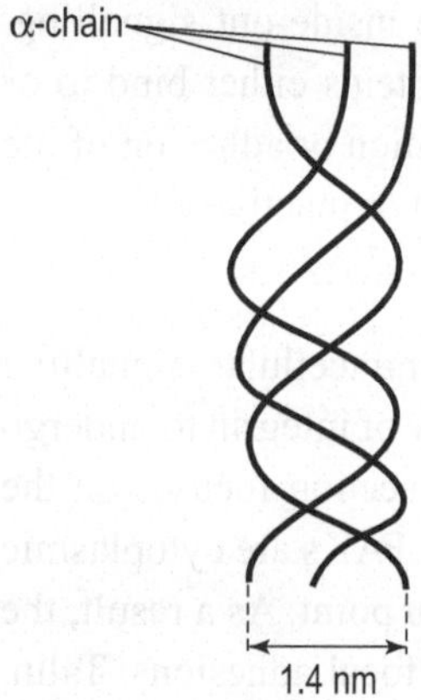

FIGURE 10.22 Structure of tropocollagen

and activates the receptor tyrosine kinase (RTK). RTK, in turn, mediates cell proliferation by influencing various intracellular signal proteins, as explained in Chapter 11. Regulation of the activities of protease enzymes and protease inhibitors is yet another function of proteoglycan. Proteoglycans regulate the proteins by blocking their action via stearic hindrance or immobilization of the proteins or by retaining their activity via prevention of proteolysis and effective presentation to their site of action. Proteoglycans can also function as co-receptors. The crucial role of proteoglycan as co-receptor is confirmed by mutation studies in Drosophilla. Proteoglycans that function as co-receptors are made of a core protein that is attached to the plasma membrane via glycosylphosphotidyl inositol anchor (GPI anchor; refer Chapter 11 for GPI anchor). These co-receptors function along with the conventional receptors to promote cell–matrix interaction and initiate cell signalling. Syndecan and betaglycan are examples for co-receptors. Syndecan is located in the epithelial and fibroblast membranes and binds to ECM protein ligands. It binds to FGF and presents it to the FGF receptor. Betaglycan binds to TGF and presents it to the TGF receptor.

10.3.2 Collagen

Collagen is a fibrous protein of ECM secreted by connective tissues. It is the most abundant protein in mammals constituting nearly 45 to 50% of the total protein. It is predominant in fibrous tissues. The basic structural unit of collagen is called tropocollagen. Each tropocollagen is a tight, right-handed triple helical structure of 1.4 nm diameter, formed by three polypeptide chains (α-chain), wound around each other (Fig. 10.22). Each polypeptide chain is made of repeated triplets denoted as X-X-Gly, where X is any amino acid residue. Approximately, 1,000 residues are present in each chain. The polypeptide constituent of collagen has approximately 33% glycine, 10% proline, 10% hydroxyproline, and 1% hydroxyl lysine. Thus, 60% of the chain is made up of Gly-Pro-X or Gly-X-HPro (pro is proline and HPro is hydroxyproline), whereas 40% is made of other amino acid residues, with every third residue as glycine, which keeps the chains tightly packed. The chain has a left-handed helix conformation, with three residues per turn. The chains differ from conventional α-helix in that they do not have intra chain hydrogen bonding. However, inter hydrogen bond between glycine's –NH group and carboxyl group or adjacent chain does exist. The pyrrolidine ring of the proline residues stabilizes the chain. Tropocollagen fibres cross link laterally to form collagen fibrils. Collagen fibrils aggregate into larger collagen fibres, which have an elongated, rigid structure. Formation of collagen from tropocollagen is shown in Fig. 10.23.

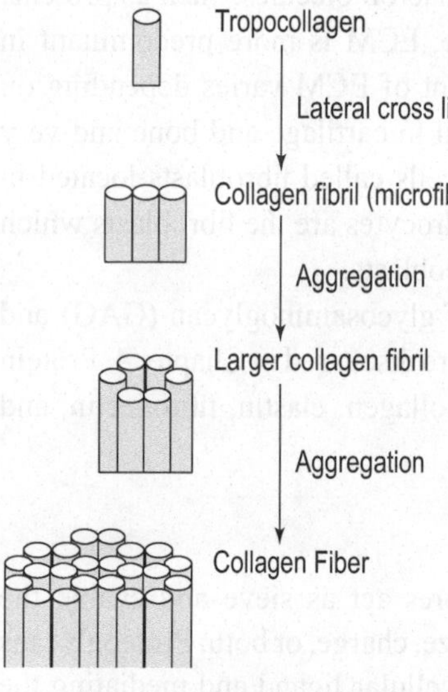

FIGURE 10.23 Organization of tropocollagen into collagen

Major types of collagen include Type I, II, III, IV, and V. Type I is the most common collagen, widely distributed in skin and bone. Approximately 90% of body's collagen belongs to Type I. Types II, III, IV, and V are distributed in cartilage, reticular fibres, basement membrane, and hair, respectively. However, other types of collagen have also been discovered, which are listed in Table 10.5.

TABLE 10.5 Types of collagen and their distribution

Type	Tissue distribution	Type	Tissue distribution
I	Tendons, skin, artery walls, myofibrils, bones, and teeth	VIII IX	Endothelial cell FACIT collagen
II	Hyaline cartilage Vitreous humour	X	Hypertropic cartilage Mineralising cartilage
III	Reticular fibre Uterus Intestine	XI	Cartilage
IV	Basal lamina Eye lens Glomerular capillary	XII	Glycosaminoglycan
V	Interstitial tissue Placenta	XIII	Basement membrane
VI	Interstitial tissue	XIV	FACIT collagen
VII	Epidermal junction		

10.3.3 Elastin

Elastin is a stretchable fibrous protein network (Fig. 10.24). Each structural unit of elastin is called tropoelastin. It is a chain of about 70 amino acid residues, highly hydrophobic. It is rich in glycine, valine, alanine, and proline, but has no hydroxyl lysine. It has a molecular weight of 6.6 kDa. It is distributed in tendons, ligaments, and walls of blood vessel. In general, tissues that function by reversible stretching and relaxing, for example, lung wall and blood vessel wall, are rich in elastin. Elastin fibres are laterally cross-linked by a tetrapeptide formed from four lysine residues. This tetrapeptide is called desmosine, which contributes to the elastic property. Actually, three of the four lysine residues in desmosine are present in modified form. They are modified into allysine by the enzyme lysyl oxidase. The integrity of the elastin itself is maintained by a microfibril sheath of glycopoteins, including fibrillin. Microfibril sheath functions as a scaffold for elastin. Deficiency of microfibril affects the function of elastin-rich tissues, causing Marfan's syndrome.

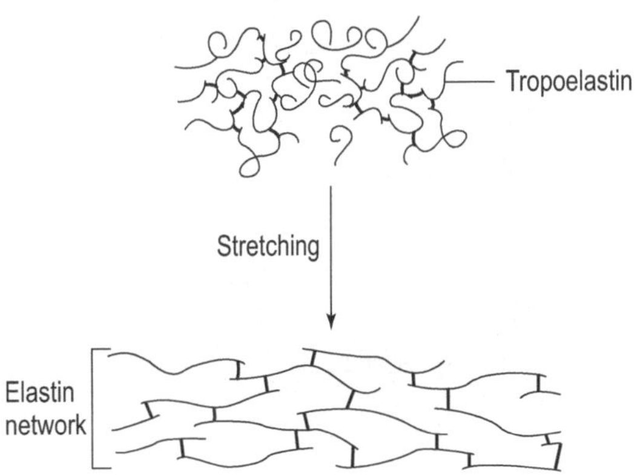

FIGURE 10.24 Structure of elastin

Elastin functions in association with collagen in the ECM. It is crucial for the function of organs that require reversible stretchability and deformability, such as lungs, skin, and blood vessels. It is very crucial for the function of arteries and aorta. It is also responsible for structural integrity of tissues. It is crucial for the renewal of dead cells in skin, especially during wound healing.

10.3.4 Fibronectin—A Matrix Glycoprotein

Fibronectin is a glycoprotein with molecular weight of approximately 440 kDa. It is present in ECM, as insoluble form, and in plasma, as soluble form. In ECM, it organizes the matrix and helps in cell adhesion. In plasma, fibronectin promotes blood clotting, wound healing, and phagocytosis.

Structure

Structurally, it is a dimer made of two long monomers (Fig. 10.25). Each monomer is approximately 2,500 residues long, both linked by a disulphide bridge near the C-terminus. Each chain has multiple domains separated by flexible peptide sequence. The domains include integrin-binding domain, heparin-binding domain, cell-binding domain, collagen-binding domain, and self-association domain. The chain also has frequently repeating domains called type III fibronectin repeats (T3FNR, Fig. 10.26). Each T3FNR is made of about 90 amino acid residues, and it is the most common protein domain in vertebrates. T3FNR has two specific sequences, namely, arginine-glycine-aspartate sequence (the RGD sequence) and synergy sequence. These sequences function as cell-binding sites. RGD sequence in fibronectin is the target for integrin receptors. That is, integrin recognizes and binds to fibronectin at the RGD sequence to promote cell adhesion. Binding of RGD sequence to integrin is enhanced by synergy site. RGD sequence is also a part of other proteins, such as fibrinogen and disintegrins (present in snake venom).

Functions

Fibronectin performs the following functions:

- As mentioned in the preceding section, the main role of fibronectin fibril is the organization of matrix (i.e., matrix assembly) and cell–ECM adhesion. To perform its

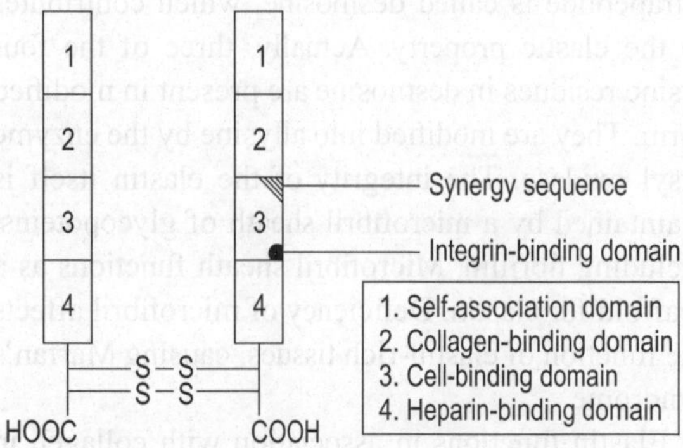

FIGURE 10.25 Structure of fibronectin dimer

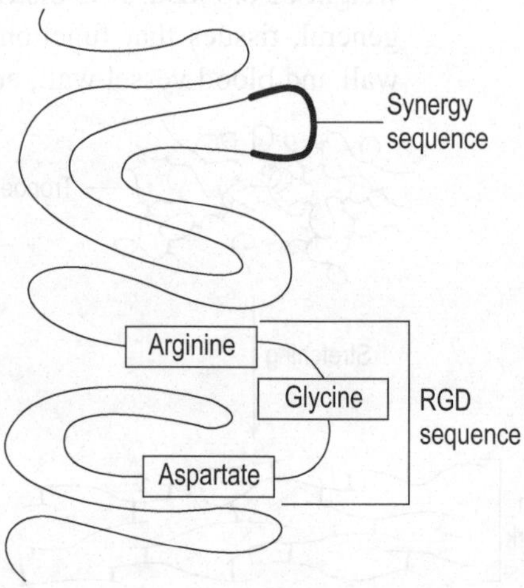

FIGURE 10.26 Structure of Type III fibronectin repeat

function, fibronectin should first be aligned into fibrils as a mosaic. Unlike collagen filbrils, fibronectin fibrils cannot form by self-assembly, as they require additional proteins, especially integrins. Fibronectin is secreted by fibroblasts. Fibronectins are then assembled into fibrils on the surface of fibroblast cells, as these cells have integrins at elongated sites called fibrillar adhesions (a form of focal adhesion). Synergy sequence binds to the α-subunit of the integrin and enhances the binding of RGD sequence

to the cleft between α and β subunits. Integrin links the fibronectin fibrils with the intracellular actin network. Actin, in turn, helps orienting the fibrils on and near the cell surface. Actin and myosin exert tension over the fibronectin fibrils. As a result, the fibrils are stretched. Stretching exposes cryptic binding site and more integrin binding sites in the fibronectin. Cryptic binding sites allow the fibrils to self-associate directly. Integrin binding sites allow the formation of fibril–integrin complex. This results in fibronectin polymerization and matrix assembly.

- Fibronectin fibrils also have functions other than matrix assembly. They are necessary to organize endothelial cells to form blood vessels. Therefore, absence of fibronectin fibrils is fatal to embryos.
- Recent research report suggests that interaction between the RGD sequence and integrin receptor mediates the attachment between the sperm cells and the human endosalpinx in the isthmic region.
- Fibronectin is responsible for cell migration during embryogenesis.
- Wound healing is also a function of fibronectin.

10.3.5 Influence of ECM on Shape, Survival and Proliferation of Cells

This section recaps the role of ECM and basal lamina to conclude that ECM influences the cell shape, survival, and proliferation. ECM is made of proteins and carbohydrates which mediate cell–matrix adhesion. Cell-matrix adhesion mediates outside-in signalling and inside-out signalling. This signalling results in various cell processes, such as growth, survival, apoptosis, or proliferation. Integrin-mediated signalling pathway results in cell survival, cell growth, cell division, cell differentiation, or apoptosis. Fibroblast growth factor (FGF) ligand binds to the proteoglycan of ECM and mediates cell proliferation by activating RTK. Elastin of ECM is responsible for the renewal of dead cells in skin, especially during wound healing. Thus, it can be stated that ECM promotes cell growth and regeneration. Fibronectin is responsible for cell migration during embryogenesis. Fibronectin also helps in the organization and assembly of ECM and maintains the shape of the cell. Thus, it is logical to say that ECM influences various cellular processes, such as growth, migration, and survival.

10.3.6 Regulation of ECM Turnover

So far, the assembly, structural organization, and functions of ECM have been discussed. In addition to that, it is also necessary to throw some light on the degradation of ECM because a balance between the synthesis and degradation of ECM is very essential for the normal functioning of the cells. In fact, even in static ECM, the components undergo slow but continuous degradation to maintain a balance between synthesis and degradation. As synthesis of ECM is important for various cellular processes discussed above, degradation of ECM is also important for various biological processes, such as cell migration, metamorphosis, and organ shrinkage (involution). Cells migrate from one place to other along the ECM way. For example, white blood cells migrate through the basal lamina of the blood vessel cells in order to reach the site of infection or injury. Cancer cells migrate from their origin to other organs for metastasis via ECM degradation. Metamorphosis is yet another process involving ECM degradation. During the conversion of tadpole larva to adult organism, the tail degenerates by the degradation of ECM. After child birth, the uterus involutes by ECM degradation.

Role of proteases

Degradation of ECM is an enzyme-dependent process. As most of the components of ECM are proteins, proteases come into action. Proteases, by name, hydrolyze the peptide bonds of proteins. Matrix metalloproteases (MMPs) and serine proteases are used in regulating the ECM turnover. MMPs are calcium- or zinc-ion dependent. Serine proteases have active serine residues to catalyze proteolysis.

Mechanism of protease action

Proteases act by three mechanisms and regulate the ECM turnover. They are as follows:

Local activation Local activation occurs in the process of dissolving blood clots. The proteases involved in local activation of ECM degradation to promote dissolution of blood clot are secreted as an inactive precursor. The precursor is plasminogen. Plasminogen is locally activated into active plasmin by plasminogen activator (a serine protease). Plasmin dissolves the blood clot by degrading the fibrin polymers.

Confinement by cell-surface receptors Some of the proteases are confined to the receptors expressed on the cell surface. As a result, the action of protease is also confined to that cell and its surroundings. Urokinase-type plasminogen activator (UPA) is confined to urokinase-type plasminogen receptor (UPR), which is expressed in migrating cancer cells. UPA cleaves the plasminogen between an ariginine and a valine residue and forms active plasminogen. Plasminogen dissolves ECM and gives way for the migration of cancer cells, and thus, promotes metastasis.

Secretion of inhibitors Inhibition of proteases is important to protect ECM which need not be degraded. For this purpose, cells secrete protease inhibitors which inhibit the catalytic activity of MMPs and serine proteases in specific areas. Tissue inhibitors of metalloproteases (TIMPs) and serine protease inhibitors (serpins) inhibit MMPs and serine proteases, respectively.

Overall, activation of proteases and inhibition of proteases are involved in the regulation of ECM degradation.

10.4 BASAL LAMINA

Basal lamina (BL) is a sheet-like layer of ECM with thickness of 60 to 120 nm secreted by the epithelial cells. The organization of BL depends on the nature of the tissue. In intestine and skin, BL forms a layer on only one surface of the cell, whereas in muscle

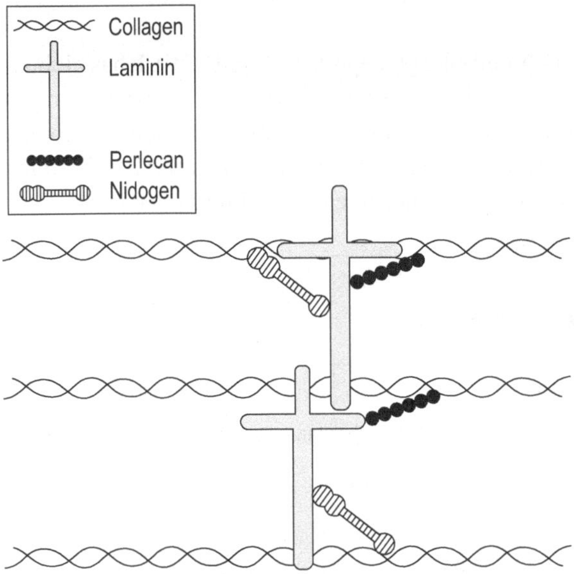

FIGURE 10.27 Components of basal lamina

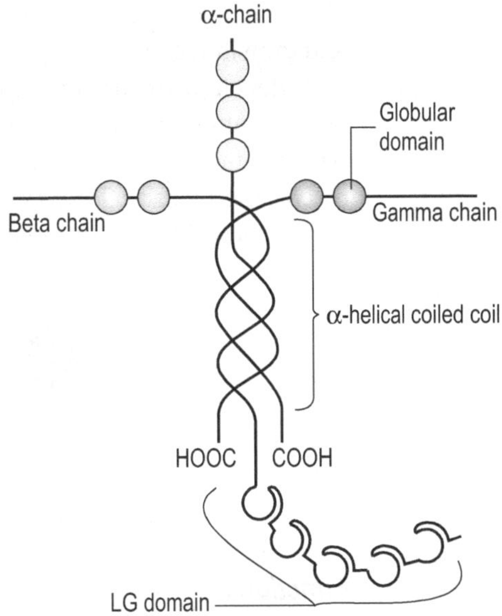

FIGURE 10.28 Structure of laminin

and adipose tissues, they surround the entire cell surface. Following are the components of the BL, which are synthesized and secreted by the cells associated with BL:

- Laminin
- Type IV collagen
- Perlecan
- Nidogen (entactin)

Structural components of BL are shown in Fig. 10.27 and discussed as follows:

Laminin is the major multi-adhesive protein of basal lamina. It is a glycoprotein with three polypeptide chains organized into a cross-shaped structure (Fig. 10.28). The three chains, named α, β, and γ, are interlinked by disulphide bonds. So far, five types of α chains, three types of β chains, and three types of γ chains have been discovered. The chains can combine in multiple ways giving rise to 45 different laminin isoforms. The laminin molecule has multiple domains for binding to different ECM ligands, including collagen, integrins, and carbohydrates. For example, the α-chain has a globular domain (laminin globular domain or LG domain) which binds to the carbohydrate moieties in cell-surface molecules, such as syndecan and dystroglycan. Laminin is the primary basal laminal ligand for many integrins. Laminin communicates internally with the cell via integrin in hemidesmosomes.

Type IV collagen is the main structural component of basal lamina. The collagen fibrils are aligned head-to-head or head-to-tail. Several such dimers then form lateral adhesion to form a sloppy sheet-like organization of the BL. Type IV collagen sheet forms a two-dimensional network with the laminin.

Perlecan is a proteoglycan. It has a core protein attached to GAG chains. The core proteins have several globular repeats, such as LG domains of laminin. This gives a pearl-like appearance to the protein, hence the name perlecan. Apart from GAG, perlecan is also attached to ECM components, including laminin and nidogen, cell-surface receptors, and growth factors.

Nidogen, also called entactin, is a dumb-bell-like glycoprotein (Fig. 10.29). It has globular ends separated by a rod-like conformation. The N-terminus has two globular domains, whereas the C-terminus has a single globular domain. The C-terminal globular domain of entactin binds to Type IV collagen and laminin. Thus, it cross links the basal laminal components. The rod-like structure has RGD sequence to enable cellular interaction.

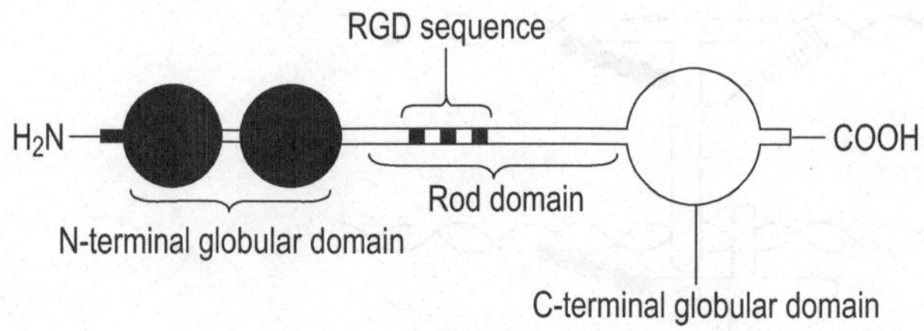

FIGURE 10.29 Structure of nidogen

Functions

Different functions of BL are mentioned as follows:

- BL mediates tissue regeneration and embryonic development. During embryogenesis, BL adheres the cells of blastula and gastrula together.
- It is also important in neural development.
- BL is the main structure that separates blood and the brain. Thus, BL functions as a blood–brain barrier.
- In kidneys, BL is much thicker with an extra layer, forming the basement membrane, which functions in filtration. Collagen and heparin sulphate proteoglycan in the BL play a role in the glomerular filtration.
- BL covering the epithelial cells are selectively permeable to certain cells, but act as a barrier for other cells. For example, fibroblasts from the connective tissues under the BL cannot get access to the epithelial cells.
- BL helps in tissue regeneration. Whenever tissues are damaged, BL supports the tissues as a scaffold and facilitates the assembly of regenerating cells. This function is well studied in the regeneration of muscle cells at the neuromuscular junction.
- BL promotes cell migration in wound healing.

10.5 PLANT CELL WALL

Cell wall is the extracellular matrix of plant cells. It is thick, strong, tough, and more rigid than the animal cell matrix. It is composed of a pectin matrix over which a network of cellulose with glycan cross links are embedded. The orientation of cell wall is determined by the intracellular microtubules. This section discusses about the composition and function of cell wall as an ECM.

10.5.1 Composition of Cell Wall

Cell wall is formed during cell division. In the cytokinesis stage of cell division, the cell plate forms a new wall as a partition. This is called the primary cell wall which is thin but tough. After the cell growth, the primary cell wall is thickened by incorporation of additional layers to form a secondary cell wall which is thick, rigid, and lignified. The ECM of animal cells is predominant in proteins and nitrogen-containing polymers, but the plant counterpart (i.e., the cell wall) is predominant in non-nitrogenous polymers, including cellulose and lignin (refer Chapter 4), which are held together by covalent and non-covalent bonds.

The primary cell wall is predominantly made of polysaccharides, including cellulose (1, 4-linked β-D-glucose residues), hemicelluloses (1, 4-linked β-D-hexosyl residues), and pectin (1, 4-linked α-D-galacturonic acid). Homogalacturonans, rhamnogalacturonans, and substituted galacturonans are pectin components of cell wall. However, minor amount of glycoproteins, enzymes, phenolic compounds, such as ferulic and coumaric acids, and minerals are also present. Lignin is the principal component of secondary cell wall. Approximately, 30% of the secondary cell wall is made up of xylulose.

10.5.2 Organization of Cell Wall

Although the composition of cell wall is known, the idea about the organization and mechanical properties of cell wall is still under debate. Four models have been proposed to describe the organization and mechanical properties (Fig. 10.30a to d). They are

(i) covalent model
(ii) tether model
(iii) diffused-layer model
(iv) stratified-layer model

Covalent model proposes that the cell wall is rich in cellulose fibres and that fibres are covalently cross linked by xyloglucan chains. This covalent linkage provides tensile strength to the cell wall. However, this model is not accepted due to lack of evidence for covalent linkage (Fig. 10.30a).

According to tether model, pectin molecules are non-covalently cross linked to form a network. Cellulose fibres are meshed in the pectin network. Cellulose fibres are linked by xyloglucan tether by hydrogen bonding (Fig. 10.30b).

As per diffused-layer model, the xyloglucan chains are just diffused on the surface of the cellulose fibres and embedded on the pectin matrix (Fig. 10.30c).

Stratified-layer model suggests that cell wall possesses cellulose fibres cross linked by xyloglucan chains to form a lamellar structure. The lamella is separated by strata of pectin network (Fig. 10.30d). Although four models were proposed, further investigation is still needed to completely elucidate the cell wall organization.

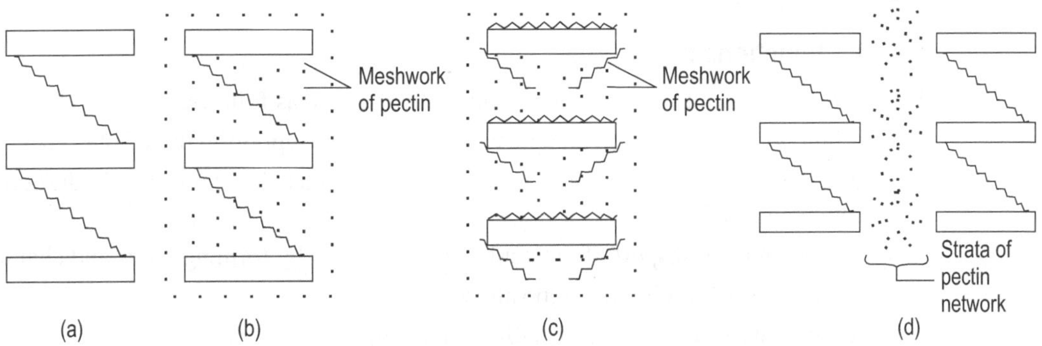

FIGURE 10.30 Proposed models for the organization of cell wall: (a) Covalent model, (b) tether model, (c) diffused-layer model, and (d) stratified-layer model

10.5.3 Cell Wall and Turgor Pressure

Turgor pressure (also called turgidity) is the pressure or force exerted by the cellular water (in vacuole) against the cell wall. The cell wall is made of fluid, which is hypotonic when compared to the intracellular compartment. This results in osmotic imbalance and inward movement of water molecule (Fig. 10.31). Consequently, an internal hydrostatic pressure, called the turgor pressure, is developed towards the cell wall. Turgor pressure is responsible for cell expansion and mechanical rigidity. This is the reason why healthy plants are mechanically rigid and wilted plants are not. The concept of turgor pressure is shown in Fig. 10.32.

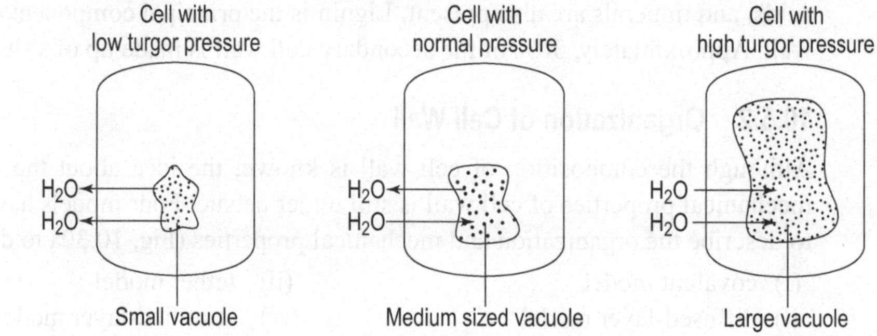

FIGURE 10.31 Osmotic imbalance in plant cells

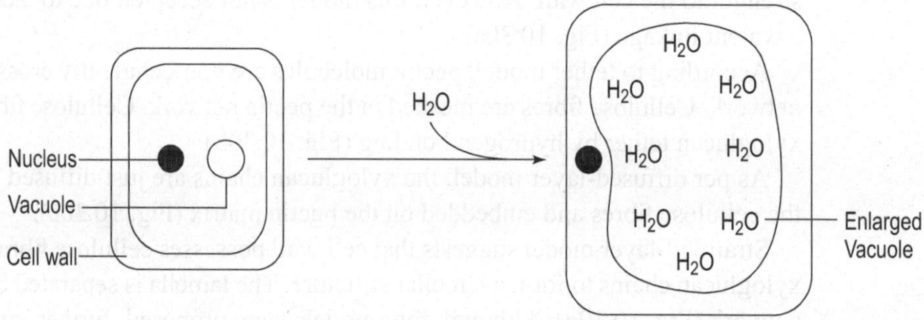

FIGURE 10.32 Turgor pressure

10.5.4 Functions of Cell Wall

Plant cell wall has multiple functions, some of which are as follows:

- It functions as a skeleton and provides mechanical support to the whole cell.
- It functions as a protective barrier and envelops each cell to prevent dehydration and pathogen attack.
- It helps in the transport of molecules between cells by forming pores and channels.
- It serves as a carbohydrate repository.
- It functions as a source of signalling molecules.
- It is responsible for determining and maintaining the shape of the cell.

- It facilitates resistance to internal turgor pressure of the cell.
- It regulates the direction and rate of cell growth.
- In some specific tissues, it becomes modified and acquires special function. For example, the cell wall is modified as a tough trichome or spike in Arabidopsis leaf. The cell wall is modified with wax and cuticle in tomato leaf to defend against pathogen.
- It facilitates cell–cell interaction.

SUMMARY

Cells vary in their shapes, sizes, and functions. When cells are organized into tissues they acquire special functions, bidirectional transport across the membrane, and signal transduction. Signal transduction regulates cell survival, proliferation, differentiation migration, and apoptosis. Cell–cell adhesion and cell–matrix adhesion plays a crucial role in facilitating all the cellular functions. Cell junctions are connections of proteins between cell to cell or cell to matrix. Occluding junction, tight junction, anchoring junctions, adheren junctions and focal adhesions, desmosomes, hemidesmosomes, and gap junctions are present in animal cells. They all help in cell–cell and cell–matrix communication, but differ in their localization and organization. Plant cells have a gap junction called plasmodesmata. Plasmodesmata facilitate communication of signals between cells, including the passage of gene regulatory proteins. The mechanism of transport function of plasmodesmata is unknown. Two families of cell-adhesion molecules (CAMs), cathedrin, and immunoglobin (Ig) superfamilies, are involved in mediating cell adhesion. CAMs are integral proteins distributed throughout the plasma membrane. Cadherin is a glycoprotein family crucial for cell adhesion in embryogenesis. Catenin is required for the function of cadherin. Selectin is a transmembrane protein involved in propelling white blood cells along the surface of the endothelial lining of the blood vessels in order to move into the tissues (to extravasate).

Yet another protein superfamily called immunoglobulin superfamily (Ig superfamily) encompasses proteins containing one or more Ig-like structural domains. Members of this superfamily are calcium-independent homophilic CAMs (e.g., neural cell adhesion molecule, N-CAM, and fasciclin III, FAS3) or heterophilic CAMs, (e.g., intercellular adhesion molecule, ICAM). N-CAM promotes fine-tuning of cell adhesion, and subsequent cell sorting, for tissue organization. FAS3 mediates the adhesion of motor neuronal junctions to the muscle cells, via synaptic connection. Cell surface adhesion receptors (CSARs) are proteins, (e.g., integrin and immunoglobulin superfamilies CAMs) that mediate cell–matrix adhesion. Integrin is a transmembrane glycoprotein. The binding of the intracellular domain to actin makes the integrin more active and provide a conformation to the integrin enabling tighter binding with its ligand. Integrin plays a role in the final step of leucocyte migration.

Extracellular matrix (ECM) is an intricate network of macromolecules, such as proteins and polysaccharides, occupying the extracellular space, secreted by the cells called fibroblasts located in the extracellular space. Through its components, such as GAGs, collagen, fibronectin, laminin, perlecan, and nidogen, ECM maintains its structural organization and performs its function. Basal lamina (BL) is a sheet-like layer of ECM with thickness of 60 to 120 nm, secreted by the epithelial cells. It performs specific functions, such as tissue regeneration, embryonic development, and formation of blood–brain barrier, molecular filtration in kidney cells, cell migration, and wound healing. A balance between the formation and degradation of ECM is very essential for normal functioning of the cells. Degradation of ECM is an enzyme-dependent process catalyzed by matrix metalloproteases (MMPs) and serine proteases.

Cell wall is the extracellular matrix of plant cells. It is thick, strong, tough, and more rigid than the animal cell matrix. It is composed of a pectin matrix over which a network of cellulose with glycan cross links are embedded. Cell walls functions as a skeleton and provide mechanical support to the whole cell. Other function of cell wall include transport, carbohydrate repository, storage of signalling molecules, regulation of cell growth, and providing resistance to turgor pressure.

TERMS TO REMEMBER

Adherens junctions Anchoring junctions of cadherin family, which interlink actin filaments of two cells.

Anchoring junctions Transmembrane proteins which connect cytoskeletal filaments of a cell to the ECM or to the cytoskeleton of the neighbouring cells.

Basal laminae A sheet-like layer of ECM with thickness of 60 to 120 nm, secreted by the epithelial cells.

Cadherin A family of glycoproteins, often called cadherin super family, crucial for cell adhesion in embryogenesis.

Cell junctions Specialized cell–cell or cell–matrix junctions which usually appear like discrete patches of protein strands.

Cell wall The extracellular matrix of plant cells composed of a pectin matrix over which a network of cellulose with glycan crosslinks is embedded.

Cell-surface-adhesion receptors (CSARs) Proteins that mediate contact between the cell and the ECM, thus establishing the cell–matrix adhesion.

Collagen The most abundant protein in mammals predominant in fibrous tissues made of triple helix

Desmosomes Desmosomes are junctions made of proteins from cadherin family and responsible for maintaining the integrity of the skin epithelial cells.

E-cadherin Cadherin found in epithelial cell.

Elastin A stretchable fibrous protein network distributed in tendons, ligaments, and blood vessel walls. It is crucial for the renewal of dead cells in skin, especially during wound healing.

Extracellular matrix (ECM) An intricate network of macromolecules, such as proteins and polysaccharides, occupying the extracellular space.

Fasciclin III (FAS3) Member of Ig superfamily responsible for adhering motor neuronal junctions to the muscle cells, thus, making a synaptic connection.

Fibronectin A glycoprotein of ECM which organizes the matrix as well as promotes cell adhesion.

Focal adhesions Anchoring junction made of transmembrane adhesion proteins of integrin family, and function in connecting actin filaments of the cytoskeleton with the ECM.

Gap junctions Small pores of 2 nm, interconnecting two cells in order to pass smaller molecules or signals between the cells.

Hemidesmosomes Junctions made from transmembrane proteins of integrin family, used to connect the intermediate filaments of the cells with the ECM.

Heterophilic adhesion Adhesion between two cells mediated by CAMs of different types.

Heterotypic adhesion CAM-mediated adhesion between cells of different types.

Homophilic adhesion Adhesion between two cells due to CAMs of same type.

Homotypic adhesion CAM-mediated adhesion between cells of same type.

Ig superfamily Protein family containing one or more Ig-like structural domains. They can function as calcium-independent homophilic CAMs or heterophilic CAMs.

Integrin A transmembrane glycoprotein that plays a role in cell signalling and in the final step of leucocyte migration.

Laminin The major multiadhesive protein of basal lamina, functioning as the primary laminal ligand for many integrins.

Matrix metalloproteases (MMPs) Enzymes that degrade ECM proteins in the presence of metals, such as calcium and zinc ions.

N-cadherin Cadherin that is predominant in nerve cell.

Neural cell-adhesion molecule (N-CAM) Member of Ig superfamily which function in fine tuning of cell adhesion and subsequent cell sorting that are required for tissue organization.

Nidogen A dumb-bell-shaped glycoprotein of basal lamina that cross links the basal laminal components – laminin, type IV collagen, and perlecan.

Occluding junctions Tight sealing junctions of epithelial cells which function as diffusion barrier or permeability barrier for specific molecules and ions.

P-cadherin Cadherin distributed in placenta and epidermis.

Perlecan A proteoglycan of basal lamina that forms network with ECM components, including laminin and nidogen, cell-surface receptors, and growth factors.

Plasmodesmata The only intercellular junction found in plants to establish direct contact between two cells.

Selectin A transmembrane protein with specificity for oligosaccharides and EGF.

Turgor pressure The pressure or force in the plant cell wall exerted by the cellular water (in vacuole).

Type IV collagen The main structural component of basal lamina that forms a two-dimensional network with the laminin.

REVIEW QUESTIONS

1. Tabulate the types and functions of cell junctions.
2. Differentiate between connexin and connexion.
3. Write the role of claudin and occludin in cell communication.
4. What are ZO proteins?
5. Define desmoglein and desmicollin.
6. How are anchor junctions classified?
7. What is shaking-B?
8. Define plasmodesmata.
9. Write the functions of plasmodesmata.
10. What are CAMs?
11. Define homotypic and homophilic cell adhesion.
12. Illustrate the mode of cis-trans mechanism in cell adhesion with a suitable diagram.
13. Describe the structure and adhesive function of cadherin.
14. Comment on the role of catenin in cell adhesion.
15. What is protocadherin?
16. Selectin and integrin help body to fight infection. Justify this statement.
17. Explain about N-CAM.
18. Draw and explain the structural moieties in integrin receptors.
19. What are the proteins used by integrin as anchors?
20. How is FAK related to integrin?
21. Write the significance of plasmodesmata.
22. Explain the structure and formation of collagen.
23. What is tropoelastin?
24. What is Marfan's syndrome?
25. Describe the structure and function of elastin.
26. Define the following: (a) Type III fibronectin repeats, (b) RGD sequence, and (c) synergy sequence.
27. Fibronectin is responsible for the organization of ECM. Explain.
28. What is the difference between basal lamina and basement membrane?
29. Around 45 isoforms of laminin exist. Give reason.
30. Draw the organization of basal laminal components.
31. What is nidogen? What is its function?
32. Enlist the functions of basal lamina.
33. What is the principle behind plant wilting?
34. Explain various models explaining the structure of plant cell wall.
35. What are the functions of cell wall?
36. What is turgidity?
37. Write about the role of MMPs in the regulation of ECM turnover.
38. Explain three mechanisms by which proteases are regulated to control the ECM turnover.

MULTIPLE-CHOICE QUESTIONS

1. Hemidesmosomes connect
   (a) actin and ECM
   (b) tubulin and ECM
   (c) intermediate filament and ECM
   (d) cell membrane and ECM
2. Which of the following is used in promoting leucocyte migration in blood vessels?
   (a) integrin
   (b) collagen
   (c) nidogen
   (d) basal lamina
3. Cadherin is a
   (a) nucleoprotein
   (b) glycoprotein
   (c) lipoprotein
   (d) synaptic junction
4. Fine-tuning of cell adhesion is facilitated by
   (a) N-CAM
   (b) catenin
   (c) plasmodesmata
   (d) selectin
5. Basal lamina is secreted by
   (a) epithelial cells
   (b) neural cells
   (c) plant cells
   (d) placental cells
6. The ECM component that is dumb-bell shaped is
   (a) perlecan
   (b) tropocollagen
   (c) nidogen
   (d) integrin

7. Turgor pressure is established in the
   (a) cell wall
   (b) nucleus
   (c) vacuole
   (b) plasmodesmata
8. Structural protein of occluding junction is
   (a) claudin
   (b) desmoglin
   (c) desmoplakin
   (d) laminin
9. Gap junction is formed by
   (a) collagen
   (b) laminin
   (c) connexin
   (d) selectin
10. Focal adhesions connect
   (a) actin and ECM
   (b) tubulin and ECM
   (c) intermediate filament and ECM
   (d) cell membrane and ECM
11. N-cadherin is predominant in
   (a) gap junction
   (b) anchor junction
   (c) adheren junction
   (d) synaptic junction
12. Perlecan connects collagen with
   (a) integrin
   (b) laminin
   (c) cadherin
   (d) anchor protein

Answers to Multiple-choice Questions

1. (c)	2. (a)	3. (b)	4. (a)	5. (a)
6. (c)	7. (c)	8. (a)	9. (c)	10. (a)
11. (c)	12. (b)			

CHAPTER 11

Cell Signalling

Tony Pawson was awarded the Kyoto Prize (also known as 'Japan's Nobel') in 2008 for proposing and proving the concept of adaptor molecules in signal transduction.

After reading this chapter, students should be able to understand and appreciate the following:

- Molecular tools involved in cell signalling
- Role of ligands, receptors, and intracellular molecules in signalling
- Types of cell signalling
- Various cellular responses produced by cell signalling
- Different types of intracellular signalling pathways
- Proteolysis-dependent signalling pathways
- Cell-signalling network and cross-talk
- Specificity of cell signalling
- Regulation of cell signalling
- Signalling in plant cells

Cells communicate with the external environment, other cells, and different intracellular compartments by transmitting and processing the signals received from these locations. This process is called *cell signalling.* When the signal is relayed from one cell to the other, it is called intercellular signalling pathway. Intercellular signalling pathway is critical for the processing of sensory information. External stimuli, such as optical and acoustic signals, stress, gradients of nutrients, and so on, are registered in sensory cells (refer Terms to Remember) and are transmitted to other cells of the organism via the intercellular pathway. If the signal is relayed within the cell, it is called an intracellular signalling pathway.

11.1 MOLECULAR TOOLS INVOLVED IN CELL SIGNALLING

The three basic components that are involved in cell signalling are *extracellular signalling molecules*, *cellular receptors*, and *intracellular signalling molecules*. These components form a cascade system and function as a signalling pathway. This section discusses the three components.

11.1.1 Extracellular Signal Molecules

Eukaryotic cells communicate with each other by means of signalling molecules, which include amino acids, proteins, nucleotides, steroids, and fatty acid derivatives. The signalling molecules are released out into the extracellular space, which in turn act on another cell, the responding cell or the target cell. As a result, the behaviour of the target cell is altered. Sometimes, the target cell may itself be the signalling cell. These two types of cell signalling are depicted in Fig. 11.1.

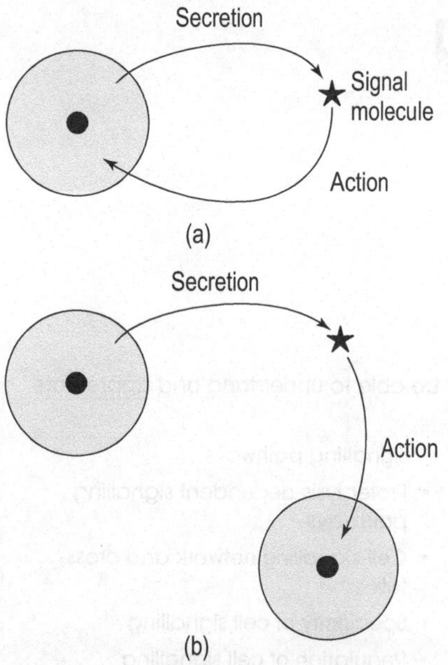

FIGURE 11.1 Two types of cell signalling: (a) intracellular signalling in which the signal molecules produced in a cell act on the same cell and (b) intercellular signalling in which the signal molecules produced by one cell act on a different target cell

TABLE 11.1 Extracellular signalling molecules that bind to cell-surface receptors for signal transduction

Family of molecule	Example
Peptides	Hormones
	Growth factors
	Neuropeptides
	Neurotropin
Neurotransmitters	Glutamic acid
	GABA
	Aspartic acid
	Glycine
	Vasopressin
	Somatostatin
	Neurotensin
	Endorphin
	Norepinephrine
	Dopamine
	Serotonin
	Acetylcholine
Eicosanoids	Prostaglandin
	Prostacyclin
	Thromboxane
	Leucotriene

Depending on the receptor molecules to which they bind, the extracellular signal molecules are divided into two types: those that bind to cell-surface receptors and those that bind to intracellular receptors. Both these types are discussed here.

Extracellular signal molecules that bind to cell-surface receptors

Various families of extracellular signal molecules act by binding to cell-surface receptors. Families and examples are summarized in Table 11.1. Peptides, neurotransmitters, eicosanoids, and leucotrienes bind to cell-surface receptors.

Peptides are the major kind of extracellular signalling molecules. They are of the following types:

- Peptide hormones (insulin, glucagon, and hypophyseal hormone)
- Neurotropin (neural growth factor, NGF)
- Neuropeptides (enkephalin and endorphin)
- Growth factors which control cell growth and differentiation, for example, epidermal growth factor (EGF), which stimulates cell proliferation in embryonic development, and platelet-derived growth factor (PDGF), which is required for blood clotting.

Neurotransmitters are chemicals that allow the transmission of signals from one neuron to the next across the synapses (the junction between two neurons). Three classes of compounds function as neurotransmitters. They are listed as follows:

- Amino acids, which include aspartate, glutamate, GABA, and glycine
- Monoamines, which include norepinephrine, dopamine, serotonin, and, acetylcholine
- Peptides, which include vasopressin, somatostatin, neurotensin, and endorphin

Among these neurotransmitters, glutamate and GABA are the primary ones. Monoamines and acetylcholine exert their action in specific structures, whereas peptides are functional in hypothalamus and other parts of the brain.

Neurotransmitters, secreted as signal molecules at the synapses in response to action potential, subsequently act on the receptor of the target cells, such as muscle cells. Guanine nucleotide binding protein (G-protein) links the neurotransmitters to the intracellular signalling cascade or intracellular response. Thus, neurotransmitter receptors are G-protein linked receptors.

Eicosanoids is a family of extracellular signalling molecules generated from arachidonic acid—which is a polyunsaturated fatty acid with 20 carbon atoms. It consists of four members: prostaglandin, prostacyclin, thromboxane, and leucotriene. Prostaglandin, prostacyclin, and thromboxane are collectively called prostanoids. Eicosanoids are excreted out of their host cells in order to exert their effect on other cells by interacting with the G-protein coupled receptors. The mechanism of action is through stimulation or inhibition of cGMP formation, or activation of phosphatidylinositol signal pathway which leads to intracellular calcium ion release.

Extracellular signalling molecules that bind to intracellular receptors

Some of the signalling molecules bind to intracellular receptors, such as cytosolic receptors and nuclear receptors. In such cases, the signalling molecules traverse through the cell membrane via passive diffusion and reach the intracellular compartment. If the signal is to be mediated through the cytosolic receptor, the signalling molecule will mediate its action within the cytoplasm. If the nuclear receptor is the target, the signalling molecule will move further along the cytoplasm and cross the nuclear membrane to reach the nuclear receptor. The signalling molecules that act on intracellular receptors are usually lipophilic molecules, such as steroid hormones and vitamin-A derivatives. Details of intracellular-receptor-mediated signalling are discussed in the section on 'Intracellular receptor for hydrophobic signal molecules'.

11.1.2 Receptors

As mentioned earlier, receptors are another important tool involved in cell signalling. A receptor may be extracellular (located in cell membrane, hence called transmembrane receptor; a part of this receptor projects from the cell-surface receptor, therefore, it is also called a cell-surface receptor) or intracellular (located in cytosol or nucleus).

On encountering the signal molecule, the receptor gets activated and transduces the external signal into an intracellular cascade or chain. In this cascade, intracellular proteins are activated (these proteins are called downstream target proteins cascade) which finally activate the target proteins leading to a change in the cell behaviour. Alternatively, some molecules may activate enzymes instead of binding to intracellular receptors. A model of pathway mediated by extracellular and intracellular receptors is shown in Fig. 11.2. Intracellular events activated during cell signalling constitute the intracellular pathway. Details of intracellular pathways are discussed later in this section.

Transmembrane receptors or cell-surface receptors for hydrophilic molecules

Transmembrane receptors are membrane-bound proteins with an extracellular domain for binding to the external signal molecule and to receive the signal from the molecule. The extracellular domain is linked to the intracellular domain via a transmembrane domain. The signal is transmitted through the transmembrane domain to reach the inner part of the cell. Thus, when the transmembrane receptor is involved in cell signalling, the external

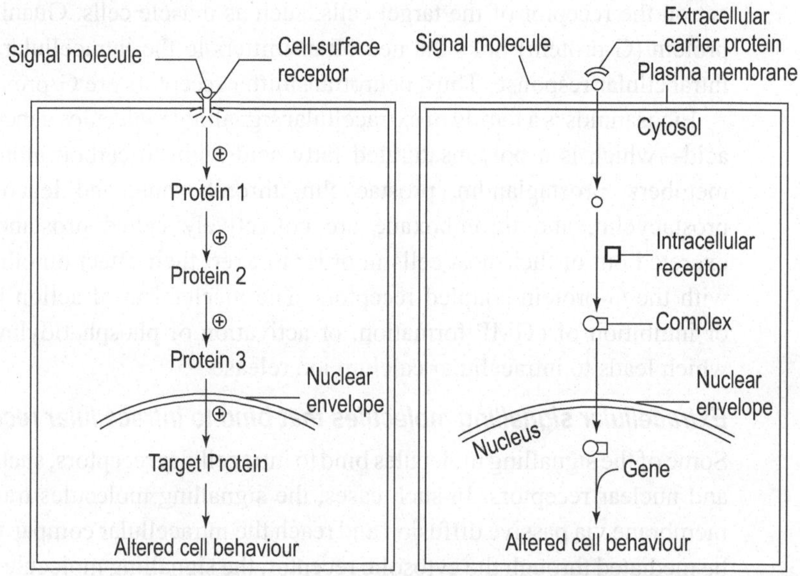

FIGURE 11.2 Models of signalling pathway mediated by extracellular and intracellular receptors

signal molecule does not have to get into the cell. According to the mechanism of transduction, these receptors are divided into three classes discussed as follows:

Ion-channel receptor or Ionotropic receptor Ion-channel-linked receptors are multipass transmembrane proteins (Fig. 11.3) which function like a channel in the membrane. They are located in electrically excitable cells. Ion channels are transiently opened or closed by neurotransmitter signal molecules, influencing the permeability of the ions across the membrane. These receptors regulate the transport of ions, such as sodium, potassium, chloride, and calcium, across the membrane. In response to neurotransmitter binding, these receptors function by two mechanisms. One is the direct mechanism in which the channel opens transiently, for the duration of some milliseconds, due to binding of the ligand. In the second mechanism, a chemical reaction occurs due to binding of the ligand. After the reaction, the channel opens for a few seconds to a few minutes.

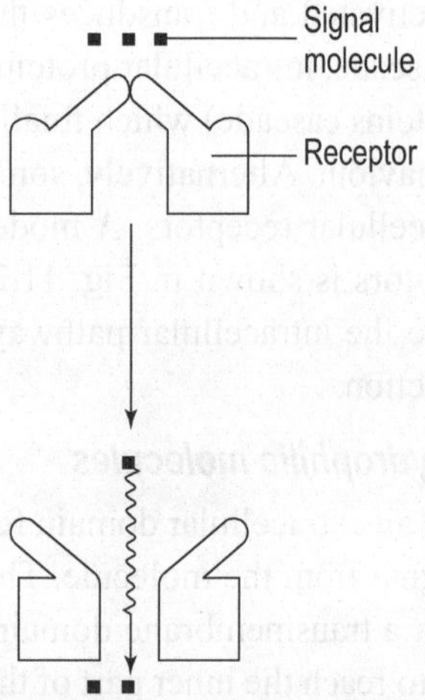

FIGURE 11.3 Ion-channel-linked receptor

G-protein-linked receptor or serpentine receptor G-protein-linked receptors are the largest family of eukaryotic cell-surface receptors that transduce signals from diverse group of signal molecules, such as drugs, amino acids, hormones, neurotransmitters, proteins, and fatty acids. These are superfamily receptors, whose members include rhodopsin receptor present in the eyes (which responds to light), olfactory receptor present in the nose, and mating receptor, which recognizes secretion of mating factor in *Saccharomyces cerevisiae*. They are seven-pass transmembrane proteins composed of a single polypeptide chain (Fig. 11.4). As they move back and forth seven times across the membrane, G-protein receptors are also called *serpentine receptors*.

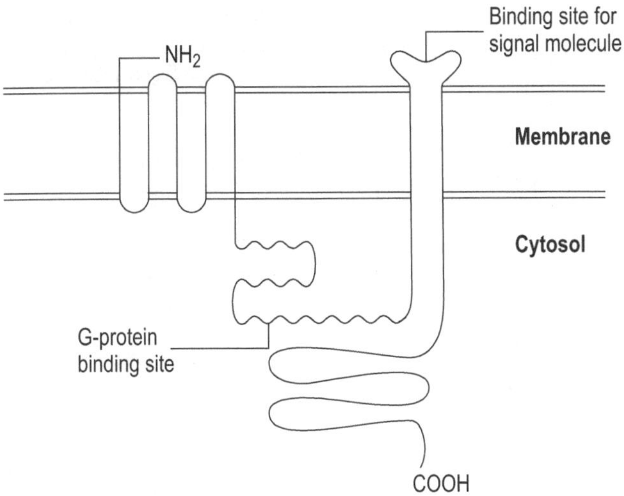

FIGURE 11.4 Structure of G-protein-linked receptor (serpentine receptor)

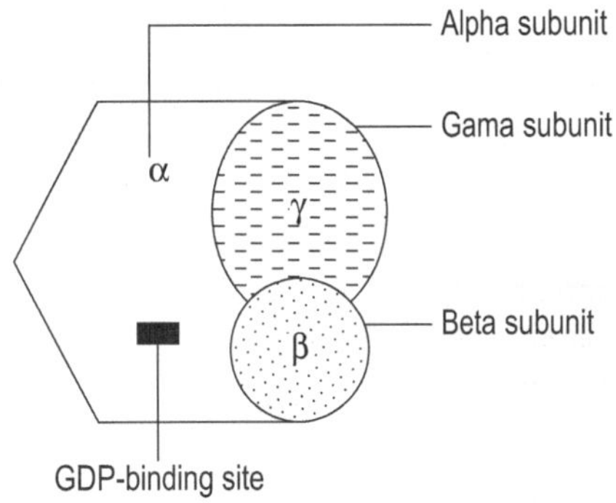

FIGURE 11.5 Structure of G-protein

Structure and function of G-protein Serpentine receptors are linked to a trimeric GTP-binding protein called G-protein. G-proteins are specific for particular serpentine receptors and for a specific group of downstream protein cascade. Structurally, G-protein is a trimer (Fig. 11.5) made of three subunits—α subunit (GTPase enzyme), β-subunit, and γ-subunit. The protein is linked to the serpentine receptor at the cytosolic face (the portion of the receptor facing the cytosol, i.e., facing the intracellular region of the membrane). In the absence of a signalling molecule, the α subunit is bound to GDP and the G-protein as such is inactive. When the signal molecule binds to the serpentine receptor, the later undergoes a conformational change which is accompanied by the activation of G-protein. In this process of G-protein activation, GDP is exchanged for GTP in α subunit with the dissociation of β- and γ-subunits. The α subunit, in turn, binds to the downstream protein forming an α-signal protein complex. This complex further binds to a second protein of the cascade and so on. The signal cascade finally interacts with the target protein which may be an enzyme or an ion channel. If the target protein is an enzyme, it results in an alteration in the intracellular mediator level. Activation of ion channel protein leads to alteration in ion permeability of the plasma membrane. After the response is produced, the GTPase activity of the α subunit cleaves the bound GTP and re-forms the α–β–γ trimer, regenerating the inactive G-protein. Thus, the G-protein cycles between the active GTP-bound monomeric form and the inactive GDP-bound trimeric form. The G-protein cycle is shown in Fig. 11.6. The mechanism of action of G-protein-linked receptor and G-protein is shown in Fig. 11.7.

Role of regulator of G-protein (RGS) in G-protein-mediated response Human genome codes for around 25 RGSs, each of which is specific for a particular G-protein. RGS is a protein which functions as an activator of α subunit GTPase of G-protein, thereby hydrolyzing the GTP to GDP. This results in the inactivation of G-protein. Because of this function, RGS is called GTPase activating protein (GAP). RGS is responsible for shutting of G-protein-mediated responses in eukaryotic cells.

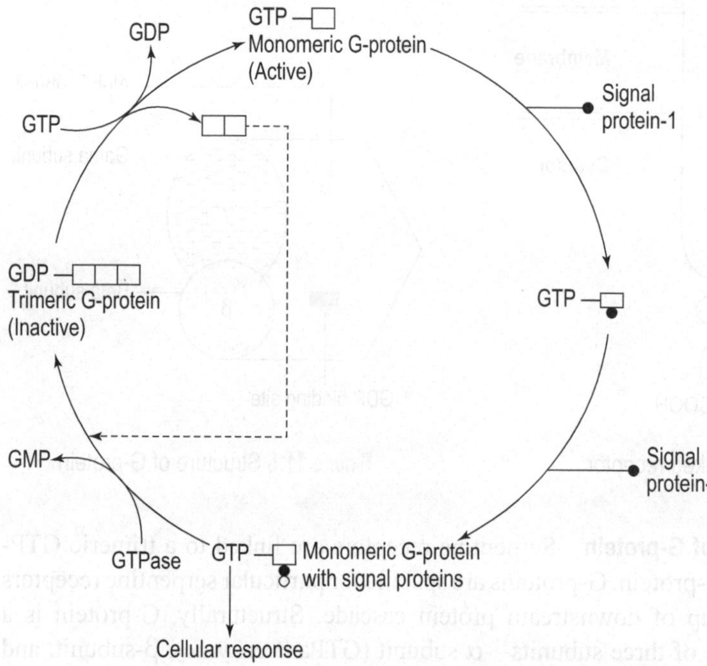

FIGURE 11.6 G-protein cycle

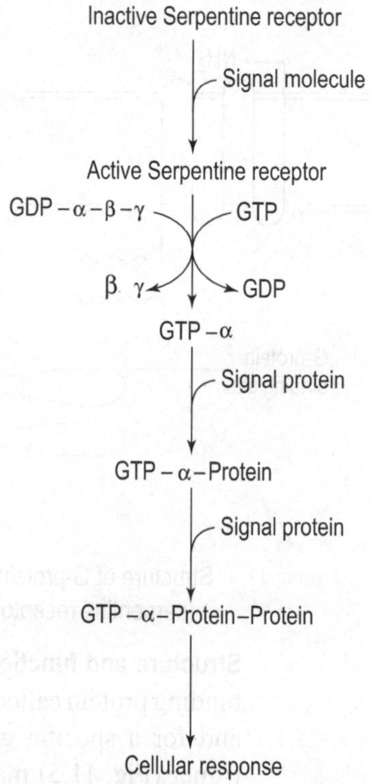

FIGURE 11.7 Mechanism of action of G-protein-linked receptor and G-protein

Enzyme-linked receptor Enzyme-linked receptors (ELRs, Fig. 11.8) are single-pass transmembrane protein with two domains— the extracellular domain that binds to the signal molecule and the intracellular domain that binds to the target enzyme. Both domains are connected by a transmembrane domain. When the signal molecule binds to the extracellular domain, the enzyme is activated to produce the response. ELR responds to the signals of growth factors. The receptors are activated by growth factors to produce the following types of response in a cell:

- Growth
- Proliferation
- Differentiation
- Survival
- Change in cell movement
- Change in cell shape

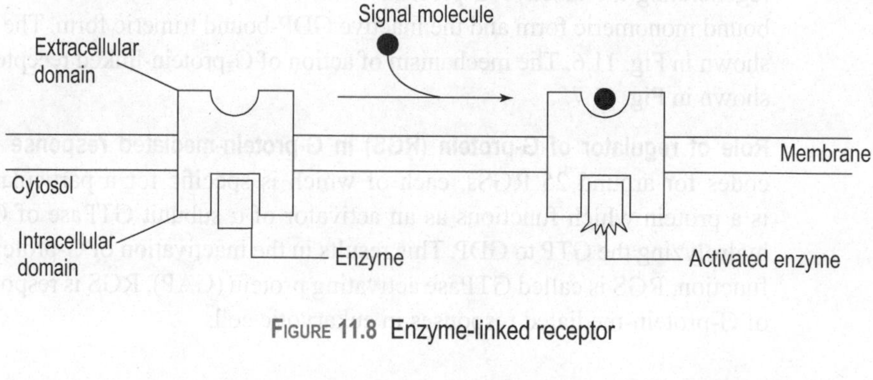

FIGURE 11.8 Enzyme-linked receptor

TABLE 11.2 Enzyme-linked receptors and their functions

Type of ELR	Function
Receptor tyrosine kinase	Phosphorylates specific tyrosine residues on the intracellular signalling proteins as well as on the receptor domain itself
Tyrosine kinase associated receptor or non-receptor tyrosine kinase	Associates with intracellular signalling proteins that has tyrosine kinase activity
Receptor serine/threonine kinase	Phosphorylates specific serine or threonine residues on intracellular latent gene regulatory proteins
Receptor guanylyl cyclase	Catalyzes the production of cyclic GMP from GTP in the cytosol
Histidine kinase associated receptor	Activates a two-component signalling pathway in which the kinase phosphorylates itself on the histidine residue and then immediately transfers the phosphate to a second intracellular protein

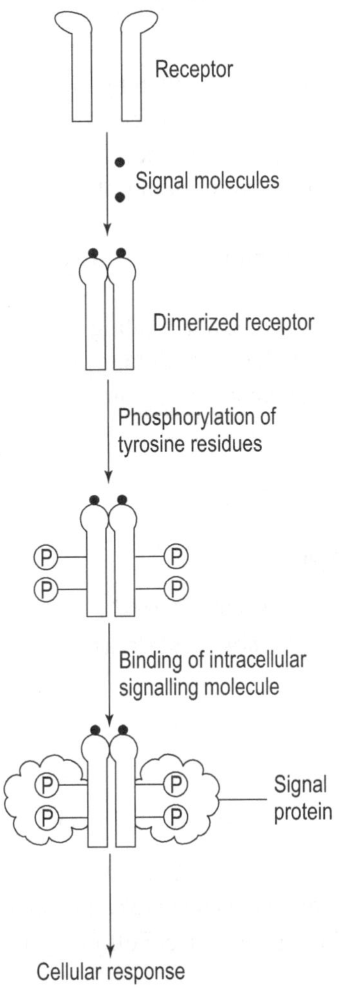

FIGURE 11.9 Signalling mediated by tyrosine receptors

Types of ELR There are five types of ELR. They are listed as follows:

(a) Receptor tyrosine kinase (RTK)
(b) Tyrosine kinase associated receptor (TKAR) or non-receptor tyrosine kinase
(c) Receptor serine/threonine kinase (RSTHK)
(d) Receptor guanylyl cyclase (RGC)
(e) Histidine kinase associated receptor (HKAR)

Functions of ELR in cell signalling are summarized in Table 11.2 and discussed in this section.

(a) Receptor tyrosine kinase As mentioned in the previous section, receptors are linked to enzymes for signalling the cell. Sometimes the receptor itself will have the enzyme activity which is activated due to binding of signal molecule. In other words, some cell-surface receptors are enzymes. Tyrosine kinase comes into this category and is referred to as tyrosine kinase receptor or receptor tyrosine kinase (RTK). RTK is the most predominant class of ELRs. RTK phosphorylates substrate proteins on tyrosine residues subsequently forming phosphotyrosine. Structurally, RTK has a double polypeptide chain. Binding of signalling molecules to these receptors results in receptor dimerization and autophosphorylation, that is, the two polypeptide chains phosphorylate each other. The resulting phosphorylated tyrosine kinase in turn binds to intracellular downstream molecules via phosphotyrosine residues. Signalling through tyrosine receptors is shown in Fig. 11.9.

Cellular responses produced by RTKs RTKs are targets for a wide range of extracellular signalling molecules. Depending on the signalling ligands that bind to it, RTK produces specific responses. Some of cellular responses produced by the activation of RTK are listed in Table 11.3.

TABLE 11.3 Cellular response elicited by RTK

Molecule	RTK type	Cellular response
Epidermal growth factor (EGF)	EGF receptor	Stimulates cell proliferation
Insulin	Insulin receptor	Utilization of glucose
Insulin growth factor IGF	IGF receptor	Promotes cell growth and survival
Neuronal growth factor (NGF)	TrkA	Promotes neuronal growth and survival
Platelet-derived growth factor (PDGF)	PDGF receptor	Induces growth, survival and proliferation of various cell types
Fibroblast growth factor (FGF)	FGF receptor	Stimulates cell proliferation
Vascular endothelial growth factor (VEGF)	VEGF receptor	Stimulates angiogenesis
Ephrins	Eph receptor	Stimulates angiogenesis and cell migration and regulates of cell adhesion

The signalling molecule binds to two chains or subunits of the RTKs, thereby inducing cross linking of receptor. Cross linking leads to cross phosphorylation of adjacent domains of receptor which in turn stimulate the kinase activity of the receptor. The active RTK domain can function as docking site for intracellular proteins. Cross linking of receptor domains occurs by two ways:

1. Clustering of monomeric signal molecules on cell surface proteoglycans; two such clusters can bind to receptors for cross linking.
2. Clustering of monomeric signal molecules in the cell membrane and binding of those clusters to the receptors in target cell membrane (intercellular cross linking).

Examples for signal molecules inducing different types of cross linking are given in Table 11.4.

TABLE 11.4 Ways of RTK cross linking induced by specific molecules

Method of cross linking	Example
Dimer binding	Platelet-derived growth factor
Clustering on proteoglycan and binding as monomer	Fibroblast-derived growth factor
Intercellular linking between membrane-bound receptor and membrane-bound ligand	Ephrin

Types of cross linking which help relay signal through RTK are shown in Fig. 11.10.

RTK-dependent signalling pathways RTK is responsible for mediating four intracellular signalling pathways. These pathways are named after the final kinases that are responsible for phosphorylation (activation) of the target protein eliciting the response. Following are the intracellular signalling pathways mediated by RTK:

- Calcium–calmodulin (CaM) kinase pathway
- Protein kinase C (PKC) pathway

- Mitogen-activated protein kinase (MAP kinase) pathway
- Protein kinase B (PKB) pathway or PI3/PKB pathway

These pathways are depicted in Fig. 11.11 and discussed in Section 11.4.

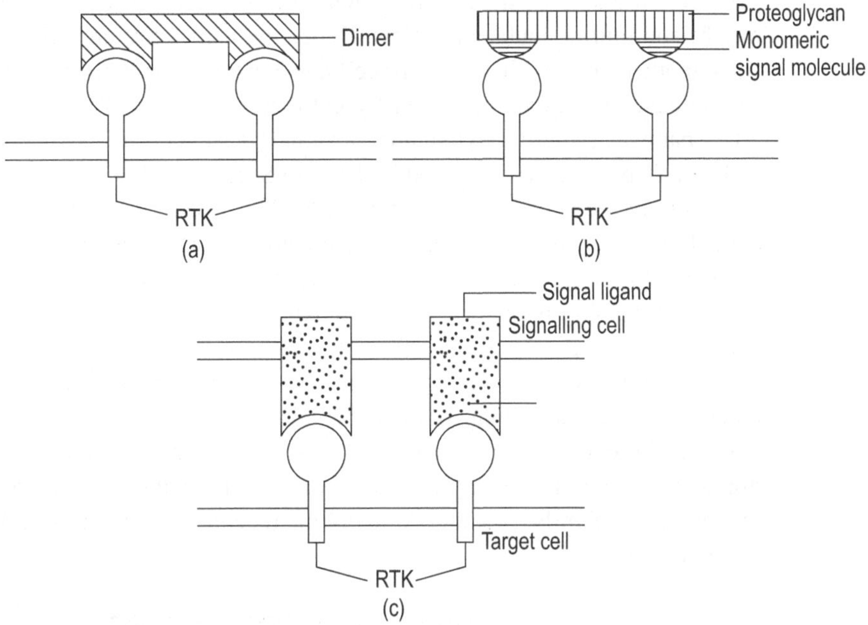

FIGURE 11.10 Cross linking of RTK induced by signalling ligand (a) Cross linking by dimer, (b) Cross linking by proteoglycan-bound clusters, (c) Intercellular cross linking

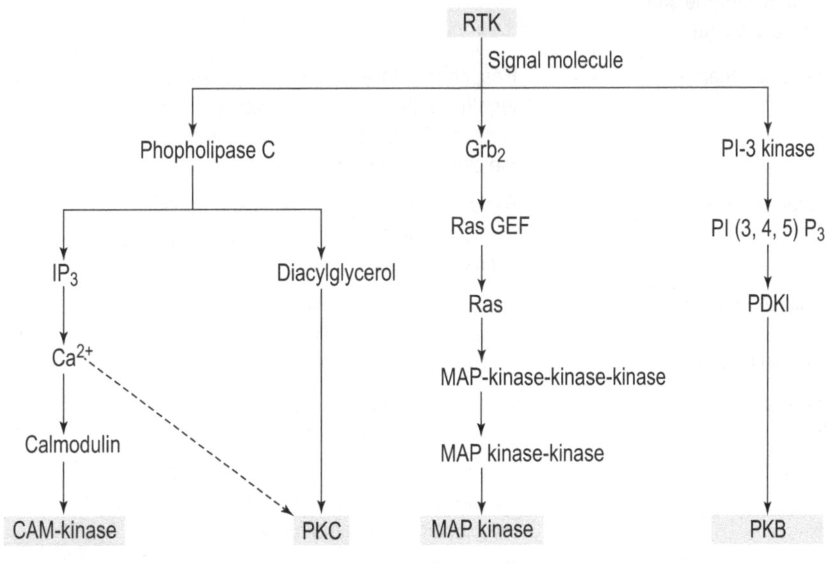

FIGURE 11.11 Pathways mediated by RTK

(b) Tyrosine kinase associated receptor or non-receptor tyrosine kinase Non-receptor tyrosine kinases (NRTKs) are enzymes that are not integral part of the receptors, unlike receptor

tyrosine kinase. They catalyse the transfer of phosphate group from ATP to a tyrosine residue in a protein. Three major classes of non-receptor tyrosine kinases have been identified—Src, Abl, and Jak. Most of these enzymes contain SH2 domains. Originally they were found to be transforming genes of viral genomes, such as Rous sarcoma virus or Abelson murine leukaemia virus. Hence the names Src and Abl. Janus kinase means 'just another kinase'. It is an intracellular non-receptor tyrosine kinase responsible for transducing the signals produced by cytokines. The signal is transduced by the JAK-STAT pathway (where STAT stands for signal transducers and activators of transcription). JAK-STAT pathway is discussed in detail in Section 11.4.

There are receptors which signal the NRTK. These receptors receive the cytokine as signalling molecule and mediate inflammatory and immune response. Because of this, these receptors are called cytokine receptor super family. Members of the cytokine receptor super family include Type I cytokine receptor, Type II cytokine receptor, tumour necrosis factor receptor, and TGF beta receptor. They are membrane-bound receptors with an extracellular domain and an intracellular domain. In Type I and Type II cytokine receptors the extra cellular domain contains a conserved motif and binds to cytokine which is the extracellular signal, while the intracellular domain is associated with SH2 domain of the JAKs. Many cell functions are regulated by members of the cytokine receptor super family. Members of cytokine receptor super family and their functions are given in Table 11.5.

TABLE 11.5 Members of cytokine receptor super family

Receptor that associates with non-receptor tyrosine kinase and responds to immune and inflammatory signal	Function of the receptor the receptor	Specific cells hosting
T-lymphocyte receptor	Responds to foreign antigens, presented together with the major histocompatibility complex (MHC). Subsequently, it regulates the clonal expansion of T cells	T lymphocytes
B-lymphocyte receptor	Responds to foreign antigens and provides first line of defence against infection by microorganisms	B lymphocytes
Interleukin-2 receptor (IL-2R) or Type I cytokine receptor	Responds to the cytokine IL-2 leading to cytolysis and immunoglobin-G secretion.	T-cells, B-cells and NK cells
Immunoglobulin receptors (IgE-R)	Induces hypersensitivity and initiates acute inflammatory responses	Mast cells and blood-borne basophils
Erythropoietin receptors	Plays an important role in the final stage of maturation of erythroid cells into mature red blood cells (haematopoiesis)	Erythroid progenitor cells
Tumour necrosis factor receptor (TNFR) or death receptor	Responds to the inflammatory cytokine TNF alpha causing immune response and necrosis of specific tumour	Expressed in most tissues
Chemokine receptor	Induces the onset of chemotaxis	Expressed in most tissues

(c) Receptor serine/threonine kinase Receptor serine/threonine kinases (RSTHKs) are ELRs that phosphorylate serine/threonine residues in intracellular proteins. They are single-pass transmembrane proteins, the cytosolic domain of which possesses serine/threonine kinase activity. Figure 11.12 shows a schematic representation of RSTHK. There are two types of RSTHK—Type 1 and Type 2. Both are homodimers and structurally similar. Ligands binding to RSTHK include members of TGF β-super family, such as TGF β, activin, and bone morphogenic protein (BMP). The ligands first bind to and activate Type 2 RSTHK, which in turn recruits and phosphorylates and activates the Type 1 RSTHK, forming an active tetrameric receptor complex. The complex then binds to and phosphorylates a latent gene regulatory protein of *smad* family, including smad 2, smad 3, and smad 5 or smad 8. Phosphorylated smad dissociates from the Type 2 RSTHK and binds to smad 4, which then complexes with any of the five smads. The smad complex then translocates into nucleus, where, in association with other regulatory proteins, it binds to and regulates the target genes—smad 6 gene and smad 7 gene. These genes encode smad inhibitory protein. The steps in RSTHK-mediated activation of smad is listed below and is shown in Fig. 11.13.

1. Signal molecule binds to and activates Type 2 RSTHK homodimer
2. Type 2 RSTHK recruits and phosphorylates Type 1 RSTHK
3. Type 1 RSTHK recruits and phosphorylates smad 2 or smad 3, 1/5/8
4. Phosphorylated smad dissociate from Type 1 receptor and bind with smad 4 to form a complex
5. Smad complex translocates into nucleus, associates with other regulatory protein and binds to SRE and initiate expression of adjacent target gene. White zone represents cytosol and shaded zone represents nuclear compartment.

(d) Receptor guanylyl kinase Receptor guanylyl kinases (RGCs) are ELR which are neither kinases nor associated kinases. They are single-pass transmembrane proteins with an extracellular domain for ligand binding and an intracellular domain facing the cytosol with guanylyl cyclase activity. Most of the RGCs are orphan receptors, that is, receptors whose specific endogeneous ligand is not defined or known so far. The well-known signal molecule that binds to RGC is the natriuretic peptide which activates guanylyl cyclase to form cGMP. The final cellular response of this signal is the dilation of blood vessels and a fall in blood pressure. cGMP activates cGMP-dependent PKs (PKG) which in turn phosphorylate serine/threonine residues in specific intracellular signal proteins. cGMP-mediated pathway is discussed in Section 11.4.

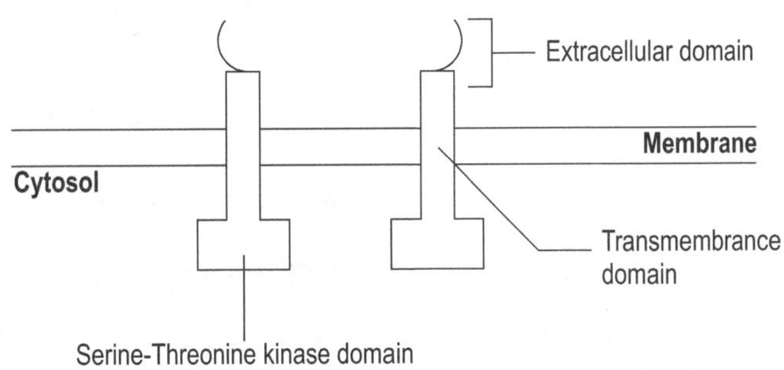

FIGURE 11.12 A receptor serine/threonine kinase

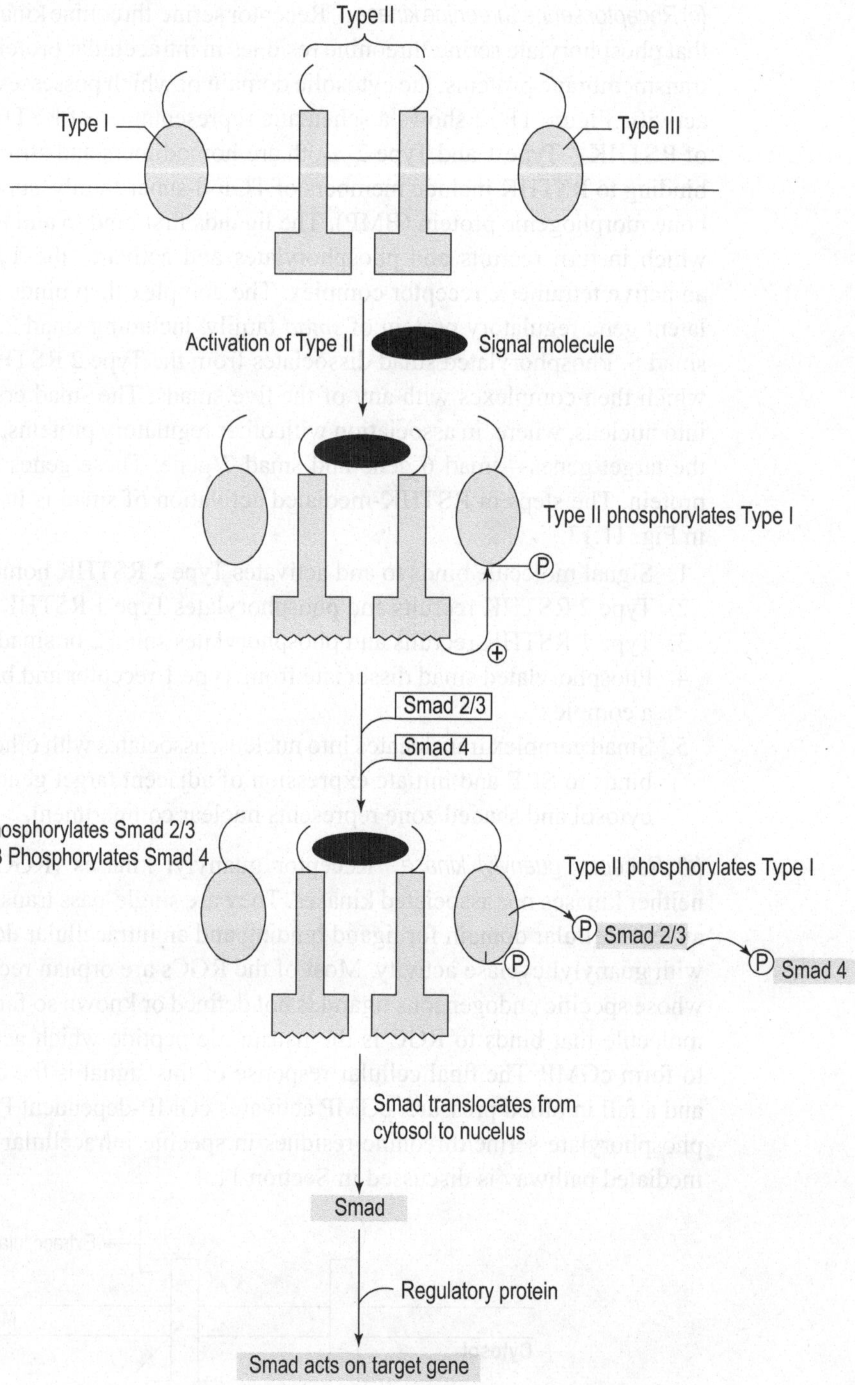

FIGURE 11.13 RSTHK-mediated smad pathway

(e) Histidine kinase associated receptor and two-component pathway Histidine kinase associated receptors (HKAR) are chemotactic receptors present in bacterial cells. They produce two types of responses, tumbling and swimming, depending on the extracellular

signalling molecules. Repellents (noxious chemicals) and attractants (nutrients, such as sugars, amino acids, or peptides) are ligands that induce tumbling and swimming, respectively. Tumbling and swimming are brought about by difference in orientation of flagella. Clockwise rotation leads to tumbling and anticlockwise rotation leads to swimming (Fig. 11.14).

As mentioned in Chapter 1, the locomotion of bacterial cell is brought about by flexible hair-like projections called flagella. The structure of flagella is described in Chapter 1. The movement of flagellum is regulated by the pathway mediated by HKAR.

HKAR is a dimeric transmembrane protein with an extracellular domain capable of binding the repellent or attractant and an intracellular domain that is bound to a signal transducing histidine kinase (CheA) via an adaptor protein (CheW). CheA and CheW couple the HKAR to the flagellum and relay the signal onward. When a repellent binds to the receptor, CheA is activated. Active CheA phosphorylates its own histidine residues in the presence of ATP and rapidly transfers the phosphate group to aspartate residue of a messenger protein CheY. Phosphorylated CheY dissociates from the receptor, diffuses in the cytosol, binds to flagella, and rotates the flagella clockwise at the expense of the energy stored in proton gradient across the membrane. Clockwise rotation results in the flagella flying apart and induces tumbling. This process is shown in Fig. 11.15.

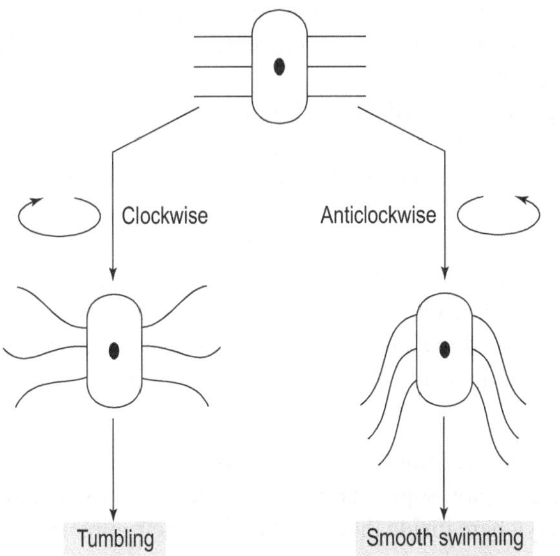

FIGURE 11.14 Tumbling and swimming responses mediated by histidine kinase associated chemotactic receptor in bacteria

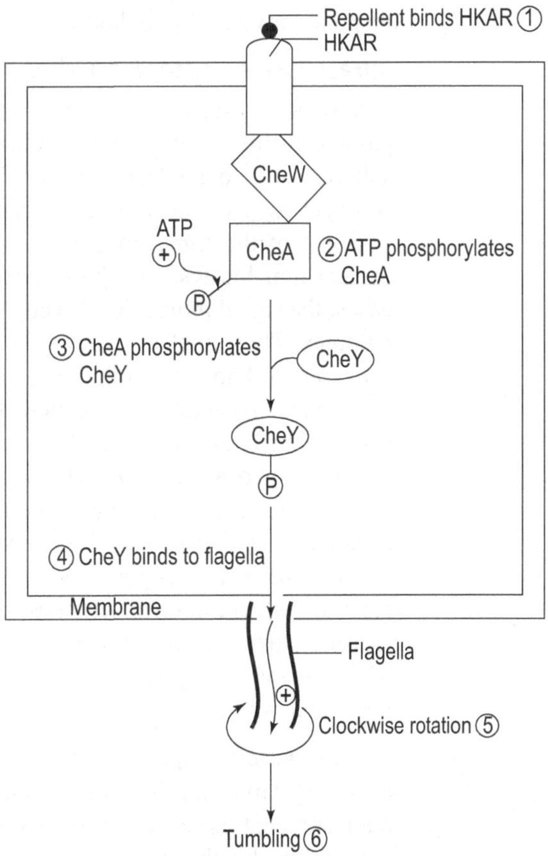

FIGURE 11.15 Mechanism of tumbling process mediated by histidine kinase associated receptor (HKAR) in bacterial cells

TABLE 11.6 Differences between nuclear-receptor-dependent signal molecules and cell-surface-receptor-dependent signal molecules

Factor	Nuclear-receptor-dependent signal molecules	Cell-surface-receptor-dependent signal molecules
Solubility	Hydrophobic	Hydrophilic
Circulatory mode	Circulate in blood by binding to carrier to enhance solubility	Circulate in blood in free or unbound form
Lifetime	Persist in blood for hours (steroid hormone) or even days (thyroid hormone)	Persist in blood for seconds or minutes, after which they are broken down, e.g., neurotransmitters are removed within seconds or milliseconds
Response time	Mediate responses that are long lasting	Mediate responses of short durations

Binding of attractant to HKAR decreases the activity of the receptor and decrease the phosphorylation of CheA and CheY causing anticlockwise rotation of flagella leading to smooth swimming. CheY has auto-dephosphorylation activity which is accelerated by another protein called CheZ. The lifespan of phosphorylated intermediates involved in HKAR-mediated flagellar motion is very short (approximately 10 s). As a result, bacteria respond very rapidly to their environmental conditions.

Intracellular receptor for hydrophobic signal molecules

Hydrophobic signal molecules from extracellular space traverse the cell membrane by passive diffusion and bind to intracellular receptors, such as cytosolic receptor, and subsequently to nuclear receptor. Nuclear-receptor-dependent signal molecules differ from cell-surface-receptor-dependent ones in some aspects as given in Table 11.6.

Some of the signal molecules bind to cytosolic receptor forming a complex which in turn translocates into the nucleus to bind with the nuclear receptor. In some other cases, the signal molecules directly move into the nucleus without forming any cytosolic complex. Thus, nuclear receptor is the target for the signal molecules. At this stage, it is necessary to know about the nuclear receptor and its family.

Nuclear receptors are called ligand-controlled transcription activators. This is because the ligand activates the receptor and the activated receptor in turn stimulates the transcription of specific DNA in the gene. Nuclear receptors are capable of directly binding to DNA sequence called hormone response element (HRE) and regulate the expression of adjacent genes; hence, these receptors are classified as transcription factors. Binding of signal molecules (ligands) to a nuclear receptor induces conformational change in the receptor, which, in turn, activates the receptor. The nuclear receptor/DNA complex then recruits other proteins that transcribe DNA downstream from the HRE into messenger RNA and eventually protein, which causes a change in cell function (cellular response to hormonal binding).

Nuclear receptor super family All nuclear receptors are structurally related and belong to a very large family called nuclear receptor super family. In the absence of ligand, nuclear receptors are accommodated anywhere inside the cell. Based on their intracellular location in the absence of ligand and the mechanism of action, nuclear receptor super family is divided into the following four subfamilies:

- Type I
- Type III
- Type II
- Type IV

Type I nuclear receptor family is accommodated in the cytosol. Only when the ligand binds, the receptor translocates to the nucleus where it binds with inverted repeats of HRE as a monomer and recruits other proteins which transcribe sequences downstream of HRE to form mRNA. The mRNA is then translated into protein that induces cellular response. Direct binding of the Type I nuclear receptor to a DNA hormone response element is referred to as transactivation. Examples of Type I nuclear receptor include sex hormone receptors, progesterone receptor, and glucocorticoid receptor.

Type II nuclear receptor family is put up in the nucleus along with a co-repressor protein. Only when the ligand binds, does the receptor release the co-repressor protein and subsequently bind to co-activator protein and DNA polymerase forming a complex. The complex binds to HRE for transcribing the gene into mRNA which is then translated into protein that induces cellular response. Examples of Type II nuclear receptor include retinoic acid receptor, retinoid X receptor, and thyroid hormone receptor.

Type III nuclear receptor family is lodged in the nucleus, which on encountering a ligand, binds to direct repeats of HRE as a homodimer to transcribe the gene into mRNA which is then translated into protein that induces cellular response. Example of Type III nuclear receptor is still not proposed clearly because the ligands that bind to these receptors are still anonymous. Because of this reason Type III receptors are also called orphan receptors.

Type IV receptor binds to a single half site of HRE either as a monomer or as a dimer and results in gene expression. Examples of Type IV receptors are found in most of the nuclear receptor subfamilies.

Signalling molecules for intracellular receptors Different classes of signal molecules that bind to intracellular receptor for eliciting response are given in Table 11.7. These molecules include steroid hormones, thyroid hormone, vitamin D, and retinoids.

(a) Steroid hormones Steroid hormones form an important class of hydrophobic signalling molecules that are intracellular-receptor dependent. They are classified into mineralocorticoids (aldosterone), glucocorticoids (cortisol), and sex hormones (oestrogen, testosterone, and progesterone). Glucocorticoids and mineralocorticoids are sometimes collectively called corticosteroids. Each of these hormones fulfils different but equally important functions in the body (Table 11.7). All these functions are responses to cellular signals.

Mineralocorticoids are largely involved in kidney functions, such as excretion of salt and water. Glucocorticoids are involved in lipid, carbohydrate, and protein metabolism. Cortisol, a glucocorticoid, binds to cytosolic receptor, forming a complex which subsequently translocates into nucleus and binds to a receptor-binding element. A straightforward reaction scheme can depict the binding of a hormone to a cytosolic receptor. This reaction is called receptor–ligand reaction scheme:

$$[H] + [K] \leftrightarrow [HR]$$

where $[H]$ = Concentration of the hormone; $[K]$ = Concentration of the receptor; and [HR] = Concentration of the hormone–receptor complex.

For the above reaction KD (the equilibrium constant for dissociation) $= \dfrac{[\mathrm{H}] \times [\mathrm{R}]}{[\mathrm{HR}]}$

Usually KD value ranges between 10^{-6} and 10^{-12} M. This formula is applicable for cytosolic receptors only.

TABLE 11.7 Hydrophobic extracellular molecules that bind to intracellular receptor for signal transduction

Family of molecule	Example and structure	Biochemical or physiological role
Steroid hormone	Oestrogen 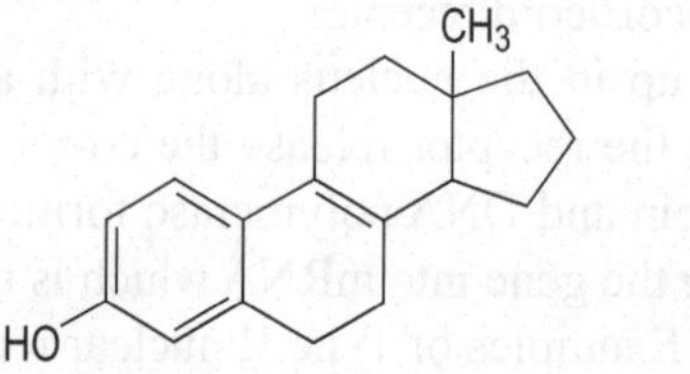	• Growth and differentiation of female reproductive system • Development of secondary sexual characteristics in females • Induces thickening of the endometrium and other aspects of regulating the menstrual cycle
	Testosterone 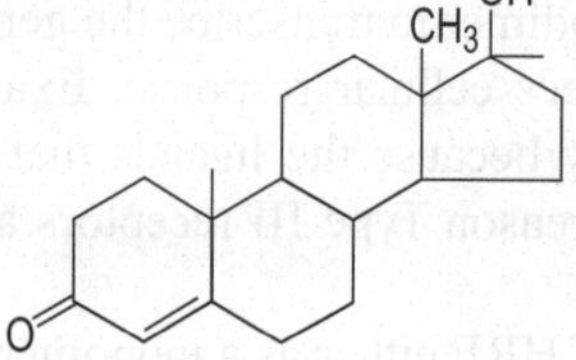	• Growth and differentiation of male reproductive system • Growth of skeletal muscle • Development of secondary sexual characteristics in males
	Progesterone 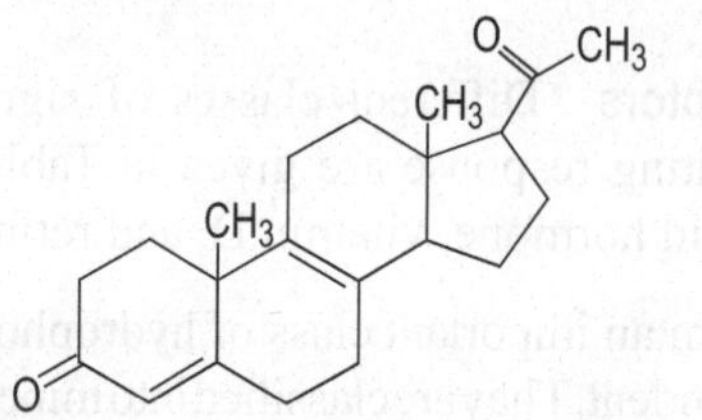	• Implantation of fertilized zygote • Maintenance of early pregnancy
	Corticosteroids 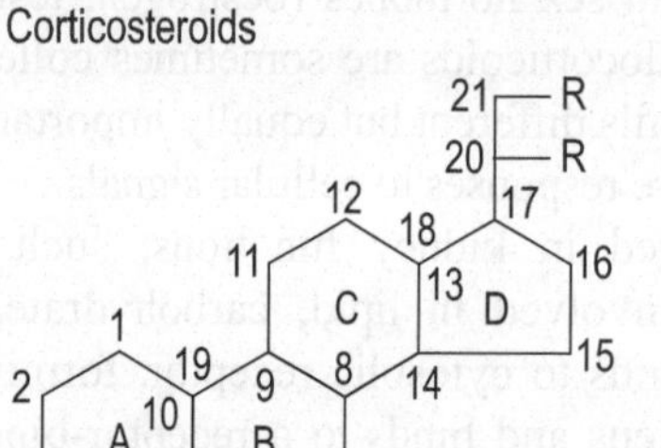	• Maintenance of water and electrolyte balance • Increase in blood sugar and storage of sugar in the liver as glycogen; aids in fat, protein, and carbohydrate metabolism; and aids in suppressing the immune system (and regulating lymphocytes)
Vitamin D	Vitamin D3 (cholecalciferol) 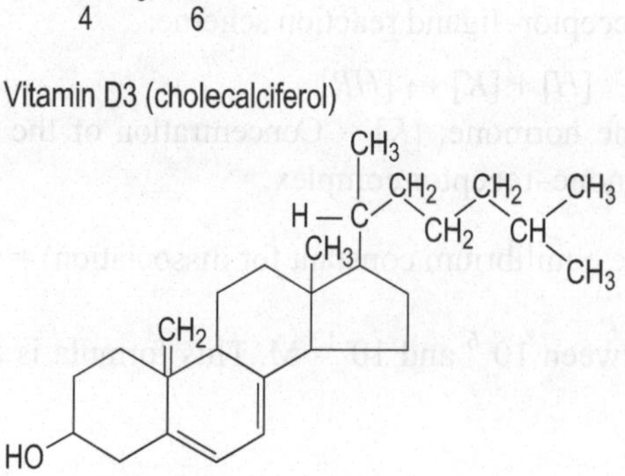	• Intestinal absorption of calcium, phosphate, and magnesium ions • Enhances calcium absorption by stimulating the expression of calcium-binding protein (calbindin) that ferries calcium across the intestinal epithelial cell

(Contd)

TABLE 11.7 (Contd)

Family of molecule	Example and structure	Biochemical or physiological role
Thyroid hormone	Thyroxine (T4) 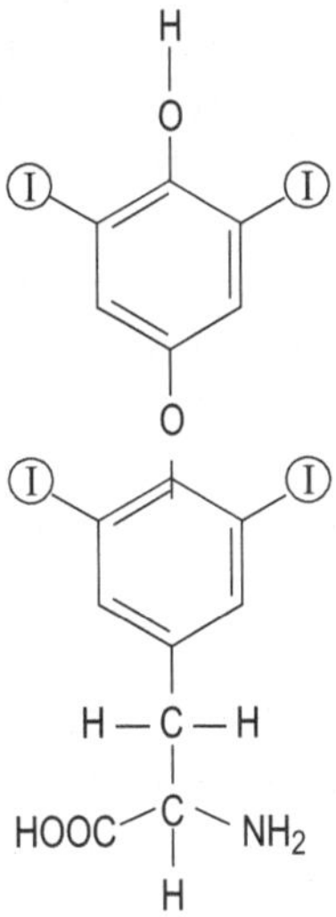	• Regulation of the metabolic processes in all cells of the body
	Triiodothyronine (T3) 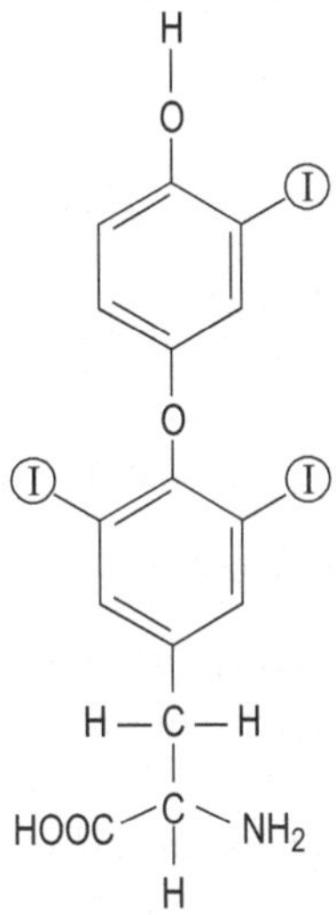	• Increases oxygen consumption, heat formation, and stimulation of glycolysis and protein synthesis
Retinoids	Retinol 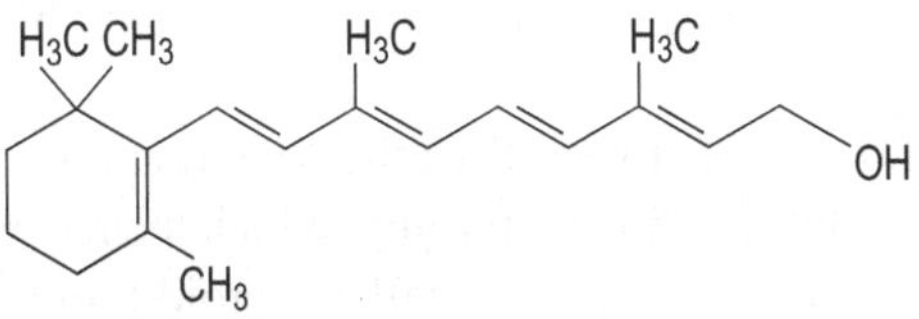	• Important in vision and bone growth • Required in the production of rhodopsin, the visual pigment used in low light levels • Influences the process of cell differentiation, growth, and development of embryos • Maintenance of epithelial cells • Synthesis of glycoprotein • Needed for normal haematopoiesis

Progesterone and oestrogen are typically linked with the female reproductive functions, such as menstrual cycles and ovulation. Testosterone is a male sex hormone responsible for development of male sex organs and secondary sexual characters in males.

Sex hormones, unlike corticoids, pass through cytosol without forming any complex and move directly to the nucleus, where the nuclear receptors are bound to the receptor-binding sequence of the gene. Thus, both types of signal molecules (cytosol-receptor-specific molecules and nuclear-receptor-specific molecules) reach the nuclear response element forming primary proteins. These proteins in turn switch off the primary response element and stimulate the transcription of another gene segment to produce a secondary response which results in the production of secondary proteins (Fig. 11.16); this leads to the final cellular response to the signal. The mechanism of action of steroid hormones is shown in Fig. 11.17.

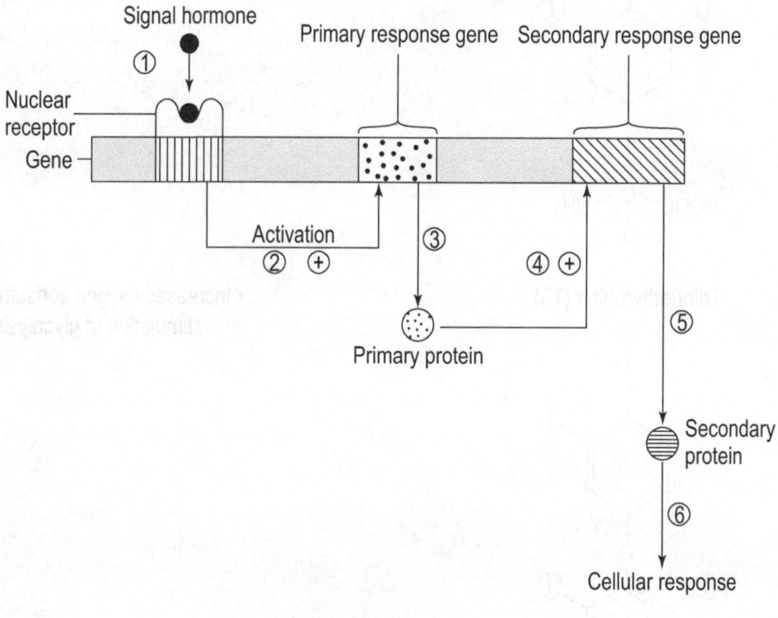

FIGURE 11.16 Response produced by nuclear receptor

Structure of steroid hormone receptors Steroid hormone receptors belong to a diverse family of ligand-activated gene regulators that share a highly conserved structure and common mechanisms affecting gene transcription. They have a molecular weight of 300 to 400 kDa. All steroid hormone receptors are made of three domains (Fig. 11.18).

- Amino terminal domain or transcription regulation domain that bestows immunogenicity and alters transcription in a gene in cell-specific manner.
- Carboxy terminal domain or steroid-binding domain that actually binds to the hormones.
- DNA-binding domain (DBD) lying between the other two domains. DBD is composed of two functionally distinct zinc fingers through which the receptor physically binds directly with the specific regions of DNA called steroid hormone responsive element (SHRE). SHRE is composed of structural region and regulatory region. Structural gene (region) encodes the specific amino acids of the protein, and the

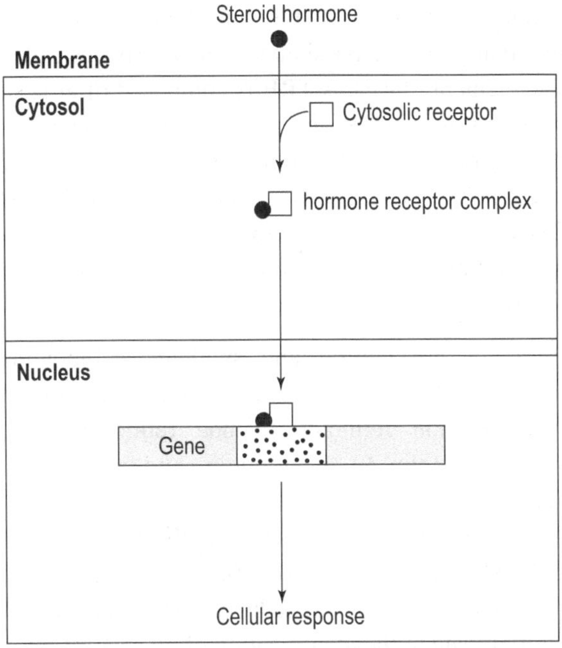

FIGURE 11.17 Mechanism of action of steroid hormones

regulatory gene (region) interacts with various proteins to control the rate of transcription.

(b) Vitamin D Vitamin D is a fat-soluble vitamin. It is of two types: vitamin D2 (ergocalciferol) and vitamin D3 (cholecalciferol). Both vitamin D2 and D3 are collectively called calciferol. Vitamin D3 is synthesized in skin on exposure of 7-dehydrocholesterol (the precursor) to sunlight and carried to liver where it is converted into 25-OH-cholecalciferolbytheenzyme25-hydroxylase. 25-OH-cholecalciferol is converted to active form of vitamin D—the 1, 25-di-OH-calciferol in kidney and macrophages by the enzyme 1-alpha-hydroxylase (Fig. 11.19). Vitamin D3 facilitates intestinal absorption of calcium, phosphate, and magnesium ions. It enhances calcium absorption by stimulating the expression of the calcium-binding protein, calbindin, which ferries calcium across the intestinal epithelial cell.

Vitamin D receptor (VDR) is also called NR111 because it belongs to nuclear receptor subfamily (subfamily 1, group I, member 1). VDR is encoded by VDR gene, and it possesses hormone-binding domain and DNA-binding domain. The VDR binds to vitamin D3 via hormone-binding domain. The receptor then forms a complex with another intracellular receptor, the retinoid-X receptor, and forms a heterodimer which in turn binds to DNA's hormone response elements via the DNA-binding domain, resulting in expression or transrepression of specific gene products.

(c) Thyroid hormones Thyroid hormones (T3 and T4) function by binding to intracellular thyroid hormone receptor. In contrast to steroid hormone receptor, thyroid hormone receptor (THR) remains bound to DNA segments even in the absence of the signalling hormone. When the hormone binds to the receptor, the later undergoes a conformational change leading to transcription of necessary gene to elicit the response. Thyroid hormone receptor proteins are encoded by alpha gene and beta gene. Both genes encode a protein which is the nuclear hormone receptor for triiodothyronine. There are four types of thyroid hormone receptors designated THRα1, THRα2, THRβ1, THRβ2. Structurally all these four types are made of three domains:

- N-terminal domain or transactivation domain
- C-terminal domain or ligand-binding and dimerization domain
- Middle domain or DNA-binding domain

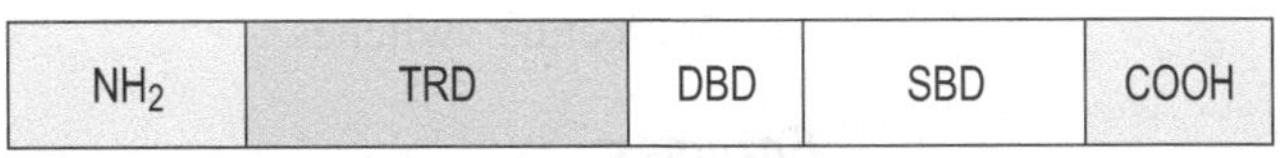

FIGURE 11.18 Structural domains of steroid hormone receptors that transmit the hormonal signal (TRD, transcription regulation domain; DBD, DNA-binding domain; SBD, steroid-binding domain; NH_2, aminoterminus; COOH, Carboxy terminus)

The DNA-binding domain of all the four types of THR is conserved. Mammalian genome has a short repeated sequence called thyroid responsive element (TRE). A TRE is composed of at least two conserved motifs of AGGTCA, separated by four nucleotides. This motif can be arranged as direct repeats, palindromes, or inverted repeats (e.g., AGGTCATGACCT). THR binds to TRE as free THR or as THR-hormone complex. Free THR binding leads to repression of transcription, whereas binding of the THR–hormone complex activates transcription.

(d) Retinoids Retinoids are natural and synthetic compounds that are chemically related to vitamin A. There are three retinoid generations—first-generation retinoids include retinol, retinal, tretinoin (retinoic acid, Retin-A), isotretinoin, and alitretinoin. Second-generation retinoids include etretinate and acitretin. Tazarotene , bexarotene, and adapalene belong to third generation.

Retinoic acid (RA) is involved in vertebrate morphogenesis, growth, cellular differentiation, and tissue homeostasis. The use of in vitro systems initially led to the identification of nuclear receptor retinoic acid receptors (RAR) and retinoid X receptor (RXR). Heterodimers are possible transducers of the RA signal. Retinoid acid binds to retinoic acid receptor. When bound to a retinoid, they act as transcription factors, altering the expression of genes with corresponding response elements. There are two basic types of retinoic acid nuclear receptors: RAR and RXR. RXR is a Type IV member, which is activated by 9-cis retinoic acid. There are three types of RXR receptors—RXR-alpha, RXR-beta, and RXR-gamma, encoded by the RXRA, RXRB, and RXR genes, respectively.

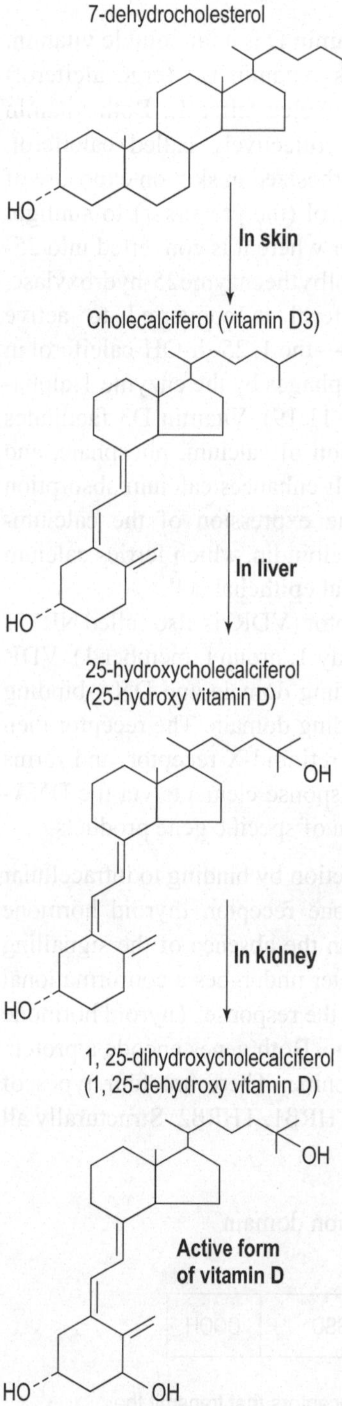

FIGURE 11.19 Biosynthesis of active form of vitamin D_3

11.1.3 Intracellular Signalling Molecules

Signals received by cell-surface receptors are transduced or passed downstream by four different types of intracellular molecules, most of which are proteins, predominantly enzymes. Molecules involved in intracellular cascade are generally classified as follows:

- Enzymes
- Small, low molecular weight effectors or second messengers
- Large intracellular mediators
- Molecular switches

Enzymes

Among the intracellular signalling proteins, enzymes are predominant and they are called signalling enzymes. They play various functions in signal transduction and regulation (Table 11.8).

TABLE 11.8 Diverse functions of signalling enzymes in signal transduction

Function	Mechanism	Example
Reversible activation or inactivation (regulation) of other proteins	By phosphorylation or dephosphorylation of proteins	PKs Protein phosphatises
Formation and degradation of second messengers	By hydrolysis and cyclization of the substrate	Adenylyl cyclases Phospholipases
Accepting and transmitting signals to downstream components	By switching between active and inactive conformations	GTPases

Protein kinases Majority of signal transduction pathways are mediated by reversible phosphorylation–dephosphorylation of enzymes and other proteins. Protein kinases play a pivotal role in this process. Protein kinases are transferases that transfer phosphate group from the ATP to a specific amino-acid residue of the acceptor protein. Depending on the type of the amino-acid acceptor protein kinases are classified into the following four types:

- Serine/threonine-specific protein kinases that add a phosphate residue to the alcohol group of the serine and threonine residues of the protein.
- Tyrosine protein kinases that transfer a phosphate group to the phenolic OH group of the tyrosine residues in the target protein.
- Histidine-specific protein kinases that transfer phosphorus group to the histidine residues forming phosphorous amide.
- Aspartate/glutamate-specific protein kinases that transfer a phosphate group to the carboxyl group of aspartate or glutamate residue of the acceptor protein forming a phosphate–carboxylate anhydride.

Reaction of the four types of protein kinases are shown in Fig. 11.20. Different types of target proteins of protein kinases are

- Enzymes
- Adaptor proteins
- Transcription factors
- Ion channels
- Transmembrane receptors
- Ribosomal proteins
- Structural proteins
- Transport proteins

Second messengers

Second messengers are small diffusible intracellular molecules. They are formed from their substrate by the action of enzymes, for example, cAMP is formed by the enzyme adenylyl cyclase. Second messengers are formed at high concentrations locally. Production of second messengers at high concentrations is responsible for signal amplification and high rate activation of further downstream signals. In response to regulatory signals, second messengers are also released from the intracellular repository, for example, calcium ions are released from endoplasmic reticulum. Second messengers bind to their specific binding sites on other proteins leading to diversification of signals. When not needed by the cell, or when the cell needs to terminate the signals produced by the second

TABLE 11.9 Intracellular messengers of signalling processes

Cytosolic messengers	Membrane-bound messengers
cAMP	Diacyl glycerol (DAG)
cGMP	Phosphatidyl inositides
Calcium ions	
Inositol phosphates	

messengers, the later can be degraded by enzymes or restored in the repository. Basically, there are two types of second messengers—cytosolic and membrane-bound. Cytosolic messengers bind to an unbound target protein and modulate the signal. Membrane-bound messengers bind to membrane-bound target or recruit the cytosolic target to the membrane for subsequent signal transduction. Major second messengers involved in signal transduction are listed in Table.11.9.

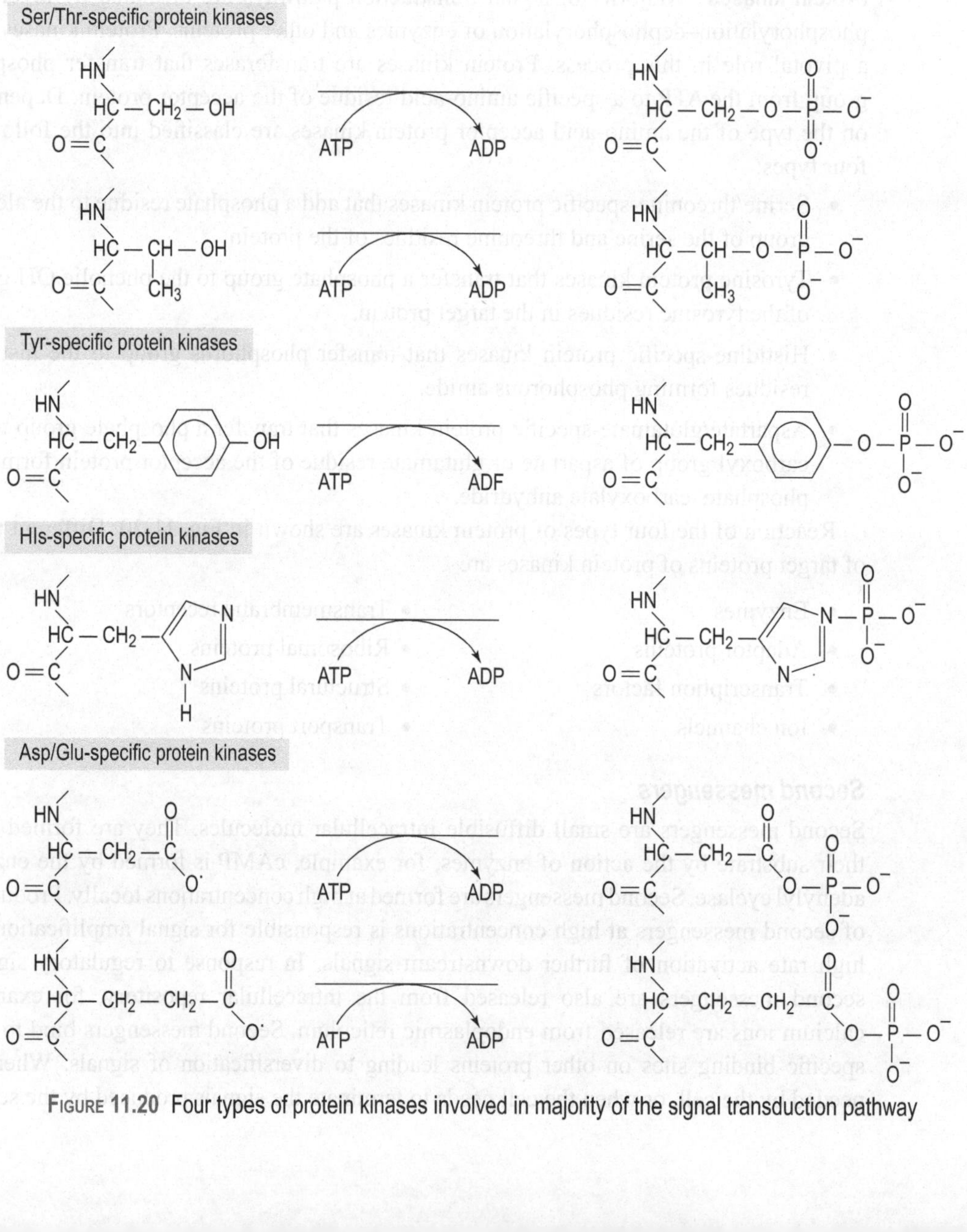

FIGURE 11.20 Four types of protein kinases involved in majority of the signal transduction pathway

Structure of these messengers are shown in Fig. 11.21 and their role in cell signalling or signal transduction is discussed in detail in Section 11.4.

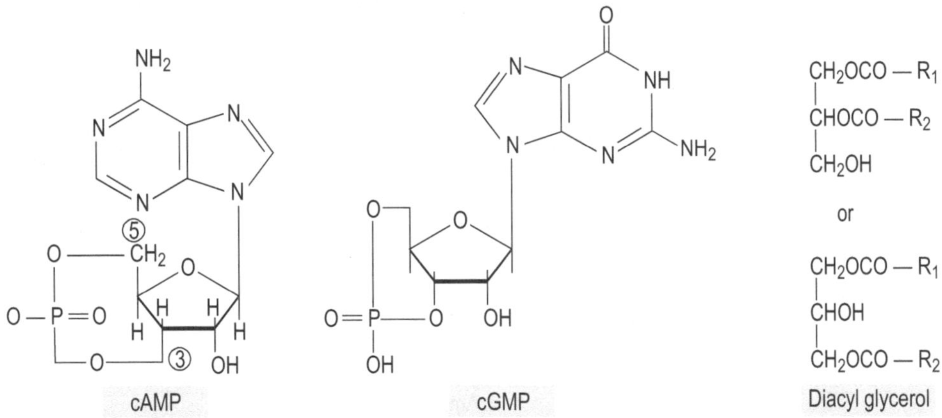

FIGURE 11.21 Structure of major second messengers

Large intracellular mediators

Intracellular mediators are large intracellular molecules. There are different types of intracellular proteins in signal transduction that are differentiated on the basis of their function. They are relay proteins, messenger proteins, adaptor proteins, scaffold proteins, bifurcation proteins, amplifier proteins, transducer proteins, integrator proteins, latent gene regulatory proteins, and anchoring proteins. Functions of large signalling proteins are given in Table 11.10 and schematized in Fig. 11.22.

Molecular switches

Depending on the cellular demand, signal molecules switch between their active and inactive conformations. These molecules are called molecular switches. Protein kinases and GTP-binding proteins are well-understood molecular switches.

TABLE 11.10 Types of intracellular proteins (large intracellular molecules)

Protein	Function
Relay proteins	Relay the message or signal to the next or neighbouring protein
Messenger proteins	Pass the signal from one part of the cell to the other compartment.
Adaptor proteins	Function as a clamp and provide docking site to connect one protein to the other for passing the signal effectively
Scaffold proteins	Function as support or substrate for the closely located signaling molecules.
Bifurcation proteins	Bifurcate the signal from one pathway to the other
Amplifier proteins	Amplify the signal by producing large amount of intracellular mediators
Transducer proteins	Convert the signal into different forms
Integrator pathway	Integrates the signals of several pathways and sends it into a single pathway
Latent gene regulatory proteins	Migrate to the molecules to induce gene transcription
Anchoring proteins	Anchor proteins at a precise location

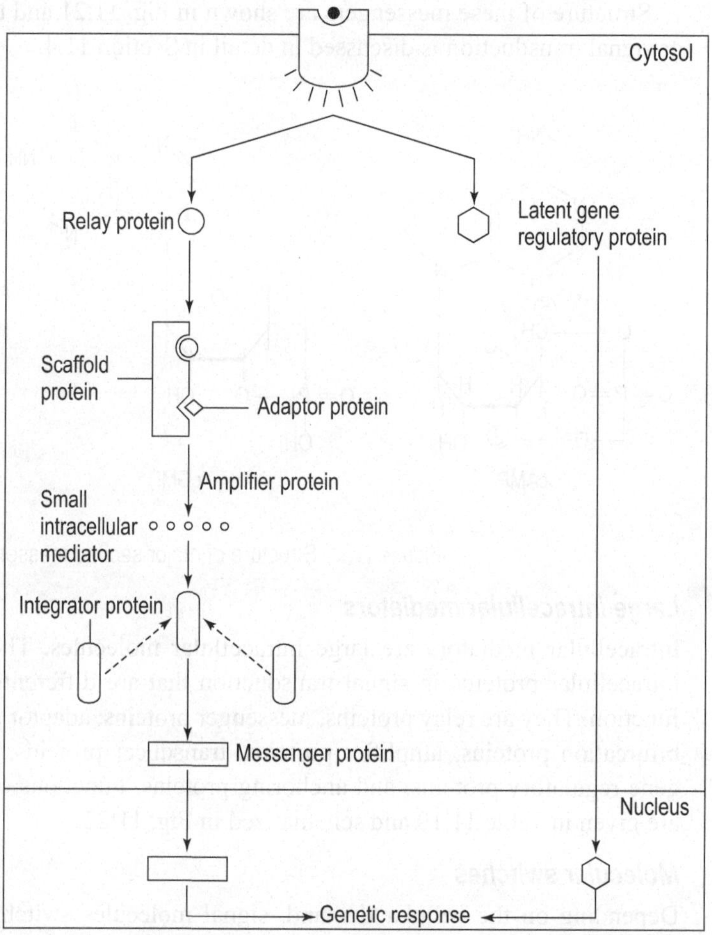

FIGURE 11.22 Types and function of large intracellular proteins (arrows represent the direction of signal transduction)

Protein kinases as molecular switches Protein kinases (PK) are class 2 enzymes, that is, transferases, which transfer phosphate group from ATP to its substrate thereby phosphorylating the substrate. PK itself is activated via phosphorylation by other PKs; or inactivated via dephosphorylation by phosphatases, depicted in the following manner:

During the signal transduction, one PK activates the next PK which in turn activates the third PK and so on as a cascade, thus leading to the relaying of signal onward (Fig. 11.23). Two major classes of PK are found in cells:

1. Serine /Threonine kinase
2. Tyrosine kinase

GTP-binding proteins as molecular switches GTP-binding proteins are a large family of heterotrimeric or monomeric proteins that bind GTP (guanosine 5′-triphosphate) as intermediaries in intracellular pathways. There are two types of GTP-binding proteins:

1. Trimeric GTP-binding protein or G-protein
2. Monomeric GTP-binding protein or monomeric-GTPases

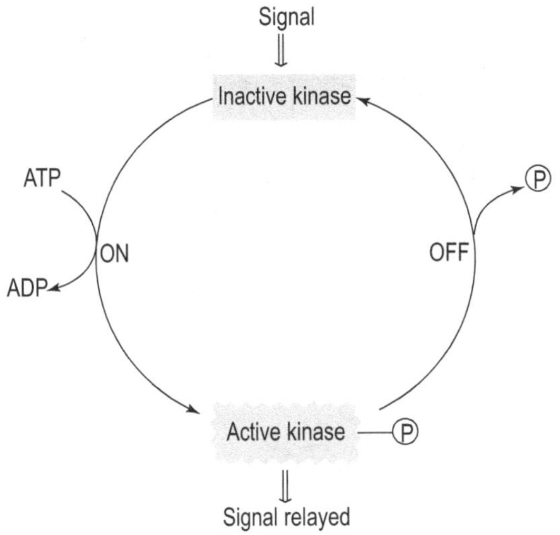

FIGURE 11.23 Protein kinase as molecular switch

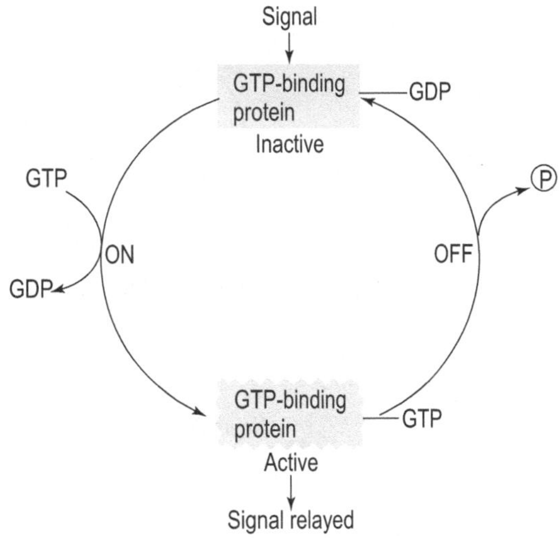

FIGURE 11.24 GTP-binding proteins as molecular switch

These proteins switch "on" and "off" by binding to GTP and GDP, respectively. On binding to GTP, the protein is activated and functions as a GTPase thereby hydrolyzing the bound GTP to GDP. In the presence of GDP, the protein remains inactive (Fig. 11.24).

11.2 TYPES OF CELL SIGNALLING

Signalling process can occur in different ways. They may be listed as follows:

1. Contact-dependent signalling
2. Paracrine signalling
3. Synaptic signalling
4. Endocrine signalling
5. Autocrine signalling
6. Gap-junction signalling

These six different types of signalling are schematically shown in Fig. 11.25. In this context, the cell that produces the molecules is called the signalling cell and the one that responds to the molecule is called the *target cell.*

Contact-dependent signalling

In this type of signalling, the signal is transported from the signalling cell to the target cell on establishment of direct contact between the plasma membrane of the two cells. This process is also called juxtacrine signalling. Contact-dependent signalling is responsible for precise control of cell differentiation during embryogenesis. Notch signalling mechanism is an example for contact-dependent signalling.

Paracrine signalling

One of the dictionary meanings of para is 'near'. In paracrine signalling, the cell releases signal molecule into the extracellular space. The molecules then act locally on nearby or neighbouring cells (target cells). The movement of paracrine-signalling molecules is hindered by the extracellular matrix; because of this, the molecules do not travel a long distance and instead act on the nearby cells. Paracrine signalling occurs in response to allergens, clotting, tissue repair, and scar tissue formation. Growth factors, clotting factors, retinoic acid, and allatostatin are examples of signal molecules that induce paracrine effect.

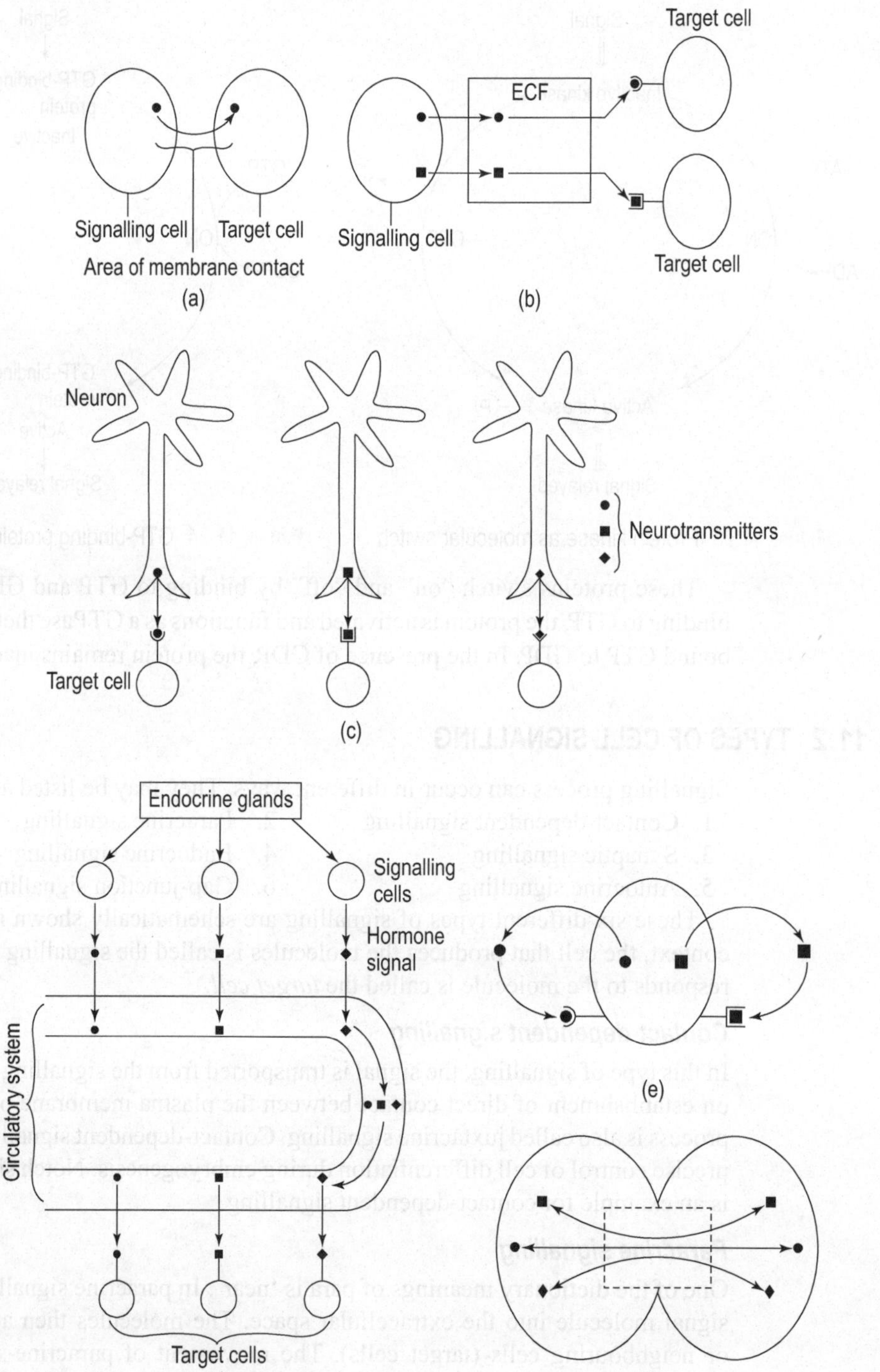

FIGURE 11.25 Different types of cell signalling (a) Contact-dependent signalling, (b) Paracrine signalling (c) Synaptic signalling, (d) Endocrine signalling, (e) Autocrine signalling, (f) Gap-junction signalling

Synaptic signalling

Synaptic signalling is common in nerve cells and it is used in the transmission of nerve impulses. Molecules involved in synaptic signalling are called neurotransmitters. In neurons, electrical signals are transmitted from the cell body along the axon and the neurotransmitter is generated at the synapse, which is the junction between the nerve cell and the target cell. Synaptic signalling occurs at a faster rate and the signalling can be transmitted over a long distance (100 m/s).

Endocrine signalling

Endocrine-signalling pathway occurs in cells of endocrine glands. These cells secrete proteins or peptide signal molecules called hormones into the blood stream. Hormones travel through the blood and reach the receptor of the target cell and elicit necessary changes. Endocrine signalling is slow when compared to synaptic signalling. Insulin is a classic example for endocrine signalling. Insulin hormone is synthesized and secreted by the pancreas into the blood stream. Blood insulin reaches the cells of muscle and adipose tissue and binds to the insulin receptor and enhances the absorption of glucose by these cells. Insulin also stimulates liver cells to store the glucose in the form of glycogen.

Autocrine signalling

When the signalling cell and the target cell are same, autocrine signalling takes place. In few cases, the target cell and the signalling cell are same group of adjacently located cells. A cell produces a signal molecule which acts on its own receptors to produce the response. This type of signalling is common during the developmental stage. Cancer cells are another example where autocrine signalling occurs helping in their proliferation and survival.

Gap-junction signalling

Gap junction refers to a nanoscale patch of size 2 to 4 nm, which is a space or gap between the membranes of two adjacent cells. Gap junctions directly connect the cytoplasm of the adjacent cells through water-filled channels. Gap junctions can mediate signalling by allowing the exchange of small intracellular signal molecules, such as calcium ions and cAMP, between the cells. These cell pairs, therefore, respond to extracellular signal in a coordinated way.

11.3 VARIOUS CELLULAR RESPONSES PRODUCED BY SIGNALLING

Irrespective of the type of signalling or the receptor, the cell will elicit a response to the signal. Some of the prominent cellular responses are survival, division, differentiation, secretion, and contraction.

Different types of signals produce different types of responses in the cell. Same types of signals may also have different effect on different cell types. For example, acetylcholine, which is a neurotransmitter signal, decreases the rate of contraction in heart muscle cells but increases the rate of contraction in skeletal muscles. The same molecule induces secretion in salivary gland cells. These differences in response are due to diversity in cell-surface-receptor molecules. If an appropriate signal is not produced, the cell will undergo suicide and this process is known as apoptosis (programmed cell death). Various cellular responses to signalling molecules are shown in Fig. 11.26.

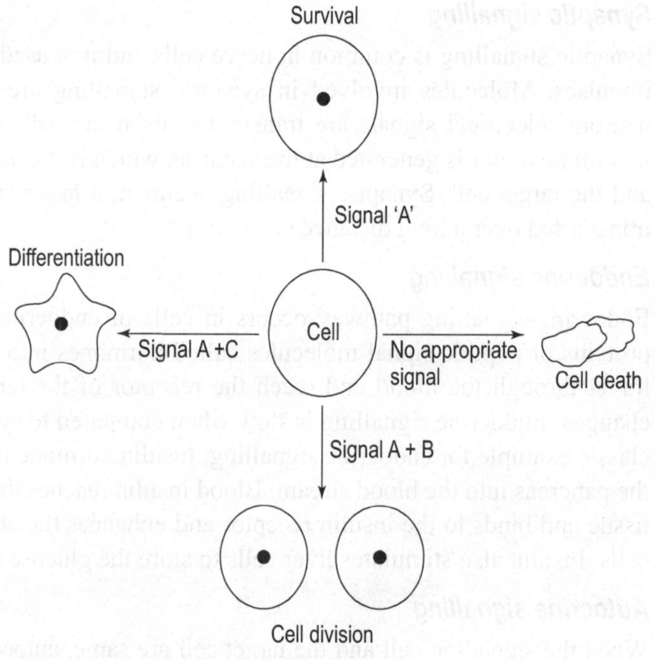

FIGURE 11.26 Cellular responses to signal molecules

11.4 INTRACELLULAR SIGNALLING PATHWAY AND SIGNAL AMPLIFICATION

Binding of extracellular signal molecules to a receptor (extracellular or intracellular) leads to transduction of the external signal into an intracellular signalling pathway or cascade. Each intracellular signalling pathway makes use of different types of cellular compounds, mostly proteins (collectively called cascade or chain), to pass the signal sequentially downstream and amplify the signal and, finally, elicit the cellular response. Cellular response is the end point of any pathway.

Depending on the amplifying signal molecules, there are diverse intracellular signalling pathways. The following are the major intracellular signalling pathways discussed in this section:

1. cAMP pathway
2. cGMP pathway
3. PLC–Ca^{2+} pathway
4. Ca^{2+}–calmodulin pathway
5. MAP kinase pathway (discussed in Section 11.9.3.)
6. Janus kinase pathway
7. NFκB pathway

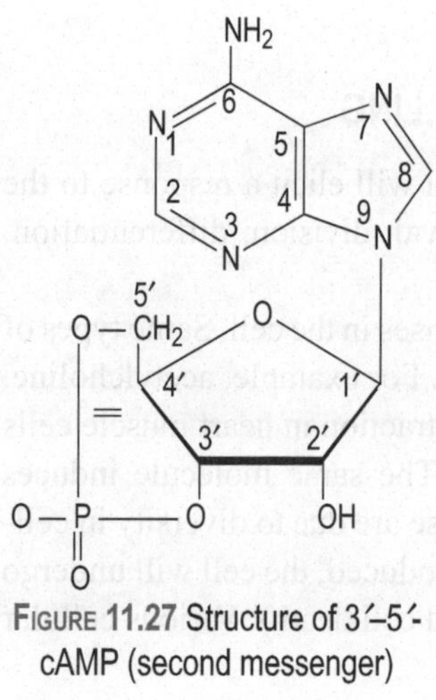

FIGURE 11.27 Structure of 3′–5′-cAMP (second messenger)

11.4.1 cAMP Pathway

The 3′–5′cyclic adenosine monophosphate (cAMP) is a cyclic nucleotide synthesized from ATP by the enzyme adenylyl cyclase (Fig. 11.27). It is a second messenger, that is, a small molecule produced intracellularly in response to binding of extracellular signal molecules to cellular receptors. It is formed from ATP by the action of adenylyl cyclase, and upon appropriate condition it is degraded by phosphodiesterase (Fig. 11.28).

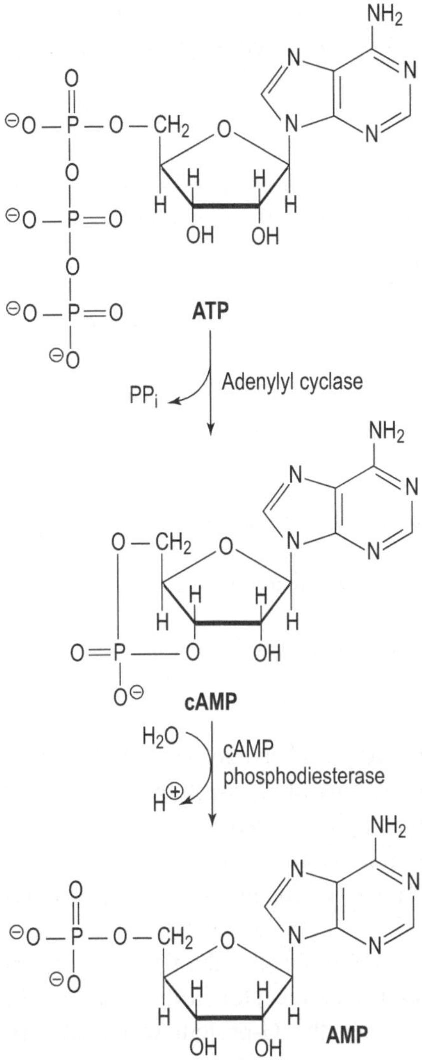

FIGURE 11.28 Enzyme-catalysed synthesis and degradation of cAMP

cAMP pathway is mediated by hormonal signals. Hormonal signals and the corresponding cAMP-mediated responses are summarized in Table 11.11.

TABLE 11.11 cAMP signalling pathway induced by hormones

Hormone signal	Target cell/organ	Cellular response
ACTH	Adrenal cortical cells	Secretion of cortisol
Epinephrine (adrenalin)	Muscle cells Heart cells	Glycogenolysis Increase in heart rate
Adrenalin	Adipocytes	Lipolysis (breaking of
ACTH		triglycerides
Glucagon		
TSH		
Glucagon	Liver	Glycogenolysis
Leutinizing hormone	Ovary	Secretion of progesterone
Parathormone	Bone	Bone resorption
TSH	Thyroid gland	Synthesis and secretion of thyroid hormones
Vasopressin	Kidney	Water resorption

One of the well-studied cAMP pathways is initiated by a hormone signal called epinephrine. Epinephrine binds to the G-protein-linked receptor and stimulates the cAMP cascade pathway. This pathway amplifies the signal and produces three types of responses in the cells:

1. Regulation of glycogen metabolism (stimulation of glycogenolysis and inhibition of glycogenesis in muscle cells)
2. Transcription of specific target gene with CRE (cAMP response element)
3. Regulation of ion channels in epithelial cells of nose (olfactory epithelium)

Among these three types of cAMP-induced responses, the first two are indirect and occur via an enzyme called protein kinase A (PKA), whereas the last response is direct and occurs without the PKA mediator.

Regulation of glycogen metabolism

Before discussing the role of cAMP in glycogen metabolism, it is a prerequisite to understand the metabolic pathway of glycogen. Polymerization of glucose molecules to form a long-chain polymer of glycogen is called glycogenesis. It is represented as follows:

UDP-glucose + (glucose) n / (Glycogen) $\rightarrow$ UDP+ (glucose) n+1

Glycogen occurs as storage polysaccharide in animals. It exists as granules in cytoplasm. It is stored in liver and muscles and is depolymerized to glucose whenever energy is needed.

The breakdown or depolymerization of glycogen into glucose units is called glycogenolysis. Glycogenesis occurs in the *fed state.* Glycogenesis pathway starts with UDP glucose – the *active form* of glucose. UDP glucose is synthesized from glucose-1-phosphate. The pathway is shown in Fig. 11.29 and is schematically represented in Fig. 11.30.

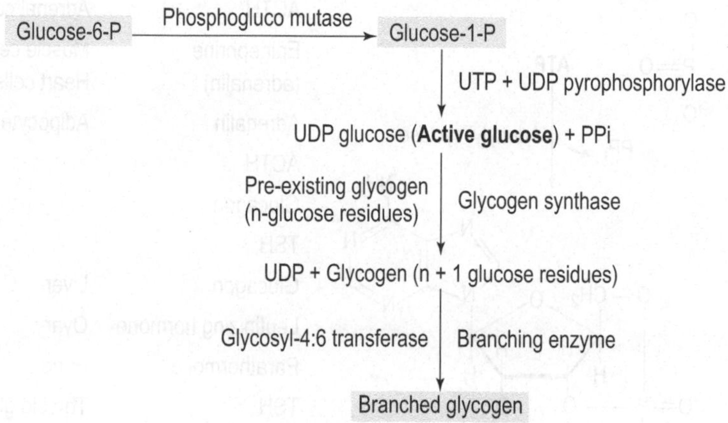

FIGURE 11.29 Glycogenesis pathway

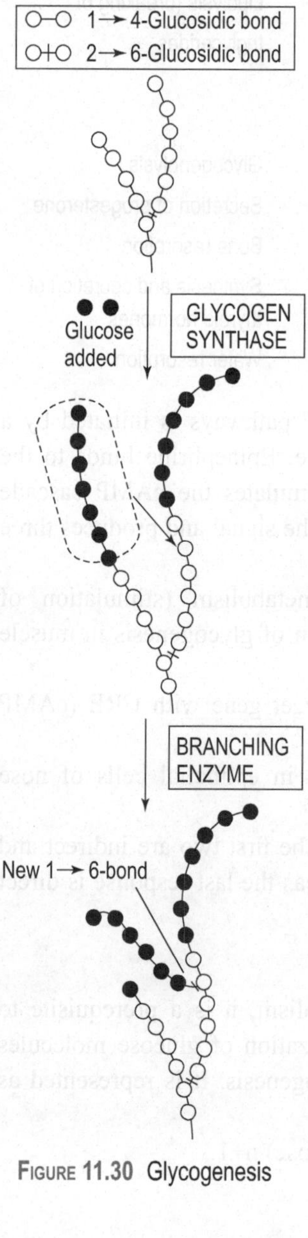

FIGURE 11.30 Glycogenesis

Glycogenolysis involves phosphorolytic cleavage of α-1-4 glycosidic linkage by phosphorylase and debranching of glycogen by *debranching enzymes* – glucosyl transferase and amylo-6-glucosidase.

$$\text{Glycogen (n-glucose residues)} + \text{Pi} \xrightarrow{\text{Glycogen phosphorylase}} \text{Glycogen (n-1 glucose residue)} + \text{Glucose-1-P}$$

Epinephrine, the extracellular signal molecule (the first messenger) binds to the G-protein-linked cell-surface receptor in the muscle cell. The hormone-bound receptor activates the G-protein which in turn stimulates adenylate cyclase. Adenylate cyclase acts on its substrate ATP and catalyzes the formation of cAMP (the second messenger).

cAMP binds to the enzyme PKA. PKA has four subunits – two regulatory subunits and two catalytic subunits. The catalytic subunits of PKA bind to two substrates which are also enzymes. PKA catalyses the phosphorylation of two enzymes—phosphorylase kinase and glycogen synthase. Phosphorylase kinase, in turn, phosphorylates glycogen phosphorylase which finally breaks glycogen to glucose-1-P.

Glycogen synthase is simultaneously inactivated on phosphorylation by PKA thereby preventing glycogen synthesis. Thus, cAMP-mediated PKA activation results in two events—glycogenolysis stimulation and glycogenesis inhibition. Depending upon the cellular need these two processes can also be reversed by the enzyme protein phosphatises, which removes phosphate group from glycogen phosphorylase and glycogen synthase.

Transcription of specific target gene with CRE (cAMP response element)

cAMP can induce transcription of genes containing regulatory sequence called cAMP response element (CRE). In this mechanism (Fig. 11.31), cAMP first activates PKA, the catalytic subunit of which enters into the nucleus and phosphorylates a transcription factor called CRE-binding protein (CREBP). Phosphorylated CREBP is active and it induces the transcription of the target genes. Thus, cAMP-induced regulation of glycogen metabolism and gene expression are both mediated by PKA.

Regulation of ion channels in epithelial cells of nose (olfactory epithelium)

Odorant receptors (Fig. 11.32) are expressed in sensory neurons of nose. These receptors are also a member of G-protein linked receptor family. Hence, these receptors are linked to an adenylate cyclase via G-protein. When an extracellular signalling molecule binds to the odorant receptor, the G-protein is stimulated to activate adenylate cyclase to produce cAMP. cAMP acts directly (without binding to PKA) to open the sodium channels in plasma membrane causing depolarization and nerve impulse.

Adenylate cyclase is a large multipass transmembrane protein with catalytic domain facing the cytosolic side of the cell. It exists in eight isoforms. Adenylate cyclase is regulated by G-proteins. There are two types of G-proteins based on their regulatory function: stimulatory G-protein (Gs) and inhibitory G-protein (Gi). Gs stimulates adenylate cyclase to raise the concentration of cAMP, whereas Gi inhibits adenylate cyclase to decrease the concentration of cAMP in the cell. Gs and Gi are targets for bacterial toxins. The following are two common examples for this statement:

Gs stimulates adenylate cyclase in intestinal epithelial cells during cholera *Vibrio cholerae* is a bacterium that is the causative organism of cholera disease. The bacterium releases a toxin called cholera toxin which is an enzyme protein. This toxin catalyses the transfer of ADP ribose moiety from intracellular NAD^+ to Gs. ADP ribosylation alters the alpha subunit of Gs; so that it cannot hydrolyze its bound GTP anymore and hence remains

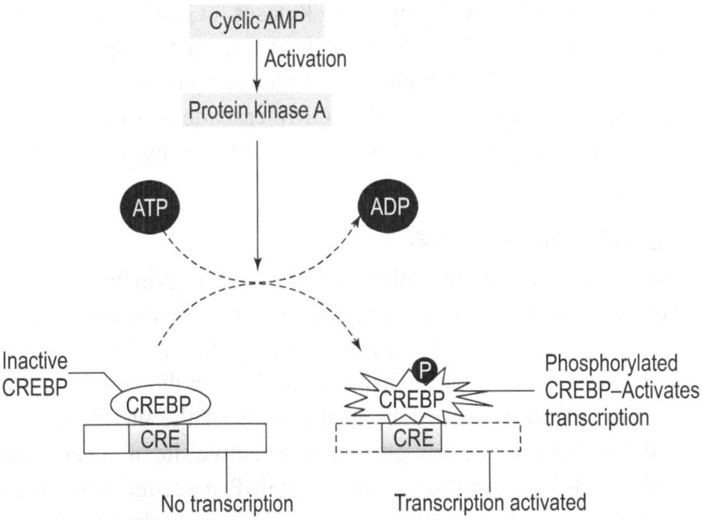

FIGURE 11.31 Role of cAMP in transcription of genes containing CRE

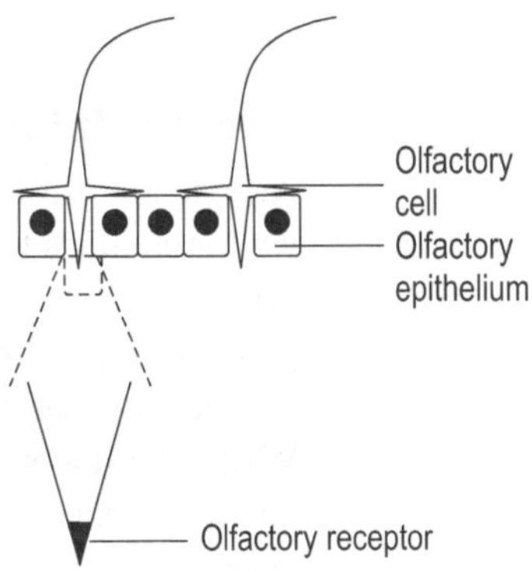

FIGURE 11.32 Olfactory epithelium

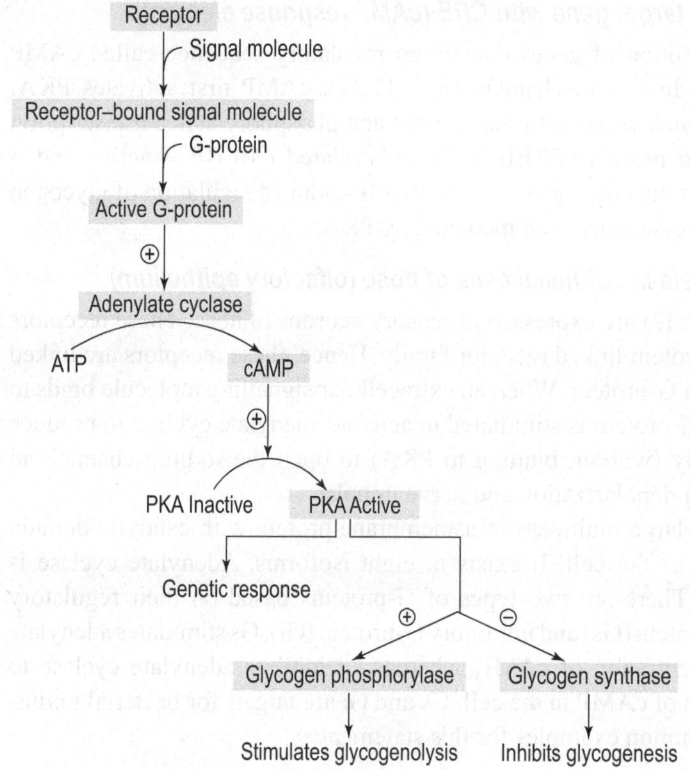

FIGURE 11.33 cAMP pathway

in an active state (instead of shuttling between GTP-bound active state and GDP-bound inactive state). As a result, adenylate cyclase is indefinitely stimulated that leads to increased cAMP; which in gastrointestinal epithelial cells causes enhanced efflux of chloride and water into the gut lumen (diarrhoea) the characteristic feature of cholera.

Gi inhibits adenylate cyclase during whooping cough *Bordtella pertussis* is a pathogen that causes pertussis, commonly called whooping cough. Gi is the target for this organism. The pathogen induces ADP ribosylation of two subunits of Gi and prevents Gi from interacting with its receptor. Gi retains its GDP-bound state and does not regulate its target protein. Various cAMP signalling pathways are shown in Fig. 11.33.

11.4.2 cGMP Signalling Pathway

cGMP (cyclic guanosine monophosphate), as the name indicates, is a cyclic nucleotide formed from GTP by the catalytic action of guanylate cyclase. cGMP is a signal-amplifying molecule or second messenger which responds to peptide ligand bound to the cell surface. cGMP usually functions by activating protein kinases. cGMP signal occurs in smooth muscle cells of blood vessels, where the final response to the signal is the dilation or relaxation of the blood vessel. cGMP pathway is initiated by a hormone called acetylcholine, which is an extracellular signal molecule. cGMP cascade is also very well characterized in rod cells of the retina. In rod cells, cGMP converts the light signal to a nerve impulse.

Role of cGMP in amplifying acetylcholine signal

Acetylcholine is a neurotransmitter secreted by the autonomic nerve endings in blood vessel wall. Ach binds to its receptor on the endothelial cell and activates the enzyme nitric oxide synthase (NOS). NOS catalyses the production of NO from arginine. NO is a hydrophobic paracrine signalling molecule capable of rapidly diffusing from the membrane of the cell to enter into the target cell—the smooth muscle cells in the walls of the blood vessels. Within the smooth muscle cell, NO binds to the iron present in the active site of the enzyme guanylyl cyclase, which catalyses the conversion of GTP to cGMP, a second messenger or intracellular molecule. cGMP relaxes the smooth muscle and enhances the blood flow through blood vessels. The cGMP pathway is schematized in Fig. 11.34.

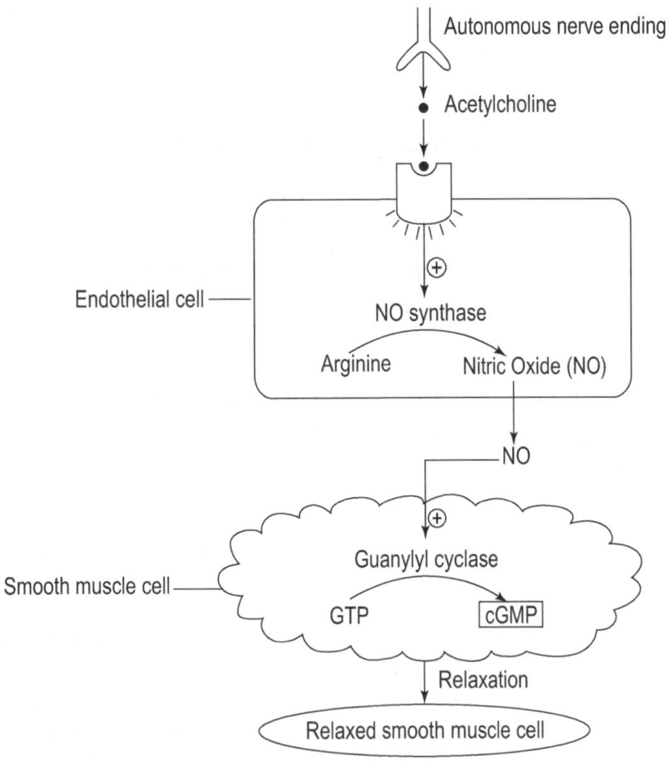

FIGURE 11.34 cGMP pathway

cGMP in rod cells

The inner layer of the eye is called the retina. It contains special types of cells called rods and cones. These are photoreceptor neurons. Rods contain a photopigment called rhodopsin. Basically, rhodopsin is made of a protein called opsin, which is conjugated to a prosthetic group called retinal, a chromophore. When light strikes the rhodopsin, 11-cis-retinal (in resting stage) is isomerized to 11-trans-retinal (active form). The trans isomer activates opsin which in turn stimulates a membrane-bound *signal-coupling protein* called transducin. Transducin activates cGMP phosphodiesterase which catalyzes the breakdown of cGMP to GMP which in turn leads to the closing of gated sodium channels. Consequently, sodium entry into the rod cell is blocked which leads to an increase in electronegativity inside the plasma membrane (–70 mV). This results in hyperpolarization of plasma membrane of the rod cells. Events involved in cGMP signalling in rod cells are shown in Fig. 11.35.

In the absence of light, that is, in dark, cGMP is not broken down and so the sodium channel gate is opened and there is an inflow of sodium with a loss in electronegativity (40mV) in the plasma membrane.

11.4.3 PLC and Inositol Phospholipid Pathway

Phospholipase is an enzyme that hydrolyzes phospholipids into fatty acids. There are four types of phospholipases—phospholipase A (PLA), phospholipase B (PLB), phospholipase C (PLC), and phospholipase D (PLD). PLA and PLB hydrolyze SN1 and SN2 acyl chains

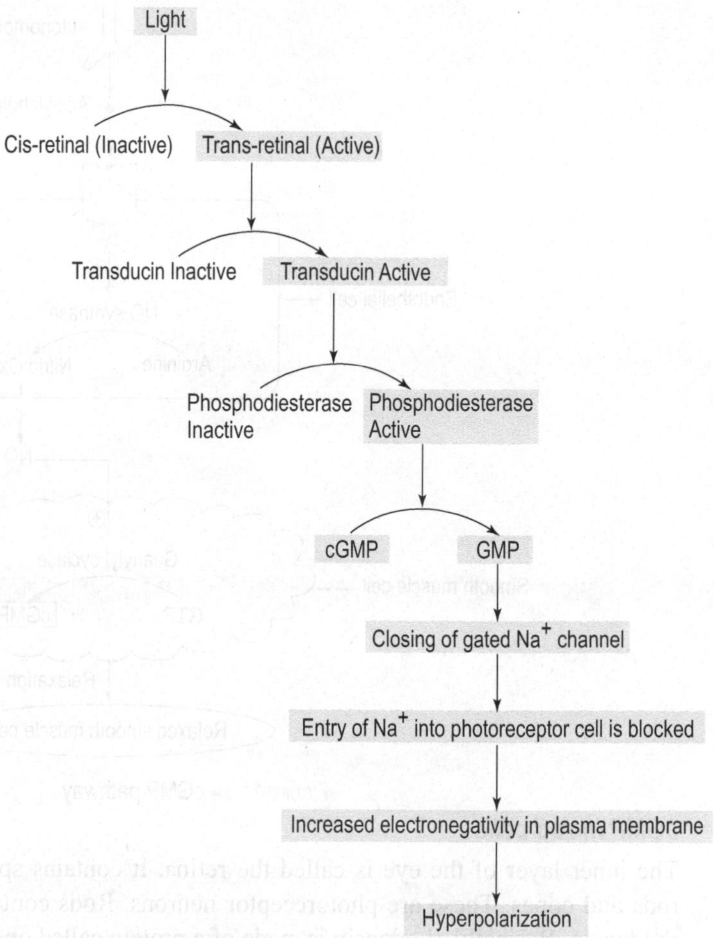

FIGURE 11.35 cGMP pathway in photoreceptors (rod cells)

releasing arachidonic acid. PLC cleaves before the phosphate releasing diacyl glycerol and a phosphate-containing group. PLD hydrolyses after the phosphate group generating phosphatidic acid and an alcohol. Among these four phospholipases, PLC has a vital role in signal transduction.

Extracellular signal molecules, such as hormones and growth factors, bind to cell-surface receptor and activate PLC. Activation of PLC produces two types of second messengers: IP3 (inositol-1, 4, 5- triphosphate) and DAG (diacylglycerol). IP3 induces calcium immobilization response, whereas DAG induces cell proliferation response.

Two types of receptors are involved in PLC-mediated signalling. G-protein linked receptors activate PLCβ, and receptor tyrosine kinase (RTK) activates PLCγ by phosphorylating the tyrosine residues of the enzyme. PLCγ contains SH2 domains. Activated PLC stimulates the hydrolysis of a phospholipid called phosphotidyl inositol-4, 5-bisphosphate present in the inner leaflet of the plasma membrane into IP3 and DAG. IP3 is a small molecule which diffuses rapidly from plasma membrane to cytosol and reaches the endoplasmic reticulum (ER) where it opens the IP3 gated calcium release

TABLE 11.12 Inositol-phospholipid-mediated signalling pathway

Extracellular sponse	Target	Cellular resignal
Acetylcholine	Pancreas	Secretion of amylase
Acetylcholine	Smooth muscle	Muscle contraction
Thrombin	Thrombocytes	Platelet aggregation
Vasopressin	Liver	Glycogenolysis

channel in ER membrane. This leads to calcium immobilization from endoplasmic reticulum reserve. The second molecule, DAG, remains in the membrane and has two fates:

- It activates PKC.
- It undergoes hydrolysis to form arachidonic acid.

PKC is a calcium dependent serine/threonine protein kinase (C in PKC represents calcium). PKC activates protein kinases or MAP kinase pathway in order to phosphorylate transcription factors. Transcription factors are proteins which switch *on* or *off* the transcription of specific gene sequences. Phosphorylation of transcription factors alters the expression of the target gene leading to cell proliferation. An example for transcription factor activated by PKC is the nuclear factor for kappa chain gene in B-lymphocytes (NF-κB).

Hydrolysis of DAG yields arachidonic acid which can function as a messenger or can be used in the synthesis of eicosanoids that produce inflammatory response. Various other cellular responses mediated by inositol phospholipid are shown in Table 11.12.

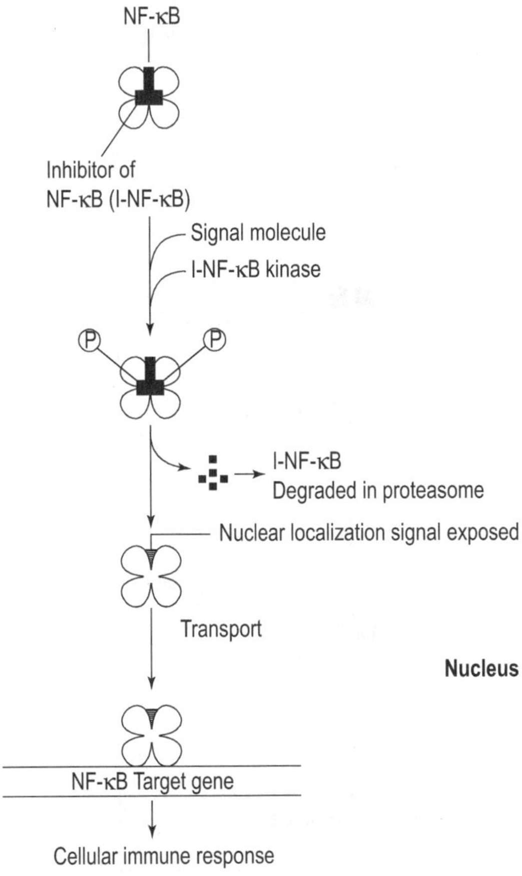

FIGURE 11.36 NF-κB pathway in B-lymphocytes

11.4.4 NF-κB pathway

NF-κB (nuclear factor kappa light-chain enhancer of activated B cells) is a protein, which functions as a transcription factor in eukaryotic cells. It binds to promoter and enhancer sequences of genes. Stress, cytokines, free radicals, ultraviolet irradiation, oxidized LDL, and antigens (such as bacteria and viruses) are stimulus or signals for NF-κB pathway. NF-κB produces immune responses and inflammatory responses following an immunological or inflammatory signal. In the absence any signal, NF-κB is located in the cytoplasm as inactive form. Inactive NF-κB is bound to an inhibitory subunit and denoted as I-NF-κB. The inhibitor hides the nuclear localization signal of the NF-κB. In the presence of signalling molecules that bind to B-cells, the I-NF-κB is phosphorylated by I-NF-κB kinase, which leads to the dissociation of the inhibitory subunit from NF-κB to form active heterodimer. The nuclear localization signal is exposed in the active and the NF-κB translocates into nucleus to switch on the gene expression to elicit immune response or inflammatory response. NF-κB pathway is shown in Fig. 11.36.

11.4.5 JAK-STAT Pathway

There are two types of tyrosine kinases existing in the cell: receptor tyrosine kinase (RTK) and non-receptor tyrosine kinase (NRTK). In RTK, the tyrosine kinase moiety is the intrinsic part of the receptor, whereas in NRTK, the enzyme is not an intrinsic part of the receptor. Cytokine signals activate cytokine receptors which in turn activate NRTK. There are two types of NRTK—Src family and Janus kinase (JAK) family. Cytokine receptors are found in association with JAK. Binding of extracellular signal molecule to cytokine receptor induces attachment of inactive STAT to JAK via SH2 domains. (STAT refers to Signal Transducers and Activators of Transcription). STAT proteins are transcription factors with SH2 domains. STAT occurs in cytosol as inactive protein in unphosphorylated form. When it binds to JAK, it is phosphorylated and becomes active. Activated STAT dimerizes and move into the nucleus where it activates the transcription factor of the target gene to induce the transcription of the target gene. JAK-STAT pathway is shown in Fig. 11.37.

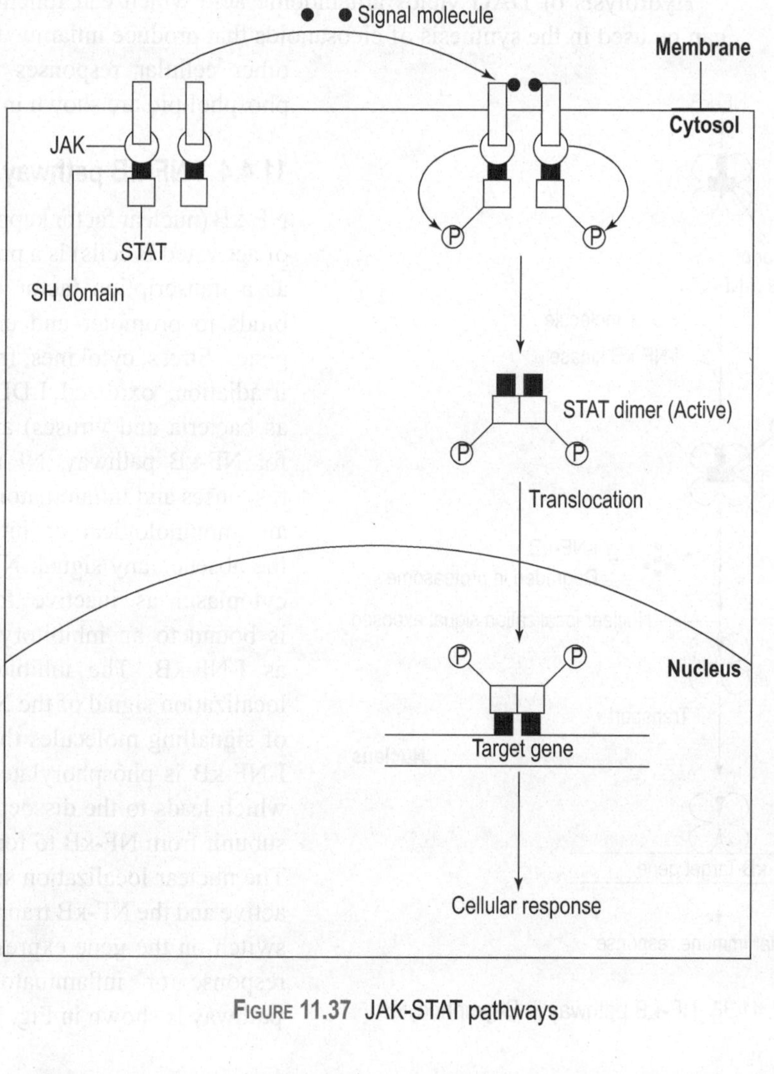

FIGURE 11.37 JAK-STAT pathways

11.5 PROTEOLYSIS-DEPENDENT PATHWAY

Proteolysis refers to hydrolytic cleavage of the peptide bond existing between specific amino acid residues in a peptide or protein. Some of the signalling pathways depend on proteolysis of at least one of the proteins in the cascade. This includes the following pathways:

- Notch pathway
- Wnt pathway
- Hedgehog pathway
- NF-κB pathway

All these pathways are crucial to animal development and are discussed in this section.

11.5.1 Notch Pathway

Notch pathway is the widely used pathway in the development of embryo. This pathway is responsible for silencing the genes responsible for neural differentiation of epithelial cells. During embryonic development epithelial cells are differentiated into neural cells. However, the neural cell should signal the other neighbouring epithelial cells to not enter the neural fate. This phenomenon by which the neural cells inhibit the further differentiation of epithelial cells into other neural cells is called lateral inhibition. Lateral inhibition is mediated by notch pathway (Fig. 11.38) which takes place in undifferentiated epithelial cells subjected to inhibition.

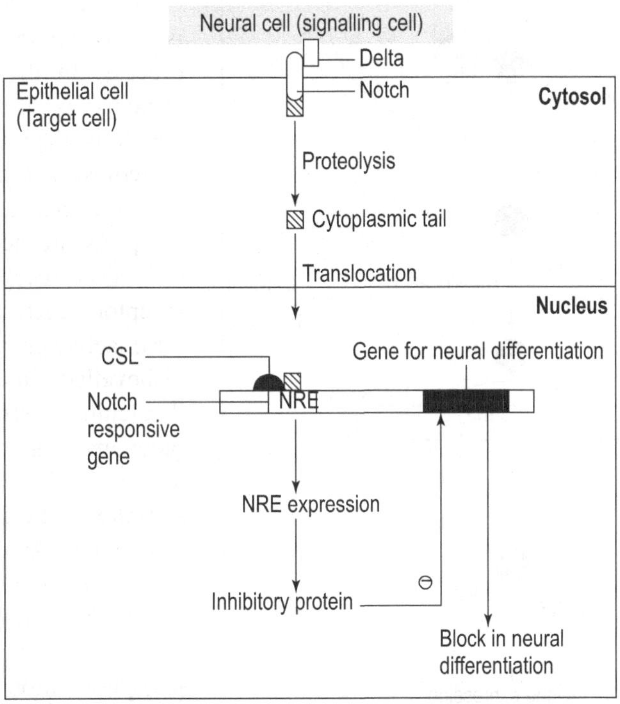

FIGURE 11.38 Notch pathway

Here the neural cell is the signalling cell and the epithelial cell is the target cell. The neural cell expresses a ligand or transmembrane signal protein called delta on its surface. The epithelial cells synthesize a protein called notch-receptor protein. Notch is processed in the Golgi vesicle by proteolytic cleavage, mediated by a protease called furin. After proteolysis, the notch becomes a mature heterodimer and reaches the cell surface with the cleavage site forming the extracellular domain and displayed as a transmembrane protein.

Delta protein binds to and activates notch, both interacting via their individual EGF-like domains. Thereafter, two other proteases cleave the notch to release the cytoplasmic tail. The tail subunit enters the nucleus, where it binds to and activates the notch responsive gene, in association with a gene-regulatory protein called CSL protein (a transcription activator, named so because it is called CBF1 in mammals, Suppressor of hairless in flies, and Lag-1 in worms). Consequently, the notch response gene produces an inhibitory protein which in turn blocks the expression of genes required for neural differentiation (in nervous system) and of various other alternate genes in other tissues.

Thus, the above discussion about notch pathway clearly indicates that proteolysis of notch is crucial in mediating the response.

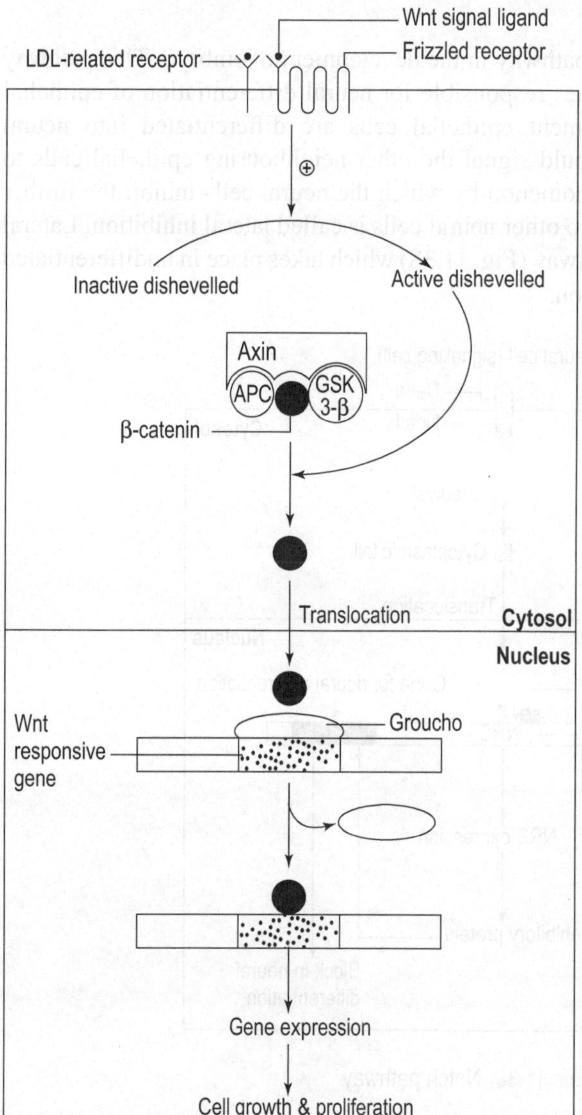

FIGURE 11.39 Wnt pathway

11.5.2 Wnt Pathway

The wnt pathway is responsible for inducing growth and proliferation via stimulating the expression of wnt responsive genes including C-myc gene. The key protein in this pathway is β-catenin that is regulated by proteolysis process. In the absence of any wnt signal proteins, β-catenin remains phosphorylated, which is then ubiquitylated, and undergoes proteolysis and degradation in proteasomes.

In the presence of ligand (wnt protein), two receptors are activated: the frizzled receptor and a co-receptor called the LDL-related receptor (LRR). The activated receptor, in turn, activates a cytoplasmic protein called dishevelled. Dishevelled acts upon β-catenin. β-catenin exists in conjugation with APC (adenomatous polyposis coli) protein and active GSK-3β. This three-component complex is held in cytoplasm by a scaffold protein called axin. Active GSK-3β maintains β-catenin in phosphorylated state. Dishevelled passes the signal to the axin-linked complex, and inactivates GSK-3β. Consequently, phosphorylation and degradation of β-catenin is inhibited. The β-catenin, therefore, accumulates in the cytosol, travels into the nucleus and dislocates a repressor protein called *Groucho*, which otherwise is bound to the wnt responsive genes to keep the genes silenced. Dislocation leads to expression of wnt responsive genes, such as the c-myc gene which produces c-myc protein, a powerful stimulator of cell growth and proliferation. β-catenin-mediated pathways in the presence and absence of wnt signals are compared in Fig. 11.39.

11.5.3 Hedgehog Pathway

Hedgehog pathway is a signalling pathway initiated by the protein encoded by the hedgehog gene (identified in drosophila). This pathway is responsible for the embryonic development as well as for maintenance of stem cell population in the adults.

Hedgehog signalling is regulated by proteolytic cleavage of a gene-regulatory protein called cubitus interruptus (Ci). In the absence of hedgehog protein signal, Ci is cleaved to form a small subunit called transcription repressor, which keeps the hedgehog protein responsive gene silenced. In the presence of signal, proteolysis of Ci is blocked and the intact (uncleaved) Ci accumulates in the nucleus and switch on the hedgehog responsive genes. Hedgehog pathway has been well studied in *Drosophila.*

Two types of receptors are involved in hedgehog pathway are Patched receptor and Smoothened receptor. Patched receptor is a 12-pass transmembrane protein and smoothened receptor is a 7-pass transmembrane protein (Fig. 11.40).

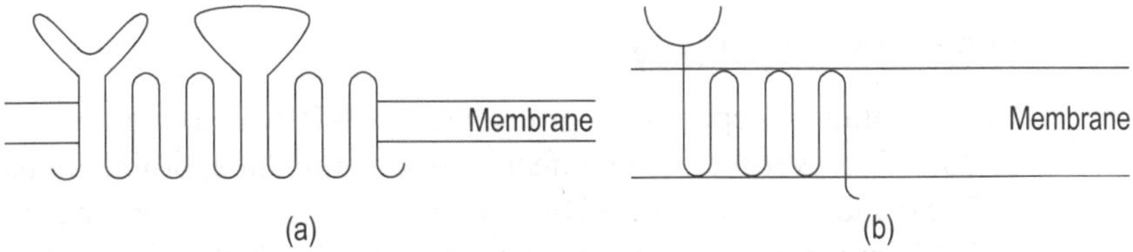

FIGURE 11.40 Receptors involved in hedgehog pathway (a) Patched receptor, (b) Smoothend receptor

In the absence of signal (polypeptide ligand-hedgehog is the signal), the patched receptor inhibits the smoothened receptor by an unknown mechanism. In the presence of a signal, patch receptor relays the signal to and activates the smoothened receptor. Smoothened receptor in turn passes the signal downstream to Ci protein which exists as a part of a complex. The complex is made of the following:

- Anchor protein called 'coastal protein'
- Serine/threonine kinase called 'fused'
- Adaptor protein called 'suppressor of fused'

Coastal protein anchors Ci to microtubule. Ci protein complex is shown in Fig. 11.41.

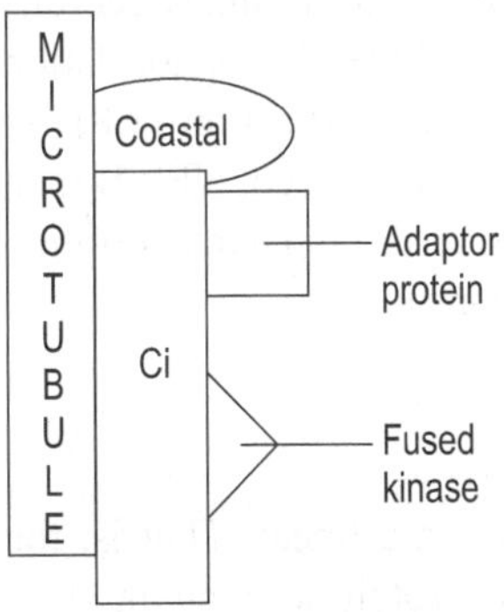

FIGURE 11.41 Cubitus interruptus (Ci) protein complex

The signal from smoothened receptor leads to the dissociation of Ci from the microtubule complex and the free Ci accumulates in nucleus. Ci, in association with a co-activator, turns on the transcription of hedgehog target genes, such as wnt wignless protein gene and patched gene (negative feedback network). Hedgehog pathway is more frequent in skin cancer (basal cell carcinoma) suggesting its proliferation response. The hedgehog pathway mediated by Ci is shown in Fig. 11.42.

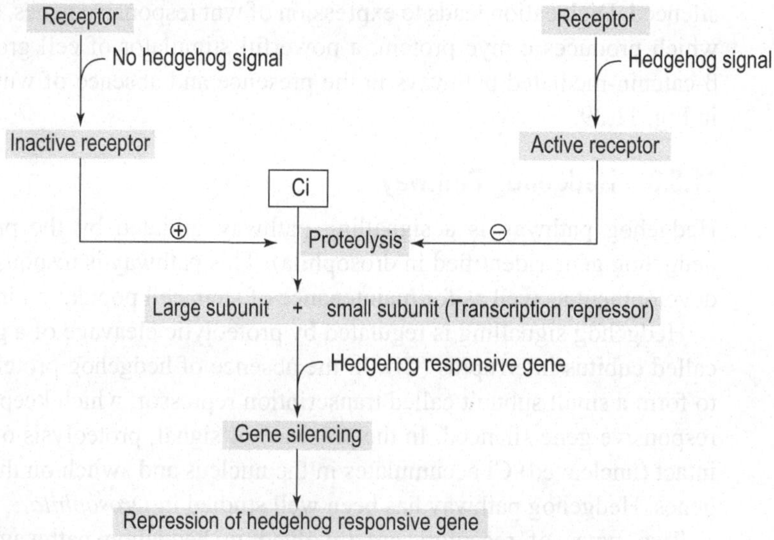

FIGURE 11.42 Hedgehog pathway mediated by Ci (components of Ci protein complex are shown in Fig. 11.41; HHRE—hedgehog responsive gene)

11.5.4 NF-κB pathway

NF-κB pathway depends on proteolysis of IκB which is associated with NF-κB (Fig. 11.36). Whenever there is an infection or inflammation, proinflammatory cytokines, such as TNFα and IL1, are released by immune cells such as macrophages. TNFα and IL1 bind to TNFα receptor and IL1 receptor, respectively. Both the receptors relay the signal by same mechanism. This section, therefore, explains about the signal transduction by TNFα receptor which also applies to IL1 receptor.

TNFα receptor is a trimer. On binding to TNFα, the trimer recruits three intracellular proteins:

1. Receptor-interacting protein (RIP)
2. TNF receptor associated death domain protein (TRADP)
3. TNF receptor associated factor 2 (TRAF2)

These three proteins recruit and activate 'IκB kinase kinase kinase' which in turn activate 'IκB kinase kinase'. IκB kinase kinase activates 'IκB kinase'. IκB kinase finally recruits NFκB–IκB complex and phosphorylates the IκB on the two serine residues. Phosphorylation marks IκB for ubiquitylation proteolysis and degradation in proteasomes (Section 11.8). As a result, free NF-κB is released and the nuclear localization signal is exposed. NF-κB then moves to the nucleus where it associates with the co-activator and turns on the transcription of NF-κB target genes which finally results in inflammatory response.

11.6 SIGNALLING NETWORK

Generally, it is assumed that most of the intracellular pathways are linear. That is, the signal passes from one protein to the subsequent downstream protein as a chain (Fig. 11.43a). In the linear cascade, the signal is passed to one protein at a time. However, there are some pathways in which at least one of the proteins pass the signal to two or more downstream proteins, and hence, the signal takes multiple routes. This situation results in branching of the signalling cascade (Fig. 11.43b).

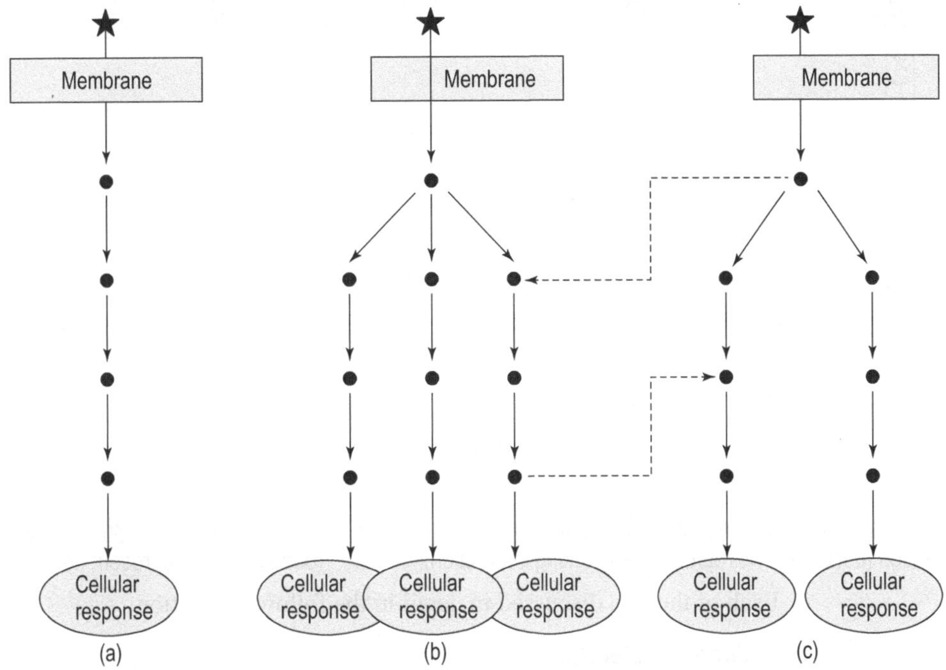

FIGURE 11.43 (a) Classic linear pathway, (b) branching pathway, and (c) crosstalk-denoted as dotted arrows (solid arrow represents the route of signal transduction. Asterisk represents the signal; closed circle represents proteins of the intracellular cascade)

11.6.1 Crosstalk

Yet another concept that occurs during signal transduction is called crosstalk. In this process, proteins of one cascade interact with that of another pathway thereby linking the two pathways. In crosstalk process, a protein of one pathway influences a molecule of another pathway either by inhibiting or activating the later. This phenomenon is shown in Fig. 11.43c. A very striking example for crosstalk is the interaction of MAPK pathway's Raf kinase protein with the protein kinases of another pathway utilizing the same enzyme.

Crosstalk helps coordinating the pathways and regulation of cellular activity. An important outcome of branching and crosstalk is the formation of network. Network may be defined as interconnections between different signalling pathways via proteins. In a network, there are two important points—junction and node. Junction is the point where two pathways integrate or converge. Node is a point where a signal diverts or splits into two or more routes.

11.6.2 Mechanism of Signalling Network

The signalling network arises due to one or all of the following four processes:

- Coincidence detection
- Gating
- Multiple diversions
- Feedback looping

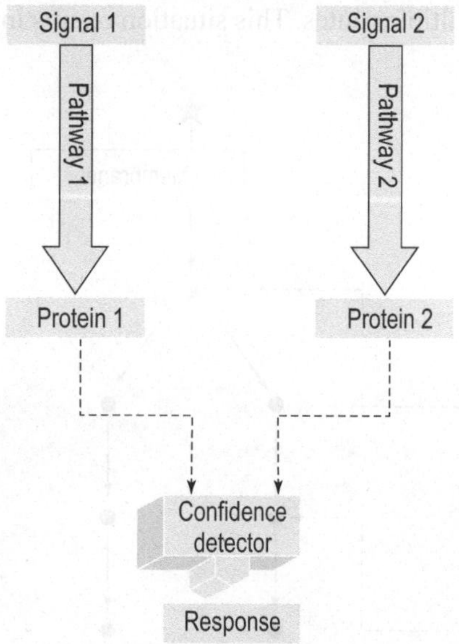

FIGURE 11.44 Cell signalling network by coincidence detection (dotted arrows represent networking interaction)

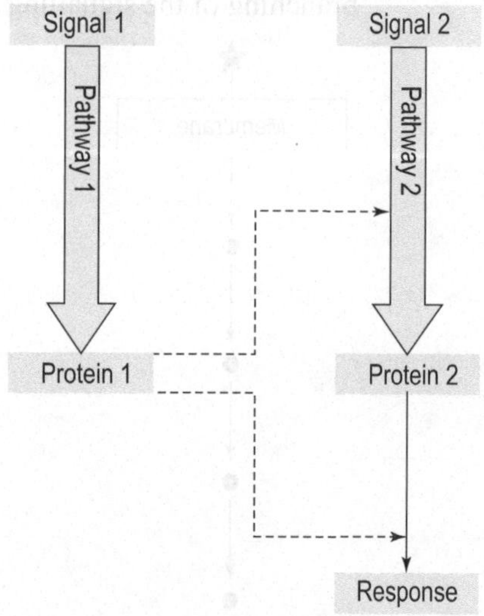

FIGURE 11.45 Cell signalling network by gating (dotted arrows represent networking interaction)

Each of these is discussed in detail in the following section.

Coincidence detection

Coincidence detector refers to a functional protein complex. Two different pathways converge upon the coincidence detector which can process the signal to produce a distinct unique response (Fig. 11.44). The response produced by this network is different from that generated singly by the converging pathways.

Gating

In gating process also two cascades are involved. One cascade called the primary cascade that produces a biological response is influenced through a gate by the other pathway called the secondary cascade (Fig. 11.45). This networking results in either inhibition or activation of the primary pathway. Thus, the signal flow through the primary pathway is regulated by the secondary pathway. Gating process just modifies the cellular response of the primary pathway instead of producing a unique response unlike coincidence detector.

Multiple diversions

Signal from a single protein can take different pathways from the node, with one pathway regulating the other (Fig. 11.46).

Feedback looping

In cell signalling, feedback refers to the regulation of a cellular response by inhibiting or activating the signal by the elements of the same pathway. Feedback control functions like a loop from downstream to the upstream of the cascade. There are three types of feedback looping that can generate signal network. They are positive feedback looping, negative feedback looping, and double-negative feedback looping. These three networks are shown in Figs 11.47a, b, and c, respectively.

In positive feedback looping, signal from one protein activates the second protein. The second protein in turn not only activates the further downstream protein, but also the immediate upstream first protein. Thus, the result of this network is the amplification and stabilization of the signal. A remarkable example for positive feedback looping occurs in Drosophila oocytes. An extracellular protein called gurken denoted as Grk passes the signal to EGF receptor on the follicle, from where it is transduced to activate the expression of rhomboid-1 (Rho-1) gene to produce rho-1 protein. Rho-1 activates another EGF receptor ligand spitz (Spi). Spi amplifies and stabilizes the EGF-receptor-linked signalling.

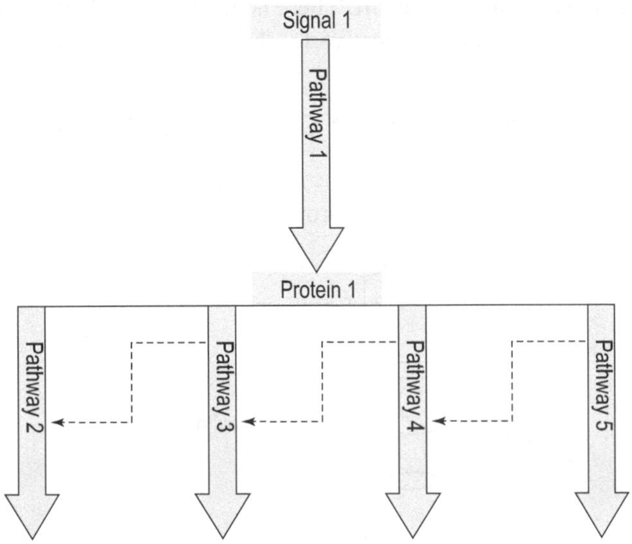

FIGURE 11.46 Cell signalling network by multiple diversions (dotted weight arrows represent networking interaction)

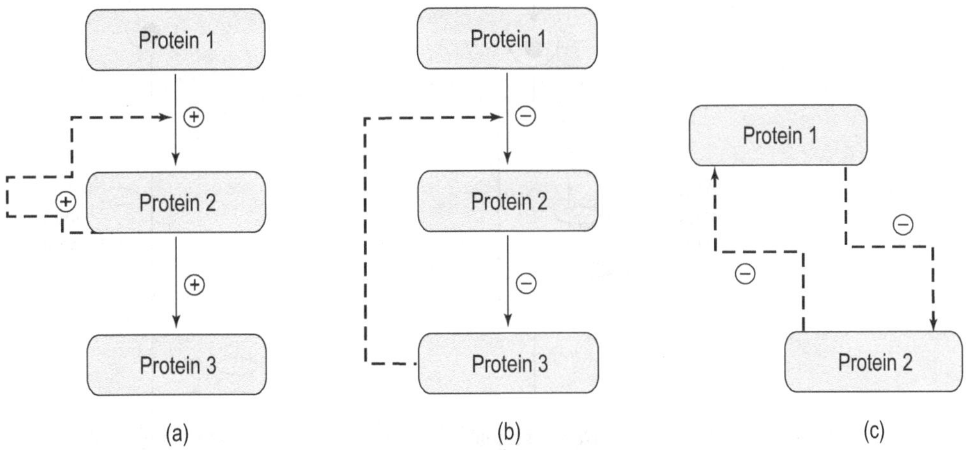

FIGURE 11.47 Feedback looping (dotted arrows represent networking interaction) (a) Positive feedback, (b) Negative feedback and (c) Double Negative feedback

In negative feedback looping, a pathway is inhibited by the product produced by the transcription activator of the same pathway. Inhibition of JAK-STAT pathway forms a negative feedback loop. As mentioned earlier, JAK-STAT pathway is initiated by the activation of cytokine receptor. In this pathway, the ligand cytokine binds to the cytokine receptor and activates it. Activated receptor takes on JAK which phosphorylates the STAT. STAT, being a transcription activator, activates the expression of SOCS proteins. SOCS moves upstream and inhibits cytokine receptor thereby inhibiting the JAK-STAT pathway. Thus, the end point of negative feedback looping is dampening and limiting of signalling

When two proteins in the pathway inhibit each other it leads to double-negative feedback looping. This type of loop toggles between two states in which only one protein is on while the other is off. There is no chance of having both the proteins on or both the proteins off. Double-negative feedback networking is found in MAPK pathway.

11.7 SPECIFICITY AND EFFICIENCY OF CELL SIGNALLING

Precision, speed, and efficiency of signalling is determined by scaffold proteins. Scaffold proteins organize the relay proteins and hold them in position forming a complex called

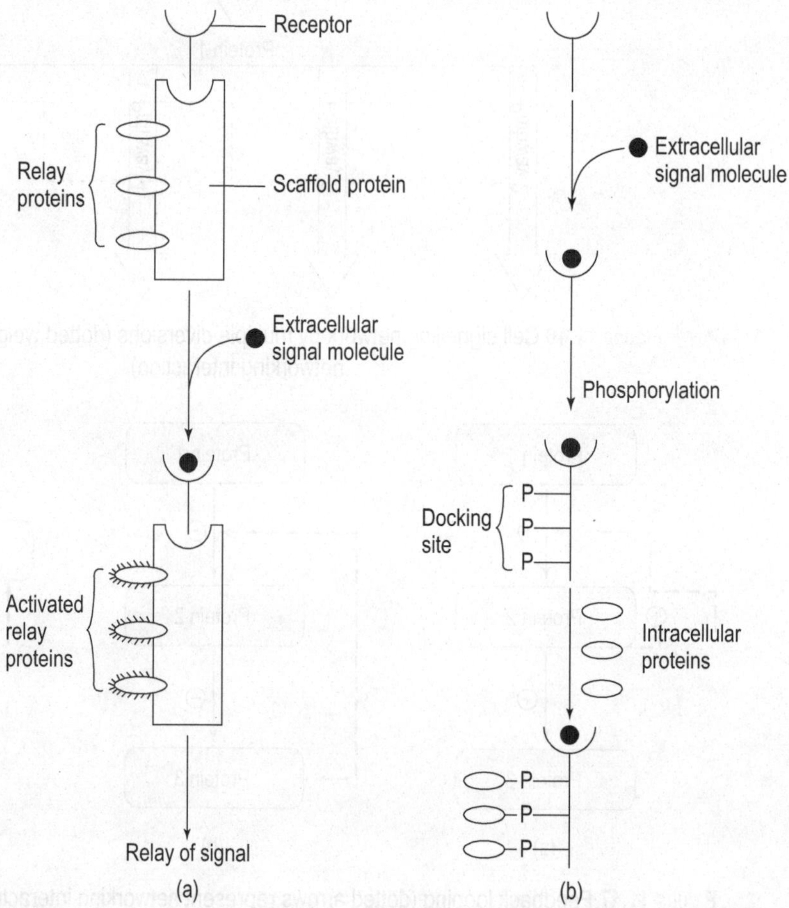

FIGURE 11.48 Intracellular signal complexes (a) Preformed complex, (b) Receptor-activated complex

signal complex. There are two types of signal complexes: preformed signal complex and receptor-activated signal complex (Fig. 11.48). The receptor is aligned with the intracellular protein which are pre-assembled and organized in a scaffold protein into a signal complex (the preformed signal complex). Scaffold protein forms a complex in response to binding of extracellular signal molecule to the receptor. On encountering an extracellular signal the receptor is phosphorylated at multiple sites. The phosphorylated residues function as docking sites for the intracellular proteins. Assembly of preformed signal complex and receptor-activated complex is shown in Fig. 11.48.

11.8 REGULATION OF CELL SIGNALLING

In order for a cell to produce a particular biological or physiological response at appropriate time, with particular intensity, the signalling pathway of that response should tightly be regulated. Therefore, regulation of signalling pathway is a noteworthy branch of cell signalling. This section discusses how signalling pathways are regulated. An overview of signalling regulation is shown in Fig. 11.49. In order to regulate the signalling pathway, the cell has to regulate certain basic components of signal transduction. Thus, the signalling enzymes and other intracellular signalling proteins are the regulatory points for appropriate cellular response.

Signal transducers are usually enzymes and proteins, operating downstream of the signalling pathway. For regulating the cell signalling, the concentration and function of proteins and enzymes should be controlled by an activation–deactivation process or by synthesis–degradation process which in turn can be accomplished by the following:

- Effector molecules
- Phosphorylation
- Ubiquitin–proteasome pathway (UbPP)
- Sumoylation
- Lipidation
- Single-site and multi-site modifications

11.8.1 Effector molecules as regulators

In many signal cascades, the second messengers function as low molecular weight effectors. They bind to target protein and regulate the flow of information through the pathway. cAMP and cGMP are best-characterized second messenger regulators of protein kinases.

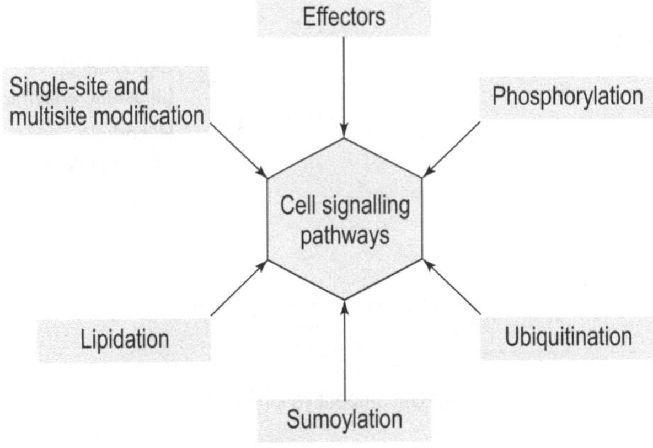

FIGURE 11.49 Regulation of cell signalling pathways

Enzyme inhibitors and enzyme activators are small effector molecules which regulate many signalling enzymes by inhibiting or activating them. Inhibitors may compete with the substrate of the enzyme and bind to the active site of the enzyme to impair the activity of the later. This process is called competitive inhibition. In some cases, the inhibitor may bind to a regulatory site commonly called the allosteric site of the enzyme and induce a conformational change in the enzyme and shift the equilibrium towards the inactive state. This is a process of non-competitive inhibition or allosteric inhibition. In some other instances, the inhibitor binds to the enzyme which has already bound to its substrate. In other words, the inhibitor binds to the enzyme–substrate complex and inhibits the enzyme. This process is called uncompetitive inhibition. In all these three types of inhibition, the enzyme kinetics (k_m and V_{max}) are altered. Types of inhibition of signalling enzymes and the kinetics profile of the inhibited enzymes are shown in Fig. 11.50 and Fig. 11.51, respectively. It is remarkable to learn that the inhibitors themselves are regulated by phosphorylation status or by degradation or by de novo synthesis.

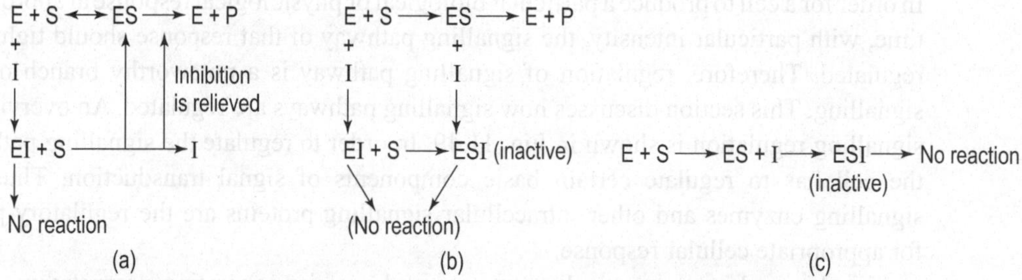

FIGURE 11.50 Types of inhibition of enzymes used for regulating pathways (a) Competitive inhibition, (b) Non-competitive inhibition (c) Uncompetitive inhibition

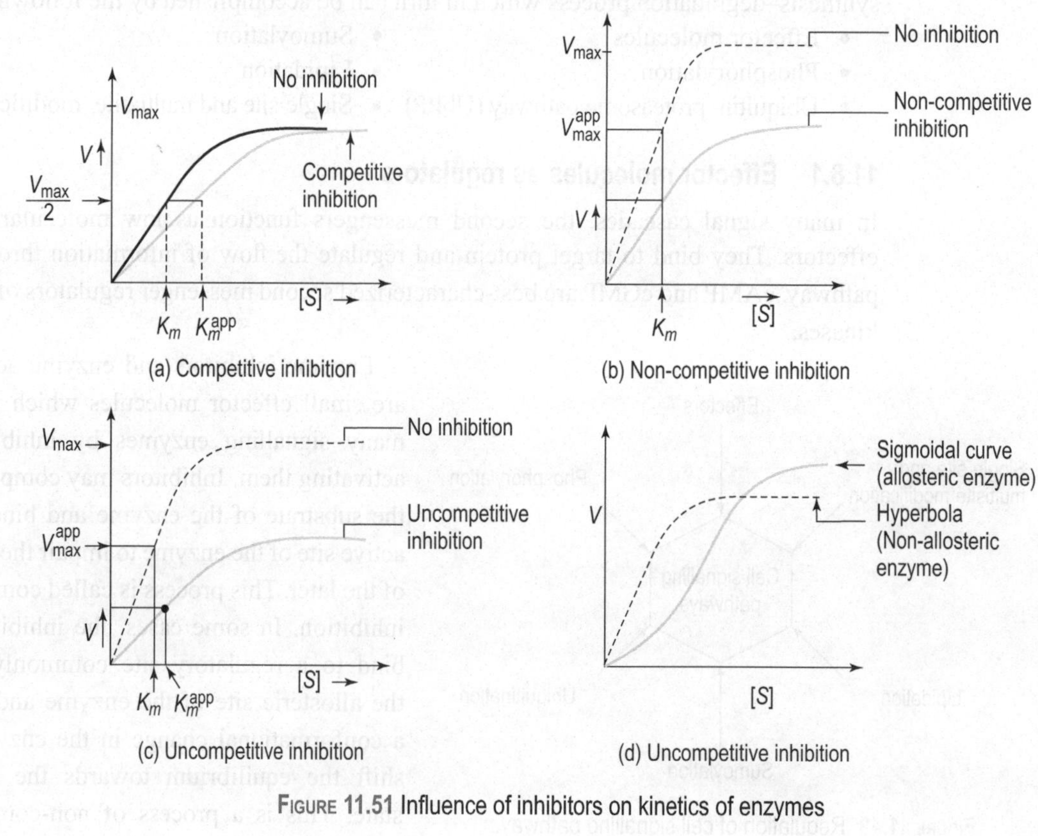

FIGURE 11.51 Influence of inhibitors on kinetics of enzymes

Activators are positive regulators of signalling enzymes. They bind to the active site of the enzyme and activate the enzyme. Ca^{2+}/calmodulin-dependent enzymes are regulated by reversible association with activators. Cyclins are best-studied examples for enzyme activators.

Metal ions also share the regulatory role in cell signalling. Ca^{2+} ions are used in the activation of protein kinase C (PKC), which in turn activates MAP kinase pathway for finally stimulating the transcription activators. Transcription activators are ultimately responsible for eliciting the final response to the signal. Thus, calcium plays a crucial role in regulating the signalling pathway, as described in Section 11.4.

11.8.2 Regulation by Reversible Phosphorylation

Phosphorylation of specific amino-acid residues of the enzymes is another way of regulation. Phosphorylation is a type of covalent modification induced by another enzyme—protein kinase. In eukaryotic cells, tyrosine, serine, or threonine residues of the enzymes are targets for protein kinases; whereas in prokaryotes, aspartate and histidine residues are often phosphorylated. Phosphorylation of enzymes can be represented as shown in Fig. 11.52.

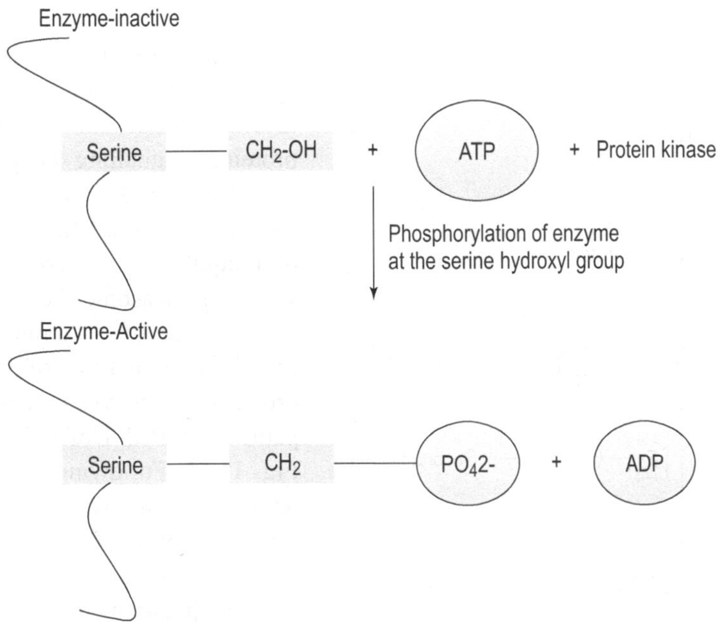

FIGURE 11.52 Regulation of enzyme by protein-kinase-mediated phosphorylation

Protein kinases transfer the phosphate group from ATP to the hydroxyl group of serine residue of the signalling enzyme. The phosphorylated enzyme bears four oxygen atoms and two negative charges and a pKa value of 6.7 at physiological pH. Phosphate group maintains the active state of the enzyme by forming a network with other regions of the enzyme via hydrogen bonds and electrostatic interactions. In helical enzymes, the phosphate group interacts with nitrogen of glycine residues. In non-helical enzymes, it reacts with the planar structure of the guanidine group of arginine residues resulting in the formation of multiple hydrogen bonds linking different regions of the signalling peptide. Electrostatic interactions occur between negative charge of phosphate group and positive charge of arginine residues. Hydrogen bonding and electrostatic interactions, both cause conformational changes in the enzyme. If the enzyme has to revert to inactive conformation, it should be dephosphorylated. The removal of phosphate group is catalyzed by another enzyme called phosphatase. Like kinases, phosphatases are also serine or threonine specific.

An example of an signalling enzyme modified by phosphorylation and subsequent allosteric conformational change is the muscle cell glycogen phosphorylase. In this enzyme, serine-14 is phosphorylated. In CDK (cyclin-dependent kinase), tyrosine is phosphorylated. Isocitrate dehydrogenase from Escherichia coli (E. coli) is unique because phosphorylation leads to only minimal conformational changes with maximal electrostatic interactions to become active.

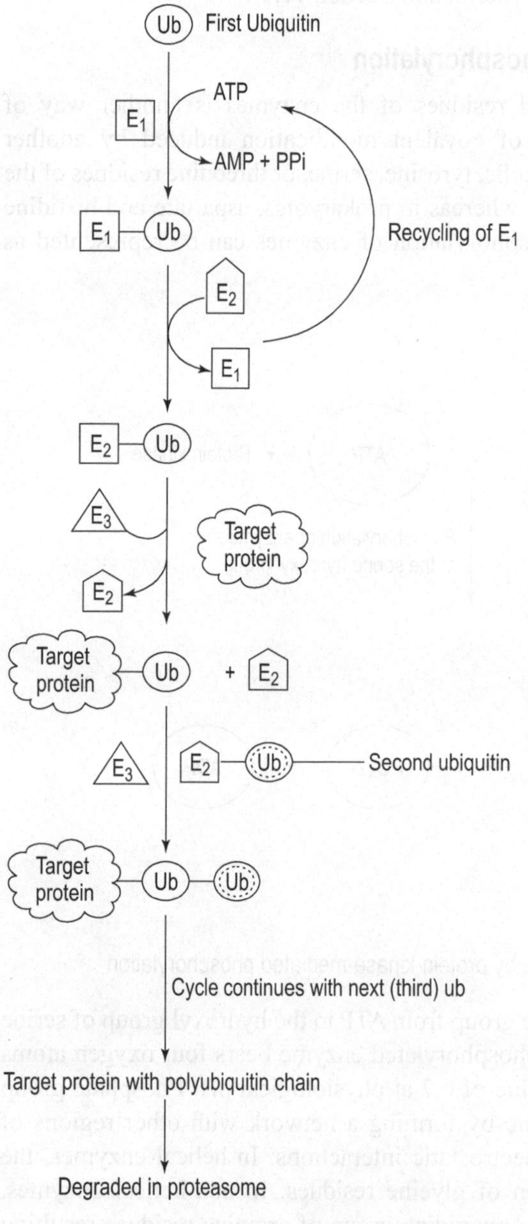

FIGURE 11.53 Degradation of proteins by ubiquitination

11.8.3 UbPP

Ubiquitin (Ub) is a protein family, members of which share a common fold, called the Ub fold. They are responsible for degrading the proteins by forming an isopeptide bond between its own glycine residue and an amino group of the target proteins. Signalling enzymes are also targets for ubiquitin family. Members of ubiquitin family include ubiquitin, NEDD8, SUMO-1, and ThiS. Role of ubiquitin in degradation of signalling enzymes within proteasome is discussed in this section. Ub is made of 76 amino acid residues. Function of Ub in regulation of signalling enzyme (target protein) occurs via repetition of enzyme-catalysed pathway called UbPP. This pathway is shown in Fig. 11.53. Following are the steps involved in the degradative pathway:

- Activation of Ub by activating enzyme—E_1.
- Conjugation of Ub to another enzyme, the conjugation enzyme—E_2.
- Transfer of Ub to the signalling enzyme (the target protein for Ub), in the presence of E_3.
- Repetition of the above three steps to form a polyubiquitin chain on the signalling enzyme.
- Transfer of poly-ub- enzyme complex to proteasome—the site of degradation.

On the whole, it can be noted that three enzymes, namely E_1, E_2, and E_3, catalyses the regulation of signalling enzymes by UbPP-mediated degradation. Initially, the first Ub molecule, denoted for convenience as Ub1 is activated by binding to E_1 in the presence of ATP. Linkage occurs between the glycine's carboxyl group of Ub and the active site cysteine's thiol group of the E_1 to form a E_1–Ub thiol ester. In the next step, Ub is transferred from E_1 to the cysteine thiol group of E_2 to form E_2–Ub. In the third step, E_3 transfers the Ub from E_2 to the target signalling enzyme. Basically, E_3 is a ligase (Ub-protein ligase) which catalyses the ligation of the C-terminal glycine of Ub to the ε-NH2 group of the lysine residue of the target via amide-isopeptide linkage. This results in the formation of Ub-signal enzyme complex. Subsequently, the same process continues again to add a second Ub to the first Ub already attached to the protein by a Ub–Ub linkage to form Ub–Ub–signal enzyme. As these processes continue, a large number of Ub will be added to form a poly-Ub chain on the signal enzyme.

E3 ligase family includes two members: HECT domain E_3 and RING domain E_3. HECT domain refers to 'homologous to E_6-AP C terminus' domain. It is used in the degradation of the tumour suppressor protein, p53. RING domain implies 'Really Interesting New Gene domain', sub members of which have conserved histidine and cyteine residues that coordinate zinc ions in a structure called the RING finger. Cb1 is a single subunit RING domain E_3 that catalyses the down regulation of tyrosine kinases (both RTK and NRTK) by binding to phosphotyrosine via SH2 domain. Cullin E_3 is another member of RING E_3. It is a multisubunit protein with a unit called cullin, attached to two RING motifs. The RING motif bound to the C-terminus of the cullin helps in recruiting the E_2 enzymes. Mechanism action of cullin is still unknown. APC, also called cyclosome, is a multisubunit RING domain E_3 which regulates cell cycle by regulating the cell cycle enzymes. It degrades proteins containing specific sequences called destruction boxes. SCF is multisubunit E_3. SCF stands for 'Skp-1-Cullin-F-box'. It has a linker protein unit, scaffold protein unit, and Rbx1/Roc1/Hrt RING domain unit. RING domain is responsible for recruiting the E_2 enzymes.

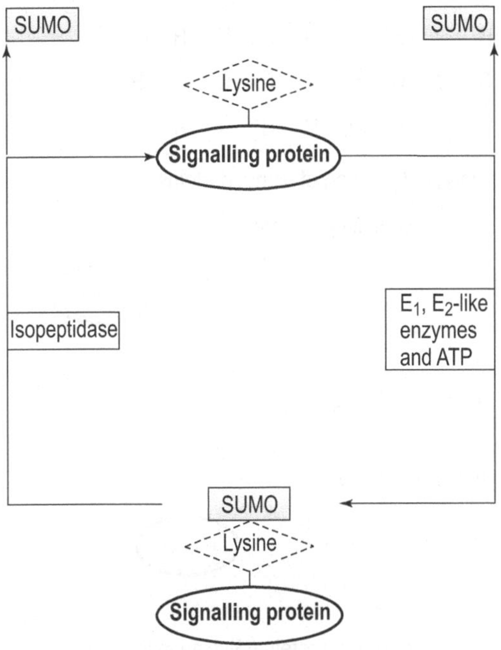

FIGURE 11.54 Schematic representation of SUMOylation

11.8.4 SUMOylation

SUMO refers to 'Small Ubiquitin-related Modifier' protein which functions in binding to and regulating other proteins. SUMOylation is one of the methods of regulation of signal proteins. SUMO binds to target proteins via lysine residues of the later in the presence of E_1- and E_2-like enzymes and ATP. SUMO binding is reversible and the target protein can be released by the enzyme isopeptidase (Fig. 11.54). SUMO is involved in the regulation of transcription factors that are responsible for producing cellular response. In the presence of co-repressors, such as histone deacetylases, SUMO binds covalently to lysine residues and inhibits transcription. One of the established roles of SUMOylation has been proved in the inhibition of IκB. When SUMO binds to IκB, the latter is stabilized and remains bound to NF-κB thereby inhibiting nuclear translocation of NF-κB and transcriptional activation. Proteins can be targeted for either mono- or poly-SUMOylation, but most proteins that have been characterized do not undergo poly-SUMOylation. A schematic representation of SUMOylation is shown in Fig. 11.54.

11.8.5 Lipidation

Some signal proteins are post-translationally modified by attachment of lipid anchors for subsequent attachment to the cell membrane or organelle membrane. Lipid anchors target signalling proteins to the membrane in order to tune the membrane-associated signalling pathways. Following are the lipid anchors used for modifying and regulating signal proteins:

- Myristoic acid anchor
- Isoprenoid anchor
  - o Farnesyl anchor
  - o Geranyl anchor
  - o Dual anchor
- Palmitic acid anchor
- Cholesterol anchor
- GPI anchor

Myristoylation

Myristic acid (n-tetradecanoic acid) is a fatty acid with 14 carbon atoms (Fig. 11.55). It is added co-translationally or post-translationally. N-myristoylation usually occurs at the consensus sequence in the N-terminal end of the protein: Gly-Cys-X-X-Ser. Presence of basic amino acids at the N-terminus signals the protein for myristoylation. Myristoyl group is added to the amino group of the glycine residue in the consensus sequence via amide bond, in an irreversible reaction catalysed by N-myristoyl transferase. Myristoylation reaction results in stable modification. The reaction is shown in Fig. 11.56. Cytoplasmic tyrosine kinases, such as Src kinases, possess myristoyl anchor.

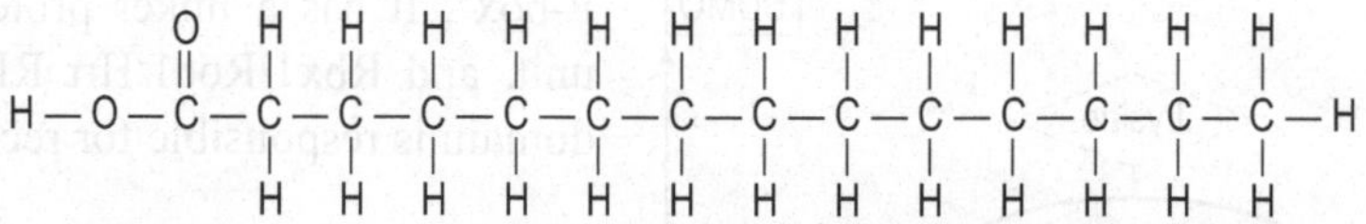

FIGURE 11.55 Myristic acid anchor used in lipidation of signal proteins

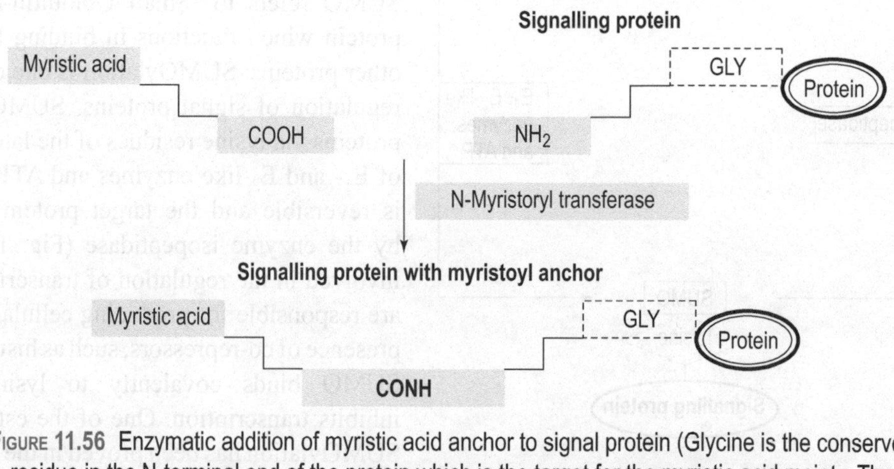

FIGURE 11.56 Enzymatic addition of myristic acid anchor to signal protein (Glycine is the conserved residue in the N-terminal end of the protein which is the target for the myristic acid moiety. The carboxyl terminal of the myristic acid interacts with the amino group of the glycine forming amide bond)

TABLE 11.13 Target signalling protein modified by palmitoylation via thiester linkage

Protein
Pleckstrin homology (PH) domain of phospholipases D1 and D2
Ras regulatory GTPases
Rho regulatory GTPases
P56 non-RTK
PSD-95 adaptor protein
RGS protein

Palmitoylation

Palmitic acid (n-hexadecanoic acid, shown in Fig. 11.57) is a fatty acid with 16 carbon atoms anchored to the signalling protein. Palmitoyl group is added to the thiol group of the cysteine residue in a process called acylation, via thioester bond. This reaction is catalysed by the protein acyl transferase. Palmitoylation reaction results in reversible modification because the thioester bond is weaker than the amide bond formed by the myristyl anchor. Sometimes, palmitoylation can also occur in the serine or threonine residues via oxy-ester bonding; or in epsilon amino group of lysine residues via amide linkage. The enzyme catalyzing the reversible reaction (i.e., deacylation) is called acyl-protein thioesterase. Targets for palmitoyl anchor are given in Table 11.13. Palmitoylation of protein by thioester linkage is shown in Fig. 11.58.

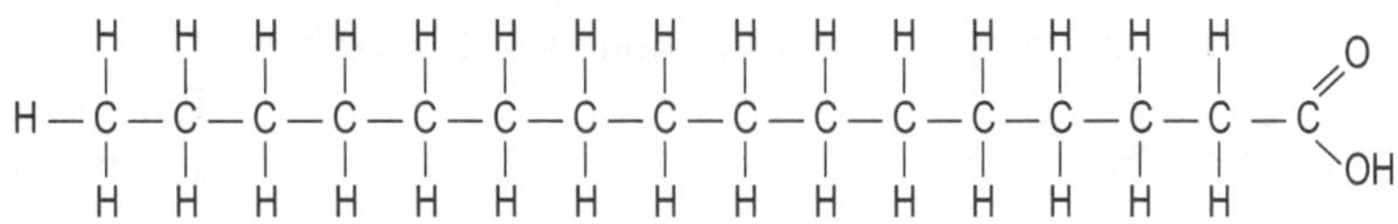

FIGURE 11.57 Structure of palmitic acid anchor used for lipidation of signalling protein

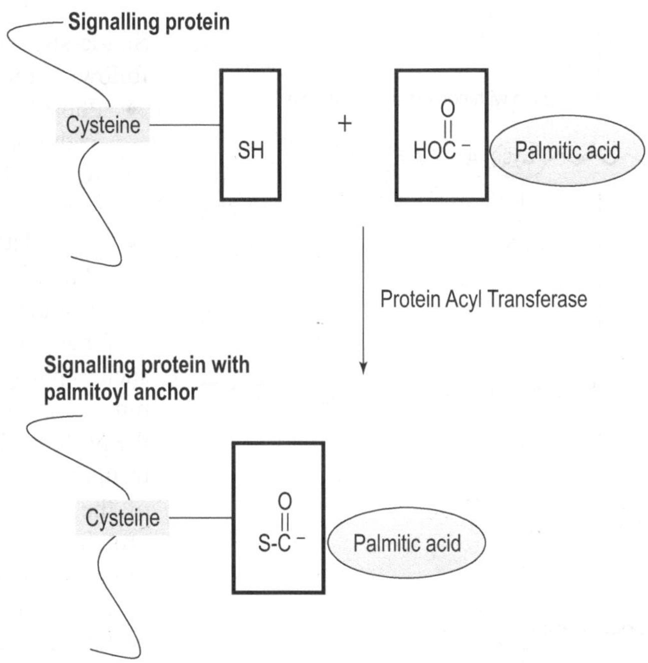

FIGURE 11.58 Enzymatic addition of palmitic acid anchor to signal protein. (Cysteine is the target residue in the protein. The carboxyl terminal of the palmitic acid interacts with the thiol group of the cysteine forming thioester bond. Groups involved in bond formation are shown in box.)

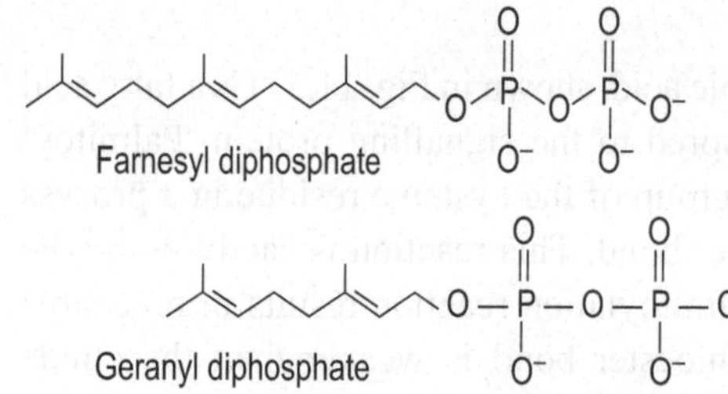

FIGURE 11.59 Farnesyl diphosphate and geranyl diphosphate; these compounds donate farnesyl and geranyl group to the signalling protein in the presence of specific transferase enzyme

Isoprenylation

Prenylation, or isoprenylation, is the enzyme-catalyzed addition of isoprenyl group to a protein. Prenyl groups used in protein modification are either a farnesyl or a geranyl–geranyl moiety to C-terminal cysteine(s) of the target protein. The signal for the isoprenylation is the presence of consensus sequence, 'Cys-A-A-X-COOH' in the carboxyl terminal of the target protein to be modified. Here, A is alanine and X is any amino acid. Farnesyl diphosphate and geranyl diphosphate (Fig. 11.59) are the donors of farnesyl and geranyl chain. They donate their prenyl group in the presence of prenyl transferases (e.g., farnesyl transferase or geranyl transferase) to the cysteine residue. Subsequently, the other three carboxyl terminal residues are cleaved by prenylation-dependent endoprotease (Rce-1) and the new COOH group of cysteine is methylated by methyl transferase. Target proteins for isoprenylation include members of Ras superfamily, G-protein subunits, and Rab protein. Farnesylation of protein is schematised in Fig. 11.60.

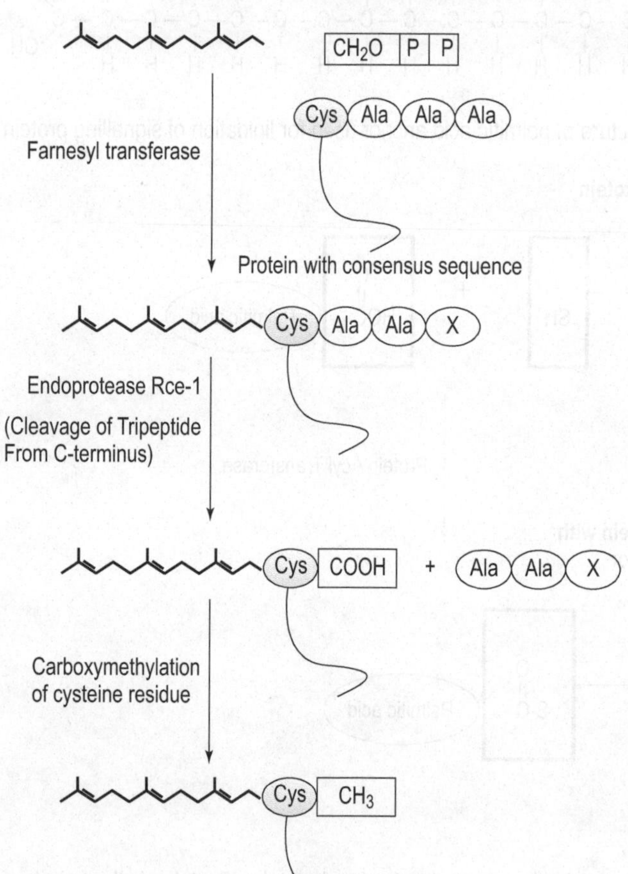

FIGURE 11.60 Enzyme-catalysed farnesylation of signal protein (the line represents the protein chain)

11.8.6 Single-site and Multi-site Regulation by Specific Functional Groups

Single-site regulation occurs by the following processes:

- Phosphorylation of serine or tyrosine or threonine residues of the protein
- Acetylation or methylation or of lysine residues
- Methylation of arginine residues
- Oxidation or nitrosylation of cysteine residues

When the regulatory functionalities, such as phosphate group, acetyl, methyl, or nitrosyl groups (Fig. 11.61), are added to the proteins they are specifically identified by interaction module or interaction domain of the downstream protein. Downstream protein specifically recognizes and binds to the regulatory groups of their upstream proteins via interaction domain. The portion of the protein that binds to the interaction domain may be a modified amino acid or another domain or a peptide sequence. Some of the interaction domains and the group of their signalling protein partner are shown in Table 11.14.

TABLE 11.14 Interaction domains and their protein partners (the portion of the protein that binds to the interaction domain may be a modified amino acid, another domain, a peptide sequence, or a phospholipid associated with a membrane-bound protein)

Interaction domain	Binding site of the protein	Nature of the binding site
Sarcoma homology 2 domain (SH2)	Phosphotyrosine	Modified amino acid residue
Phosphotyrosine binding domain (PTB)	Phosphotyrosine	Modified amino acid residue
forkhead-associated domain (FHA)	Phosphotyrosine	Modified amino acid residue
MAD homology domain (MH2)	Phosphoserine	Modified amino acid residue
Chromo	Methyl lysine	Modified amino acid residue
Bromo	Acetyl lysine	Modified amino acid residue
*PDZ	PDZ	Specific domain
Sterile alpha motif domain (SAM)	SAM	Specific domain
Caspase recruitment domain (CARD)	CARD	Specific domain
Death-effector domain (DED)	DED	Specific domain
Sarcoma homology 3 domain (SH3)	Proline rich motif (PXXP)	Peptide sequence
Enabled/VASP homology 1 domain (EVH1)	FPPPP (one phenylalanine residue followed by four proline residues)	Peptide sequence
WW	PPXY motif (P = proline, x = any amino acid, and Y = tyrosine)	Peptide sequence
PTB	NPXY (a domain of low density lipoprotein receptor)	Peptide sequence
PDZ	Val-COOH	Peptide sequence
Pleckstrin homology dmain (PH)	Phosphoinositide-3 (PIP3)	Phospholipid
*FYVE	Phosphoinositide-2 (PIP2)	Phospholipid
Diacyl glycerol(DAG) domain	Diacyl glycerol (DAG)	Phospholipid
*FERM	PIP2	Phospholipid
Ca2+ binding domain (C2)	Phospatidic acid/phospatidyl serine (PA/PS)	Phospholipid
Epsin N-terminal homology domain (ENTH)	PIP3 PIP2	Phospholipid

* PDZ is an acronym representing the first letters of three proteins namely, post synaptic density protein (PSD95), Drosophila disc large tumor suppressor (Dlg1), and zonula occludens-1 protein (zo-1). These proteins were first discovered to share the domain

FYVE zinc finger domain represents first letters of four cysteine-rich proteins: Fab 1 (yeast orthologue of PIKfyve), YOTB, Vac 1 (vesicle transport protein), and EEA1, in which it has been found.

FERM domain is the abbreviation of 4.1 protein, ezrin, radixin and moesin.

SH2 domain is a well-known example for interaction domain which binds specifically to the phosphorylated tyrosine residues of the other protein. MH2 domain is specific for phosphoserine, the bromo domain interacts with acetylated lysine. Lipid rafts are micro-domains in lipid the bilayer, which are rich in cholesterol and glycolipid. Lipid rafts act as scaffold for clustered intracellular proteins and cell-surface receptors. In a lipid raft, the phospholipid associated with the protein is the site of binding for the interaction domain. SH2 domains and PTB domain bind to phosphorylated tyrosine residues which are present in activated receptor domain (e.g., RTK) or any of the intracellular proteins of the cascade.

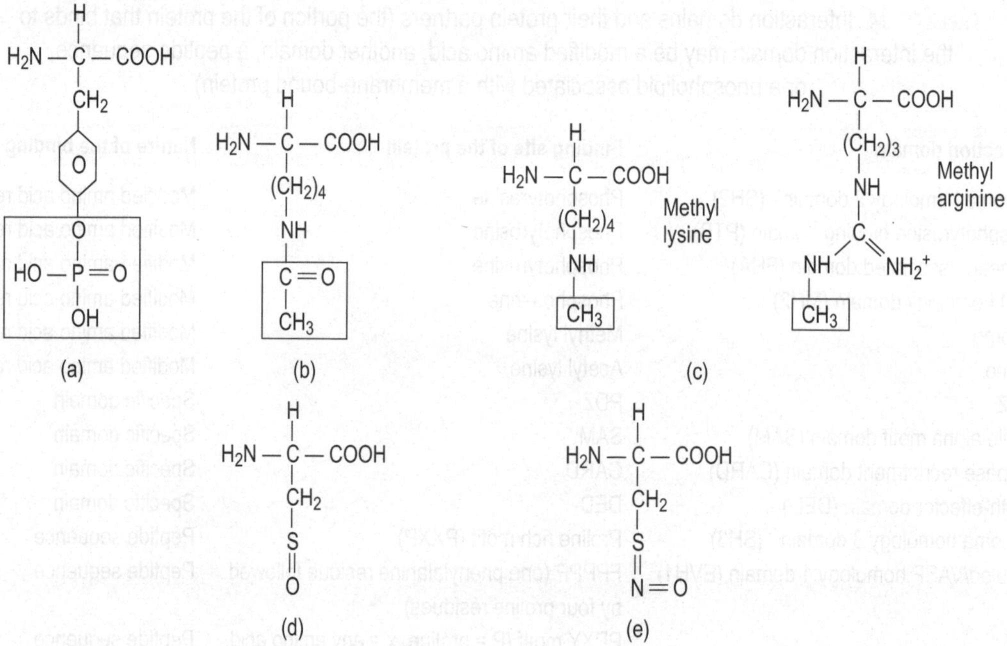

FIGURE 11.61 Regulatory groups of signal proteins (functional group attached to the aminoacid residue of the protein is shown in box) (a) Phosphate group—Example phosphotyrosine, (b) Acetyl group —Example accetyl lysine, (c) Methyl group—Example methyl lysine and methylarginine, (d) Oxide group—Example oxycysleine, (e) Nitrosyl group—Example Nitrosyl cysteine

PH domain (pleckstrin homology domain) binds to phosphorylated inositol phospholipid in membrane to form complex. Thereafter, the protein of the complex interacts with other signalling proteins of the transduction assembly. SH3 domain binds to proline stretches. PDZ domain binds to specific motifs in receptor or signalling protein.

Multi-site modification occurs by attachment of same groups in different amino acid residues of the same protein, or by attachment of different functional groups in same protein. The best illustrated multi-site modified signal protein is the tumour suppressor protein p53. It is phosphorylated, acetylated, SUMOylated, and ubiquitinated on several sites. Sometimes different modifying groups are added to the same type of amino acid residues. One such example is the lysine residue which can be acetylated, methylated, needylated, SUMOylated, and ubiquitinated.

11.9 SIGNALLING IN PLANT CELLS

Although little information is available about plant cell signalling, it is known that plant cells mainly depend on ELR of RSTHK type for cell signalling. Response to light signals by photoproteins and two component signalling pathway of ethylene have been studied in plant cells.

11.9.1 Cell-surface Receptors in Plant Cells

Similar to animal cells, plant cells depend on cell-surface receptors for signalling. The largest class of ELR in plant is RSTHK. The extracellular domain of RSTHK in plant cells

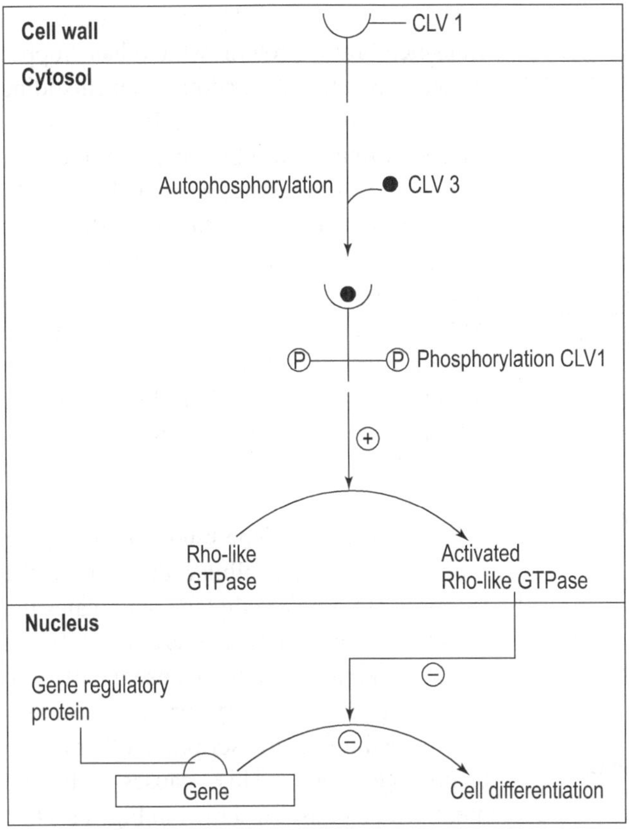

FIGURE 11.62 Cell differentiation response induced by CLV 3 and CLV 1

has leucine rich repeats (LRR). Signalling mediated by LRR receptor is well studied in Arabidopsis. The best-studied LRR receptor is CLAVATA 1 (CLV 1). The signalling ligand that binds to CLV 1 is secreted by the neighbouring cell. Binding of CLV 3 to CLV 1 produces cell differentiation in shoot meristem. Although the exact mechanism of the signalling response is unknown, it is thought to occur as follows: On binding of ligand, the receptor phosphorylates itself at serine/threonine residues and gets activated. Active receptor in turn activates rho-like GTPase which in turn inhibits a gene regulatory protein that is an inhibitor of cell differentiation. Thus, CLV 3 induces cell differentiation (Fig. 11.62).

11.9.2 Photoproteins-mediated Signalling in Plant Cells

Response of plant cells to light is mediated by receptors called photoproteins; they are of three types:

- Phytochromes, which respond to red light and far-red light
- Cryptochromes, which respond to blue light
- Phototropin, which responds to blue light

Phytochrome is a dimer with RSTHK and RHK activity. It is a cytosolic receptor. RSTHK is responsible for signalling. When cells are exposed to red light, inactive phytochrome is activated and it is phosphorylated at its serine/threonine residues. Phosphorylated phytochrome is thought to relay the signal, by any one of the following mechanisms:

1. Directly translocating into the nucleus, and activating a nuclear gene regulatory protein to initiate transcription of light responsive gene.
2. Activating a cytosolic gene regulatory protein which, in turn, moves into nucleus for initiating the transcription of the light responsive gene.
3. Triggering signalling cascade in the cytosol, which produces cell response without the involvement of nuclear elements.

Phytochrome-mediated response to light is shown in Fig. 11.63.

11.9.3 Ethylene-mediated Two-component Signalling Pathway in Plant Cells

Ethylene is one of the growth regulators in plant cells, others being auxin, cytokinin, giberellin, and abscisic acid. Ethylene functions as a signal molecule by binding to

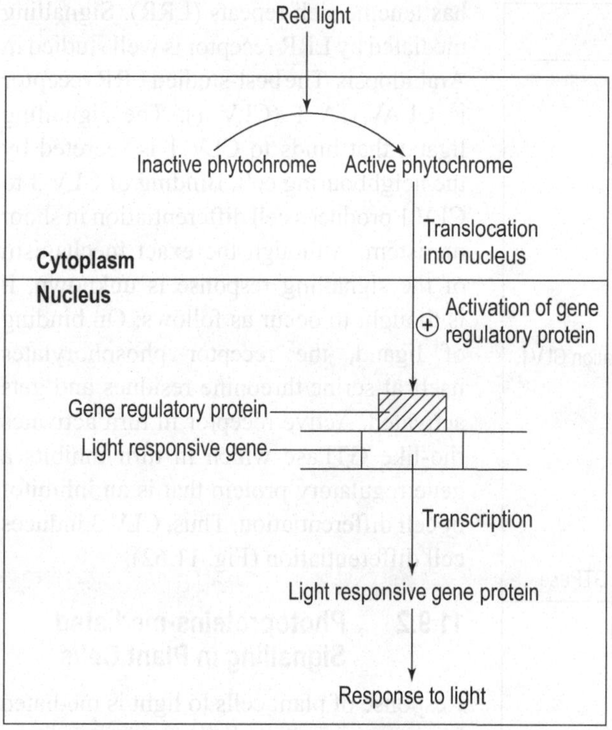

FIGURE 11.63 Phytochrome-mediated response to light in plant cells

ethylene receptor. Ethylene receptor is a transmembrane protein which has copper atoms at its extracellular domain and histidine kinase activity at its intracellular domain. In the absence of ethylene, the receptor remains in active state, so that the kinases phosphorylate its own histidine residues. The phosphate moiety is immediately transferred sequentially to MAP kinase kinase kinase, MAP kinase kinase and MAP kinase. These three kinases together form a MAP kinase module in the cell. MAP kinase finally inhibits the nuclear gene regulatory protein which is responsible for transcription of ethylene responsive gene. This response is reversed when ethylene binds to the copper atoms in the extracellular domain of the receptor. Binding of ethylene inactivates the histidine kinase domain thereby inhibiting the MAP kinase module. Consequently, the nuclear gene regulatory protein becomes active leading to transcription of ethylene responsive gene. This causes ethylene mediated response as shown in Fig. 11.64.

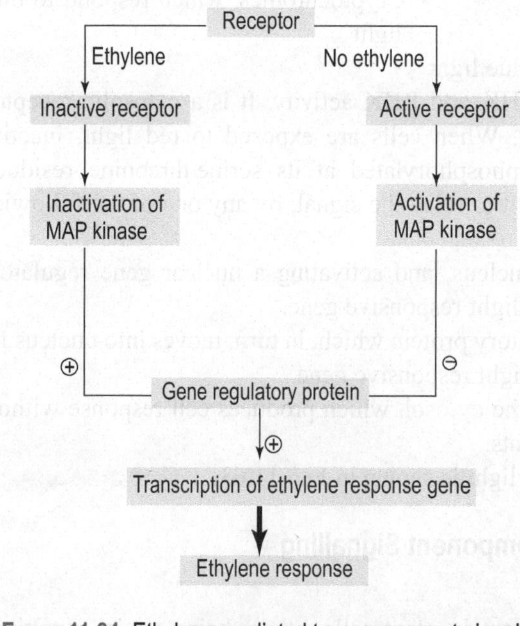

FIGURE 11.64 Ethylene-mediated two-component signalling response in plant cells

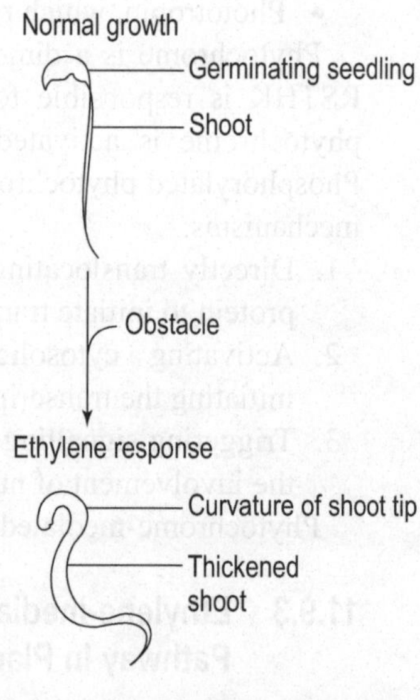

FIGURE 11.65 Ethylene-mediated triple response in shoot cells

Ethylene response occurs whenever the cell encounters a stress. For example, when plant cells in a seedling encounter a hurdle, such as a piece of stone or gravel, while growing, it produces a triple response (Fig. 11.65) which is controlled by ethylene. The triple responses include the following:

- Thickening of the shoot
- Curvature of shoot tip to form a shield
- Growth of the shoot away from the direction of gravity to avoid the obstacle

SUMMARY

Cell signalling is a process by which cells communicate with one another and with the environment in order to produce a specific response. Membrane-bound receptors, intracellular receptors, intracellular molecules, such as enzymes, second messengers, large intracellular mediators, and molecular switches are molecular tools which enable cell signalling. Cell signalling can occur via autocrine mode, paracrine mode, endocrine mode, or gap-junction mode. Major responses produced by cell signalling include survival, division, differentiation, secretion, and contraction. Cell signalling occurs as a cascade or pathway mediated by signal proteins and enzymes.

Intracellular signalling pathways include cAMP-signalling pathway, cGMP-signalling pathway, PLC and inositol phospholipid signalling pathway, NF-κB signalling pathway, and JAK-STAT signalling pathway. Some pathways are turned on or off by the proteolysis of the key proteins involved in that pathway. Such pathways are called proteolysis-dependent signalling pathway which include notch pathway, wnt pathway, and hedgehog pathway. Signalling pathways with the cell are interconnected to form a network.

Cell signalling should be tightly regulated for proper functioning of the cell. Cell signalling is regulated by modification of signalling proteins by SUMOylation, myristoylation, palmitoylation, isoprenylation, or by degradation of the protein by ubiquitin in proteasomes. Signalling in plant cells is poorly understood. Photoproteins and ethylene are two signalling molecules in plant cells whose functions are well understood.

TERMS TO REMEMBER

3'5'-cyclic AMP It is a second messenger, that is, a small molecule produced intracellularly in response to binding of extracellular signal molecules to cellular receptors. It is formed from ATP by the action of adenylyl cyclase.

Autocrine signalling A process in which a signal molecule produced by one cell acts on its own receptors to produce the response.

Cell signalling Communication of cells with the external environment, with other cells, and with different intracellular compartments by transmitting and processing the signals received from these locations.

Cholera A severe bacterial infection caused by the bacteria *Vibrio cholerae*, which primarily affects the small intestine. The main symptoms include profuse watery diarrhoea and vomiting.

Contact-dependent signalling Signalling in which the signal is transported from the signalling cell to the target cell on establishment of direct contact between the plasma membrane of the two cells.

Crosstalk A process of cell signalling in which proteins of one cascade interact with that of another pathway thereby linking the two pathways. Crosstalk helps in coordinating the pathways and in regulation of cellular activity.

Endocrine signalling A process by which the signals produced by the endocrine glands move through the blood vessel and reach the target organ to elicit the effect.

Epidermal growth factor (EGF) A growth factor that plays an important role in the regulation of cell growth, proliferation, and differentiation by binding to its receptor EGFR.

Fibroblast growth factors (FGFs) A family of growth factors involved in angiogenesis, wound healing, and embryonic development.

Gap junction A nanoscale patch of 2 to 4 nm size which is a space or gap between the membranes of two adjacent cells. Gap junctions can mediate signalling by allowing the exchange of small intracellular signal molecules like calcium ions and cAMP between the cells.

Isoprenylation The enzyme-catalysed addition of isoprenyl group to a protein with consensus sequence 'Cys-A-A-X-COOH' in the carboxyl terminal.

Janus kinase A family of intracellular non-receptor tyrosine kinases that transduce cytokine-mediated signals via the JAK-STAT pathway.

Myristoylation The addition of myristic acid (n-tetradecanoic acid) to a protein co-translationally or post-translationally. This process is involved in the regulation of signal proteins.

Nerve growth factor (NGF) A small secreted protein which is important for the growth, maintenance, and survival of certain target neurons (nerve cells). It also functions as a signalling molecule.

Neurotrophins A family of proteins that induce the survival, development, and function of neurons.

Palmitoylation The addition of palmitic acid (n-hexadecanoic acid) to the thiol group of the cysteine residue of the signalling protein via thioester bond.

Paracrine signalling The action of signal molecules in the extracellular space on the neighbouring cells (target cells).

Phototropins Photoreceptor proteins (specifically, flavoproteins) that mediate phototropism responses in higher plants. Along with cryptochromes and phytochromes they allow plants to respond and alter their growth in response to the light environment.

Phytochromes A photoreceptor pigment used by plants for the detection of light. It is sensitive to light in the red and far-red region of the visible spectrum.

Platelet-derived growth factor (PGDF) One of the numerous growth factors, or proteins, that regulate cell growth and division. In particular, it plays a significant role in blood vessel formation (angiogenesis) and the growth of blood vessels from already-existing blood vessel tissue.

Second messengers Molecules that relay signals from receptors on the cell surface to target molecules inside the cell, in the cytoplasm or nucleus. They relay the signals of hormones, such as epinephrine (adrenalin), growth factors, and others, and cause some kind of change in the activity of the cell.

Sensory cells Cells that can sense the information, such as sounds, light, touch, smell, taste, and temperature) through receptors on their surface.

Steroid hormones An important class of hydrophobic signalling molecules that are intracellular-receptor dependent. They are classified into mineralocorticoids (aldosterone), glucocorticoids (cortisol), and sex hormones (oestrogens, testosterone, and progesterones).

SUMO Small Ubiquitin-related Modifier protein which functions in binding to and regulating other proteins. SUMOYlation is one of the methods of regulation of signal proteins.

Synaptic signalling Transmission of signal along the neurons.

Transmembrane receptors Membrane-bound proteins with an extracellular domain, transmembrane domain, and an intracellular domain.

Ubiquitin A protein family, members of which are responsible for degrading the proteins by forming an isopeptide bond between its own glycine residue and an amino group of the target proteins.

Vascular endothelial growth factor (VEGF) A chemical signal produced by cells that stimulates the growth of new blood vessels. It is part of the system that restores the oxygen supply to tissues when blood circulation is inadequate.

Whooping cough A highly contagious disease caused by the bacterium *Bordetella pertussis*. It is known to last for approximately 6 weeks before subsiding.

REVIEW QUESTIONS

1. What are the processes that are controlled by cell signalling?
2. List the types of extracellular signalling molecules.
3. Differentiate between ionotropic receptor and serpentine receptor.
4. What is RGS?
5. Enlist the functions of different types of enzyme-linked receptors.
6. Define RTK and write the responses produced by RTK.
7. What is JAK?
8. What is Smad?
9. What do you mean by molecular switches? Give an example.

10. In which type of signalling does the signal molecule act remotely from its origin?
11. Define crosstalk with reference to cell signalling.
12. How is cellular signalling process regulated?
13. Explain the role of cGMP-mediated signalling pathway in rod cells.
14. Describe the role of patched receptor and smoothened receptor in cell signalling.
15. How is ethylene used in signalling plant cells?

MULTIPLE-CHOICE QUESTIONS

1. Enzyme catalyzing conversion of ATP to 3′5′-cyclic AMP is
   (a) cyclic AMP synthase
   (b) cyclic AMP synthetase
   (c) adenylyl cyclase
   (d) ATP hydrolase
2. Signalling proteins that induce the growth and development of brain cells are called
   (a) neurokinin
   (b) neurotrophin
   (c) growth factor
   (d) survival factor
3. The major component that induces angiogenesis is
   (a) epidermal growth factor
   (b) fibroblast growth factor
   (c) platelet-derived growth factor
   (d) neuronal growth factor
4. Janus kinase is a
   (a) non-receptor tyrosine kinase
   (b) receptor tyrosine kinase
   (c) vascular endothelial growth factor
   (d) secondary messenger
5. In autocrine signalling, the signal molecule produced by one cell acts on
   (a) the adjacent cell
   (b) the same cell
   (c) a remote cell
   (d) a similar group of cell
6. Synaptic signalling is more common in
   (a) epidermal cells
   (b) neurons
   (c) muscle cells
   (d) liver cells
7. The causative agent for whooping cough is
   (a) *Wuchereria bancrofti*
   (b) *Bordetella pertussis*
   (c) *Vibrio cholera*
   (d) *Pseudomonas aureginosa*
8. During ubiquitination, isopeptide bond is formed between
   (a) glycine and amino group
   (b) glycine and carboxyl group
   (c) lysine and amino group
   (d) lysine and carboxyl group
9. The common name for n-tetradecanoic acid is
   (a) myristic acid
   (b) palmitic acid
   (c) isoprenoic acid
   (d) stearic acid
10. Signalling proteins are regulated by phosphorylation at the
   (a) glycine residue
   (b) lysine residue
   (c) tryptophan residue
   (d) tyrosine residue
11. Which of the following is an isoprenyl anchor?
   (a) GPI anchor
   (b) palmitoyl anchor
   (c) geranyl anchor
   (d) myristoyl anchor
12. Which of the following is not a second messenger?
   (a) cAMP
   (b) cGMP
   (c) calcium ions
   (d) G protein

Answers to Multiple-choice Questions

1. (c)	2. (b)	3. (c)	4. (c)	5. (c)
6. (b)	7. (b)	8. (a)	9. (b)	10. (a)
11. (c)	12. (d)			

CHAPTER 12

Cell Differentiation and Cancer Biology

Harald zur Hausen, received the Nobel Prize in 2008 for discovering the role of papilloma viruses in cervical cancer.

After reading this chapter, students should be able to understand and appreciate the following:

- Process of cell differentiation
- Cell differentiation in unicellular and multicellular organisms
- Determinants of cell differentiation
- Causes, prevention, diagnosis, and treatment of cancer
- Stages in carcinogenesis
- Host–tumour interaction

12.1 CELL DIFFERENTIATION

Cell differentiation is the process of transformation of less specialized cells into more specialized cells, accompanied by a change in the morphology, membrane potential, metabolism, and signal transduction process. The development of a zygote into different organs to form a foetus and the subsequent development into an adult occur due to cell differentiation. Cells that can differentiate into all types of cells, forming the whole organism are called *totipotent* (e.g., zygote), those that can differentiate into many kinds of cells are called *pluripotent* (e.g., inner cell mass of the blastula—refer Chapter 13), and those that can differentiate into cells with specific functions, such as haematopoiesis, and immunity are called *multipotent* (e.g., haematopoietic stem cells—refer Chapters 3 and 5).

12.1.1 General Characteristics of Cell Differentiation

Cell differentiation process has the following four characteristics:

1. Stability
2. Inducibility
3. Morphological difference
4. Controllability by genetic factors

Stability

The phenomenon of cell differentiation is highly stable. Once the differentiation process starts, the process continues for a long period of time. Thus, the differentiation continues for several cell generations. For example, stability of neurons is so remarkable that they remain stable throughout the lifetime of an organism.

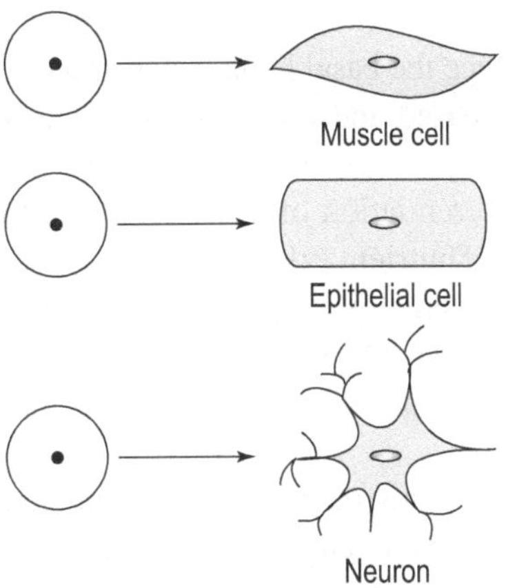

FIGURE 12.1 Morphological differences produced by cell differentiation

Inducibility

Differentiation is induced and initiated by a stimulating factor which signals the cells to differentiate (refer Chapter 11). For example, epidermal growth factor induces proliferation of the embryo. Once the differentiation process is induced, the process can continue without the need for further stimulation. Oestrogen induces differentiation of female reproductive organs and cytokines induce the differentiation of haematopoietic stem cells into blood cells and immune cells (refer Chapters 3 and 5).

Morphological difference

Cell differentiation always produces morphological differences in the cell or cell organelles (Fig. 12.1). For example, during lens development, the lens fibre cell differentiation is accompanied by the shrinkage of the nucleus, condensation of chromatins, and movement of the chromatins towards the periphery of the nucleus. The nuclear pore complexes form aggregates and get associated with the chromatin. Neural differentiation is yet another example. It leads to neurite growth and elongation of the cells.

Controllability by genetic factors

Cell differentiation is controlled by genetic factors, which are coordinated by regulatory proteins. In order to induce cell differentiation, the stimulating signal ultimately *turns on* the gene encoding the protein that controls or regulates the differentiation. The gene is translated into proteins which are accumulated in the cytoplasm. For example, the notch responsive gene is turned on to inhibit differentiation of epithelial cells into neural cells (refer Chapter 11). When neuron specific genes and epithelial specific genes are expressed, neurons and epithelial cells are produced, respectively (Fig. 12.2).

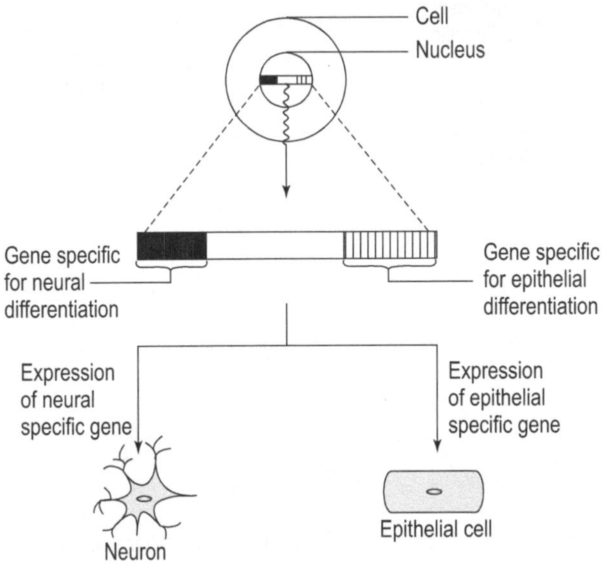

FIGURE 12.2 Control of differentiation by genetic factors

12.1.2 Differentiation in Unicellular and Multicellular Organisms

The differentiation of unicellular organisms is not as complicated as that of multicellular organisms. Unicellular differentiation is understood in *Salmonella* species, *Bacillus subtilis*, and *Rhizobium* bacteria. Formation of bacterial flagellum, endospore and cyst, and cell elongation are good examples of cell differentiation in unicellular organisms. All these processes are driven by proper synthesis and assembly of cellular macromolecules. Proteins play a major role in this process, as the specific genes are transcribed and translated.

In *Salmonella* species, the basal body hook is first constructed, during which the fla gene

(gene encoding the flagellum protein) is silenced by the inactivation of the corresponding transcription factor (sigma 28) by the antisigma factor. Once the basal hook is constructed, the cellular concentration of the antisigma factor is decreased and the sigma 28 gene is expressed leading to the flagellum synthesis, leading to morphological change (Fig. 12.3).

Bacillus subtilis is an example to understand the differentiation of bacterial cell into dormant spores (Fig. 12.4). Sporulation occurs as a result of nutrient exhaustion. Whenever there is nutrition demand, the cells enter into a dormant stage called spore, which lasts for around eight hours. During this period, the cells divide into two compartments or chambers —the mother cell chamber and the forespore chamber. The latter develops into a mature spore and functions as a germ line cell.

Rhizobium bacteria reside in the root nodules of leguminous plants. The bacteria replicate their DNA without undergoing cytokinesis. The cell's genome is amplified and the cells undergo elongation with an increase in the membrane permeability.

In multicellular organisms, the zygote develops into gastrula which develops the three germ layers. The differentiation process occurs from the three germ layers—ectoderm, mesoderm, and endoderm. The ectoderm differentiates into nervous system and sense organs (skin, eye, nose, throat, ear); the endoderm differentiates into endocrine glands, digestive glands, stomach, intestine, and the lungs and the mesoderm differentiates into

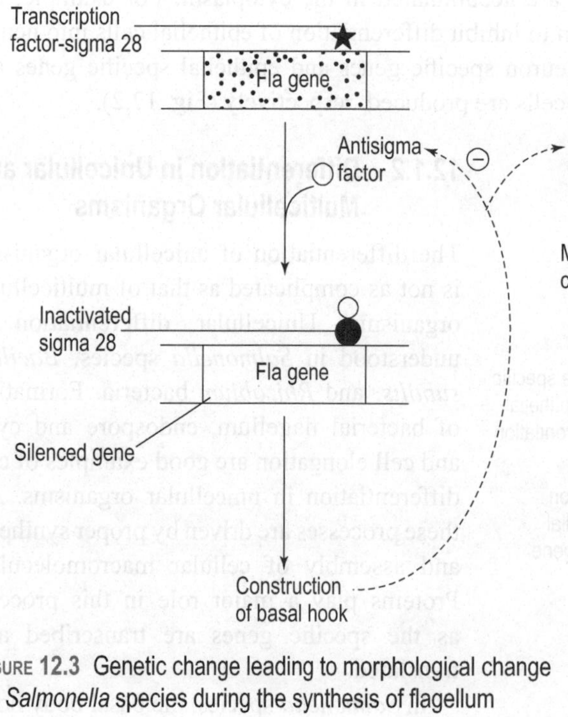

FIGURE 12.3 Genetic change leading to morphological change in *Salmonella* species during the synthesis of flagellum

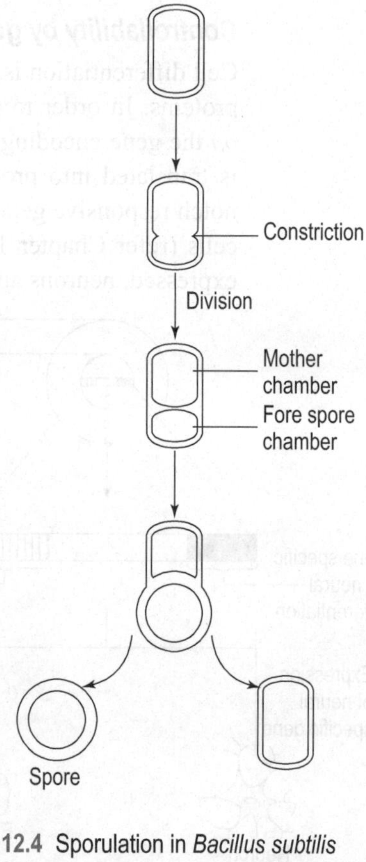

FIGURE 12.4 Sporulation in *Bacillus subtilis*

the organ systems, including skeletal system, muscular system, reproductive system, and excretory system, thus, a constituting a complete organism. In addition to the organs derived from the three germ layers, the body contains cavities, including pleural cavity, peritoneal cavity, and pericardial cavity. The three-dimensional differentiation process in multicellular organisms was studied using a multicellular test-bed known as the Evolutionary Developmental System (EDS). The EDS is a model of biological development. In EDS-based studies, the multicellular morphology has been designed from a single cell via asymmetric division, which reveals that differential gene expression following asymmetric division is responsible for diverse cell fates. The fate of a cell in multicellular organisms is determined by the presence or absence of a signal molecule, which produces an intracellular cascade event that leads to gene expression, producing a specific protein which induces the specific fate (refer Chapter 11). Apart from the functional organs, the body contain stem cells, which are pluripotent and differentiate into specific types of cells or organs. This is discussed in detail in Chapter 13.

12.1.3 Cytoplasmic Determinants

The determination of a cell type is defined as its progression into a particular fate, that is, differentiation follows determination. Once the cells are determined the fate becomes stable, and does not change. Determination is influenced by regulatory molecules called determinants.

Cytoplasmic determinants are regulators of cell differentiation present in the cytoplasm. They are mRNAs of transcription factors or are transcription factors themselves. They are distributed in the cells in an asymmetric fashion, mostly concentrated at the periphery. Owing to asymmetric distribution, the mRNA will be segregated to only one of the two daughter cells. This results in differential gene expression and difference in the cell fate (Fig. 12.5).

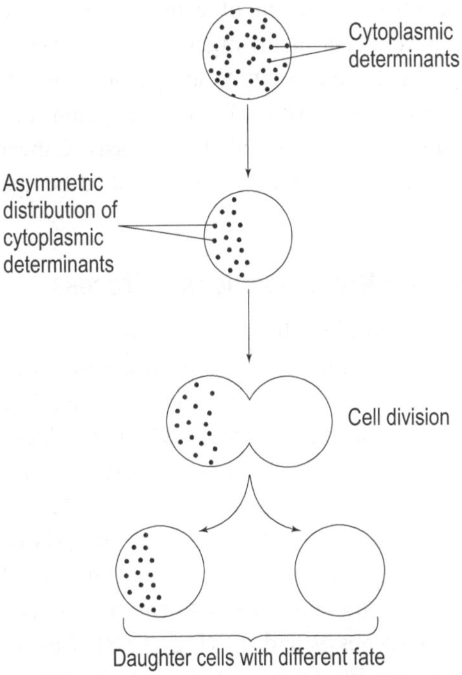

FIGURE 12.5 Asymmetric distribution and segregation of cytoplasmic determinants during cell determination

P-granules of the germ cells are good examples for cytoplasmic determinants. P-granules are dispersed in the embryo of the nematode *Caenorhabditis elegans* (Fig. 12.6). During the complete zygote formation, and just before the cell division starts, the granules move towards the periphery of the zygote. During the four-celled stage, the granules are present in the single cell present in the posterior end. Finally, during the formation of larva, the granules are present in the primordial germ cells.

Asymmetric cell division leads to asymmetric segregation of cell-fate determinants, leading to differentiation of cell into a specific fate. This was studied in *Caenorhabditis elegans*. During fertilization, and the one-celled stage, the centrosome of the sperm induces asymmetric distribution (polarized distribution) of PAR proteins. Followed by this, cell determinants are asymmetrically segregated between the anterior and the

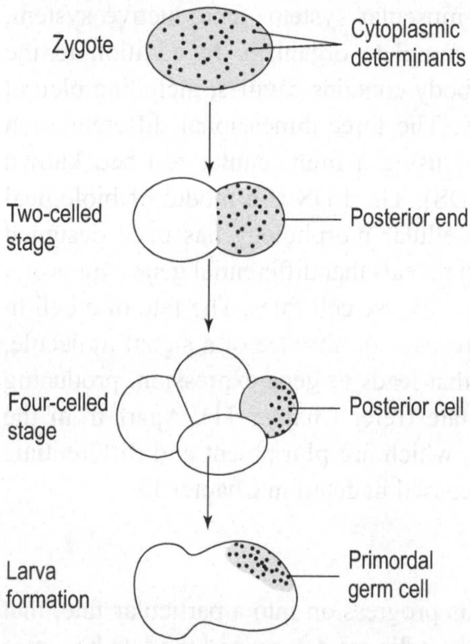

FIGURE 12.6 Asymmetric distribution of P-granules in germ cells of nematode

posterior blastomeres. The microtubules are positioned by a pulling force generated by a G-protein-dependent pathway. In the two-celled stage, the dorso-ventral axis is established. In the four-celled stage, the left–right axis (or anterior–posterior axis) is established.

12.1.4 Nucleoplasmic Interaction

Differential gene expression caused by nucleoplasmic protein interaction influences the differentiation process. Differentiation occurs due to differential gene expression caused by changes in chromatin proteins, chromatin-modifying enzymes, retinoblastoma protein, and transcription factors and repressors. Laminin, the major protein of nuclear envelope, and its interaction partner LAP-α (laminin-associated peptide-α) induces the changes associated with chromosomes and influence the cell differentiation process. The laminin complex organizes the chromatin in the nuclear periphery and aids cellular differentiation.

12.2 CANCER

Cancer is an uncontrolled growth of cells and abnormal cell division which can occur in any organ of the body (Fig. 12.7). Cancer cells have the capacity to invade other tissues and spread throughout the body via blood and lymph; this process of spreading of cancer is called metastasis. Cancer which does not spread to the other organs is called benign tumour.

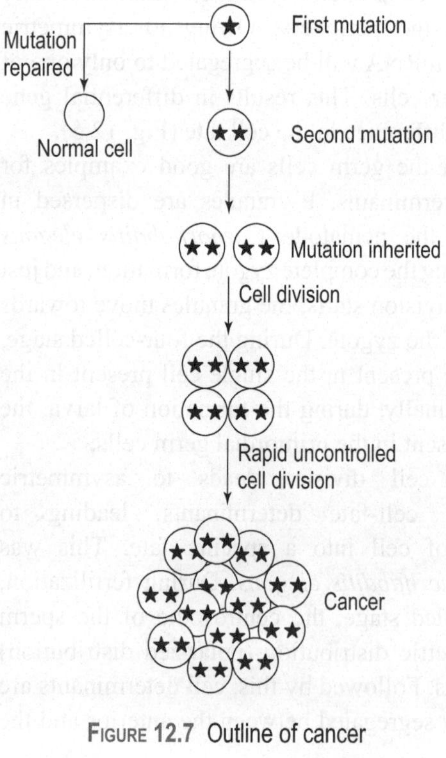

FIGURE 12.7 Outline of cancer

12.2.1 Cancer as a Microevolutionary Process

Cancer arises from a single cell mutation and progresses via repeated rounds of mutation and proliferation; this indicates that cancer cells are genetically unstable. The mutated cell forms a primary tumour, which produces progeny that undergo further changes and grow into a huge cancer. The evolution of cancer from a single cell aberration is evident from the chromosomal analysis of chronic myelogenous leukaemia. It was found that in all forms of leukaemia, cells show a translocation between the arms of chromosomes 9 and 22 (Fig. 12.8). Cancer cell inherits its abnormality to the subsequent generation, due to genetic change. After the exposure to a causative agent, the cells undergo a long incubation period to evolve

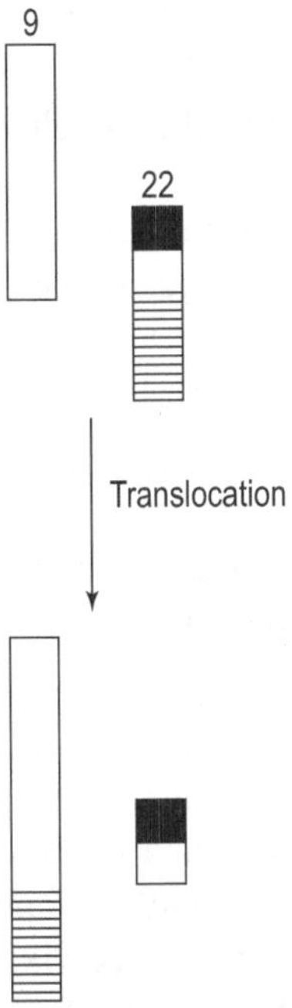

FIGURE 12.8 Leukaemia cells showing the chromosomal translocation

into cancer. The incubation period for smoke-induced lung cancer is around 10 or 20 years and for leukaemia is 5 years. For chemical carcinogen, the incubation period is more than 20 years.

Evolution of carcinoma in the epithelium is well understood in the cervix. The cervical carcinoma evolves as a low-grade neoplasia, progresses to a high-grade neoplasia, finally resulting in malignancy (Fig. 12.9). In normal cervix, the epithelium is organized as a stratified (multilayered) squamous epithelium, which mimics the epidermis of the skin or the lining of the buccal cavity. Within the stratified epithelium, the basal layer starts proliferating, and produces more cells. These cells move to the surface layer, and differentiate into flat cells which are sloughed off. In low-grade cancer (or low-grade neoplasia), the proliferation occurs in the lower third of the epithelium, and differentiation occurs in the upper layer in a disordered fashion. In the later stage, that is, in high-grade neoplasia, proliferation occurs in all the layers, without differentiation. In the next stage, the proliferating cells migrate through the basal lamina, and invade the underlying connective tissue, leading to malignancy. The rate of cancer progression is inversely related to the rate of programmed cell death.

The cells of the malignant cancer detach from the original site, invade the capillary, pass through the basal lamina and endothelium, and enter blood circulation. From the circulation, the cells extravasate the blood vessel and then escape to some other organ and survive in that alien environment. The body has certain proteins called metastasis suppressors in order to resist metastasis (refer Section 12.2.10). Only a tiny portion of the malignant cells metastasize. Metastatic cells are resistant to apoptosis. The molecular mechanism behind metastasis is still not clear. Fig. 12.10 shows the metastasis process.

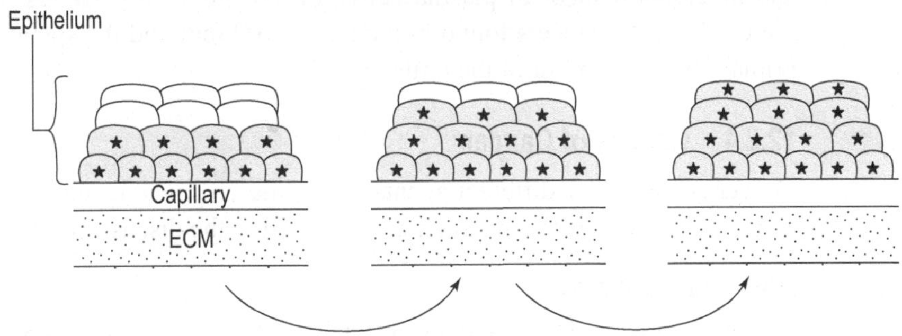

FIGURE 12.9 Evolution of cancer in the cervical epithelium (cancer cells are represented as asterisk)

12.2.2 Hallmarks of Cancer

Following are the hallmarks of cancer cells:

1. They undergo uncontrollable cell proliferation.
2. They are resistant to apoptosis.
3. They are capable of escaping replicative senescence, and hence, they undergo rapid replication.

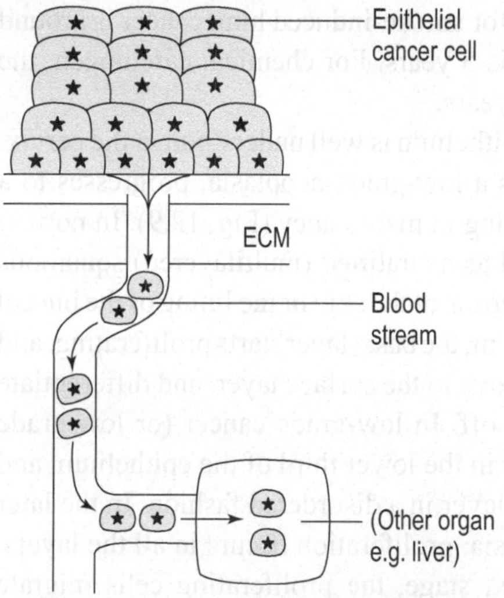

FIGURE 12.10 Process of metastasis (asterisk indicates cancer cells; arrow indicates migration of cancer cells in blood stream)

4. They have potential to avoid differentiation.
5. They show genetic instability.
6. They are capable of invading the neighbouring tissues.
7. They are capable of detaching from their origin and hence invade the remote tissues by metastasis.

12.2.3 Organ-specific Cancer

Depending on the organ of origin, cancer is divided into the following five types:

1. Carcinoma
2. Sarcoma
3. Leukaemia
4. Lymphoma and myeloma
5. Cancer of the CNS

Carcinoma refers to the malignant tumour of epithelial tissues of the skin, or those lining the internal organs. It is of three types—squamous cell carcinoma is the cancer of squamous epithelium in skin, lips, mouth, oesophagus, urinary bladder, prostate, lungs, vagina, and cervix; basal cell carcinoma is the common form of skin cancer and adenomatous cancer is the cancer of epithelium in the glandular organs.

Sarcoma is the name given to growth of cancer cells in the connective tissues of mesodermal origin, such as bone and cartilage, or other tissues, such as muscles and blood vessels. Sarcoma of muscles and blood vessels is called soft tissue carcinoma. Sarcoma of bone and cartilage is called osteosarcoma and chondrosarcoma, respectively. Leukaemia (cancer of blood-forming cells), lymphoma (cancer of the lymphatic system), and myeloma (cancer of plasma cells) are discussed in Chapters 3 and 5. Cancers of the CNS are the cancers found in the tissues of brain and the spinal cord. Glioma is the primary benign tumour of the brain.

12.2.4 Causes of Cancer

Cancer is caused by different agents, including chemical carcinogen, viruses, radiation, and tumour promoters. This section is devoted to the discussion of these causes.

Chemical carcinogen

Chemical carcinogens are chemicals which induce mutation in a cell, and initiate cancer. Many chemicals are proved carcinogens, whereas others are suspected carcinogens. Types of chemical carcinogens are listed in Table 12.1. Chemical carcinogens enter into the body via inhalation, ingestion, or through skin, and affect lungs, liver, kidney, gonads, and even brain. The influence of carcinogen depends on the duration of exposure and the presence of cocarcinogenic factors, such as smoking, alcohol consumption, high fat diet, and drug intake. Most of the carcinogens act directly on DNA (genotoxic carcinogen) to induce mutation, for example, alkylating agents. Some carcinogens gets metabolized to active metabolite (indirect carcinogens), which in turn induce DNA damage, for example, polycyclic aromatic hydrocarbons, nitrosoamines, and aromatic amines.

TABLE 12.1 Types of chemical carcinogens

Carcinogen type	Example	Carcinogen type	Example
Epigenetic carcinogen		Hydrazine	1, 2-dimethylhydrazine hydrochloride
Aromatic hydrocarbons	Benzene	Miscellaneous compounds	Thiourea
Benz[a]anthracene			Thioacetamide
Benzo[a]pyrene			Formaldehyde
Aromatic amine	Benzidine		Acetaldehyde
	Aniline		1, 4-dioxane
	o-Anisidine		Ethyl carbamate
	o-Toluidine		Styrene
Nitroso compounds	Nitrosamines		2-Nitropropane
	Nitrosodimethylamine		Arsenic compounds
Epoxides	Ethylene oxide		Chromium compounds
	Propylene oxide		Thorium dioxide
Organohalogens	Vinyl chloride		Beryllium compounds
	Carbon tetrachloride		Cadmium compounds
	Chloroform		Lead compounds
	Hexachlorobenzene		Nickel compounds
	Trichloroethylene		Selenium sulphide

Viruses

Hepatitis virus, human papilloma virus (HPV), Epstein–Barr virus (EBV), and human T-cell lymphocytic virus 1 (HTLV-1) are known to cause cancer. Hepatitis B virus causes hepatocellular carcinoma (a type of liver cancer). HPV is the causative agent for cervical cancer, vaginal cancer, vulvar cancer, oropharyngeal cancer, anal cancer, and penile cancer. EBV causes Burkitt lymphoma, non-Hodgkin lymphoma, Hodgkin lymphoma, and nasopharyngeal carcinoma. Human herpes virus and Kaposis sarcoma-associated herpes virus cause Kaposi's sarcoma. Retrovirus is the causative organism for hair cell leukaemia.

Although it is known that the virus integrates its genome in the host cell, the exact mechanism by which each type of virus causes the specific cancer is still under investigation. It was reported that the EBV inactivates the p16 gene, which is responsible for inhibiting the cell division. However, two general mechanisms were proposed to be the mechanism of virus action. The first is a direct mechanism in which the viral genome is expressed in the host cell to produce products that enhance the growth and survival of the host. Growth enhancement is inherited to the subsequent generations resulting in cancer. The second is an indirect mechanism in which the virus acts as a cofactor without taking direct part in inducing cancer. HIV is known to function by this mechanism. HIV suppresses the immune system and enhances the infection of other cancer-causing viruses, such as EBV and herpes virus, leading to cancer.

Apart from virus, organisms such as bacteria and parasites also cause cancer. Helicobacter pylori induces stomach cancer, bladder cancer and cholangiocarcinoma are caused by parasites—schistosomes and liver fluke.

Ultraviolet light

Ultraviolet (UV) radiation penetrates the layers of skin causing skin cancer. UV radiation is absorbed by DNA, which results in C→T and CC→TT transformations, leading to mutation. UV radiation can also generate free radicals, which induce cancer, such as cutaneous melanoma.

Ionizing radiation (gamma rays)

Ionizing radiation is a well-known cause of cancer. Ionizing radiations transfer their energy to the atoms of the cellular biomolecules. This leads to generation of free radicals and damage to the vital molecules including DNA. Although the damage can be repaired at low-dose exposure, high-dose exposure leads to irreparable damage. The sensitivity and the susceptibility to damage of cells towards ionizing radiation are in the following order:

Haematopoietic organs > reproductive organs > skin > bone and teeth > muscle > nervous system

Tumour initiators and tumour promoters

Tumour initiators are compounds that are transformed into chemically reactive carcinogens (e.g., epoxide derivatives) or are reactive themselves (e.g., alkylating agents). These compounds cause latent genetic damage so that the cells develop a greater chance for cancerous growth, subsequent to further exposure of the carcinogen. Thus, tumour initiators just initiate cancer. Tumour promoters are compounds which do not necessarily induce DNA damage. Instead, they induce free radical generations in the membrane. They do not have intrinsic carcinogenic activity, but indirectly promote carcinogenesis in tissues previously exposed to tumour initiator. O-tetradecanoylphorbol-13-acetate is a tumour promoter. The role of a tumour initiator and promoter is shown in Fig. 12.11.

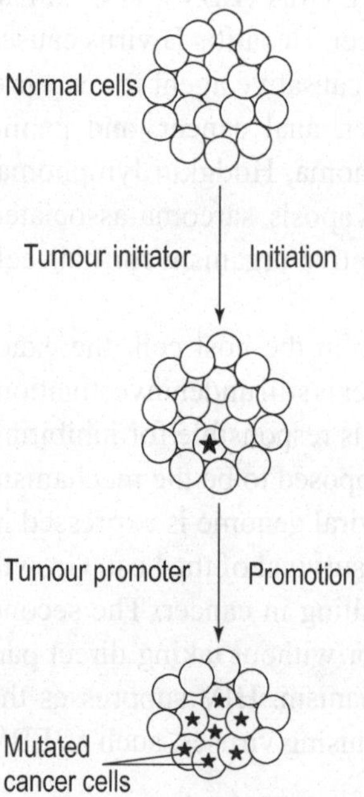

FIGURE 12.11 Role of tumour initiator and promoter (asterisk indicates cancer cells)

12.2.5 Prevention of Cancer

Cancer vaccines play a preventive role against cancer. Cancer vaccines are compounds that boost up or modify the immune system to fight against cell proliferation. Cancer vaccines are of two types—prophylactic or preventive vaccine and therapeutic or treatment vaccine. The US Food and Drug Administration (FDA) has approved two cancer preventive vaccines—hepatitis B virus vaccine to prevent liver cancer and human papilloma virus vaccine to prevent cervical cancer. The preventive vaccine targets the antigen that causes the cancer and inactivates it or kills it. Cancer preventive vaccines are produced by using the epitope or the antigen of the causative organism, thus, avoiding the risk of the disease caused by the whole organism. The vaccines can also be synthesized in vitro to mimic the natural antigen, and also modified to be more effective than the vaccine obtained from the original antigen. Side effects of cancer vaccines include inflammatory responses at the site of injection, such as erythema, pain, oedema, fever, itchiness, and skin rash.

Cancer can be prevented by antioxidants. This is due to the fact that free radicals are one of the causative agents for cancer. Free radicals are highly reactive atoms with unpaired electrons. They have the ability to damage biomolecules, including DNA. These results in DNA mutation, leading to cancer (refer Fig. 12.1). Compounds that can neutralize the free radicals are called antioxidants. Therefore, antioxidants have the capacity to prevent cancer. The preventive role of antioxidants against cancer is proved by research studies. β carotene, vitamin E, and selenium supplements can prevent gastric cancer, and β carotene and retinol are preventive against lung cancer.

A cohort study by the American Cancer Society reveals that high intake of dietary calcium and calcium supplements decreases the risk of colorectal cancer. The mechanism of action of calcium in the prevention of cancer is still unclear. However, it is proposed that calcium prevents colon cancer by three mechanisms: (1) by forming insoluble complexes with bile acids and fatty acids in the gastrointestinal tract, and thereby preventing their ability to induce proliferation, (2) by directly reducing the cell proliferation in the lining of colon, and (3) by mediating cell signalling to promote differentiation and apoptosis of the cancer cells.

Some studies have proved the role of vitamin D in reducing the risk of cancer. Epidemiological studies reveal that vitamin D reduces the incidence of cancer and cancer death. In vitro studies suggest that vitamin D induces apoptosis and prevents cell proliferation. Regular intake of vitamin D reduces the risk of colon cancer in a randomized clinical trial. Vitamin D decreases the risk of prostate cancer and pancreatic cancer. The mechanism of action of vitamin D in cancer prevention is attributed to the gene expression mediated by vitamin D receptor. Vitamin D binds to the vitamin D surface receptors (VDR). VDR interacts with the retinoid X receptor (refer Chapter 11) and passes the signal to the nucleus and transcribes the gene coding for calbindin, and CYP3A4. Calbindin enhances the uptake of calcium by the cells and prevents cell proliferation. CYP3A4 expression leads to detoxification of lithocholic acid, a stimulator of cell proliferation. Vitamin D is also reported to activate the detoxifying enzymes, in order to prevent cell proliferation.

Statin is an inhibitor of HMG CoA reductase. Statin is found to prevent colon cancer in cell lines and rodents. Physical activities, such as strenuous muscular exercises, prevent breast and colorectal cancers. A balanced, low-fat diet rich in fruits, vitamin D, and calcium is known to prevent the risk of cancer.

12.2.6 Diagnosis and Treatment of Cancer

Generally, cancer is diagnosed by signs, symptoms, screening tests, and medical tests such as blood tests, X-rays, computed tomography (CT) scans, and endoscopy. Above all, the confirmatory evidence for cancer should be obtained by histopathological examination of the suspected tissue, excised by biopsy or surgery. Histopathological diagnosis reveals the type of proliferating cells and the stage of cancer. Molecular changes of cancer cells can be revealed by immunohistochemistry (refer Chapter 14).

We shall now focus on the methods of diagnosis for specific cancer type. In women, the Papanicolaou (Pap) test and mammography are the screening tests conducted to diagnose cervical cancer and breast cancer. Checking for occult blood is used for testing colorectal cancer. Once the screening test is done, the next stage of diagnosis is confirmation of

TABLE 12.2 Cancer-specific diagnostic markers

Marker	Cancer
Alpha fetoprotein	Colon cancer
Hepatocellular carcinoma	
Beta chorionic gonadotropin	Placental cancer
Testicular cancer	
Beta microglobulin	Multiple myeloma and other blood cancers
Calcitonin	Medullary thyroid cancer
Carbohydrate antigen 125	Ovarian cancer
Carbohydrate antigen 19-9	Pancreatic cancer
Carbohydrate antigen 27.29	Breast cancer
Carcinoembryonic antigen	Colon cancer
Prostate-specific antigen	Prostate cancer
Thyroglobulin	Benign thyroid tumour

tumour by computed tomography scan, X-ray scan, magnetic resonance imaging, and microscopic imaging of a biopsy sample. Following this confirmation, specific tumour markers are measured to monitor the severity. Table 12.2 lists some of the cancer-specific markers used for the diagnosis of cancer.

Nanotechnology-based diagnosis

Recently, nanotechnology-based techniques have been developed for the highly sensitive and specific detection of cancer markers. A live cancer cell detector (explained in Chapter 14) is a good example for nanotechnology-based cancer diagnosis. For example, prostate-specific antigen (PSA) present in the serum is used as a confirmation for prostate cancer. Alpha fetoprotein and chorionic gonadotropin are markers for the diagnosis of hepatocellular carcinoma and gonadal cancer.

Biobarcode assay is used to detect prostate cancer, hepatocellular carcinoma and testicular cancer. In this technique, the gold nanoparticle is coated with an antibody specific for the cancer-specific antigen marker. A magnetic microbead is also coated with the antibody to recognize the antigen. The antigen is sandwiched between the two antibodies. The nanoparticle is pre-labelled with a DNA probe which acts as a biobarcode. The sandwich is detected by the hybridization of the biobarcode and subsequent detection of the biobarcode using fluorophore labelled complementary DNA.

Cancer cells can also be detected using functionalized quantum dots. Core shell quantum dots made from cadmium selenide and zinc sulphide are functionalized with targeting antibodies and allowed to bind with the cancer cell antigens. They are then irradiated; the quantum dots absorb the radiation and emit radiation which can be visualized through microscope and the tumour can be imaged.

Treatment

The following are some techniques used for the treatment of cancer.

Chemotherapy In cancer biology, chemotherapy refers to the treatment of cancer by killing cancer cells using any compound. The compound may be a cytotoxic drug or an antineoplastic drug; it functions by interrupting the mitosis or by facilitating the apoptosis process. Generally, alkylating agents, antimetabolites, plant alkaloids, antineoplastics, and topoisomerase inhibitors are used as chemotherapeutic agents. Examples of chemotherapeutic agents are given in Table 12.3.

Radiation therapy Radiation therapy is the killing of cancer cells by irradiating the cells with high-energy radiations. Depending on the mode of passage of radiation, there are three types of radiation therapy.

1. External beam radiation therapy, in which the radiation is passed from an external machine.

TABLE 12.3 Chemotherapeutic agents

Type	Example	Mechanism of action
Alkylating agent	Mechlorethamine Cyclophosphamide Chlorambucil Ifosfamide Cisplatin Carboplatin Oxaliplatin	Impairs cell function by forming covalent groups bonds with the functional of of biologically important molecules
Antimetabolites	Azathioprine Aercaptopurine	Prevent the synthesis of DNA in the S phase
Plant alkaloids	Vincristine Vinblastine Vinorelbine Vindesine Etoposide Taxol	Prevent microtubule formation
Topoisomerase inhibitor	Camptothecins Irinotecan Topotecan Amsacrine Etoposide Etoposide phosphate Teniposide	Disturbs DNA topology and prevents replication and transcription of DNA
Antineoplastic agent	Doxorubicin Epirubicin Bleomycin	Intercalates with DNA and inhibits its replication

2. Internal radiation therapy, in which the radioactive material capable of generating radiation is placed near the tumour.
3. Systemic radiation therapy, in which the radioactive material is injected into the circulation, to reach the tumour.

Radiations kill cancer cells by direct DNA damage or by generating free radicals.

Vaccine therapy Cancer treatment vaccines are known to treat existing cancer. Treatment vaccines are designed in three ways: (1) They can be developed from the cancer-causing antigens themselves, which include proteins, carbohydrates, glycoproteins, glycopeptides, and glycolipids. (2) Sometimes, the vaccines are also developed from the antigens expressed in the cancer cells. These antigens are obtained from the patient under treatment (aulogous vaccine) or from another individual possessing the same cancer (allogenic vaccine). (3) They can also be developed by using the DNA or RNA that encodes the cancer-associated antigens. The DNA or RNA is injected into the patient to produce more antigens and to boost the immune system. The efficacy of the treatment vaccine is increased by using an additional compound called adjuvant, which is capable of increasing the immune response. Cytokines are used as adjuvants for cancer treatment vaccines. The number of treatment vaccines available is less due to the limitations posed by the FDA. So far, the FDA has approved two types of cancer treatment vaccines—vaccines against hepatitis B virus and human papilloma virus [these vaccines function by activating the immune system and facilitate the killing of cancer cells (refer Chapter 3)], and another vaccine called sipuleucel-T, developed by isolating the antigen presenting cell (refer Chapter 3) is also approved by FDA for the treatment of prostate cancer.

Surgery Cancer surgery is done by physical removal of cancer tissue. This is a mode of treating cancer. This may be coupled with other treatments, such as chemotherapy or radiation therapy. Apart from treatment, surgery is also used for staging (to find out the stage of cancer, i.e., how advanced is the cancer), diagnosis, debulking (removing a part of tumour to increase the efficacy of radiation therapy), relieving the side effects and symptoms of cancer, and prevention (removal of tissue from the suspected tissue in advance).

Gene therapy Gene therapy is used to delete or replace the defective gene causing cancer. Details of gene therapy are discussed in Chapter 14.

Nanotechnology-based treatment Experimental studies have proved the use of nanocarriers for anticancer drug delivery, and for tumour killing. Dendrimer, a branched core-shell nanoparticle can be labelled with a targeting antibody specific for the cancer cell antigens. The dendrimers are also loaded with boron atoms. The carriers are delivered into the body, which can target the cancer cell. The particles are then irradiated to decompose the boron into lithium and high-energy radiation. The radiation kills the tumour cells. This type of treatment is called boron neutron capture therapy (BNCT). It was successful in rodents.

12.2.7 Genes Related to Cancer

As cancer is a genetic disease, it is necessary to know about the genes that are involved in the tumourigenesis process. These genes are called cancer-critical genes. There are two types of cancer-critical genes—proto-oncogenes and tumour-suppressor genes. Mutation of proto-oncogenes leading to gain in function (over-activity of gene) results in cell proliferation. Mutation of tumour-suppressor gene leading to loss of function (under-activity of gene) results in cell proliferation. Thus, cancer is induced by over-activity of proto-oncogene and under-activity of tumour-suppressor gene (Fig. 12.12). This section discusses the two categories of cancer-critical genes contributing to cancer.

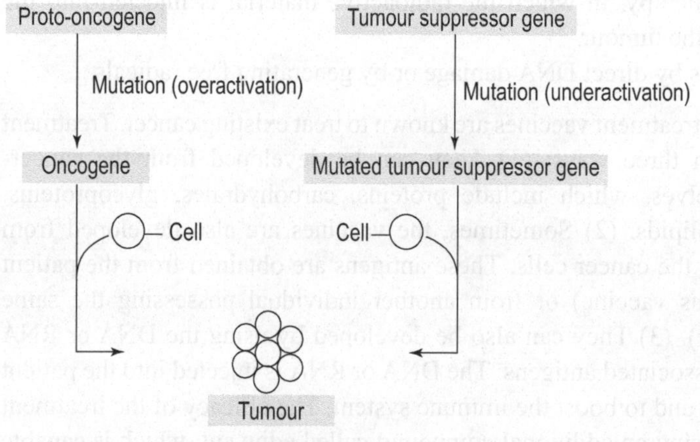

FIGURE 12.12 Role of cancer-critical genes (oncogene and antioncogene)

Proto-oncogenes and oncogenes

Proto-oncogene is a normal gene which produces proteins that regulate cell division, cell growth, and differentiation. Ras gene, myc gene, and wnt genes (refer Chapter 11) are examples for proto-oncogenes. On activation, proto-oncogene is converted to an oncogene. Oncogene induces cancer by increasing cell proliferation and inhibiting apoptosis. Factors leading to the activation of proto-oncogene to oncogene include retroviral transduction or retroviral integration, point mutations, insertion mutations, gene amplification, chromosomal translocation, and/or protein–protein interactions.

Tumour-suppressor genes—Rb gene and p53 gene

A tumour-suppressor gene is the gene that encodes the product responsible for preventing the cell from cancer. It is also called antioncogene. The retinoblastoma gene (Rb gene) is a tumour-suppressor gene. The Rb gene encodes retinoblastoma protein, which plays a crucial role in the regulation of cell cycle (refer Chapter 1). The protein prevents the cell from entering into division. Thus, mutation in Rb gene allows the cell to divide and produce cancer. The mutation occurs by a 'two-hit hypothesis'. According to this hypothesis, mutation is successful only when both the copies of the gene are mutated. If only a single copy is mutated, the gene can still produce the correct protein to suppress tumour. When both the copies are mutated it results in the tumour progression. The genes undergo deletion mutation (under-activity mutation) to cause retinoblastoma, and the cancer of lung, breast, and bladder. Another example of a tumour-suppressor gene is the p53 gene, which when mutated leads to colon cancer, breast cancer, and lung cancer.

12.2.8 Telomere and Cancer

Telomere is a repetitive DNA sequence, located as a cap, at the end of the chromosome (Fig. 12.13). It is made of a repeated sequence of AATCCC. It puts off the degradation DNA sequence, fraying of chromosome, and unravelling of genes at the end of the chromosome. Thus, it maintains the stability of the chromosomes, every time the cell divides. The length of the telomere varies from 600 base pairs in yeast to several kilobases in humans. The telomeres get shortened after every cell division, consequently limiting the life span of the cell, leading to ageing.

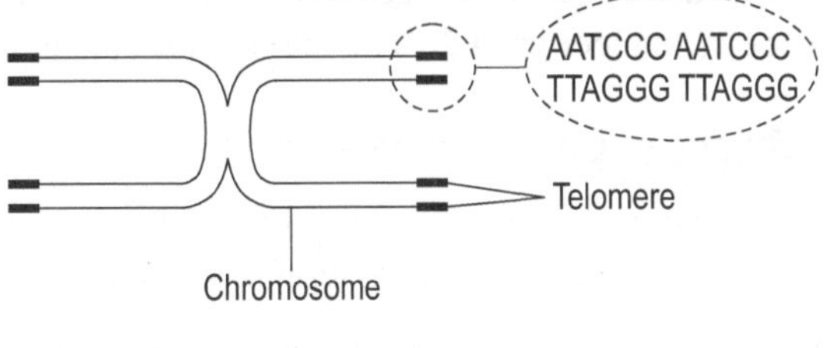

FIGURE 12.13 Telomere

The length of the telomere can however be increased by an enzyme, telomerase. Telomerase is a ribonucleoprotein with reverse transcriptase activity. The RNA component of telomerase provides AAUCCC sequence as a template to synthesize TTAGGG in the telomere. Telomerase is found only in embryonic stem cells, adult stem cells, progenitor cells, and cancer cells, and not in normal cells. The cancer cells can therefore activate their telomerase and maintain the length of the chromosomes, after every division. This results in further cell division and growth leading to progression of cancer. Thus, telomerase

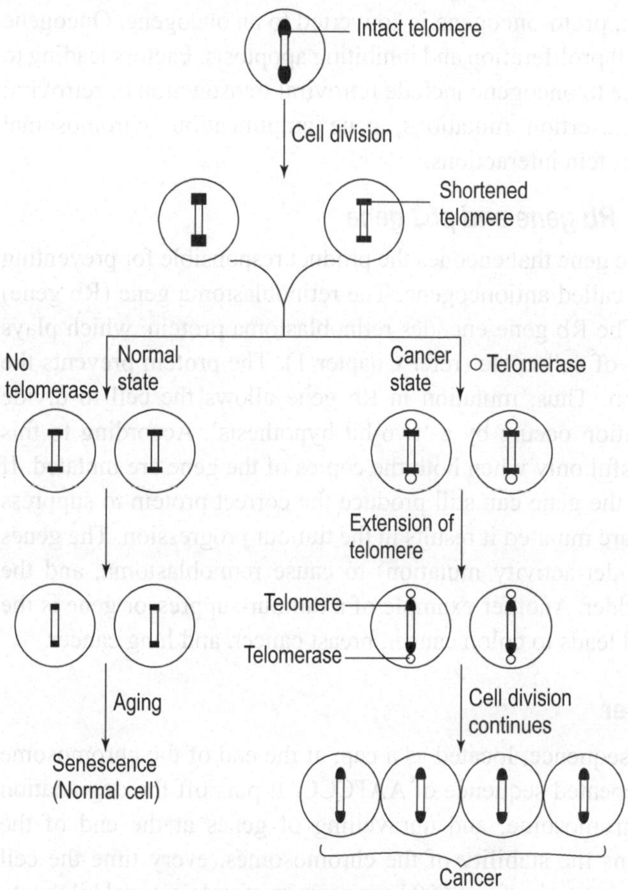

FIGURE 12.14 Telomerase in cancer cell (compared to a normal cell)

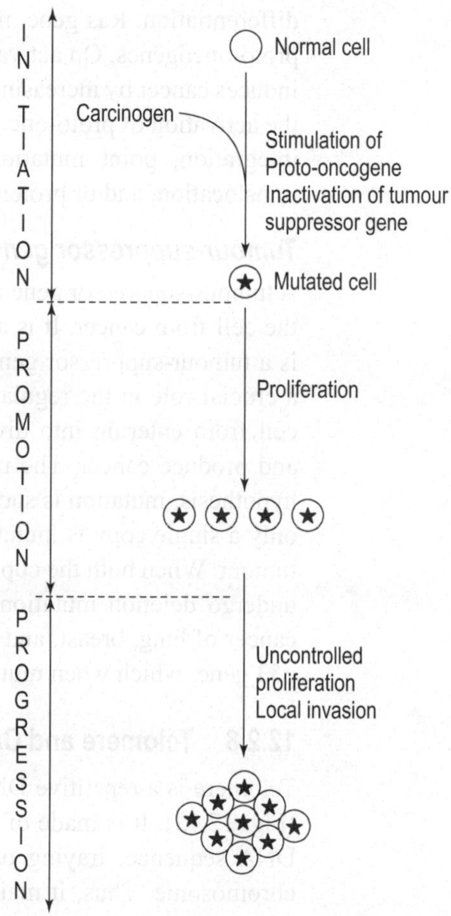

FIGURE 12.15 Stages in carcinogenesis

activity gives idea about the progression of cancer and blocking the activity may provide a way to control cancer growth. The telomerase concept is shown in Fig. 12.14.

12.2.9 Stages in Carcinogenesis

Stages in carcinogenesis are shown in Fig.12.15. Carcinogenesis occurs in three stages, leading to cancer—initiation, promotion, and progression.

Initiation The normal cell is exposed to a carcinogen. Carcinogen results in irreversible mutation of the cell, thus activating the proto-oncogene and inactivating the tumour-suppressor gene. As a result, a mutated cell is formed, in which signal transductions occur to promote proliferation and survival and to inhibit apoptosis (refer Chapter 11).

Promotion The mutated cell undergoes further growth and cell division, in the presence of tumour promoter, and gets inherited to the other cell generations. This results in the formation of malignant cells.

Progression The malignant cells undergo uncontrolled proliferation, and cause local invasion of neighbouring cells, forming a large mass called a malignant tumour.

12.2.10 Functional Proteins in Metastasis and Angiogenesis

Metastasis is the spread of cancer from its origin to other organs of the body via blood or lymph or both. One of the essential criteria for metastasis to occur is the formation of a new network of blood vessels to nourish the rapidly growing cells. This process of formation of new blood vessels is called angiogenesis.

The process of metastasis is resisted by the body via protein factors called metastasis suppressors. Metastasis is activated by another class of protein factors called metastasis activators. Understanding about these factors helps to plan for the treatment for metastasis and angiogenesis as both processes are inter-related. So far, very few metastasis suppressors have been identified. They function by different mechanisms. Mechanism of action of metastasis tumour is given in Table 12.4.

Just as some proteins are suppressors, certain proteins are activators of metastasis and angiogenesis. Metastasis-associated protein-1 (MTA-1) is responsible for increasing the rate of metastasis and angiogenesis. MTA-1 upregulates the expression of a protein called hypoxia inducible factor-1alpha (HIF-1alpha). It is a vascular endothelial growth factor. These two proteins cause metastasis and angiogenesis.

Angiogenesis is controlled by angiogenesis inhibitors. They may be of endogeneous or exogeneous origin. Examples of inhibitors of endogeneous origin include vascular endothelial growth factor receptor, angiopoietin-2, thrombospondin, angiostatin, endostatin, vasostatin, calreticulin, platelet factor-4, tissue inhibitors of metalloproteinases (TIMPs), and interferons α and β. These factors function by one or more of the following mechanisms:

- By inhibiting cell migration
- By inhibiting proliferation
- By inhibiting adhesion
- By inhibiting endothelial cell survival
- By stimulating apoptosis

Examples of exogeneous angiogenesis inhibitors include the following drugs: bevacizumab, thalidomide, interleukin 12, batimastat, and marimastat.

Angiogenesis is stimulated by angiogenic factors. Examples include fibroblast growth factor (promoter of proliferation and differentiation of smooth muscle cells and fibroblasts), vascular endothelial growth factor (stimulator of vasculogenesis), angiopoietin (stabilizer of blood vessels), transformation growth factor (stimulator of ECM production), plasminogen activator inhibitor-1 (blood vessel stabilizer), ephrin (stimulator of arteries and veins).

TABLE 12.4 Factors (malignant suppressors) that affect metastasis

Factor	Mechanism of action	Cancer type
NM23	Inhibits the activity of kinases which promote cell division	Melanoma, breast, and colon cancers
MKK4	Promotes apoptosis of cancer cells	Prostate cancer Ovarian cancer
KAI1	Binds to integrin and interferes with the cell detachment and migration of tumour cells	Prostate cancer and breast cancer
BRMS1	Promotes gap junction to inhibit metastatic growth	Colon cancer
RHOGD12	Inhibits the proteins that promote cell migration	Bladder cancer
VDUP1	Inhibits the proteins that promote cell proliferation	Melanoma

12.2.11 Tumour–host Interaction

Tumour cells reside in a surrounding environment (Fig. 12.16) comprising the following:

1. Non-cancerous cells—fibroblasts, endothelial cells
2. Connective tissue
3. Extracellular matrix
4. Collagen
5. ECM components
6. Macrophage infiltrates
7. Lymphocyte infiltrate
8. Growth factors
9. Proteases

These components are collectively called tumour microenvironment (also called host). The conditions of the tumour microenvironment are completely different from that in the normal environment. Tumour microenvironment exhibits hypoxia, low pH, and low sugar level. Hypoxia leads to free radicals generation, further inducing mutations and enhancing carcinogenesis. Hypoxia also attracts macrophages towards the tumour site. Macrophages adhere to tumours, hence the name, tumour-associated macrophages (TAMs). TAMs themselves secrete free radicals and enhance further mutation and promote cancer. TAMs can carry the associated tumour cells, circulate in the blood, and result in spreading (metastasis). Fibroblasts secrete alpha-smooth muscle actin, which allows the formation of myofibroblasts. Myofibroblasts form a dense proliferative network with collagen, forming desmoplasia, which induces angiogenesis, thus promoting cancer. The ECM contains matrix metalloproteases (MMPs, refer Chapter 10). MMPs help in tumour–surrounding interaction. It is also known to promote angiogenesis, initiation, and metastasis.

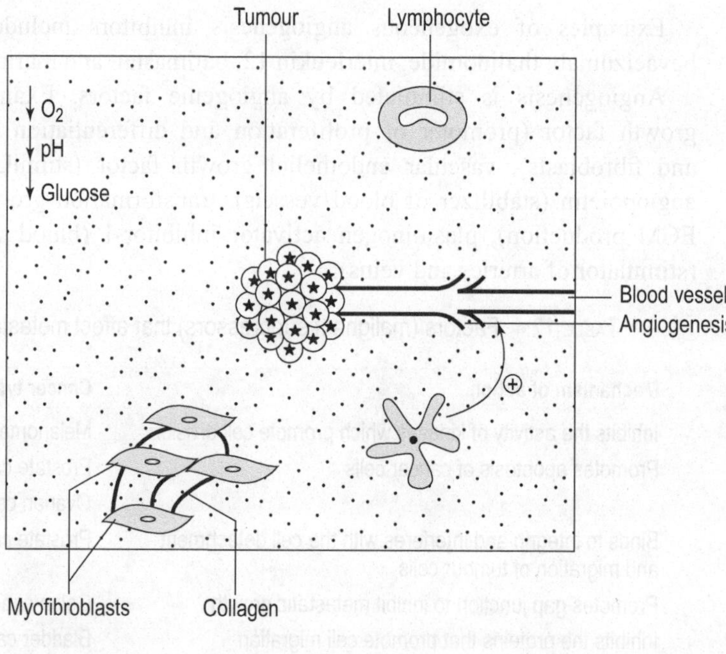

FIGURE 12.16 Tumour microenvironment (↓ indicates low level)

SUMMARY

Transformation of less specialized cells into more specialized cells of specific function is called cell differentiation. Differentiation plays a crucial role in the development of zygote into a foetus. Although cells are differentiated into different fates, there are few general characteristics of cell differentiation. They are stability, inducibility, morphological changes, and controllability by genetic factors. Differentiation process is more complicated in eukaryotes than in prokaryotes due to higher level of organization. Formation of flagellum and endospores in prokaryotes occur via differentiation. In multicellular organisms, the germ layers —ectoderm, endoderm, and mesoderm—differentiate into different organs and organ system. Formation of blood cells and immune cells occur via differentiation of haematopoietic stem cells.

Cell differentiation is determined by cytoplasmic determinants and nucleoplasmic interactions. Asymmetry of cytoplasmic determinants results in differential gene expression, thus leading to different fates.

Cancer is a mass of tissue formed due to uncontrolled and rapid cell division and growth. Cancer confined to a single location in a body is called benign tumour. The spreading of cancer through the body organs via blood and lymph is called metastasis. Cancer evolves from a single mutation induced by some tumour initiator followed by inheritance of mutation via a tumour promoter. Evolution of cervical carcinoma is well understood. The cancer evolves in the lower third of the stratified squamous epithelium, and then spreads to the upper layer and enters the blood stream and spread by metastasis. Uncontrollable cell division, resistance to apoptosis, genetic instability and capacity to spread to other tissues are hallmarks of cancer. Cancer is caused by any one of the following agents: chemical carcinogens, UV radiation, ionizing radiation, and tumour initiators and promoters. Cancer can be prevented by preventive vaccines, supplementation of vitamin D or calcium, antioxidants, HMG CoA reductase inhibitors, regular physical activity, and a low-fat diet.

Cancer is usually diagnosed by screening test, diagnostic test, computed tomography scanning, and X-ray scanning. Diagnosis of cancer by nanotechnology methods, such as biobarcode assay and quantum dot imaging, were successful in animal experiments. Cancer is treated using vaccines, chemotherapy, radiotherapy, surgery, and gene therapy. Nanoparticles-based drug delivery using dendrimers and boron neutron capture therapy will be used in human cancer treatment in future.

Cancer is caused by mutations in proto-oncogenes and tumour-suppressor genes. Telomere is a cap-like repetitive DNA sequence located at the end of the chromosome. It protects the DNA. The telomeres get shortened after repeated cell division. In normal tissue, the cell division stops at the older age and attains senescence, but in cancer cells, an enzyme called telomerase is activated which regenerates the telomere and promotes further cell division, resulting in tumour. Thus, telomerase plays a crucial role in cancer.

REVIEW QUESTIONS

1. Zygote develops into a foetus by cell differentiation. Justify this statement.
2. Give evidence for the stability of the differentiation process.
3. What are the morphological changes induced by the differentiation of eye lens fibre cells?
4. Explain the genetic changes associated with the formation of basal hook and flagellum in Salmonella species.
5. Define mother chamber and fore spore chamber.
6. Write a note on cytoplasmic determinants and asymmetric segregation.
7. List out the hallmarks of cancer.
8. Write the role of virus in causing cancer.
9. Write the mechanism of action of calcium and vitamin D in the prevention of cancer.
10. Write a note on chemotherapy and radiotherapy.

MULTIPLE-CHOICE QUESTIONS

1. Inner cell mass is
   (a) multipotent
   (b) pluripotent
   (c) cancerous
   (d) haematopoietic
2. Oestrogen stimulates the differentiation of
   (a) female reproductive system

(b) male reproductive system
(c) neuron
(d) mesoderm

3. Sigma 28 is the
(a) transcription factor for fla gene
(b) transcription factor for oncogene
(c) transcription inhibitor for fla gene
(d) transcription inhibitor for oncogene

4. Desmoplasia is associated with
(a) angiogenesis
(b) metastasis
(c) carcinogenesis
(d) apoptosis

5. P granule is a
(a) carcinogen
(b) cytoplasmic determinant
(c) plant alkaloid
(d) cancer vaccine

6. Cervical carcinoma arises in the
(a) stratified squamous epithelium
(b) myofibroblasts
(c) inner cell mass
(d) stem cells

7. Vincristine is a
(a) chemical carcinogen
(b) plant alkaloid
(c) tumour initiator
(d) tumour promoter

8. Malignant tumour of epithelium is called
(a) carcinoma
(b) sarcoma
(c) lymphoma
(d) myeloma

9. Cancer of the connective tissue is called
(a) leukaemia
(b) lymphoma
(c) carcinoma
(d) sarcoma

10. Biobarcode assay makes use of
(a) gold nanoparticles
(b) cadmium selenide quantum dots
(c) zinc sulphide quantum dots
(d) dendrimers

11. O-toluidine is a chemical carcinogen that belongs to the class of
(a) aromatic amine
(b) alkaloid
(c) antimetabolite
(d) topoisomerase inhibitor

12. Human papilloma virus causes
(a) cervical cancer
(b) hepatocellular carcinoma
(c) prostate cancer
(d) testicular cancer

13. Burkitt lymphoma is a cancer caused by
(a) Epstein barr virus
(b) human papilloma virus
(c) herpes virus
(d) bacillus subtilis

14. Statin is
(a) inhibitor of cancer
(b) inhibitor of HMG CoA reductase
(c) promoter of cancer
(d) activator of HMG CoA reductase

15. UV light induces
(a) C→T transition
(b) T→C transition
(c) T→A transition
(d) G→A transition

16. O-tetradecanoyl phorbol-13-acetate is a
(a) tumour promoter
(b) tumour initiator
(c) cancer preventive vaccine
(d) cancer therapeutic vaccine

Answers to Multiple-choice Questions

1. (b)	2. (a)	3. (a)	4. (a)	5. (b)
6. (a)	7. (b)	8. (a)	9. (d)	10. (a)
11. (a)	12. (a)	13. (a)	14. (b)	15. (a)
16. (a)				

CHAPTER 13

Stem Cell Biology

Oliver Smithies (left), Mario R. Capecchi (centre), and Sir Martin J. Evans (right) were awarded the Nobel Prize in Physiology and Medicine in 2007 for their discoveries of principles for introducing specific gene modifications in mice by the use of embryonic stem cells.

After reading this chapter, students should be able to understand and appreciate the following:

- Definition, characteristics, and types of stem cells
- Derivation and applications of embryonic stem cells
- Derivation, applications, and types of adult stem cells
- Functions of amniotic fluid stem cells and cord blood stem cells
- Differentiation of neural stem cells and skeletal muscle stem cells

13.1 STEM CELL NOTION

A stem cell is defined as a cell capable of generating multiple different cell types (multipotency), self renewal, and creating more stem cells (clonality). Thus, we can say that multipotency, self renewal, and clonality are the three characteristic features of stem cells as represented in Fig. 13.1. Stem cells can be found in different human tissues at all stages of life before and after birth.

13.1.1 Properties of Stem Cells

As mentioned in the definition of stem cell, self renewal and differentiation are important characteristic features of stem cells.

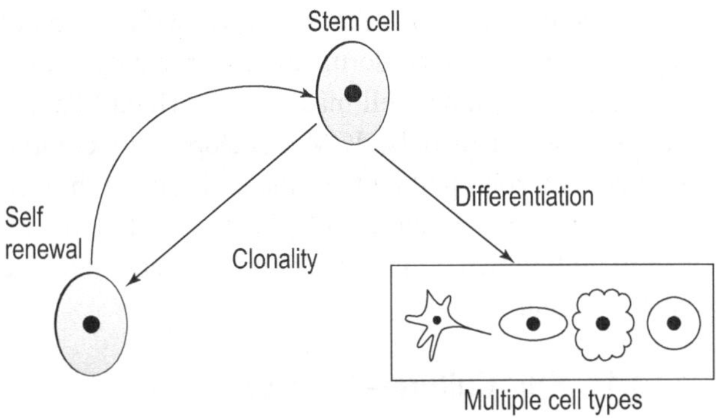

FIGURE 13.1 Characteristic features of stem cells

Self renewal

Stem cells are unspecialized cells capable of self-renewal, that is, they are capable of renewing or replicating or proliferating themselves through cell division a large number of times. Other cells, such as muscle cells, blood cells, or nerve cells, lack this feature. If the daughter cell obtained by self renewal of a stem cell is also unspecialized, the process is called long-term self renewal.

Differentiation

Under specific conditions, stem cells can be induced to become a specific cell type with a specific function. This process of converting an unspecialized stem cell population to a specialized functional cell is called differentiation. Stem cell differentiation is controlled by two types of signals—intracellular signal, which usually depends on genes in the cell, and extracellular signal, which consists of molecules secreted by the stem cells themselves or the neighbouring cells or contributed by the microenvironment. These signals induce cell differentiation by causing the DNA of the cells to acquire epigenetic markers that restrict gene expression. How specific cell signals induce differentiation into specific cell types is still unclear.

13.2 TYPES OF STEM CELLS

There are two types of stem cells: embryonic stem cells (ESC) and adult stem cells (ASC). The differences between these two types are summarized in Table 13.1.

TABLE 13.1 Differences between embryonic stem cells and adult stem cells

Embryonic stem cell	Adult stem cell
Derived from 3- to 5-day -old embryo (blastocyst)	Derived from multiple tissues (adult tissues)
Show more pluripotency	Non-pluripotent
Can differentiate into all cell types of the body	Can differentiate into cell type of the tissue where they reside
Can form many types of cell	Can form limited types of cells
Can be easily isolated	Isolation is difficult due to their scarcity in the tissue
Can be easily grown in culture	Culturing is difficult
Likely to initiate rejection after transplantation	Less likely to initiate rejection after transplantation

13.3 EMBRYONIC STEM CELLS

In vitro fertilization of egg and sperm produces an embryo. A 3- to 5-day -old embryo, called blastocyst (Fig. 13.2), is the origin of embryonic stem cells (ESCs). A blastocyst is a microscopic ball-like structure with an outer layer called trophectoderm, an inner cavity called blastocoel, and inner cell mass (ICM) which fills one end of the blastocoels. ICM develops into embryo proper. The zona pellucida of the embryo is removed by the enzyme pronase, and the trophectoderm by rabbit antihuman antibody. The ICM can then be liberated out.

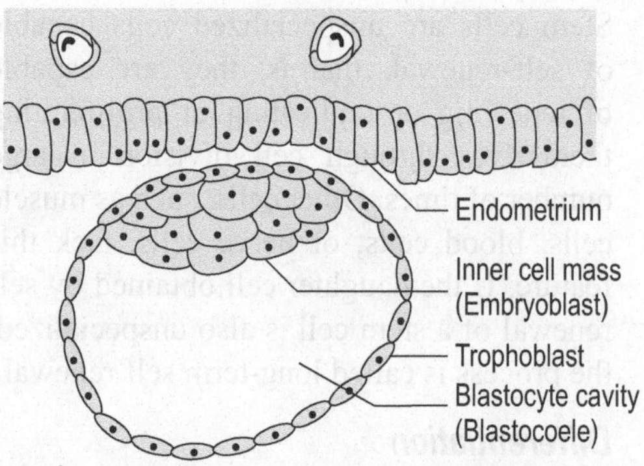

FIGURE 13.2 Structure of blastocyst showing the inner cell mass—the stem cell source

13.3.1 ESC Culture—Passage

As mentioned in the preceding discussion, the zona pellucida of the embryo is removed by the enzyme pronase, and the trophectoderm by rabbit antihuman antibody. The ICM can then be liberated out. ESCs can be cultured using a small amount of ICM as a starter. Thereafter, they are propagated and expanded in vitro, without losing their stem cell identity. The culture dish is coated with a layer of mouse embryonic stem cells, which is referred to as the feeder layer. The feeder layer serves the following two purposes:

- It functions as a sticky layer to anchor the stem cells.
- It releases nutrients into the medium.

ICM survives in the culture condition, and undergoes self renewal generating a large number of similar cells, filling the culture plate. A small portion of this cell mass can be gently removed and plated to a second culture disc or multiple culture discs and recultured. This process of replating and reculturing can be repeated several times for months. Each cycle of replating and reculturing is called a passage. Several passages may establish an ESC line which is genetically normal and pluripotent in nature. The pluripotent cell line can be cryopreserved for further uses in research, laboratory experiments, or medicines.

Basic fibroblast growth factor (B-FGF) and ECM components, such as activin A, are used as supplements for propagation and maintenance of undifferentiated human embryonic stem cell (hESC) colonies. These undifferentiated cells are characterized by (a) specific embryonic antigen (stage-specific embryonic antigen SSEA-3 and SSEA-4), and (b) keratin sulphate-associated antigen (TRA1-60 and TRA1-81; TRA is for transcription factor).

13.3.2 Characteristics of ESC

An ESC possesses several fundamental but specific characteristics. All these characteristics are explained in this section and can be used as markers to confirm ESC in the laboratory.

Long-term self-renewal capacity

ESC is capable of replicating even when cultured repeatedly several times, even for several months. The new generations remain undifferentiated or unspecialized and healthy. Even after freezing, thawing and replating, they can regrow, maintaining their integrity.

Expression of specific transcription factors—Nanog and Octamer binding transcription factor (oct4)

ESC expresses certain transcription factors which are characteristic of undifferentiated cells. Nanog (derived from Tir nan Og, the mythologic Celtic land of the ever young) and oct4 are two remarkable transcription factors that are specific for undifferentiated ESCs. Both these factors control the gene sequence responsible for differentiation and self-renewal.

Expression of specific cell-surface markers

ESC display cell-surface markers specific for undifferentiated cells; these include SSEA-3, SSEA-4, TRA-l-60, TRA-1-81, and alkaline phosphatase.

Intact chromosome structure

Chromosomes present in ESC are intact, that is, they are not damaged, even after several passages. The number of chromosomes is also not altered after several passages. The integrity of the chromosome can be monitored using a microscope.

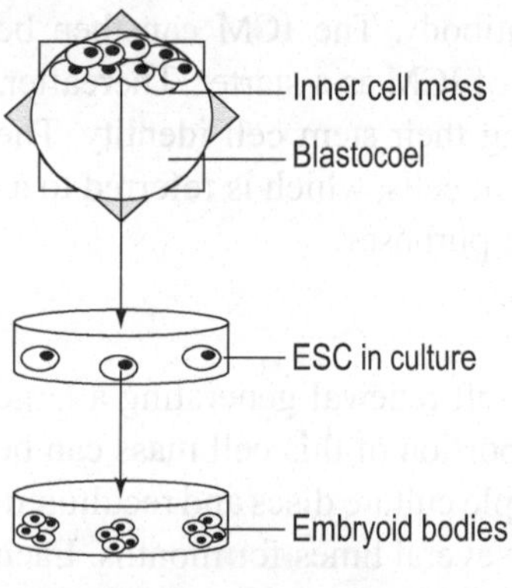

FIGURE 13.3 Embryonic stem cells and embryoid bodies

Spontaneous differentiation

Although ESCs remain undifferentiated in cell line, they can develop the potential to differentiate when they are allowed to clump together. Clumps of ESCs are called embryoid bodies that can differentiate spontaneously to form multiple cells, such as muscle cells, nerve cells, and so on (Fig. 13.3). This spontaneous differentiation indicates that the cells are healthy.

Pluripotency

ESCs are capable of developing into most of the cell types (multiple cells) under specific conditions. They can develop into a benign tumour called teratoma which contains a mixture of many differentiated cells. This is the confirmation for the pluripotent nature and this test can be done in immunosuppressed mice.

Directed differentiation

Directed differentiation is the process of directing an ESC population to differentiate into a specific cell type. This process is different from embryoid differentiation or spontaneous differentiation discussed earlier. Directed differentiation can be performed in the following three ways:

1. by altering the chemical/biochemical composition of the growth medium
2. by altering the surface of the culture plate
3. by modifying the cells by inserting specific gene into the cells

Nanog, oct4, signal transducer and activator of transcription 3 (stat3), Hex (homeobox gene product produced during haematopoiesis), and sirt1 are intracellular signalling molecules implicated in the growth and differentiation of ESCs. Sirt1 stands for sirtuin (silent mating type information regulation 2 homolog) 1. Sirt1 is a deacetylase that regulates Nanog expression by controlling the subcellular localization of p53 tumour suppressor.

13.3.3 Potential Applications of ESC

Specific cell types obtained by direct differentiation of ESC can be used to treat certain diseases by transplantation or tissue regeneration, for example, Parkinson's disease, trauma, spinal cord injury, diabetes, Dunchennes muscular dystrophy, hearing loss, and heart damage. ESCs are also used in the treatment of retinal diseases associated with retinal cell types, such as retinal pigment epithelium, rods, and cones.

13.3.4 Derivation of Mouse ESC

Mouse ESC can be derived from delayed blastocysts, which are obtained by the following procedure:

A mouse couple is mated overnight and the female mouse is observed for the next two days. On the third day, the oestrogen activity is removed by using 10 mg tamoxifen (antiestrogenic compound) followed by 1 mg depo-provera. A 100X stock solution of tamoxifen in ethanol is diluted in sesame oil to a concentration of 10 mg and injected intraperitoneally to the female mouse. Depo-provera is given subcutaneously. After the injection, the mice can be sacrificed and the blastocyst flushed from the uterus between day 6 and 8.

The blastocyst is placed in a mitotically inactivated feeder cells or leukaemia inhibitory factor (LIF). These factors help in keeping the stem cell in undifferentiated state. Fibroblast feeder is used for the maintenance of the totipotent (possessing the ability to differentiate into all cell types to form a complete organism) ESC in culture. The feeder layer should be pre-incubated in ESC medium and observed for four days. At this stage, the medium is removed and the cells washed twice with PBS with 10^{-4} M EGTA and then with phosphate-buffered saline/ethylene glycol tetraacetic acid (PBS/EGTA) containing 0.125% trypsin. The cells are then aspirated using a pipette and then incubated at 37°C. Then, using a dissecting microscope, the complete ESC medium is flooded over the ICM area which is then drawn out using a Pasteur pipette. ICM is then blown out over the culture dish containing pre-incubated feeder layer. The dish is incubated for 7 to 10 days. The medium should be replaced on the fifth day. After this process, ESC colonies can be observed, which can be passaged regularly. At the time of fourth to sixth passage, lot of ESCs will be obtained. Population doubling time of 15 to 20 h indicates a healthy culture.

The medium used for ESC culture is a 1:1 combination of DMEM: Ham's F12M medium. This combination gives a better yield when compared to the individual medium. DMEM medium is formulated to have high glucose, low pyruvate, non-essential amino acids, and nucleosides. The medium is a serum-containing medium for which foetal calf serum (10%) can be used.

13.3.5 Derivation of Human ESC (hESC)

Derivation of hESC starts with a frozen blastocyst produced by in vitro fertilization. Blastocyst can be obtained after getting an informed consent from the donor. The blastocyst is treated with acid Tyrode solution to remove the zona pellucida. The trophectoderm, which protects the ICM, is degraded by incubating the blastocyst with anti-human RBC antibody and subsequently with complement system (refer Chapter 3). The ICM is then exposed and removed. Mouse embryo fibroblast (MEF) feeder is mitotically inactivated and seeded on gelatine-coated culture vessel with a density of approximately 50,000 cells/cm2. Gelatin prevents the 'rolling up' of feeder layer. MEF competes with ESCs for media nutrient and space. Hence, feeder layer should be mitotically inactivated before hESCs are plated on the MEF layer. Mitomycin-C or gamma radiation treatment is used for the mitotic inactivation. The feeder layer is flooded with hES derivation medium.

ICM is transferred to the culture vessel and incubated at 37°C, undisturbed for at least 48 h. As a result, the ICM attaches to the feeder layer and produces outgrowth. hESCs can be interpreted visually or macroscopically. The medium colour indicates the culture quality. Normal fresh medium is orange or straw coloured. Change in the colour of the medium to yellow indicates nutrition depletion or acidic pH shift and overcrowding of cells. Appearance of pink colour in the medium indicates the basic pH. Following the visual observation, the medium can be changed accordingly. After the macroscopic observation, the culture can be viewed on low power microscope, with 2x or 5x magnification.

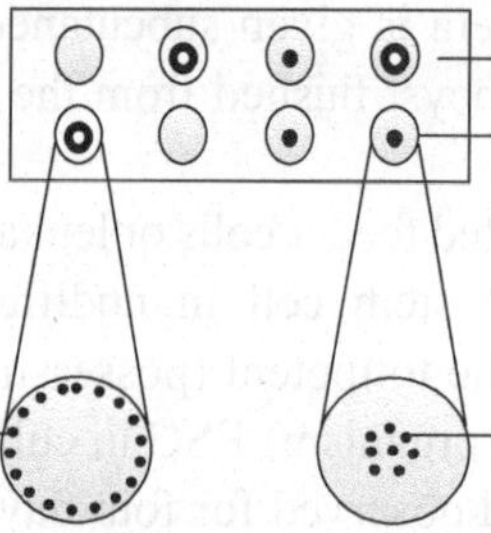

FIGURE 13.4 Differentiated colonies present in the undifferentiated colonies of the culture well

Any one of the following observations can be made—less dense undifferentiated colonies, highly dense undifferentiated colonies, or differentiated colonies. It is better to use hot stage microscope to maintain the temperature. If the colonies observed under microscope are not dense, no passaging is required, but feeding is. Hence, the medium should be replaced with a fresh medium. This is done in a sterile biological safety cabinet, where the spent medium depleted with nutrition is aspirated with a pipette without touching the sides or bottom of the well, in order to avoid air bubble formation as well as contamination. (Touching the sides and bottom may also prevent aspiration). Then, using a fresh pipette, fresh warm hESC medium is added. Once the new medium is added, the dish is transferred from the cabinet to the incubator maintained at 37°C and 5% CO2. When the undifferentiated colonies are dense and appear closely packed, covering the entire well, it indicates the necessity for passaging. Appearance of differentiated colonies insists the need for their removal to maintain the stem cell identity. A picking tool with blunt and curved end is sterilized using UV light and used to remove the differentiated colonies which are usually present along the edges or as a spot in the centre of the well (Fig. 13.4). The differentiated colonies leave the bottom of the plate and float in medium. The plate is then transferred into the safety cabinet and the differentiated cells removed along with the medium. The well is filled with fresh hESC medium.

13.3.6 Basic Characteristics of hESC

Basic characteristics of hESC are high nucleus-to-cytoplasm ratio, prominent nucleoli, and distinct colony morphology.

High telomerase activity is an important feature of hESCs. Role of telomerase is discussed in Chapter 12. Telomerase function in extending the chromosomal end (the telomere) after replication and cell division, thus, maintaining the chromosomal length. High telomerase activity contributes to immortality of the cell lines.

hESC express surface markers SSEA3, SSEA4, Tra-1-60, Tra-1-81, ALP, but not SSEA1, unlike mouse ESC. High levels of transcription factor oct4 and Nanog are also expressed in hESCs.

In appropriate culture conditions, hESC differentiates into all the three germ layers—ectoderm, mesoderm, and endoderm, both in vivo and in vitro, which can then form all types of adult cells (Table 13.2). Differentiation potential can be confirmed in vivo by the formation of teratoma in SCID mouse. In vitro, the formation of embryoid bodies by hESCs is the confirmation of differentiation potential.

TABLE 13.2 Tissues derived from the germ layers of human embryonic stem cells

Germ layer	Tissues or cell derived
Ectoderm	Gut epithelium
Mesoderm	Cartilage
	Bone
	Smooth muscle
Endoderm	Neural epithelium
	Embryonic ganglia
	Stratified squamous epithelium

Mouse ESC can proliferate and remain undifferentiated in the presence of LIF, but in the absence of fibroblast feeder layer. In contrast, LIF is insufficient to maintain the self-renewal capacity of hESC. Hence, hESCs are grown on a mouse embryonic fibroblast feeder layer (MEFs). MEFs can also be replaced with mouse embryonic fibroblast cell line, which is easy to maintain and propagate. MEF medium is a high-glucose DMEM, containing 6 mL penicillin or streptomycin, 6 mL glutamax-1, and 50 mL foetal bovine serum (FBS).

To maintain a healthy, replicating, undifferentiated hESC culture, the culture medium should be changed every day. This is needed to replenish lost nutrients and avoid unwanted differentiation factor. Still a small level of differentiation occurs during hESC cell culture, and hence, the differentiated areas should be removed manually. This step is very essential for maintaining a healthy population of stem cell.

13.3.7 Adaptations of hESC

In vitro, hESC undergoes two types of adaptations: (1) epigenetic change, which is due to modification in DNA methylation (to determine which part of genome is expressed) and (2) karyotypic change in individual chromosome. These changes provide selective advantages for long-term proliferation of pluripotent cells in vitro. They commonly occur in relatively high passage 'adopted' hESC cell lines.

hESC can be differentiated into ocular lineages by coculturing them with mesenchymal bone marrow cells (PA6 stromal cells). This phenomenon is called stromal cell-derived inducing activity (SDIA).

13.3.8 Human ESC in the Ontogenesis of Retinal Pigment Epithelium

Retina is the photosensitive component of central nervous system located in the eye. Age-related muscular degemeratopm (AMD) and retinitis pigmentosa (RP) are diseases associated with retina. In both these diseases, retinal pigment epithelium rods and cones are irreversibly damaged. hESC can be used in the retinal repair and to treat AMD and RP.

Gastrula differentiates into fore brain in the presence of bone morphogenetic protein (BMP) antagonist, noggin, and Wnt signalling antagonist, dickkopf-1. Fore brain gives rise to telencephalon and eye field under the antagonism of Wnt signalling. Eye field transcription factor induces the specification of eye field and, subsequently, insulin-like growth factor signalling pathway causes the induction of eye field. Eye tissue is formed from the eye field by the notch signalling pathway. Following this, optic vesicles arise bilaterally from the fore brain and expand into the mesodermal tissue of the head. Optical vesicle develops into optical cup and lens vesicle. The neurectodermal tissue of optical cup forms neural retina, retinal pigment epithelium (RPE), iris, and ciliary body. The proximal part of optical cup differentiates into RPE in response to TGF-β superfamily signalling molecules, such as BMP and/or activin. BMP is responsible for proper development of RPE, and activin is responsible for maintaining the RPE-specific gene expression. RPE cells gain pigmentation during development. Mitf is responsible for acquisition of pigmentation.

Rods and cones are specialized photoreceptive neurons present in retina. These cells are involved in transduction of energy from captured photons into electrical signals. These cells contain membrane-bound G-protein coupled proteins. Proteins of rods are called rhodopsins which respond to dim light. Proteins of cones are called S-opsin, M-opsin, and L-opsin which respond to bright light. These cells are known to hyperpolarize in response to light stimuli. Proper development of rods and cones is due to various components— Crx genes, neural retina leucine zipper (Nr1) transcription factor, and Nr2e3 (nuclear receptor).

13.4 AMNIOTIC FLUID STEM CELLS

Amniotic fluid is a mixture of urine and peritoneal fluid from the foetus and ultrafiltrate from the plasma of the mother entering the placenta. It contains a novel type of stem cells which can be maintained in an undifferentiated state in culture for long period and induced to differentiate into many cell types. It can be isolated by a technique called amniocentesis. For this, a thin needle is inserted into the uterus transabdominally at 16 weeks of gestation and a sample of amniotic fluid surrounding the baby is withdrawn and centrifuged. The supernatant contains the cells. These cells are from the foetus, which are mostly sloughed-off cells from the epithelium, digestive tract, urinary tract, and the amnion of the foetus. The cells are resuspended in the culture medium and seeded in the presence of 5% CO_2 at 37°C for 3 to 4 weeks and cultured. (The mitotic phase cells can be isolated from the culture and the chromosome can be characterized.) The cultured cells are heterogeneous population, displaying a diverse morphology from round to squamous in shape and 6 to 50 μm in diameter. Most cells are terminally differentiated along epithelial lineages and have limited proliferative and differentiation capability. Approximately, 0.8 to 14% of the amniotic fluid cells consist of the pluripotent progenitor cell population. These cells express a receptor tyrosine kinase (RTK, refer Chapter 11) called c-kit on their membrane surface. Hence, this population can be isolated from the heterogeneous population by a positive c-kit selection, using fluorescent-activated cell sorting.

Isolated progenitor cells when cultured are spherical for one week and show low proliferating capacity. After a week, they show high proliferative capacity, high confluence (80%), high self-renewal capacity. More than 300 doublings (which is much higher than Hay flicks limit—it is the number of times a normal cell population will divide before it stops presumably, because the telomeres reach a critical length). When expanded, these cells maintain a similar proportion in growth potential. These proliferated cells exhibit normal karyotype and also G1 and G2 cell cycle check points. They exhibit telomere conservation in undifferentiation and telomerase activity in late passages. They display the following two markers:

1. Human embryonic stage-specific marker
2. The stem cell marker oct4

They lack the following markers—SSEA1, SSEA3, CD4, CD8, CD34, CD133, C-met, ABCG2, NCAM, BMP4, TRA1-60, and TRA1-81.

This marker profile shows that the amniotic fluid stem cells (ASFCs) express the key markers of ESCs.

AFSCs show pluripotency and they can be in vitro differentiated into osteogenic, adipogenic, myogenic, neurogenic, endothelial, and hepatic phenotype (Fig. 13.5).

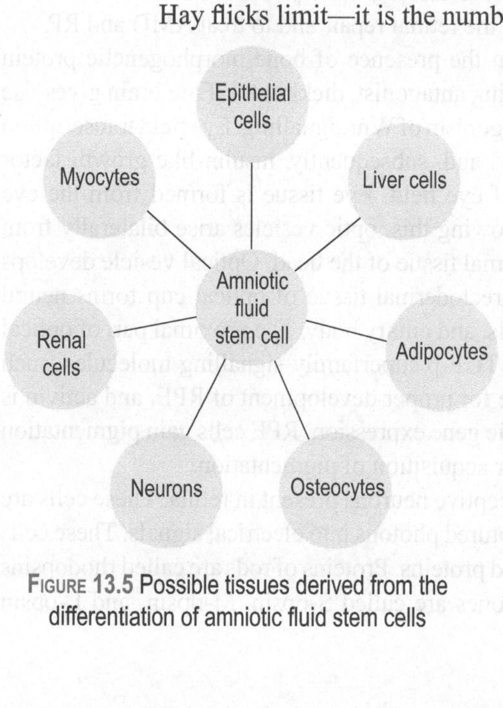

FIGURE 13.5 Possible tissues derived from the differentiation of amniotic fluid stem cells

TABLE 13.3 Characteristics of adipogenic cells differentiated from AFSCs

Phenotype	AFSCs	Adipogenic induced cells
Lipoprotein lipase – 8 days	–	+
Lipoprotein lipase – 16 days	–	+
Transcription factor – 8 days	–	+
Transcription factor – 16 days	–	+

13.4.1 Differentiation of AFSC into Adipogenic Cells

AFSCs undergo adipogenic differentiation in the presence of the following compounds as adipogenic supplements—dexamethasone, 3-isobutyl-1-methyl-xanthine, insulin, and indomethacin. This differentiation occurs within 8 days. On the eighth day, the AFSC is converted from an elongated shape to a spherical shape. Spherical cells contain intracellular lipid droplets. In these cells, the genes for lipoprotein lipase and a transcription factor, ppasγ2, are expressed, but these genes are silenced in the undifferentiated AFSCs (Table 13.3). Thus, the adipogenic cells express the functional and biochemical characteristics of adipocytes.

13.4.2 Differentiation of AFSC into endothelial cells

When AFSC is cultured in a gelatine-coated vessel containing endothelial basal medium, it is differentiated into endothelial cells and the new phenotype can be recognized in 8 to 10 days. The new phenotype (Table 13.4) in the endothelial cells includes the expression of an endothelial cell-surface marker (P1H2), transcription factor VIII (FVIII), presence of 'kinase insert domain'—containing receptors, such as CD31 and VCAM. These phenotypes cannot be recognized in control AFSCs which are not grown in endothelial basal medium.

13.4.3 Differentiation of AFSC into Hepatic Cells

For hepatic differentiation, AFSC is grown in a culture vessel coated with material or collagen for 45 days. Differentiation is induced in the presence of the following components:

- Hepatocyte growth factor
- Insulin
- Fibroblast growth factor 4
- Dexamethasone
- Monothioglycerol
- Oncostatin M

TABLE 13.4 Comparison of phenotypes in AFSC and differentiated endothelial cells

Phenotype	AFSCs	Endothelial cells
Endothelial cell-surface marker (P1H2)	–	+
Transcription factor VIII	–	+
Kinase insert domain containing receptor	–	+
CD31	–	+
VCAM	–	+

TABLE 13.5 Markers for hepatocytes differentiated from AFSCs

Marker	AFSCs	Hepatocytes
Albumin	–	+
AFP	–	+
MDR membrane transporter	–	+
Transcription factor HNT4α	–	+
C-met receptor	–	+

After 8 days of differentiation, the cells undergo a change in shape from elongated shape to cobble-stone-like appearance. These cells express various proteins which are negative in undifferentiated AFSC (Table 13.5). The proteins expressed in hepatic cells at the 45th day are albumin, alpha fetoprotein (AFP), multidrug resistance (MDR) membrane transporter, transcription factor HNT4α, and C-met receptor.

13.4.4 Differentiation of AFSC into Osteogenic Cells

In the presence of growth medium enriched with dexamethasone, beta-glycerophosphate, and ascorbic acid-2-phosphate, AFSCs can be differentiated into osteocytes. This is accompanied by conversion of elongated cell into osteoblast-like cells with finger-like excavations in the cytoplasm. On the 8th day, the cells express core-binding factor alpha-1 (cbfal) and osteocalcin proteins, which are characteristics of bone cells. Gradually, the cells aggregate and on the 16th day, they show lamellar bone-like morphology. These cells functionally differ from progenitor cells in that they synthesize alkaline phosphatase enzyme and accumulate calcium.

13.4.5 Differentiation of AFSCs into Neurons

AFSCs can be differentiated into neurons in the presence of neurogenic medium containing specific compounds, including DMSO, butyrilated hydroxyanisole, and neural growth factor. The phenotype changes start within a day. The AFSCs will change into branched cells which express all neural specific proteins.

13.4.6 Differentiation of AFSCs into Myocytes

AFSC can be differentiated into myocytes when induced in a matrigel-coated vessel in the presence of 5-azacytidine, horse serum, and chick embryo extract. The differentiation involves formation of bundles of multinucleated cells which express tropomyosin and desmin, which are not present in the progenitors.

13.5 UMBILICAL CORD BLOOD STEM CELLS

Umbilical cord and placenta are tissues present in human foetus, which support the developing foetus during pregnancy. These two tissues come along with the baby during the delivery. After the delivery, they are usually discarded. Still, umbilical cord is harvested, frozen, and stored in cord blood bank and cryopreserved in laboratories. The preserved umbilical cord is yet another source of haematopoietic stem cells (HSCs, also called neonatal stem cells). Umbilical cord blood stem cells (UCBSCs) are used in the treatment of Fanconi anaemia and leukaemia. They are likely to contain cells that can be multipotent (i.e., capable of developing into multiple germ layers or capable of developing into all the three germ layers—ectoderm, mesoderm, and endoderm). UCBSCs respond rapidly to proliferation-inducing signals (growth factors) namely, stem cell factor (SCF), fms-like tyrosine kinase receptor-3 (Flt3) ligand, and colony stimulating factor (CSF). Members of signalling factor family are shown in the Table 13.6. In the presence of the growth signals, they proliferate into specific cell types. SCF and Flt ligand bind to their corresponding RTK namely c-kit and Flt3, respectively, and induce proliferation of UCBSC. SCF and CSF enhance the frequency of the stem cells. IL-20 enhances the proliferation of the cord

TABLE 13.6 Signalling factors for cord blood stem cells

Factor	Name
GM-CSF	Granulocyte macrophage colony stimulating factor
M-CSF	Macrophage colony stimulating factor
G-CSF	Granulocyte colony stimulating factor
IL-3	Interleukin-3
EPO	Erythropoietin
TPO	Thrombopoietin
SCF	Cytokine stem cell factor or steel factor
FL	Flt3 ligand

blood cells in the presence of SLF and EPO in vitro. Cysteine motif chemokine (CXC 112) and stromal cell derived factor-1 (sdf1) increase the survival and *ex vivo* expansion of haematopoietic progenitor (HPG) cell.

Due to this proliferative capacity, UCBSC is superior to BMHSC for engraftment. The homing and engrafting capacity of the stem cells can be increased by decreasing the DPPIV (dipeptidyl peptidase IV) activity of CD26 or by eliminating the CD26 on the stem cells. For this purpose, inhibitors of DPPIV, such as diprotin A or Val-pyr, are used. Inhibition of DPPIV blocks the truncating effect of CD26 on sdf1/CXC112.

The disadvantage of UCBSC is that very limited number of cells can be harvested and there is a delayed immune reconstitution following UCBSC transplant. However, UCBSCs do possess the following advantages:

- They are easily available.
- They are easy to harvest.
- There is a reduced risk of graft versus host disease.
- They possess greater proliferative capacity than adult BMHSC.
- They are abundant and easy to obtain. Unlike bone marrow cells, they do not cause graft versus host rejection (GVHR).

13.6 FOETAL HAEMATOPOIETIC SYSTEM

HSCs appear in a 7-day-old foetus as blood islands in the yolk sac. HSC is also reported to exist in the AGM region (the region where aorta, gonads, and foetal kidney originate) of 10 to 11-day-old embryo. In AGM, HSCs migrate to the liver, where they redivide and again migrate to spleen and thymus. This migration occurs during the development of foetus. At the birth time of the foetus, these HSCs migrate to the bone marrow. During migration of HSCs from liver to bone marrow, the number of CD34+ decreases and the expression of cell adhesion molecules increases in these cells. Thus, the amount of CD34+ and cell adhesion molecule is the factor for differentiating adult bone marrow stem cells from foetal liver stem cell.

13.7 ADULT STEM CELLS OR SOMATIC STEM CELLS

Adult stem cells (ASCs) are undifferentiated cells found among differentiated cells of adult organs or tissues. Like ESCs, ASCs are capable of self-renewal. They can also differentiate into all of the major cell types. ASC is also called somatic stem cell (SSC). As ASCs are located among non-stem cell population of the organ, they are capable of repairing or regenerating the tissues of that organ.

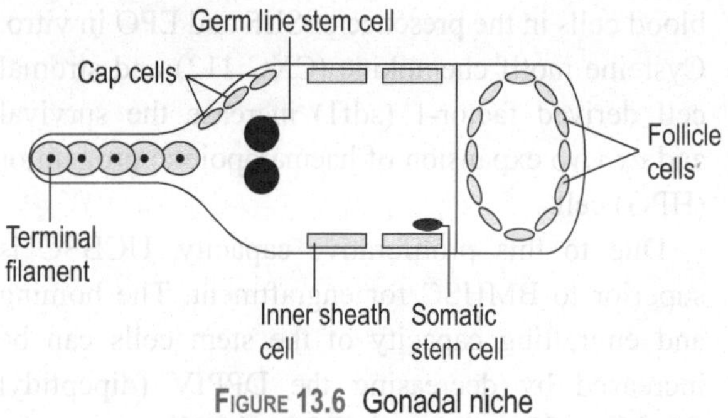

FIGURE 13.6 Gonadal niche

13.7.1 Origin of the ASC—The Stem Cell Niche

Unlike ESC, which originates from the blastocyst, the origin of ASCs cannot be defined as they are found in different organs. It is a known concept that ASCs are found in many organs, including bone marrow, brain, muscle, blood vessels, skin, teeth, heart, gut, liver, and gonads. The area of the particular organ where the ASCs reside is called the stem cell niche. Within the niche, the stem cells maintain their identity. When they come out of the niche, they lose their stem cell characteristics. Different types of stem cell niche are known. They are as follows:

Gonadal niche

Gonadal niche can be understood in drosophila. The gonadal niche is a part in the gonad, adjacent to the somatic signalling centres. It is a zone (Fig. 13.6) made of terminal filament and cap cells at the posterior end and follicle cells at the anterior end, with inner sheath cell in between. The niche accommodates two populations of stem cells, namely, germline stem cells (GSC), which give rise to the gametes, and somatic stem cells (SSCs) whose progeny form support cells that are essential for gametogenesis.

HSC bone marrow niche

Bone marrow is the niche for the haematopoietic stem cells. This niche is composed of bone marrow stromal cells (reticular cells), subendosteal osteoblasts, osteoclasts, fibroblasts sinusoidal endothelial cells, and adipocytes (Fig. 13.7).

Epithelial stem cell niche

In intestine, the subepithelial fibroblasts and the myofibroblasts form a network around the crypts. This network is the niche for the intestinal stem cells (Fig. 13.8). Limbal epithelial stem cell niche is found in the cornea of eye. The cells in this niche express putative limbal stem-cell markers—p63-alpha and ATP-binding cassette superfamily G member 2 (ABCG2).

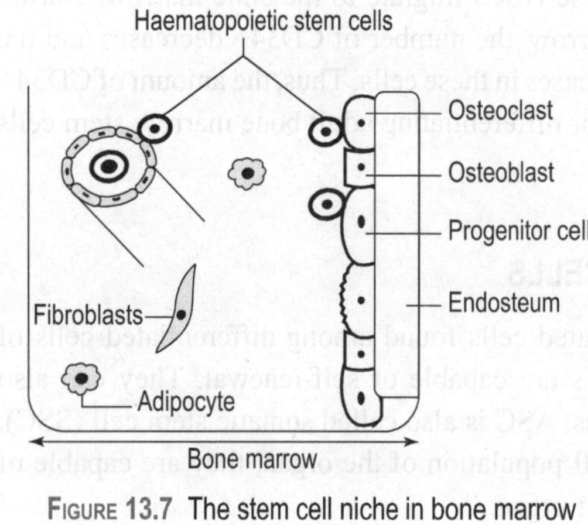

FIGURE 13.7 The stem cell niche in bone marrow

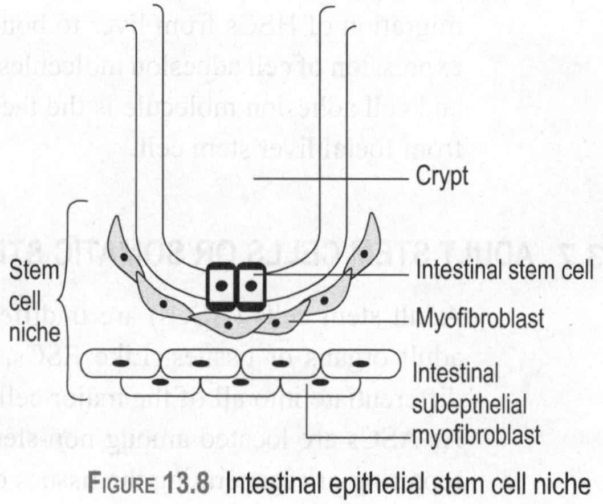

FIGURE 13.8 Intestinal epithelial stem cell niche

Neural stem cell niche

Neural stem cell niche is the zone where the stem cell resides to produce new cells of nervous system after embryonic development. Neural niche represents the subependymal zone of lateral ventricle and subgranular zone of himpocamus region (Fig. 13.9). The niche is constituted by astrocytes, transit amplifying precursor, neuroblast, and ependymal cell

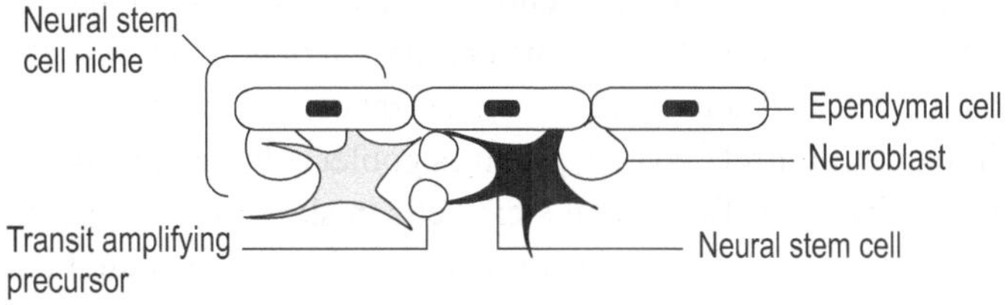

FIGURE 13.9 Zone of neural stem cell niche

13.7.2 Identification of ASC

ASCs are usually identified by labelling. The cells in the living tissue are labelled with molecular markers. After differentiation of the ASC, specialized cells are identified using the label. In another method, the potential of ASC from one animal can be identified in another animal. ASCs are isolated from a living animal, labelled in culture, and transplanted into another animal. The labelled cell can then be identified in the second animal and the ability of the differentiation of the ASC checked.

13.7.3 Differentiation of ASC

Stem cells remain quiescent until they are activated by the signal given by the organ in requirement. For example, when an organ has to be repaired during injury, the ASCs of that organ can be signalled by the neighbouring cells to differentiate and repair the tissue injury or to regenerate the damaged area. Recently, scientists have successfully regenerated bone cells using stromal stem cells, pancreatic cells, and heart cells. Since the ratio of ASC in an organ is very low when compared to non-stem cell population, it is difficult to isolate and culture them efficiently.

ASC undergoes the following three types of differentiation (Fig. 13.10):

1. Normal differentiation 2. Transdifferentiation 3. Reprogramming

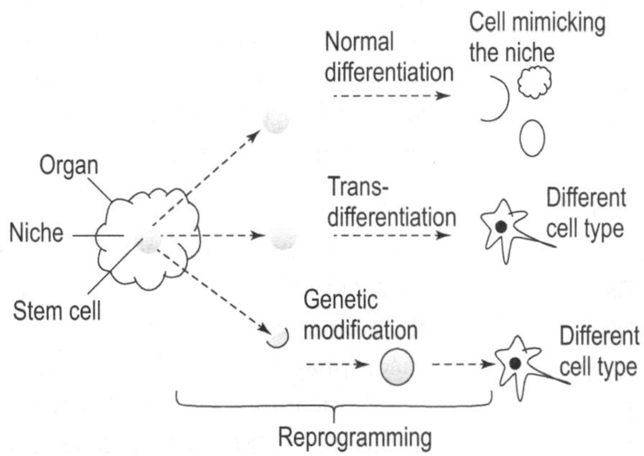

FIGURE 13.10 Three types of adult stem cell differentiation

Usually, depending on the cells' need, ASC can get the signals to differentiate into specific cells which are morphologically, structurally, and functionally similar to the tissue surrounding the niche. This phenomenon is called normal ASC differentiation. Most or all of the ASCs undergo normal differentiation both in vivo and in vitro. For example, HSC can undergo normal differentiation to form all types of blood cells. Mesenchymal stem cells of the bone can differentiate into osteocytes, chondrocytes, adipocytes, and fibrous connective tissues. Normal differentiation of neural stem cells produces oligodendrocytes, astrocytes, and neurons. Epithelial stem cells present in the crypt of gut lining differentiate into digestion-related cells, such as goblet cells, absorptive cells, paneth cells, and entero-endocrine cells. Epidermal stem cells in skin differentiate into keratinocytes.

Sometimes, ASCs from an organ can be differentiated into cells which have functions of a different organ. This phenomenon is called transdifferentiation. For example, brain stem cells can differentiate into blood cells. Blood-forming cells differentiate into cardiac cells. Transdifferentiation phenomenon was observed in some vertebrate species but its occurrence in human beings has not been proved so far.

Reprogramming is a type of induced transdifferentiation. In this process, one cell type can be differentiated into another cell type by genetic modification. The genetically modified cells are called reprogrammed cells. Reprogramming is done in exocrine cells to convert them into insulin-producing endocrine cells. This reprogrammed cell can be used to substitute damaged insulin-producing cells of the diabetic pancreas. Adult pancreatic exocrine cells can be reprogrammed by expression of three critical β-cell (endocrine cell) genes to form a 'β-cell-like cell' capable of secreting insulin. These cells mimic endocrine β-cells in shape, appearance, size, and function, thereby producing insulin. Thus, reprogramming can be used to restore blood sugar level.

TABLE.13.7 Surface markers of haematopoietic stem cells

Origin	Markers
Mouse	Sca-1
	CD27
	CD34
	CD38
	CD43
	CD90.1
	CD117 (c-kit)
	CD150
	AA4.1
	MHC class I
Human	CD34
	CD43
	CD45RD
	CD45RA
	CD59
	CD90
	CD109
	CD117
	CD133

13.7.4 Types of ASC

Different types of adult stem cells have been discovered so far. They are discussed as follows:

Bone marrow stem cells

Bone marrow stroma functions as a structural framework for the progress of haematopoiesis. In addition to this role, it also serves as a source of stem cells. There are two types of bone marrow stem cells: (i) haematopoietic stem cell (HSC) and (ii) non-haematopoietic stem cells.

Haematopoietic stem cell HSC is very easy to collect from the bone marrow of all ages and they are characterized by the presence of specific cell-surface protein markers. Markers are present in differentiated haematopoietic lineages, such as RBC lineage and macrophage lineage. These markers are collectively referred to as 'Lin' (meaning lineage). Some of the HSC markers are shown in Table 13.7. HSC differentiates into different types of blood cells. HSCs can be detected in vitro by the presence of cobblestone morphology or appearance of haematopoietic progenitor cells after culture. In vivo, HSCs are detected by colony forming unit (CFU) assay. CFU assay is used to assess the proliferative capacity of the bone marrow, and to study the haematopoiesis.

The HSC lineage is shown in Fig. 13.11. HSC can give rise to two types of progenitors—myeloid progenitor and lymphoid progenitor, both stimulated by an intricate network of growth factors and cytokines. Myeloid progenitor produces proerythroblast, megakaryocytes, and white blood cells (granulocytes and monocytes). Proerythroblast gives rise to erythrocyte, whereas megakaryocytes produce platelets. The lymphoid progenitor generates T-lymphocytes, B-lymphocytes, and natural killer cells.

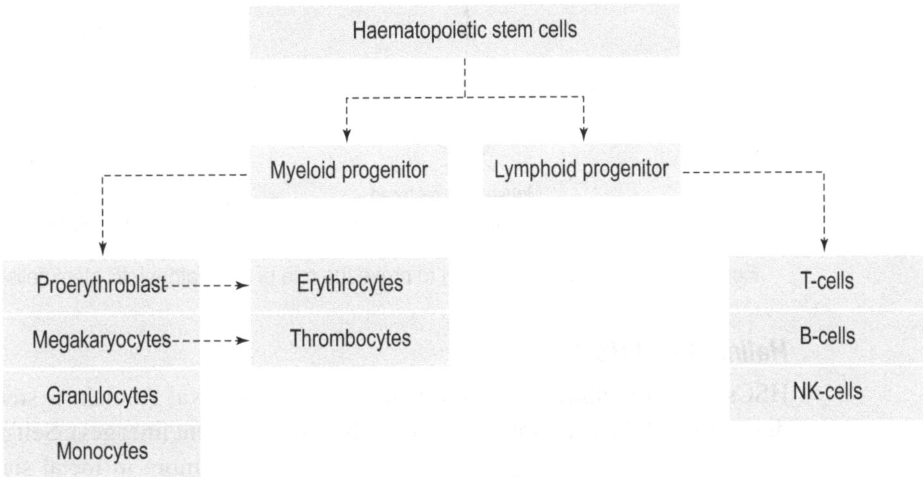

FIGURE 13.11 Haematopoietic stem cell lineage

Cytokines and their receptors that induce proliferation of HSC are the following:

- SCF/c-kit
- FL/FLT3
- Notch ligand-notch
- Wnt-3a-Frizzled

The intracellular signalling molecules implicated in HSC proliferation are the following:

- P21cip1/Waf1
- P27 kip1
- P18-Inc4c
- Hoxb4-Pbx1
- Stat3
- Stat5

Thrombopoietin, oncostatin M, and interleukin-20 are cytokines implicated in the ex vivo expansion of haematopoietic stem cells.

HSC plays a crucial role in the replenishment of blood cells, which are usually short lived, yet perform important functions in vivo. The role of HSC in haematopoiesis is well understood by a simple experiment illustrated in Fig. 13.12. A normal mouse is irradiated to kill blood-producing cells, resulting in a mouse devoid of blood cells. Following this, bone marrow cells are injected to the mouse. This results in the restoration of haematopoietic system over a period of several months. These bone marrow cells are nothing but HSCs. They are capable of self-renewal. There are also some HSCs which can restore haematopoietic system, but incapable of self-renewal. They are called short-term progenitor or precursor cells. HSC can be purified from other cells using magnetic enrichment or fluorescent-activated cell sorter (FACS). In magnetic enrichment, (Fig. 13.13), antibodies to CD34 antigen are labelled with iron particles and passed through the mixed cell population. The antibodies bind to the HSCs and can be isolated using magnetic field. In FACS method, fluorescent-labelled antibodies are used to identify and isolate HSCs.

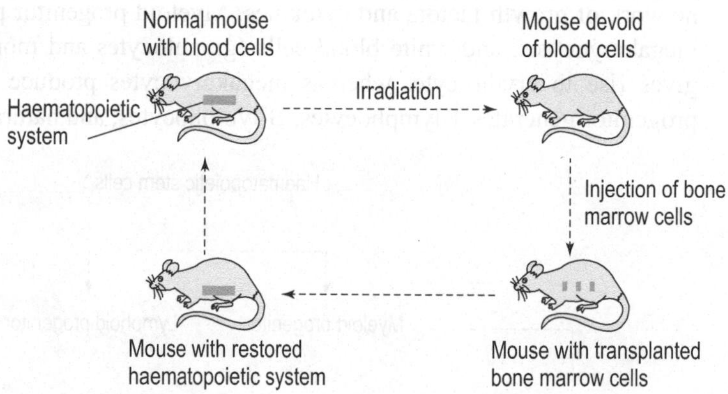

FIGURE 13.12 Experimental model to prove the role of haematopoietic stem cells

Hallmarks of HSC

HSCs have the ability to choose between self-renewal (remain a stem cell after cell division) or differentiation (maturing to become different lineages). Self-renewal capacity of HSC is more in foetal stage than in adult stage of the life. For example, HSCs undergo faster replication in foetal liver, whereas show slower rate of replication in adult stage. It was found that two cytokines, namely stem cell factor and thrombopoietin are responsible for self-renewing potential of HSCs. A signalling molecule, gp130, also plays a critical role in survival and proliferation of mouse HSC in vitro. Another factor influencing the self-renewal capacity of HSC is the telomerase enzyme. Telomerase is responsible for extending the telomeres, which are DNA sequence at the end of the chromosomes. Telomerase activity is essential for cell proliferation. Slower rate of self-renewal of stem cells may be due to decrease in telomerase which is associated with shortened telomeres. Differentiation of progenitor cells into blood cells or immune cell lineages is governed and coordinated by complex interaction among key growth factors and cytokines.

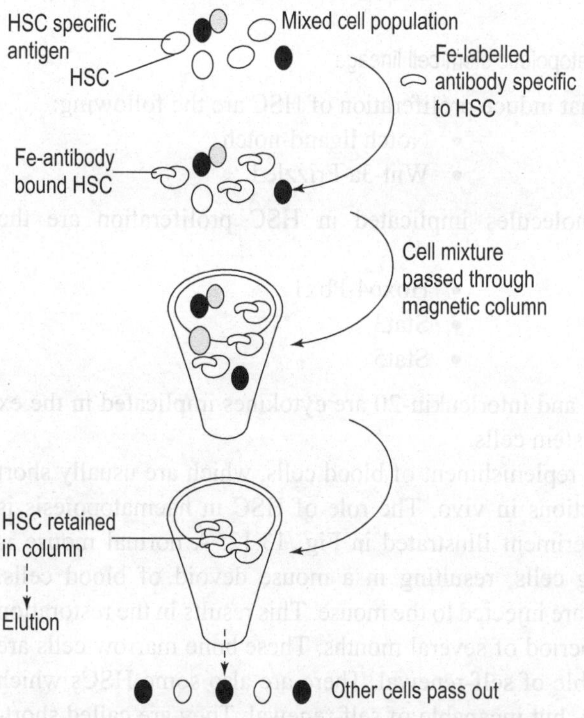

FIGURE 13.13 Isolation of HSC by magnetic enrichment

HSCs possess cell adhesion molecules on their surface with which they are immobilized in the stroma of the bone marrow. In the presence of cytokine, they are mobilized and the mobile HSCs expedite out of the bone marrow and reach peripheral blood circulatory spleen and other tissues. In peripheral blood, HSCs remain in a quiescent non-dividing stage.

When not needed, stem cells proliferation and differentiation stops by the process of self-destruction of the cells called apoptopsis. In culture, absence of growth factor in medium leads to apoptosis of HSCs. In vivo, absence of life-sustaining signals from bone marrow stroma leads to apoptosis. Adhesion between HSC and stroma is disturbed during apoptosis.

Applications of HSC

HSC has a large number of of medical applications and is used in the treatment of many diseases, including cancer and autoimmune diseases. The medical applications of HSCs are discussed as follows:

Use of allogenic HSC in the treatment of blood cancers Allogenic HSC is used in the treatment of blood cancers, such as leukaemia, lymphoma, lymphoblastic leukaemia, acute myeloblastic leukaemia, chronic myelogenous leukaemia, Hodgkins disease, multiple myeloma, and non-Hodgkins lymphoma. The use of HSC in treating leukaemia and lymphoma is widely studied. Leukaemia and lymphoma are blood cancers caused due to unlimited proliferation of white blood cells in the blood. This cancer can be cured by transplantation of HSC from peripheral blood of a donor to the patient. All leucocytes express a specific antigen on their surface called 'HLA'. HLA (human leukocytic antigen) is the major histocompatibility complex on the cell surface. It possesses immune function. Antigens are presented to the cells on these complexes. For a successful transplantation, the HLA patient and donor should be similar. Thus, the brother or the sister of the patient may be a donor. The patients' own white blood cells are first destroyed by irradiation or chemotherapy. The killed cells are replaced by the bone marrow transplant or HSC transplant.

HSC transplant is also used in the treatment of various blood disorders and inborn errors of metabolism as listed in Table 13.8.

TABLE.13.8 Blood disorders and inborn errors of metabolism treated by using HSC

Blood disorder	Inborn error of metabolism
Aplastic anaemia	Osteoporosis
Sickle cell anaemia	Hunter's syndrome
β-Thalessemia	Hurless syndrome
Severe combined immunodeficiency syndrome (SCID)	Lesch–Nyan syndrome
Globoid cell leukodystropy	
Black fan diamond syndrome	
Wiskott–Aldrich syndrome	
X-linked lymphoproliferative syndrome	

TABLE 13.9 Autoimmune diseases and their symptoms

Autoimmune disease	Major symptoms
Rheumatoid arthritis	Gout
Type-1 diabetes	Polyuria Polydypsia Polyphagia
Sjogren's syndrome	Inflammation of salivary gland leading to mouth sores and gum disease Dry eyes Dry mouth
Lesch–Nyhan syndrome	Behavioural disturbance Neurological dysfunction Gout
Multiple sclerosis	Numbness Muscle weakness

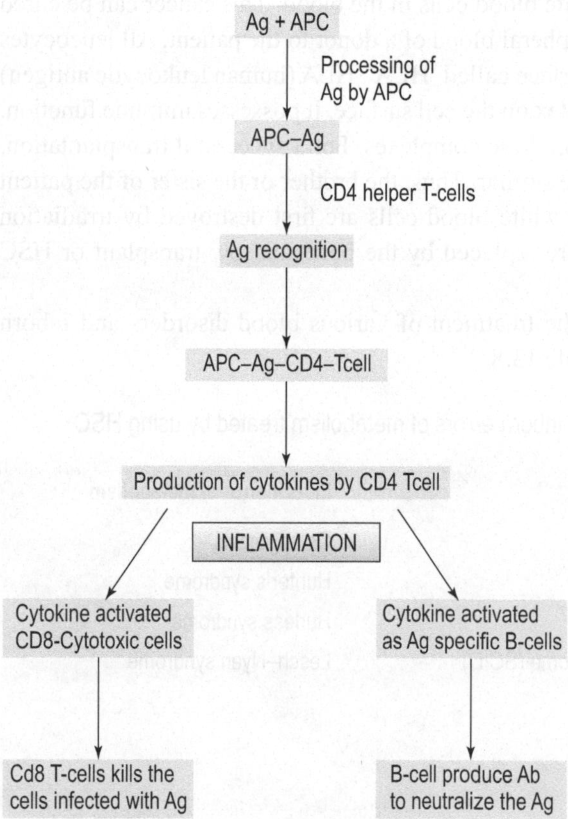

FIGURE 13.14 Steps in immune response

Use of autologous HSC to combat the side effects of chemotherapy Destruction of HSC is the major side effect of chemotherapy. Chemotherapy or radiation kills rapidly dividing cells, such as cancer cells. HSCs are also rapidly dividing, hence they are also killed during chemotherapy. As a result, the destroyed HSCs have to be replaced. For this purpose, autologous HSC can be mobilized from patient's own peripheral blood, and the CD34+ and Thy++ cells are isolated and reintroduced. This marker-specific isolation is necessary to avoid the possible contamination by cancer cells.

Allogenic peripheral blood HSC in increasing the efficacy of the cancer treatment Even after chemotherapy, some of the cancer cells may still remain in the patient's body. Therefore, in order to kill the remaining cells, allogenic peripheral blood HSC is transplanted into the cancer patient. The WBCs from the HSCs kill the tumour cell as it is recognized as foreign. Here, the WBC is called a graft, and the process is called graft versus tumour (GVT) treatment or GVT effect. GVT effect is used in the treatment of chronic lymphocytic leukaemia, advanced colon cancer, and solid malignancy.

Treatment of autoimmune disease Immune system of the body recognizes and differentiates self from non-self components (antigens). If the system recognizes the self as an antigen and attacks its own tissue, it results in autoimmune diseases of varying symptoms (Table 13.9). It is necessary to mention about two concepts to understand autoimmune disease: (i) steps in immune response and (ii) tolerance induction. Steps in immune response of a body towards the self or foreign antigens are summarized in Fig. 13.14 and discussed in the following text.

The antigen-presenting cells (APCs) process the antigen (Ag), and express it on the MHC/HLA on the surface of APC, forming an APC–Ag complex. CD4 helper T cells recognize the antigen on the APC and produce cytokines.

Cytokines then induce inflammation by the following two processes—cytokines activate CD8 in cytotoxic T cell which, in turn, kills the cells infected with the antigen. Cytokines also activate the antigen specific B-cells, which in turn, produce antibodies to neutralize the antigen. Thus, it is emphasized that the aberrant inflammatory immune response is the characteristic feature of autoimmune diseases.

Immune response involves participation of different cell types, all of which are derived from the HSC, via lymphoid progenitors. T cells and B cells are major immune cells. T cells, after their production, reach the thymus, where they get matured and become functional. Functionally, a cell can recognize a vast array of antigen. Thymus provides an environment to inactivate the T cells which are capable of recognizing the system's own antigen. This process is called tolerance induction; using this, the T cell tolerates the system's own protein (without recognizing them as non-self or foreign antigen). When this tolerance falls, the T cell recognizes self protein as foreign antigen, thus giving rise to autoimmune disease.

When cells present in the joints of the body are attacked by T cells, it leads to rheumatoid arthritis, whereas the destruction of pancreatic beta islet cell produces type-1 diabetes. Destruction of myelin covering of nerves causes multiple sclerosis. Failure of tolerance may be due to genetic, environmental, or hormonal factors, but the exact reason is unknown.

Now, let us discuss the role of stem cells in treating autoimmune disease. Stem cell transplants can introduce new T cells and re-establish the 'tolerance'. Researchers have found that HSC transplant is useful in treating lupus, in which multiple tissues, such as skin, muscle, joints, blood vessels, and nerves are attacked by immune cells. In this experiment, scientists have replaced the immune cells (hence, the name replacement therapy). According to the research report, the patients are injected with growth factors for coaxing the HSC to release them into peripheral blood circulation. HSCs harvested from peripheral blood circulation are purified and isolated from mature immune cells and stored. The patient's immune cells are destroyed and eliminated using radiation or cytotoxic drug. The stored cells are transfused back to the patient, where they migrate to the bone marrow. As the person is made immunodeficient previously, the HSC differentiate into immune cells to restore the immune system. This type of therapy has a risk (disadvantage) because the patient suffers a course of immune deficiency with increased susceptibility to many infections.

HSCs, when allowed to express specific genes, can treat diseases, such as type-1 diabetes, multiple sclerosis, and rheumatoid arthritis. The names and functions of the genes expressed in stem cells and the corresponding diseases that can be cured are shown in Table 13.10.

Application of stem cell in treating ALS Adult brain contains neural stem cells which mimic the developing foetal cells that give rise to brain and spinal cord. Three major types of brains cells are formed which include neurons (the message carrier), oligodendrocytes, and astrocytes (support cells). This regenerative capacity can be used in treating neurodegenerative diseases, such as Parkinson's disease, amylotropic laterial sclerosis (ALS), or Lou Gehrigs disease and spinal cord injury in stroke and trauma.

TABLE 13.10 Genes expressed in stem cells for curing diseases

Gene expressed in stem cell	Function of the gene	Disease cured	Experimental model
Gene encoding for the decoy receptor for interferon gamma	Binds to cytokine gamma interferon	Lupus	Mouse
Modified interleukin-2 (I-12) gene	Blocks the inflammatory activity of natural I-12 (inflammatory cytokine)	Type-1 diabetes	Mouse
Gene for anti-inflammatory cytokine IL-2	Resumes the tolerance of CD4 helper T cells to self protein Blocks bone resorption	Rheumatoid arthritis	Mouse
Gene encoding decoy receptor for the inflammatory cytokine—TNFα	Inhibits inflammation	Rheumatoid arthritis	Mouse
Gene coding for inhibitor of nuclear factor-κB*	Inhibits inflammation	Rheumatoid arthritis	Mouse

* nuclear factor kappa-B is a protein which stimulates the production of inflammatory cytokines.

Researchers have used rat model to study the use of stem cells in treating ALS. ALS is a neurodegenerative disease characterized by damaged motor neurons of spinal cord. These motor neurons are responsible for controlling movements. ALS results in impairment of muscle movement, muscle weakness causing paralysis, and finally leads to death. Sinabis virus is used to induce ALS in rats; the actual cause of ALS is still unknown. Human embryonic germ cells, when differentiated into neuronal lineages, express nestin and neuron-specific enolase, both of which are specific neuronal markers. These cells, when injected into the injured region of the spinal cord, regenerate the nerve resulting in the recovery of muscle movement and treatment of paralysis within three months of administering the injection. Thus, it is evident that stem cells based therapy can be used to treat neurodegenerative diseases, such as ALS.

Tissue repair Tissue repair refers to the regeneration of diseased or damaged tissues resulting in new tissues. Stem cells can grow into a specific tissue and repair the degenerated tissues.

Vision Repair In vivo study in mice has proved that bone marrow stem cells can repair damaged retina. The retina of the eye has a central region called macula. Macula degenerates due to ageing, consequently leading to vision loss. Bone marrow stem cells have been shown to repair the damaged retina, and the results show that the stem cells can be used in vision repair.

Muscle repair Dunchenes muscular degeneration is a muscular disease caused due to degeneration of the skeletal muscles. So far, there is no treatment available for this disease. Researchers in Harvard University have proved that stem cells from normal adult induced the regeneration of the skeletal muscle.

Stem cell engineering Stem cell engineering refers to the construction of body organs using stem cells. For example, cartilage, bone, blood vessels, and even heart can be engineered using stem cells. Stem cell researchers in Germany have proved that autologous stem cells can engineer a new heart!

Alternative sources of HSCs

Although bone marrow is a classic source of HSC, the following alternative sources are also available:

- Peripheral blood
- Umbilical cord blood
- Foetal haematopoietic system

Peripheral blood is the pool of blood in systemic circulation, remote from the heart. Peripheral blood cells are those in the circulating peripheral blood pool and not sequestered within lymphatic system, spleen, liver, or bone marrow. Cytokine mobilization releases HSCs into the blood leading to the peripheral blood source. Harvesting HSCs from the peripheral blood is easier than harvesting from bone marrow.

A large number of cells from the bone marrow can be allowed to migrate to peripheral blood, following an injection of a cytokine, such as granulocyte colony stimulating factor (GCSF), to the donor. GCSF is given to the donor for four to five days, before harvesting the cells. After 5 days, an IV tube is injected into the blood stream through vein, and peripheral blood collected and passed through a filtration system. The system pulls out CD34+ WBCs and returns the RBCs to the donor. Peripheral blood HSC (PBHSC) can be used for autologous and allogenic bone marrow transplantation. Peripheral blood HSC harvest yield is greater than bone marrow HSC (BMHSC). Moreover, PBHSCs engraft faster than BMHSCs, that is, the recipients develop the HSC lineages from the PBHSCS faster than that from BMHSCs. PBHSCs with CD34+ and Thy-1+ engraft faster. Another important difference between the mobilized PBHSCs and BMHSCs is their efficiency in taking up foreign genes. BMHSCs show slow gene transfection rate.

Non-haematopoietic stem cell

Non-haematopoietic stem cells, characterized by the lack of haematopoietic surface antigens, such as CD4, CD34, D11b, and CD114, are responsible for the maintenance of non-haematopoietic elements of the blood. They are also called multipotent marrow stromal cells or mesenchymal stem cells (MSCs). MSCs are fibroblast-like cells which, when cultured, can generate colony-forming unit—fibroblasts. They are capable of osteogenic differentiation, forming bone.

MSCs are required in large quantities for tissue repair and they can be easily obtained. It is easy to culture MSC under low oxygen tension and propagate them. MSCs cultured in vitro are sudan black positive, ALP positive, esterase negative, collagen IV positive, and fibronectin positive. Cultured MSCs can differentiate into multiple lineages, including bone, cartilage, fat, fibrous and connective tissues, dermaocytes, adipocytes, and tendons. Thus, the dogma that ASCs can differentiate into cell lineages of their own origin is not applicable for MSC. MSC lineage is shown in Fig. 13.15. MSCs can be preserved for a long time without losing their multipotency. MSCs can be genetically modified by introducing desired genes.

Sources for MSCs other than bone marrow include cord blood, peripheral blood, foetal liver and lung, adipose tissue, skeletal muscle, amniotic fluid, and synovium.

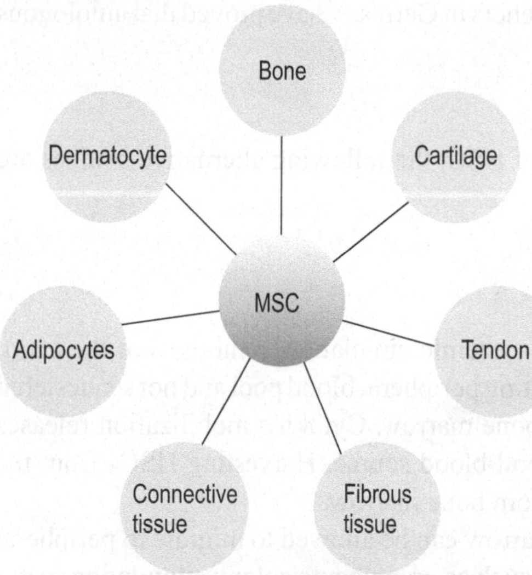

FIGURE 13.15 Mesenchymal stem cell lineage

Applications of MSCs (MSC therapy)

MSCs are used in the treatment of many diseases, including autoimmune disease and multiorgan dysfunction. Following are the medical applications of MSC.

- MSCs derived from cord blood were reported to differentiate in vitro into neuronal, hepatic, osteoblastic, and cardiac tissues. Hence, they can be used in the regeneration of brain, liver, bone, and heart. They are also shown to possess anti-inflammatory and immunomodulatory activity, and hence, they can be used in the treatment of autoimmune diseases
- Autologous and allogenic MSCs are used in treating multiorgan dysfunction associated with systemic lupus erythmatosis.
- MSCs are used to repair skeletal muscle defect.
- MSCs can enhance the graftment of transplanted haematopoietic stem cells.
- MSCs are reported to induce immunomodulatory and antiproliferative effects on T-cells during cell therapy.
- MSCs are used in the treatment of osteogenesis imperfecta in children.

13.8 HAIR FOLLICLE STEM CELLS

The structure of a hair follicle is given in Fig. 13.16.

The bulge region of hair follicle is rich in stem cells. Hair follicle stem cells (HFSC) are multipotent and capable of differentiating into all cell types of hair shaft, epidermis, outer root sheaths, inner root sheaths, and sebaceous gland. This was proved by Oshima, et al. using a transgenic mouse containing LacZ reporter gene.

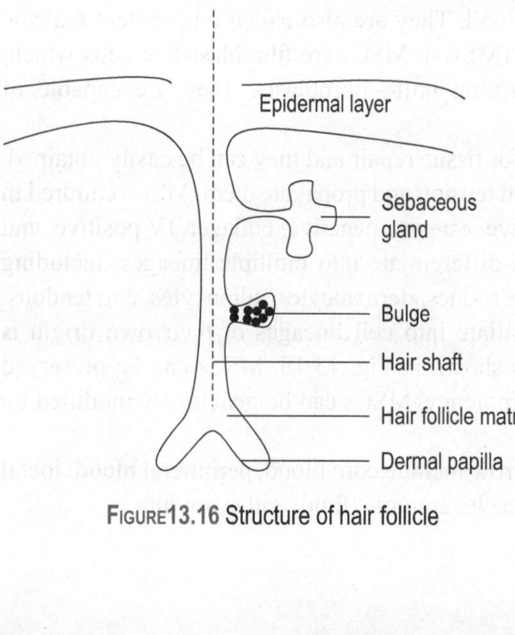

FIGURE13.16 Structure of hair follicle

Bulge region from the transgenic mice, when transplanted into a wild type embryo, stains positive for the beta galactosidase showing blue colour throughout the graft and the epidermis. This suggests that the stem cells from the bulge region form the entire follicle, including the outer root sheath, inner root sheath, and hair shaft. This study also suggests that the bulge cells possess stem cell characteristics, generating all the elements of epidermis and sebaceous gland.

Hair follicle generates a keratinized product called hair shaft. Hair shaft degenerates and sheds eventually. Again, on receiving signal, the follicle regenerates the hair shaft. The process of degeneration and regeneration occurs as a cycle.

13.8.1 Characteristic Features of HFSCs

HFSCs are slow cycling cells. Because of their slow cycling capacity, they conserve their proliferative potential and minimize DNA replication-related errors. Follicle stem cells divide into two cells, one of which is a stem cell, whereas the other is a transient amplifying cell (TAC) with limited proliferative potential. Like other adult stem cells, HFSC cells retain labels (stem cells which divide rarely and reside within their niche retain the labelled DNA precursor, such as tritiated thymidine). However, unlike stem cells, TACs lose most of their labels.

HFSCs express a variety of integrins which are responsible for the maintenance of stem cell characteristics and their position in the stem cell niche. Recently, researchers have found that a small cluster of cells called hair germ, found at the base of the bulge, originate from the bulge and are molecularly similar to bulge. The germ also responds to hair growth signal and functions like bulge stem cells.

Hair follicle stem cells (HFSC) are characterized by the presence of markers, such as b1 integrin, keratin-197, a6 integrin, CD71 (a transferin receptor), p63 (transcription factor), and CD34. HFSCs are positive for CD34 (stem cell marker) and negative for k15 (keratinocyte marker), suggesting their undifferentiated state. HFSCs expressing CD34 markers can be isolated by using anti-CD34 antibodies via FACS, as mentioned earlier.

Nestin, which is a marker for neural stem cells, is also a marker for HFSC. It is a regulatory protein responsible for the expression of green fluorescent protein (GFP). GFP is also a marker for HFSC.

13.8.2 Applications of HFSC

HFSCs have potential application in the treatment of skin loss, skin disease, and hair loss. HFSCs can be used in the treatment of burn wounds, chronic wounds, and ulcers. Bulge stem cells, cultured in vitro, produce autologous epithelium sheets, which re-epithelialize the wounds. HFSCs are also used to prepare composite skin substitute.

13.9 ADULT NEURAL STEM CELLS

Before discussing neural stem cells, it is a prerequisite to understand the neural development and differentiation pattern. During the gastrulation stage of embryogenesis, the outer layer ectoderm, which is also called epiblast, forms the dural plate on the dorsal region. The cells of the neural plate, that is, the neural epithelial cells (NEC) form a tube along the anterior and posterior axis of the embryo that acts as a progenitor and subsequently differentiates into three types of cells:

- Neurons
- Oligodendrocytes
- Supporting astroglia

These three cell types comprise the central nervous system. The neural epithelial cells are called neural progenitors or neural stem cells. Neurogenesis from embryonic stem cells takes place by embryonic induction which induces the specific cells to take a particular fate. In the gastrulation stage of embryogenesis, the ectoderm has epidermal fate in the presence of BMP signal. However, the dorsal ectoderm can be induced to neural fate by inhibitors of BMP, such as chordin, noggin, follistatin, and cerebrus.

Other signal molecules involved in neural induction are Sry-related HMG box-3 (sox3), early response to neural induction (ERNI), insulin-like growth factor (IGF), and wnt signals (wnt is a catenation of 'wingless' and 'int-1' oncogenes). Initially, the neural plate is formed which is then converted to neural tube. Neural patterning occurs in the four domains of neural tube, namely anterior, posterior, dorsal, and ventral domains. Anterior patterning occurs by default and posterior neural patterning is determined by sonic hedge hog (SHH), fibroblast growth factor (FGF), wnt, and retinoid signals. BMPs, wnt, and members of TGF-β family [growth and differentiation factor (GDF) and activin] are signals for dorsal patterning. Sonic hedgehog homolog (SHH) secreted by the notochord and floor plate determines the dorsal and ventral identity. Patterned neural cells (precursor cells) subsequently undergo differentiation into neurons first, followed by astrocytes, and finally, into oligodendrocytes.

Neural stem cells can be obtained from developing brain and developed adult brain. These neural stem cells show self-renewal and multilineage differentiation into neurons, astrocytes, and oligodendrocytes. Neural stem cells cannot give rise to all types of neurons, but only to two types of neurons—GABA neurons and glutamate neurons.

13.9.1 Culture

Neural stem cells can be isolated from brain, based on cell-surface markers, such as AC133 and Lex1. The ability of these cells to function as the stem cell (i.e., the stem cell identity) can be tested by neurosphere formation. Neurospheres are non-adherent free-floating clusters of neural progenitor cells, formed under in vitro culture conditions. Neurospheres can differentiate into brain cells. Neurosphere formation is signalled by the addition of EGF, FGF2, and leukaemia inhibitory factor (LIF). Two other methods for neural stem culture are monolayer culture and radial glia-like neural stem cell progeny culture. In monolayer culture, neural stem cells are grown in a matrix of fibronectin or laminin in the presence of FGF2. In neural stem cell progeny method, the radial glia-like cells are proliferated in the presence of FGF2 and EGF.

13.9.2 Neural Differentiation from ESC

ESCs can be differentiated into neural cells by neural induction. There are two types of neural induction—embryoid-based neural induction and stromal feeder-based neural induction.

Embryoid-based neural induction

Embryoid bodies (EB) are aggregates or clumps of ESC in suspension culture. EB can be differentiated into neural progeny by four methods:

1. 4–/4+ protocol
2. Hepa G2 protocol
3. Nestin+ protocol
4. SFEB (serum-free embryoid body) protocol

4–/4+ protocol EBs are grown for four days in the absence of retinoic acid. For the next four days, retinoic acid exposure is done (hence, the name 4–/4+ protocol). Retinoic acid induces neural progeny and exerts a caudalizing effect on anterior–posterior patterning by activating Hox gene cascade.

HepaG2 protocol HepaG2 is a hepatocarcinoma cell line. EBs are exposed to Hepa G2-conditioned medium. Conditioned medium is a sterile filtered medium in which cells have already been cultured for a period of time. This medium is enriched with growth factor, cytokines, and other mediator substances secreted by the cells. Hence, they are used for the culture of fastidious cells, such as embryonic stem cells. For example, hepGe conditioned medium is obtained by filtering the medium containing the epithelial cells from hepatocellular carcinoma. This results in the direct induction of neurectodermal fate, without expressing endoderm and mesoderm. Active principle of HepaG2 responsible for neural induction is unknown.

Nestin+ protocol In this strategy, EBs are exposed to a serum-free medium containing insulin, transferin, and selenite (so-called ITS medium). In the ITS medium, most of the EB-derived cells do not survive and consequently, a distinct population of immature cells are formed. These cells express higher amounts of nestin, an intermediate filament. Therefore, these cells are called nestin+ precursors (nestin positive precursors or NPP). NPP undergo neural induction to differentiate into neuronal and glial cells when exposed to a combination of three factors called triple factor. Triple factor is a combination of patterning prompting factor, survival factor, and lineage factor.

SFEB protocol SFEB protocol, as the name indicates, makes use of a medium free of serum and other inducing signal factors as used in the previously described methods. The EB aggregates differentiate into cortical cells in a serum-free medium by default.

Stromal feeder-based neural induction

Stromal cell lines have the ability to induce neural induction in stem cells. Two types of stromal lines can be co-cultured with EB for neural induction. They are stromal cells at pre-adipocytic stage, isolated from the bone marrow; and stromal cell lines isolated from aorta-gonad-mesonephros (AGM) region. Stromal line co-culture results in differentiation of EB into midbrain dopamine neurons. However, other neural fates can be achieved by using extrinsic signalling factors.

In addition to these methods of neural induction, default differentiation is possible, which is based on the default hypothesis that absence of signals in primitive ectodermal cells will lead to neural differentiation.

13.9.3 Neuronal Subtypes Derived from ESC

Differentiation of ESC into a specific neuronal subtype depends on the mode of neuronal induction. The following types of neurons can be derived from mouse ESC—dopamine neuron, seretonergic neuron motor neuron, GABA neuron, and glutamate neuron.

Dopamine neuron

Dopamine neurons are used as transplants in the treatment of Parkinson's disease. Therefore, differentiation of ESC into dopamine neuron deserves significance. As mentioned earlier, dopamine neuron can be derived from EB using stromal feeder-based neural induction. This method can generate 16% of the dopamine neuron. However, evidence exists to confirm the role of other factors in dopamine neuron generation (Table 13.11) with high yield. FGF8 and SHH factors yield 34% of dopamine neurons which express tyrosine hydroxylase, a rate-limiting enzyme of the dopamine synthesis. Dopamine neuron yield can be raised to 80% in the presence of Nurr1. Lmx1a and Fox2

TABLE 13.11 Factors involved in the generation of neuron subtypes

Neuron	Factors
Dopamine neuron	FGF8
	SHH
	Nurrl
	Lmx1a
	Fox2
Seretogenic neuron	FGF4
	FGF8
	Shh
	Lmx1
Motor neuron	RA
	Shh
GABA neuron	FGF2
	Shh
	FGF8

were recently reported to induce the generation of dopamine neuron. Seretogenic neurons can be obtained by stromal feeder-based EB induction in the presence of FGF4, in addition to FGF8 and SHH. Transcription factor Lmx1b and Nkx22 have a role in seretonergic differentiation. Motor neuron can be obtained by induction using retinoic acid. Increased efficiency in motor neuron differentiation can be achieved by using stromal feeder protocol in the presence of both RA and SHH factor. GABA neurons can be achieved by 4–/4+ protocol or stromal feeder protocol or SFEB protocol or even by default differentiation. FGF2, SHH, and FGF8 are the factors responsible for GABA neuronal differentiation. Glutamate neuron can be derived by any method using SHH factors.

Neural crest stem cells

Neural crest refers to a collection embryonic cell, which pinches off during the development of neural tube, and thereafter migrates to different locations where they differentiate into diverse structures, that are associated with the nervous system. Thus, neural crest is a source of multipotent adult stem cells. They are used in disease modelling and regenerative medicine. Neural crest stem cell can be obtained from both mouse and human ESC system by exposing to BMP alone or in combination with Wnt and FGF signals. Neural crest stem cells can be differentiated into neuronal cells, glial cells, or mesenchymal cells.

13.10 SKELETAL MUSCLE STEM CELLS

Skeletal muscle stem cells are called satellite cells. Initially, it was thought that satellite cells are committed unilineal stem cells, capable of differentiation into cells of myofibers. However, recent experiments have proved that satellite cells have plasticity, that is, they can be converted into primary myoblast which when treated with BMP can differentiate into osteocytes.

Satellite cells are quiescent mononucleated muscle stem cells located between the sarcolemma and the basement membrane of terminally differentiated muscle fibres. Satellite cells are capable of mediating the repair and regeneration of adult skeletal muscle cells. They are quiescent in healthy muscles, but activated during muscle injury and degeneration after which they proliferate and differentiate. Quiescent satellite cells express certain specific genetic markers or transcription factors—Pax7 and Pax3. Pax7 is expressed in quiescent and activated cells but not in differentiating satellite cells. During differentiation, Pax7 is down regulated.

Activated muscle satellite cells express myogenic regulatory factor, a family that is composed of transcription factor that mediates myogenic differentiation. Myf5, myf3 (myoD), myf6, and myogenin are the members of MRF family.

Pw1 is a gene in satellite cell that encodes a large zinc finger protein that contains 12 zinc finger motifs and a proline-rich repeat motif. It is expressed in prenatal stage, but declines in adult stage. In vitro, pw1 protein is shown to mediate TNFα-induced blocking of muscle cell differentiation. Thus, pw1 is a gene involved in the regulation of satellite cell expression.

13.10.1 Evidences to Show that Satellite Cells have Stem Cell Identity

Satellite cells possess a stem cell character and are capable of differentiation, proliferation, and tissue regeneration. Following are the experimental evidences which prove the stem cell identity of the satellite cells:

1. Studies using tritium-labelled thymidine suggest that satellite cells proliferate and differentiate to repair muscles during muscle damage.
2. Satellite cells show a hypertropic response following muscle stress, such as exercise.
3. In mouse model of Duchenne muscular dystrophy (DMD), satellite cells show high level of differentiation. DMD is due to point mutation in the dystropin gene, which causes a block in translation process of this protein, thereby producing a truncated protein or no dystropin. The function of dystropin is to provide a structural link between muscle cytoskeleton and ECM to maintain muscle integrity. In DMD mouse model, a small portion of muscle fibres still show dystropin staining and such fibres are called revertant fibres. These revertant fibres are generated due to differentiation of satellite cells.
4. Satellite stem cells differentiate into myoblast. Myoblasts undergo fusion to form multinucleated fibres called myotubes.
5. A knockout mouse that lacks Pax7 gene dies in first few months of life, due to severe impairment of postnatal growth and regeneration. Pax7 is the transcription factor which is a marker of the satellite cell. Mouse that lack Pax7 gene shows complete absence of satellite cells which impairs muscle growth. Knockout mice for pw1 gene also show impairment in post natal muscle growth.

All these evidences show that muscle satellite cells are largely responsible for the regeneration and the maintenance of the skeletal muscle. Thus, satellite cells can function as muscle stem cells.

13.10.2 Satellite Cell Niche

Satellite cell niche refers to the collection of satellite cells, cell receptors, secreted factors, the basal lamina (surrounding matrix), and the microenvironment in which the cell resides. Satellite cell forms the integral part of the niche. In addition to specific genes and regulatory factors, satellite cell niche is also responsible for regulating the activity of the satellite cell

SDF-1

Stromal cell derived factor (SDF) is a migration factor secreted by the myofibre. It binds to cxcR4 receptor expressed by the satellite cells and exerts a quiescent effect.

Proteinaceous matrix called basal lamina is the structural network of laminin, fibronectin, entactin type-II collagen, and proteoglycan. It functions as a capsule covering the satellite cell. The component proteins of basal lamina regulate quiescence and proliferation.

Proteoglycan and entactin are activated as a result of injury and prime the niche for the satellite cell response. Satellite cells are closer to the microvasculature within the muscle. The endothelial cell secretes vascular endothelial growth factor (VEGF) enzymes and other growth factors which induce proliferation and expansion of satellite cells.

In vitro studies indicate the following facts about stem cell regulation:

1. TGF beta depresses proliferation and differentiation.
2. FGF stimulates proliferation and depressed differentiation.
3. IGF stimulates proliferation and differentiation.
4. Alteration in the concentration of differentiation factors can make the satellite cells to differentiate or to remain quiescent.

13.11 ADULT PLURIPOTENT STEM CELLS

Adult pluripotent stem cells are scarce and occur in very small amounts. Umbilical cord is the reported source of adult pluripotent stem cells (refer Section 13.5). However, adult stem cells (ASCs) are specific for a particular tissue (refer Section 13.12). However, adult stem cells can be reprogrammed by inducing the genes and transcription factors to form ESC-like cells. These cells are shown to exhibit pluripotency and other properties of ESCs, including expression of cell-surface markers and differentiation into teratogenic tumour. Such cells are called induced pluripotent stem cells (iPSCs). iPSCs were reported in both mice and human beings. They are used in drug development and modelling of diseases. Oct4, Sox2, C_myc, and Kruppel-like factor (KLF) allow for the reprogramming of ASC to iPSCs.

13.12 STEM CELLS WITH TISSUE SPECIFICITY

Stem cells with tissue specificity are the cells which are concentrated on a particular organ or tissue of an adult. Adult stem cells are tissue specific and help in the regeneration of that tissue. Examples of tissue-specific stem cells include those present in the liver, gut, skeletal muscle, and cardiac muscle. This section discusses the stem cells of specific tissues or organs.

13.12.1 Stem Cells from Liver

Research reports reveal that the liver has resident stem cells called oval cells. Oval cells are small elliptical cells of 10 μm size with unusually high nucleus–cytoplasm ratio (Fig. 13.17). They are located in the terminal bile duct, in the canal of Hering. Following are the surface molecules of oval cells:

- A6 and G7 surface antigens
- Thy-1
- α-Fetoprotein
- Albumin
- Cytokeratin-19

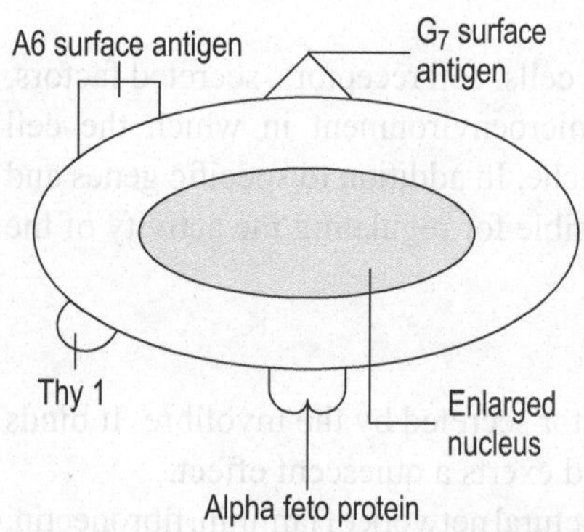

FIGURE 13.17 Oval cells in the liver

Similar to HSCs, the oval cells help in regenerating the liver following an injury. Oval cells respond to certain cytokines and differentiate into hepatocytes and biliary cells. HSCs can also produce the same lineage as oval cells. Oval cells can regenerate liver even in the absence of hepatocytes. Thus, it is clear that oval cells function as stem cells. However, the relation between the HSCs and oval cells is still unclear. The mechanism of liver regeneration by oval cells is shown in Fig. 13.18. During liver injury, the canal of Hering produces oval cells. Oval cells respond to the following cytokines:

- Tumour necrosis factor from inflammatory cells
- Interferon γ from inflammatory cells
- Tumour necrosis factor from Kupffer cells
- Interleukin-6 from Kupffer cells
- Epidermal growth factor from stellate cells (cells with star-like appearance, e.g., neurons)
- Transforming growth factor alpha and beta from stellate cells

As a result, the oval cells differentiate into hepatocytes and biliary cells, leading to the regeneration of the liver.

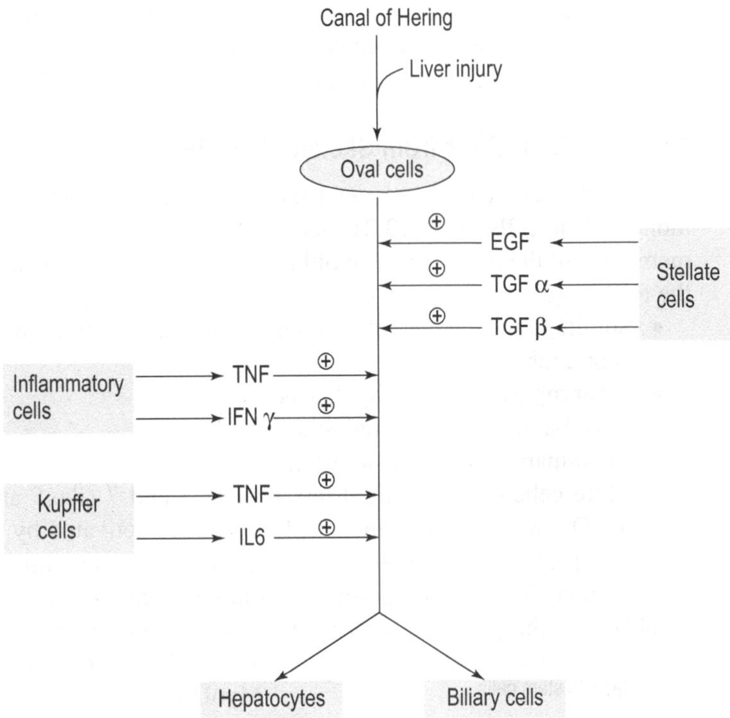

FIGURE 13.18 Role of oval cells in the regeneration of liver (TNF-Tumor necrosis factor; IFN-Interferon; IL-Interleukin; EGF-Epidermal growth factor; TGF-Transforming growth factor)

13.12.2 Stem Cells from Gut

The cells on the walls of the small intestine and the colon have invaginations called crypt. The walls of these two regions of the gut differ in the fact that the small intestine cells have protrusions called villi, which are absent in the colon (Fig. 13.19). The crypt has paneth cells at its base. Adjacent to the paneth cells, gut stem cells (GSCs) are found.

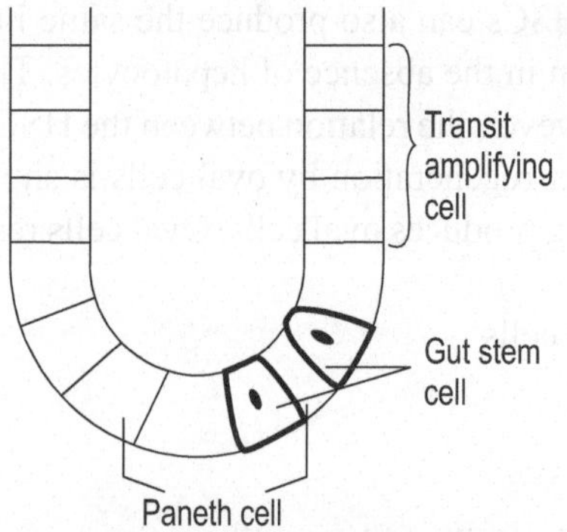

FIGURE 13.19 Gut stem cells in the crypt of the small intestine and the colon

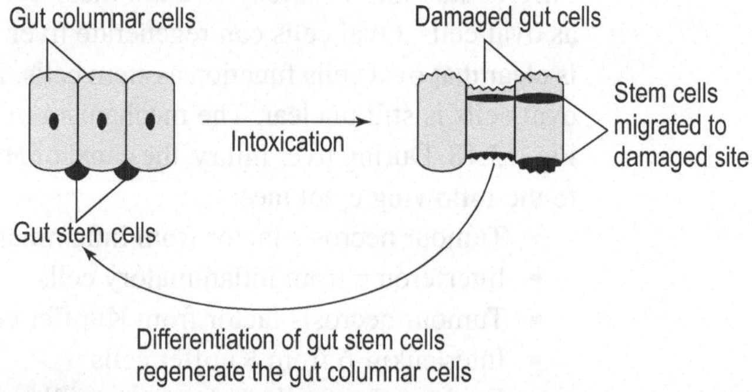

FIGURE 13.20 Function of gut stem cell in the regeneration of gut cells

GSCs are capable of differentiating into transient amplifying cells, which are committed to differentiation into different cells of the gut, including goblet cells, entero-endocrine cells, and absorptive cells. The stem cell function is proved by intoxication studies, which show that when the normal cells are affected by intoxication with toxic chemicals, gut stem cells differentiate to regenerate the cells, as shown in Fig. 13.20.

13.12.3 Stem Cells from Skeletal Muscle

Stem cells present in the skeletal muscle are called satellite cells. They are quiescent mononuclear cells (Fig. 13.21) located in between the basal lamina and the plasma membrane of the muse cells (myofibrils). Embryonic precursors for satellite cells include the following:

- Somite (transient epithelial spheres that pinch off from the mesodermal lining of the neural tube)
- Embryonic and foetal myoblasts
- Vessel-associated mesangioblasts
- Bone-marrow-derived stem cells

Satellite cells express paired box protein 5 and 7 (Pax5 and Pax7) as their surface markers. On muscle injury, the satellite cells are activated by notch signalling pathway (refer Chapter 11). During activation, the myogenic determinant genes Myo-D, myf-5, and the myoblast marker desmin are upregulated. The satellite cells proliferate into myoblasts in the presence of growth factors, such as FGF, insulin-like growth factor-1, and HGF/SF. Proliferating myoblasts are Myf-5/MyoD-positive, desmin-positive, and wnt 5a, 5b-positive. Proliferated myoblasts differentiate into myocytes, forming new myofibril. Differentiation is accompanied by down regulation of Pax5 and upregulation of myogenin. Skeletal muscle stem cell lineage is shown in Fig. 13.22.

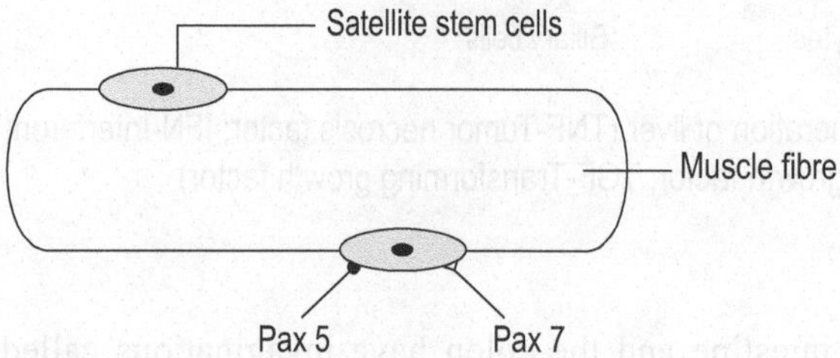

FIGURE 13.21 Satellite stem cells in skeletal muscle fibre

13.12.4 Cardiovascular Progenitor Cells

Cardiovascular progenitor cells are Flk-1+ cells that can give rise to different types of heart cell—

cardiomuyocytes, cardiovascular smooth muscle cells, and endothelial cells. (Flk-1 is the kinase insert domain protein receptor, also known as Kdr.) These cells were reported to regenerate heart cells following heart damage. Lin-cKit+ (CD34) and Sca-1+ cells are also known to produce the cardiac progeny. However, other details about these progenitors are still not known.

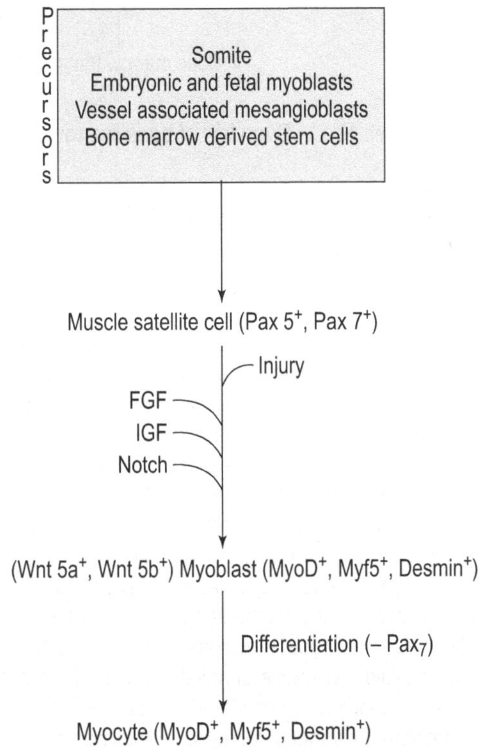

FIGURE 13.22 Satellite stem cells lineage

13.13 BONE MARROW TRANSPLANTATION

Bone marrow transplantation (BMT) is the process of transporting the bone marrow stem cells from a donor to a recipient (Fig. 13.23). There are two types of BMT—autologous transplantation and allogenic transplantation. In autologous transplantation, the donor and the recipient are the same person. In allogenic transplantation, the cells from a healthy individual are transported to the diseased individual. However, the donor and the recipient should have compatible HLA (human leucocytic antigen, refer Chapter 3). Usually, the donor is chosen to be a sibling of the recipient with tissue compatibility. The bone marrow of an individual are collected from the top of the hip bone and filtered. The individual is subjected to chemotherapy to kill the unhealthy cells, and the bone marrow stem cells are transfused intravenously via IV line. After transfusion, the cells naturally fill in the cavities of the bone marrow and replace the old bone marrow. BMT is usually performed for treating life-threatening diseases or conditions, such as congenital neutropenia, severe combined immunodeficiency syndrome (SCID), aplastic anaemia, sickle-cell disease, neuroblastoma, lymphoma, aggressive cancer treatments (chemotherapy or radiation therapy), and thalassemia.

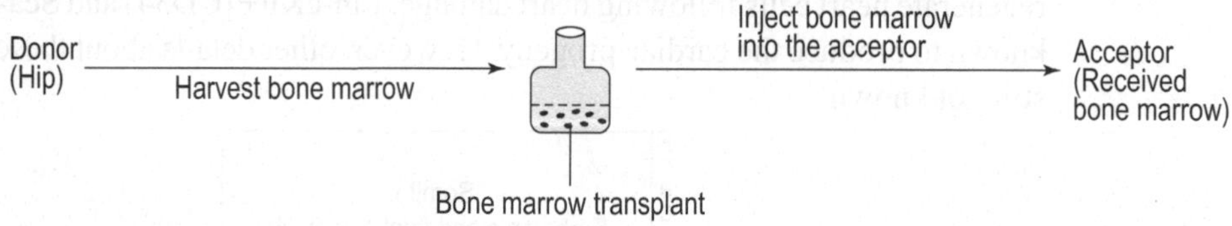

FIGURE 13.23 Overview of bone marrow transplantation

SUMMARY

A stem cell is defined as a cell which has multipotency, self-renewal capacity, and clonality. The two major types of stem cells are embryonic stem cells and adult stem cells. An embryo of 3 to 5 days old, termed blastocyst, is the origin of ESC. ESCs have long-term self-renewal capacity and express specific transcription factors—Nanong and Oct4 —specific for undifferentiated cells. They also possess intact chromosomes. They have the ability to undergo spontaneous differentiation. ESC can be used to treat certain diseases by transplantation or tissue regeneration. The basic characteristic of hESC is high nucleus-to-cytoplasm ratio, prominent nucleoli and distinct colony morphology, high telomerase activity, and expression of surface markers, and transcription factors. Amniotic fluid is a source of novel type of stem cells which can remain undifferentiated for long period and can be induced to differentiate into osteogenic, adipogenic, myogenic, neurogenic, endothelial, and hepatic phenotype in vitro. Umbilical cord is likely to contain cells that can be multipotent (i.e., capable of developing into multiple germ layer) or pluripotent (i. e., capable of developing into all three germ layers—ectoderm, mesoderm, and endoderm). HSCs can be formed in the 7-day old foetus, and it appears like blood islands in the yolk sac.

Adult stem cells (ASC) occur among the cells of adult organs and tissues, and remain in undifferentiated state. ASCs can undergo self renewal and are capable of differentiating into many types of cells. Bone marrow stroma functions as a structural frame work for the progress of haematopoiesis. In addition to this role, it also serves as a source of stem cells. There are two types of bone marrow stem cells—haematopoietic stem cells and non-haematopoietic stem cells. HSCs are used in the treatment of blood cancers, to combat the side effects of chemotherapy, to increase the efficacy of the cancer treatment, treatment of autoimmune disease. Non-haematopoietic stem cells are devoid of haematopoietic surface antigens, such as CD$_4$, CD$_{34}$, D11b, and CD$_{114}$ mesenchymal stem cells are fibroblast-like cells capable of generating colony forming units—fibroblasts. It can undergo osteogenesis to form bone. The bulge region of hair follicle is also rich in stem cells. HFSC, being multipotent are capable of differentiating into hair shaft, epidermis, outer root sheaths, inner root sheaths, and sebaceous gland. HFSC can be used in the treatment of skin loss, skin disease, hair loss, burn wounds, chronic wounds, and ulcers.

ESC can undergo neural induction to differentiate into neural cells. Following types of neurons can be derived from mouse ESC—dopamine neuron, seretonergic neuron motor neuron, GABA neuron, and glutamate neuron. Skeletal muscle stem cells are called satellite cells. Satellite cells can be differentiated into myoblasts. BMP can convert myoblasts into osteocytes.

TERMS TO REMEMBER

Allogenic transplant A process where the stem cell of one person is transmitted to another person with genetic similarity. This is done for treatment of diseases.

Autologous transplant A process where the stem cells are used for a patient from his own body.

Blastocyst The pre-implantation stage of embryogenesis, at which the embryo has an monolayer of cells called trophectoderm and an inner cell mass.

CD26 DPPIV An enzyme that truncates and inactivates certain chemokines, including sdf1/cxzc112 and selective growth factors (e.g., cytokines) thereby blocking the activity of these molecules.

Chemokine Cytokine proteins that induce chemotaxis.

Cytokine A group of signalling molecules, usually, proteins, peptides or glycoproteins, secreted by immune cells and glial cells of the nervous system.

Embryonic stem cells These are stem cells derived from the pre-implantation embryo with an unrestricted capacity for self renewal that has the potential to form all types of adult cells, including germ cells. Thus, for, such cells have only been obtained from certain strains of mice.

Epiblast stem cells Cells obtained from early post implantation mouse embryo that resemble the embryonic stem cells from other species in terms of the conditions required for maintaining pluripotency and inducing differentiation.

Epigenetic factors Factors influencing gene expression during the development without affecting DNA sequence.

Graft-versus-host-disease (GVHD) A transplantation complication in which the white blood cells from the donor attack tissues of the recipient.

Haematopoietic progenitor cell The progeny of a stem cell that has a little or no self-renewal capacity, but is committed to produce mature blood cell types of one or more lineages, for example, granulocytes, macrophages, erythrocytes, megakaryocytes, and lymphocytes.

Haematopoietic stem cell Cell that can reproduce itself (self renew) to give rise to all the blood forming tissues.

Hay flick's limit The number of times a normal cell population will divide before it stops presumably because the telomeres reach a critical length.

Induced pluripotent stem cells Cells that are induced to return to a pluripotent state by manipulating them to express genes associated with pluripotency.

Multipotency The ability of the stem cells to generate multiple cell types that belongs to a particular germ layer.

Niche The local environment that supports stem cell maintenance and self-renewal.

Passaging The process of repeated replating and reculturing the stem cell.

Pluripotency The potential of a cell to differentiate into most or all types of specialized cells; to be distinguished from totipotency, this is the ability to form a complete embryo capable of giving rise to offspring.

Stem cell plasticity The ability of adult stem cell to cross lineage barriers and adopt the expression profile and functional phenotypes of cells that are unique to other tissues.

Tophectoderm The outer layer of blastocyst. It is the sole precursor to the trophoblast lineage of the placenta, but do not contribute cells to the embryo proper.

Totipotency The ability of the cells to generate a whole animal autonomously.

Trophoblast A general term used to describe all cell types of the developing and mature placenta derived from trophectoderm. In mouse, the trophoblast lineage would include extraembryonic ectoderm, ectoplacental cone, trophoblast giant cells, chorionic ectoderm, labyrinthine trophoblast, and spongiotrophoblast.

REVIEW QUESTIONS

1. List the properties of stem cells.
2. Compare embryonic stem cells and adult stem cells.
3. Define pluripotency, totipotency, and multipotency.
4. What is a niche?
5. What is a trophoblast?
6. Define epigenetic factors.
7. Comment on stem cell plasticity.
8. Write the principle of amniocentesis.
9. Give reason for the addition of mitomycin-C to the feeder layer used in HESC culture.
10. Define the following: GVHD and Hay flick's limit.
11. Compare totipotency, multipotency, and pluripotency.
12. Describe the stem cell database.
13. What is a knock-out mice? What is its role in stem cell biology?
14. Write about the satellite cell niche.
15. Discuss the protocols used in the transformation of embryoid bodies into neural progeny.
16. Give a detailed account on the applications of HSC.

MULTIPLE-CHOICE QUESTIONS

1. Proliferation of stem cell is termed
   (a) differentiation
   (b) transformation
   (c) self renewal
   (d) stem cell therapy
2. Conversion of stem cell into a specific cell type with specific function is called
   (a) pluripotency
   (b) differentiation
   (c) self renewal
   (d) transplantation
3. Which of the following statements is true?
   (a) ASC is derived from blastocyst
   (b) ASC culturing is easy
   (c) ESC is derived from blastocyst
   (d) ESC culturing is difficult
4. Reculturing of a stem cell type is called
   (a) differentiation
   (b) transplantation
   (c) passaging
   (d) isolation
5. The supplement required for the propagation of ESC is
   (a) bone morphogenic protein
   (b) myogenin
   (c) activin A
   (d) oncastatin
6. Stem cells are maintained in their undifferentiated state by
   (a) activin A
   (b) leukaemia inhibitory factor
   (c) interleukin
   (d) transforming growth factor
7. The mixture of foetal peritoneal fluid and mother's plasma is called
   (a) niche
   (b) amniotic fluid
   (c) bone marrow transplant
   (d) ultrafiltrate
8. The factor that facilitates the differentiation of AFSC into liver is
   (a) oncastatin
   (b) interferon beta
   (c) mitogen
   (d) interferon gamma
9. Bone marrow is the niche for
   (a) ESCs
   (b) HSCs
   (c) ASCs
   (d) AFSCs
10. The intestinal epithelial stem cells reside in the
    (a) myofibroblast network
    (b) adipocyte network
    (c) inner sheath cell
    (d) cap cell
11. Terminal filament is a part of
    (a) neural niche
    (b) bone marrow niche
    (c) gonadal niche
    (d) epithelial niche

Answers to Multiple-choice Questions

1. (c)	2. (b)	3. (c)	4. (c)	5. (c)
6. (b)	7. (b)	8. (a)	9. (b)	10. (a)
11. (c)				

CHAPTER 14

Techniques in Cell Biology

Gerd Bining, Cal Quate, and Christoph Gerber discovered the atomic force microscope (AFM) in 1985.

After reading this chapter, students should be able to understand and appreciate the following:

- Preliminary protocols for cell biology
- Principle, instrumentation, and applications of centrifugation technique
- Working protocol and applications of chromatography technique
- Types, classification, and applications of electrophoresis
- Radioactivity and its applications in cell biology
- Electrochemical technique and its applications

Cells are used for various biochemical investigations. Sometimes, they are also used as models for investigation of diseases. In some other cases, cellular biomolecules are analysed to derive information about diseased status, diagnose diseases, or monitor the treatment of diseases. During tissue or cell damage of a specific organ, certain markers leak out of the cells and accumulate in the circulation. Monitoring the level of these markers by biochemical techniques gives an idea about the damage and the disease of the cells. Research laboratories utilize cells and their macromolecules for various analyses. In all these investigations, several biochemical techniques are used to learn about cells and their biomolecules. Microscopy is widely used for analysing changes within the cells. This chapter focuses on the principle, methodology, and the applications of various techniques, including microscopy, used in cell biology. However, before starting with the techniques, it is necessary to understand some of the preliminary methods (Table 14.1) which are fundamental for any further investigations on cells and their constituents. Therefore, the first section of this chapter deals with the preliminary methods used in cell biology.

TABLE 14.1 Preliminary protocols used for cell biology investigations

Method	Principle
In vivo approach	Use of animal or plant models as the source of cells or tissues
In vitro approach	Use of cell cultures and established cell lines
Homogenization	Disruption of cell architecture to liberate the cell organelles and biomolecules
Microscopy	Processing of tissue slices to study the cellular structures
Cell sorting	Sorting of a cell population from a mixed population based on some physical properties
Cell counting	Counting of number of cells in a given suspension
Cell preservation	Storage and maintenance of cells for further investigation

Biochemical techniques, such as centrifugation, chromatography, and electrophoresis, used for further analysis are discussed in subsequent sections.

14.1 PRELIMINARY METHODS USED IN CELL BIOLOGY

To work with tissues or cells, we need to culture or harvest them. This can be done by in vivo and in vitro approaches. For microscopic investigations and histopathological studies, tissues should be processed and fixed. In some other cases, cell sorting should be done to isolate a specific cell population. Cell counting is done to get diagnostic information. Above all, the cells should be preserved and stored under appropriate conditions to maintain the viability for any further investigation. This section discusses all the preliminary protocols, namely, in vivo and in vitro approaches, tissue processing for microscopy, cell sorting, cell counting, and cell preservation.

14.1.1 In Vivo Approach

In vivo technique refers to investigations on a whole organism or on a tissue within the biological system. Animals and plants are commonly used as in vivo models. Animal models such as rats, mice, guinea pigs, and rabbits are used mainly for in situ metabolic investigations. In this technique, the animal is anaesthetized and perfused with the artery that carries blood to the organs such as liver and kidney. Subsequently, the blood is drawn from the veins coming from the organs. Biochemical investigations (enzyme assay, protein estimation, DNA estimation, lipid estimation, and sugar level analysis) are carried out in blood. Tissues can be excised from the animal model and homogenized (see Section 14.3) and analysed for various biomolecules. Even the membrane of the cells can be pelleted to analyse the membrane fluidity. Cells can be isolated from the bone marrow and genotoxicity studies can be done. Whole plants are used for investigating photosynthesis, respiration, nitrogen assimilation, pytotoxicity, micropropagation, xenobiotic testing, plant pathology, plant breeding, production of secondary metabolites, and metabolism.

14.1.2 In Vitro Approach

In vitro approach is the investigation done outside a living system. Cell culture, homogenates (disturbed cells), isolated erythrocytes, puffy coats platelets, membranes, and tissue slices are widely used as in vitro models. Cell culture is the widely used in vitro approach, although it requires stringent experimental and laboratory conditions.

Cell culture

Cell culture is the growth of excised biological material in a sterilized nutrient medium outside the living system, that is, in the laboratory. Cell culture is of three types—primary cell culture, secondary cell culture, and established cell line culture.

Primary cell culture The excised tissue is disaggregated enzymatically and incubated in a nutrient medium and grown into a primary cell culture. Primary cultures have chromosomes similar to that of the parent tissue.

Secondary cell culture Secondary cell culture is formed when primary culture is incubated and grown in a fresh medium.

Established cell line culture When cells are recultured for numerous generations, the cells become 'established' and 'adapted' to indefinite growth. The cells are called established cell lines established cell lines are characterized by an increase in chromosome number.

Cell culture technique

Cell culture is a basic technique in cell biology. As an initial step, the nutrient media is prepared and sterilized by autoclaving at 121°C, 15 lb for 15 min. The equipment used for the culture, such as the culture plate, pipettes, and flasks, are also autoclaved. The primary culture is taken as a starter and added to the medium maintained in a sterile laminar flow hood. The culture is maintained and the viability checked before every use. Viability is checked to assure that the cells are normal and alive. The cells are maintained for future use by various preservation techniques discussed in Section 14.1.6. Let us now discuss about the techniques of culturing microbial cells, animal cells, and plant cells.

For the growth of microbial cells, two types of nutrient media are used:

1. Natural or complex or non-synthetic media, in which extracts of plant and animal tissues are used in the medium. The extract supplies sugars, amino acids, and mineral salts to the cells.
2. Synthetic or defined media, in which a blend of precise amount of carbon source, nitrogen source, energy source, amino acids, vitamins, and mineral salts are used. Examples include LB (Luria Bertani) broth media and LB solid media. The components of these two media are shown in Table 14.2.

TABLE **14.2** Ingredients of microbial media

Media	Compound	Amount	Function
LB broth	Yeast extract	5 g	Source of vitamins and minerals
	Tryptone	10 g	Source of C, N, and energy
	NaCl	10 g	Maintains osmotic pressure
	0.1 M NaOH	1 ml	Maintains pH
LB solid media	Yeast extract	5 g	Source of vitamins and minerals
	Tryptone	10 g	Source of C, N, and energy
	NaCl	10 g	Maintains osmotic pressure
	0.2 M NaOH	1 ml	Maintains pH
	Agar	20 g	Solidifying agent; provides a supporting surface

The nutrient media is sterilized in an autoclave at 121°C under 15 pounds pressure for 15 min. A primary culture (starter culture or the inoculum) is inoculated into the media in a Petri plate. This can be done using pour plate method, spread plate method, or quadrant streak method (Fig. 14.1; see Terms to Remember). In pour plate method, the inoculums are taken in a tube and the tube is tilted to transfer the inoculums into a petri plate; in spread plate method, an 'L' rod is dipped into the inoculum and the medium stuck to the rod is spread over the medium, and in quadrant streak method, a platinum loop is dipped in the inoculum, and the loop is streaked on different areas of the plate. Inoculated petri

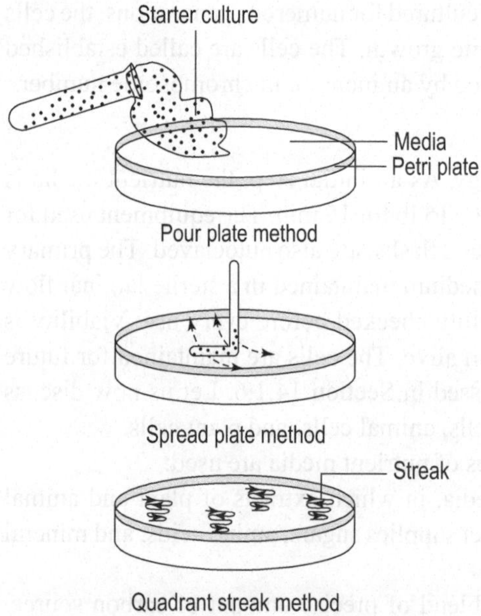

FIGURE 14.1 Inoculation method for primary culture of bacteria

plate is closed and incubated in a 37°C incubator overnight. Bacterial culture can be observed the next day. The culture is stored in glycerol storage buffer.

Bacterial culture is used in biodegradation (degradation of toxic materials), bioconversion (conversion of a toxic element into a non-toxic and useful compound), industrial production of organic acids, alcohols and peptides, genetic engineering for the production of antibiotics, interferon, amino acids, single cell protein and secondary metabolites; and mutation studies.

Plant cell culture

Plant cell culture is the in vitro culture of whole plant or plant parts from explants (a small mass of a tissue) under defined experimental conditions. A complete plant culture media should contain inorganic micro and macroelements, organic energy source, vitamins, complex extract, and growth factors. Growth of plants in vivo is promoted and regulated by organic molecules called phytohormones, for example, cytokinin, auxin, and gibberellin. Growth of plants in vitro can be promoted and regulated by some molecules which functionally mimic phytohormones. These molecules are called growth regulators, for example, indole acetic acid, kinetin, and zeatin. Murashige and Skoog media (MSM) is a commercially available plant culture media. It has the following constituents:

- Salts of Mg, Zn, Cu, Ca
- Ammonium sulphate
- Glucose
- Urea
- Casein hydrolysate
- Antioxidants
- Polyvinyl pyrolidine
- Activated charcoal
- Growth regulators

MSM is dissolved in sterile distilled water and the pH is adjusted to 5–7 using 0.1 N NaOH or HCl. The solution is then blended with agar and dispensed in culture tubes or culture flasks and autoclaved at 121°C at 15 lb for 15 min.

There are three types of plant cell cultures:

- Callus culture, in which a small fragment of the plant, called the callus is allowed to grow in a medium.
- Suspension culture, where the cells are grown in a suspension rather than in a solid medium.
- Organ culture (organogenesis), where the cells are grown into different organs of the plants, such as shoot or root by varying the hormone ratio. For example, high auxin:cytokinin ratio induces shoot formation, whereas low auxin:cytokinin ratio induces root formation.

Figure 14.2 represents the scheme of these cultures.

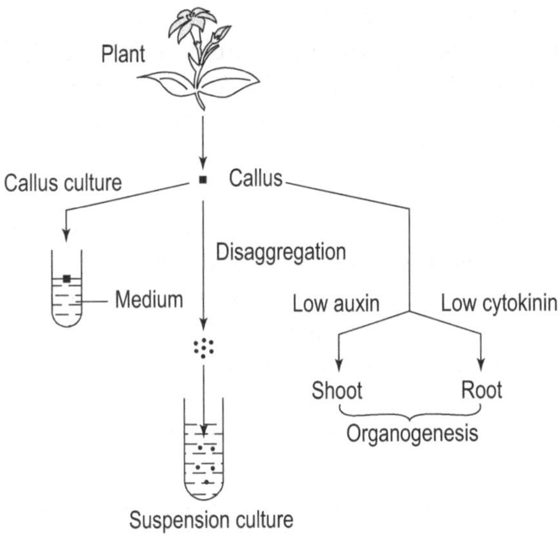

FIGURE 14.2 Plant cell culture techniques

Incubation of explants in MSM incorporated with growth factors results in the formation of an unorganized, undifferentiated mass of parenchymatous cells called as callus. When callus is disaggregated by a mechanical rotary shaker or by enzyme digestion, a cell suspension is produced. The suspension can be stored in a sterile culture flask and subcultured in MSM with appropriate manipulation of the growth factor to develop into tissues such as roots, stem, leaves, or even whole plant. This phenomenon is called totipotency. For example, high ratio of auxin:cytokinin induces root development (rhizogenesis), whereas low ratio induces shoot development (caulogenesis). Plant cell culture is used:

- to separate encapsulated seeds.
- to separate disease-resistant plants.
- to separate stress-resistant plants.
- for somatic hybridization.
- for protoplast formation.
- for genetic manipulation of plant genomes.
- for producing hybrid plants.
- to investigate cell division, cell differentiation, cell expansion, and intermediary metabolism.
- to produce pharmacologically significant metabolites such as caffeine, diosgenin, papain, vinblastin, aromas, benzoqinone, flavonoids, xanthotoxin, alkaloids, reserpine, morphine, nicotine, and digoxin.

Animal cell culture

Animal cells can be cultured in natural or synthetic media. Natural media contains crude tissue extract, ascitic fluid, placental blood, cord blood, and serum or plasma incorporated with antibiotic. Eagle defined medium (EDM, also called as synthetic medium) contains mineral salts, glucose, essential amino acids, glutamine (energy source), vitamins, and peptide growth factor (from serum).

The following two techniques are used to culture animal cells:

1. Monolayer culture or anchorage-dependent culture in which the cells are grown on a supporting surface, for example, teflon, glass, metal, plastic surface. Examples of such culture are fibroblast cells and epithelial cells.
2. Suspension culture or anchorage-independent culture in which the cells are suspended and dispersed free in a liquid medium without any supporting surface, for example, human tumour cells.

A scheme of animal cell culture is shown in Fig. 14.3. Any animal tissue of interest is first disaggregated, either by using enzymes such as collagenase, elastase, trypsin

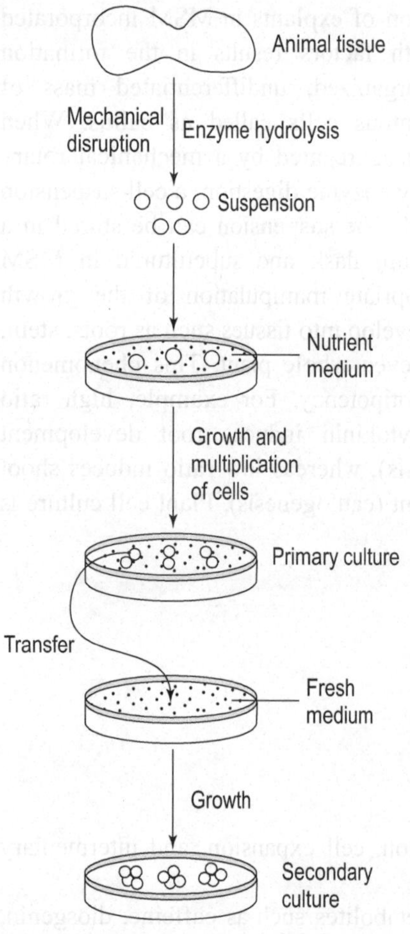

FIGURE 14.3 Animal cell culture techniques

and papain, or by mechanical disruption, and suspended in a medium in the presence of proper proportion of O2 and CO2 (by using CO2 incubator). This leads to a primary culture, which is then transferred to a free medium to form a secondary culture.

Animal cell culture is applied:

1. to culture lymphocytes, polymorphonuclear leucocytes, HeLa cells, macrophages, and hybridoma.
2. to produce commercially and pharmacologically important products, including enzymes, hormones, vaccines, interferon, growth hormones, blood factors, and monoclonal antibodies.
3. as in vitro model system for biochemical, physiological, and pharmacological investigations.
4. to investigate drug and peptide absorption across blood–brain barrier.

14.1.3 Homogenization Technique

Homogenization or cell disruption is defined as the disorganization of cell architecture in the presence of a buffer to release the cellular contents to form an extract (called as cell lysate or homogenate). The homogenate can be further used for biochemical analysis. The molar concentration and the pH value of the buffer are the important factors that determine the subsequent analysis. Usually, Tris HCl buffer or phosphate buffer (0.1–0.2 M, pH 7–8) incorporated with a carbon source (sucrose/sorbitol) is used as homogenization buffer. Certain additional constituents (Table 14.3) that play a specific role in disruption are sometimes incorporated into the buffer.

TABLE 14.3 Ingredients of tissue homogenizing buffer

Compound type	Name of the compound	Function
Antioxidants	Dithiothrietol	Prevention of protein oxidation
	β-mercaptoethanol glutathione	Prevention of disulphide bonding in proteins
Protease inhibitors	Serine protease inhibitor	Inhibition of lysosomal proteases
Chelating agents	EDTA	Prevent thiol groups from the formation of mercaptide
Phenolic inhibitors	Polyvinyl pyrolidine	Prevent the conjugation of plant phenolic compounds with enzymes
Metal ions	Magnesium	Maintenance of the integrity of ribosomes and nucleic acids
Antimicrobial agents	Sodium azide	Prevent contamination by microbes
DNAse inhibitors	Citrate	Prevention of DNA hydrolysis
Organic solvents	Chloroform, ethanol	To isolate chloroplast
Trihydroxy alcohol	Glycerol	To isolate plastids

Homogenization can be done by using the following devices:

Blender Blender is a small chamber equipped with a set of blades of different shapes. The chamber is filled with the specimen and the blades rotated electronically to exert a shearing force to disturb the cell architecture and release the homogenate. Plant cells and mammalian cells may be homogenized by blender.

FIGURE 14.4 Mortar and pestle

Mortar and pestle Mortar and pestle is made of heavy marble. It is composed of a small bowl (mortar) and a blunt rod (pestle) (Fig. 14.4). A small amount of the tissue sample is ground in a mortar using the pestle in the presence of abrasives, such as fine glass pieces, and buffer to form a homogenate. Plant leaves, liver, kidney, and other organs can be homogenized on a small scale by this method

Dynomill or bead mill The components of a dynomill are shown in Fig. 14.5. The sample or the cell suspension and the glass beads (0.5 μm diameter) are put in the chamber. The impeller disc is rotated at a high speed of 10 m/s. The beads collide with the disc and the cells are trapped between the beads and the disc leading to cell disruption.

High-pressure homogenizer The homogenizing press (Fig. 14.6) has an upper chamber and a lower chamber, both connected through a small orifice. The cell suspension or the cell paste is loaded into the upper chamber and subjected to a high pressure created by a mechanically driven piston. The cells are thus mechanically pushed through the orifice into the lower chamber where the pressure drops abruptly. As a result, the cells burst due to turbulence and shearing force. The lysate is collected via the outlet. Plant cells and microbial cells can be disturbed by this method. Commercially available homogenizers, such as French press and Huges press, could disturb large volume of cells.

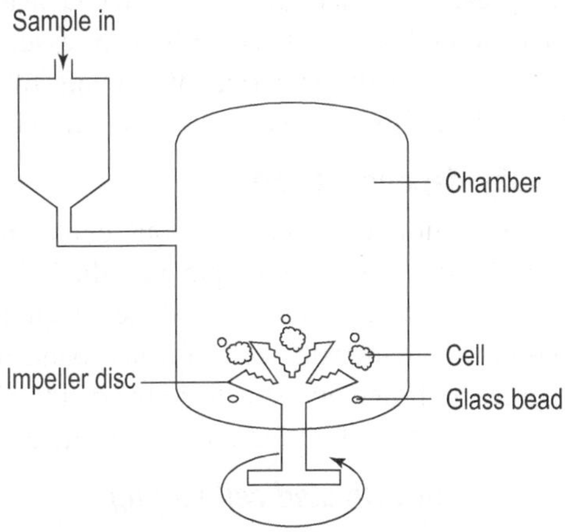

FIGURE 14.5 Dynomill

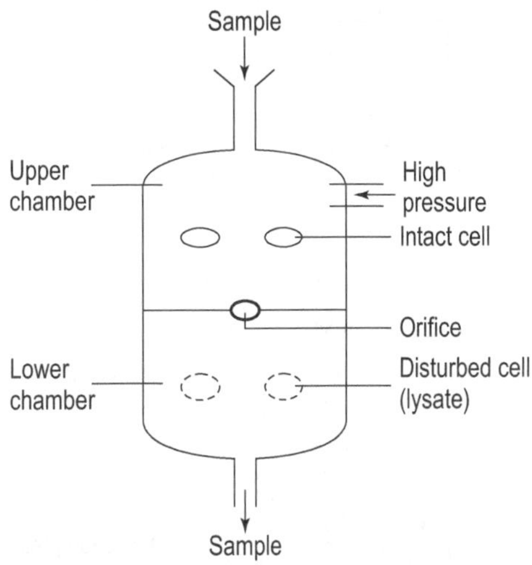

FIGURE 14.6 Homogenizing press

Osmotic shock Animal cells can be lysed by osmotic shock by placing them in a buffer of high osmotic pressure (e.g., buffer with 20% sucrose) for half an hour to one hour. As a result, water molecules diffuse out of the cells causing the cells to shrink. The shrunken cells are then transferred to a buffer of low osmotic pressure or to distilled water to cause diffusion of water molecules into the cell, causing the cells to swell and rupture.

Enzymes Specific enzymes are used to disrupt a wide range of cells. For example, lysozymes are used to disturb the integrity of gram-positive bacteria, whereas zymolase/lyticase is used for breaking yeast cell wall. For fungal cell wall, chitinase is used. EDTA or SDS is used for gram-negative bacterial cell disruption.

Detergents and chemicals Detergents (such as polysorbate, sodium lauryl sulphate, and Triton), solvents (such as toluene and acetone), and chaotropic agents (such as urea and guanidine) are used in the large-scale disruption of microbial cells.

Sonicator Sonicator is a device that generates sound waves capable of disturbing cells. Cells are exposed to high-frequency sound waves of greater than 20 Hz for 30 to 60 s in the presence of ice. Sound waves build up shearing force to disturb the cell architecture.

14.1.4 Cell Sorting

While working with cells, we may be interested in certain specific cell types and not as much in other types. In such cases, cell types can initially be separated based on some properties which usually differ among different cell populations. These properties include sedimentation rate, density, charge, surface-specific antigens, avidin–biotin interaction, and centrifugal elutriation. Thus, cell sorting is the separation of a specific cell type from a mixed population based on a specific property.

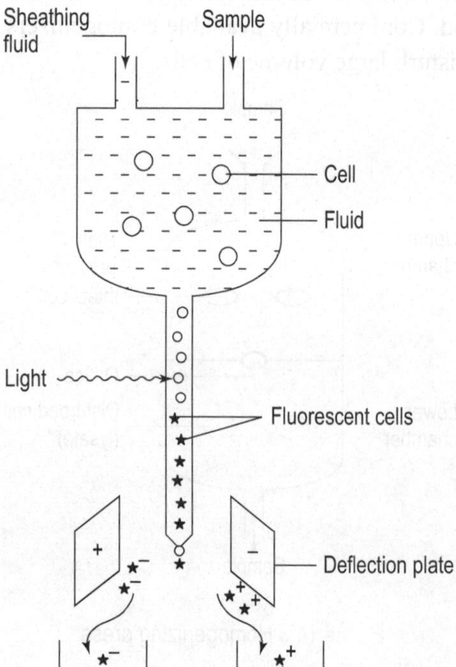

FIGURE 14.7 Cell sorting by flow cytometer

Size-dependent sorting

Centrifugation is used to separate cells and cell organelles differing in their size, based on their sedimentation rate. Cell suspension is centrifuged in a sucrose medium. Cells with greater sedimentation rate pellet out at a faster rate and get separated, whereas the cells with lesser sedimentation rate remain in the supernatant. Details of centrifugation and sedimentation rate are discussed in Section 14.3.

Density-dependent sorting

Centrifugation can also sort out cells on the basis of their density. This technique is called density gradient centrifugation. Separation is done in an isotonic Ficoll medium. Low-density cells form a band at the surface, whereas high-density cells form band at the bottom. Blood cells can be sorted out on the basis of their density.

Fluorescent-activated cell sorting

Cells bearing net positive or negative surface charges can be sorted out by an instrument called flow cytometer or fluorescent activated cell sorter (Fig. 14.7). The cells are mixed with a sheathing fluid in a chamber and passed through a nozzle to emerge out as a stream of charged droplets. This stream passes through an electrostatic field existing between two high-voltage deflecting plates. Charged droplets are attracted towards oppositely charged plates and collected separately while neutral droplets move undeflected.

Sorting on the basis of surface antigen

Certain types of cells, such as cancer cells, express specific antigens on their surface. Monoclonal antibodies (Ab) specific for the surface antigen (Ag) are conjugated with magnetic spheres (dynabeads) to form a dynabead–Ab conjugate, as shown in Fig. 14.8. The conjugate is added to the cell suspension. The cells with specific Ag alone bind to the conjugate, whereas the remaining cell types remain unbound. After washing the unbound cells, the bound cells are isolated by a magnetic field. The FAb region of the dynabead-bound Ab is then made to interact with polyclonal antibodies in order to weaken the affinity of Ab to the cells, thus leading the cells to be sorted out.

Sorting on the basis of avidin–biotin interaction

Cell sorting done on the basis of avidin–biotin interaction is shown in Fig. 14.9. Avidin is a glycoprotein with high affinity for biotin (Vitamin H). The cell mixture is allowed to react with a specific Ab conjugated to biotin. Cells with specific Ag are bound to the conjugate, whereas the non-specific cells remain unbound. The mixture is passed through a column of avidin and washed. The unbound cells pass out of the column while the Ab-bound cells are retained in the column and eluted out subsequently.

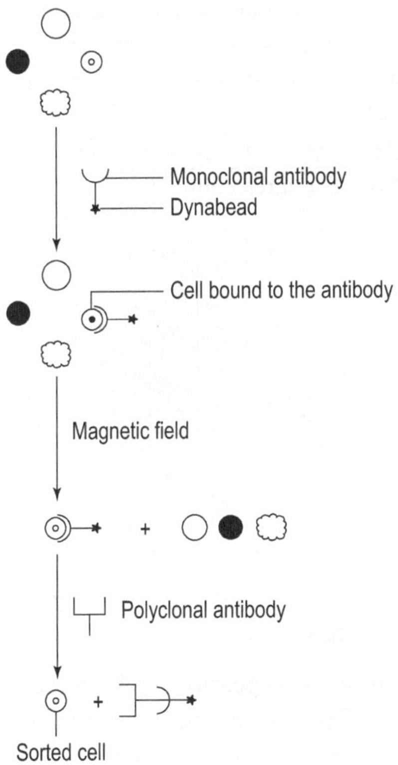

FIGURE 14.8 Cell sorting using dynabeads

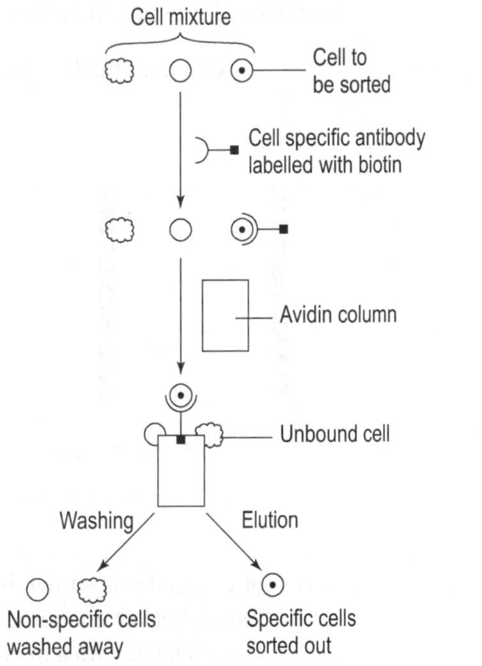

FIGURE 14.9 Cell sorting based in avidin–biotin interaction

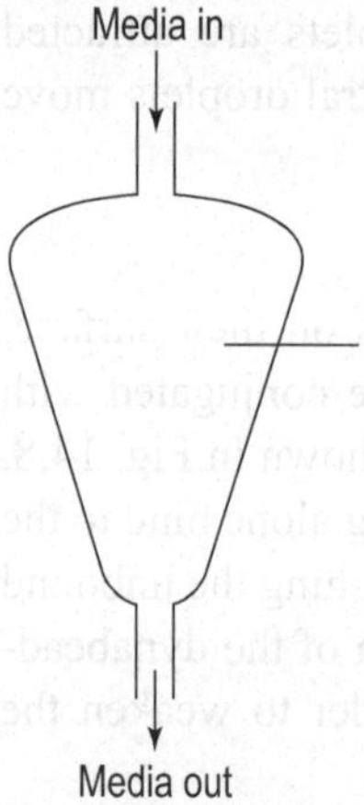

FIGURE 14.10 Centrifugal elutriation

Sorting by centrifugal elutriation

Centrifugal elutriation (counter stream centrifugation) is done in a conical chamber called centrifugal elutriator. Cell suspension is pumped under gentle pressure into the centrifugal elutriator (Fig. 14.10). The cells are then subjected to two opposing forces: (i) the media flow and (ii) the applied centripetal force. These opposing forces allow the cells to separate rapidly depending on size. This technique is used to sort out monolayer cells, cells in tissue homogenate, cells in body fluids, cells liberated from solid tumour, and cells in different phases of cell cycle.

14.1.5 Cell Counting Techniques

Cell counting is done for the assessment of growth and viability of cultured cells. Cells can be counted by three methods: direct microscopy, haemocytometer, and electronic coulter counter.

In direct microscopy, as the name indicates, the cells are magnified in a microscope and counted directly. This is used for counting microbial cell colonies. Haemocytometer is used to count blood cells. Blood cell count is used as the marker for various infections and diseases. A haemocytometer (Fig. 14.11a) contains two chambers, each of 9 mm^2 area and 1 mm depth. The chamber (Fig. 14.11b) is separated into nine squares of 1 mm^2 each. The central square is again redivided into 25 (5 × 5) small squares. Each of the 25 squares is further divided into 16 tiny squares. The four corner squares are just graved into 16 smaller squares. The chamber is filled with the cell suspension and sealed with a cover slip. Cells in the four corner squares and the central squares are counted and substituted in the following formula:

$$\text{Total no. of cells/mm}^3 = \frac{\text{No. of cells counted} \times \text{Dilution} \times 1000}{\text{No. of squares counted}}$$

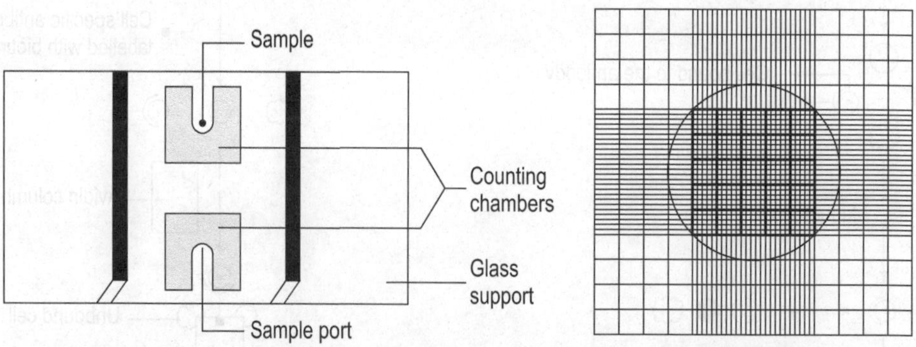

FIGURE 14.11 Haemocytometer and its counting chamber (a) Haemocytometer device (b) Counting chamber

A coutler counter is shown in Fig. 14.12. The cells suspension in physiological saline is filled into a beaker containing an orifice tube. The orifice tube with a tiny aperture of 0.1 μm is associated with a manometer. One platinum electrode is dipped directly into the beaker while a second platinum electrode is placed inside the orifice tube. Current flows from one electrode to the other via the orifice. The manometer draws a fixed volume

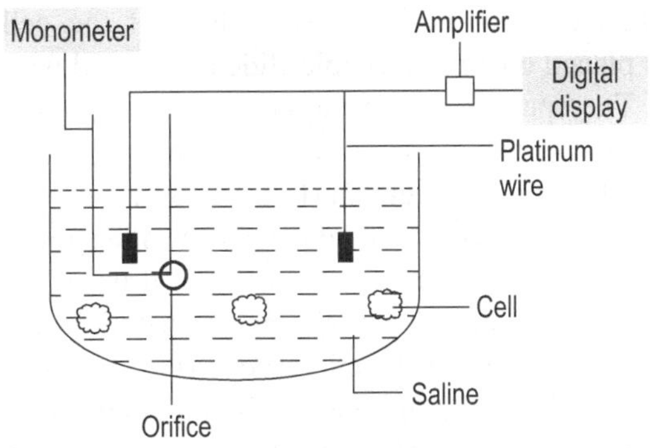

FIGURE 14.12 Coutler counter

of the cell suspension via the orifice and the cells displace their own volume of saline. This leads to a modification in (a drop in) the conductivity, which is converted into a voltage change, amplified, and recorded. The voltage change is directly proportional to the number of particles (cells) in a fixed volume of suspension.

14.1.6 Maintenance and Preservation of Cells

Cells can be stored and preserved by four methods—cryopreservation, freeze drying, liquid drying, and bead desiccation.

Cryopreservation

Cryopreservation refers to the storage of biological materials in a frozen yet viable form at ultra low temperature (–196°C). The cells in suspension are separated by centrifugation and mixed with foetal calf serum containing 10% glycerol and dispensed in sterile cryoampoules and tightly screwed. The ampoules are then arranged in a cryofreezing container and stored at –80°C. After 24 h, the container is transferred to a liquid nitrogen freezer (–196°C) and stored. For example, semen and fish sperm can be cryopreserved.

Lyophilization

Lyophilization is the long-term storage of cells in vacuum, below their freezing point. In this method, the cells are suspended in a storage solution or protective agents, such as albumin, dimethylsulphoxide, milk protein or sucrose, and dispensed in sterile storage vials. The vials are then frozen at –30°C to –80°C and sublimated to evaporate the water in order to render the cells metabolically inert. The cells are then dried in vacuo and sealed under vacuum or inert gas. Lyophilization is useful in marketing, antibodies diagnostic reagent, enzymes, hormones, and vaccines.

Liquid drying

Liquid drying is superior to freeze drying for preserving delicate samples. Delicate anaerobic microbes are sensitive to freezing and freeze drying. Hence, they are preserved by liquid drying. In this method, cells are suspended in protective agents, such as mesoinositol, honey, glutamate, or raffinose and adsorbed on to activated charcoal disc. The disc is placed in air tight glass vials and dried at 20°C.

Bead desiccation

Bead desiccation method is used to preserve cells in a simple laboratory that lacks deep freezer and liquid nitrogen freezer. In this method, small volume of cell suspension is dispersed over large surface of glass beads or porcelain beads and dried on a desiccant. Bead desiccation is suitable for storing spore-forming bacterial cells and fungal cells.

14.2 MICROSCOPY

Microscopy refers to the magnification and examination of live cells or fixed tissues using microscope. Live cell suspension can be viewed under microscope by the following three methods:

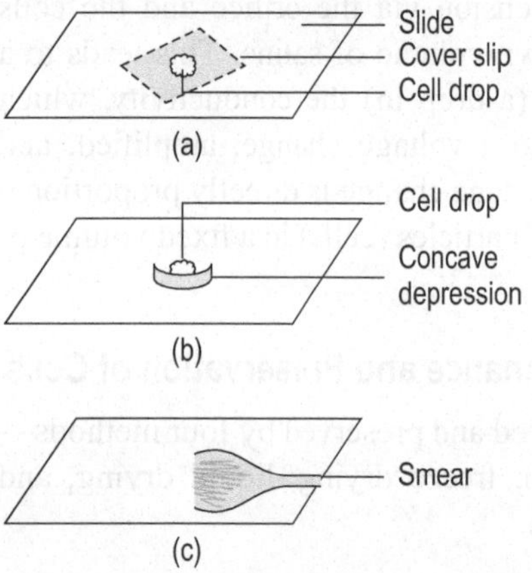

FIGURE 14.13 Methods for viewing live cell culture suspension (a) Wet mount method (b) Hanging drop method (c) Smear method

1. Wet mount method—In this method, a drop of the cell suspension is placed on a microscopic slide and sealed with a cover slip. The space between the edge of the slip and the slide is sealed with petroleum jelly and imaged under microscope for further investigation (Fig. 14.13a).
2. Hanging drop method—This method makes use of a special microscope slide with a concave depression. A drop of the cell suspension is kept on a cover slip and inverted on the depression and imaged under the microscope (Fig. 14.13b).
3. Smear method—The cell suspension is dropped on a microscopic slide using a platinum loop and the drop dragged or spread using another slide and magnified under the microscope (Fig. 14.13c).

Processed tissues

The tissue samples excised from an experimental animal are processed stepwise, stained specifically, and finally viewed under the microscope. The tissue is first fixed using fixative chemicals, such as alcohol, formaldehyde, and mercuric chloride. Acid fixatives are used to fix nucleus and chromosomes. Acetone, formaldehyde, and glutaraldehyde are used for fixing enzymes. During fixation, the cells die and the cellular macromolecules are cross linked and precipitated. The tissues gain more stiffness and stability, and develop more affinity for dyes.

The fixed specimen is dehydrated by alcohol and embedded with embedding agents such as molten paraffin, durcopan, or araldite. The tissues are then sliced into thin sections (using microtome, razor, or diamond) and kept on a mounting support such as a slide or a disc. Finally, the specimen in the slide is stained with a single dye (simple staining) or with two or more dyes (differential staining) and magnified. Differential staining is used to identify different cells and cell components, for example, Gram staining. Different types of dye used for staining are shown in Table 14.4. The stained specimen is magnified and imaged under a microscope.

TABLE 14.4 Types of staining dyes

Name of the dye	Specificity of the dye	Example
Acidic dye	Binds to basic molecules	Eosin Fast green
Basic dye	Binds to acidic molecules	Haematoxycilin
Vital dye	Binds to intracellular structures	Methylene blue Osmium tetroxide
Cytochemical dye	Specific for a particular molecule (such as DNA or lipid) or for a particular cell (such as mast cell)	Toluidine blue Alcian blue

14.2.1 Types of Microscopy

On the basis of magnification and resolution power, the microscope is basically divided into two types—light microscope and electron microscope which are compared in

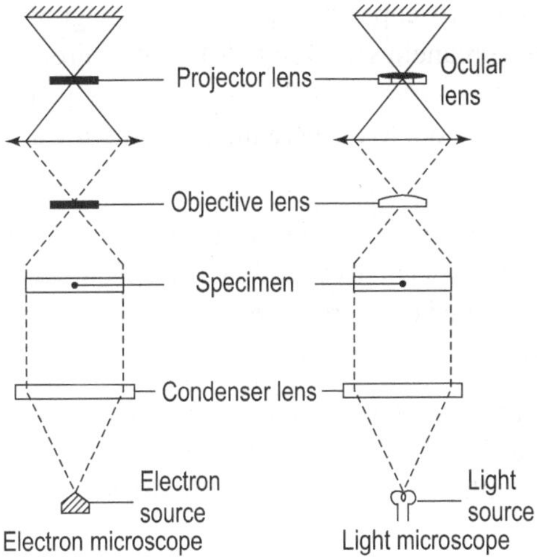

FIGURE 14.14 Light and electron microscopy

Table.14.5 and Fig. 14.14. Earlier, these two microscopes were widely used. Subsequently, several improved versions of microscopes were discovered which include scanning tunnelling microscope, atomic force microscope, scanning electron microscope, and transmission electron microscope.

Light microscopy is subclassified into five types—dark field, bright field, phase contrast, fluorescence, and polarizing microscopy. Electron microscopy is further divided into two types—transmission electron microscopy (TEM) and scanning electron microscopy (SEM).

Light microscopy

Light microscopy is the technique used to obtain microscale images of elements, including cells. It makes use of visible light and lens system to magnify the images.

Different types of light microscopy are discussed as follows:

Dark field microscopy In dark field microscope, the specimen looks brilliantly intense against a dark background. This is used for investigating unstained live cells in suspension.

Bright field microscopy Under bright field microscope, the specimen looks very dark against a bright background. This is used for imaging of stained cells.

TABLE.14.5 Comparison of light and electron microscopy

Parameter	Light microscopy	Electron microscopy
Factor responsible for magnification of the sample	Light beam and glass lens	Electron beam and electromagnetic lens
Material used for embedding the sample	Paraffin	Durcopan, araldite glass/diamond, 10–100 nm
Device used for mounting the sliced tissues	Glass slide	Perforated metal disc
Dyes used for staining the samples	Haematoxycilin, eosin, Gram stain	Heavy metal salts—osmium tetroxide, lead acetate, uranium acetate
Region for viewing the sample	The sample is magnified between the condensing lens and the objective lens and the image is viewed in a phosphorescent screen	The sample is magnified between the condensing lens and the objective lens and the image is viewed in an ocular lens
Subtypes	Bright field microscopy	Scanning electron microscopy
	Dark field microscopy	Transmission electron microscopy
	Phase contrast microscopy	
	Polarizing microscopy	
	Fluorescence microscopy	

Phase contrast microscopy In phase contrast microscope, the specimen is magnified in such a way that different cell components (organelles) exhibit different degree of brightness. The difference in brightness (i.e., the contrast) is due to variation in the refractive index of the cell organelles. This is a good tool to observe the cell structure and mitosis of live cells.

Fluorescence microscopy In fluorescence microscopy, the cells to be magnified are labelled with specific antibodies bound to a fluorescent dye. When the specimen is illuminated, the cells fluoresce. As a result, the labelled cells are exposed. This is used for the identification of a single microbial cell, and helps in diagnosis.

Polarizing microscopy Under polarizing microscope, the sample is illuminated with polarized light to produce a bright image on a dark background. This is used to scan highly ordered materials such as bundles of parallel filaments as in cytoskeleton (refer Chapter 9).

Microscopy at nanoscale

Scanning electron microscopy (SEM), transmission electron microscopy (TEM), scanning tunnelling microscopy (STM) is widely used nowadays in cell biology. In fact, they have replaced the conventional light optical microscopy. The subsequent sections describe the working principle and applications of SEM, TEM, and STM in greater details.

14.2.2 Scanning Electron Microscopy

Scanning electron microscopy (SEM) images the sample surface by scanning it with a high-energy beam of electrons in a raster scan pattern. The electrons interact with the atoms that make up the sample producing detectable signals. The signals provide us with information about the sample's surface topography, composition, external morphology (texture), chemical composition, crystalline structure, orientation of materials constituting the sample, and other properties such as electrical conductivity.

Instrumentation

A schematic representation of SEM is shown in Fig. 14.15. SEM contains the following components:

- electron gun (shown separately in Fig. 14.16)
- cathode cap or cathode shield or Wehnelt cylinder
- anode grounded to the microscopic column
- magnetic lens and condenser lens

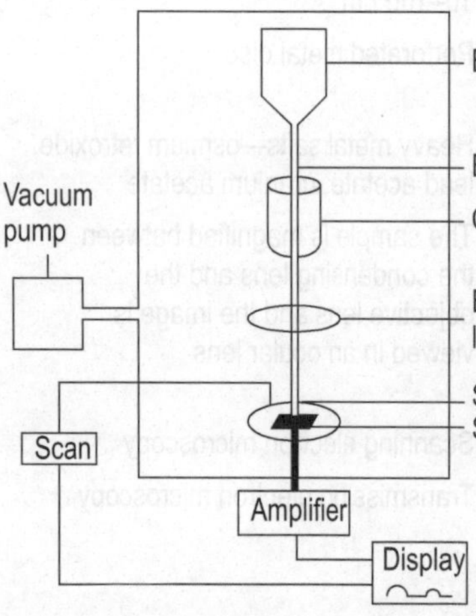

FIGURE 14.15 Schematic representation of SEM

Working

Electron beam is thermionically emitted from an electron gun fitted to a tungsten filament which is held at a negative high voltage [accelerating voltage of 1 to 30 kV; this voltage is used to accelerate the electron beam (also called primary electrons)]. Wehnelt cylinder enclosing the filament is more negative than the filament and repulses electrons. Anode grounded to the microscopic column pulls the electrons and

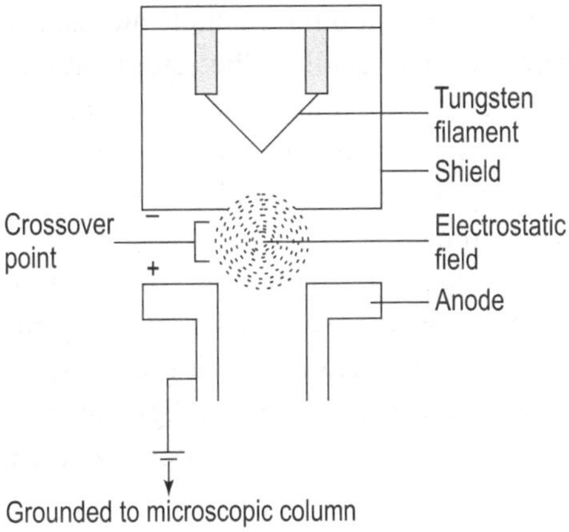

FIGURE 14.16 SEM electron gun

accelerates them down the column. The area between the shield and the anode is approximately 50 μm in diameter and is called the crossover point. Electrons pass through the crossover point to reach down to the condenser lens and electromagnetic lens following a helical trajectory. After emerging from the lens, they converge at a focal point and then hit the sample surface.

Various signals are produced due to interaction of electron beam with the sample (Table 14.6 and Fig. 14.17). Signals produced by incident electron beam include secondary electrons, backscattered electrons, characteristic X-rays, and light (cathodoluminescence).

Secondary electrons are ejected from k-orbitals of the sample atoms by inelastic scattering (scattering of electrons in which the electron loses kinetic energy, but changes trajectory only minimally) of the beam. Using secondary electron signals, SEM produces very high resolution images of 1 to 5 nm of sample surface. Such images yield a characteristic 3-D appearance useful for understanding the surface structure of a sample.

Back-scattered electrons (BSE) are a beam of electrons that are reflected from the sample by elastic scattering. Elastic scattering refers to electron scattering where little kinetic energy is lost, but the trajectory of the electron is substantially changed. The intensity of BSE signal is strongly related to the atomic number of the sample. Therefore, SEM images from BSE signal yield information about distribution of element in the sample. For example, BSE signal had been used to image gold nanoparticles of 5 to 10 nm size. BSE signal can image colloidal gold immune-labels of 5 to 10 nm which cannot be imaged using secondary electron signals.

TABLE 14.6 Signals produced due to interaction of electron beam with the sample in SEM

Signal	Production of the signal	Application
Secondary sample sample	Ejected from k-orbitals of the sample atoms by inelastic scattering of the interacting beam	To get high-resolution images of a sample surface revealing morphology and topography at about 1–5 nm To get characteristic 3-D appearance useful for understanding the surface structure of a sample
Backscattered electrons	Produced due to reflection of electron beam from the sample by elastic scattering	To provide information about the distribution of different elements in sample Most valuable for illustrating contrasts in composition in multiphase samples (i.e., for rapid phase discrimination)
Characteristic X-rays	Emitted when the electron beam removes an inner shell electron from the sample atoms, causing a higher energy electron to fill the shell and release energy	To identify the composition of the sample To measure the abundance of elements in the sample
Light (cathodoluminescence)	Produced when the sample atoms, excited by the incident high-energy electrons, return to their ground state	To get an emission spectrum of the sample in colour

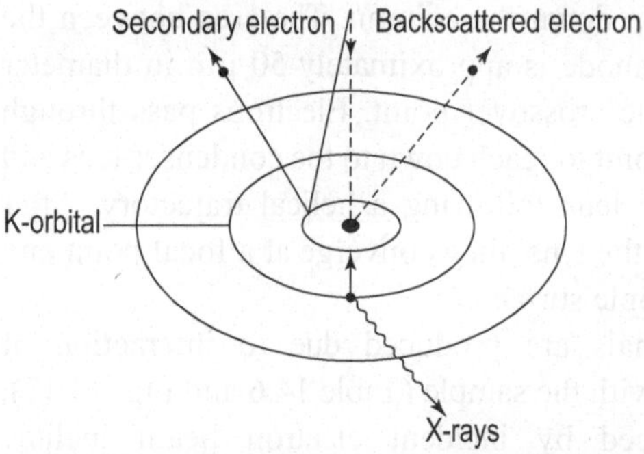

FIGURE 14.17 Signals produced in SEM

The electron beam when hit on the sample atoms, can eject electron from the inner shell of the sample atoms. The vacant shell is refilled by a higher energy electron. In this process, X rays are emitted. X-ray signal helps to determine the composition and the abundance of the elements in a given sample.

Cathodoluminescence (CL, emission of photons or light) occurs when the sample atoms excited by the incident high-energy electron return to their ground state. In SEM, CL detectors collect all the emitted light, analyse its wavelengths, and display an emission spectrum in real colour.

The ejected electrons are captured by a detector and the information is converted into an electric signal, amplified, and digitalized. The electrons that are backscattered from the sample, as well as X-rays produced in the process should be well shielded for safety.

Sample preparation

In order to be observed with a SEM, the specimen should fulfil two criteria:

1. First, the objects should be made conductive for current. This is done by coating them with an extremely thin layer (1.5–3.0 nm) of gold or gold–palladium.
2. Objects must be able to sustain high vacuum and should not alter the vacuum, for example, by losing water molecules or gases. While using biological samples, all water present must be removed from the samples because it would vaporize in vacuum. Biological materials are subjected to prefixation with cold slush nitrogen (cryofixation) or with chemical compounds.

Biological samples and soft samples The following steps are used in sample preparation:

Extraction Extraction refers to removal of a part from the specimen. If the part to be extracted is soft, manual dissection can be done by cutting and pulling the part. However, if the part required is hard, such as the exoskeleton, shell of snails, or teeth or cartilage, chemical extraction can be done in which the chemicals dissolve the soft body material leaving the hard parts.

Fixation Fixation is done to preserve the cellular structure. Fixation is carried out using hexamethyldisilazane. The fixation process results in the enhancement of mechanical strength and the stability of the tissues. The biochemical reactions of the fixed tissues get terminated, and the tissues are prevented from decay.

Cleaning In order to view the specimen under high magnification, it should be thoroughly clean. Cleaning can be done by three ways:

- Manual cleaning—Dirt is picked up using fine pins or thin forceps.
- Vibrational cleaning—The grime can be removed using a sonicator which sends a high-frequency vibration through the specimen, shaking off the dirt. This method is ideal for hard samples.
- Chemical cleaning—Various chemicals are used to dissolve the dirt and remove the wax layer.

Drying SEM works under vacuum and thus the specimen must be dried. Drying prevents collapsing and blowing up of samples in the vacuum chamber. Drying can be done by the following three methods:

- Air drying by exposing to air
- Critical point drying by replacing the liquid in the cells with gas
- Chemical dehydration using alcohol

Mounting After completing all the steps of sample preparation listed so far, the sample should be mounted on an aluminium stub. A stub is a small, flat, round piece of metal that has a stem; it looks like a flattened mushroom. The specimen is glued to the stub. Mounting stabilizes the specimen and reduces the extent of handling.

Gold sputtering Before working with SEM, the mounted sample should be coated with gold using a gold sputter coater. Gold coating is done because gold is not transparent to electron beam used in SEM (i.e., gold is electron opaque). If the sample is not sputtered with gold, the electron beam would travel right through the specimen, creating no image and probably destroying the specimen. Conductive samples, such as metals, do not require any prior processing. However, non-conductive samples should be covered with a conductive layer by using a sputter coater. The sample to be coated is placed in a vacuum chamber. Argon gas is passed into the chamber and an electric field is applied. The argon atoms get ionized and migrate to a negatively charged gold foil. As the argon ions bombard the gold foil, gold atoms are knocked out, which can be deposited o n the sample to be coated.

Cryo SEM

Cryo SEM refers to the imaging of frozen samples, such as tissues and cells. Cryo SEM is also used to image fractures in samples.

Applications

SEM is used

- to generate high-resolution images of shapes of objects and to show spatial variations in chemical compositions
- to identify phases based on qualitative chemical analysis and/or crystalline structure
- for precise measurement of very small features and objects down to 50 nm in size
- for rapid discrimination of phases in multiphase samples by using backscattered electron images

14.2.3 Transmission Electron Microscopy

Transmission electron microscope (TEM) functions like a slide projector. A projector projects a beam of light which transmits through the slide. The patterns present on the slide accept few beams and transmit most of them. The transmitted beam is the replica of the patterns on the slide forming an enlarged image of the slide when it falls on the screen. TEMs work in a manner similar to a slide projector, except that they shine a beam of electrons over the sample instead of light. When a beam of electron is passed through an object, the beam transmits through the sample and gets impinged on a phosphor

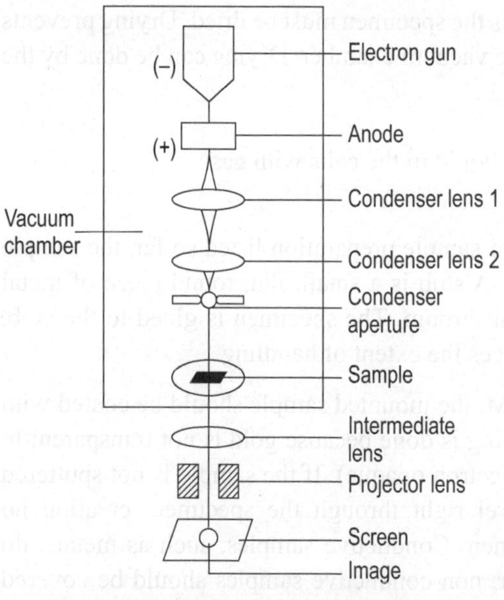

FIGURE 14.18 Components of TEM

screen to show the image of the sample. Proportion of transmission depends on properties of the sample, such as density and composition. For example, porous material will allow more electrons to pass through, whereas a dense material will allow a few. As a result, a specimen with a non-uniform density can be examined by this technique. Using TEM, ultrathin sections of tissues and intact samples, such as viruses and macromolecular aggregates, can be clearly imaged.

Instrumentation

Figure 14.18 represents the components present in a TEM, which may be listed as follows:

- Electron gun
- First and second condenser lens
- Condenser aperture
- Sample holder
- Objective lens
- Objective area aperture (optional)
- Selected area aperture (optional)
- Intermediate and projector lens
- Phosphor screen

Working

- The electron gun at the top produces a stream of monochromatic electrons.
- The light is focussed into a fine coherent beam by a couple of condenser lens.
- In the next stage, the condenser aperture knocks out the high-angle electrons and forms a restricted beam.
- The restricted beam hits the sample and is transmitted out.
- The objective lens converts the beam into an image.
- The intermediate and the projector lenses magnify the image.
- The image is allowed to strike the phosphor screen and viewed.

TEM images are found to have two contrast areas, dark and light. The darker regions are formed due to fewer electron transmissions, whereas the lighter regions are formed due to transmission by more number of electrons. In addition, the dark region represents the thicker or denser part of the sample, whereas lighter areas represent the less dense area of the sample. Before TEM analysis, biological samples are fixed by using staining material, such as uranyl acetate, or by plastic embedding. Biological samples can also be embedded in vitreous ice and held at liquid nitrogen. The tissue samples should be coated with a thin layer (several nanometers) of conducting material such as carbon. Coating can be done by sputtering.

Applications of TEM

- TEM is used to magnify biological and non-biological samples with a good resolution of up to 600,000X.
- It is used to obtain ultrastructural information about animal cells and human cells. Structural information of Golgi complex, mitochondria, endoplasmic reticulum,

lysosomes, vesicles, membranes, nucleus, cytoskeleton, extracellular matrix (collagen), and intercellular junctions can be obtained via TEM.

- It is used to identify the morphology of viruses. Thus, TEM is useful in diagnosis of small pox as it images pox virus. It is also used to identify the morphology of hepatitis B, bacteriophages, and papovirus B19.
- It is used to elucidate the structure and function of many bacterial structures such as flagellae and fimbriae.

14.2.4 Scanning Probe Microscopy

Scanning probe microscope (SPM) has fine probes which can function at nanoscale. Hence, it has wider biological applications. SPM, as the name indicates, probes the samples including biological samples at nanoscale. The first SPM discovered was scanning tunnelling microscope (STM), which was succeeded by atomic force microscope (AFM). Principle and applications of STM and AFM are discussed here.

Scanning tunnelling microscopy

A schematic representation of STM is shown in Fig. 14.19. It has the following components:

- Piezoelectric tube with a tip and electrodes
- Fine needle tip for scanning the sample surface
- Tunnelling voltage source to tunnel electrons between the sample and the tip
- Tunnelling current amplifier
- Distance control unit to control the distance between the tip and the sample
- Data processing cum display unit to display the topography of the sample
- Vacuum chamber for isolating STM from vibrations and protection against contamination

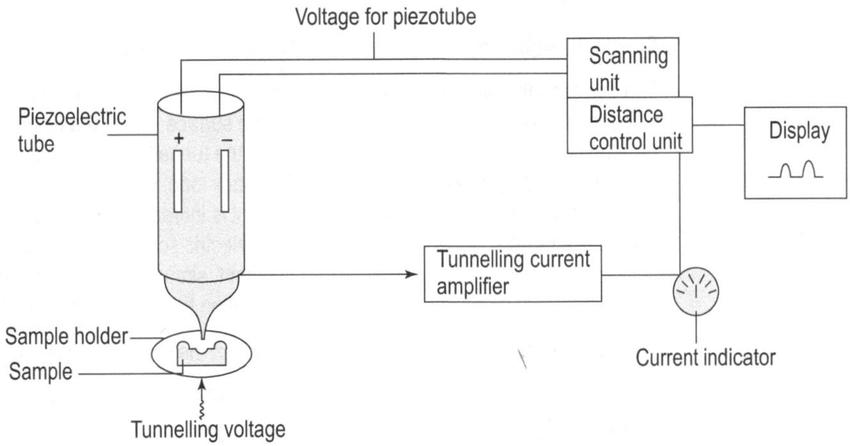

FIGURE 14.19 Schematic representation of a scanning tunnelling microscope

The STM has a metal needle that scans a sample by moving back and forth over it, gathering information about the curvature of the surface and following the smallest changes in the contours of a sample. This concept is similar to closing our eyes and

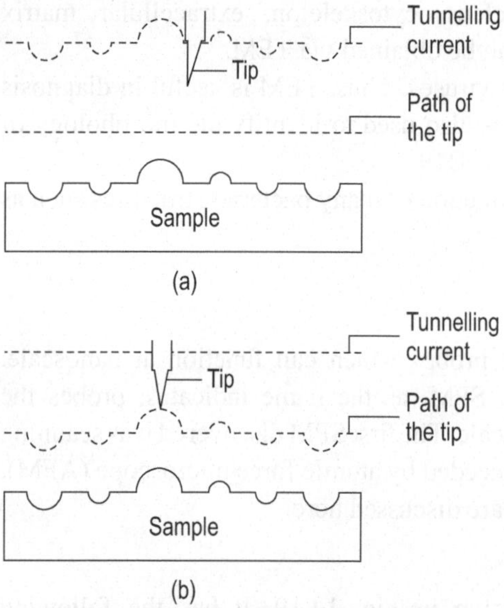

FIGURE 14.20 Working mode in STM (a) Constant height mode (b) Constant current mode

running our finger along the top of a row of books for identifying the changes in height. The needle scans the sample at a distance, approximately equivalent to two atoms above the surface.

Working of STM—The tunnel effect A voltage is applied between the tip of the probe and the sample. The tip scans the surface of the sample. Electrons tunnel between the tip and the sample, generating an electric current. If the sample has a protrusion, the tip encounters the protrusion, and the tunnelling current increases. At this stage, the piezotube retracts. Again, if the tip scans a depression, the piezotube extends. A digital computer system records the movement of the tip and the height of the tip during each scan and displays the topology of the sample. The computer can display the information as a grayscale image or topographical map. STM should be vibrationally isolated from the building. Various methods for vibrational shielding include building the STM on bedrock, using an air table to cushion the STM, or hanging the STM from bungie cords. A small vibration, even a sound, could smash the tip and the sample together. A single dust particle, for example, could damage the needle.

STMs can be designed to work in either of the following two modes (Fig. 14.20)—constant-height mode or constant distance mode and constant-current mode. These two modes are compared in Table 14.7.

TABLE 14.7 Working mode in STM

Factor compared	Constant-height mode	Constant-current mode
Principle	The tip scans the sample surface in a uniform horizontal pattern. Depending on the topography and the electronic properties of the scanned surface, the tunnelling current is altered.	The tip scans the sample, in line with the corrugations on the surface. Thus, the height of the tip is altered while the tunnelling current is kept constant by a feedback loop. For example, whenever the tunnelling current is increased, the voltage applied to the piezoelectric scanner is adjusted. This keeps the current constant, simultaneously increasing the distance of separation between the tip and the sample surface.
Parameter measured	Tunnelling current	Motion of the scanner (distance moved)
Source of topographic image	Tunnelling current measured at each location on the sample surface constitutes the data forming the corresponding image	Motion of the scanner constitutes the data forming the image
Speed	Fast	Slow
Samples preferred	Sample with smooth surface	Sample with irregular surface

Applications of STM STM has a wide range of applications in imaging biological samples, material science samples, and liquid–solid interface. It is also used in atomic manipulation. The applications of STM are discussed as follows.

Measuring thickness of biological specimens STM is a very useful tool for measuring the thickness of biological objects. For example, STM has been used to determine the thickness and to get the structural information about bacterial cells. The purple membrane from *Halobacterium halobium* and the surface protein of *Deinococcus radiodurans* were investigated under STM. Purple membrane is a membrane crystal containing a protein called bacteriorhodopsin. STM study reveals that the highly ordered hexagonal lattice has a lattice constant of 6.3 nm. The purple membrane is a very compact structure with thickness of 4.7 nm and a surface corrugation of <0.7 nm. Patches of purple membrane are typically 0.5 to 1 m in diameter.

Deinococcus radiodurans possesses a surface protein with hexagonally packed intermediate (HPI) layer. STM reveals that HPI layer is a hexamer of a single polypeptide, with lattice constant of 18 nm, and it forms a highly corrugated network.

Atomic-scale imaging at the liquid–solid interface Biomolecules such as nucleotides and proteins in aqueous solutions can be imaged with STM. This process involves prior immobilization or anchoring of the molecule to a flat surface without changing its functions. A self-assembled monolayer of thiol formed on atomically flat gold surfaces, for example, Au (111), is used. Thiols, the organic molecules with a thiol functional group -SH, can form a strong S-Au bond on gold surfaces [approximately 1eV (1 electron volt)], leading to self-assembled monolayers. The other functional groups in that molecule, such as –OH, $–CH_3$, –CHO, can provide anchors to the sample molecules of interest. The mono layers (S), anchor (X), and the sample molecules(Y) are schematically shown in Fig. 14.21.

Alkanethiols $C_nH_{2n+1}SH$ are the most commonly used thiols to form self-assembled monolayers. Organic molecules containing carboxyl (–COOH) and amine (–NH2) functional groups are important for immobilization of biological macromolecules. Among the thiols, cysteine is of particular interest. A self-assembled monolayer of cysteine on gold surface provides both the carboxyl group and the amine group for the immobilization of biological macromolecules, especially proteins.

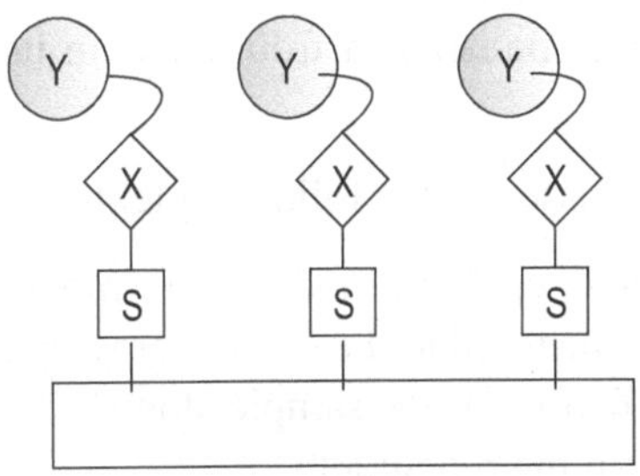

FIGURE 14.21 Self-assembled monolayer of thiol group (S) providing anchor (X) to immobilize biomolecules (Y)

Atomic force microscopy

Atomic force microscopy (AFM) is the most widely used SPM. It is a powerful technique for investigating surfaces by scanning the surface using a sharp tip provided by a cantilever. The local attractive or repulsive force between the tip and the sample is converted into a bending or deflection of the cantilever. Depending on whether the interaction at the tip is attractive or repulsive, the cantilever deflects towards or away from the surface, respectively.

This cantilever deflection must be detected in some way and converted into an electrical signal to produce the images. A laser beam and a detector are used for this purpose. A laser beam is hit on

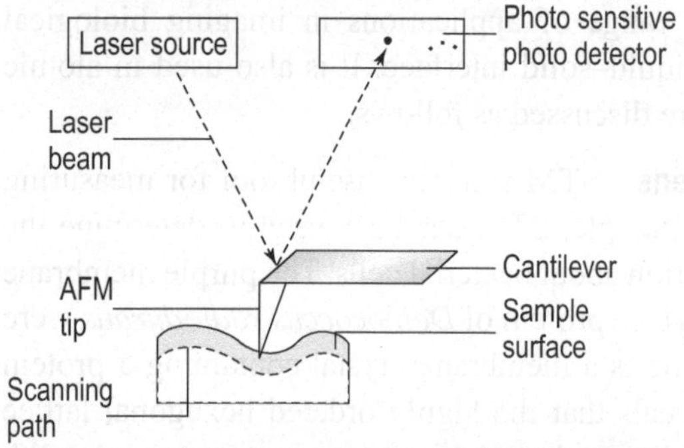

FIGURE 14.22 Schematic representation of AFM

the back of the cantilever and reflected onto a position sensitive photodetector. Whenever the cantilever moves, depending on the surface topography, the deflection is altered, and hence the position of the beam in the PSPD is also displaced. The displacement is measurable by the PSPD. The displacement gives a measure of the cantilever deflection, which in turn gives the topographic data of the sample surface

The cantilever can function in two modes: constant height mode and constant force mode. In constant height mode, the height of the scanner is fixed, and the cantilever deflection gives information about the topography. This mode of working is useful for atomic-scale imaging. In constant force mode, the deflection is used to move the scanner via feedback loop. The movement of scanner produces the image. On the whole, the cantilever deflection and the force applied to the sample are held constant, and the height of the scanner is varied.

Instrument A schematic illustration of AFM is shown in Fig. 14.22. The components are discussed as follows:

Scanner stage in which the sample is placed The most fundamental component of the AFM and the heart of the microscope is the 3-D piezoelectric scanner. Piezocrystals are ceramic materials that expand or contract in the presence of voltage gradient, and conversely, they develop an electrical potential in response to mechanical pressure. In this way, movements in x, y, and z directions are possible. The scanner functions in two ways. It can either move the sample itself (if the sample is little), or it can move the tip over the sample (if the sample is large).

Cantilever Cantilever with an extremely sharp (10–100 Å in radius) probing tip is used for scanning the sample. AFM tips and cantilevers are microfabricated from Si or Si_3N_4. Cantilevers typically range from 100 to 200 μm in length (l), 10 to 40 μm in width (w), and 0.3 to 2μm in thickness (t). They are characterized by their force constant and resonant frequency, which have to be chosen according to the sample to be studied.

Components of an AFM also include the following:

- A laser light to illuminate the back side of the tip
- A photodiode which is sensitive to laser light deflected at the tip

To achieve the highest resolution, the microscope must be isolated from noise in its surroundings and floor vibrations.

Working of AFM An AFM works under three modes discussed in the following text:

Contact mode Contact AFM mode (C-AFM) is also called repulsive mode. In this mode, the AFM is fabricated from silicon or silicon nitride and anchored to the cantilever with a spring constant lower than the effective spring constant that holds the sample atoms. The tip rasters the sample surface, and gives information about the topography.

Non-contact mode (scanning done without touching the surface) In non-contact AFM (NC-AFM), the cantilever tip is prevented from contacting the surface. For this purpose, a stiff cantilever is used which may not be pulled into contact with the surface. In NC-AFM,

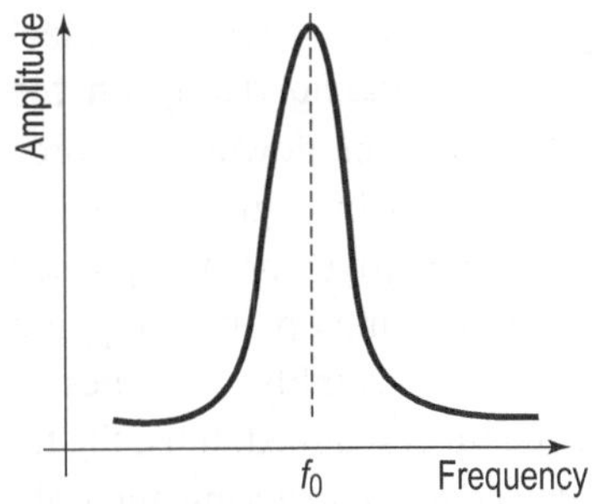

FIGURE 14.23 Resonant frequency

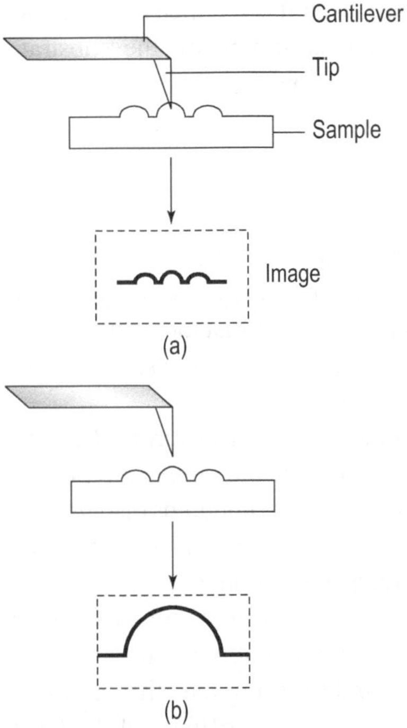

FIGURE 14.24 Comparison of C-AFM and NC-AFM (a) Contact mode (b) Non-contact mode

the tip vibrates above the sample surface with resonant frequency (frequency that gives maximum amplitude, as shown in Fig. 14.23). The tip–sample distance is 10 to 100 Å. As a result, the force between the tip and sample is low (approximately 10–12 N). Hence, the force is difficult to measure unlike C-AFM. Two factors of AFM—the stiff cantilever and the low force value—make the NC-AFM signal small and therefore difficult to measure.

The stiff cantilever vibrates with its resonant frequency (typically 100–400 kHz). Thus, variation in resonant frequency of the cantilever reflects the changes in tip–sample spacing or topography. The events that occur during scanning in NC-AFM are as follows:

- Because of the attractive force between the probe tip and the surface atoms, the cantilever vibration at its resonant frequency near the sample surface experiences a shift in spring constant from its intrinsic spring constant (k_0). This is called the effective spring constant (k_{eff}), and the following equation holds:

$$k_{\text{eff}} = k_0 - F'$$

  where F = force gradient
- The cantilever's resonant frequency varies as the square root of its spring constant.
- The spring constant varies with the force gradient experienced by the cantilever.
- Finally, the force gradient varies with tip–sample separation distance.

Thus, the variation in resonant frequency of the cantilever reflects the changes in sample spacing or topography.

In NC-AFM, the resonant frequency and hence the tip-sample distance/force is monitored and kept constant with the aid of a feedback system that moves the scanner up and down. Then, the motion of the scanner is used to generate the dataset to give the topographic image. Thus, NC-AFM operates in constant force mode.

C-AFM and NC-AFM are compared in Table 14.8 and Fig. 14.24.

TABLE 14.8 Comparison of contact and non-contact AFMs

Property	C-AFM	NC-AFM
Force between tip and sample	High	Low
Type of sample scanned	Soft and elastic samples cannot be scanned	Soft and elastic samples can be scanned
Force measurement	Easy	Difficult
Cantilever	Flexible cantilever	Stiff cantilever
Tip and sample degradation	Frequent degradation	No degradation
Sample with water layer	Can be equally scanned due to penetration of the tip	Cannot be scanned

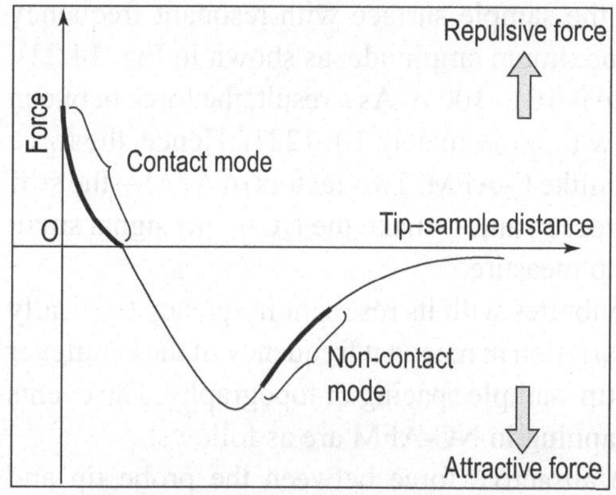

FIGURE 14.25 Force–distance curve

Tapping mode (scanning done in with occasional contact with the surface intermittently tapping gently on the sample) In C-AFM, the constant downward force on the tip often damages the soft biological samples. Moreover, the sample is often destroyed or even pushed out of the field of view by the rastering tip. In the tapping mode, the AFM tip–cantilever assembly oscillates at the sample surface while the tip is scanned; thus, the tip lightly taps the sample surface while rastering and only touches the sample at the bottom of each oscillation. This prevents damage to soft specimens and avoids the 'pushing' of specimens around on the substrate. By using constant oscillation amplitude, a constant tip–sample distance is maintained until the scan is complete. Tapping mode AFM can be performed on both wet and dry sample surfaces.

Interatomic forces versus distance curve

Several forces typically contribute to the deflection of an AFM cantilever. The force most commonly associated with atomic force microscopy is an interatomic force called the van der Waals force. The relation of the van der Waals force with the distance between the tip and the sample is shown in Figure 14.25. The figure shows two regimes: the contact regime and the non-contact regime.

In the contact regime, the cantilever is held less than a few angstroms away from the sample surface, and the interatomic force between the cantilever and the sample is repulsive. In the non-contact regime, the cantilever is held on the order of tens to hundreds of angstroms away from the sample surface, and the interatomic force between the cantilever and sample is attractive (largely a result of the long-range van der Waals interactions).

Because the atomic force microscope relies on the forces between the tip and sample, knowing these forces is important for proper imaging. The force is not measured directly, but calculated by measuring the deflection of the lever, and knowing the stiffness of the cantilever. Hook's law gives $F = -kz$, where F is the force, k is the stiffness of the lever, and z is the distance the lever is bent.

Applications of AFM

Following are some applications of AFM:

DNA manipulation DNA can be manipulated using AFM. This process has two steps:

- Nanoscale incision of DNA using the tip to create fragments (nanoshaving).
- Repositioning of the fragment to predetermined position using the side of the tip.

To determine the rupture force of DNA

- Rupture force is the force that ruptures and disorganizes a DNA helix. AFM can be used to find the rupture force. In this application, the DNA is held between the probe tip and a surface. The probe tip and surface comes into contact initially. The force at which the contact occurs determines the rupture force. It was revealed that the rupture force needed to separate a strand from the DNA duplex in buffer solution is 44.3±0.4 pN.

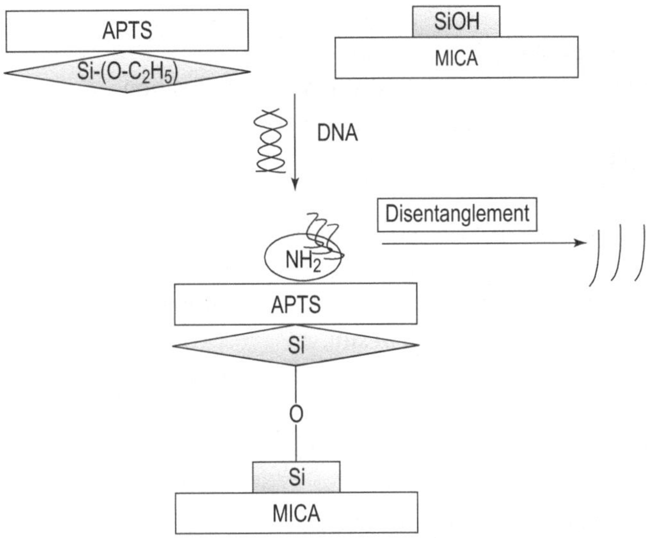

FIGURE 14.26 Immobilization of DNA on silanated mica for AFM imaging

Imaging of DNA For AFM imaging, biomolecules should be immobilized on an atomically flat surface without impairing its biological activity. Gold surface with a self-assembled monolayer of thiol derivatives are used. One of the best substrates is silanated mica. The general formula for minerals of the mica group is $XY_{2-3}Z_4O_{10}$ $(OH, F)_2$ with X = K, Na, Ba, Ca, Cs, (H_3O), (NH_4); Y = Al, Mg, Fe^{2+}, Li, Cr, Mn, V, Zn; and Z = Si, Al, Fe^{3+}, Be, Ti.

Biomolecules do not adsorb on native mica surfaces. For this purpose, the mica is modified by a process called silanization (treating mica with a silanation agent). Aminopropyl triethoxysilane (NH_3–$(CH_2)_3$–Si–$(O$–$C_2H_5)_3$), (APTS) is used as a silanation agent. Here, an ethyl group (–C_2H_5) of APTS reacts with a hydroxyl group (–OH) of mica due to which a Si-O-Si bond is formed. The amine group (–NH2) is exposed and enables the adsorption of biomolecules (Fig. 14.26).

The bonding of DNA on APTS-treated mica is too strong and hence the DNA molecules are often entangled. Disentanglement is done using a process called 'molecule combing'. In this process, DNA is subjected to fluid flow using buffer fluid, parallel to the surface, and the DNA molecules are aligned to the direction of the flow, forming a more-or-less regular pattern, thus, effectively disentangled. This results in effective imaging of DNA.

Analysis of DNA kinking Kinking is the localized abrupt bending of DNA bending, frequently existing in protein-DNA complexes. In fact, kinking is a type of conformational change in DNA. DNA kinking is induced by the bound protein or changes in the physiological environment, such as pH, ion concentration, or temperature. Kinking plays a role in the control of gene expression, replication, recombination, and packaging in the nucleus. Kinking can be imaged and analysed at a spatial resolution on the order of at least 1 nm. An AFM is used for in situ imaging of conformational changes such as kinking in the DNA molecules.

Other applications The following are some additional applications of AFM:

- It is generally used to measure surface characteristics including topography, elasticity, friction, adhesion, magnetic fields, and electrical fields.
- AFM cantilever is so thin and sensitive that it can sense minute surface forces, such as van der Waals forces, magnetic forces, and electrostatic forces.
- It is used to gather information about the topography of a sample.
- It is used for investigation of stratified lipid layers.
- It is used to measure the elasticity, or softness, of a sample.
- It is used to scan biological samples.

- It is used to image DNA.
- It is used to image individual proteins.
- It is used to image DNA/protein complexes rapidly and without fixation.
- It is used for the characterization of the geometry and spatial relationship of several protein-nucleic acid complexes.
- It is used to image living cells.
- It is used to study physical properties of bacteria living in simple biofilm communities on a glass surface.
- It is used to image the virus particles, measure their size and distribution.
- It is used for visualization of nucleus structure in situ.
- It is used for high resolution analysis of 3-D organization of human metaphase chromosomes.
- It is used for high resolution imaging of bacteriorhodopsin.
- It is used to observe oxidative stress in yeast cells.
- It is used for the analysis of human fibroblasts.
- It is used for the analysis of corneal tissues.
- It is used for the study of surface structural changes in spermatozoa.
- It is used for the calculation of cuticle step heights from AFM image of outer surface of human hair.
- It is used for imaging lung chondrocyte surface structure.
- It is used for evaluating demineralization and mechanical properties of human dentin.
- It is used for imaging bacterial shape, surface, and appendages.
- AFM is used to image live cells in order to study the effect of drugs and other parameters on cell morphology and behaviour.
- It is used for imaging cells infected with virus and parasite.
- It is used for imaging of cytoskeletal structure and membrane surface can be done with high resolution.
- It is used for studying membrane proteins and ion channels.

14.3 CENTRIFUGATION

We now move on to study the different biochemical techniques used for getting specific information about cells, tissues, and cellular biomolecules. First, we start with centrifugation technique.

Centrifugation is a separation technique, where cells, cellular organelles, or cellular biomolecules are separated under the influence of applied force. The biological sample is suspended in a suitable medium (solvent) and subjected to a force and spun, larger particles present in the sample sediments first, and the smaller particles sediments later. That is, the sedimentation rate is greater for the larger particles than the smaller particles. The force applied is called centrifugal force (extragravitational force). The instrument used to hold the sample and generate the centrifugal force is called a centrifuge. Table 14.9 summarizes the use of centrifugation techniques.

TABLE 14.9 Applications of centrifugation technique

Type of centrifugation technique	Application in cell biology
Preparative centrifugation	Preparation of tissue homogenate for all biochemical investigations including enzyme assay
	Separation of blood cells
Differential centrifugation	Sub cellular fractionation (separation of subcellular organelles)
Density gradient centrifugation	Separation of lipoproteins
Continuous linear gradient centrifugation	To separate ribosomal subunits, poly ribosomes, and viruses
Rate zonal centrifugation	Separation of enzymes, hormones, and RNA–DNA hybrids
Isopycnic centrifugation	Separation of Golgi bodies, mitochondria and peroxisome
	To determine the buoyant density of the organelles
	To separate mixtures of DNA, RNA, and protein
	To separate heterogeneous mixture of DNA fragments
Analytical ultra centrifugation	Determination of molecular weight of macromolecules
	Assessment of conformational changes in DNA
	To determine the purity of macromolecules

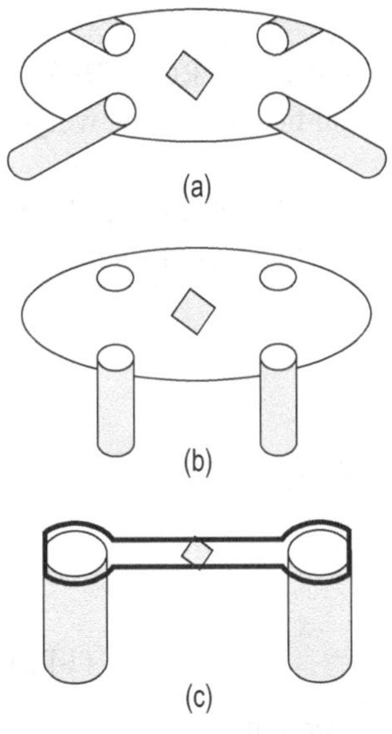

FIGURE 14.27 Types of rotors (a) Fixed angle rotor (b) Verticle rotor (c) Swinging bucket rotor

Rotor is the main part of a centrifuge. A rotor can provide room for the sample tube and spin along its own axis at different speeds to generate the required centrifugal force. The speed with which the rotor spins is denoted as rotation per minute or rpm. On the basis of the rpm, centrifuges are classified into four types: low-speed centrifuge, high-speed centrifuge, analytical centrifuge, and airfuge. Different types of centrifuges are compared in Table 14.10. Low-speed centrifuge is used to obtain crude tissue homogenate and blood cells. High-speed centrifuge is used to isolate protein precipitates, and larger sub cellular organelles. Analytical ultracentrifuge is used to characterize cellular macromolecules such as DNA and proteins. For example, the molecular weight and purity of proteins and the conformational changes in DNA samples can be determined. Airfuge is used for the analysis of amino acids and for ligand-binding assay and receptor assay.

Rotors are divided into three types—vertical tube rotors, fixed angle rotors, and swinging bucket rotors (Fig. 14.27). In vertical tube rotors or zero angle rotors, the tubes containing the sample are held vertically parallel to the axis of rotation. In fixed angle rotors, the sample tube is held at an angle of 30° to the horizontal. In swinging bucket rotors, the sample tubes are kept within buckets, which are *perpendicular* to the axis of rotation.

TABLE 14.10 Types of centrifuge and their applications in cell biology

Type of centrifuge	Rotor speed (rpm)	Temperature control	Applications
Low speed	5000	Absent tissue sample	Preparation of crude Preparation of tissue homogenate Separation of blood cells
High speed	25,000	Present	To isolate larger subcellular organelles To precipitate proteins To isolate cell membranes
Ultra	70,000	Present	To characterize proteins and nucleic acids To determine molecular weight of proteins and nucleic acids To find the conformational changes in DNA To assess the purity of macromolecules
Airfuge	100,000	Present	To analyse amino acids For ligand-binding assay For receptor assay

Before discussing the applications of centrifugation technique, it is relevant to understand about two basic concepts related to this technique—relative centrifugal force and sedimentation coefficient.

14.3.1 Relative Centrifugal Force

Centrifugal force is generally expressed as the relative centrifugal force (RCF); RCF is the ratio of the weight of the sedimenting particle in the centrifugal field (F) to the weight of the same particle in the gravitational field (g).

When a sample solution containing the particles to be separated is centrifuged, the particles experience a *centrifugal force*, *F*, and move radially from the axis of rotation.

The centrifugal force can be written as

$$F = \omega^2 r \tag{14.1}$$

where ω = angular velocity of the rotor

r = radial distance moved by the particle from the axis of rotation

ω can be represented in terms of rotor speed, rpm.

1 rpm = 2π radians

$2\pi \times$ rpm

Therefore, $$\omega = [2\pi \times \text{rpm} / 60] \text{ radians/s} \tag{14.2}$$

Substituting the value of ω from Eqn (14.2) in Eqn (14.1), we obtain

$$F = [2\pi \times \text{rpm} / 60]^2\, r$$

$$= [4\,\pi^2\,(\text{rpm})^2 / 3600]\, r \tag{14.3}$$

By definition, RCF is the ratio of the weight of the sedimenting particle in the centrifugal field (F) to the weight of the same particle in the gravitational field (g). That is

$$\text{RCF} = F / (g) \qquad (14.4)$$

but,

$$g = 980 \text{ cm/s} \qquad (14.5)$$

Substituting the value of F and g from Eqns (14.3) and (14.5), respectively in Eqn (14.4), we get

$$\text{RCF} = [4\,\pi^2\,(\text{rpm})^2 / \frac{3600}{980}]\,\text{r} \times 1$$

$$= 1.118 \times 10^{-5}\ \text{rpm}^2 \times r \qquad (14.6)$$

14.3.2 Sedimentation Coefficient

Sedimentation coefficient (S) of a molecule is its sedimentation rate per unit centrifugal field, that is

$$S = V / \omega^2 r \qquad (14.7)$$

The particles are sedimented in different medium at different temperatures. The density of the medium and variation in temperature influences the value of S. Therefore, S is always corrected to a sedimentation constant obtainable in water at 20°C. This is called standard sedimentation coefficient, Sω20.

$$S\omega^{20} = (\eta T / \eta_{20}) \times (\eta / \eta 0) \times (1 - \nu\rho_{20}\,\omega / 1 - \nu\rho T) \qquad (14.8)$$

where

S = experimentally measured sedimentation coefficient

$\eta T/\eta_{20}$ = relative viscosity of water at temperature T compared with that at 20°C

η/η_0 = relative viscosity of the solution to that of medium solvent

ν = partial specific volume of the solute

ρ_{20} = density of water at 20°C

ρT = density of the medium at experimental temperature T°C

The basic unit of sedimentation coefficient is *Swedberg unit* (S), which is equal to 1×10^{-13} s. For example, sedimentation coefficient of rRNA = 5×10^{13} s, that is, 5 S.

14.3.3 Centrifugation Techniques

Centrifugation techniques are of two types—preparative centrifugation and analytical centrifugation. Preparative centrifugation is used for large volume of samples. Particles of different morphology and density are separated and purified using this technique. In analytical centrifugation, small volume of samples is used and pure particles are characterized.

Preparative centrifugation

Preparative centrifugation is divided into three types:

1. Differential centrifugation
2. Density gradient centrifugation (rate zonal and isopycnic centrifugations)
3. Continuous flow centrifugation

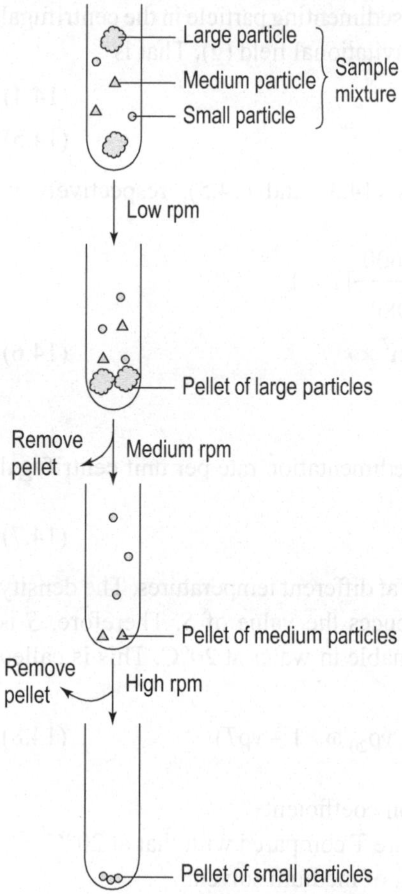

FIGURE 14.28 Steps in differential centrifugation

Differential centrifugation The tissue sample is homogenized in a mortar and pestle (described in Section 14.1.3) and suspended in a medium of uniform density. The homogenate is subjected to a stepwise increase in centrifugal field. Under each centrifugal field, particles are sequentially separated based on their sedimentation rate as shown in Fig. 14.28.

Applications Differential centrifugation is used for sub cellular fractionation, that is, for isolation of sub cellular organelles from tissue homogenate. In this technique, approximately 10 g of the tissue is homogenized using 0.25 M sucrose and the homogenate is subjected to differential centrifugation by raising the rotor speed (centrifugal field) in a stepwise manner as shown in Fig. 14.28. Cell organelles fractionated by differential centrifugation could be further purified using density gradient centrifugation, and the purity could be tested by marker enzyme assay.

Density gradient centrifugation Density gradient is a liquid column whose density increases towards the bottom (Fig. 14.29). The gradient stabilizes the liquid column, thus preventing the mixing of separated particles and improving the resolution of separated particles.

Following are the criteria for a density gradient material:

- It should be stable, non-toxic, non-inflammable, and sterile.
- It should not absorb light during monitoring.
- It should have negligible osmotic pressure.
- It should provide minimum change in pH, ionic strength, and viscosity

Caesium chloride, rubidium chloride, ficoll, and metrizamide are examples for ideal density gradient materials.

Two types of density gradient can be prepared for density gradient centrifugation. They are step gradient and continuous gradient.

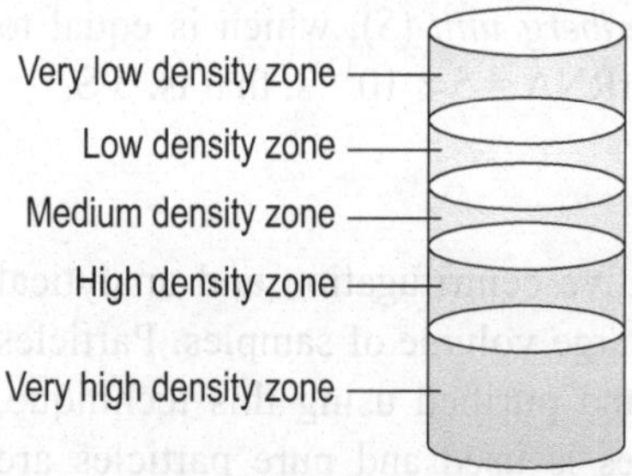

FIGURE 14.29 Density gradient

Step gradient Step gradient is prepared using glycerol or sucrose as shown in Fig. 14.30. A serial concentration of the density gradient solution (10%, 20%, 30%, 40%, 50%, 60%, 70%, and 80%) is prepared. One millilitre of each solution is pipetted out into a centrifuge tube with the lowest concentration (10% solution) at the top and the highest concentration (80% solution) at the bottom. Step gradient is used to separate serum lipoprotein fractions, such as VLDL, LDL, and HDL.

Continuous linear gradient A solution of the lowest and highest concentrations (say, 10% and 80%) is taken in a special device called gradient former (Fig. 14.31). It has a donor chamber loaded with the low density (10%) solution and a mixing chamber loaded with the high density (80%) solution. The chamber is also equipped with a stirrer and an outlet.

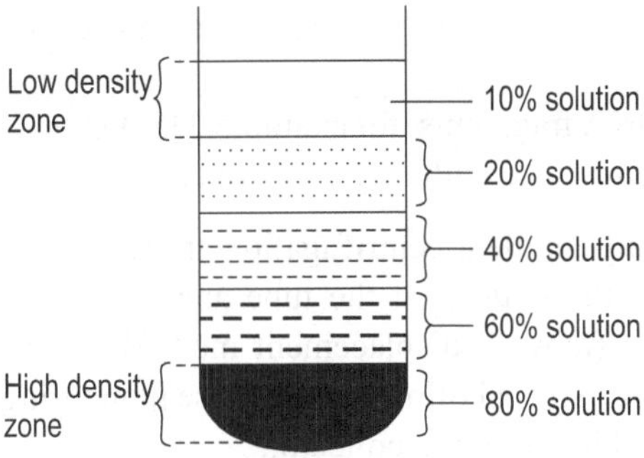

FIGURE 14.30 Step gradient

One millilitre of the high density solution is collected into a centrifuge tube through the outlet and an equal volume of the low density solution is allowed to pass from the donor chamber into the mixing chamber and stirred, so that the density of the solution in the mixing chamber falls a little. One millilitre of this solution is immediately collected in the centrifuge tube. This process of (i) liquid entry from the donor chamber, (ii) mixing in the mixing chamber, and (iii) collection in the tube is repeated in order to form a continuous linear gradient in the tube.

Once the density gradient is formed, about 0.5 to 1 cm^3 of the sample is pipetted along the sides of the centrifuge tube slowly. The pipette should be held 2 to 3 cm above the column at 45° to prevent the formation of air bubbles.

Applications Continuous gradient is used for separating ribosomal subunits, poly ribosomes, and viruses.

Types of density gradient centrifugation Density gradient centrifugation is of two types—rate zonal centrifugation and isopycnic centrifugation.

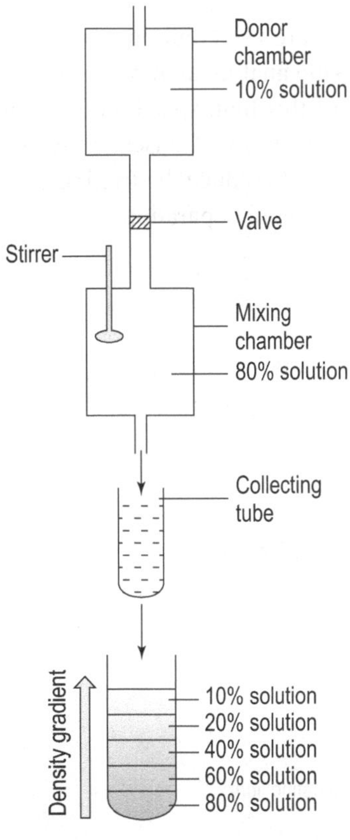

FIGURE 14.31 Formation of continuous linear gradient

Rate zonal centrifugation Rate zonal centrifugation requires a preformed step gradient. The maximum density of the step gradient should always be smaller than the density of the 'densest particle' present in the sample. The sample is placed on the top of the gradient and centrifuged in a swing bucket rotor, at a low rpm for shorter duration. During centrifugation, the particles are separated based upon their sedimentation coefficient to form distinct zones. Particles of higher density and larger size will be at the bottom zone, whereas tiny particles with smaller density will be at the top zone. This technique is used to separate enzymes, hormones, and RNA–DNA hybrids.

Isopycnic centrifugation The gradient forming solution is taken in a tube and mixed with the sample to form a homogeneous mixture. The maximum density of the gradient should be higher than the density of the densest particle present in the sample. The tube is sealed and centrifuged at high speed for several hours. The particles will sediment till the isopycnic position is reached. Isopycninc position is that region of the gradient where the buoyant density of the particle and the gradient density are same. At the isopycnic position, bands are formed.

This technique is used

- to separate subcellular organelles with different buoyant densities, such as Golgi bodies, mitochondria, and peroxisome.
- to find the buoyant density of the organelles.

- to segregate mixtures of DNA, RNA, and protein into different bands on caesium chloride linear gradient.
- for separating heterogeneous mixture of DNA fragments, for example, DNA strands with high G:C content can be separated from those with high A:T content.

Recovery of zones Zones formed during density gradient centrifugation are recovered by (i) pumping 60 to 70% sucrose solution to the bottom of the tube and displacing the zone upwards to recover them sequentially (upward displacement method) or (ii) puncturing the base of the tube to collect the zone using a hollow needle (base puncturing method). The recovered samples can be collected in a fraction collector.

Continuous flow centrifugation Continuous flow centrifuge has a mechanically operated hollow bowl or basket, equipped with a discharge valve and an outlet. This type of centrifuge can generate higher gravitational force to centrifuge excessive volume of crude sample extract at industrial scale. The crude extract is continuously pumped into the bowl and centrifuged and the supernatant is simultaneously collected via the outlet. The pellets are intermittently collected through the discharge valve. Approximately 100 litres of samples can be continuously centrifuged per hour. Continuous flow centrifugation is used for the large-scale separation of cell and cell debris from the crude tissue extract.

Analytical centrifugation

In analytical centrifugation, very small volumes of samples are used and pure particles are characterized. Analytical centrifugation is performed using analytical ultracentrifuge. It functions at a very high speed of 70,000 rpm. Because of this high speed the friction between the rotor and air becomes elevated, leading to heat generation. Consequently, the analytical ultracentrifuge is refrigerated and evacuated in order to reduce the friction. It is used for the characterization of biological samples rather than their separation.

It consists of two major components:

1. An armoured chamber
2. An optical system (Fig. 14.32)

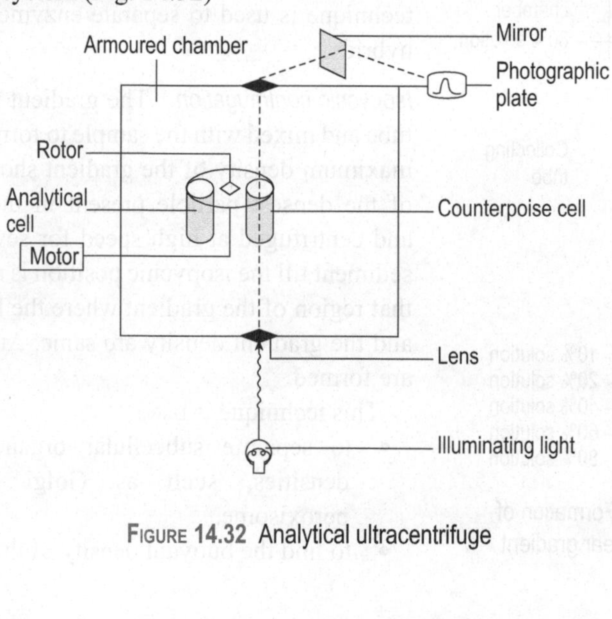

FIGURE 14.32 Analytical ultracentrifuge

The armoured chamber is evacuated and refrigerated and it accommodates a rotor consisting of analytical cell and counterpoise cell, and a thermistor, for measuring the temperature. The optical system is made of a light source, a lens system, a mirror, and a photographic plate.

The sample to be characterized is suspended in a suitable medium (solvent) and filled in the analytical cell and counter balanced by the counterpoise cell. The centrifuge is then operated at a high speed to sediment the particles. A parallel beam of light strikes the sedimenting particle. The light gets refracted at the sample solution, and monitors the sedimentation in the form of peaks. The area of the peak is proportional to the concentration of the sample. A shift in the peak or the movement of the peak gives information about the sedimentation velocity of the particles.

Three types of optical systems are present in the analytical ultra centrifuge. They are discussed as follows:

1. UV light absorption system—The sample is irradiated by UV light and the intensity of the transmitted light is determined and recorded as peak in the photographic plate
2. Schleiren optical system—When samples are separated into density zones, the samples are irradiated so that the light gets refracted in the zone. The refractive index is plotted against the distance of the analytical cell and shown as peak.
3. Rayleigh interference system—This system makes use of two slits. One slit is located behind the sample solution and the other behind the reference solution. Both the solutions are centrifuged and irradiated. Light gets refracted in both the liquids, and the refraction is monitored as interference fringes in the photographic plate.

Applications of analytical ultracentrifuge Analytical ultracentrifuge is used to characterize the molecular weight and conformational changes of macromolecules.

Determination of molecular weight using the sedimentation coefficient The sample of unknown molecular weight is centrifuged in the analytical cell of the rotor. During centrifugation, the sample molecules move radially outwards and create a boundary, which is irradiated and monitored as peak by UV light or Schleiren optical system. The movement of the peak is proportional to the sedimentation rate (v) at a given time (t). The sedimentation coefficient (S) is determined using Eqn (14.7). S is then substituted in the following to obtain the molecular weight:

$$M = [RTS / D\,(1 - v)\,\rho] \tag{14.9}$$

where

M = anhydrous molecular weight of the molecule
R = molar gas constant
T = absolute temperature in Kelvin
S = sedimentation coefficient
D = diffusion coefficient of the molecule
v = partial specific volume of the molecule
ρ = density of the medium at 20°C

Determination of molecular weight by sedimentation equilibrium method Sample solution and a suitable reference solvent are placed in a double sector analytical cell and centrifuged at high speed until there is no net sedimentation of the solute throughout the length of the cell. The centrifugation event is monitored by Rayleigh interferometer system. The system generates fringes. From the fringes, the concentrations C_1 and C_2 of the sample at a distance r_1 and r_2 from the centre of rotation is calculated. The molecular weight is then determined using the following formula:

$$M = 2RT \ln (C_1 - C_2) / \omega^2 r (1 - v\rho) (r_2^2 - r_1^2) \qquad (14.10)$$

Detection of purity of macromolecules The macromolecule solution is centrifuged and the sedimentation is followed by the Schleiren optical system and the peak is observed. Appearance of a single sharp peak represents the purity of the macromolecule. Appearance of an additional peak, shouldering of the main peak, and asymmetric peak are evidences of impurity.

Determination of conformational changes in macromolecules The sedimentation rate is higher for native macromolecules. However, when there is any conformational change in the native molecule, it becomes disorganized leading to a fall in sedimentation rate. Therefore, difference in sedimentation rate before and after the conformational change gives an idea about the extent of conformational changes. For example denaturation of DNA and enzymes (e.g., aspartate transcarbamoylase) can be detected.

14.4 CHROMATOGRAPHY

Chromatography is a separation technique used for the isolation of a wide range of compounds depending on their difference in distribution between two immiscible phases, namely the *stationary phase* and the *mobile phase.* Biomolecules such as amino acids, carbohydrates, fatty acids, hormones vitamins, lipids, mineral ions, plant pigments, and cell organelles can be separated by chromatography. Chromatography can be done in the paper mode (paper chromatography), thin layer mode (thin layer chromatography), or column mode (column chromatography). In paper mode, separation is based on partition of the test compounds between two phases—the mobile phase and the stationary phase (Section 14.4.2). In thin layer chromatography, separation is based on partition principle or adsorption of the compounds on a stationary phase. In column chromatography, particles are separated based on various principles, such as liquid–liquid partition, gas–liquid partition, adsorption, ion exchange, biological affinity, or molecular exclusion (Section 14.4.2). The stationary phase, the mobile phase, and the separation principle of different types of chromatography are summarized in Table 14.11. Specific applications of each type of chromatograph are mentioned under corresponding sections discussed subsequently (Table 14.12).

14.4.1 Basic Principles of Chromatography

In chromatography, compounds are separated based on their distribution between two immiscible phases called the stationary phase (a liquid or a solid) and the mobile phase (a liquid or a gas). The ratio of the concentration of the compound present in one phase to that present in the other phase is called as the *distribution coefficient* (k_d) or *partition coefficient* of that compound.

$$k_d = \frac{\text{Concentration of the compound in one phase}}{\text{Concentration of the compound in the other phase}}$$

TABLE 14.11 Overview of chromatography

Type of chromatography	Nature of the stationary phase	Nature of the mobile phase	Principle of separation
Paper	Water entrapped in cellulose paper	Organic solvent:water mixture	Liquid–liquid partition
Thin layer	Water entrapped in silica layer	Organic solvent:water mixture	
	Water present in alumina layer	Adsorption	Partition between two liquids
Column			
Normal phase partition chromatography	Polar liquid associated with silica	Non-polar liquid	Partition between two liquids
Reverse phase partition chromatography	Non-polar liquid associated with silica	Polar liquid	Partition between two liquids
Gas–liquid chromatography	High boiling point organic liquid, bonded to silanized celite	Gas	Partition between gas and liquid phases
Adsorption chromatography	Alumina or silica	Organic solvents, such as acetone, alcohol, or hexane	Adsorption of test compounds via electrostatic or van der Waal's interaction or hydrogen bond
Ion exchange chromatography	Charged matrix of agarose, sepharose, polyacrylamide or polystyrene	Buffer containing metal ions	Exchange of ion between another ion and a matrix
Gel filtration chromatography	Porous glass beads Porous glass granules	Buffer	Molecular exclusion or molecular sieving through pores of gel beads
Affinity chromatography	Matrix of agarose or polyacrylamide or polystyrene, provided with specific ligands	Buffer with metal ions	Biological affinity between the ligand and the target molecules

TABLE 14.12 Applications of chromatography in cell biology

Chromatography	Applications
Adsorption chromatography	Used to separate triglycerides, parathyroid hormone, amino acids, and vitamins
Hydrophobic interaction chromatography	Used to separate nucleic acids and proteins from cell extract
Hydroxyapatite chromatography	Used to resolve double-stranded and single-stranded DNAs
Partition chromatography	Used to resolve fats and proteins
Chiral chromatography	Cell fractionation
	Cell organelle fractionation
	Isolation of membrane receptors
	Separation of acidic and basic drugs in pharmaceutical industries

(*Contd*)

TABLE 14.12 (Contd)

Chromatography	Applications
Counter current chromatography	Used to separate hormones, nucleic acids, antibiotics, vitamins, various tRNAs
Ion exchange chromatography	Used to – separate amino acids, polysaccharides, nucleotides, vitamins, proteins and enzymes, metal ions – prepare deionized water
Gel filtration chromatography	Used to – resolve compounds of different molecular weights – determine the molecular weight of proteins
Affinity chromatography	Used to separate proteins, enzymes, immunoglobins, nucleotides, and nucleic acids Used for Fractionation of cell mixture into homogenous population
Lectin affinity chromatography	Used to separate glycoproteins and membrane receptors
Immunoaffinity chromatography	Used to separate viruses
Dye–ligand chromatography	Used to separate kinases, dehydrogenases, and DNA polymerases
Covalent chromatography	Used to separate thiol-containing proteins
HPLC chromatography	Used to – identify and resolve unknown fatty acids present in fat hydrolysate – calculate the retention index of a compound – analyse polyhalogenated compounds – analyse nitrogen- and phosphorus-containing compounds – identify functional groups of compounds – determine the position of double bond in unsaturated organic compounds

There is another related term called the effective distribution coefficient. It is given as the product of kd with the ratio of the volume of the two phases. In general, the compound to be separated is called the solute, whereas the mobile phase is called the solvent.

14.4.2 Types of Chromatography

Chromatography is divided into the following three types:

1. Paper chromatography (works on the partition principle)
2. Thin layer chromatography (separates compounds on the basis of the adsorption principle)
3. Column chromatography (separation is performed via partition, adsorption, ion exchange, molecular exclusion, or affinity principles)

Paper chromatography

In paper chromatography, the stationary phase and the mobile phase are both liquids. The stationary phase is water present around the cellulose granules of the filter paper, and the mobile phase is an organic solvent saturated with water. The solvents used in paper chromatography are listed in Table 14.13. The sample mixture (the solute) is separated on the basis of difference in the partition coefficients.

TABLE 14.13 Solvents used in paper chromatography

Solutes separated	Solvent mixture	Ratio of the solvent
Amino acids	*n*-butanol:acetic acid:water	40:10:50
	n-butanol:pyridine:water	33:33:33
	methanol:pyridine:water	25:12:63
Monosaccharides	*n*-butanol:pyridine:water	50:28:32
Disaccharides	*n*-butanol:ethanol:water	52:35:15
Chlorophyll	*n*-propanol:petroleum ether	4:96
Carotenoids	Chloroform:petroleum ether	30:70

Technique A rectangular piece of Whatman No. 1 filter paper of 15 to 30 cm length and 1 to 5 cm width is used as the stationary phase. A horizontal line is marked at 5 cm from the bottom edge of the paper as shown in Fig. 14.33. Cross marks of 1.5 cm apart are drawn over the line. The number of cross marks is equivalent to the number of sample compounds to be resolved. Approximately 50 μl of the sample solution is loaded in a capillary tube or micro syringe applied on the cross marks as a spot and the spots are dried. The spotted paper is further developed by (a) ascending development, (b) descending development, or (c) disc development techniques.

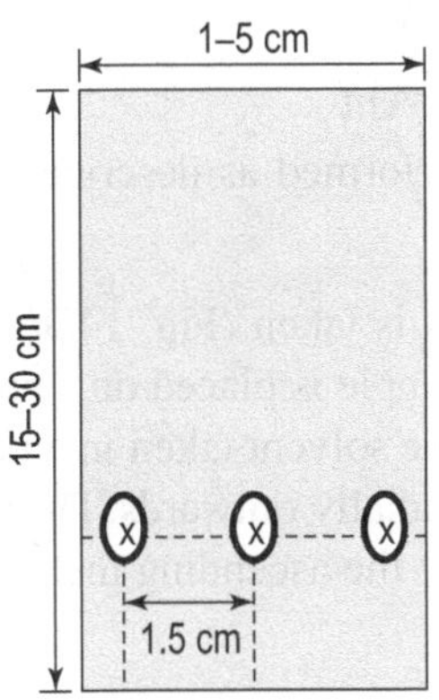

FIGURE 14.33 Paper chromatography

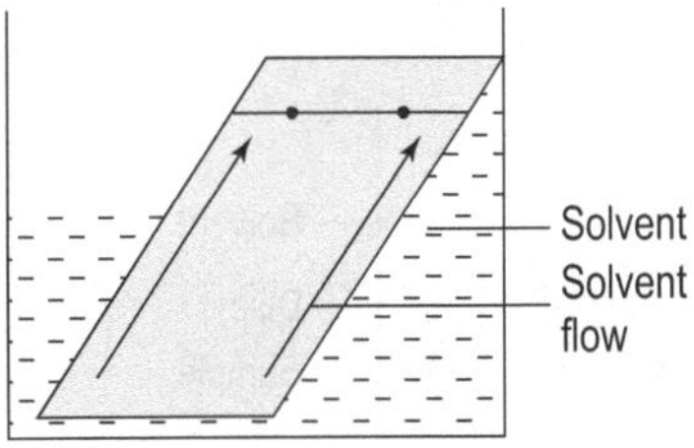

FIGURE 14.34 Ascending development

Ascending development In ascending development method, a solvent is taken in a tank. Solvents used for separating different solutes are shown in Table 14.13. The top of the paper is supported by a clip, and the other end is vertically immersed into the solvent with the spots lying above the solvent level (Fig. 14.34). The solvent moves vertically upwards due to capillary action. A propelling force is developed that pulls the sample along the solvent flow. The compounds that are highly soluble in the solvent move faster than the compounds with lesser solubility. Difference in partition of substances between the mobile phase and the stationary phase can also lead to separation. The distance moved by the solvent (approximately 2 cm from the top edge) is called solvent front. The paper is dried in a dry air oven or air drier. The dry paper containing the solvent front is known as a chromatogram. Compounds separated on the paper can be located and identified by different methods depending on the nature of the compounds.

Hydrophobic compounds are identified as waxy spots by spraying water. Fluorescent compounds are seen as dark spots by exposing the chromatogram to the UV radiation. Unsaturated compounds can be recognized as brown spots by exposing the chromatogram to iodine vapours. Amino acids are located as purple spots after spraying with 0.1% ninhydrin in acetone and heating the chromatogram at 100°C for

3 to 10 min. Carbohydrates can be identified by plunging the chromatogram in alkaline silver nitrate solution. Radiolabelled samples can be seen as dark areas by exposing to X-rays. After the location, the distance moved by the solute (the solute front) is measured from the origin.

The compounds located on the paper are determined by calculating their retardation factor (*Rf*) value and comparing with the Rf value of a standard compound (*Rf* value is constant for a given compound).

$$Rf = \frac{\text{Distance travelled by the solute}}{\text{Distance travelled by the solvent}}$$

The spots in the chromatogram can be quantified by the following two methods:

1. On-paper quantification—In this method, the intensity of the spot is measured using a densitometer. The intensity is directly *proportional* to the amount of the compound distributed in the spot.
2. Off-paper quantification—In this method, the spot is scraped from the chromatogram and dissolved in a suitable solvent in a flask and the compound is eluted. The eluant is then centrifuged and quantified by colorimetry or fluorimetry (refer Section 14.7).

Descending development The spotted paper is immersed in a trough of solvent present in the upper chamber of a tank. As a result, the solvent migrates vertically downwards under gravity and capillary action, as represented in Fig. 14.35. Once the solvent reaches the end of the paper, the Rf value is calculated using the following formula:

$$Rf = \frac{\text{Distance travelled by the solute}}{\text{Distance travelled by the solvent}}$$

Location, identification, and quantification protocols are performed as described for the ascending method.

Disc development In this method, a circular paper (or a disc) is taken (Fig. 14.36). A wick is cut from the edge towards the centre of the disc. The sample is placed on a cross mark drawn in the centre of the disc. The disc is plunged into the solvent taken in a petri plate. The solvent flows along the wick and drags the sample radially outwards. Drying, location, identification, and quantification can be done similar to the ascending method.

Uses Paper chromatography is used:

- to resolve carbohydrates and amino acid mixture.
- for qualitative analysis of amino acids.
- to separate plant pigments, such as chlorophyll and carotenoids.

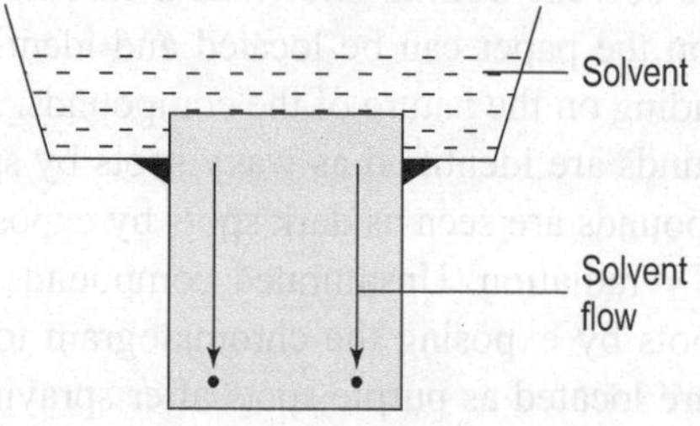

FIGURE 14.35 Descending development

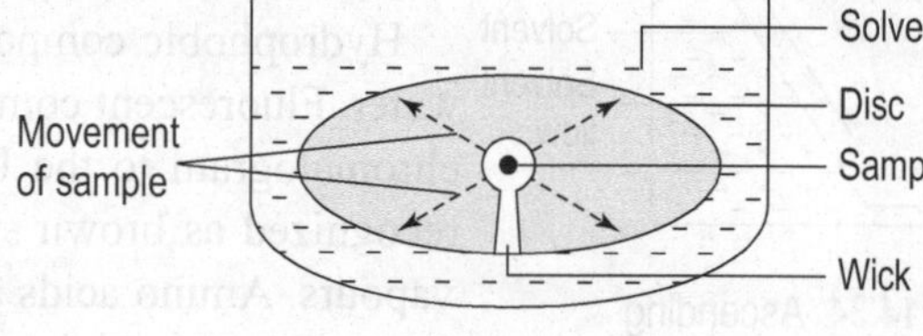

FIGURE 14.36 Disc development

Thin layer chromatography

Thin layer chromatography (TLC) can be carried out by using a thin layer of alumina (adsorption principle) or silica gel (partition principle). The stationary phase is spread as thin layer on a glass plate or metal or plastic foil. TLC is superior to paper chromatography since it gives rapid separation, higher sensitivity, and good resolution. It separates a wide range of compounds.

Technique

The adsorbent is blended with 10 to 15% gypsum (binding agent) and water in a flask and shaken vigorously to form a thick, homogenous slurry. The choice of the adsorbent is determined by the nature of the compounds to be separated (Table 14.14).

TABLE **14.14** Stationary phases used in TLC

Stationary phase/Adsorbent	Compounds separated
Silica gel	Amino acids, fatty acids, steroids, phospholipids
Hydroxyapatite	Protein
Kieselghur G	Monosaccharides, disaccharides, and triglycerides
Cellulose powder	Nucleotides

Two glass plates or foils are dipped in the slurry taken out slowly in a vertical position, to form a thin layer. The plate is then activated by drying in an oven at 110°C for two hours. Activated plates are called chromatoplates. The thin layer should be 2.5 mm for preparative separation and 0.25 mm for analytical separation. Alternatively, chromatoplates can be prepared by spreading the slurry via applicator or glass rod or pipette. Commercially available poly β-caprolactum plates are used for running the standards and the unknown compounds simultaneously.

Loading of sample Baseline and cross marks are drawn on the plate as explained for paper chromatography. The sample is dissolved in 0.5 to 1 ml of volatile solvent and placed in the plate.

Development Chromatoplates are developed by ascending or descending methods. The solvent is allowed to travel a distance of 10 to 12 cm above the origin for about one hour. Nature of the compounds to be separated determines the nature of the mobile phase to be used (Table 14.15).

TABLE **14.15** Mobile phases used in TLC

Mobile phase	Ratio	Compounds separated
n-butanol:acetic acid:water	8:2:2	Amino acids
Chloroform:methanol:water	2:2:1	Amino acids
n-butanol: acetone:water	4:5:1	Mono, di, tri, and tetrasaccharides
n-butanol:ethanol:water	5:3:2	Polysaccharides
Hexane:diethylether:acetic acid	80:20:1	Lipids
Petroleum ether:diethylether:acetone	90:10:1	Triglycerides

As the mobile phase moves, the compounds which are strongly adsorbed by the stationary phase remain nearer to the point of origin. Compounds that are weakly adsorbed by the stationary phase are carried by the mobile phase along the plate and separated. Alternatively, compounds which are more soluble in the solvent (mobile phase) are carried faster than the compounds with poor solubility, thus leading to separation.

Compounds separated in the TLC are identified and the Rf value calculated as described for paper chromatography.

Uses TLC is used:

- to separate nucleic acids.
- to identify vitamin A.
- for the qualitative, quantitative, small scale, and preparative separation of trace elements, amino acids, steroids, terpenoids, nucleic acids, and carbohydrates.
- for detecting the presence of impurities in any biological preparations.
- to resolve plant pigments such as chlorophyll and carotenoids
- to identify C^{14} labelled compounds in tracer technique

Column chromatography

In column chromatography, a column of fine granular material is used for separation. Compounds are separated based on any one of the following principles: adsorption, partition, ion exchange, molecular exclusion, or biological affinity (Table 14.11). Depending on the mode of separation, the mobile phase and the stationary phase column chromatography is divided into seven types (Table 14.11). The column packing protocol, the method for sample application, recovery, detection, and collection of the separated compounds are all generally similar for all types of chromatography.

General protocol for column preparation A thin-walled pyrex glass column of length 15 to 20 cm and width 1 cm, closed by a cotton wool plug at the lower end is used for chromatography. Column preparation has the following four steps:

Column packing The column is packed with a slurry made by mixing the stationary phase (porous granular materials, such as agarose, cellulose, dextran, polyacrylamide, polystyrene, or silica) and the mobile phase (wet packing, alternatively, the stationary phase alone is packed by suction under gravitation dry packing).

Packed column is then equilibrated using the mobile phase without forming air bubbles. The column surface is always covered with the mobile phase and provided with filter paper disc, nylon gauze, rayon gauze, or a plunger. The column is connected to a detector, recorder, and a fraction collector by capillary tubing as illustrated in Fig. 14.37.

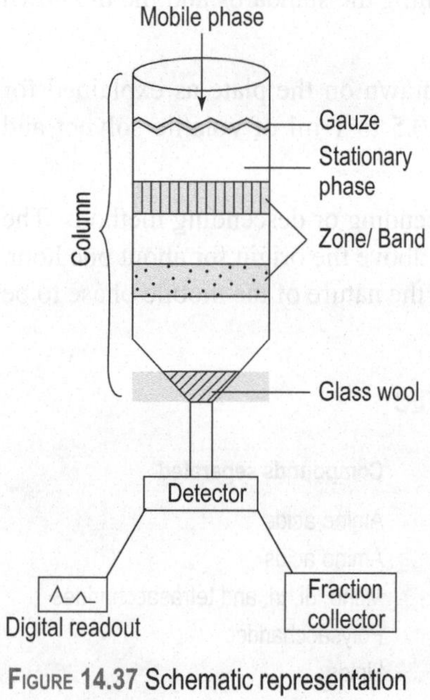

FIGURE 14.37 Schematic representation of column chromatography

Sampling The sample to be resolved is dissolved in a small volume of the solvent (mobile phase) and placed over the column surface using a capillary tube or microsyringe or a micropipette.

The mobile phase is constantly percolated into the column at a steady rate using a low-pressure peristaltic pump. As a result, the sample mixture is separated within the column and distributed in the form of zones or bands.

Recovery of bands The zones are recovered by eluting the column with suitable liquid called eluting agent. (Usually, the solvent itself can be used as an eluting agent.) If the pH, ionic strength, and polarity of the eluting solvent are constant, it is called isocratic elution. If the pH, ionic strength, or polarity of the solvent is continuously altered during the process of elution, it is called gradient elution. Gradient elution is done by adding acetic acid, ammonium hydroxide, or 1 M NaCl to the solvent. For separating macromolecules, the column is eluted by adding a second type of compound for which the column has greater affinity than the compound to be separated. For separating enzyme, the column is eluted with a high concentration of the substrate or the reversible inhibitor. Chromatography can be performed under a pressure of 5 bar, (low-pressure liquid chromatography [LPLC]) or under a high pressure of greater than 50 bars, (high-pressure liquid chromatography [HPLC]).

Detection UV spectrometer, fluorescent spectrometer, colorimeter, refractive index monitor, flame ionization detector, thermionic detector, or electrochemical detector detect the eluted compounds and produce a specific signal. This signal is registered in a recorder. The signals are optical density (for spectrometer), refractive index (for refractive index monitor), electrons generated (for flame ionization detector), and current drop (for electron capture detector). A plot of the signal versus time or volume generates a series of peaks (chromatogram). Areas under the peak is proportional to the quantity of each compound. A fraction collector segregates the detected compounds into small fractions of 1 to 10 cm^3.

During the separation process, the time taken by the compound to emerge from the column is called the retention time, *TR*. The volume of the solvent required for eluting the compound under standard condition is called the retention volume *VR*. The following equation relates *TR* and *VR*:

$$VR = TRF$$

where *F* is the flow rate of the solvent.

In the following sections, we focus on the working principle and the applications of different types of column chromatography.

Adsorption chromatography

Adsorption chromatography is used to separate compounds based on the difference in their adsorption to the stationary phase (the adsorbent).

Technique Adsorbents are finely divided porous granular material capable of adsorbing particles via electrostatic interaction, hydrogen bonding, van der Waal's forces, or dipole interaction. The column is packed with the adsorbent (silica, charcoal, or alumina). Acidic dilanol group (Si–OH) present in the silica is ideal for separating basic compounds. Alumina is used for separating acidic compounds due to the presence of a basic group. The sample is mixed with the mobile phase and placed on the column surface. Choice of the mobile phase depends on the functional group present in the sample compound. Alcohol is the mobile phase for separating compounds with hydroxyl group. Acetone or ester is used for separating compounds containing carbonyl group, whereas hexane, heptane, or toluene can be used for compounds possessing non-polar groups. The sample components are resolved into zones, and eluted by gradient elution or affinity elution, and detected by UV or visible spectrophotometer or colorimeter.

Applications Adsorption chromatography is used:

- to separate triglycerides, parathyroid hormone, amino acids, and vitamins.
- to separate nucleic acids and proteins from cell extract.
- to resolve single-stranded, double-stranded, linear, circular, and super helical forms of DNA. This technique is called hydrophobic interaction chromatography (HIC).
- to resolve double-stranded and single-stranded DNAs. This technique is called as hydroxyapatite chromatography (HAC).

Hydroxyapatite chromatography

A column of hydroxyapatite [Ca_{10} (PO_4) OH_2] is used as an adsorbent to separate DNAs. The adsorbent is placed on agarose matrix column. Cell homogenate (sample) prepared in 20 mM phosphate buffer is placed on the column surface. At 20 mM ionic strength, the adsorption of DNA is greater than that of protein and RNA. Therefore, DNA is adsorbed in the calcium and phosphate ions of the hydroxyapatite column, whereas protein and RNA can be washed out using more buffer. Subsequently, single-stranded DNA is desorbed using a buffer of slightly higher concentration. Finally, double-stranded DNA is desorbed using the buffer of still higher concentration.

Hydrophobic interaction chromatography

Hydrophobic interaction chromatography (HIC) is useful to exclusively separate and purify proteins. Separation occurs due to difference in interaction between the surface-specific hydrophobic amino acids of the protein and the adsorbent.

Proteins in the cell extract are precipitated using ammonium sulphate in order to expose their hydrophobic amino acid residues. The precipitate is dissolved in buffer (whose pH is equal to the isoelectric point of the protein) and placed on the column of hexyl, octyl, or phenyl groups supported on agarose column. The pH of the buffer should be equal to the isoelectric point of the protein. As each protein has specific hydrophobic amino acids their adsorption also differs. Thus, protein can be separated due to difference in their hydrophobic interaction with the adsorbent.

The proteins are eluted with the buffer of decreasing the ionic strength or increasing pH (gradient elution). Alternatively, non-ionic detergent with higher affinity for the adsorbent, such as Tween 20 or Triton X-100 can be passed through the column to displace the protein.

Partition chromatography (liquid–liquid chromatography)

Compounds of different polarities are separated by partition chromatography. Separation is based on partition between a liquid stationary phase and a liquid mobile phase The liquid stationary phase is held in a column of finely divided granular material As the stationary phase and the mobile phase are both liquid, partition chromatography is also known as liquid–liquid chromatography.

Partition chromatography may be a normal phase liquid chromatography (when the stationary phase is a polar liquid, and the mobile phase is a non-polar) or reverse phase liquid chromatography (where the stationary phase is a non-polar liquid, and the mobile phase is a polar liquid).

Technique Table 14.16 represents the stationary and mobile phases used for partition chromatography. The stationary phase is chemically bonded to a silica column. The sample containing compounds of different polarity is placed on the column surface, and then the mobile phase is passed through. The column is then eluted with dioxane or methanol of increasing polarity. In normal phase chromatography, non-polar compounds are eluted first, whereas the most polar compounds come out. In reverse phase, highly polar compounds come first, whereas the least polar compounds come out last.

TABLE 14.16 Stationary and mobile phases used for partition chromatography

Partition mode	Stationary phase		Mobile phase	
	Polar	Non polar	Polar	Non polar
Normal	Alkyl amine	–	–	Hexane Heptane Dichloromethane Ethyl acetate
Reverse	–	Butyl silane Octyl silane	Water Aqueous buffer Acetonitrile Methanol	–

Applications Partition chromatography is used to separate fats, proteins, and chiral compounds (chiral chromatography).

Chiral chromatography

Chiral chromatography resolves the enantiomeric forms of chiral compounds. Functional groups in the enantiomers of chiral compounds are derivatized using chiral derivatizing agent (CDA), and resolved into diastereomers. Diastereomers form a transient diastereomeric complex with CDA given as follows:

$$(R + S) + CDA\text{-}R' \text{ (Enantiomers)} \longrightarrow R\text{–}R' + S\text{–}R' \text{ (Diastereomers)}$$

CDA functions as a stationary phase.

Technique Compounds such as phenyl glycine or triacetyl cellulose or cyclodextrin are used as CDA, and bonded to a silica column. The enantiomeric sample is placed on the surface of the column. The rectus (R) and the sinister (S) form of the enantiomer (sample) have chiral centres and possess their functional groups with different spatial arrangement. CDA is used to derivatize the functional groups and resolve them sample into diastereomers.

CDA possess an open truncated conical structure with the secondary hydroxyl group present on its rim, thus, forming a chiral cage or chiral cavity. Diastereomers interact with the cage via hydrogen bonding or van der Waals force forming a transient diastereomeric

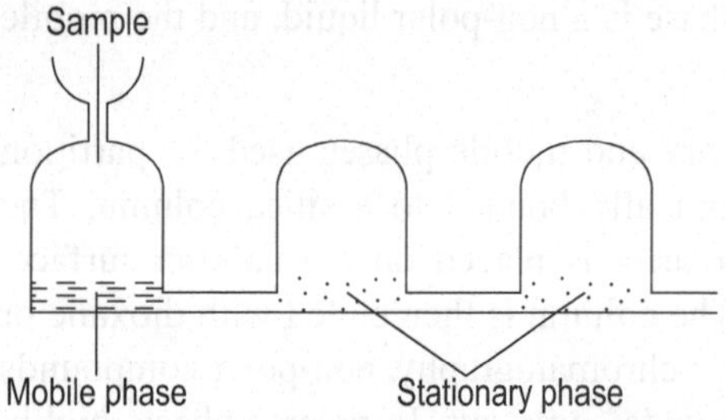

FIGURE 14.38 Craig's countercurrent apparatus

complexes, R–R′ and S–R′. The complex is eluted by reverse phase gradient elution and detected by UV spectrometer.

Applications Chiral chromatography is used for cell fractionation, cell organelle fractional ion, and isolation of membrane receptors.

Countercurrent chromatography

Counter current chromatography (CCC) is a partition chromatography that separates particles on the basis of difference in partition coefficient between two immiscible phases filled in a special type of apparatus called Craig's counter current apparatus. (Fig. 14.38)

The apparatus is made of 30 to 100 H-shaped vessels, interconnected to each other. The first vessel is filled with the mobile phase, whereas the other vessels are filled with the stationary phase. The mobile phase is equilibrated with the sample by rocking the vessel through 90°. The equilibrated mobile phase is then allowed to enter the next and mixed with the stationary phase. Then, the first vessel attains the original partition, and a fresh mobile phase is filled in. The process is repeated to progressively transfer the mobile phase along the series of stationary phase. Compounds with different partition coefficient get accumulated in a specific vessel.

Applications CCC is used for the isolation and purification of

- hormones
- nucleic acids
- antibiotics
- vitamins
- various tRNAs

Ion exchange chromatography

Ion exchange chromatography is used to separate biomolecules with net positive or negative charge based on attraction between the opposite charges present in the sample and the stationary phase. The stationary phase is the solid matrix called an ion exchanger.

Ion exchanger Ion exchanger is an inert matrix with covalently bound ions that are electrostatically linked to exchangeable counter ions. A matrix that has a covalently bound positive ion is called a cationic exchanger (Fig. 14.39a). Compounds that are stable below their isoelectric point are negatively charged. They are separated by a cationic exchanger. A matrix that has a covalently bound negative ion is called an anionic exchanger (Fig. 14.39b). Compounds which are stable above their isoelectric point are positively charged, and they can be separated by an anionic exchanger.

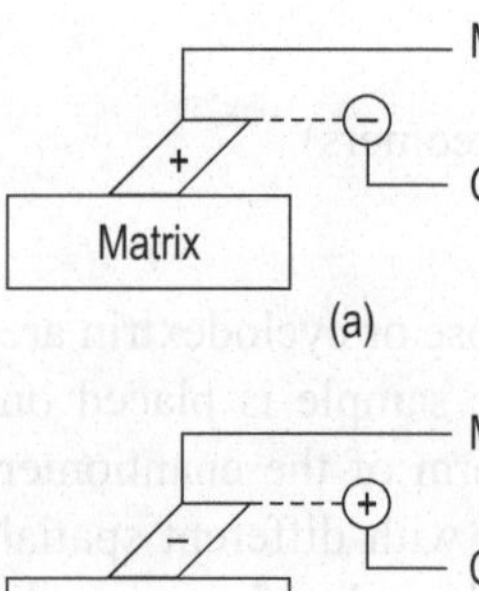

FIGURE 14.39 Ion exchangers (a) Cationic exchanger (b) Anionic exchanger

The number of exchangeable counter ions present in one gram of dry exchanger is the total exchange capacity of the exchanger. Ion exchangers such as polystyrene, agarose, and cellulose are used under low-pressure, whereas the bonded phase porous silica and hydrophilic polyether are used for

HPLC. Depending on the functional groups, ion exchangers are divided into strong cationic, weak cationic, strong anionic, and weak anionic (Table 14.17).

TABLE 14.17 Types of ion exchangers

	Cationic exchanger		Anionic exchanger	
	Strong	Weak	Strong	Weak
Example	Cellulose	Agarose	Cellulose	Agarose
	Dextrose	Cellulose	Dextrose	Cellulose
	Polystyrene	Dextrose	Polystyrene	Agarose
Functional group	$-SO^{3-}$	–OH	$(C_2H_5)_3$–N′–C_2H_5–	(C_2H_5)2–NH–C_2H_5–
	COO-		$(CH_3)_3$–N+–CH_2–	
	C_6H_5OH			
	CH_2COO–			

Separation strategy When a compound is stable at a pH greater than its pI, anion exchanger is used. When a compound is stable at a pH less than its pI, cation exchanger is used. This separation strategy is represented in Table 14.18.

TABLE 14.18 Separation strategy in ion exchange chromatography

Resin type	Charge in the resin	Charge of the target compound to be separated	pH condition
Cationic exchanger	–	+	< pI of target compound
Anionic exchanger	+	–	> pI of target compound

Technique Ion exchangers are allowed to swell by treating with 0.5 N NaOH and 0.5 N HCl and suspended in water to expose the functional group. They are then equilibrated with a suitable counter ion and packed into a column. Cationic exchanger is equilibrated with anionic buffer, such as acetate, barbiturate, or phosphate. Anionic exchanger is equilibrated with cationic buffers such as tris, pyridine or alkyl amine are used for anionic exchanger.

The mobile phase used for ion exchange chromatography is a buffer with pH one unit above or below the isoionic point (the pH at which the compound exist as zwitter ion) of the compound to be separated. The sample is dissolved in the mobile phase and diffused into the column at a rate of about 20 to 30 ml/h. At this stage, the charged group of the sample compound is exchanged for the counter ion, so that the compound gets reversibly adsorbed to the column.

The sample compound bound to the column is desorbed and eluted by percolating the column with the buffer of eluting pH. Eluting pH is the pH at which all the bound samples are completely desorbed from the exchanger without loss of activity. Binding pH and eluting pH are predetermined. Gradient elution is ideal for the separation of amino acids from protein hydrolysate. Acidic amino acids, neutral amino acids, and basic amino acids

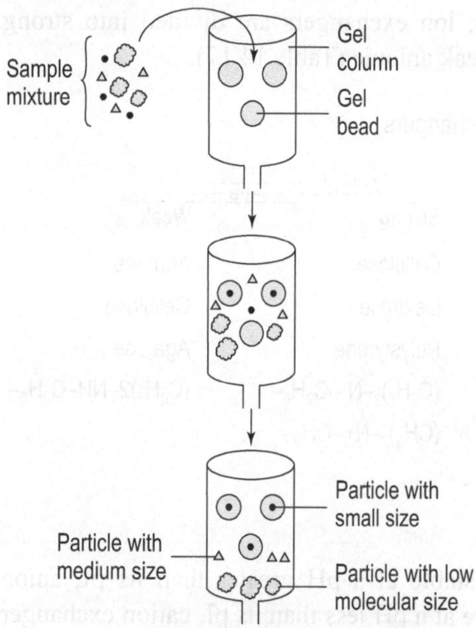

FIGURE 14.40 Principle of gel filtration chromatography

are eluted sequentially by using a buffer of increasing pH and ionic strength. Eluted compounds are transformed into coloured compounds via chemical reaction and detected using colorimeter or fluorescence spectrometer.

Applications Ion exchange chromatography is used:

- to separate amino acids, polysaccharides, nucleotides, and vitamins.
- to purify proteins and enzymes.
- to prepare deionized water.

Gel filtration chromatography

Gel filtration chromatography (also called gel permeation chromatography, gel exclusion chromatography or molecular sieve chromatography) is used to separate compounds of different molecular weight or molecular size. This technique is also useful in the determination of molecular weight of proteins.

A column of porous gel beads (stationary phase) is equilibrated with a liquid (mobile phase) that is distributed inside and outside the gel pores (sieves). As the sample mixture is passed into the gel filtration column, the compounds of high molecular weight are totally expelled from the sieve and move down the column at a faster rate. On the other hand, compounds of low molecular weight are fully retained in the mobile phase present in the sieve and move down the column at a slower rate. Compounds with intermediate molecular weight bypass the interstitial space between the gel beads and travel down at moderate speed. The principle behind gel filtration chromatography is shown in Fig. 14.40.

Porous beads Porous glass granules or porous gel beads made from polymeric organic compounds constitutes the stationary phase of gel filtration chromatography (Table 14.19). Pores of gel beads create a three-dimensional cross-linked network that possesses molecular sieving property.

Packing of gel column The column material is first degassed under vacuum and allowed to swell up by adding 0.1 M NaCl, thus, forming a thick, semisolid mass called slurry. The slurry is then filled into a column without air bubbles and settled under gravity. Longer column with smaller diameter is ideal for better resolution. Following are the parameters related to gel filtration column:

- Total bed volume, V_t (The total length of the column containing the gel beads)
- Void volume, V_o (the external volume of the gel bead)
- Matrix volume, V_m (the volume of the gel surface)
- Internal volume, V_i (the volume of the gel pores)
- Total liquid volume of the gel bed, V (the volume of the liquid required to elute low-molecular weight compounds that are equally distributed in V_i and V_o)

TABLE 14.19 Stationary phases used in gel filtration chromatography

Chemical name	Commercial name
Agarose	Sepharose
	Biogel-A
Polyacrylamide	Biogel-P
Dextran	Sephadex
	Sephacryl
Polystyrene	Biobead-S

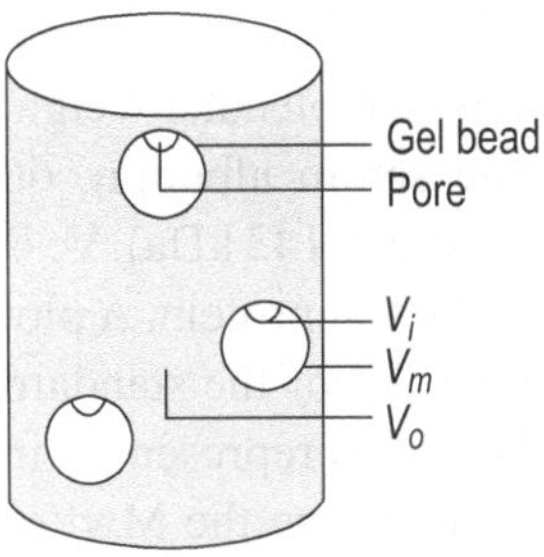

FIGURE 14.41 Parameters of gel column

The quantities V_t, V_o, V_m, V_i, and V are represented in Fig. 14.41. These parameters are related as per the following equations:

$$V_t = V_o + V_m + V_i \tag{14.11}$$

$$V = V_o + V_i = V_t - V_m \tag{14.12}$$

V_o can be calculated experimentally by percolating a substance that is completely excluded by all types of gel beads (i.e., a substance with K_d value distribution coefficient = 0). For example, blue dextran has K_d value of '0'. V_m can be calculated by percolating a substance which is equally distributed between V_i and V_o, (i.e., a substance with K_d value of 1).

A gel column is used

- for separating particles of different molecular weights.
- to determine the molecular weight of an unknown compound, for example, protein.

Application of gel filtration in resolving particles of different molecular weights There are two different ways of expressing protein size: molecular weight and molecular mass. Both quantities are functionally equivalent. Molecular weight is the ratio of the mass of a protein to 1/12 of the mass of carbon 12. It is unitless. Molecular mass refers to the atomic mass of the protein. It is expressed in daltons (Da). Sometimes molecular weight values are also suffixed with kilodaltons as unit. For example, the molecular weight of bovine serum albumin is 66 kDa. Gel filtration chromatography is used to resolve particles of different molecular weight. For this purpose, the gel column is equilibrated with 2 to 3 bed volumes of the mobile phase (bed volume is the total volume of the column packed with the gel). The column is loaded with a small volume of the sample and percolated with excess of mobile phase. The mobile phase is distributed inside and outside the gel pores. As a result, the sample compounds distribute between the inner and the outer mobile phase. Higher molecular weight compounds are completely distributed in the outer mobile phase, with the distribution coefficient value equal to zero ($K_d = 0$). Low-molecular weight compounds distributed in the inner mobile phase with the distribution coefficient value equal to one ($K_d = 1$). Intermediate molecular weight compounds acquire partial access to the inner mobile phase with the distribution coefficient value between zero to one ($K_d = 0$ to 1).

Therefore, compounds of different molecular weight have different K_d values and can be resolved using gel filtration chromatography. The resolved compounds can finally be eluted using different elution volume and detected using a suitable detector. The volume of the buffer or the mobile phase required for eluting a particular compound with a K_d value between 0 and 1 is called elution volume, V_e. V_e is represented by the following equation:

$$V_e = V_o + K_d \times V_i \tag{14.13}$$

V_e value for two compounds of dissimilar molecular weight and hence different Kd values say K_d' and K_d'' is represented by the following equation:

$$V_e = (K_d' \; Kd'')V_i \tag{14.14}$$

Application of gel filtration for the determination of molecular mass of a protein Molecular weight of the protein separated by gel filtration is calculated based on the linear relation between log molecular mass and V_e. The equilibrated column is percolated with standard protein of known molecular weight (e.g., β-galactosidase, 116 kDa; bovine serum albumin, 66 kDa; carbonic anhydrase, 29 kDa; or lysozyme, 143 kDa). Ve is calculated for the standard proteins and the test protein. A plot of log molecular weight (log Mwt) versus Ve of the standard protein is drawn to obtain a linear relation, as represented in Fig. 14.42. The plot is extrapolated to determine the Mwt of the test protein.

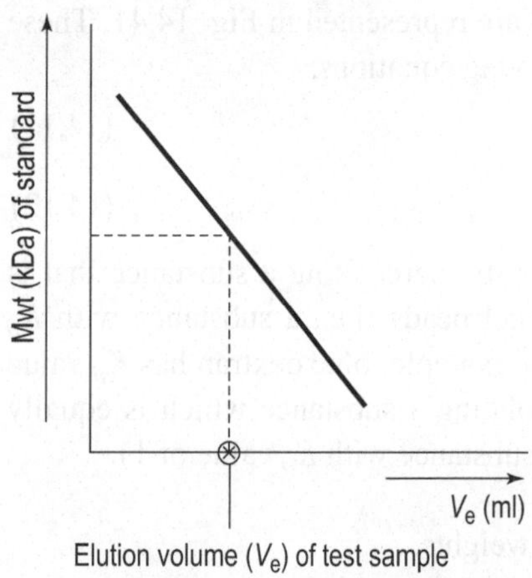

FIGURE 14.42 Determination of molecular weight of protein using gel filtration chromatography

Affinity chromatography

Affinity chromatography makes use of a stationary phase containing immobilized ligands. Macromolecules, whole cells, and cell fragments are separated based on their specific and reversible biological affinity towards the immobilized ligands.

Affinity column Stationary phases such as agarose, dextran, polystyrene, and polyacrylamide are first activated by cyanogen bromide. Ligands with chemical groups such as $-NH_2$, $-COOH$, $-SH$, or $-OH$ are immobilized on the stationary phase via spacer arms. Spacer arm is a hydrophobic or hydrophilic compounds with 6 to 10 carbon atoms, for example, 1, 6-diaminohexane or 6-amino hexanoic acid. The stationary phase with the ligand is then packed into a column called the affinity column (Fig. 14.43).

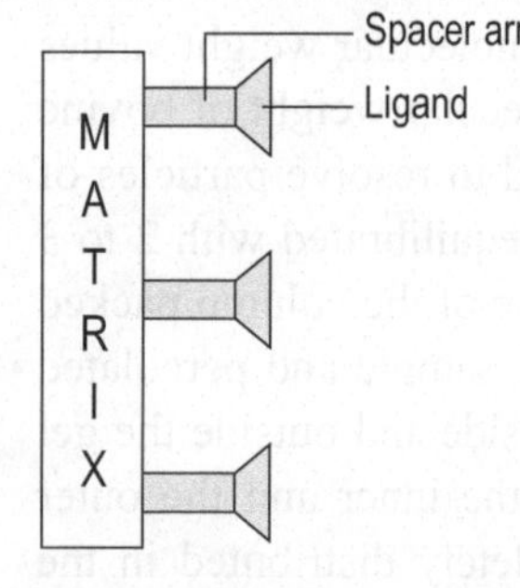

FIGURE 14.43 Affinity column

Two types of ligands are used in the affinity column: (i) absolute specificity ligand with affinity for a single specific compound and (ii) group specificity ligand with affinity for a number of closely related compounds. Examples for ligands are given in Table 14.20.

TABLE 14.20 Matrix-bound ligands used in affinity chromatography

Ligand type	Example	Applications
Absolute specificity ligand	Lysine	Separation of ribosomal RNA
	Tryptophan	Separation of alpha chymotrypsin
Group specificity ligand	AvidinHeparin	Separation of biotin containing enzymes
		Separation of coagulation factors

Protocol The sample is mixed with the buffer containing metal ions and layered on the affinity column. As the column is percolated with mobile phase (buffer), metal ions promote the interaction between the sample and the ligand. Non-specific molecules are washed out with excess buffer. Enzyme samples are eluted using substrate or reversible inhibitor of the enzyme. Non-enzymatic compounds are eluted by affinity elution using a second ligand for which the matrix has greater affinity than the first ligand. Proteins can be eluted by gradient elution using buffers of varying pH or ionic strength. Eluted

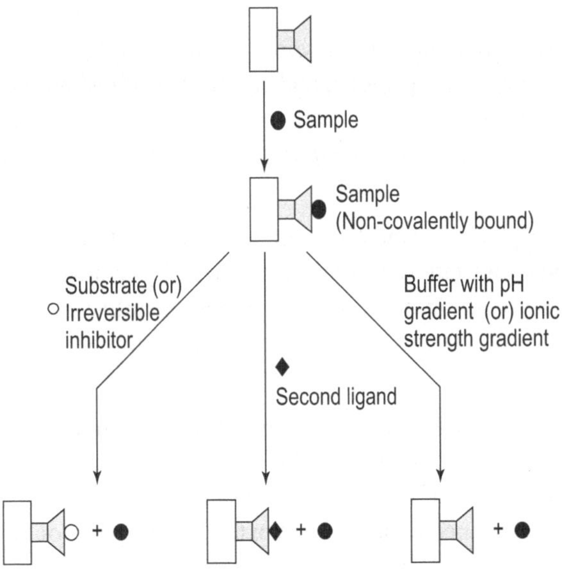

FIGURE 14.44 Scheme of affinity chromatography

compounds can be detected by UV detector or fluorescent detector and collected in a fraction collector for further analysis. The principle of affinity chromatography is schematized in Fig. 14.44.

Applications Affinity chromatography is used

- to separate proteins, enzymes, immunoglobins, nucleotides, and nucleic acids.
- for fractionation of cell mixture into homogenous population.
- to separate and purify specific biomolecules using specific ligands.
- Depending on the specific ligand used for separating a specific molecule, the affinity chromatography can be named as follows:
  - Lectin affinity chromatography that uses lectin as a ligand to separate glycoproteins and membrane receptors.
  - Immunoaffinity chromatography that uses immunoglobulin to separate virus.
  - Dye–ligand chromatography that uses cibacron blue dye to separate enzymes, such as kinases, dehydrogenases, and DNA polymerases.
  - Covalent chromatography that uses disulphide-2′-pyridyl group to separate thiol-containing proteins.

Gas–liquid chromatography

In gas–liquid chromatography (GLC), low-polarity volatile compounds are separated using a gaseous mobile phase. GLC gives high sensitivity, reproducibility, and good resolution.

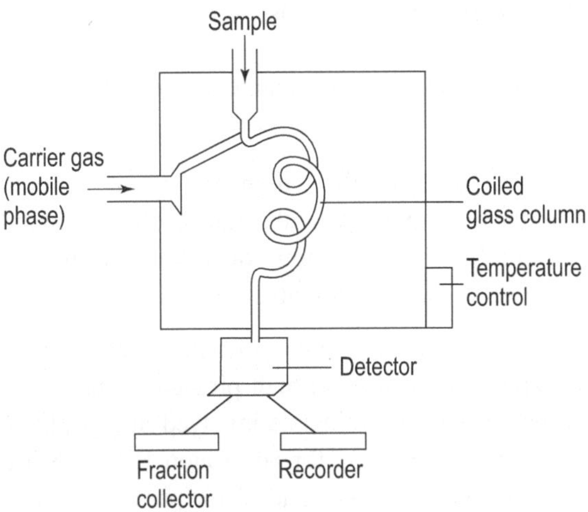

FIGURE 14.45 Scheme of GLC apparatus

GLC column Figure 14.45 shows the schematic representation of a GLC apparatus. It consists of a thin coiled capillary column of glass, steel, or metal. The column is usually 1 to 3 m long and 2 to 4 mm diameter width. Two types of GLC column are known—wall-coated open tubular (WCOT) column, where the stationary phase is directly coated on the wall of the column. Porous layer open tubular (PLOT) column, where the stationary phase is coated to an inert support such as celite (diatomaceous silica) already bonded to the wall of the column. The –OH group of the celite adsorbs polar organic molecule and impair resolution. This can be circumvented by deactivating the –OH group by treating with hexamethyl silazane (Fig. 14.46) (a process called silanization).

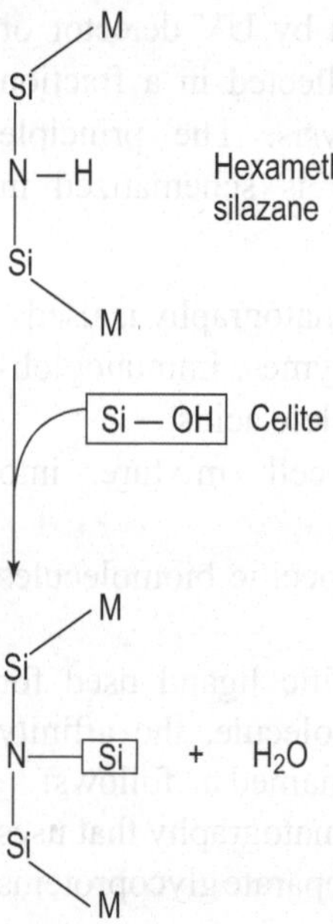

FIGURE 14.46 Silanization

Organic compounds with high boiling point are used as stationary phase for GLC. Examples include polyethylene glycol, methyl–vinyl silicone gums, methyl–phenyl silicone gums, apiezone, and esters of succinic, adipic, and phthalic acids. The stationary phase is coated to the GLC column and activated by heating for 24 to 28 h. Activated column is flushed with the mobile phase (carrier gas).

Sample Polar samples are processed before loading in the GLC column. Carbohydrates, amino acids, and fatty acids are derivatized by silanization, methylation, or perfluoro acylation in order to enhance the volatility and the effective distribution coefficient. Derivatization is not essential for non-polar samples.

Technique The sample is dissolved in acetone, heptane, or methanol and a volume of 0.1 to 10 mm^3 (0.1 to 10 μl) is introduced into the column. The temperature of the column is elevated to volatilize the sample. Nevertheless, the column temperature is maintained to be lesser than the boiling point of the stationary phase.

Carrier gas such as nitrogen, helium, or argon is injected into the column at a flow rate of 40 to 80 cm^3/min. Being a mobile phase, the gas carries the sample through the column and separate the sample compounds based on difference in partition coefficient. Flame ionization detectors are use for detecting organic compounds, nitrogen phosphorus detectors are used for detecting nitrogen- and phosphorus-containing compounds, whereas electrochemical detectors are used for polychlorinated pesticide samples. Signals produced by the detector are monitored in the recorder and the sample collected in a fraction collector.

Applications GLC is used

- to recognize and resolve unknown fatty acids present in fat hydrolysate.
- to calculate the retention index of a compound.
- to analyse polyhalogenated compounds.
- to analyse nitrogen- and phosphorus-containing compounds.
- to identify functional groups of compounds.
- to determine the position of double bond in unsaturated organic compounds.

High-pressure liquid chromatography

High-pressure liquid chromatography (HPLC), also called high-performance liquid chromatography, separates compounds under high pressure with high resolution, based on adsorption, partition, ion exchange, molecular exclusion, or affinity principle. Figure 14.47 shows a schematic representation of HPLC apparatus.

HPLC column HPLC column is made of stainless steel and it has a length of 20 to 50 cm and a diameter of 1 to 4 mm. It has the capacity to withstand high pressure of up to 5.5 × 107 Pa. Stationary phase is chosen on the basis of the principle involved in separation. Usually, microporous alumina or microporous silica is used for adsorption HPLC. Silica support chemically linked to alkyl amine or octadecyl silane is ideal for partition HPLC. Polystyrene resin chemically bonded to silicone network is preferred for ion exchange HPLC.

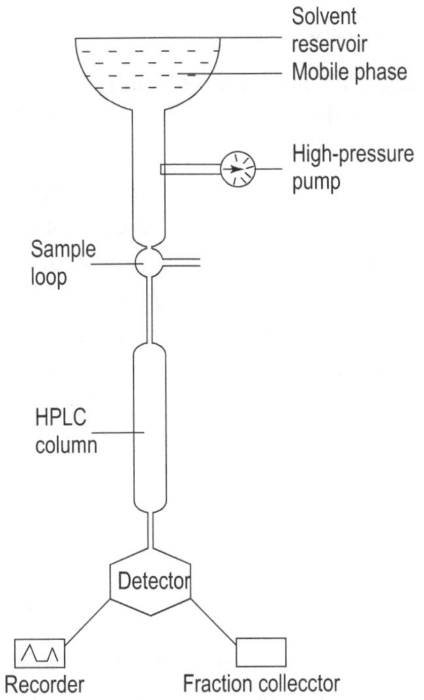

FIGURE 14.47 HPLC apparatus

Porous glass beads or porous polyvinyl acetate beads is best for gel filtration HPLC. Ligand-bound polystyrene can be used for affinity HPLC. Stationary phases used in HPLC are summarized in Table 14.21.

TABLE 14.21 Stationary phases used in HPLC

Principle of separation	Stationary phase
Adsorption	Microporous alumina/silica
Partition	Silica bonded to amine/silane
Ion exchange	Polystyrene resin
Gel filtration	Glass beads, polyvinyl acetate beads

Technique Degassed mobile phase is mixed with the stationary phase to form slurry. The slurry is pumped rapidly into the HPLC column under high pressure. A metal sample loop situated on the top of the column is used for loading the sample. When biological samples such as blood, urine, plasma, and serum are frequently used, there may be impairment in column resolution. To circumvent this, a tiny column of 2 cm, called the guard column, is kept at the top of the main column. Guard column is made from the same material as the main column and can be swapped with new column to maintain the resolution.

The HPLC solvent is purified in a micro filter and degassed by warming, evacuation, vigorous stirring, or ultrasonic vibration. Thereafter, the solvent is filled in a solvent reservoir that is linked to a high pressure pump, which in turn, is attached to the sample loop.

The HPLC pump consists of a chamber, a piston, an entry valve, and an exit valve. As the piston is pulled, the pulling stroke opens the entry valve forcing the solvent into the chamber. The piston is then compressed, and the compression stroke opens the exit valve. Thus, the solvent exits the chamber and reaches the sample loop. Finally, the sample is flushed into the column and the compounds present in the sample are separated based on adsorption, partition, ion exchange, molecular exclusion, or ligand affinity principle. Separated compounds come out from the column to reach the detector (UV–visible spectrophotometer fluorimeter, refractive index monitor or electrochemical detector). The detector produces a signal which is recorded in a recorder data system, simultaneously the compound is collected in a fraction collector.

Applications HPLC is used to

- isolate peptides, vitamins, polyphenols, and steroids.
- separate highly polar compounds, such as organic acids, amino acids, and catecholamines.
- separate and purify biological compounds present in blood, urine, serum, and plasma.
- isolate proteins from cell extract and oligopeptides.
- separate drugs and their metabolites.
- isolate aldolase, transferin, cytochrome C, and thyroglobulin.

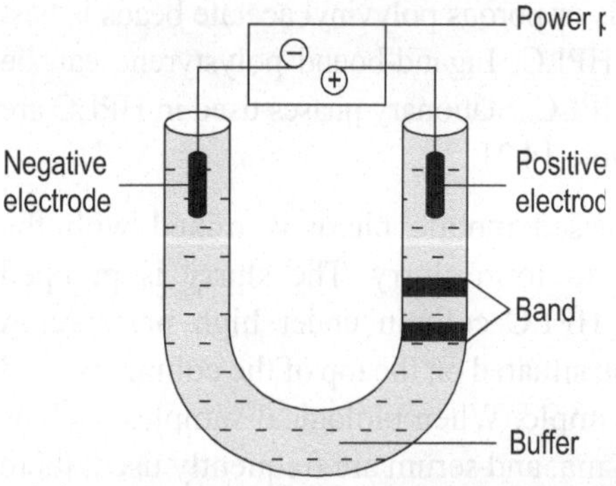

FIGURE 14.48 Moving boundary electrophoresis

14.5 ELECTROPHORESIS

Electrophoresis is a technique used for separating charged molecules such as amino acids, proteins, and nucleic acids, under the influence of an electric field. Separation is based on difference in migration rate of the molecules through a buffer towards cathode or anode. The separated compounds form discrete zones or bands that are detected by staining, autoradiography, UV absorption, or enzymatic colour reactions. The rate of migration of the particle per unit electrical field is called electrophoretic mobility. When electrophoresis is performed in a free solution, it is called moving boundary electrophoresis. If it is carried out in a stable supporting medium, it is called zone electrophoresis.

14.5.1 Moving Boundary Electrophoresis

The apparatus for moving boundary electrophoresis (MBE) consists of a U tube filled with a buffer solution that bathes the electrodes (Fig. 14.48). The sample is kept over the liquid surface and an electrical field is applied. The sample compounds obtain ionic charges depending on the pH of the buffer and travel towards the respective electrodes; during migration, the compounds are separated into zones. The boundary formed between the zone and the pure buffer alters the refractive index of the solution. The variation in the refractive index is monitored by Schleiren optical system (Section 14.3.3.) to show the electrophoretic pattern (peak). Peak gives information about the direction of migration and rate of migration of the compounds. Area under the peak is proportional to the quantity of the compound.

Though MBE is used for large-scale separation of protein and purification of blood plasma, zone broadening and zone diffusion is caused by convection current and gravity field. This problem can, however, be prevented by using a stable porous mechanical support called the supporting medium, and this technique is called zone electrophoresis.

14.5.2 Zone Electrophoresis

Zone electrophoresis can be performed using a sheet of Whatman filter (paper electrophoresis), a strip of cellulose acetate (cellulose acetate electrophoresis), or a slab of porous gel (gel electrophoresis). At present, gel electrophoresis is commonly used.

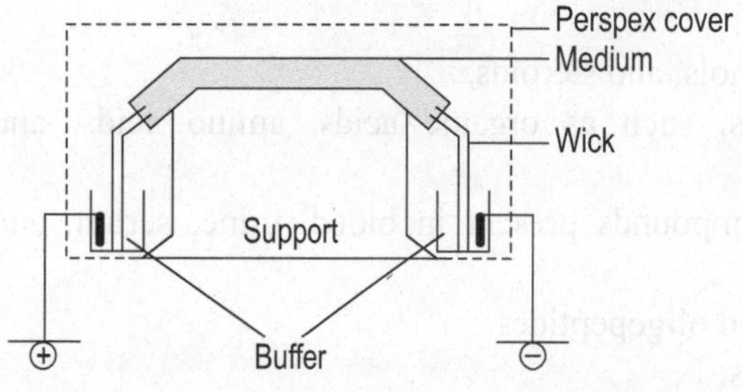

FIGURE 14.49 Components for zone electrophoresis

Following are the basic components required for zone electrophoresis (Fig. 14.49):

- Supporting medium
- Buffer chamber (to hold buffer of appropriate ionic strength and pH)
- Power pack and electrode
- Sample dissolved in a mixture of buffer and tracking dye
- Perspex cover to encase the entire electrophoretic unit

General procedure for zone electrophoresis

The sample to be resolved is dissolved in acetate, pyridine, barbitone, Tris, or citrate buffer, and a small amount of a tracking dye. The sample is spotted on the supporting medium held between the electrodes, and the power pack is switched on. As the voltage is applied, the current is generated to carry the buffer ions along with the sample. The sample compounds migrate along the supporting medium with different electrophoretic mobility. The rate of mobility is determined by the charge, size, and molecular weight of the compound. The specimen front (the movement of the sample) can be tracked by the tracking dye. Voltage is applied till dye reaches the other side of the medium (approximately 2 cm away from the other end of the medium). After this, the sample is resolved into distinct zones/bands. The medium is then removed, and stained with a dye to visualize the bands.

14.5.3 Factors Influencing Zone Electrophoresis

Factors influencing the rate and the resolution of zone electrophoresis are the following:

Supporting medium

Earlier, paper and cellulose acetate strips were used to separate small molecules. However, due to the difficulty of adsorption and electroendosmosis (EEO), these media have been replaced by gel, which can give a better resolution.

Buffer

pH and ionic strength of the buffer determines the resolution of electrophoresis. An optimum ionic strength of 0.05 to 0.1 M and an alkaline pH is ideal for achieving a maximum resolution.

Electric current

Mobility of the sample molecule is directly proportional to the applied voltage (volts) and current (amperes) but inversely proportional to the resistance.

Sample

Rate of migration of the sample molecules is determined by their size and the shape. Tiny molecules encounter lesser frictional force and show greater mobility than larger molecules. Spherical molecules are faster than non-spherical molecules.

Heat

Heat dissipation occurs during electrophoresis, and this may lead to the following problems:

- Generation of convection current
- Diffusion of zones
- Denaturation of samples like DNA and protein
- Loss of enzyme activity
- Alteration in buffer viscosity

To overcome these setbacks a cooling system is used.

Electroendosmosis

The supporting medium contains ionizable groups. For example, paper possesses carboxyl group, agarose contains sulphate group, and the capillary wall is made of silanol group. At alkaline pH (which is suitable for separation of biomolecules), these groups are ionized into negatively charged anionic site. These anionic sites attract the positive ions of the buffer forming an electrical double layer.

Under the influence of applied voltage, the buffer ions carry the current and travel towards the cathode. As the sample usually moves towards the anode at alkaline pH, the sample will come across a buffer counter flow. This leads to a decrease in the migration rate of the sample. This phenomenon is called as electroendosmosis (EEO).

14.5.4 Paper Electrophoresis

Paper electrophoresis is primarily used for separation of proteins (at low voltage), amino acids, and nucleotides (at high voltage).

Low voltage electrophoresis

Approximately 1 μl of the protein sample is mixed with bromophenol blue (the tracking dye) and applied on the centre of a filter paper strip saturated with barbitone buffer of pH 8.6. The ends of the paper are then plunged into the buffer taken in the buffer chambers. At alkaline pH (pH 8.6), the protein molecules have a net negative charge and travel towards the anode. As a voltage of 200 V is applied for 5 to 6 h, the proteins are separated into five different bands based on electrophoretic mobility. Albumin has the highest mobility and form bands closer to anode and γ globulins move slowly and form a zone near the cathode. The amount of molecules in the bands can be quantified using densitometry.

High voltage electrophoresis

The major drawbacks of paper electrophoresis are diffusion, adsorption, and EEO. These factors hinder the migration of molecules. To overcome this, a high voltage of 10,000 V can be used. This results in a better resolution and faster separation. Small molecules such as amino acids and nucleotides can be separated with good resolution using high voltage electrophoresis.

Cellulose electrophoresis can be performed using low or high voltages, similar to the paper electrophoresis. Still, cellulose acetate electrophoresis has replaced the paper electrophoresis because of the following advantages:

- Cellulose acetate gives a better resolution in shorter time.
- It displays negligible adsorption and EEO.
- It is suitable for fractionation of polysaccharides and lipoproteins also.
- It produces easily detectable bands.

14.5.5 Agarose Gel Electrophoresis

Agarose is a polymer of agarobiose (galactose + 3, 6-anhydro galactose). Agarose solution is prepared by boiling the agarose powder in aqueous buffer. The clear solution is then cast into a horizontal slab, and cooled to form a porous rigid agarose gel with anti-conventional properties. The pores of the gel function as a molecular sieve. The pore size of the gel is inversely related to the percentage of the agarose. As the molecular weight of DNA is relatively high, the penetration of DNA into polyacrylamide gel is hindered. Therefore, agarose gel electrophoresis is the most ideal technique for the separation of nucleic acids; 0.8% agarose is conventionally used for separating DNA.

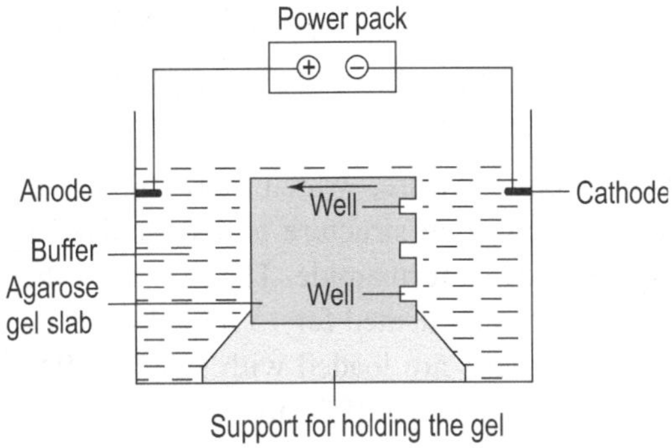

FIGURE 14.50 Agarose gel electrophoresis (Arrow indicates movement of sample)

Principle

The phospho-sugar backbone of DNA contributes negative charge to the DNA molecule. Thus, DNA functions as a polyanion and moves towards the anode. The rate of migration depends on the size and the molecular weight. Low molecular weight DNA fragments migrate faster than the high molecular weight fragments.

Separation of DNA

Agarose gel is cast on a horizontal glass plate to form a slab of 25 cm length and 12 cm width, and wells are drilled using tiny combs. The DNA sample (of unknown molecular weight) and the standard markers (of known molecular weight) are mixed with bromophenol blue and glycerol and loaded into separate wells. Glycerol increases the density of the medium and allows the samples to sink into the well. The gel is then submerged into the buffer tank of the electrophoresis apparatus as illustrated in Fig. 14.50. The tank is filled with Tris acetate (0.04 M) or Tris borate (0.08 M) buffer containing EDTA, and coupled to a power pack. A direct current of 200 mA at 200 Vs is applied for 6 to 8 h. The DNA molecule moves towards the anode. The rate of migration is based on size and molecular weight. Low molecular weight DNA fragments travel faster than the high molecular weight ones. The voltage is applied until the dye has reached distance, which approximately 2 cm from the anodic end of the gel.

Development of gel

The gel is removed carefully and stained with ethidium bromide solution (0.5 μg/cm^3) for 30 to 45 min. As a result, the cyclic planar structure of the ethidium bromide intercalates between stacked base pairs of DNA. The gel is then washed repeatedly with distilled water till the molecules are developed into visible bands.

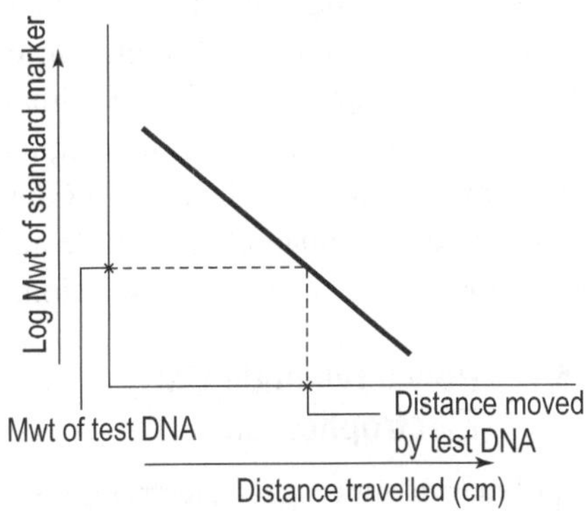

FIGURE 14.51 Determination of molecular weight of DNA

Visualization of DNA bands

The developed gel is allowed to dry and visualized in a UV transilluminator set at 300 nm. At this wavelength, the DNA-bound dye can be observed as fluorescent orange red bands. The bands can be cut, melted, and diluted with buffer and treated with ethanol to precipitate the DNA. The DNA precipitate is purified by chromatography.

Applications

Agarose gel electrophoresis is used for determining the molecular weight of DNA. The sample DNA of unknown molecular weight and standard markers of known molecular weight are run simultaneously.

A plot of the distance travelled by the DNA versus log molecular weight of several is used for determining the molecular weight of the test fragment (Fig. 14.51). Agarose gel electrophoresis is also used for DNA sequencing, to assess the purity of DNA, and to analyse recombinant DNA molecules.

Separation of RNA

RNA is stabilized by the secondary stem loop structure, which may hinder the electrophoretic mobility. To overcome this, the stem loop structure is disorganized by mixing with denaturing agents such as formaldehyde or formamide. The sample is then mixed with the buffer, glycerol, and the tracking dye and heated for 15 min at 65°C and placed on the sample well. Simultaneously, other wells are loaded with standard RNA markers. The gel is submerged into the buffer and a voltage of 75 to 100 V is applied for 3 to 4 h. The RNA fragments are separated into bands based on their molecular size. The gel is then incubated with acridine orange to stain the band and visualized in UV transilluminator set at 260 nm. Molecular weight can be determined using calibration curve as explained for DNA.

14.5.6 Pulsed Field Gel Electrophoresis

Pulsed field gel electrophoresis (PFGE) is used to separate DNA by varying the strength and the direction of electrical field between electrodes. (Recall that in the conventional agarose gel electrophoresis a constant electric field is used). PFGE can separate DNA of several mega bases (conventional agarose electrophoresis can separate DNA up to 500 kb). In PFGE, the DNA sample loaded to the gel is first exposed to the electric field in one direction. This leads to a conformational change so that the coiled molecule is stretched in the direction of the applied electric field. Next, another electric field of different strength is applied in a second direction. This leads to the orientation and realignment of the molecules in the new direction. This process of changing the electric field is repeated. Reorientation and migration of DNA are functions of molecular size. Small molecules realign rapidly and migrate quicker than the larger molecules. PFGE is used to separate human chromosome and yeast chromosome with good resolution.

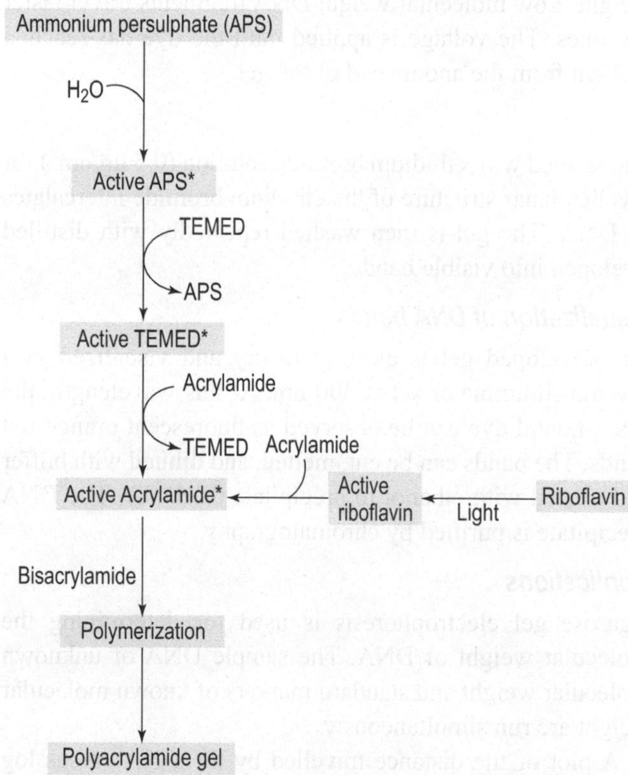

FIGURE 14.52 Polymerization reaction for the formation of polyacrylamide gel

14.5.7 Polyacrylamide Gel Electrophoresis

In polyacrylamide gel electrophoresis (PAGE), polyacrylamide gels are used for separating protein fractions in the presence of sodium dodecyl sulphate

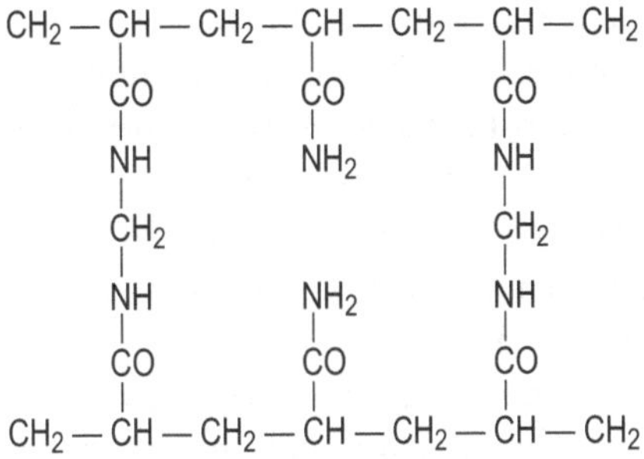

FIGURE 14.53 Polyacrylamide gelmatrix

(SDS), hence, the name SDS-PAGE. Polyacrylamide gel is prepared by polymerization and cross-linking of acrylamide with bisacrylamide, under the catalytic action of free radicals generated from either ammonium persulphate (APS) or riboflavin. The polymerization reaction is shown in Fig. 14.52. APS decomposed to a free radical from (APS*), and mixed with TEMED. Alternatively, riboflavin (RF) can be converted to free radical form (RF*) by exposing to the fluorescent light. The free radical from either TEMED* or RF* is used to activate acrylamide monomer, the acrylamide is then polymerized and cross linked with bisacrylamide to form a matrix of polyacrylamide gel (Fig. 14.53).

Vertical PAGE

Polyacrylamide can be cast as a tube (for small volume of samples) or a vertical slab (for large volume of samples). Vertical slab PAGE apparatus is shown in Fig. 14.54. The pore size of the gel is determined by the acrylamide concentration. Higher concentration of acrylamide gives smaller pore size. Gels with large pore size are ideal for separating proteins with low molecular weight, whereas those with smaller pore size are preferred for separating proteins with high molecular weight. The vertical PAGE gel has two distinct regions of different pore size and different pH—upper gel of 1 cm and larger pore size (the stacking gel) and the lower gel of 10 cm with smaller pore size (the separating gel). Separating gel prepared from 15% acrylamide and 0.5 M Tris buffer of pH 8 is first cast between two glass plates using stacking gel prepared from 4% acrylamide, and 0.5 M Tris buffer of pH 6.8 is then cast on the top of the separating gel using. Water saturated with butanol is placed on the surface of the stacking gel to avoid oxygen, which may otherwise impair the polymerization reaction. Sample wells are drilled on the stacking gel surface using gel combs. The sample protein and the standard markers are prepared in Tris buffer with pH 6.8, and mixed with bromophenol blue, SDS, and urea (denaturing agent).

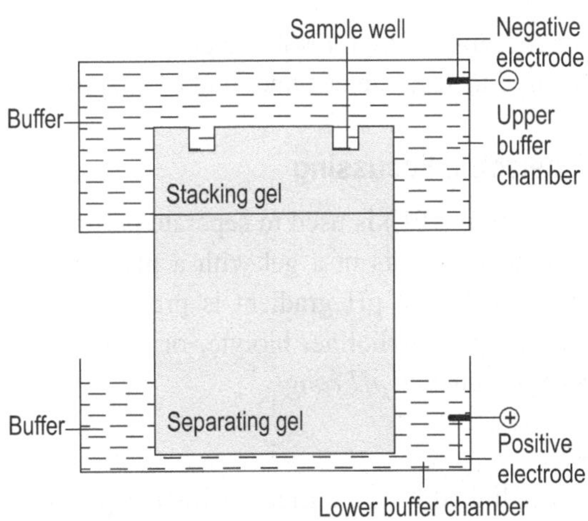

FIGURE 14.54 Vertical PAGE apparatus

Role of denaturing agents

The secondary structure of protein is highly stable due to hydrophobic interaction, hydrogen bond, and disulphide bond (S–S). This stability affects the electrophoretic mobility of the protein. The secondary structure should, therefore, be disturbed by using SDS, β-mercaptoethanol, and urea. SDS is an anionic detergent that binds to two amino acid residues and breaks the hydrophobic interaction, producing a random coil configuration of identical charge:mass ratio. β-mercaptoethanol and urea denatures the protein by breaking the disulphide hydrogen bonds.

Separation

The sample well is filled with the sample solution and the gel is placed in the electrophoretic unit. 0.5 M Tris glycinate buffer with pH 8.8 is poured into the buffer tank present above and below the gel slab. Electrodes coupled with the power pack are dipped into the buffer, and a voltage of 80 V is applied. Tris glycine buffer enters into the stacking gel. As the pH of the stacking gel (6.8) is almost equal to the isoelectric point of glycinate ion (6.0), the glycine molecules undergo poor ionization and move slowly. The chloride ions, however, undergo complete ionization and migrate quickly down the gel. The proteins get tightly stacked as sharp discs between the $\bar{Cl}$ ions (leading ions), and the glycinate ions (trailing ions) migrate down into the separating gel.

A pH of 8.8 in the separating gel enhances the ionization of glycine the glycinate, in turn overtakes the protein disc. The protein disc is then separated within the gel due to molecular sieving property of the gel. Proteins with low molecular weight pass through the gel pores quickly, whereas those with high molecular weight are retained due to frictional resistance and migrate at a slower rate. Once the tracking dye has reached a distance, approximately 5 cm above the lower end of the separating gel, the electrical field can be withdrawn.

Staining and development of gel

The gel is stained by soaking the gel in coomassie brilliant blue (CBB) solution or alkaline silver nitrate solution, containing formaldehyde. Blue bands and metallic bands appear in CBB staining and silver staining, respectively. Occasionally, zinc stain and sypro orange stain are also used to visualize protein bands.

Destaining

The stained gel is destained with a mixture of methanol, water, and acetic acid (destaining solution). Destaining results in removal of excess dye and formation of distinct bands. Destained gel is then dried using an air dryer.

Applications SDS-PAGE is used

- to determine molecular weight of proteins by constructing a plot of log molecular weight versus distance travelled in the gel. Molecular weight is calculated based on inverse relation between molecular weight and distance travelled as described for DNA.
- to fractionate protein subunits.
- to assess the purity of protein samples (because of high resolving power).
- PAGE without SDS (native PAGE) is used to separate enzymes and isoenzymes.

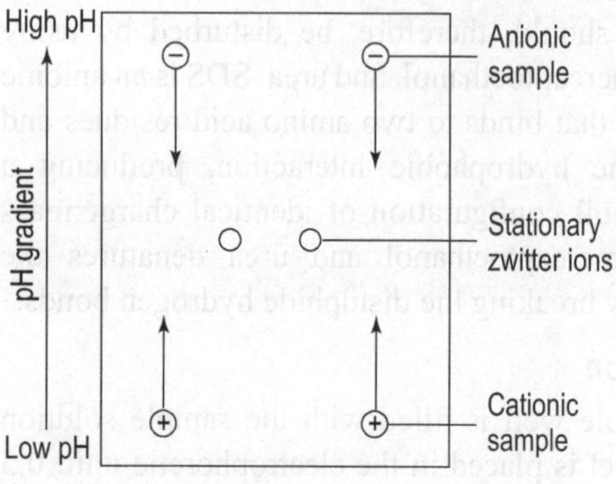

FIGURE 14.55 Isoelectric focussing

14.5.8 Isoelectric Focussing

Isoelectric focusing (IEF) is used to separate particles of different isoelectric points in a gel with a pH gradient. Horizontal IEF gel with pH gradient is prepared using ampholytes such as ampholine, biolyte, or pharmalyte. IEF gel has a *preselected pH range*.

Separation

The IEF gel is linked to the buffer via filter paper wick (Fig. 14.55) and the required voltage is applied to establish

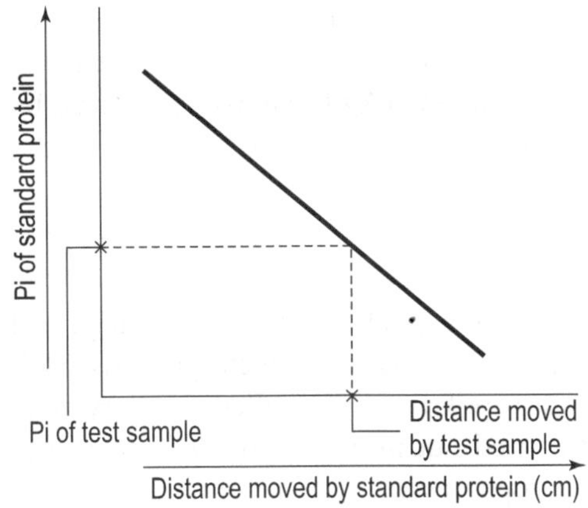

FIGURE 14.56 Determination of molecular weight using IEF

a pH gradient. Once the gradient is established, the power pack is switched off and the sample is loaded into the sample well and the voltage (2,500 V) is applied and a temperature of 10°C is maintained. Compounds with a pH less than the isoelectric point obtain a net negative charge and migrate towards the cathodes, whereas molecules with a pH greater than the isoelectric point obtain a net positive charge and travel towards the cathode. Molecules become stationary and exist as zwitter ions, as they encounter a pH region in the gel that is equal to their isoelectric point. The gel is then developed using CBB staining or silver staining, as explained for SDS-PAGE.

Applications

IEF is used to:

- determine the isoelectric point (pI) of a protein. Sample and the standard markers of known isoelectric point are run in an IEF gel. A plot of distance moved versus isoelectric point is extrapolated to determine the isoelectric point of the test protein (Fig. 14.56).
- separate isoenzymes.
- fractionate proteins with higher resolution.
- separate all amphoteric substances.

14.5.9 Two-dimensional Electrophoresis

In two-dimensional electrophoresis (2D-PAGE), the molecules are separated consecutively by two methods—by IEF and by SDS-PAGE. Molecules are separated in a pH gradient gel column, based on their pI. The gel is then extruded from the tube and incubated with SDS for 30 min to denature the protein. The gel is placed—in conjunction with the stacking gel—for SDS-PAGE. When a voltage is applied, the denatured protein subunits are separated on the basis of size and molecular weight as described under SDS-PAGE. 2D-PAGE is used for resolving proteins and enzymes with high resolution as well as to analyse the purity of proteins.

14.5.10 Immunoelectrophoresis (Rocket Electrophoresis)

Immunoelectrophoresis is used to quantify antigens by combining the sensitivity of gel electrophoresis with the specificity of immune reaction. Molten agar solution saturated with suitable antibody (Ab) specific for the test antigen to be quantified is cast on a horizontal plate, and wells are drilled. One of the well is loaded with about 1 mm^3 of the test antigen (Ag t), and other wells are loaded with equal volume of serial concentrations of standard antigens (Ag1, Ag2, Ag3, Ag4, Ag5). The gel is immersed into the buffer of alkaline pH, and a direct current is applied. At the alkaline pH, the Ag acquires a net negative charge, and travels towards the anode

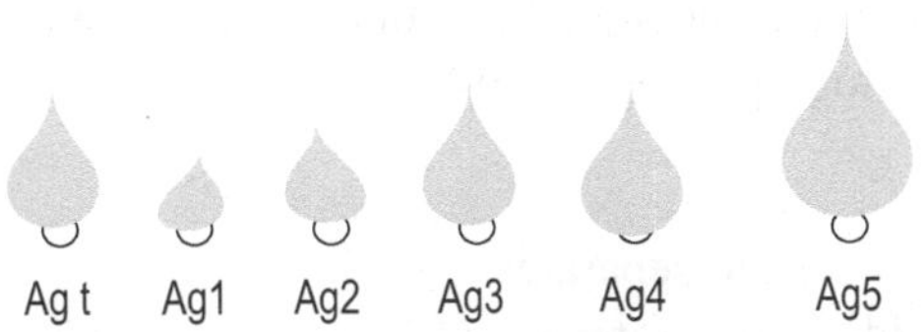

Ag1, Ag2, Ag3, Ag4, Ag5- standard antigen concentrations
Ag t-test concentration

FIGURE 14.57 Rocket electrophoresis

and interacts with the Ab to form Ag–Ab complex called as immunoprecipitates. The gel is then stained with CBB to expose the immunoprecipitates as rocket-shaped arcs as shown in Fig. 14.57. The concentration of the test antigen (Ag t) can be determined by constructing a plot of height of the rocket arc versus the corresponding standard antigen concentration.

14.5.11 Capillary Electrophoresis

Capillary electrophoresis is performed within a capillary tube under high voltage and high pressure to achieve better resolution in shorter time. Agarose electrophoresis, SDS electrophoresis, and IEF can be carried out in a capillary mode (called capillary zone electrophoresis). Electrophoresis can also be carried out in a free solution taken in a capillary tube (called free solution capillary electrophoresis).

Capillary electrophoresis apparatus (Fig. 14.58) is made of a long capillary tube of 50 to 100 cm long and with 50 μm internal diameter. The tube is loaded with the sample solution and the ends of the tube dipped into the buffer tank. The buffer is connected to the power pack via electrodes, and a voltage of 20 kV is applied.

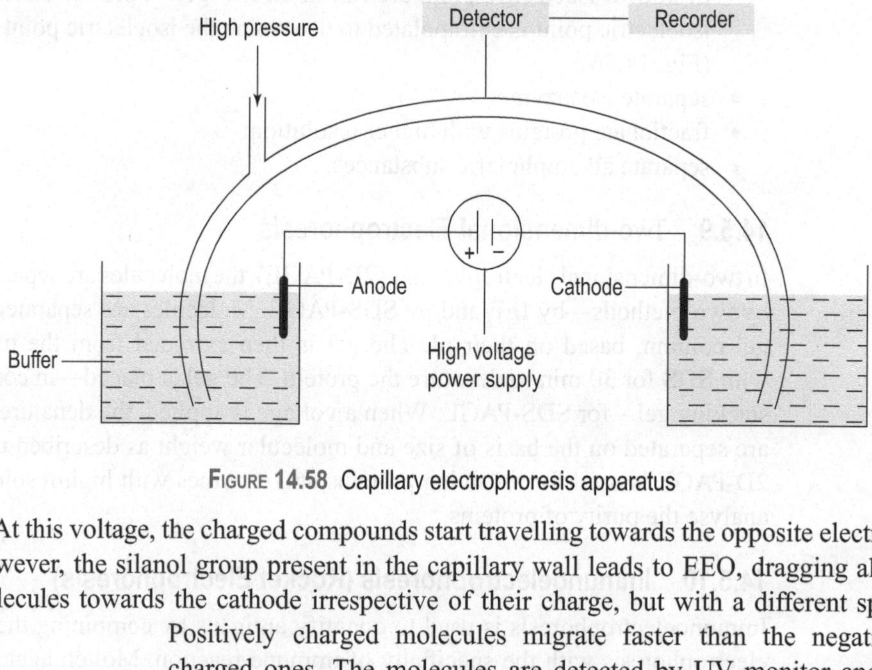

FIGURE 14.58 Capillary electrophoresis apparatus

At this voltage, the charged compounds start travelling towards the opposite electrode. However, the silanol group present in the capillary wall leads to EEO, dragging all the molecules towards the cathode irrespective of their charge, but with a different speed. Positively charged molecules migrate faster than the negatively charged ones. The molecules are detected by UV monitor and the signal is recorded in a recorder. The signal is plotted against time to get a capillary electrophoretograph (Fig. 14.59).

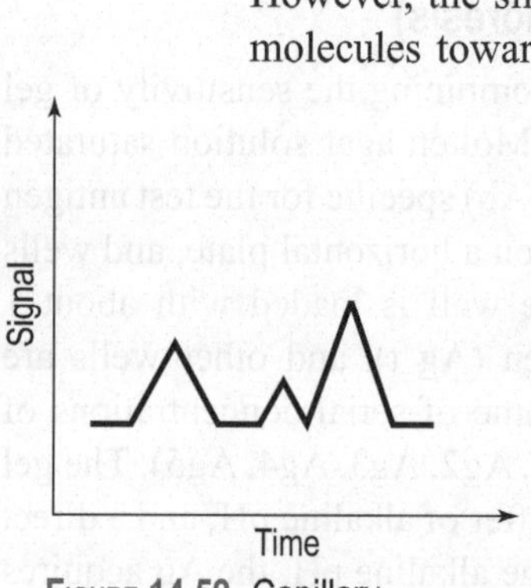

FIGURE 14.59 Capillary electrophoretogram

Applications

Capillary electrophoresis is used to separate:

- amino acids, peptides, and proteins
- oligonucleotides and nucleic acids
- metal ions
- enantiomers

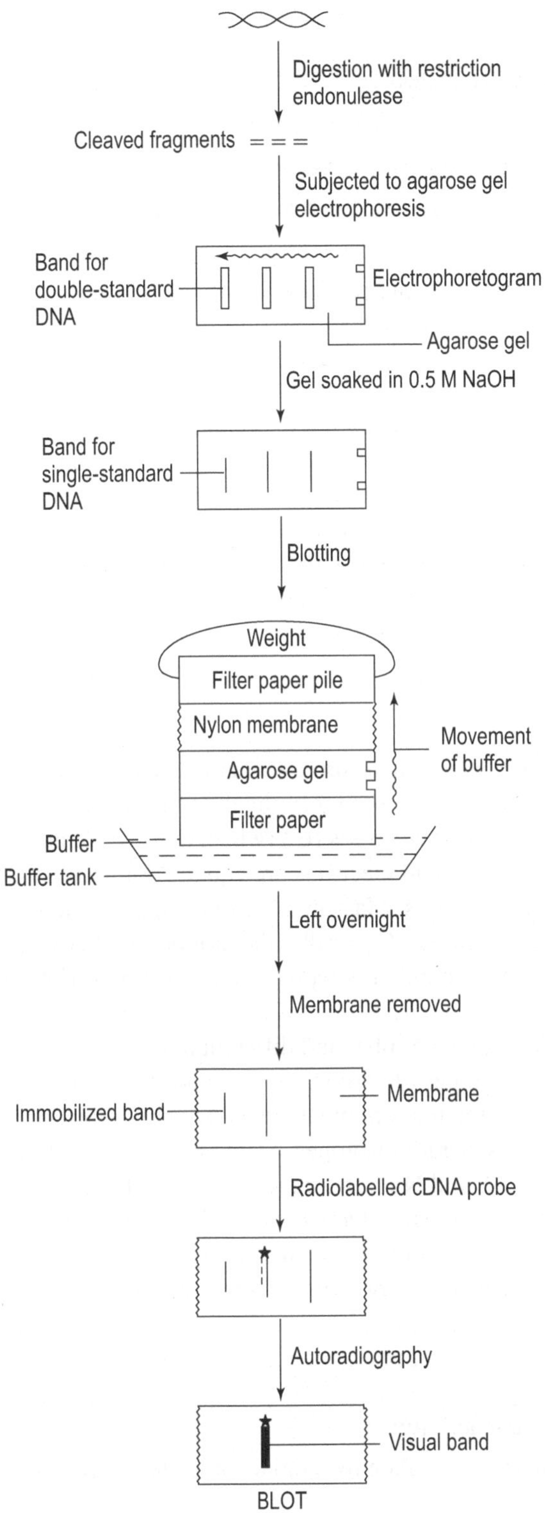

FIGURE 14.60 Protocol for Southern blotting

14.5.12 Blotting Techniques

Blotting refers to the transfer of the fractionated molecules from the gel to the nitrocellulose or nylon membrane. The molecules are immobilized in the membrane by hydrophobic interaction. The membrane with the immobilized molecules is called as a blot. Molecules on the blot can be probed, located, and identified. E.M. Southern first identified DNA in a blot. Hence, the transfer and immobilization of DNA is called Southern blotting. In later stage, RNA and proteins were blotted and the technique is called by the jargon terms—Northern blotting and Western blotting, respectively.

Southern blotting

The protocol of Southern blotting is schematized in Fig. 14.60. Blotting protocol involves the following steps:

- Agarose gel containing the DNA band is denatured into single-stranded fragments by using alkali.
- Then the gel is sandwiched between a filter paper at the bottom and nitrocellulose membrane at the top. The filter paper should saturate with the buffer.
- A pile of dry filter papers is kept over the membrane and pressed with an iron plate.
- The entire blotting unit is left undisturbed for overnight.
- As a result, the buffer moves vertically upward from the filter paper through gel and carries the DNA fragments to the nylon membrane.
- DNA molecules are immobilized in the membrane by hydrophobic interaction and looks like a replica of the parent agarose gel.
- The DNA is immobilized permanently to the membrane by baking at 80°C; the membrane is then known as a blot.
- The blot is incubated with radiolabelled mRNA (mRNA*) that is complementary to the specific DNA fragment to be identified, for example, radiolabelled insulin c DNA* hybridized with DNA fragment coding for insulin. The fragment is then located and identified by autoradiography (refer Section 14.6).

TABLE 14.22 Labels used for probing and visualization of proteins in Western blotting

Label	Example	Method for visualization
Enzyme	Horse radish peroxidase Acid phosphatase	A substrate that gives coloured product on reacting with the enzyme is added, and the colour visualized
Radioactive label	I125	Radioactivity detected by autoradiography and densitometry
Fluorescent label	Fluorescein isothiocyanate	Fluorescent band visualized by irradiating with UV light
Affinity label	Biotin	Addition of avidin–enzyme conjugate, followed by addition of substrate to give coloured product

Northern and Western blotting

Northern blotting and Western blotting are performed similar to Southern blotting, except that the sample is RNA and protein, respectively. Western blots are however probed by immunochemical methods. In this method, the blot is incubated with a primary antibody that is specific for the protein to be probed. In the next step, a secondary antibody conjugated with a suitable label is allowed to a bind with the primary antibody, and visualized. Labels used for conjugating the secondary antibody and methods used for visualization are shown in Table 14.22.

14.6 RADIOACTIVITY

Matter is composed of atoms. An atom is made of positively charged particles called protons (P), uncharged particles called neutrons (N) (both forming the central core called nucleus), and negatively charged particles called electrons (e-) (which revolve round the nucleus). For any atom, the number of protons and electrons are equal. Atomic number of an atom is the number of its protons or electrons. Mass number of an atom is the sum of the number of protons and number of neutrons. Generally, the atomic number of an element is denoted as a subscript and the mass number is represented as a superscript for the symbol of that element.

Atoms of a given element with similar atomic number and different mass numbers are known as isotopes. For example, the two isotopes of carbon are ${}_6C^{14}$ and ${}_6C^{12}$, where '6' is the atomic number and '14' and '12' are the mass numbers. Isotopes are divided into (i) stable isotopes and (ii) unstable isotopes or radio isotopes. Stable isotopes are atoms with atomic number less than or equal to 20, and N/P ratio equal to 1. Unstable isotopes (radioisotopes) are atoms with atomic number greater than 20, and N/P ratio greater than or lesser than 1. Excess of neutron or proton contributes to the instability of radioisotopes. Radioisotopes compensate their excess neutrons or protons by spontaneously emitting energy particles such as alpha (α), beta (β), and gamma (γ) rays. The emission continues until a stable N/P ratio is obtained. This spontaneous phenomenon of emission of radiation is called *radioactivity* or radioactive decay.

Following are some terms related to radioactivity:

Half life period—Half life period ($t½$) is the time taken by a radioisotope to decay to one half of its original volume.

$$t½ = 0.693/k, \text{ where } k = \text{decay constant}$$

Radioactive decay energy—The energy emitted during radioactive decay is called radioactive decay energy.

Units of radioactivity—curie (Ci), rutherford (Rd), and becquerel (Bq) are the units used to express radioactivity.

14.6.1 Types of Radioactive Decay

There are five types of radioactive decay undergone by radioisotopes to reach the stable ratio N/P = 1. They are positron emission, negatron emission, alpha ray emission, gamma ray emission, and X-ray emission.

Positron emission

Positrons are positively charged beta particles (β^+) produced by radioisotope containing excess of protons (i.e., N/P < 1). During positron emission, each excess proton is converted to a neutron. Positron emission is represented as follows:

Proton + Electron → Neutron + β^+

β^+ emission decreases the atomic number by 1 without influencing the mass number. For example, sodium emits positron in the following manner:

$$_{11}Na^{22} \longrightarrow {}_{10}Nc^{22} + \beta^+$$

Negatron emission

Negatrons are negatively charged beta particles (β^-) produced by a radioisotope having excess of neutrons (i.e., N/P > 1). Excess neutron is converted to proton with the emission of negatrons. Negatron emission can be represented as follows:

$$\text{Neutron} \longrightarrow \text{Proton} + \beta^-$$

β^- emission increases the atomic number by 1, whereas the mass number remains constant. Decay of carbon is an example for β^- emission.

$$_6C^{14} \longrightarrow {}_7N^{14} + \beta^-$$

Alpha emission

Alpha particles (α) are helium atoms ($_2He^4$) emitted by isotopes with higher atomic number. α emission results in a reduction in atomic number by two and a fall in mass number by four. For example

$$_{88}Ra^{226} \longrightarrow {}_{86}Rn^{222} + He \longrightarrow Pb$$

Gamma emission

Gamma rays (γ) are electromagnetic radiations of short wavelength emitted by a radioisotope that has neutron deficiency, (i.e., N/P < 1). During gamma emission, the proton captures an electron from the innermost K-shell to form a neutron. γ emission can be denoted as follows:

K-shell electron + Proton (in nucleus) ⟶ Neutron (nucleus) + γ

This phenomenon by which electron is captured from the K shell is called K-capture

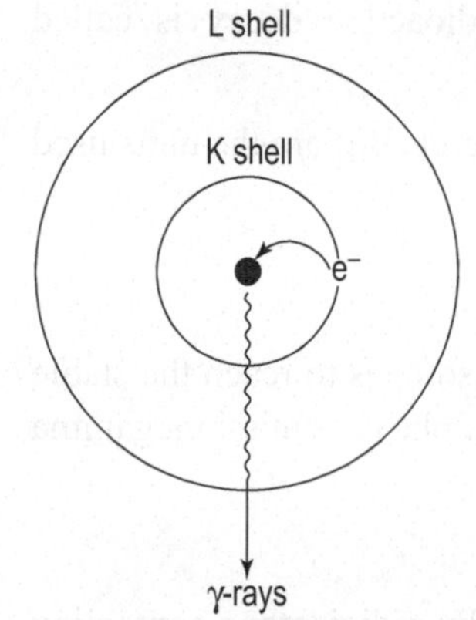

FIGURE 14.61 K capture and X-ray emission

(Fig. 14.61). K capture leads to loss of energy in the form of γ rays. Iodine decays by γ emission in the following manner:

$$_{63}I^{131} \longrightarrow {}_{64}Xe^{131} + \gamma$$

X-ray emission

X rays are electromagnetic radiations with long wavelength and high penetrating power. K-capture generates a vacancy in the K-shell. This vacancy is compensated by transfer of electron from L-shell to K-shell, generating a vacancy in L-shell, which in turn is compensated by the capture of electron from M-shell (Fig. 14.61) accompanied by the emission of electromagnetic radiation in the form of X rays. The following are some examples for X-ray emission:

$$_{25}Mn^{54} \longrightarrow {}_{24}Cr^{54} + \text{X rays}$$

$$_{53}I^{125} \longrightarrow {}_{52}Te^{125} + \text{X rays}$$

14.6.2 Detection of Radioactivity

Radioactivity has wide applications in the fields of medicine and biological research. In order to understand these applications, we must study the methods of radioactivity detection.

Radioactivity can be detected by any one of the following five methods:

- Geiger–Muller counter (GM counter) method
- Scintillation counter method
- Bubble chamber method
- Cloud chamber method
- Autoradiography technique

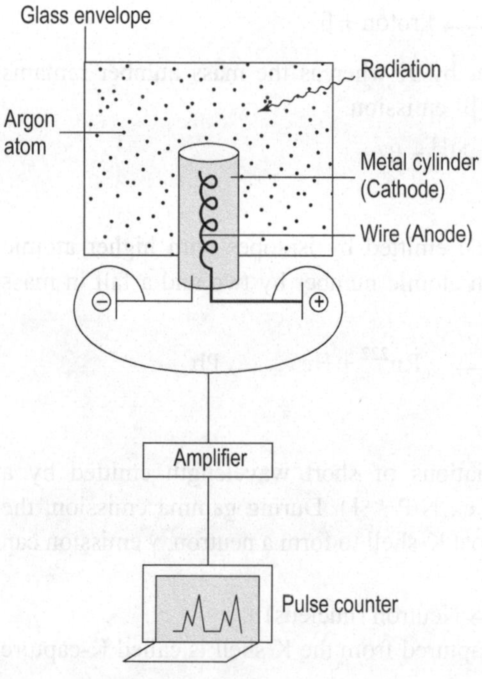

FIGURE 14.62 GM counter

GM counter method

GM counter works on the principle of ionization of atoms. Radiations induce ionization of gas atoms into cations and electrons. Cations migrate to cathode and the electrons move to the anode resulting in a current flow. The magnitude of the current is proportional to the radioactivity.

A schematic representation of GM counter is shown in Fig. 14.62. It consists of a metal cylinder (cathode) and a fine wire (anode), both enclosed in a glass mantle filled with argon gas. A voltage of 1000 V is applied between the electrodes. The counter is connected to a pulse counter via an amplifier. Radiations emitted by the radioisotope are penetrated through the glass envelope. Radiations ionize the argon atoms into argon ions (cations) and electrons (anions).

$$\text{Ar} \longrightarrow \text{Ar}^{2+} + 2e$$

Cations migrate towards the cathode, and electrons travel towards the anode resulting in electrical conductivity.

At 1000 V, electrons collide with inert gas atoms and induce further ionization (secondary ionization), generating a stream of electrons. This phenomenon is known as gas amplification, which creates a strong current pulse. The current is amplified by an amplifier and fed into a pulse counter. The pulse counter measures the electrical conductivity that is equivalent to the radioactivity of the radiation penetrated into the GM counter.

The extent of gas amplification differs at different voltages. For example, at 1000 V the gas amplification is maximum and constant over time, this is called GM counter region. At voltages below 1000 V, other regions, such as ionization chamber region, proportional counter region, and continuous discharge region, also exist in the chamber shown in Fig. 14.63. This figure shows the different regions produced by gas amplification at different voltages.

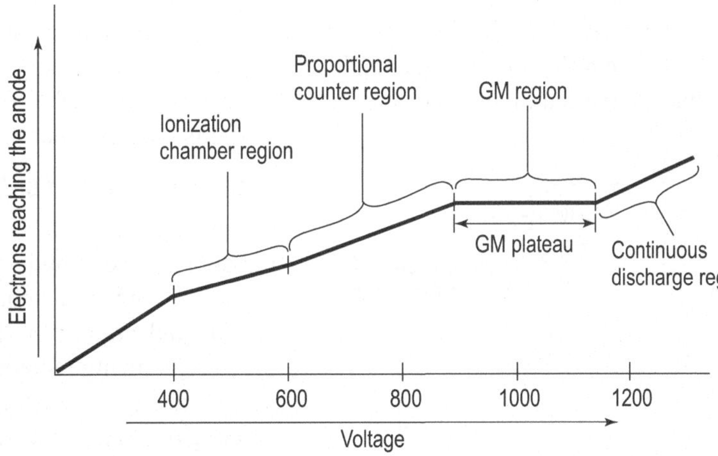

FIGURE 14.63 Effect of voltage on gas amplification in GM counter

Scintillation counter method

Scintillation counter allows the collision of radiation with atoms of certain solids or liquids (fluors) in order to excite the electrons from ground state to higher energy state. When the excited electrons come back to their ground state, tiny flashes of light are emitted. This phenomenon is called scintillation. The light energy produced by scintillation is transformed into electrical pulse, and the magnitude of the pulse is proportional to the intensity of the radiation.

Scintillation counters are of two types, solid scintillation counter and liquid scintillation counter (Fig. 14.64). In solid scintillation counter, the radioactive sample is in contact with a solid fluor (crystal) that is held adjacent to a photomultiplier. In liquid scintillation counter, the radioactive sample is dissolved in a mixture of a solvent, primary fluor, and secondary fluor, called scintillation cocktail. The mixture is held adjacent to a photomultiplier. The working principle in solid and liquid scintillation counters is schematized in Fig. 14.65.

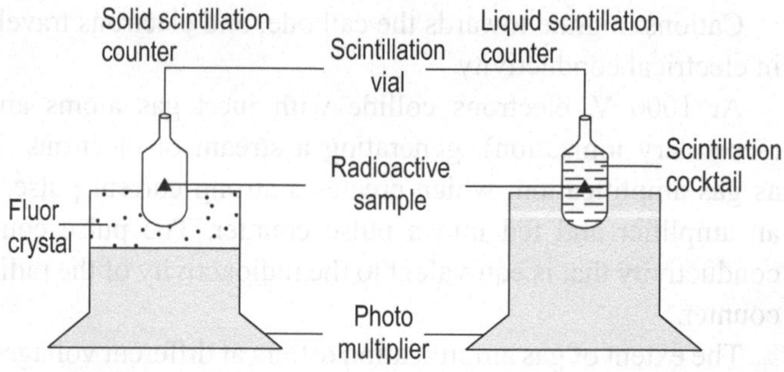

FIGURE 14.64 Solid and liquid scintillation counter

In solid scintillation counters, the radiation interacts with the crystal and produces scintillation. The photomultiplier converts the scintillation into electrical energy and the resultant electrical pulse is recorded. The magnitude of the electrical pulse is proportional to the intensity of the radiation produced by the radioactive sample. In liquid scintillation counters, the radiation interacts with the solvent, producing scintillation. The light energy of the scintillation is sequentially transferred via the primary fluor and the secondary fluor to the scintillation cocktail and converted into detectable light. The photomultiplier converts the light energy into electrical energy.

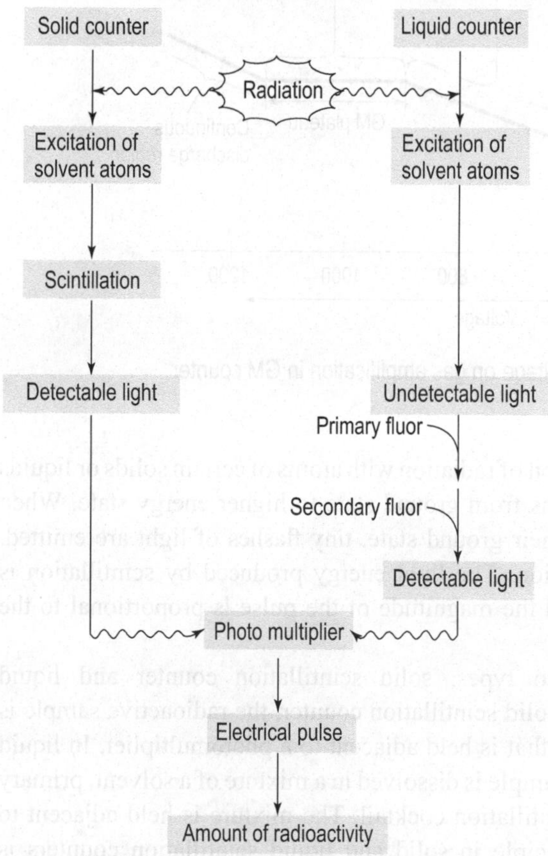

FIGURE 14.65 Principle of working in solid and liquid scintillation counters

Cloud chamber and bubble chamber methods

Cloud chamber detects the nature of ionizing radiations. The cloud chamber, provided with a camber and a piston, is filled with alcohol vapour and argon gas (Fig. 14.66). Under low pressure, ionizing radiations from the radioactive sample are sent into the chamber to ionize the argon gas atoms into argons ions. Argon ions bring about the condensation of alcohol vapour into liquid droplets. The drops are aligned into tracks, illuminated by a light flash, and viewed by the camera. Natures of the tracks provide information about the nature of the radiation. Short, broad, densely packed track is indicative of β radiation. Thin, beaded, tortuous track is a sign of α radiation.

The working principle of the bubble is similar to that of the cloud chamber except that a superheated liquid is ionized to form bubble tracks which can be illuminated and viewed to find out the nature of the radiation.

Autoradiography technique

Autoradiography is a technique by which the radioactivity present in the electrophoretogram, chromatogram, tissue slice, or whole animal or

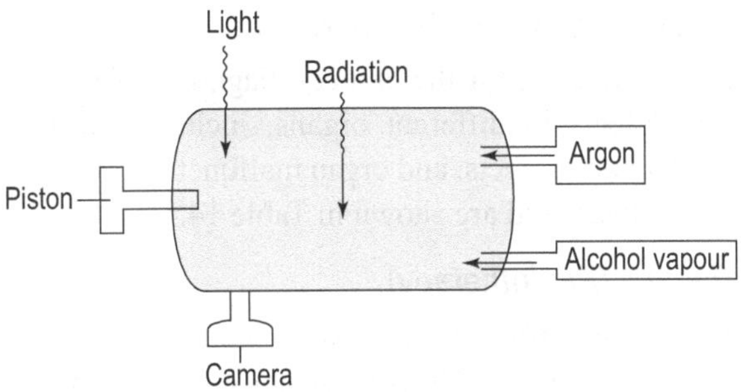

FIGURE 14.66 Cloud chamber

plant can detected and quantified using silver halide emulsion.

When the radioactive sample is made to come in close contact with an X-ray film, coated with silver halide (AgCl) emulsion, the radiation penetrates into the film and reduces silver halide into metallic silver to form a latent image. The intensity of the image is proportional to the radioactivity of the sample.

Steps in autoradiography The following five steps are involved in autoradiography:

1. Preflashing
2. Exposure
3. Development
4. Fixation
5. Detection

All the five steps are demonstrated in Fig. 14.67.

Preflashing—An X-ray film is coated with AgCl emulsion and exposed to light flash for few seconds for activating the film. This process is called preflashing.

Exposure—Preflashed film is held in close contact with the radioactive sample. The radiation dissipates from the sample to the AgCl crystals. This process is called exposure.

Development—The exposed film is submerged in a developer solution composed of a metal + Na2CO3 + KBr + Na2SO3. Consequently, the activated (exposed) AgCl is reduced to metallic silver, and converted into a latent image. However, the unexposed AgCl remains unreduced.

Fixation—The developed film bearing the latent image is washed with water and dipped into the fixer solution (hypo), for fixing the latent image on the film, the unexposed AgCl is converted into insoluble complex salts which are easily washable.

Detection—The intensity of the developed image is detected by a densitometer. The intensity is directly proportional to the radioactivity of the sample.

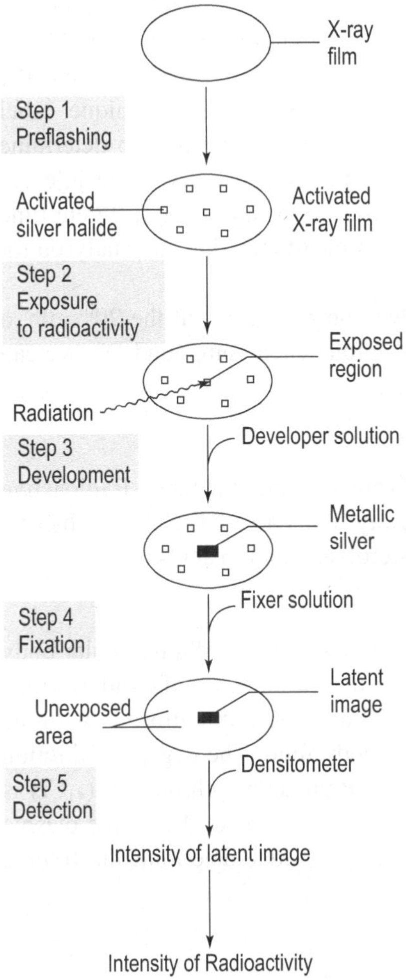

FIGURE 14.67 Steps in autoradiography

Applications In cell biology, autoradiography is used to:

- locate and study the structure and function of intracellular components.
- determine the precise site of synthesis of DNA, RNA, and protein.

14.6.3 Applications of Radioactivity

Radioactivity is used in medicine for diagnosing diseases and for treating cancer. It is used in analysing metabolic pathways, and radiocarbon dating. Various applications of radioactivity are discussed in this section.

Table 14.23 Applications of radioactive isotopes in organ diagnosis

Name of the isotope	Organ/disease diagnosed
$_{131}$I	Thyroid cancer Hyperthyroidism Hypothyroidism Kidney malfunction
$_{90}$Sr	Osteoblastoma (Bone cancer)
$_{99}$Te	Brain seizure Liver abscess Pancreatic malignancy Pancreatic inflammation
$_{51}$Tc	Ectopic spleen Spleen infarcts
$_{201}$Th	Myocardial infarction
$_{133}$Xe	Lung defect

Diagnosis of diseases of different organs

Radioisotopes are used for the clinical diagnosis of various diseases or defects in different organs, including cancer, hormone imbalance defects, and organ malfunction. Isotopes and the organ diagnosed are shown in Table 14.23.

Cancer therapy (Radiotherapy)

Cancer refers to abnormal cell division and uncontrolled cell growth. Radiations are used to control cancer by inducing

- direct DNA damage
- local blood vessel damage
- necrosis
- impairment of blood and oxygen supply to the cells

Radioactive tracer technique

From Chapter 7, it is clear that several biochemical pathways occur simultaneously within the cells. Therefore, it becomes difficult to trace the metabolic pathway of specific compound. However, radioactive tracer technique label specific compound to trace its metabolic pathway. For example, in order to determine the metabolic pathway of a compound, say 'A', A is radiolabelled (A*) and injected into the experimental animal. Blood samples collected from the animal at different time intervals, say 10th min, 20th min, and 30th min from the time of injection, are analysed for radioactivity.

If at the 10th min, blood possesses a radiolabelled compound B*; at the 20th min, a radiolabelled compound C*; and at the 30th min, a radiolabelled compound D*, we can predict the metabolic pathway of A as

$$A \longrightarrow B \longrightarrow C \longrightarrow D$$

That is, we can say that A is metabolized to D via B and C as intermediates. Radioactive tracer technique is used to trace the metabolic pathway of acetic acid, glycine, glucose, chromium, vitamin B_{12}, calcium, TCA cycle, cholesterol, and amino acids.

Isotopic dilution technique

Isotopic dilution technique is a method to find out the total volume of a particular body fluid, by determining the dilution of a radioisotope in that fluid. A specific radioisotope of known radioactivity is injected into the body, and allowed to mix uniformly with the particular body fluid. The isotopes get diluted in the body fluid. The degree of dilution is proportional to the total volume of the body fluid. Radioactive chromium ($_{51}$Cr) is used to determine blood volume, radioactive iodine ($_{131}$I) is used to determine plasma volume, and radioactive sodium (^{24}Na) and tritium ($_3$H) are used to determine the volume of extracellular fluid and total body water.

Radio immunoassay

Radio immunoassay (RIA) is a technique used to quantify a specific antigen based on competition between labelled and unlabelled Ag to bind to a specific antibody (Ab).

A constant volume of the radiolabelled antigen (Ag*) and its Ab is allowed to react with serially varying amount of standard unlabelled Ag. A fixed amount of unlabelled test Ag is also added to the reaction mixture. A competition exists between the labelled and unlabelled Ag to bind to the Ab. The unlabelled Ag reduces the binding of labelled Ag by an extent equivalent to the concentration of unlabelled Ag present in the sample. Thus, the quantity of labelled Ag conjugated to the Ab is inversely proportional to the concentration of the Ag.

Radiocarbon dating in plant cells

Radiocarbon dating is a method used to estimate the age of a piece of wood or fossil using the ratio of two isotopes of carbon (^{14}C:^{12}C ratio). Atmospheric CO_2 is composed of a mixture of radioactive ^{14}C and ^{12}C in a fixed ratio. When the atmospheric CO_2 is assimilated by living plants to prepare cellulose (wood), 14C disintegrates into nitrogen and positron as follows: $^{14}C \rightarrow N + \beta-$. Nevertheless, the loss is compensated by the absorption of atmospheric CO_2, thus, maintaining a constant $_{14}C$:$_{12}C$ ratio in the wood. However, when the plant dies, ^{14}C disintegration is not compensated by the uptake of atmospheric CO_2, causing a decrease in ^{14}C:^{12}C ratio. Using the $_{14}C$:$_{12}C$ ratio and the half life of ^{14}C, the age of the wood can be calculated by applying the following formula:

$$\text{Age} = [2.303T/0.693] \times \log [^{14}C:^{12}C \text{ in fresh wood}/^{14}C:^{12}C \text{ in dead wood}]$$

where T = half life of ^{14}C = 5760 years.

Study of reaction mechanism

The mechanism of reaction of a compound can be studied by radiolabelling any atom of that compound. For example, oxygen-labelled water (H_2O*) is used to trace the mechanism of hydrolysis of ester.

$$\underset{\text{Ester}}{R\text{—}\overset{\overset{O}{\|}}{C}\text{—}R'} \quad + \quad H_2O^* \quad \rightarrow \quad \underset{\text{Acid}}{R\text{—}\overset{\overset{O}{\|}}{C}\text{—}O^*H} \quad + \quad \underset{\text{Alcohol}}{R'\,OH}$$

The above reaction reveals that the oxygen atom of the water is incorporated into acid, whereas the oxygen atom of the ester is incorporated into alcohol.

Other applications

Radioisotopes are also used:

- to investigate the cellular accumulation, metabolism, mechanism of action, site of action, and the metabolic products of drugs.
- to study the absorption and distribution of phosphate fertilizers in plant cells by using ^{32}P.
- in DNA sequencing and blotting.
- to determine blood cholesterol.
- to investigate plant respiration.
- to study the production and average life of erythrocytes in the body.
- in autoradiography, to study about biological samples.

14.6.4 Hazards of Radioactivity

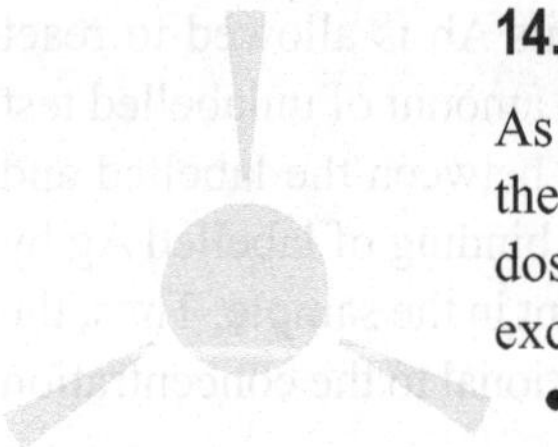

FIGURE 14.68 Radioactive alarm sign

As the application of radioactivity is immense, it is important to understand the hazards of radioactivity to tissues and cells of living organism. At a high dose, radiations affect cells and cellular biomolecules by inducing ionization and excitation resulting in following biological hazards:

- DNA mutation
- Enzyme inactivation
- Metabolic degeneration
- Cancer
- mRNA damage
- Leukaemia
- Nausea
- Diarrhoea
- Hair loss
- Mouth ulcer
- Skin blemishes
- Haemorrhage
- Cataract
- Infertility

Radiation hazards can however be prevented by

- accumulating the radioactive samples in thick-walled lead containers to prevent the radiation from coming out.
- using lead aprons and lead gloves.
- using remote control robots to handle radioactive sample.
- avoiding mouth pipetting.
- monitoring the exposed dose using film badges (refer Terms to Remember for film badge).
- proper disposal of radioactive wastes in a remote place.
- avoiding eating and drinking in the radioactivity laboratory.
- labelling the instruments and glass wares with radioactive sign as shown in Fig. 14.68.

14.7 SPECTROSCOPY

Spectroscopy is a technique used for the quantification, characterization, and structural analysis of a substance (matter) based on interaction of light with that substance. Interaction of light with a substance/sample may result in different phenomena and generate a specific analytical signal. The signal is detected by a detector and recorded by a digital recorder in the form of a spectrum. The spectrum can be correlated with the amount, structure, property, and functional group of the sample.

Understanding about light is essential to get a clear idea about spectroscopy. Light is a combination of electrical and magnetic waves, hence the name electromagnetic radiation (EMR).

14.7.1 Electromagnetic Radiation

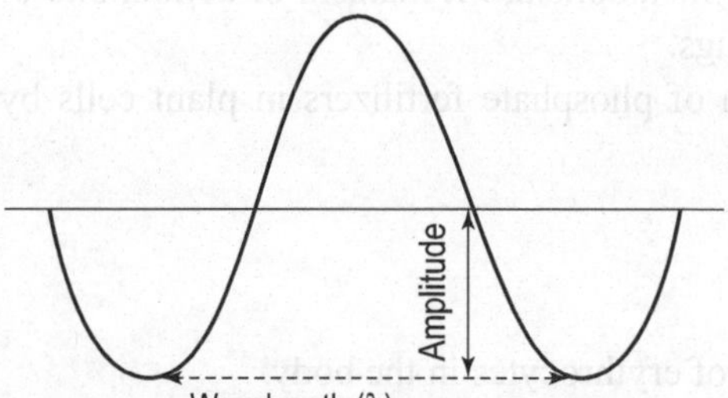

FIGURE 14.69 Electromagnetic radiation

Electromagnetic radiation (EMR) consists made of electrical and magnetic waves oriented perpendicular to each other. The wave nature of EMR is shown in Fig. 14.69.

Let us consider that an EMR is passed through a medium, then various parameters of the EMR, such as frequency, wave number, velocity, and wavelength, can be defined. Frequency (ν) of the radiation is the number of times it oscillates in the medium per second. It is expressed as Hertz (Hz). Wave number (n) is the number of waves propagating per centimetre of the medium. Velocity (c) is the distance travelled by the light wave per second. The unit of velocity is expressed in cm s1. Wavelength (λ) is the distance between two successive peaks of a light wave, expressed as nanometer (1 nm = 10 Å = 10^{-9} m). The colour or visibility of a light wave is determined by its wavelength, for example, light possessing wavelength between 380 and 750 nm is visible to human eye as violet, indigo, blue, green, yellow, orange, red (VIBGYOR) colours, and therefore, called visible light or white light. Radiations with wavelength below 380 nm and above 750 nm are not visible to the eye. Thus, EMR constitutes radiation of different wavelengths, hence, the name polychromatic radiation.

EMR is a source of energy packets called photons. The energy (E) of EMR is directly related to its frequency and velocity but inversely related to its wavelength. This relation can be represented as

$$E \propto \nu$$

But $$\nu = c/\lambda$$

Therefore $$E \propto c/\lambda$$

or $$E = hc/\lambda$$

where h = Planck's constant.

14.7.2 Spectrum

When a radiation from an incandescent source is passed through a prism, it is dispersed into orderly pattern of different wavelengths called a spectrum (Fig. 14.70).

Types of spectrum

Spectrum may be classified as shown in Fig. 14.71.

An emission spectrum is produced when a radiation is directly examined by a spectrometer, whereas an absorption spectrum is formed when a radiation passing through an absorbing medium is examined. For example, when sodium atoms are excited in a flame, light energy is emitted to produce emission spectrum (Fig 14.72a). On the other hand, when sodium atoms are irradiated with a light from a carbon arc, absorption spectrum is produced (Fig. 14.72b). Emission and absorption spectra may be either a line spectrum or a band spectrum. The spectrum produced by atoms in gaseous state appears as lines, and is therefore called a line spectrum or atomic spectrum. The spectrum produced by molecules in solution or crystal appears as bands, and is therefore called a band spectrum or molecular spectrum. Figure 14.73 represents line and band spectra. Table 14.24 summarizes the differences between emission and absorption spectra.

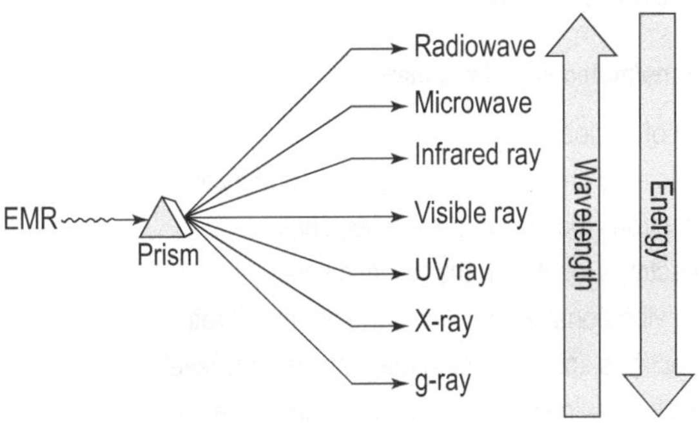

FIGURE 14.70 Spectrum of EMR

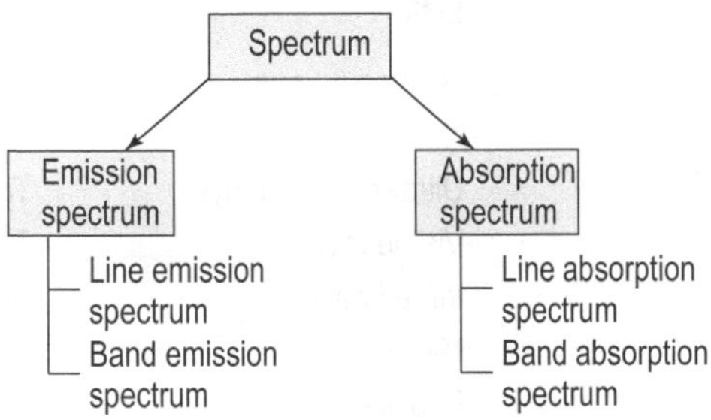

FIGURE 14.71 Types of spectrum

FIGURE 14.72 (a) Emission spectrum and (b) absorption spectrum of sodium

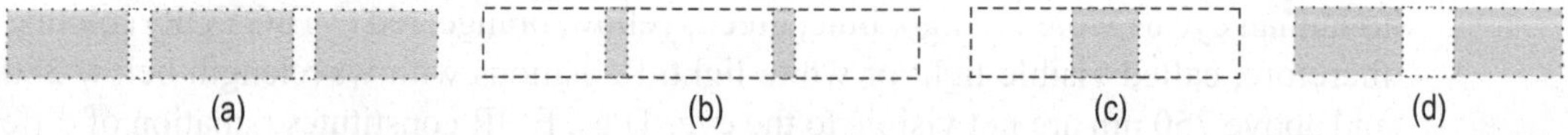

FIGURE 14.73 Line (atomic) spectrum and band (molecular) spectrum (a) Line emission spectrum (b) Line absorption spectrum (c) Band emission spectrum (d) Band absorption spectrum

TABLE 14.24 Differences between emission and absorption spectra

Emission spectrum	Absorption spectrum
Attributed to the emitting atoms or molecules in excited state.	Attributed to the irradiated absorbing atom or molecule in the ground state.
Line emission spectrum consists of fine bright lines on a dark background. For example, sodium atoms excited by flame gives line emission spectrum.	Line absorption spectrum consists of dark lines on a bright back ground. For example, sodium atoms irradiated by white light produce line absorption spectrum.
Band emission spectrum consists of broad bright bands. For example, heating Ca salts in a flame produces band emission spectrum.	Band absorption spectrum consists of broad dark bands. For example, iodine vapour irradiated with white light produces band absorption spectrum.

14.7.3 Interaction of Light with Matter

When radiations of different wavelengths are incident on matter, various changes take place in the matter as listed in Table 14.25.

Interaction of light with atoms of matter can be described by Beer–Lambert's law. Beer's law states that the quantity of light absorbed by a solution is directly proportional to the concentration of the solution. If A is the absorbance and C is the concentration of the in the solution, $A \propto C$. Lambert's law states that the amount of light absorbed by a solution is proportional to the depth through which light passes in the solution. If L is the depth through which light passes in the solution, then

$$A \propto L$$

TABLE 14.25 Effects of EMR on matter

EMR	Changes in atoms/molecules of the matter
Gamma (γ) rays	Rearrangement of nucleus
X rays	Transition of electrons present in the inner shell of the atom
Ultraviolet (UV) rays	Transition of electrons located in the valence shell
Visible rays	Transition of electrons located in the valence shell
Infrred rays	Alteration in the vibrational energy level and atomic vibration
Microwave	Transition of electrons spinning in the magnetic energy level
Radiowave	Transition of nucleus spinning in the magnetic energy level

The two laws can be integrated to obtain to Beer–Lambert's law. We get, A ∝ C × L or A = K × C × L, where K is a numerical constant. If A_1 is the absorbance of the test solution and C_1 its concentration, and if A_2 is the absorbance of the standard solution and C_2 its concentration, then we get

$$\frac{A_1}{A_2} = \frac{K \times C_1 \times \text{L}}{K \times C_2 \times L} = \frac{C_1}{C_2}$$

$$C_1 = \frac{A_1}{A_2} \times C_2$$

$$\text{Concentration of test solution} = \frac{\text{Absorbance of test}}{\text{Absorbance of standard}} \times \text{concentration of standard}$$

where L = length of the cuvette.

If this value of concentration of test is present in X ml of sample taken for analysis then 100 ml sample contains

$$\frac{\text{A of test CS}}{\text{100 units A of Std. X}}$$

14.7.4 Types of Spectroscopy

Spectroscopy is basically divided into atomic spectroscopy and molecular spectroscopy (Fig. 14.74). Atomic spectroscopy is subdivided into atomic emission spectroscopy and atomic absorption spectroscopy. Molecular spectroscopy is divided into UV-visible spectroscopy, IR and Raman spectroscopy, NMR spectroscopy, ESR spectroscopy, luminometry, CD spectroscopy, nephlometry, turbidometry, and X-ray spectroscopy. Figure 14.74 represents different types of spectroscopy.

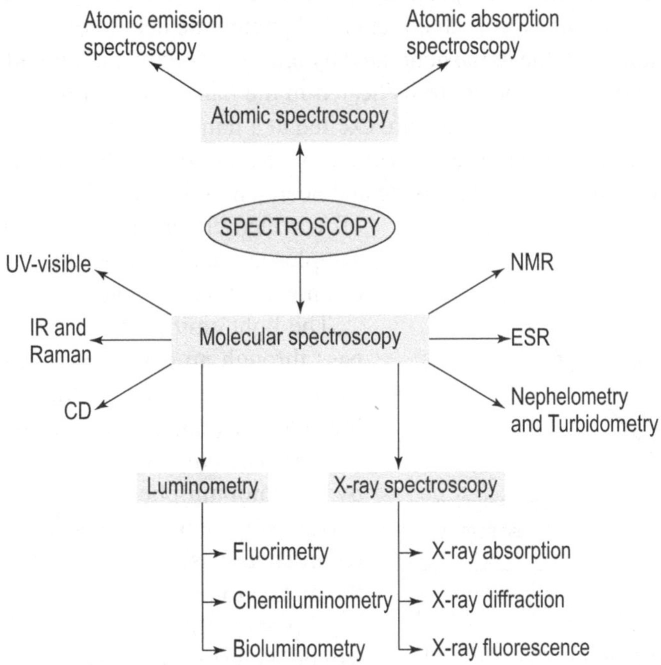

FIGURE 14.74 Types of spectroscopy

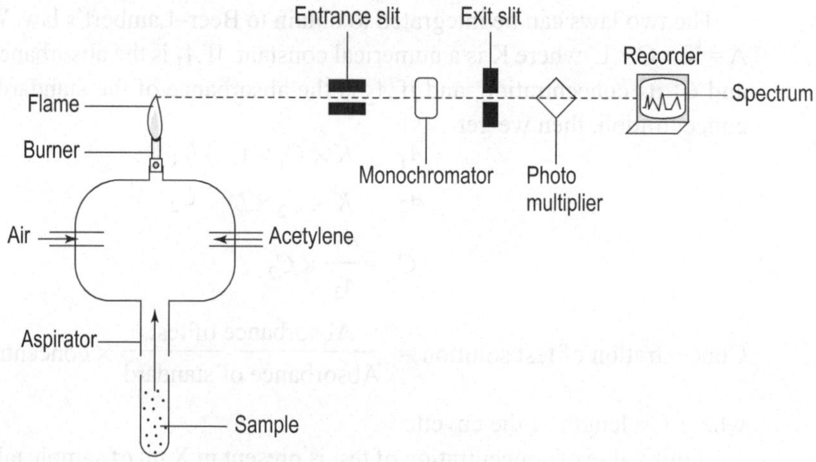

FIGURE 14.75 Atomic emission spectrophotometer

Atomic emission spectroscopy

Atomic emission spectroscopy, also called flame emission spectroscopy, is used to quantify the amount of the metal in solution. When a metal solution is heated in a flame, the atoms are vaporized and excited to emit coloured light. The nature of the metal is determined by the colour and the wavelength of the emitted light. The intensity of the flame is proportional to the quantity of the metal. For example, yellow colour flame is indicative of sodium, and red colour flame is indicative of calcium.

Instrument and working Atomic emission spectrophotometer is the instrument used in atomic emission spectroscopy (Fig. 14.75). The sample solution containing metal atoms is aspirated into a chamber called pneumatic nebulizer which sprays the solution as a fine mist over the flame generated by igniting a mixture of air and natural gas in a Mercker type burner. The metal atoms heated in the flame are volatilized, atomized, and electronically excited. Alkali metals are excited at a flame temperature of 2200 K, but for other metals, higher temperature is essential, which is produced by igniting air and acetylene mixture (2450 K) or nitrous oxide and acetylene mixture (3200 K). The excited atoms emit light of wavelength characteristic of the metal element. Metal samples are analysed simultaneously with standard solutions of known concentration.

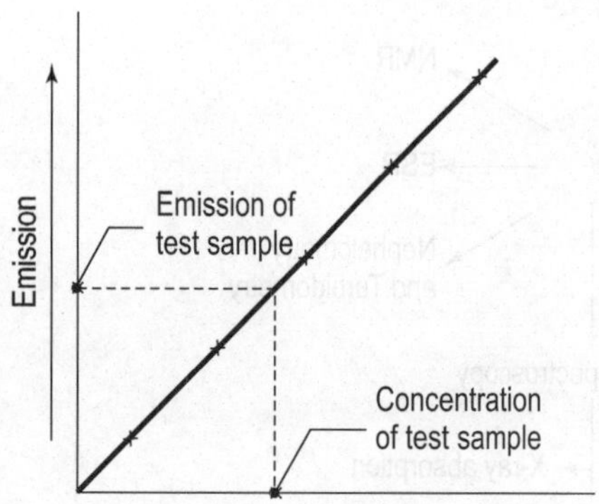

FIGURE 14.76 Calibration curve for atomic emission spectrophotometer

The light emitted by the sample and the standard solution pass through an entrance slit and a sintered glass filter. The filter selects the specific wavelength emitted by the specific metal atoms (e.g., 589 nm for Na, 767 nm for K, and 443 nm for Ca). The emitted light enters an exit slit and it is detected by a photomultiplier or a photocell. A pattern of emission versus wavelength (emission spectrum) is recorded in a recorder. Emission is directly proportional to the amount of the metal atoms.

A plot of emission versus concentration is created and extrapolated (Fig. 14.76) to determine the concentration of the test solution.

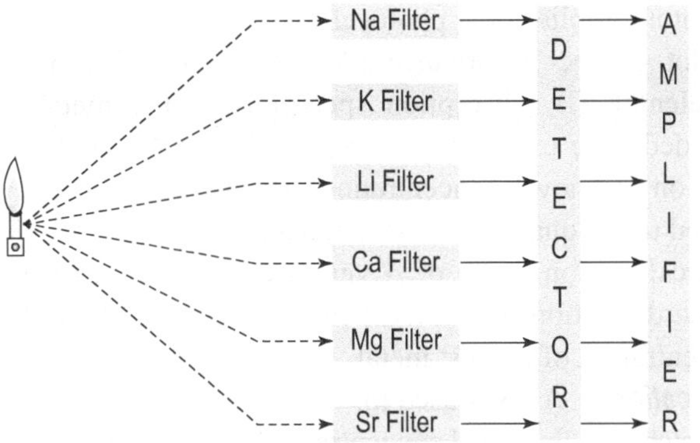

FIGURE 14.77 Multichannel flame emission spectrophotometer

Multichannel flame emission spectrophotometer Multichannel flame emission spectrophotometer consists of a multichannel monochromator (Figure 14.77), capable of simultaneously selecting the wavelength of the light emitted by six different elements at the same time.

Applications Flame emission spectroscopy is used to measure the amount of metal atoms, such as Na, K, Li, Ca, Mg, Sr, and Ba, in body fluids. It can also quantify trace elements, such as Cu, Fe, and Mg, in biological samples.

Atomic absorption spectroscopy

In atomic absorption spectroscopy (AAS), neutral metal atoms in ground state are vaporized and irradiated with white light. The atoms absorb the light, and the amount of the absorbed light (i.e., absorbance A) is proportional to the amount of the metal atoms. The wavelength of the absorbed light is characteristic of a specific metal.

The instrument for (atomic absorption spectrophotometer) is shown in Fig. 14.78. The instrument is similar to atomic emission spectrophotometer except that it has a white light source for irradiating the sample.

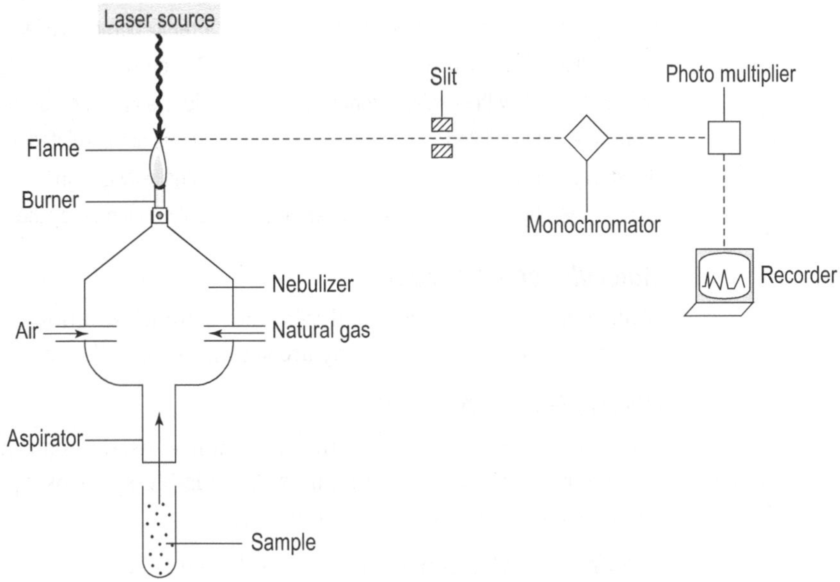

FIGURE 14.78 Atomic absorption spectroscopy

A mixture of air, O_2, and N_2O, acetylene and natural gas is ignited to produce a ribbon-shaped flame on a slot type burner or premix burner. The test solution is aspirated into a nebulizer, from where it is sprayed as a fine mist of aerosol over the flame. The metal atoms are vaporized and atomized in the flame. A white light from a hollow cathode lamp or a laser source is incident on the atoms. The metal atoms absorb the light and

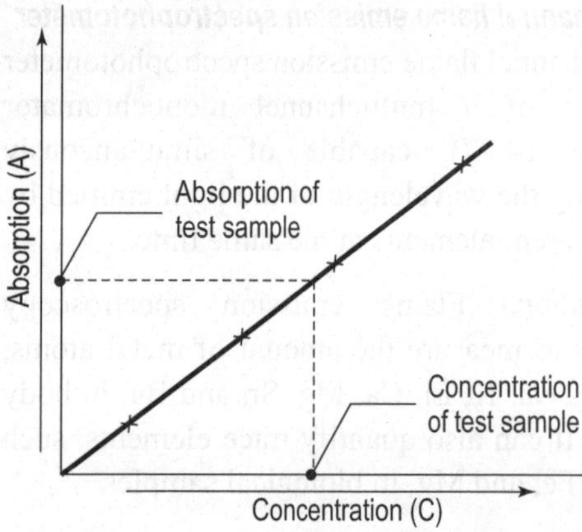

FIGURE 14.79 Calibration curve for AAS

a photomultiplier or a photocell detects the absorbance. A plot of the absorbance (A) versus the absorption wavelength (l) (absorption spectrum) of the metal is recorded. The experiment is repeated with the standard solution of known concentration. Absorbance is directly related to the quantity of the metal atoms.

A calibration curve of A versus concentration of the standard solution (Fig. 14.79) can be used to find out the concentration of the test metal.

Applications AAS is used to:

- determine the metal concentration in body fluids, such as blood and urine, using calibration curve.
- quantify the amount of alkali metal atoms and transition elements in homogenates of animal and plant tissues.

Differences between atomic (flame) emission and atomic absorption spectroscopies are tabulated in Table 14.26.

TABLE 14.26 Differences between atomic (flame) emission and atomic absorption spectroscopies

Atomic (flame) emission spectroscopy	Atomic absorption spectroscopy
Mercker burner is used to produce flame	Premix burner is used to produce flame
Circular flame is produced	Ribbon shaped flame is produced
Light is emitted by the excited atoms	Light is absorbed by the atoms in ground state
Atoms are not irradiated	Atoms are irradiated with a light
Sensitivity is low	Sensitivity is high
Light emitted by the metal atom is determined	Light absorbed by the metal atom is determined

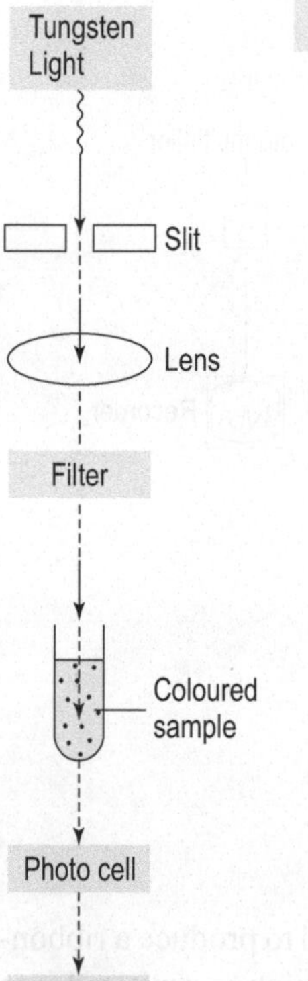

FIGURE 14.80 Schematic representation of a colorimeter

Molecular spectroscopy

Molecular spectroscopy analyses the sample solution without atomization. Different types of spectroscopy are summarized in Table 14.27.

UV-visible spectroscopy

UV-visible spectroscopy is the commonly used technique to quantify and characterize biological compounds. UV-visible spectroscopy is divided into (1) colorimetry and (2) spectrophotometry.

Colorimetry Colorimetry is a method used to quantify coloured compounds in solution using a colorimeter. A colorimeter works by irradiating with visible light of 400 to 700 nm wavelength, and measuring the absorbance. The schematic representation of a colorimeter is given in Fig. 14.80.

Polychromatic visible light from a tungsten filament lamp enters a slit and a condensing lens and exits as a parallel beam. This beam is interrupted by a filter, to filter out a beam of specific wavelength. The light is then passed into a coloured solution taken in a glass cuvette. As a result, the valence electron in the sample

TABLE 14.27 Types of spectroscopy

Type of spectroscopy	Light source used for irradiating sample	Phenomenon exhibited by the sample due to interaction with light	Signal detected	Detector	Major applications
UV spectroscopy	Deuterium lamp	Transition of valence electron	Radiation absorbed	Photocell	Estimation of unsaturated compounds, compounds with sigma and pi bonds, and colourless compounds
Visible spectroscopy	Tungsten filament valence electron	Transition of	Radiation absorbed	Photomultiplier	Estimation of coloured compounds
IR spectroscopy	Sintered silicon carbide rods or Nichrome alloy wire lamp	Molecular vibration and change in dipole moment of the molecule	Absorbed radiation	Golay cell	To detect the functional group of a compound
Raman spectroscopy	Tungsten or deuterium lamp	Molecular vibration and change in polarizability of the molecule	Scattered radiation	Photomultiplier at 90° angle	To detect the functional group of a compound
NMR spectroscopy	Radiofrequency	Transition of spinning nucleus to higher magnetic energy level in the presence of external magnetic field (nuclear resonance)	Absorbed radiation	Oscilloscope	Quantification and structural determination of hydrogen-containing compounds oscillator
ESR spectroscopy	Klystron oscillator generating microwave	Transition of spinning electron to higher magnetic energy level in the presence of an external magnetic field (electronic resonance)	Absorbed radiation	Silicon crystal	Quantification and structural analysis of paramagnetic substances and free radicals
CD spectroscopy	UV/visible plane polarized light	Differential absorption of left and right circularly polarized light. Differential transmission of elliptically polarized light	Ellipticity of transmitted radiation	Photomultiplier	Quantification and characterization of optically active compounds
Luminometry	UV or visible light source	Electronic excitation	Emitted light		To determine the amount of analyte in a reaction
Nephlometry	Mercury arc lamp	Scattering of light by suspended particles	Scattered light	Photomultiplier	To quantify bacterial suspension and paste
Turbidometry	Tungsten lamp ordueterium lamp	Transmission of light by turbid particles	Transmitted light	Photocell	To quantify bacterial suspension and paste
X-ray	Coolidge tube	Excitation of inner shell electrons	Absorbed radiation; diffracted radiation; fluorescent radiation	GM counter	To detect bone fractures and haemor rhage spectroscopy To detect block in blood vessels

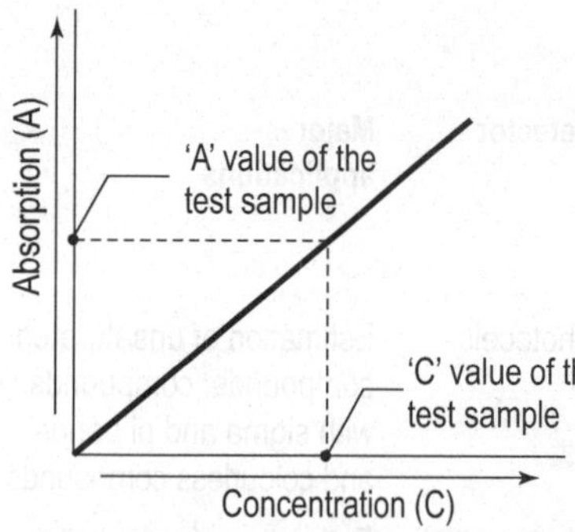

FIGURE 14.81 Calibration curve of A vs C

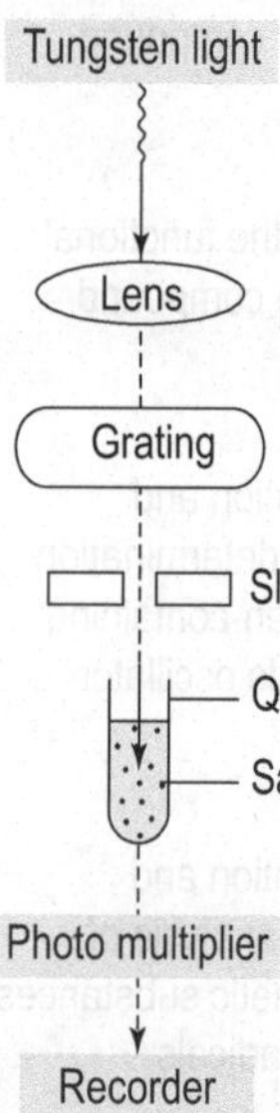

FIGURE 14.82 Schematic representation of a UV-visible spectrophotometer

is shifted to higher energy level. The absorbed beam which induces electronic transition is detected by a photocell detector. Absorbance of a series of standard solution of known concentration is also recorded. A standard calibration curve is constructed by plotting absorbance versus concentration of the standard solution (Fig. 14.81). The concentration of the test solution can be estimated from the curve.

Spectrophotometry Spectrophotometry is a technique used for the quantification and characterization of colourless compounds with the help of spectrophotometer. The sample is irradiated with UV or visible light and the absorbance measured. The schematic representation of a UV-visible spectrophotometer is given in Fig. 14.82.

UV light emitted by a deuterium lamp, hydrogen lamp, mercury arc, or xenon discharge lamp is transformed into a monochromatic beam by quartz. The beam is made parallel by a slit. The parallel monochromatic beam is then incident on the sample solution taken in a silica or quartz cuvette. The UV light is absorbed by the functional groups of the test compound, leading to excitation of bonding electrons (n-electrons and π-electrons). The absorbance is detected by a photomultiplier, and recorded against the wavelength of the UV light by a recorder, to generate the UV-visible spectrum. The intensity of the peak is directly proportional to the concentration of the compound. As usual, a standard solution of known concentration is also irradiated and the absorbance is plotted against the concentration to generate a calibration curve. From the curve, the concentration of the test solution can be measured.

The absorption spectrum (peak) of the test compound shows the λ_{max} (wavelength sowing maximum absorption) of the compound. However, change in the functional group of the compound and solvent environment may shift the λ_{max} and the intensity of the peak as shown in Fig. 14.83. The λ and A (or intensity) change may be defined under different terms as shown Table 14.28.

TABLE 14.28 Terms related to absorption spectrum

Term	Change in absorption peak	Factor inducing the change
Bathochromic shift (Red shift)	Shift of λ_{max} to longer wavelength	NH_2 and OH groups; decreased polarity of the solvent
Hypsochromic shift (Blue shift)	Shift of λ_{max} to shorter wavelength	Deconjugation
Hyperchromic shift	Increase in intensity absorption	Addition of new group, e.g., CH_3 group
Hypochromic shift	Decrease in intensity of absorption	Geometrical distortion of compound by CH_3 group

TABLE 14.29 Differences between colorimetry and spectrophotometry

Parameter	Colorimeter	Spectrophotometer
Light source	Tungsten filament	Deuterium or hydrogen lamp
Monochromator	Glass filter	Prism or grating
Detector	Photocell	Photomultiplier
Application	Coloured compounds are quantified	Colourless compounds are quantified

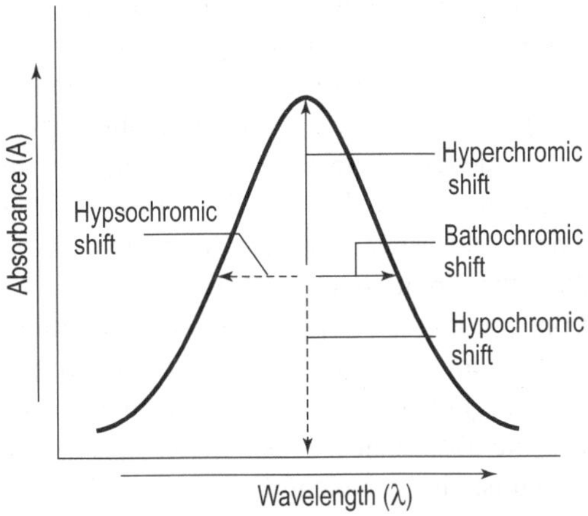

FIGURE 14.83 Absorption spectrum

Differences between colorimetry and spectrophotometry are summarized in Table 14.29.

Applications UV-visible spectroscopy is used to

- quantify protein, carbohydrates, NADPH, haemoglobin, chlorophyll, and carotenoids in biological compounds (using calibration curve and peak intensity).
- detect impurities in biological compounds by analysing the additional peaks in the spectrum.
- investigate geometrical isomers by comparing the λ_{max} of the cis form and the trans form. (Cis isomers show greater λ_{max} than trans isomers).
- determine the functional group of a compound by calculating the λ_{max} and comparing with the standard λ_{max} values, for example, λ_{max} for C–N is 165 nm and for C=C is 175 nm.
- validate the presence of conjugated double bonds, aromatic ring and keto-enol tautomeric forms in biological compounds.
- identify redox pairs by finding out the difference in λ_{max} values between the oxidized and the reduced forms. Examples include ubiquinone:ubiquinol and Cyt $C_{(oxd)}$:Cyt $C_{(red)}$.
- determine the specific activity of enzymes belonging to oxidoreductases and isomerases.

Nowadays, sophisticated variants of spectrophotometers have been developed and applied for specific purposes as shown in Table 14.30.

TABLE 14.30 Variants of UV-visible spectrophotometers

Spectrophotometer	Applications
Dual beam spectrophotometer	To determine the absorbance of test and blank simultaneously
Multibeam spectrophotometer	To record absorbance of different samples at the same time To find the λ_{max} value of a sample
Low temperature spectrophotometer	To determine the concentration of heat-sensitive compounds
Reflectance spectrophotometer	To determine the concentration of paste and suspension of microorganism
Microscale spectrophotometer	To find out the absorbance of a single living cell
Recording spectrophotometer	To record the change in absorbance with time at fixed λ To follow the enzyme-catalysed reaction

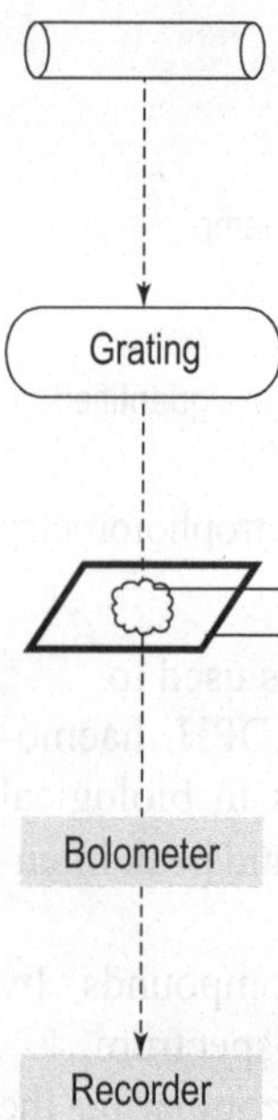

FIGURE 14.84 Instrumentation of IR spectroscopy

Infrared spectroscopy

Asymmetric molecules possessing slight positive and negative charges exhibit a permanent dipole moment. Irradiating such molecules with infrared (IR) radiation induces molecular rotation and vibration leading to an alteration in the dipole moment. The amount of IR light absorbed is proportional to the alteration in the dipole moment, which in turn is characteristic of a specific functional group.

Instrumentation and working Instrumentation schematic of an IR spectrometer is shown in Fig. 14.84. Polychromatic IR light of 800 to 150 nm wavelength emitted by a sintered silicon carbide rod or a zirconium oxide rod is passed into a grating. The emerging monochromatic beams possess frequency similar to the frequency of vibration of the sample with permanent dipole moment. The sample is usually mixed with mineral oil and placed in a rock salt plate. The IR ray brings about vibration and rotation of the sample molecules and alters the dipole moment. The possible modes of vibration (scissoring vibration, rocking vibration, stretching vibration, wagging vibration, twisting vibration, and bending vibration) of the sample atoms are represented in Fig. 14.85. The alteration in dipole moment of the molecule is directly related to the absorbance. A bolometer detects the absorbance, which is recorded against the wavelength in a recorder and shown as an absorption spectrum. The absorption maximum is the characteristic of a specific functional group of the test molecule.

Applications IR spectroscopy is used to:

- identify the functional group of a compound.
- determine the dipole moment of a molecule.
- find out the bond length and the bond angle of a molecule.
- study about photosynthesis, respiration, CO_2 metabolism, and nitrogen fixation.
- detect the secondary structure of macromolecules.
- determine isomeric mixtures.
- sense the presence of hydrogen bonds in a compound. Hydrogen bonds are shown as broadening of peaks.

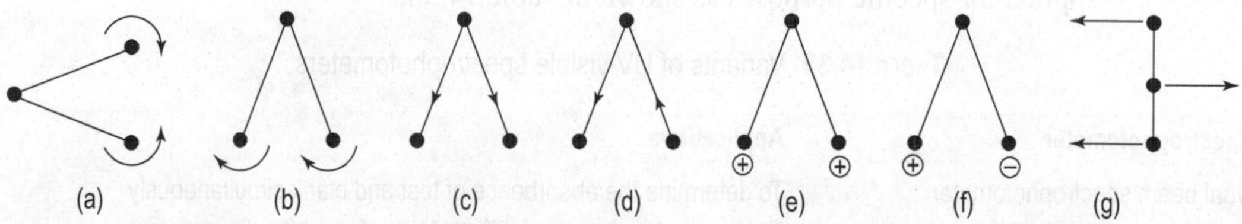

FIGURE 14.85 Molecular vibrations induced in IR spectroscopy (a) Scissoring vibration (b) Rocking vibration (c) Symmetric streching vibration (d) Asymmetric stretching vibration (e) Wagging vibration (f) Twisting vibration (g) Perpendicular bending vibration

Raman spectroscopy

Raman spectroscopy works on the following principle—a beam of light hits a molecule without permanent dipole moment, and changes its polarizability. The molecule vibrates and scatters the light. The frequency of the scattered light (vs) differs from the frequency of the incident light (vi). This phenomenon of change in frequency is known as Raman effect and the spectrum of the scattered light is known as Raman spectrum. The difference in frequency, that is, vs – vi, is called Raman frequency, vR; vR is characteristic of the vibrating functional group. The intensity of the Raman line is proportional to the change in the polarizability.

Instrumentation and working Raman spectrometer has a Toronto lamp or a helium–neon laser to emit radiation. The radiation enters a nickel oxide filter and the emerging beam is incident on the sample. The sample may be loaded in a melting point capillary or adsorbed onto roughened metal surface or colloids to get strong spectra. The light scattered by the sample molecule is detected by a photomultiplier and the spectrum of the scattered light is recorded in a recorder as Raman lines (Stokes or anti-Stokes lines). Raman lines are generated by two phenomena (Fig. 14.86) discussed as follows:

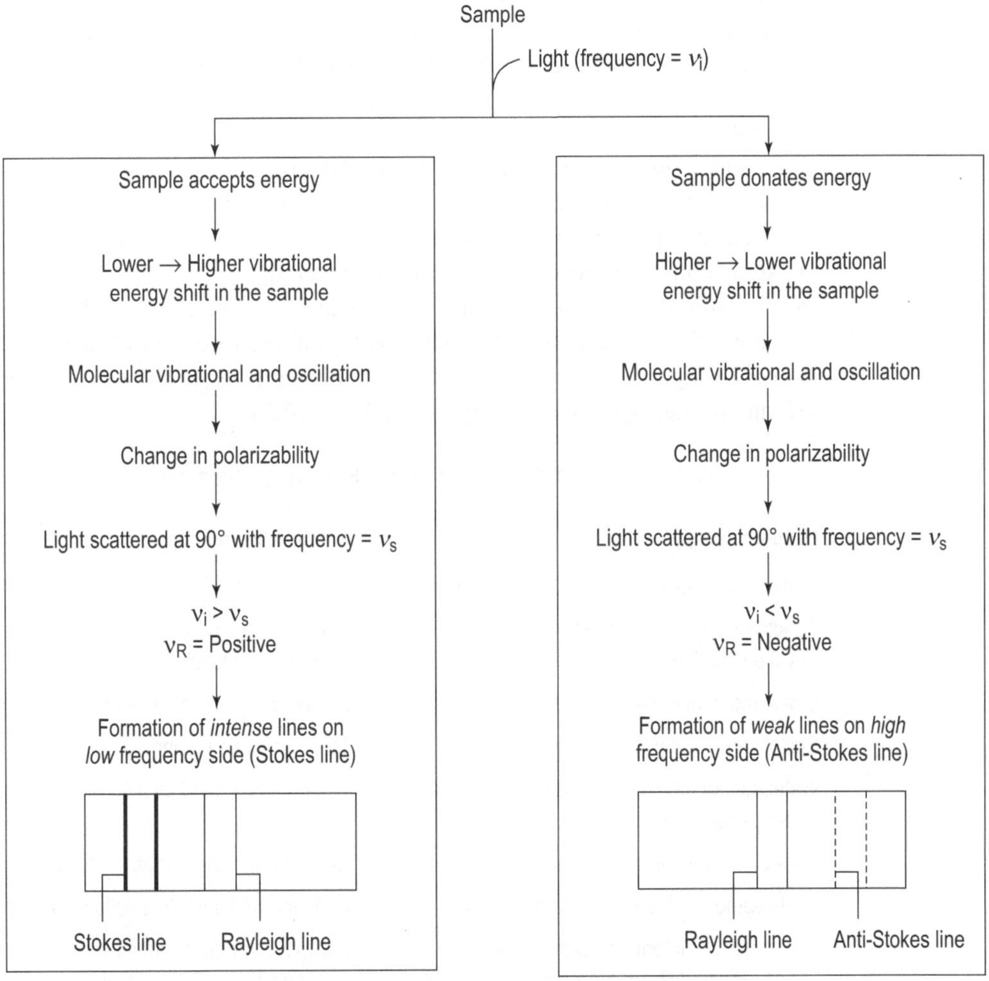

FIGURE **14.86** Phenomenon behind Raman lines

Phenomenon 1 The incident beam may donate energy to the molecule. In such case, the molecule is elevated to higher vibrational energy level, the polarizability changes, and the light gets scattered. In this phenomenon, the scattered light will have a lesser frequency less than the incident light, that is, $\nu_i > \nu_s$. The spectrum of the scattered light has intense lines called Stokes lines at the low frequency side of the incident spectrum (Rayleigh line). The difference in frequency between the incident light and the scattered light is called Raman frequency, ν_R. In phenomenon 1, ν_R is positive.

Phenomenon 2 The incident light may absorb energy from the molecule. In such case, the molecule is shifted to lower vibrational energy level, the polarizability changes, and the light is scattered. In this phenomenon, $\nu_i < \nu_s$. The spectrum of the scattered light shows weak lines, called anti-Stokes lines on the higher frequency side of the Rayleigh line. In this phenomenon, ν_R is negative. ν_R is characteristic of a particular functional group. Stokes and anti-Stokes lines are together called Raman lines.

Applications Raman spectrum is used to:

- determine the *vibrating functional group* of a molecule on the basis of ν_R. For example, ν_R of OH group is 350 cm and that of CH group is 1500 cm.
- determine the concentration of test compounds on the basis of height and intensity of Raman lines.
- elucidate linear, triangular, and pyramidal structure of molecules.
- detect the presence of double and triple bonds present in a compound.
- determine cis-trans isomers and tautomers.
- study about electrolytic and hydrolytic reactions.
- determine amorphous and crystalline nature of any compounds (amorphous form shows diffused Raman lines, whereas crystalline form shows sharp lines).
- sense impurities in dyes (impurities are displayed as extra peaks or lines).

IR and Raman spectra are compared in Table 14.31.

TABLE 14.31 Comparison of IR and Raman spectroscopy

IR spectroscopy	Raman spectroscopy
IR radiation is emitted from silicon carbide rod	Any UV/visible light source may be used
Spectrum is produced due to absorption of light by the vibrating molecule	Spectrum is produced due to scattering of light by the vibrating molecule
Dipole moment of the molecule is altered	Polarizability of the molecule is altered
Signal cannot be photographed	The signal can be photographed
Water cannot be used as a solvent because it is opaque to IR rays	Water can be used as a solvent
Rock salt plate holds the sample	Capillary or metal surface holds the sample
Dilute sample solutions are preferred	Concentrated sample solutions are preferred
Homonuclear diatomic molecules such as O_2, H_2, and N_2 are IR inactive (i.e., these molecules do not give IR spectrum)	Homonuclear diatomic molecules such as O_2, H_2, and N_2 are Raman active (i.e., these molecules give Raman spectrum)

TABLE 14.32 Correlation of I with atomic and mass numbers

Atomic number	Mass number	I value
Even	Even	0
Even	Odd	1/2, 3/2, 5/2, ...
Odd	Odd	1/2, 3/2, 5/2, ...
Odd	Even	1, 2, 3, ...

Nuclear magnetic resonance spectroscopy

Spinning (denoted as I) is an important property of nucleus. The value of I is determined by the atomic number and the mass number of the element as represented in Table 14.32.

Compounds with $I = 1/2$ produce a magnetic field known as nuclear magnetic field (NMF). An external magnetic field (EMF) can split NMF into two energy levels:

1. Lower energy level, which is parallel to EMF
2. Higher energy level, which is anti parallel to EMF

In NMR spectroscopy, a compound with $I = 1/2$ is exposed simultaneously to EMF and radiowave. At a specific combination of EMF strength (H_o) and radio wave frequency (rwf), the nucleus is excited to a higher magnetic energy level, causing resonance. Such a resonance signal is called nuclear magnetic resonance (NMR) signal, and is recorded as NMR peak. NMR peak provides structural information about the compound.

Instrumentation and working As hydrogen itself produces NMR signal, the sample for NMR analysis is dissolved in a heavy solvent containing deuterium in the place of hydrogen, For example, deuterium oxide is (D_2O) used. In NMR spectrometer, the sample taken in an inert glass tube is kept between two permanent electromagnets containing Helmholtz coil as shown in Fig. 14.87. A radio frequency oscillator coil is wound around the sample tube. The magnet creates a homogenous magnetic field (Ho), whereas the frequency oscillator coil generates the radiowave.

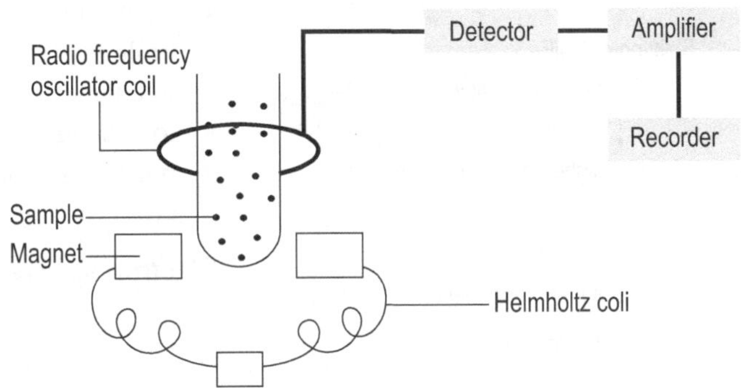

FIGURE 14.87 NMR spectrometer

A particular combination of the external magnetic field (H_o) and radiowave frequency (rwf) shifts the nucleus to a higher magnetic energy level leading to resonance. The combination of H_o and rwf may be achieved by two methods:

- Field sweep method, where H_o is varied and the rwf is kept constant.
- Frequency sweep method, where the rwf is varied and H_o is kept constant.

The frequency at which the nuclear resonance takes place is known as resonance frequency. The nuclear resonance is sensed by a detector, amplified by an amplifier, and recorded as an NMR signal or NMR peak (i.e., a plot of absorbance versus magnetic field). NMR spectroscopy is summarized in Fig. 14.88. The information derived from the NMR peak is summarized in Table 14.33.

TABLE 14.33 Information derived from NMR peak

Peak character	Information derived
Number of peaks	Number of peaks is proportional to the number of equivalent protons (i.e., no. of protons with same environment) present in the compound, for example, ethane (C_2H_6) shows 1 peak; ethyl chloride (CH_3 CH_2 Cl) shows 2 peaks
Area of the peak	Area of the peak is proportional to the number of protons giving rise to the peak
Intensity of the peak	Intensity of the peak is equivalent to the number of protons
Splitting of the peak (hyperfine splitting)	Splitting is related to the environment of the proton absorbing the wave, with respect to the neighbouring proton
Position of the peak position	Position of the peak gives an idea about the electronic environment of each proton

Applications NMR is used:

- to identify the active site amino acid residue of enzymes.
- to elucidate the structure of antibiotics, complement proteins, interleukin, growth factor, and nucleic acids.

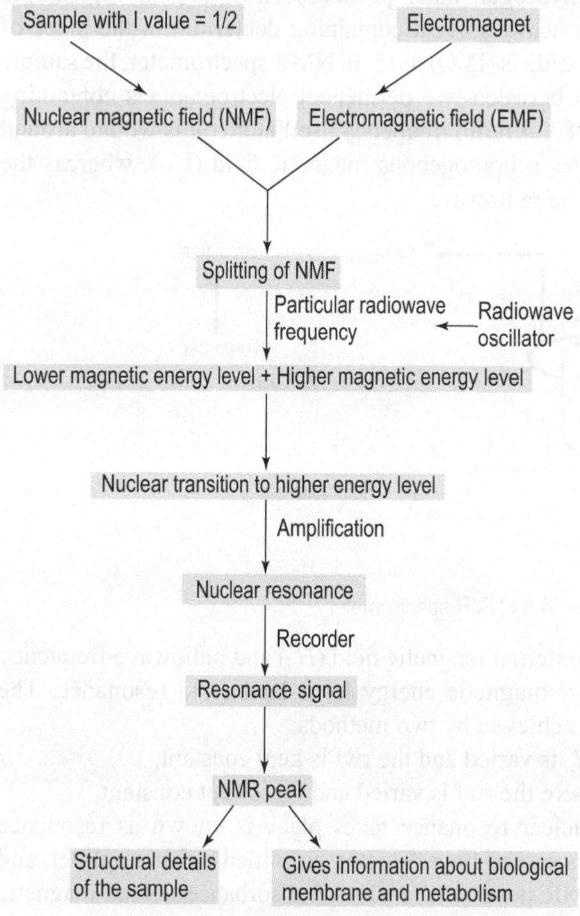

FIGURE 14.88 Outline of NMR spectroscopy

- to examine protein folding.
- for investigating the kinetics of kinases, proteases, and phosphatases.
- for investigating tissue biopsy samples.
- to analyse drug metabolism.
- for analysing the conformations of macromolecules.
- to monitor tumour mass and polyps in biological systems.
- to characterize membrane dynamics.
- to determine the characteristics of polymers, such as branching, crystalinity, cross linking.

Electron spin resonance spectroscopy

Electron spin resonance (ESR) spectroscopy is always exhibited by paramagnetic substances with spinning electrons, therefore, it is also called electroparamagnetic resonance (EPR) spectroscopy. The principle of ESR spectroscopy is similar to NMR spectroscopy except that the spinning component is electron. The paramagnetic substances containing unpaired electron spin, I = ½, generate a magnetic field. The substance is exposed to a microwave radiation along with an external magnetic field. At a particular combination of magnetic field strength and microwave frequency, the electron is excited to a higher

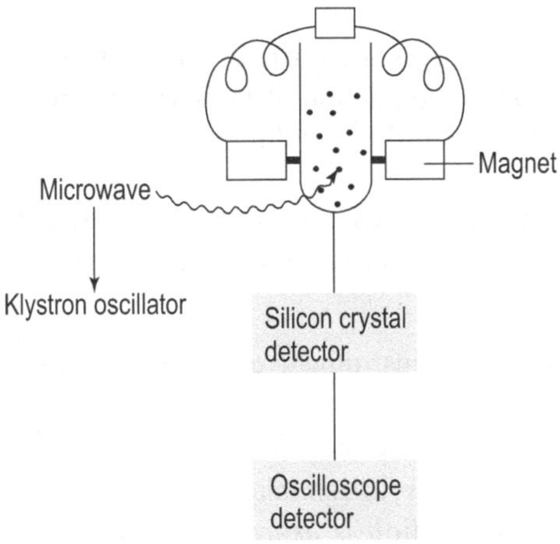

FIGURE 14.89 Components of ESR spectrometer

magnetic energy level, causing electron resonance. The resonance signal is detected, and a plot of absorbance versus magnetic field strength (ESR spectrum) is generated.

Instrumentation and working The instrument required for ESR spectroscopy is known as ESR spectrometer (Fig. 14.89).

The paramagnetic substance is taken in a sample holder held between a pair of electromagnets. The microwave produced by a Klystron oscillator is passed into the sample (thus, ESR is a type of absorption spectroscopy). At a specific combination of microwave frequency and magnetic field strength, the electrons show resonance and excitation to the higher magnetic energy level. A silicon crystal detector detects the resonance signal and an oscilloscope records the ESR signal [a plot of absorbance versus magnetic field (A vs Ho)]. The intensity of ESR peak is related to the total number of unpaired electrons present in the sample.

Simultaneous resonance of an electron with microwave and of nucleus with a radiowave improves the resolution of ESR signal. This technique is called electron nucleus double resonance, ENDOR. Simultaneous irradiation of sample with microwaves of two different frequencies to induce resonance is called electron double resonance, ELDOR.

Applications ESR is used:

- for investigating the molecular structure of paramagnetic substances.
- to study about free radicals.
- to find out the oxidation state of metals, for example, Cu I and Cu II.
- to investigate the electron exchange reactions.
- for the diagnosis of diseased tissues based on the conformation of free radicals.

Luminometry

Luminometry works on the principle of luminescence. Luminescence is the phenomenon where light is emitted when an excited electron in the higher energy level comes back to the ground level.

Luminescence is of the following types:

1. Chemiluminescence—electrons are excited by a chemical reaction.
2. Bioluminescence—electrons are excited by an enzyme-catalysed biological reaction.
3. Fluorescence—luminescence stops as the light source inducing excitation is cut off.

Chemiluminometry and Bioluminometry Chemiluminometry and bioluminometry are techniques used for measuring the chemiluminescence and bioluminescence, respectively. In chemiluminometry, the light emitted during the chemical reaction is measured. The intensity of chemiluminescence is proportional to the reaction rate and the amount of the reactant or the product involved in the reaction. For example, the oxidation reaction of luminol in the following reaction is determined by chemiluminometry.

The intensity of the light emitted by the reaction is proportional to the rate of oxidation of luminol.

In bioluminometry, the intensity of the bioluminescence is proportional to (i) the concentration of the reactant utilized or the product generated, (ii) the activity of the enzyme, and (iii) the rate of the reaction. For example, oxidation of luciferin by fire fly luciferase in the presence of ATP (in the following reaction) is a bioluminescent reaction, which can be analysed by bioluminometry.

The intensity of the light emitted by the bioluminescent reaction is proportional to the (i) the concentration of the ATP, (ii) the activity of the luciferase enzyme, and (iii) the rate of oxidation of the luciferin.

Instrumentation Luminometry is performed using an instrument called luminometer (Fig. 14.90). It contains a light-resistant reaction chamber into which the reactants are introduced using a syringe. Since EMR is not used in luminometer, light source and monochromator are present. Under standard conditions, reaction takes place within the reaction chamber and the light emitted during the reaction is examined directly by a photomultiplier tube. The intensity of emission (luminescence) is recorded against the reaction time to generate a luminescence spectrum.

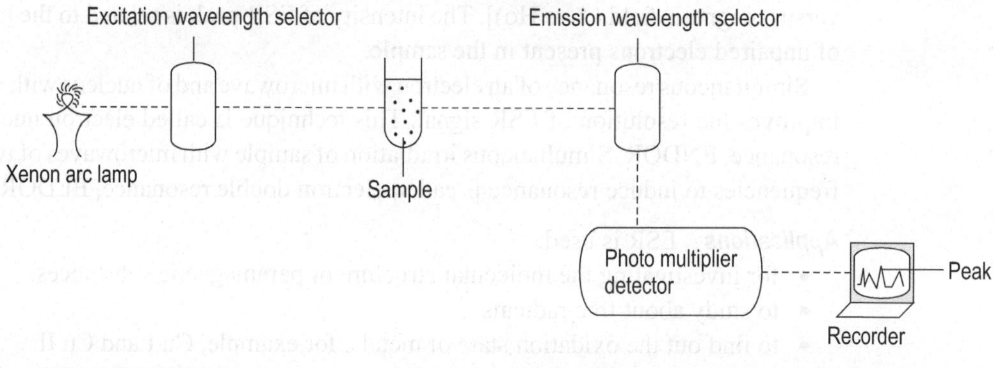

FIGURE 14.90 Components of a fluorimeter

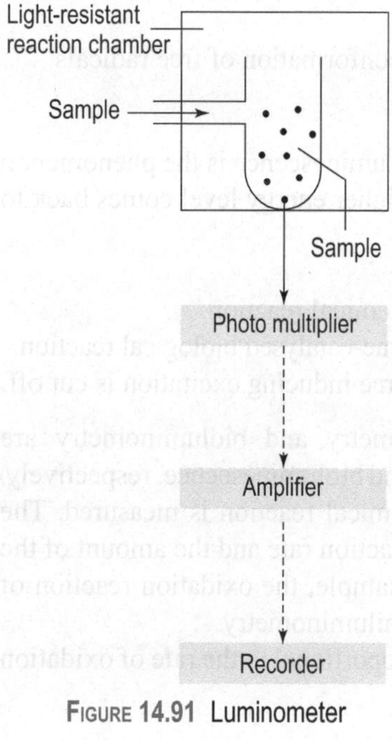

FIGURE 14.91 Luminometer

Applications Luminometry is used:

- to determine metal ions resolved by ion exchange chromatography.
- to determine the concentration of analytes and activity of the enzymes involved in a chemical or a biochemical reaction.
- for analysing DNA in DNA finger printing, DNA sequencing, and Southern blotting.

Fluorimetry

Fluorimetry is a technique used to measure fluorescence. When a light of higher energy and shorter wavelength is incident on certain substances, the electron is excited from ground state to excited state. When the electron is transferred back to the ground state, the substance glows by emitting light of lesser energy and longer wavelength than the incident light. This phenomenon is known as fluorescence and the substances emitting light are referred to as fluors. Fluorescence stops immediately after the excitation is arrested. Measurement of fluorescence intensity is used for the quantification and characterization of fluorescent compounds.

Instrumentation and working Fluorimetry is performed using fluorimeter, which is schematically represented in Fig. 14.91.

Polychromatic light generated by a xenon arc lamp falls on a wavelength selector (the primary monochromator). This results in the filtration of UV radiation only. The radiation is then penetrated through the sample. Compounds possessing double bonds and electron donating groups, such as amino group and hydroxyl group, fluoresce. The fluoresced light is selected by another wavelength selector (the secondary monochromator) and the light is detected by a photomultiplier and recorded as a fluorescence emission spectrum (intensity versus emission). The intensity of fluorescence is proportional to the concentration of the sample.

Applications Fluorescence measurement is used:

- for quantifying compounds such as amino acids and peptides by labelling with an extrinsic fluor, acridine orange.
- for detecting enzymes in cells and metals in metalloproteins.
- for determining the conformations of enzymes proteins and nucleic acids.
- to sense single-stranded and double-stranded DNA and 3-D structure of tRNA.
- to determine the radioactivity using scintillation counter.
- to perform fluorescent immunoassay and blotting techniques.
- to locate malignant cells (acridine orange labelled malignant cells fluoresce more intensely than normal cells).
- to detect compounds separated by chromatography (fluorescence assay).
- to analyse membrane structure and fluidity.
- in cell sorting and cell counting.
- for in situ imaging of cells and tissues.

Circular dichroism spectroscopy (CD spectroscopy)

CD spectroscopy makes use of a plane polarized light for quantifying a chiral compound (optically active compound). A plane polarized light is made up of two components—left circularly polarized component and right circularly polarized component. When a plane polarized light is passed through a chiral compound, the compound absorbs the two components differentially due to its asymmetric nature. Consequently, the resultant transmitted light will be elliptically polarized, and the ellipticity is directly proportional to the quantity of the chiral compound.

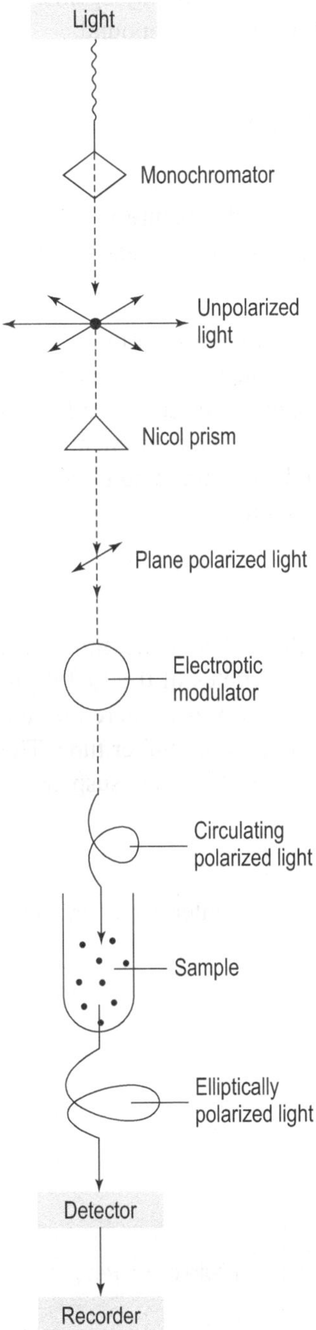

FIGURE 14.92 CD spectrometer

Instrumentation A CD spectrometer is schematized in Fig. 14.92. A beam of monochromatic light is transformed into a plane polarized light by a Nicol prism. The plane polarized light in turn is transformed into a circularly polarized light by an electro optic modulator. The circularly polarized light is penetrated through an optically active sample. The sample absorbs and transmits the radiation as an elliptically polarized beam. A photomultiplier detects the ellipticity and a recorder records the ellipticity and generates a CD spectrum by plotting the ellipticity against the wavelength. The ellipticity is related to the concentration of the chiral compound.

Applications CD spectroscopy is used:

- to determine the percentage of secondary structure in a protein.
- to identify the denaturation of proteins.
- for studying the conformation of proteins, enzymes, and nucleic acids.
- to determine the structure and nature (single- or double-stranded nature) of DNA.
- to identify the structure of optically active molecules such as amino acids and peptides.

Nephelometry and turbidometry

Nephelometry and turbidometry are used to analyse suspensions and turbid solutions such as bacterial suspension. In nephelometry, an instrument called nephelometer measures the intensity of the light scattered by a suspension. The intensity of the scattered light is proportional to the concentration of the suspension. In turbidometry, an instrument called turbidometer determines the intensity of the light transmitted by a turbid solution. The intensity is proportional to the concentration of the turbid solution.

Instrumentation Figure 14.93 represents the components of a nephelometer and a turbidometer.

Nephelometer Radiation produced by a mercury arc is allowed to enter through a lens and a filter and passed through a suspension containing particles of 0.1 to 1.0 μm diameter. The suspension is held in a rectangular cell. The light ray is scattered by the sample maximally at 90°. The scattered light is detected by a photomultiplier tube. The intensity of the scattered light is proportional to the concentration (C) of the suspension as represented by the following equation:

$$I_s = k_s \times I_i \times C$$

where I_s = intensity of scattered light, k_s = empirical constant, I_i = intensity of incident light, and C = concentration of the sample.

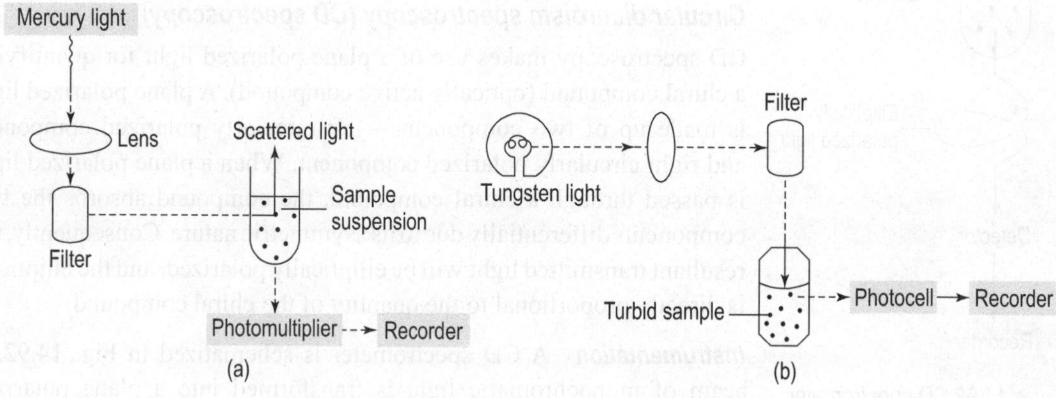

FIGURE 14.93 (a) Nephelometer (b) Turbidometer

TABLE 14.34 Differences between nephelometry and turbidometry

Nephelometry	Turbidometry
Mercury arc lamp is used to irradiate the sample	Tungsten/deuterium lamp is used to irradiate the sample
Sample is held in a rectangular cuvette	Sample is held in a semi octagonal cuvette
Scattered light is measured	Transmitted light is measured
Measurement is done at 90°	Measurement is done in straight line
Photomultiplier detects the scattered light	Photocell detects the transmitted light

Turbidometer In a turbidometer, a beam of UV light or visible light from a tungsten lamp or deuterium lamp, respectively, is passed through a condensing lens and a filter. The filtered beam is incident on the turbid solution taken in a semi octagonal glass cell or quartz cell. The solution absorbs and transmits the light. The intensity of the transmitted light, T is detected by a photocell and recorded in a recorder. Transmittance is related to the turbidity (S) and concentration (C) by the following equation:

$$\log T = S = kbC$$

where k = turbidity coefficient and b = path length.

The transmittance of light by known concentration of formazin polymer can be used as standard and a calibration curve can be created. Differences between nephelometry and turbidometry are shown in Table 14.34.

Applications Nephelometry and turbidometry are used:

- for determining the concentration of bacterial suspension and paste.
- to determine the amount of amino acids, proteins, vitamins, glycogen, and antibiotics in blood.
- to calculate the molecular weight of biomolecules using the following formula—

$$M = T/HC$$

where M = molecular weight of the biomolecule, T = turbidity of the solution of the biomolecule, H = a constant, and C = concentration of the macromolecule in g ml^{-1}.

X-ray spectroscopy

X-rays are electromagnetic radiations with shorter wavelength and high penetrating power. Although, X-rays are a component of EMR spectrum, they can also be generated in a Coolidge tube by impinging fast-moving electrons on a metal surface of high melting point.

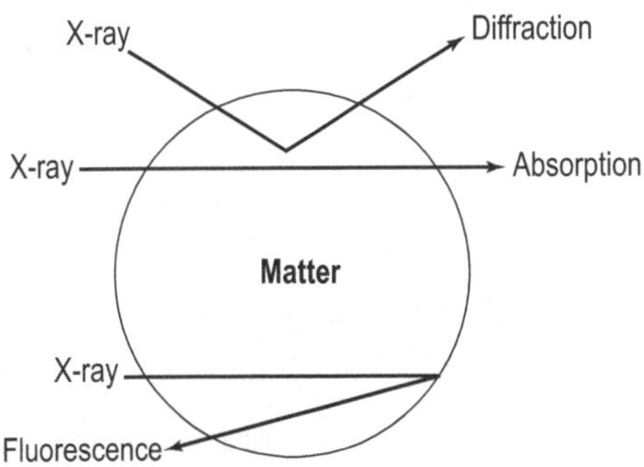

FIGURE 14.94 Phenomena due to interaction of X-rays with matter

Interaction of X-rays with matter can give rise to any one of the following three phenomena (Fig. 14.94):

- X-ray absorption
- X-ray diffraction
- X-ray fluorescence

X-ray absorption When X-rays are absorbed by a sample, the corresponding absorption spectrum is characteristic of the sample. The intensity of the absorption spectrum is related to the concentration of the element in the sample.

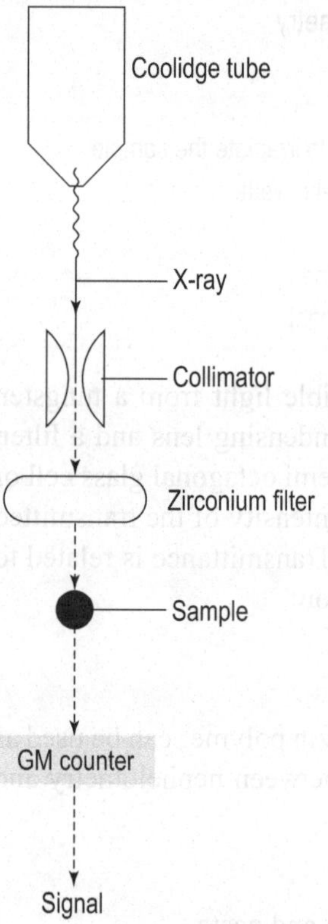

FIGURE 14.95 X-ray absorption spectrometer

Instrumentation X-ray absorption is determined by X-ray absorption spectrometer (Fig. 14.95). It contains a Coolidge tube, which generates X-rays. The X-ray is collimated into a narrow pencil of beam by a collimator. A zirconium filter converts the narrow beam into a monochromatic beam. The monochromatic beam is absorbed by the sample. Absorption is detected by a GM counter and the absorbance is amplified and recorded as an absorption spectrum. The absorption spectrum is characteristic of the sample.

Applications X-ray absorption is widely used in medicine for detecting breaks, fractures, and cracks in bone. It is also used to locate blocks and deposits in blood vessels. It is also used to determine the phosphorus uptake by plant cells.

X-ray diffraction Crystalline solids, such as metals, polymers, and ceramics, diffract the X-rays. The diffraction pattern determined by an X-ray diffractometer is used for characterization and structural analysis of crystalline solids. This technique has limited applications in cell biology.

Instrumentation The X-ray diffractometer is of two types—single crystal diffractometer and powder diffractometer (Figs. 14.96 and 14.97). In both types, a Coolidge tube produces polychromatic X-ray, which is converted into a narrow monochromatic beam by a collimator and a monochromator, respectively. The monochromatic beam is incident either on a solid rotating crystal sample (in single crystal diffractometer) or on a crystal powder (in powder diffractometer). In either case, the beam is diffracted. A camera set parallel along the diffracted beam and axis of rotation of the sample is used to display the diffraction pattern and the diffraction intensity is displayed in a photographic film. The diffraction pattern is used for characterization and structural analysis of the crystalline samples.

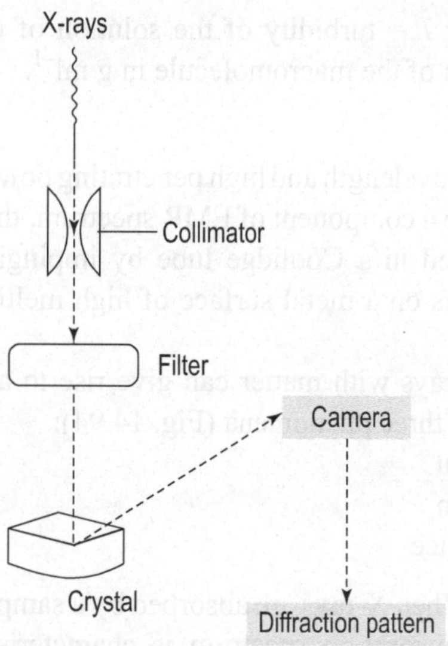

FIGURE 14.96 Single crystal diffractometer

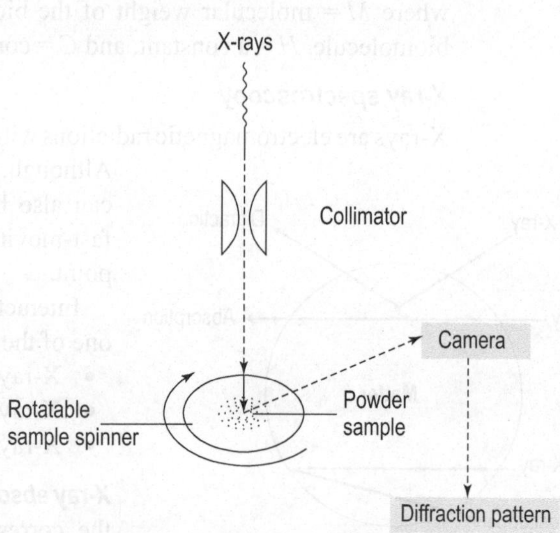

FIGURE 14.97 Powder diffractometer

Applications X-ray diffraction is used to:

- detect the degree of crystallinity of polymers.
- differentiate allotropes of any substance.
- find out the symmetry, dimension, and lattice structure of crystals.
- assess the quality of metals (non-biological applications).

X-ray fluorescence X-ray fluorescence is a phenomenon by which an element absorbs an X-ray (the incident ray or the primary X-ray) of a specific wavelength, undergoes electronic excitation, and releases an X-ray beam (the fluorescent ray or the secondary X-ray) of a different wavelength. The wavelength of the fluorescent X-ray is characteristic of the element, whereas the intensity of the fluorescent X-ray is related to the quantity of that element.

Applications X-ray fluorescence is used:

- to determine the sulphur content of protein samples.
- for the qualitative and quantitative analysis of elements in blood, tissues, and bone.
- to detect trace elements, insecticides, and pesticides in fruits and vegetables.

14.8 ELECTROCHEMICAL TECHNIQUES

Electrochemistry is the study of relationship between a chemical reaction and electricity and its practical application. In fact, biochemical reactions done in vitro can also be linked with electrochemical technique. Electrodes (anode and cathode) are basic components used in this technique. A system that consists of an element in contact with the solution of its ion is called an electrode or a half cell. On changing the nature of the electrodes used, this technique can be used to derive information on various components of the cell and body fluids. For example, calomel electrode and glass electrode are used to measure the pH of any body fluid, whereas silver–silver chloride electrode and Clark's electrode are used for measuring oxygen, and pCO_2 electrode is used for measuring carbon dioxide content of the blood. A system in which anode and cathode are dipped in an ionic solution and interconnected by a salt bridge is called an electrochemical cell or galvanic cell. It is used to measure oxygen-consuming enzyme system.

14.8.1 Galvanic Cell and Measurement of Enzyme in Oxygen-Consuming Reaction

Galvanic cell is used for understanding the basic principle of electrochemical technique. A galvanic cell consists of two electrodes of different metals, interlinked by a salt bridge. In the presence of applied voltage, a redox reaction occurs. That is, oxidation takes place at the positive electrode (anode) and reduction takes place at the negative electrode (cathode), leading to the generation of electricity. A galvanic cell, thus, converts chemical energy into electrical energy. This electrochemical process can be easily understood by studying the principle of a Daniel cell (Fig. 14.98). It consists of two half cells, one half cell is composed of a zinc rod immersed in zinc sulphate solution, and connected to the positive terminal of the battery. The other half cell consists a copper rod immersed in copper sulphate solution, and connected to the negative terminal of the battery. A KCl salt bridge connects both the half cells. A galvanic cell may be represented as:

$$Zn / Zn^{2+} SO_4^{2-} \parallel Cu^{2+}SO_4^{2-} / Cu$$

Anode (half cell) Cathode (half cell)

When a voltage is applied between the anode and the cathode, oxidation (loss of electrons) occurs at the anode to form zinc ions, that is, $Zn \rightarrow Zn^{2+} + 2e^-$. These electrons flow through the external circuit (shown by arrow in Fig. 14.98) and reach the cathode, where, it is accepted by the Cu^{2+} ions to form copper atoms. That is:

$$Cu^{2+} + 2e^- \rightarrow Cu$$

Thus, reduction takes place at the cathode.

The tendency of an electrode to reduce or oxidize itself is called single electrode potential. In a galvanic cell, the tendency of the Zn anode to donate electron is called the oxidation potential. The tendency of the Cu cathode to accept electron is called the reduction potential. The sum of the oxidation potential and the reduction potential of the galvanic cell is called the cell potential or electromotive force (EMF) or E_{cell}.

EMF = E_{cell} = Oxidative potential of the anode + Reduction potential of the cathode

The oxidation or reduction potential can be calculated by integrating the anode or the cathode with a standard reference electrode; the EMF can then be determined. A reference electrode is of known potential, insensitive to current passage and unknown solution. Standard hydrogen electrode (SHE, Fig. 14.99) is an example of a reference electrode. SHE consists of an inert platinum wire dipped in 1N HCl, through which purified hydrogen gas is bubbled at 1 atm pressure. The potential of the standard hydrogen electrode is assigned as 0.00 V.

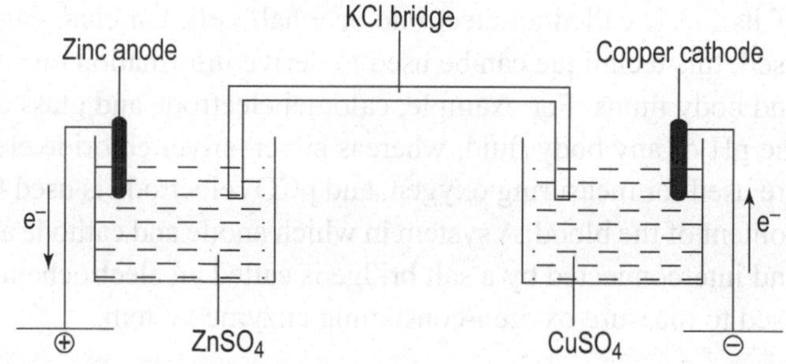

FIGURE 14.98 Galvanic cell (Arrow indicates the movement of electrons)

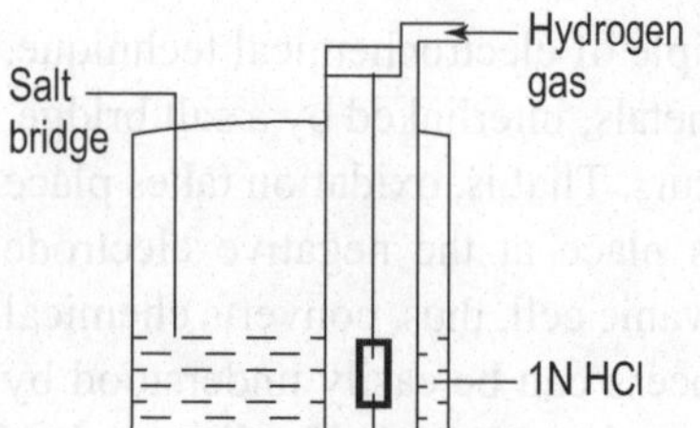

FIGURE 14.99 Standard hydrogen electrode

To measure the oxidation potential, the Zn anode is coupled with the SHE and a voltage is applied. The EMF is measured using a potentiometer. As the potential of SHE is assigned 0.00 V, the EMF is equal to the oxidation potential of the zinc electrode. The EMF for zinc anode is reported to be +0.76 V. Similarly, the reduction potential of Cu cathode is estimated to be +0.34 V. Therefore, EMF of the Daniel cell is the sum of oxidation potential of zinc anode and the reduction potential of copper cathode. That is,

$$EMF_{Daniel\ cell} = (0.76) + (0.34) = 1.1\ V.$$

Galvanic cell is used in the measurement of changes in oxygen-consuming systems which consume less oxygen than what can be measured by conventional manometry. Tyrosinase is a copper-containing enzyme of plant and animal cells. As discussed in Chapter 7, it catalyses the formation of melanin from tyrosine. Activity of tyrosinase can be measured by measuring the oxygen changes using Galvanic cell. Galvanic cell has also proved to be efficient in measuring the reactions which occur too faster. For example, oxidation of DOPA by tyrosinase and oxidation of L-leucine by leucine oxidase can be investigated by Galvanic cell.

14.8.2 Electrodes and Their Applications

This section discusses different types of electrodes and their components along with their applications. Electrodes are incorporated into a small device called pH meter which is mainly used in the measurement of pH, an essential parameter for all the biology related experiments and research. For example, assay of enzyme activity in a given tissue requires a buffer solution of optimum pH, quantification of cellular DNA from any tissues, or even isolation of plasmid DNA from bacterial cell needs optimum pH to be maintained during the experimental conditions. Experiments involving isolation, fractionation, and purification of tissue proteins and blood proteins also depend on pH measurement. We can, therefore, say that electrochemical technique is used in cell biology research and laboratory experiments. Measurement of pH using the pH electrodes is discussed in Section 14.8.2(d). Apart from this, another electrode called oxygen electrode has wider biological applications as mentioned in the following discussion.

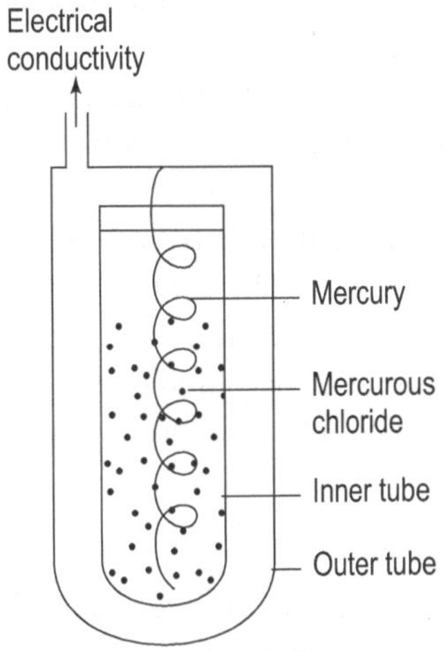

FIGURE 14.100 Calomel electrode

(a) Calomel electrode

Calomel electrode refers to a mercury electrode dipped in mercurous chloride (calomel). It is highly stable and used as a reference electrode in pH measurement. Figure 14.100 shows the components of a calomel electrode. It has a tube provided with a glass wool at the bottom and packed with a paste of mercury and mercurous chloride dispersed in a saturated solution of potassium chloride. This tube placed inside another tube, containing a frit at its bottom. The function of the frit is to allow electrical contact with the outside solution.

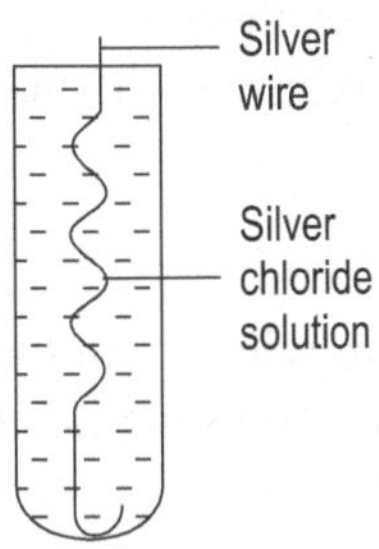

FIGURE 14.101 Silver–silver chloride electrode

(b) Silver–silver chloride electrode

Silver–silver chloride (Ag–AgCl) electrode consists of a silver wire immersed in silver chloride solution (Fig. 14.101). It is used as an internal reference electrode for pH meter.

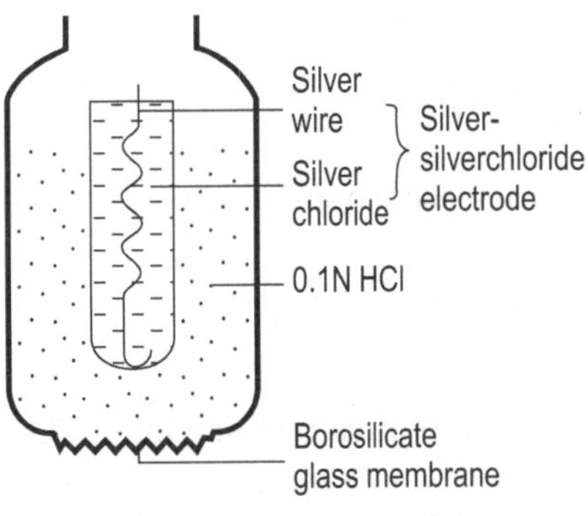

FIGURE 14.102 Glass electrode

(c) Glass electrode

Glass electrode is composed of a silver–silver chloride electrode dipped into a thick glass tube filled with 0.1N HCl, and provided with a thin borosilicate glass membrane at its bottom (Fig. 14.102).

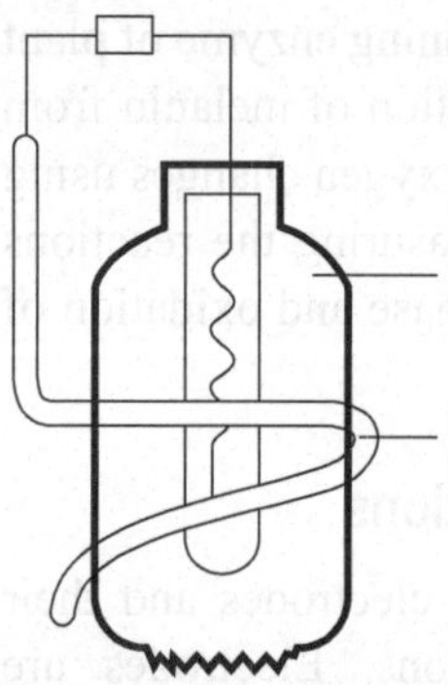

FIGURE 14.103 pH electrode

The borosilicate membrane is selectively permeable to protons (hydrogen ions). This electrode is used for determining the pH of biological fluids or biological solutions.

(d) pH electrode

pH electrode is made up of a glass electrode and a calomel electrode (Fig. 14.103). The electrode is incorporated into a small device called pH meter. The pH meter is used for measuring the pH of solutions and thus assumes greater biochemical significance.

The pH meter is a pH electrode connected to a potentiometer. The electrode is immersed in a solution of unknown pH, and a voltage is applied. As the calomel electrode is insensitive to the unknown solution it is the reference electrode and it possess a constant potential of –0.250 V. The glass electrode is highly sensitive to the unknown solution. Depending on the hydrogen ion concentration (pH) of the test solution, the glass electrode establishes a potential, resulting in a current flow. The magnitude of the current is measured by the pH meter, which is directly proportional to the ion concentration and inversely proportional to the pH of the test solution.

(e) Oxygen electrode

An oxygen electrode consists of a hydrogen electrode and a silver–silver chloride electrode (reference electrode), both dipped in an electrolyte into which oxygen can diffuse.

The electrode assembly is kept in a chamber containing the test solution of unknown oxygen content. An oxygen-permeable membrane is present between the electrode system and the test solution. In the presence of applied voltage, silver is oxidized with the loss of electrons as follows:

$$Ag \rightarrow Ag^{2+} + 2e^-$$

These electrons move along the circuit and reach the platinum wire of the hydrogen electrode. The oxygen molecule diffuses to the platinum wire and accepts the electron. As a result, oxygen is reduced to water as follows:

$$O_2 + 4\,e^- + 4H^+ \rightarrow 2H_2O$$

This redox reaction and electron flow in the oxygen electrode generates current that is proportional to the partial pressure of the oxygen (pO_2).

Oxygen electrode is used to:

- measure the partial pressure of oxygen in the human blood.
- study mitochondrial respiration.
- study the properties of enzymes such as glucose oxidase, amino acid oxidase and catalase.
- determine the oxygen evolution during photosynthesis in plant cells.

(f) pCO_2 electrode

The pCO_2 electrode is used for measuring the partial pressure of carbon dioxide (pCO_2) in blood samples. It is a combination of a calomel electrode and a glass electrode, both enclosed in a CO_2 permeable rubber membrane case (Fig. 14.104). A thin film of water is located between the rubber membrane and the pH electrode. The electrode system is kept

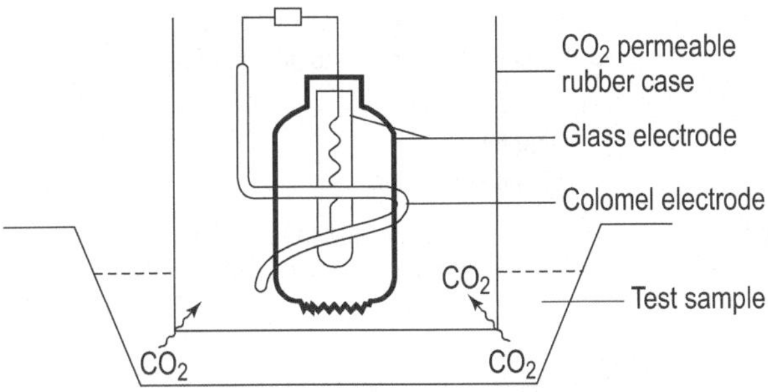

FIGURE 14.104 pCO_2 electrode

in the test solution. The CO_2 molecules from the test solution diffuse via the rubber membrane and dissolve with water film. The pH electrode component measures the pH of the water film. pCO_2 value is then interpreted on the basis of the linear relationship between pH and pCO_2. Knichwitz et al. have suggested that when there is a gastrointestinal disorder, metabolic and histologic changes occur. This results in an alteration in the intramucosal pH and pCO_2. In such conditions, pCO_2 measurement can be done using this electrode.

14.9 OTHER TECHNIQUES RELATED TO CELL BIOLOGY

The previous sections have discussed about the major separation techniques, characterization techniques, and imaging techniques used in cell biology. This chapter concludes by briefing to the readers about other techniques of indirect relevance to cell biology.

14.9.1 DNA Fingerprinting (DNA profiling, DNA typing)

DNA fingerprinting is the creation of a pattern of an individual's DNA fragments. This pattern is unique for an individual.

There are different methods for carrying out DNA fingerprinting:

- Polymerase chain reaction (PCR)
- Restriction fragment length polymorphism (RFLP)
- DNA foot printing

Polymerase chain reaction

PCR is used in DNA fingerprinting. It makes numerous DNA copies and amplifies a desired copy. PCR technique is explained in the following text. Steps involved in PCR are schematized in Fig. 14.105. The components needed for PCR are:

- Primer
- Target DNA (Template)
- Taq polymerase
- Magnesium ion
- Deoxynucleotide substrates
- Ligase
- PCR equipment

In PCR-dependent fingerprinting, the primers are designed to flank the variable number of tandem repeats (VNTR) locus. Thus, the size of the PCR product is proportional to the number of repeats. This creates a unique finger print.

PCR (polymerized chain reaction) is a technique used to amplify DNA molecules. It is a stepwise process that makes use of various enzymes, coenzymes, and metal cofactors. PCR involves four steps—denaturation, renaturation or annealing, polymerization or synthesis, and cycle repetition. The first three steps constitute one complete PCR cycle. Several cycles are repeated for amplification, which is considered as the fourth step for convenience.

Denaturation Denaturation is the first step in PCR. DNA strands to be amplified are separated by denaturation which is brought about by increasing the temperature of the reaction mixture to 95°C. The two strands are called templates.

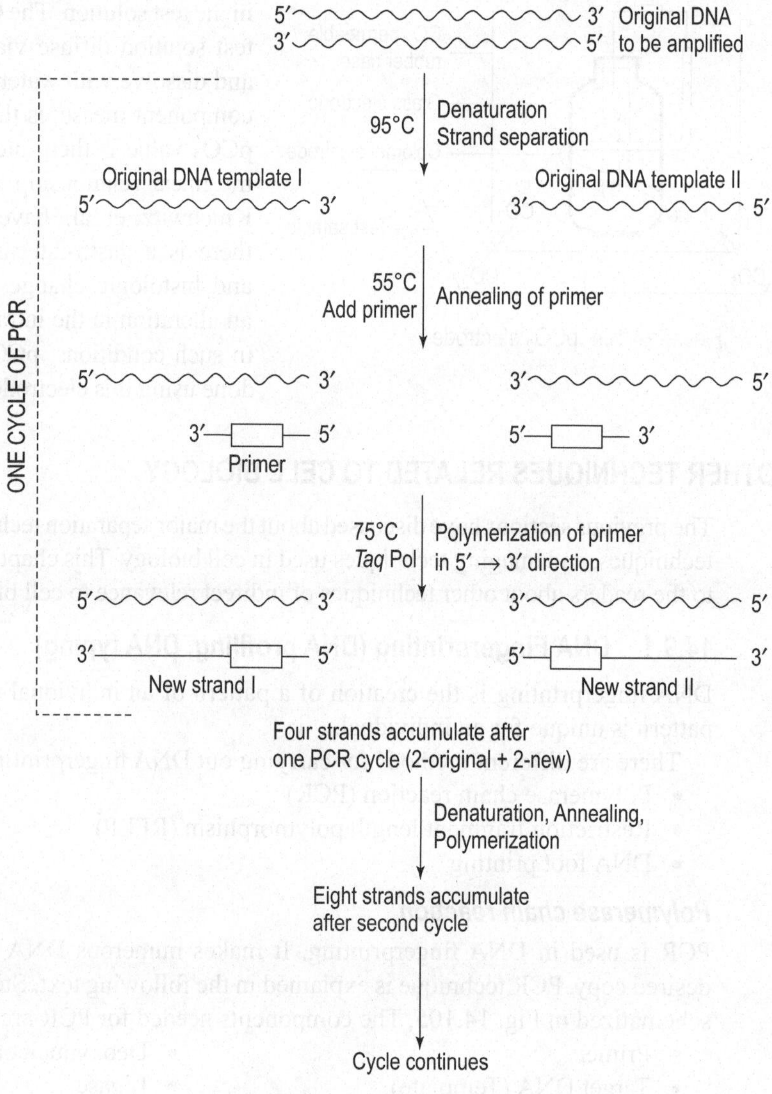

FIGURE 14.105 Polymerase chain reaction

Renaturation or annealing After strand separation, the temperature of the mixture is slowly decreased to 55°C; this process is called annealing or renaturation. At this stage, the primer strand base pairs with the respective complementary template.

Polymerization or synthesis Polymerization is the last step of PCR cycle and is usually done at a temperature of 75°C. In this step, DNA synthesis initiates at the 3′OH site of the two primers. This results in extension or polymerization of primers along the length of the template. Polymerization process is catalysed by an enzyme called Taq polymerase (Taq Pol) essentially in presence of magnesium ion as a cofactor. Polymerization process of PCR mimics leading strand replication. It ends up in four strands—two templates and two newly synthesized strands.

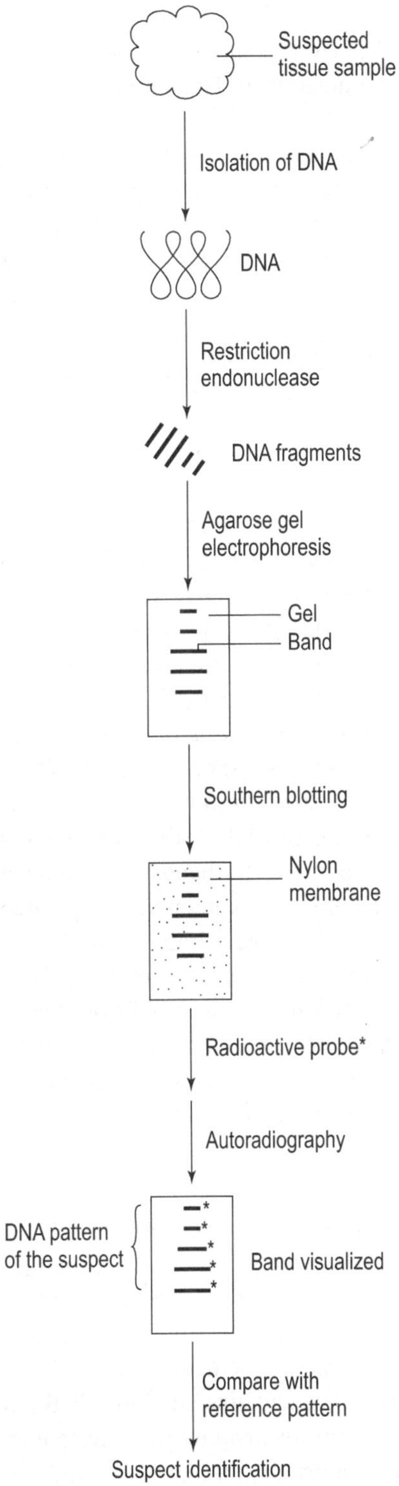

FIGURE 14.106 Restriction fragment length polymorphism

Cycle repetition The three processes mentioned above constitute a full cycle of PCR. This cycle is repeated several times, usually more than thirty times. The next cycle always starts by using the total number of strands obtained in the previous cycle as individual templates. Thus, the second cycle starts with four templates which resulted from the first cycle. Hence, at the end of second cycle, eight strands would be obtained, four template strands plus the corresponding four new strands. At the end of the 32nd cycle, large amount of target DNA accumulates.

The number of copies obtained after each cycle can be summarized as follows:

1st cycle $= 2^2$ copies	7th cycle $= 2^8$ copies
2nd cycle $= 2^3$ copies	8th cycle $= 2^9$ copies
3rd cycle $= 2^4$ copies	9th cycle $= 2^{10}$ copies
4th cycle $= 2^5$ copies	10th cycle $= 2^{11}$ copies
5th cycle $= 2^6$ copies	35th cycle $= 2^{36}$ copies
6th cycle $= 2^7$ copies	= 68 billion copies

Restriction fragment length polymorphism

Restriction fragment length polymorphism (RFLP) technique produces the DNA profile of an individual. DNAs are isolated from tissue samples, such as skin, semen, hair, or blood. The buccal swab can also be used as a sample. RFLP is done as follows:

- The DNA is fragmented using restriction endonucleases.
- The fragments are then subjected to agarose gel electrophoresis (Section 14.5.6).
- The fragments are separated based on their size, forming distinct bands.
- The bands are blotted onto a nylon membrane (Southern blotting—refer Section 14.5.13).
- Radioactive probe is added to the blotted bands to form DNA fingerprints.
- The probe is identified by autoradiography.
- The fingerprints are unique for each individual.
- The suspected fingerprint can be compared with a reference fingerprint (obtained from another individual).

RFLP protocol is shown in Fig. 14.106.

Applications of RFLP Genetic mutations abolish or alter or induce recognition sites. Therefore, RFLP is used to diagnose genetic diseases such as Huntington's disease, cystic fibrosis, sickle cell anaemia, retinoblastoma, and Alzheimer's disease.

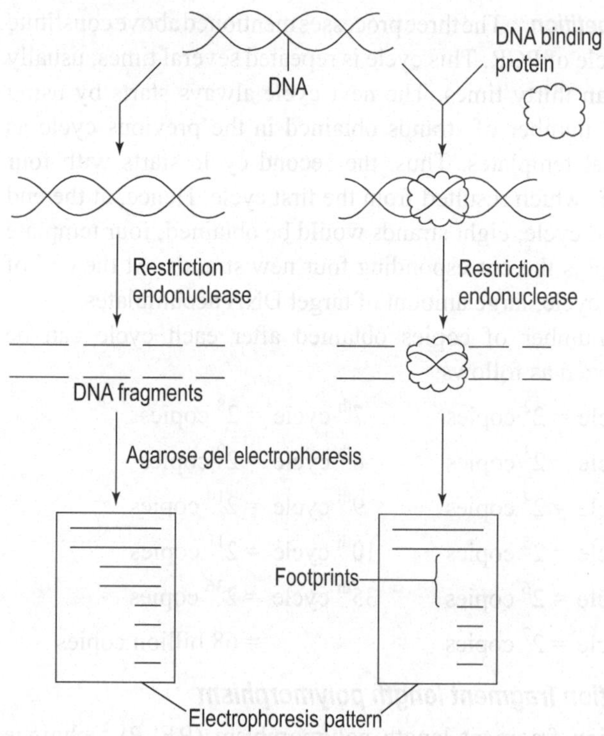

FIGURE 14.107 DNA footprinting

VNTR refers to variable number of tandem repeats. VNTR is a short DNA stretch with 14 to 400 nucleotides. They are clustered as tandem repeats. They are repeated for 4 to 40 times. The number of times the VNTR is repeated varies between individuals. Hence, the fingerprints produced by RFLP will also vary. Therefore, RFLP is also used to identify individuality in forensic sciences.

DNA footprinting

DNA foot printing refers to the identification of protein-bound DNA as a 'footprint' in the electrophoresis gel. It is used to study about DNA–protein interaction in cells and in vitro. Cells can be grown in artificial conditions to enable DNA–protein binding. The DNA is then broken by DNAse or by passing UV radiation to permeate through the cell membrane. The cells are then lysed, and DNA extracted. In vitro, the DNA suspected to have a bound protein is radiolabelled. In both the above cases, the protein-bound DNA sequence is blocked from lysis. The DNA is run on denaturing PAGE. Simultaneously, a control PAGE is run, where the radiolabelled DNA without any protein is electrophoresed. After electrophoresis, the bands are visualized by autoradiography (Section 14.6.2). In the control gel, the DNA fragments are separated into bands in the form of continuous ladder, whereas in the protein-bound sample, the ladder has a break or a hole called the 'footprint' indicating the presence of protein. Thus, the protein bound sequence is identified. Protocol of DNA footprinting is illustrated in Fig. 14.107. DNA footprinting is used to identify DNA–protein interaction in the cells. For example, binding of transcription factors can be studied and gene expression can be investigated. It is also used to find the binding of regulatory enzymes to DNA.

14.9.2 Cell Imaging by Nanoparticles

Cell imaging is defined as the process of locating a diseased cell or cancerous cell for further treatment. Cell imaging is the best method for early detection of cancer or tumour. Recently, nanoparticles have proved to be a good tool for imaging cancer cells. Quantum dots (QD) are nanocrystals with 2 to 8 nm diameter. They are made of atoms from groups II and VI elements (e.g., CdSe and CdTe) or groups III and V elements (e.g., InP and InAs) of the periodic table. QD with cadmium selenide core and zinc sulphide shell (so called core shell QD, Fig. 14.108) were proved to be useful for imaging prostate cancer. The imaging is due to the luminescent property of QD. On irradiation, QDs emit light of different colours depending on their size. This property is used as diagnostic tool. For the diagnosis of prostate cancer, monoclonal antibodies (against prostate specific antigen,

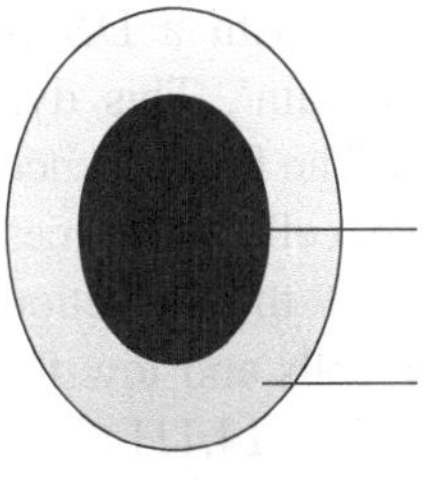

FIGURE 14.108 Core shell quantum dots

PSA) are produced by hybridoma technology, as discussed in Section 14.9.3. The protocol for cell imaging by QD is shown in Fig. 14.109. The QDs are tagged with the antiprostrate antibodies and polyethylene glycol (PEG). PEG is used for enhancing the circulatory time of the QDs. The QDs are injected into the experimental animal, say rat or mice, with prostate cancer. The QDs will be targeted to the prostate cancer cells as they over express the PSA. The animal is then irradiated. QDs absorb the radiation and glow. The glowing light can be viewed through fluorescent microscope and the cancer cell can be imaged.

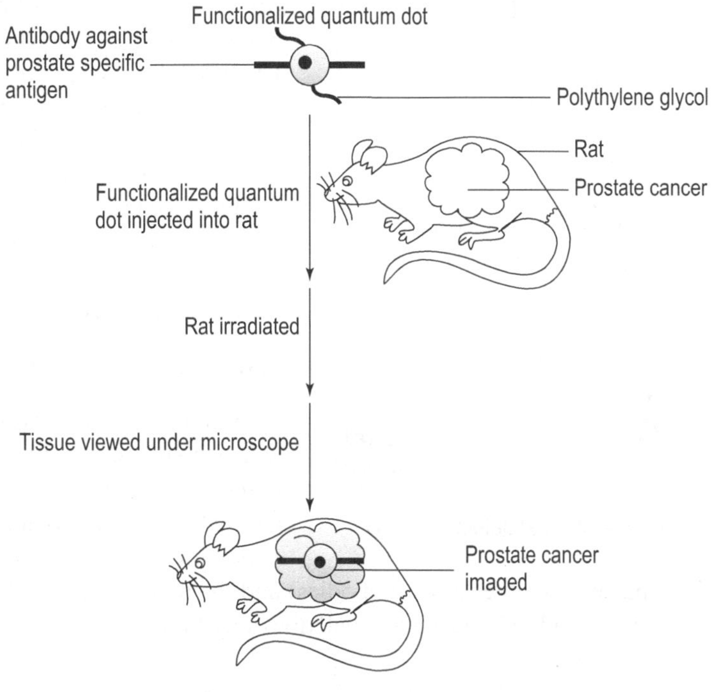

FIGURE 14.109 Imaging of cancer cells by quantum dots

Live cancer cell imaging by gold nanoparticles

Gold nanoparticles are used as a live cancer cell detector system (LCCD; Fig. 14.110). LCCD detects epidermal growth factor receptor (EGFR) which is over expressed in epidermal cancer cells. Surface of a siliconized glass slide (SGS) is modified by treating with 3-aminopropyltrimethoxysilane. This results in the amination (addition of amine group) of the SGS. Aminated SGS is gently stirred with gold nanoparticles (AuNp) solution. This results in immobilization gold to the SGS surface, resulting in a wine red colour solution. Cetuximab (CET) is a monoclonal antibody raised against EGFR. AUNp–SGS is soaked with CET to form CET–AuNp–SGS. CET–AuNp–SGS is called LCCD which is used to sense cells over expressing EGFR amidst non-cancerous populations which lack these receptors.

Cells with EGFR are considered as a positive cell line. EGFR deficient cells are considered as negative cell line. The cells are incubated with LCCD in a 12-well plate for 30 min.

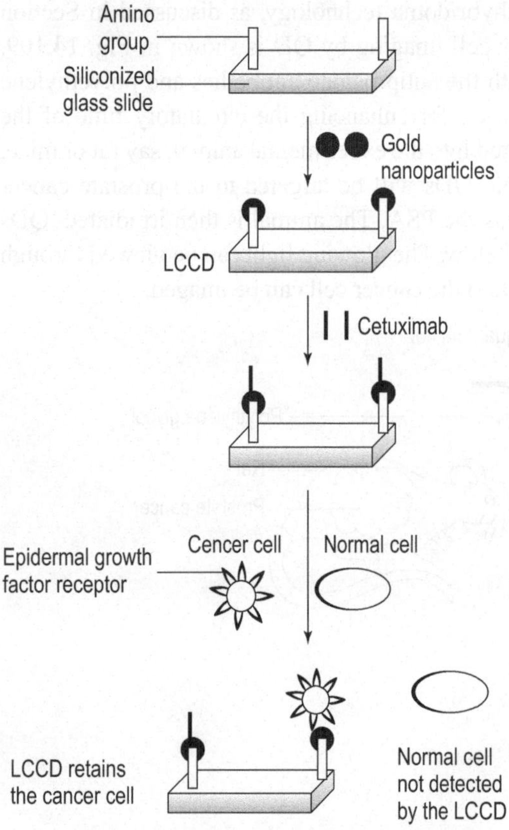

FIGURE 14.110 Live cancer cell detector (LCCD)

The incubation mixture is treated with a DNA-specific fluorescent stain 'Hoechst stain'. This dye has λ_{ex} = 350 nm and λ_{em} = 461 nm. The fluorescence intensity is very high for positive cells (12 times higher) than the negative cells. So the detection efficiency is thus high for positive cells and low for negative cells. This is shown in Fig. 14.111. The cells in the culture plate can also be viewed under fluorescent microscope. The CET–AuNP–SGS treated positive cells will show more fluorescent spots when compared to negative cells. CET–AuNP–SGS treated negative cells, AuNp-SGS + positive and AuNp-SGS + negative cells show negligible spots. Thus, the dye specifically interacts with the antibody-bound cancer cells. This is illustrated in Fig. 14.112.

Fluorescence spectra and fluorescence image show that LCCD has high cellular affinity and high specificity to live cancer cell with EGFR, and hence, it can be used to detect epithelial cancer cells.

14.9.3 Hybridoma Technology

Hybridoma technology is the production of immortal cells which are capable of producing monoclonal antibodies (antibodies specific for a single antigen). The hybrid cell is obtained by fusing two types of cells with two specific properties. One is the B-cell (capable of synthesizing antibodies) and the other is the myeloma cell (capable of being immortal and capable of rapid cell division). Following are the steps used in the hybridoma technology:

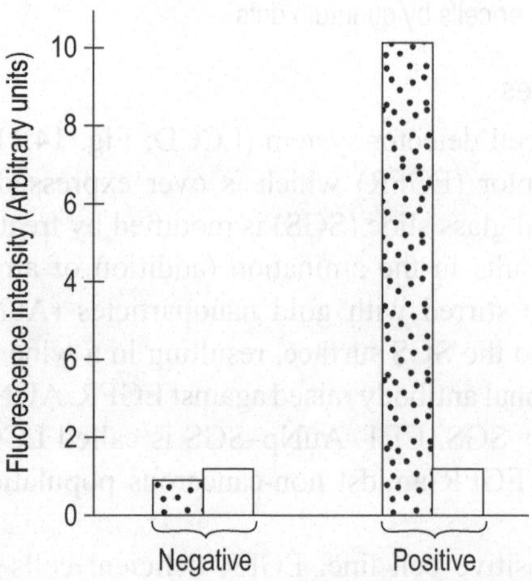

FIGURE 14.111 Fluorescence intensity of negative and positive cells

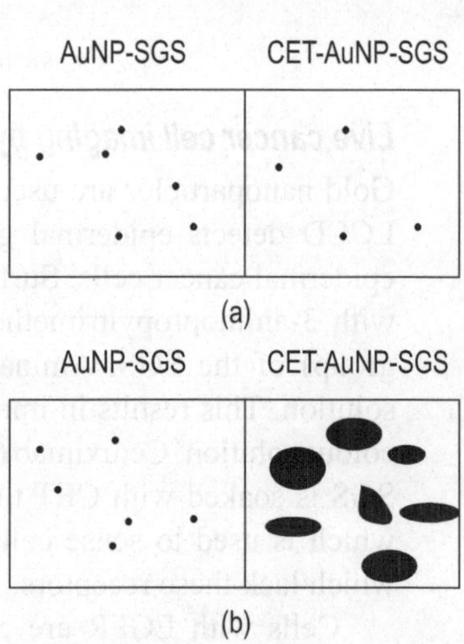

FIGURE 14.112 Fluorescent spot profile for positive and negative cells (a) Negative cell line (b) Positive cell line

- The antigen that can elicit the antibody of interest is injected into a mouse or rat for several weeks.
- The spleen of the mouse is dissected and the B-cell isolated. B-cell is denoted as 'HGPRT++Ab+. This means that B-cells possess HGPRT gene and have the ability to synthesize antibodies.
- Myeloma cells are then selected with two criteria. The cells should be deficient in HGPRT gene (gene coding for the nucleotide-synthesizing enzyme, HGPRT), and they should not have the ability to synthesize antibodies.
- B-cells and myeloma cells are fused in HAT medium to form hybrid cells or hybridoma. Hybrids alone will survive in the HAT medium. Cells without HGPRT gene are sensitive to HAT medium (hypoxanthine-aminopterin-thymidine medium), so, unfused myeloma cells will die. Unfused B-cells will also die due to shorter life span.
- The hybrids will divide rapidly (as myeloma cells) and synthesize antibodies for the specific antigen (as B-cells).
- Hybrids can be screened by adding a secondary antibody labelled with enzyme, followed by the addition of a chromogenic substrate. Production of a coloured product indicates the presence of hybrids.
- Hybrids can be cultured and cryopreserved and used for producing large quantities of specific monoclonal antibodies.

Steps involved in the hybridoma technology protocol are summarized in Fig. 14.113.

Applications of hybridoma technology

Hybridoma technology is widely used in the diagnosis of disease and delivery of drugs. For example, monoclonal antibodies to prostrate specific antigen (PSA) are used to image and diagnose prostrate cancer. Monoclonal antibodies are used for targeting specific cells. For example, they are used in targeting cancer cells by binding to cancer specific antigens present on the surface of the cancer cells.

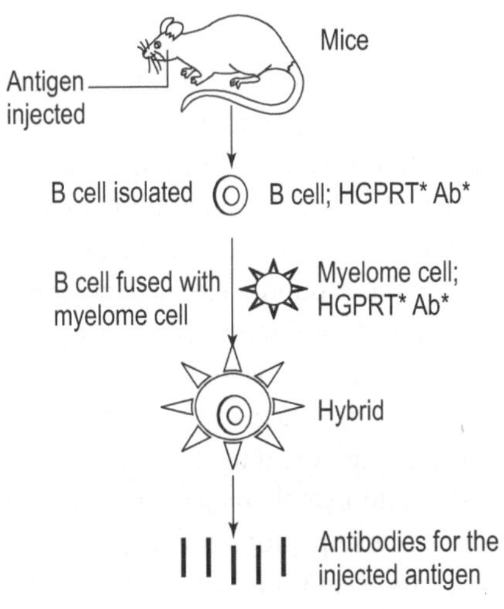

FIGURE 14.113 Protocol for hybridoma technology

14.9.4 GFP Tagging

Green fluorescent protein (GFP) is a fluorophore with 238 amino acids and molecular weight 27 kDa, first isolated from the jelly fish Aequorea victoria. It emits green light at 509 nm. GFP has gradually replaced the use of FITC in cell biology. This is because GFP is less toxic, whereas FITC is toxic to live cells. GFP is used as a fluorescent tag for tagging proteins and genes which can be probed through fluorescent microscopy. GFP gene can be incorporated into vector and fused with the recombinant gene. Thereafter, GFP fluorescence can be used as a marker for the expression of recombinant gene. Thus, GFP is used in the identification of recombinant hosts in recombinant DNA technology.

14.9.5 In Situ Hybridization

In situ means 'in original position'; in situ hybridization (ISH) is the hybridization of a DNA or RNA in a tissue (in situ) by a labelled complementary DNA or RNA probe. This is done to locate the tissue DNA or RNA by the probe. Radiolabels or fluorescent labels are used for ISH. The probe that is complementary to the target is tagged with radiolabel, fluorescent compound, antibodies, or biotin. The probe is then allowed to hybridize with the target. The probe target hybrid is identified by autoradiography, fluorescence microscope, or enzyme-based colour reaction. ISH is used in blotting (Section 14.5.13).

14.9.6 Gene Therapy

Gene therapy is the technique used to treat genetic diseases by introducing, replacing, or deleting a gene present in a tissue. Gene insertion is used to treat single-gene defects, such as cystic fibrosis, haemophilia, muscular dystrophy, and sickle cell anaemia. Gene therapy by gene deletion can be done by knocking out the defective gene using endonucleases. For example, knocking out of CCR5 gene from T-cells is used to control HIV infection.

Gene therapy can be done in somatic cells (somatic gene therapy) or germ cells (germ line gene therapy). However, germ cell gene therapy faces some ethical issues. In somatic cells, the therapeutic gene is introduced by the following methods:

- By viral vectors using reterovirus, adenovirus, adeno-associated virus, or Herpes virus
- By physical methods such as electroporation, sonoporation, gene gun targeting, and magnetofection
- By chemical methods using dendrimers

Gene therapy can be used to treat:

- inherited blindness
- cancer
- myeloid disorder
- Parkinson's disease
- deafness
- sickle cell anaemia
- cystic fibrosis
- thalassemia
- severe combined immunodeficiency syndrome

14.9.7 Immunohistochemistry

Immunohistochemistry is a technique used for detecting a specific protein (the antigen) in the cells of a tissue by using antibodies. The antibodies are labelled to detect the signal produced by the immune reaction. This technique is used to find abnormal tissues such as cancer. The tissue is sectioned into a thin slice using microtome and mounted on a slide. Monoclonal antibodies specific for the target protein are allowed to bind with the antigen. The antibody is labelled with a probe, such as enzyme or fluorophore. The chromogenic substrate is added to the enzyme to produce colour, or the fluorescence emitted by the fluorophore is measured. Alkaline phosphatase, horseradish peroxidase, fluorescein, or rhodamines are the commonly used probes. Alternatively, the probes may be labelled to a second antibody (the secondary antibody) which is complementary to the first (the primary antibody). This method is called indirect immunohistochemistry. In indirect method, the secondary antibody is labelled with biotin. Subsequently, enzyme-labelled streptavidin is added. The enzyme-avidin conjugate binds to avidin with high affinity. The enzyme is then detected by adding the chromogenic substrate. Principle of direct and indirect immunohistochemistry is shown in Fig. 14.114.

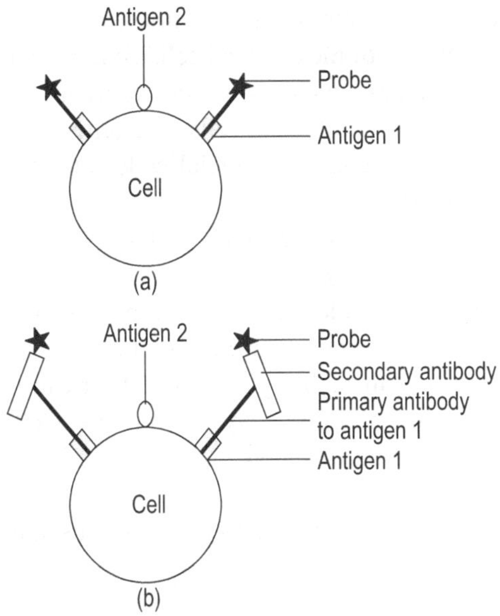

FIGURE 14.114 Immunohistochemistry (a) Direct method (b) Indirect method

14.9.8 Transgenic Animals

Transgenic animals are genetically modified animals. They contain a new or exogenous DNA introduced into the original genome, thus, expressing a new character or property. The phenomenon of producing a transgenic animal is called transgenesis. Transgenic animals have new or improved genetic characteristics and they are used to study gene expression, developmental process, investigate disease state, and improve the quality and quantity of milk, meat, and wool.

Transgenic animals (e.g., transgenic mice) are produced using embryonic stem cell technology. The blastocyst of a mouse is taken and the inner mass isolated to get the pluripotent stem cells. A new gene of interest (called the transgene) is introduced into the pluripotent embryonic stem cell via electroporation or microinjection. The stem cells are introduced into another blastocyst (the recipient blastocyst), which in turn is implanted into a foster mother. The mother mouse produces lot of baby mice, called the transgenic founder, which express the new genetic character expressed by the transgene.

Transgenic animals can be engineered to bear a human disease of interest. Such animals can be used for the development of therapeutic drugs and gene therapy trial. For example, transgenic animals with sickle cell anaemia can be created by inserting beta globin gene of sickle cell haemoglobin, and different aspects of sickle cell anaemia can be investigated. Similarly, HGPRT gene can be mutated to create a Lesch Nyhan syndrome model. Animals with Kaposis sarcoma, X-linked muscular dystrophy, and malignancy can also be engineered and the corresponding diseases can be investigated.

14.9.9 Transgenic Plants

Transgenic plants are genetically modified to express a new character such as improved resistance to stress, pests, diseases, or produce pharmaceutical components or nutrients. The desired gene (the transgene) is introduced into the plant cells by biolistic method or particle gun method. In this method, the transgene is coated on the surface of gold or tungsten particles and loaded in a gun and bombarded into the cells under high force. The gene enters the plant cell and integrates with the host genome and expressed. Alternatively, *Agrobacterium tumifaciens* can be modified by replacing the T-gene with the desired gene. It is then allowed to infect the plants. This organism has the natural ability to highly infect the plants and insert its genome.

14.9.10 Artificial Cells

Artificial cells are man-made microscopic structures containing biologically active materials, such as lipid covering (the membrane), proteins, and so on. They perform certain biological functions similar to those performed by natural cells. The first artificial cell containing a polymersome membrane enclosing haemoglobin was first engineered by Thomas Chang at McGill University. The membrane of the artificial cell is engineered using liposomes (vesicles of lipid). Macromolecules, such as genes, proteins, or enzymes, are introduced into the vesicle. Microbial cells, biological cells, and genetically engineered cells can also be accommodated inside liposomes. Artificial cells can have macro dimensions running down to molecular dimensions.

Artificial cells have wider applications. They can be engineered to accommodate drugs like proteins and thereafter used to treat diseases and disorders.

The cells can be encapsulated with other cells, for example, islets for treating diabetes, hepatocytes for treating liver failure, and kidney cells to produce erythropoietin and treat anaemia. If the cells are engineered to encapsulate enzymes, they can be used for treating enzyme deficiency disorders including phenylketonuria and Lesch Nyan syndrome. Cells containing charcoal and resins are used in detoxification of blood. Artificial red blood cells (respirocytes) are used to increase the oxygen-carrying capacity of the blood. Cells containing the cholesterol metabolizing organism Pseudomonas pictorum can be used to treat hypercholesterolemia and associated risks.

14.9.11 Cells and Cell Ghosts as Drug Carriers

Cell ghosts are cells without their internal mass but still containing the outer membrane and maintaining their morphology. Cells ghosts are used in drug delivery. They are created by lysis of cells and release of cytoplasmic contents. For example, after haemolysis, the red blood cells are called red cell ghosts. Different types of cells can be lysed to produce the cell ghost and subsequently used as drug carriers for drug delivery. Bacterial cells can be lysed by disturbing the membrane to release the cytoplasmic content, and the resulting bacterial ghost suspensions are incubated with drugs, vaccines, or genetic material. Once the materials are loaded, the membrane is sealed by fusion with membrane vesicles. The resultant resealed cells are called drug loaded bacterial ghosts. They can efficiently release anticancer drugs. Erythrocyte ghosts can load drugs, enzymes, and peptides and release them with high efficiency to the reticuloendothelial system of liver and spleen. Drug delivery efficacy of polymorphonuclear leukocyte (PMN) ghosts was proved in vitro. When incubated with high concentration of antibiotic, the PMN can upload the drug and deliver it to the infected cells. Dendritic cells were reported to deliver the anti-tumour drug O-galactosylceramide with high efficacy.

14.9.12 Cells as Nanofactories

DNA can be manipulated into two- and three-dimensional nanostructures, which in turn can be used as template to engineer other nanostructures as well as a scaffold for organizing nanoparticles and proteins. Hao Yan from Biodesign Institute at Arizona State University has used cells as factories producing DNA nanostructures. He used the four basic building blocks of DNA—A, G, T, and C. Bacterial cells were manipulated into a nanofactory using phages and such factory can produce DNA nanostructures.

In this technique, a single-stranded DNA of 1.7 kb is folded into an octahedron (nanostructure) in the presence of helper DNA strands. Initially, the DNA is introduced into a phagemid vector and the vector is transferred into E. coli (host). Following transformation, the host is infected with helper phage which helps folding the phagemid DNA into a nanoscaffold structure and secretes them into the culture media. The production of DNA nanostructure can be increased by repeated culturing of the host cell. Thus, the bacterial cells function as a nanofactory, producing DNA nanostructure.

SUMMARY

Microscopy, cell sorting, cell counting, and cryopreservation are basic protocols for any investigations and studies on cell biology. Live and dead cells can be processed, stained and magnified for further investigations either by light microscopy or electron microscopy. Isolation of a specific group of cells from a mixed population is called cell sorting. Cells can be sorted on the basis of several aspects such as sedimentation rate, density, charge, surface specific antigen, avidin–biotin interaction, and centrifugal elutriation. Number of cells in a population can be counted using Coutler counter or dye exclusion principle. Cryopreservation (storage in liquid nitrogen) and lyophilization (storage in vacuo below freezing pint) are the common cell preservation methods.

Centrifugation is a separation technique by which particles of different shape, size, and density are separated on the basis of their sedimentation rate by spinning under the influence of centrifugal force (extra gravitational force). An instrument called centrifuge generates centrifugal force. Differential centrifugation is used for subcellular fractionation. Density gradient centrifugation is used for the separation of lipoprotein fraction, hormones, enzymes and RNA–DNA hybrids. Analytical centrifugation is used for the determination of molecular weight, purity and conformational changes of macromolecules.

Chromatography is a technique used for the separation of a wide range of compounds based on their distribution between two immiscible phases, namely the stationary phase and the mobile phase. Compounds such as amino acids, carbohydrates, lipids, fatty acids, vitamins, hormones, mineral ions, drugs and their metabolites, plant pigments, pesticides, cell organelles, and gases can be resolved by chromatography. Chromatography can be performed in paper mode (paper chromatography), thin layer mode (thin layer chromatography), or in a column mode (column chromatography).

Electrophoresis is a technique in which charged molecules, such as amino acids, proteins, nucleic acids, and enzymes, are separated into zones or bands, based on the difference in their migration towards the electrodes under the influence of electric field. Agarose gel is used to separate DNA and RNA and to quantify antigen concentration using immunoprecipitation (immunoelectrophoresis). Polyacrylamide gel is used to resolve denatured protein subunits (SDS-PAGE) on the basis of their molecular weight; to separate amphoteric compounds on the basis of their isoelectric point (isoelectric focusing); to separate proteins and enzymes by isoelectric focusing in one dimension and SDS-PAGE in the second dimension (2-D PAGE). Apart from routinely used slab mode, agarose and polyacrylamide gels are also used in capillary mode under high pressure and high voltage and this separation technique is called capillary zone electrophoresis (CZE). CZE is used to resolve drugs, metal ions, oligonucleotides, amino acids, peptides, and diastereomers. Transferring the bands from the gel to a nitrocellulose paper for further analysis can identify specific DNA, RNA, and protein fragments. This technique is called blotting.

Radioactivity can be measured by using some peculiar properties of radiations such as ionizing capacity, excitation capacity, condensation capacity, and capacity to expose photographic emulsion. GM counter, scintillation counter, silver chloride (AgCl) coated X-ray film, cloud chamber, and bubble chamber are some detectors used to detect and measure radioactivity. Besides their extensive use in medicine, research, agriculture, and industry, radioisotopes are also known for their biological hazards. However, these hazards can be prevented by stringently following some safety aspects. Spectroscopy is a technique which is used for quantification, characterization, and structural analysis of a substance (matter) based on interaction of light with that substance. Interaction of light with a substance/sample may result in different phenomena and generate a specific analytical signal. The signal is detected by a detector and recorded by a digital recorder in the form of a spectrum. The spectrum can be correlated with the amount, structure, property, and functional group of the sample.

Study of the relationship between chemical reaction and electrical energy is called electrochemical technique. This technique is used to determine the pH of a solution which is crucial for any biochemical investigation, including enzyme assay. It is also used to determine the oxygen-consuming enzyme reactions, and partial pressure of oxygen and carbon dioxide in the blood.

TERMS TO REMEMBER

2D-PAGE A technique in which molecules are subjected to two different types of separation. First, molecules are separated in a cylindrical gel by isoelectric focusing. Later, the molecules are separated by SDS-PAGE in a second gel.

Abscess Localized collection of pus formed by disintegration of tissues.

Absolute specificity ligand Ligand which binds exclusively to a particular compound in affinity column chromatography. For example, tryptophan (ligand) binds to chymotrypsin.

Absorbance (A) When light passes through a medium, the ratio of log of intensity of incident light to the intensity of transmitted light is called absorbance.

Adsorption Adhesion of compounds to the surface of a granular solid (called adsorbent) by means of electrostatic bond, hydrogen bond, or van der Waals forces.

Affinity elution A method for eluting compounds from the column. This is done by percolating a second compound for which the column has greater affinity than the separated compound.

Agar A dried polysaccharide extract of red algae used to solidify media.

Ampholytes Substances which exist as anion at higher pH, as cation at lower pH, and as zwitter ions at neutral pH.

Analytical ultracentrifugation A technique in which biomolecules are characterized by sedimenting under very high centrifugal field in a refrigerated condition.

Anionic exchanger A matrix with covalently bound negative ions. This is used as stationary phase in ion exchange chromatography.

Anode Positive electrode in an electrochemical cell, at which oxidation occur.

Atomic number Number of protons present in an atom. It is always equal to the number of electrons.

Atomization Decomposition of elements into free atoms using flame (in flame photometry).

Auto radiography A method of detecting radioactivity. Radiations from the radioactive element expose photographic film by reducing AgCl to metallic Ag. This leads to formation of latent image. The intensity of the image is proportional to the radioactivity.

Autoclave Equipment for sterilizing objects under steam pressure.

Bed volume (Vt) The total volume occupied by the gel bead column used for gel filtration chromatography.

Binding pH The pH of the eluting buffer at which the charged compounds bind to an ion exchanger in exchange for counter ions. This should be predetermined before starting the experiment.

Bioluminescence Luminescence produced during an enzyme-catalysed biological reaction.

Blot Nitrocellulose membrane in which (DNA/RNA/protein) molecules are immobilized by hydrophobic interaction.

Blotting Molecules (bands) resolved in the gel are transferred to a nitrocellulose/nylon membrane, and therein immobilized by hydrophobic interaction. This process is called blotting.

Blue dextran Cibacron blue covalently coupled to a high molecular weight dextran. Since blue dextran is completely excluded by all types of gel beads with K_d value of 0, it is used to find the interstitial volume (void volume), V_0, of the gel filtration column.

Blue shift Shift of λ_{max} of a compound to shorter wavelength.

Bright field microscope Illuminates and magnifies the dark image of the specimen against a bright back ground.

Broth A liquid medium used for microbial cell culture.

Callus An unorganized, undifferentiated mass of parenchymatous cells.

Capillary electrophoresis Separation of molecules in a capillary tube in the presence of high voltage and high pressure.

Carrier gas Gases such as H_2, He, and Ar can be used as a mobile phase in gas liquid chromatography.

Cathode Negative electrode in an electrochemical cell, at which reduction occur.

Cationic exchanger A matrix with covalently bound positive ions. This is used as stationary phase in ion exchange chromatography. Cation exchangers bind electrostatically to negative counter ions.

Caulogenesis Generation of shoots from callus using cytokinin rich medium is called caulogenesis.

Cell counting Assessment of number of viable cells in a population.

Cell potential (EMF) The sum of the oxidation potential of anode and the reduction potential of the cathode.

Cell sorting Isolation of a specific cell type from a mixed population.

Centrifugation A separation technique in which particles of different shape, size, and density are separated (based on sedimentation rate) under the influence of centrifugal field.

Chemiluminescence A phenomenon by which light is emitted due to chemical reaction between the radioactive sample and the scintillation cocktail. It is a disadvantage of scintillation counter.

Chemiluminescence Luminescence produced during a chemical reaction.

Chiral chromatography A type of liquid–liquid partition column chromatography exclusively used to separate chiral compounds with asymmetric carbon atom.

Chromatogram Compounds resolved by column chromatography is detected by a detector and shown as a signal. A plot of signal versus time or volume is called a chromatogram. A thin layer plate or a paper showing the spot of the separated compounds can also be called a chromatogram.

Chromatography A technique by which particles are separated on the basis of distribution between two immiscible phases, the mobile phase and the stationary phase.

Chromatoplate A thin layer plate with marked solvent front and spots of separated compounds.

Chromophore Specific functional group of a compound which is responsible for absorbing light.

Cloud chamber An instrument used to measure radioactivity. Gas atoms are converted into ions by bombardment with radiation. Ions thus generated induce condensation of alcohol vapour to form liquid droplets. The tracks of the droplets give idea about the radiations.

Continuous buffer system An electrophoretic system in which the tank buffer (running buffer) and the gel buffer are the same.

Covalent chromatography A type of affinity chromatography in which disulphide group is used as a ligand to separate thiol-containing proteins.

Cryopreservation Preservation of biological material in a viable form at ultra low temperature (–196°C).

Culture Cell population grown in vitro in a nutrient medium.

Cuvette A small rectangular or cylindrical containers used to hold samples used in spectroscopy. It is also called a sample cell, for example, glass cuvettes and quartz cuvettes.

Degassing Removal of air bubbles from the mobile phase used in high-pressure liquid chromatography. This can be done by warming, evacuation, vigorous stirring, or ultrasonic vibration.

Densitometer A device used to measure the intensity of the latent image in autoradiography.

Density gradient centrifugation A type of preparative centrifugation in which particles are separated in a density gradient under uniform centrifugal field.

Density gradient A liquid column whose density increases from top towards the bottom.

Desalting Process of removal of low molecular weight compounds and impurities from macromolecules. For example, removal of nucleic acids and ammonium sulphate from proteins using gel filtration chromatography.

Detector A device used to detect the signal produced due to interaction of light with matter (e.g., photocell and photomultiplier tube).

Detector An accessory device used for detecting the presence of compounds separated in column chromatography. For example, UV/visible spectrophotometer, colorimeter, fluorimeter, radioactivity detector, flame ionization detector, electrochemical detector, nitrogen phosphorous detector, and so on.

Differential centrifugation A type of preparative centrifugation, in which particles are separated in a uniform density medium under different centrifugal field.

Discontinuous buffer system In an electrophoretic system, if the tank buffer (running buffer) and the gel buffer are different (e.g., in SDS-PAGE the tank buffer is Tris glycinate, whereas the gel buffer is Tris HCl).

Dye–ligand chromatography A type of affinity chromatography in which cibacron blue is used as a ligand to separate albumin and enzymes.

ELDOR (Electron double resonance) Simultaneous irradiation of a paramagnetic sample with microwave of two different frequencies (during ESR spectroscopy).

Electrode A metal rod or a wire dipped in a solution of its ion. It is also called half cell.

Electroendosmosis (EEO) A phenomenon in which the charged groups present in the supporting medium is ionized into anionic site, which in turn associates with the positive buffer ions and creates a buffer counter flow against the sample migration. This leads to slower migration of the sample. It is the major disadvantage of paper electrophoresis.

Electrophoresis Separation of charged biomolecules under the influence of electric field on the basis of molecular weight, size, or isoelectric point.

Electrophoretic mobility The velocity with which the charged molecules move towards oppositely charged electrodes during electrophoresis.

Electrophoretogram A developed gel containing the resolved macromolecules (usually as bands).

Elution pH The pH of the eluting buffer at which the charged compounds bound to the ion exchanger are completely desorbed and eluted without loss of activity.

Elution volume (Ve) The volume of the liquid required to elute a compound with Kd value between 0 and 1.

Elution Collection of separated compounds from the column using a suitable buffer or mobile phase.

ENDOR (Electron nucleus double resonance) Simultaneous resonation of electron of a molecule with microwave and nucleus of the same molecule with radiowave is called ENDOR. It is used to improve ESR spectral resolution.

Equivalent protons Group of protons with same environment.

Established cell line Cells that are established and adapted to indefinite growth, after repeated culture for numerous generations.

Film badge When a person works in a radioactive laboratory, the radiation dose received by him can be continuously monitored by a small device called film badge.

Fluorescence A phenomenon by which the luminescence induced by a light source stops as soon as the light is cut off.

Fraction collector A device used for dividing the compounds separated by column chromatography into smaller fractions for further analysis.

Gas amplification Secondary ionization of inert gas atoms in GM counter at 1000 V to produce a torrent of electrons.

GM counter An instrument used to measure radioactivity on the basis of ionization of gases by radiations.

Gradient elution Elution of compounds from the column by percolating a buffer of constantly varying pH or ionic strength or polarity.

Gradient former A special device with two chambers linked by a tube and used for forming a continuous linear gradient column.

Gradient gel When a gel slab incorporated with ampholytes is subjected to an electrical field, a pH gradient is established along the gel. Such a gel is called as gradient gel and it is used for isoelectric focusing.

Group specificity ligand Ligand that binds to a group of closely related compounds in affinity column chromatography.

Growth regulators Synthetic compounds added to the medium in order to govern the growth of the plant in vivo. Examples include indole acetic acid, kinetin, zeatin, and so on.

Guard column Repeated use of biological samples such as urine, blood, plasma, and serum may impair the resolution of HPLC column. To avoid this problem, a short column of length 2 cm packed with the same material as the main column is installed on the top of the main column. This is called guard column.

Half life Time taken by a radioisotope to decay into one half of its original volume.

Homogenate Cell content/cell lysate (lysate) that are released after homogenization.

Homogenization Disturbance of cell architecture to release the cell content by grinding.

Hydrophobic interaction chromatography A type of adsorption column chromatography exclusively used to separate proteins based on differences in specific hydrophobic amino acid residues on the protein.

Hydroxyapatite chromatography A type of adsorption column chromatography. It is exclusively used to separate nucleic acids from cell extract using a column of hydroxyapatite.

Hyperchromic shift Increase in intensity of absorption by a compound due to addition of a new chromophore.

Hypochromic shift Decrease in intensity of absorption by a compound due to geometrical distortion.

Immunoaffinity chromatography A type of affinity chromatography in which antibodies are used as ligand to separate viral proteins

In situ Experiment that is done in an organ that is still in its original place in the body.

In vitro An experiment that is done outside the living system, that is, in the laboratory equipment.

In vivo An experiment that is carried out within a living organism.

Infarction Localized ischemic necrosis produced by occlusion of blood vessels.

Inflammation Response of tissue to injury or distraction.

Inoculation Addition of a biological material onto a medium under aseptic condition to initiate growth.

Internal volume (v_i) The volume of the pores (sieves) of the gel beads used in gel filtration chromatography.

Ion exchanger Cross linked inert matrix made of agarose, dextrose, or polystyrene. It has covalently bound ionic groups which in turn is electrostatically bound to oppositely charged counter ions. It is used as stationary phase in ion exchange chromatography.

Isocratic elution Elution of compounds from the column by percolating a buffer of uniform properties.

Isoelectric focusing (IEF) An electrophoretic technique in which compounds are run through a pH gradient gel and separated on the basis of their isoelectric point.

Isoelectric pH The pH at which a substance exists as zwitter ions.

Isopycnic centrifugation A type of density gradient centrifugation in which particles are separated solely on the basis of their buoyant density.

Isotope dilution technique Determination of the total volume of a particular body fluid by measuring the dilution of radioisotope in that fluid.

Isotopes Atoms of same element with similar atomic number but different mass number (e.g., ${}_{6}C^{14}$ and ${}_{6}C^{12}$).

K-capture A process by which the proton of a radioisotope with N/P < 1 captures an electron from the nearest K shell to form a neutron with the emission of γ rays.

Lectin A tetrameric protein with affinity for carbohydrates. It is produced by plant seeds. It is used as a ligand to separate glycoproteins, receptors, and lymphocytes by affinity chromatography.

Light Electromagnetic waves or electromagnetic radiations made up of components of different wavelength and frequency.

Luminescence Emission of light by an atom due to electronic excitation.

Lyophilization Preservation of biological material in a viable form at frozen state under vacuum or inert gas.

Mass number Mass number of an atom = Number of protons (P) + Number of neutrons (N).

Matrix volume (V_m) The surface volume of the gel bead used in gel filtration column.

Media A nutrient mixture used to promote the growth of cells in vitro.

Metal affinity chromatography A type of affinity chromatography in which, metal ions are used as ligand. It is used to separate fibrinogen, SOD, and non-histone nuclear proteins.

Molecular sieve Porous gel beads are used as stationary phases in gel filtration chromatography. Pores of these beads form a three-dimensional cross-linked network called sieve. These sieves are called so because they retain low molecular weight compounds and exclude high molecular weight compounds.

Monochromatic light Light of a single wavelength.

Monochromator A device which selects a light of single wavelength and excludes all other wavelength.

Monolayer culture Cells grown on a solid support (substrate/anchor) as a single layer.

Nebulization Conversion of a solution of an element into a fine spray; it is used in flame emission spectroscopy.

Necrosis Morphological changes in the tissues. It is an index of cell death.

Negatron Negatively charged beta particles (β^-).

Oxidation potential Tendency of anode to oxidize.

Oxidation Loss of electrons at anode (e.g., $Zn \rightarrow Zn^{2+} + 2e^-$).

Partition coefficient Concentration the compound (to be separated) in one phase divided by the concentration of the same compound in the other phase is called partition coefficient. It is also called distribution coefficient.

Partition Distribution of a compound in definite ratio between two immiscible phases.

Path length Internal cross length of the cuvette through which light passes is called path length; usually it is 1 cm.

Phospholuminescence A phenomenon by which the pigmented samples and the scintillation vial absorb and re-emit light. It is a disadvantage of scintillation counter.

Phosphorescence A phenomenon by which the luminescence induced by a light source continues even after the light is cut off.

Photo multiplier A device in a scintillation counter used to convert light energy into electrical energy.

Photons Light is a source of energy which exists as discrete packets of quanta called photons.

Plane polarized light Light that has electromagnetic vibrations in only one plane. When light passes through a Nicol prism plane polarized light is obtained.

PLOT It implies Porous Layer Open Tubular column. Here the stationary phase is coated on a celite support that is chemically bonded to the wall of the column. It is used for GLC.

Positron Positively charged beta particles (β^+).

Potentiometric titration Determination of the end point of a titration, by using the change in EMF. Here, the indicator electrode solution is titrated against a suitable reagent called titrant and the resulting change in EMF is measured.

Quenching Interference in the transfer of light energy from the radioactive sample to the secondary fluor within a scintillation counter. It is a disadvantage of scintillation counter.

Radioactivity Isotopes with N/P ratio greater than or lesser than one are called unstable isotopes or radioisotopes. Radioisotopes undergo spontaneous decay by emitting radiations until they are converted into stable isotopes. This phenomenon is called radioactivity.

Radiocarbon dating Determination of the age of a material using 14*C: 12*C ratio.

Radioimmunoassay Quantification of hormones, steroids, and drugs by using radiolabelled antigens.

Radioisotopes Isotopes with N/P ratio greater than or lesser than one are called unstable isotopes or radioisotopes.

Raman effect Consider a UV or visible light incident on a molecule without permanent dipole. The molecules undergo a change in polarizability and scatter the light. The scattered light will have a different wavelength and frequency from that of the incident light. This phenomenon is called Raman effect.

Rate zonal centrifugation A type of density gradient centrifugation in which particles are separated on the basis of their sedimentation coefficient.

Recorder An accessory device used for recording the signal produced by the detector in response to the presence of a compound resolved by column chromatography.

Red shift Shift of λ_{max} of a compound to longer wavelength.

Reduction potential Tendency of cathode to undergo reduction.

Reduction Gain of electrons at the cathode (e.g., $Cu2^+ + 2\,e^- \rightarrow Cu$).

Reference electrode An electrode whose potential is constant, accurate, reproducible, and insensitive to unknown solution. For example, standard hydrogen electrode used to determine the potential of other electrode.

Relative centrifugal force (RCF) The ratio of the weight of a particle sedimenting in a centrifugal field to the weight of the same particle in a gravitational field.

Rf value (relative flow or ratio of front) Ratio of the distance travelled by the solute (sample compound) to the distance travelled by the solvent in paper chromatography or thin layer chromatography.

Rhizogenesis Generation of roots from callus using an auxin-rich medium.

Salt bridge A 'U' shaped tube containing gelatin saturated with KCl. It couples the two half cells and allows intermigration of ions between the two half cells.

Scintillation cocktail Scintillation cocktail is a mixture of solvent, primary fluor, and secondary fluor. It is used for dissolving radioactive sample in scintillation counter.

Scintillation counter An instrument used to measure radioactivity based on excitation of electrons present in a solid or liquid.

Scintillation When electrons are excited from ground state to higher energy level, light is emitted. This phenomenon is called scintillation.

Sedimentation coefficient Sedimentation rate per unit centrifugal field.

Sedimentation rate The velocity with which the particles sediment during centrifugation.

Silanization In GLC, the liquid stationary phase is coated to a celite support. The OH group of the celite impairs resolution. So, the OH group is deactivated by treating hexamethyl silazane. This process is called silanization.

Solvent front Distance travelled by the solvent (mobile phase) from the sample origin to the opposite edge of a paper chromatography.

Spacer arm A hydrophobic or hydrophilic compound of 6 to 10 carbon atoms. It is used to attach ligand to affinity column matrix (e.g., 1, 6-diamino hexane).

Spectrum An orderly pattern of wavelength of a light. In spectroscopy, the wavelength is usually plotted against the intensity of the light.

Spin–spin coupling Interaction between magnetic moment of a spinning electron with that of a nucleus in the same molecule.

Stable isotopes Isotopes with N/P ratio equal to one. They do not emit radiations or decay spontaneously.

Stacking gel and separating gel Two types of gels are used in vertical SDS-PAGE.

Stacking gel has larger pore size with pH 6.8 and it is approximately 1 cm in length. It forms the top portion of the vertical gel slab.

Separating gel has smaller pore size with pH 8.8 and it is approximately 10 cm in length, it forms the lower portion of the vertical gel slab.

Standard hydrogen electrode (SHE) A system in which pure hydrogen gas is bubbled through a platinum wire dipped in 1N HCl.

Supporting media A stable, porous, mechanical support with anticonventional property. It is used to perform electrophoresis.

Suspension culture Cells that are grown and dispersed on a liquid medium.

Total exchange capacity This term is used in ion exchange chromatography. It is the number of counter ions present per gram of dry ion exchanger.

Total liquid volume (V_d) Volume of the liquid (mobile phase or buffer) required to elute low molecular weight compounds with $K_d = 1$, which is equally distributed in V_i and V_o in gel filtration chromatography. Adenine can be used to find V of all gel filtration columns.

Totipotency This is a phenomenon by which roots, stem, leaves or even whole plants are generated from a callus, using appropriate growth medium.

Tracer technique Tracing the metabolic pathway of a compound by tagging that compound with a radiolabel is called tracer technique.

Tracking dye A dye usually mixed with the sample in order to monitor the extent of migration during gel electrophoresis.

Transmittance When light passes through a medium, the ratio of intensity of the transmitted light to the intensity of incident light is called transmittance.

Visible light Light of wavelength 370 to 780 nm which is visible to naked eye; it is also called white light.

Void volume (V_o) Volume external to the gel bead in a gel filtration column.

WCOT WCOT means wall coated tubular column. In WCOT, the liquid stationary phase is coated directly on the wall to the column and used for GLC.

X-rays Invisible electromagnetic radiations of longer wavelength and higher penetrating power.

α rays Helium nucleus ($_2He^4$) is called an alpha ray.

λ_{max} The wavelength at which a light is maximally absorbed by a substance is called λ_{max} of that substance, for example, λ_{max} for protein = 280 nm.

REVIEW QUESTIONS

1. Define LB broth and LB and media.
2. Explain the protocol for plant cell culture.
3. What is the role of sodium azide, glycerol, and DNAse in tissue homogenization?
4. Name the dyes used for staining the DNA and intracellular structures for microscopic examination.
5. Explain the principle and working of TEM.
6. Describe the principle and applications of AFM.
7. How are dynabeads used in cell sorting?
8. How will you count the blood cells using haemocytometer?
9. Write a note on cryopreservation.
10. Derive the formula for relative centrifugal field (RCF).
11. How will you determine the molecular weight of a macromolecule using the analytical ultracentrifuge?
12. Describe the technique of hydrophobic interaction chromatography and hydroxyapatite chromatography.
13. Give reason for running protein electrophoresis at alkaline pH.
14. Mention the disadvantages of heat generated during electrophoresis.
15. Agarose gel is preferred for DNA separation, whereas polyacrylamide gel is preferred for protein separation. Justify this statement.
16. Write the role of glycerol in agarose gel electrophoresis.
17. Explain the protocol for resolving a mixture of different molecular weights.
18. Write about Craig's countercurrent apparatus and its applications.
19. Define silanization.
20. Define electroendosmosis.
21. What is PFGE?
22. Explain the southern blotting technique.
23. What is the function of Coolidge tube?
24. Define Stokes and anti-Stokes lines.
25. Compare Raman and IR spectroscopies.
26. What is a galvanic cell?
27. How will you determine the pH of a buffer to be used for tissue homogenization?

MULTIPLE-CHOICE QUESTIONS

1. X-rays are produced from
   (a) Galvanic cell
   (b) Coolidge tube
   (c) radiofrequency oscillator
   (d) Golay cell
2. Decrease in absorption of a compound is called
   (a) red shift
   (b) blue shift
   (c) hyperchromic shift
   (d) hypochromic shift

3. In the production of DNA nanostructures which of the following is used as helper?
   (a) phage DNA
   (b) E. coli DNA
   (c) liposome
   (d) octahedron
4. Taq polymerase is a
   (a) extremozyme
   (b) endonuclease
   (c) ligase
   (d) primase
5. Which of the following is associated with RFLP?
   (a) agarose electrophoresis
   (b) differential centrifugation
   (c) gas liquid chromatography
   (d) NMR spectroscopy
6. Helium nucleus is called
   (a) alpha ray
   (b) beta ray
   (c) gamma ray
   (d) X-ray
7. Scintillation cocktail is a mixture of
   (a) solvent and primary flour
   (b) solvent and secondary flour
   (c) solvent, primary flour, and secondary flour
   (d) primary flour and secondary flour
8. Hydroxyapatite chromatography is used exclusively for the separation of
   (a) proteins
   (b) enzymes
   (c) nucleic acids
   (d) lipids
9. Spacer arm is used in
   (a) HPLC
   (b) affinity chromatography
   (c) gel filtration chromatography
   (d) ion exchange chromatography
10. In standard hydrogen electrode
    (a) platinum wire is dipped in 1N hydrochloric acid
    (b) platinum wire is dipped in 1N sulphuric acid
    (c) silver wire is dipped in 1N hydrochloric acid
    (d) silver wire is dipped in 1N sulphuric acid
11. Silicon carbide rod is a source of
    (a) UV radiation
    (b) visible radiation
    (c) infra red radiation
    (d) radiofrequency radiation

Answers to Multiple-choice Questions

1. (b)	2. (d)	3. (a)	4. (a)	5. (a)
6. (a)	7. (c)	8. (c)	9. (b)	10. (a)
11. (c)				

Appendix

TYPES OF CELLS IN HUMAN BODY

Cell type	Example	Cell type	Example
Neural cells	Neurons	Gland cells	Serous cells
	Astrocytes		Islet cells
	Oligodendrocytes		Leydig cells
Muscle cells	Cardiocytes		Follicular cells
	Myocytes	Blood cells	Erythrocytes
	Smooth muscle cells		Leucocytes
Bone cells	Osteoblasts		Thrombocytes
	Osteocytes		
	Osteoclasts		
Reproductive cells	Sperm cells	Stem cells	Hematopoietic stem cells
	Egg cells		Cord Blood stem cells
			Placental stem cells
			Hair follicular stem cells
			Embryonic stem cells

CANCER RELATED GENES

Gene type	Associated Cancer	Function of the gene	Example
Tumour suppressor gene	Familial adenomatous polyposis	Regulation of cell cycle	APC
	Breast cancer	Apoptopsis	BRCA
	Ovarian cancer	Inhibition of Cyclin dependent kinase	P^{53}
	Prostate cancer	DNA repair	RB
	Colonic neoplasm		
	Pancreatic neoplasm		
	Leukemia		
	Multiple myeloma		
	Retinoblastoma		
Oncogene	Lymphoma	Inhibition of Apoptosis	BCL2
	Ovarian neoplasm	Stimulation of angiogenesis	HER2
	Burkit's lymphoma	Stimulation of interaction of proteins with various cellular receptors	MYC
	Bladder neoplasm kidney neoplasm	Stimulates growth factor	RAS
	Lung neoplasms		

GLOSSARY OF TERMS

Quantity Quantity refers to the amount of any substance (solute) present in a solvent/solution, irrespective of the volume of the solution. *Example* Consider 10 g of NaOH dissolved in 100 ml or 200 ml of solvent. The quantity of NaOH in both 100 ml solution and 200 ml solution is 10 g.

Concentration Concentration refers to the amount of solute present in a specific volume of solution or solvent. *Example* When 10 g of NaOH is dissolved in 100 ml of water the concentration of the resulting solution is 10%.

Atomic number (A) It is the number of protons present in the nucleus of an atom. It is equal to number of electrons present in the same atom. *Example* Carbon atom has 6 protons and 6 electrons. Atomic number of carbon = 6.

Mass number (z)/atomic weight It is the sum of number of protons and neutrons present in an atom. *Example* Carbon has 6 protons and 6 neutrons. Atomic weight of carbon = 6 + 6 = 12

Molecular weight Sum of atomic weights of all the atoms in a molecule is its molecular weight. *Example* Molecular weight of glucose is 180.

$C_6 H_{12} O_6$:

$$\begin{aligned} &(C)\ 6 \times 12 = 72 \\ &(H)\ 12 \times 1 = 12 \\ &(O)\ 6 \times 16 = 96 \\ \hline &\quad 180 \end{aligned}$$

Equivalent weight Equivalent weight of a compound is its molecular weight divided by the number of replaceable hydrogen ion or hydroxyl ion present in that compound. *Example* Equivalent weight of H_2SO_4:
M. wt of H_2SO_4 = 98.1
No. of replaceable hydrogen in H_2SO_4 = 2

$$\text{Equivalent weight of } H_2SO_4 = \frac{98.1}{2} = 49.05$$

Mole (mol) Molecular weight of a compound in grams is called mole. It is also called gram molecular weight.

Example 1
1 mol/L of NaOH:
M. wt of NaOH = 40
1 mole of NaOH = 40 g
1 mole/L of NaOH is 40 g of NaOH dissolved in water and made up to 1 litre.

Example 2
1 mole/L of H_2SO_4:
M. wt of H_2SO_4 = 98.08
1 mole of H_2SO_4 = 98.08 g
1 mole/L of H_2SO_4 is 98.08 g of H_2SO_4 made up to 1 litre with the solvent.

Molarity (m) Molarity of a solution is the number of moles present in 1 litre of the solution divided by molecular weight.

Example 1
Molarity of H_2SO_4:
Concentration of commercial H_2SO_4 = 1730g/L
Mol. wt. of H_2SO_4 = 98.08

$$\therefore \text{ Molarity of } H_2SO_4 = \frac{1730}{98.08} = 17.6$$

Example 2
Molarity of HCl:
Concentration of commercial HCl = 440g/L
M. wt. of HCl = 36.5

$$\text{Molarity of HCl} = \frac{440}{36.5} = 12.05$$

Molality Molality is the number of moles dissolved in 1 kg of the solvent.

Normality Equivalent weight of a compound present in 1 litre of the solution is called the normality of the solution.

Example 1
Normality of H_2SO_4:
Concentration of commercial H_2SO_4 = 1730g/L

Equivalent wt. of $H_2SO_4 = 49.05$

$$\text{Normality of } H_2SO_4 = \frac{1730}{49.05} = 35.2$$

Example 2
Normality of CH_3COOH:
Concentration of commercial CH_3COOH = 1060 g/L
M.wt. of CH_3COOH = 60

$$\text{Normality of } CH_3COOH = \frac{1060}{60} = 17.6$$

Molar solution (M) One molar solution of a compound is the molecular weight in grams in one litre of solution.

Example 1 1 M solution of glucose:
M. wt of glucose = 180
1 M solution of glucose is 180 g of glucose dissolved in water/solvent and made up to 1 litre solution

Example 2
1 mM solution of glucose
1 mM solution of glucose is 18 g of glucose dissolved in water / solvent and made upto 1 litre solution.

Normal solution One normal solution is the equivalent weight of the compound in grams present in 1 litre of the solution

Example 1N H_2SO_4: 49 g of H_2SO_4 in 1 litre of solution is 1N H_2SO_4

Percentage solution It is the volume of a substance present in 100 parts by volume of solution. *Example* 20% acetic acid is:

20 ml acetic acid + 80 ml H_2O (v/v)
(or)
20 g acetic acid + 80 ml H_2O (w/v)

Parts per million (ppm) It is the milligram of a substance present in 1 litre of a solution. *Example* 1 ppm NaCl is 1 mg NaCl in 1 litre of solution.

Parts per billion (ppb) It is the microgram of a substance present in 1 litre of the solution *Example* 2 ppb NaCl solution is 2 mg of NaCl in 1 litre of solution.

Accuracy Accuracy is the degree of agreement of the 'determination' of an analysis with the true value.

Precision Precision is the 'reproducibility' of the biochemical analysis. It is expressed in terms of variation among several determinations on the same sample.

Reliability Reliability is the measure of the congruence of accuracy and precision.

Sensitivity It is the percentage of positive results of diseased patients.
Lower sensitivity occurs in the early stages of many diseases in contrast to the higher sensitivity in well established or severe stages.

Specificity It is the percentage of negative results among people who do not have the disease.

Predictive value It is the percentage of positive results that are true positives.

Strong acid A strong acid is one that ionises almost 100% in aqueous solution.

Example HCl is a strong acid as it is essentially 100% ionised in solution as:

$$HCl + H_2O \rightarrow H_3O^+ + Cl^-$$

Strong base A strong base is one that ionise in solution to yield OH– ions. *Example* NaOH is a strong base as it is ionized as:

$$NaOH \rightarrow Na^+ + OH^-$$

pH and pOH pH is the short way of representing the hydrogen ion activity of a solution. It is the negative log of H^+ ion activity. It is inversely proportional to the acidity of a solution. pOH is the short way of representing the hydroxyl ion activity of a solution. It is the negative log of OH^- ion activity. It is inversely proportional to alkalinity of a solution.

BIOCHEMICAL REFERENCE VALUES

Particulars	System	Values
Total Protein	Serum	6 – 8 g/dl
Albumin	Serum	3.2 – 4.5 g/dl
Globulin	Serum	2.3 – 3.5 g/dl
A/G ratio	Serum	≥ 1
Fibrinogen	Plasma	0.2 – 0.6 g/dl
Reducing sugar (Fasting)	Whole blood	80 – 120 mg/dl
Glucose (Fasting)	Serum or Plasma	65 – 110 mg/dl
Total cholesterol	Serum	150 – 220 mg/dl
Total lipid	Serum	400 – 800 mg/dl
Phospholipid	Serum	150 – 380 mg/dl
Chloride	Plasma	96 – 106 m Eq/L
Bicarbonate	Plasma	21 – 28 m Eq/L
Urea	Whole blood	15 – 45 mg/dl
Uric acid	Serum	2.5 – 7.0 mg/dl
Creatinine	Serum	0.6 – 1.2 mg/dl
Ceruloplasmin	Serum	23 – 50 mg/dl
Glutathione	Whole blood	24 – 37 mg/dl
Transferin	Serum	200 – 400 mg/dl
Ascorbic acid	Plasma	0.4 – 1.5 mg/dl
β-carotene	Serum	50 – 300 mg/dl
Vitamin A	Serum	15 – 60 mg/dl
Vitamin B12	Serum	> 200 pg/ml
Vitamin D	Serum	
25-OH-cholecalciferol		8 – 55 ng/ml
1, 25-di OH-cholecalciferol		26 – 65 pg/ml
24, 25-di OH-cholecalciferol		1 – 5 ng/ml
Bilirubin	Serum	
Total		0.2 – 1.2 mg/dl
Direct conjugated		0.1 – 0.4 mg/dl
Indirect		0.2 – 0.7 mg/dl
Sodium	Serum	135 – 145 m Eq/L
Potassium	Serum	3.5 – 5.0 m Eq/L
Calcium	Serum	8.5 – 10.3 mg/dl
Magnesium	Serum	1.8 – 3.0 mg/dl
Copper	Serum	70 – 150 mg/dl
Zinc	Serum	50 – 150 mg/dl
Iron	Serum	60 – 150 mg/dl
Lead	Whole blood	0 – 50 mg/dl
pH	Whole blood	7.45 – 7.55
pO_2	Whole blood	95 – 100 mm Hg
pCO_2	Whole blood	30 – 45 mm Hg

pI VALUES OF AMINO ACIDS

Amino acids	M. wt	pI values	Amino acids	M. wt	pI values
a alanine	89.1	6.02	Hydroxy proline	131.1	5.83
b alanine	89.1	69.0	Isoleucine	131.2	6.02
Arginine	174.2	10.76	Leucine	131.2	5.98
Asparagine	132.1	5.41	Lysine	146.2	9.74
Aspartate	133.1	2.98	Methionine	149.2	5.75
Citrulline	175.2	5.92	Ornithine	132.2	9.70
Cysteine	121.2	5.02	Phenylalanine	165.2	5.48
Cystine	240.3	5.06	Proline	115.1	6.30
Glutamine	146.1	5.65	Serine	105.1	5.68
Glutamate	147.1	3.22	Taurine	125.1	5.12
Glycine	75.1	5.97	Threonine	119.1	6.53
Histidine	155.2	7.58	Tryptophan	204.2	5.88
Homocysteine	135.2	5.54	Tyrosine	181.2	5.65
Homocystine	268.3	5.53	Valine	117.1	5.97
Hydroxylysine	162.2	9.15			

MOLECULAR WEIGHT OF PROTEINS

Protein	M. wt
Bovine serum albumin	65,400
Egg albumin	45,000
Bovine catalase	250,000
Cytochrome C	13,000
Bovine hemoglobin	64,500
Bovine Insulin	5700
Horse radish peroxidase	40,000
Bovine trypsin	23,800
Egg lysozyme	14,600

SPECIFIC ROTATION OF SOME CARBOHYDRATES

Name	*	■	Name	*	■
β D arabinose	–175	–103	β D glucose	+18.7	+52.7
α L arabinose	+55.4	+105	α L glucose	– 95.5	–51.4
β L arabinose	+190.6	+104.5	α D mannose	+29.3	+14.5
β D fructose	–133.5	–92	β D mannose	–16.3	+14.5
α D galactose	+150.7	+80.2	α L Rhamnose	– 8.6	+8.2
β D galactose	+52.8	+80.2	α D Xylose	+9.36	+18.8
α D glucose	+112	+52.7			

* Specific rotation of original form

■ Specific rotation of the equilibrium mixture of a and b forms after mutarotation

ABSORPTION MAXIMA AND ABSORPTION COEFFICIENT OF SOME BIOMOLECULES

Compound	l_{max} (nm)	Am × 10^{-3}	Compound	λ_{max} (nm)	Am × 10^{-3}
Adenine	260.5	13.3	Guanine	275.5	8.1
Adenosine, AMP,	259	15.4		246	10.7
ADP, ATP			Guanosine, GMP,	252	13.7
Ascorbic acid	265	7	GDP, GTP		
Cytosine	267	6.1	NAD^+, $NADP^+$	259	18
CMP, CDP, CTP	271	9.1	NADH, $NADPH^+$	339	6.22
Coenzyme A	260	15	Nicotinamide	260	4.6
Coenzyme Q	275	16.8	Riboflavin	450	12.2
Cytochrome C	550	30.2		375	10.6
FMN	450	12.2		260	27.7
	375	10.4	Thymine	264	7.9
	260	27.1	Uracil	259.5	8.2
FAD	450	11.3	Uridine, UMP, UDP,	262	10.0
	375	9.3	UTP		
	260	37	Vitamin A	325	52.4

References

Alberts, Bruce, Alexandar Johnson, Jullian Lewis, Martin Raff, Keith Roberts, and Peter Walter (2002), *Molecular Biology of the Cell*, 4th edition, Garland Science, Taylor and Francis Group, New York,.

Baron, M.H. (2004), *Developmental Haematopoiesis, Methods and Protocols*, Humana Press, New Jersey.

Barry R. Zirkin (1998), *Germ Cell Development, Division, Disruption and Death*, Springer, United Kingdom.

Berg, J.M., John L. Tymoczko, and Lubert Stryer (2002), *Biochemistry*, 5th edition, W. H. Freeman and Company, New York.

Bershadsky, A.D. and Vasiliev, J.M. (1988), *Cytoskeleton (Cellular Organelles)*, Springer, UK.

Bhagvan, N.V. (2002), *Medical Biochemistry*, 4th edition, Harcourt/Academic Press, Massachusetts.

Chris Wylie (2000), 'Germ Cells', *Current Opinion in Genetics & Development*, vol. 10, pp. 410–413.

Colin Russell Short and R.V. Austin (1980), *Germ Cells and Fertilization*, University Press, USA.

Conn, Erice E., Parul K, Stumpf, George Bruening, and Roy H. Doi (2002), *Outlines of Biochemistry*, 5th edition, John Wiley and Sons, Singapore.

De Robertis, E. D. P., and D.M.F. De Robertis, Jr. (2003), *Cell and Molecular Biology*, 8th edition, Lippincott Williams and Wilkins, Philadelphia.

Devasena T., Lalitha S., and Padma K., 'Lipid peroxidation, osmotic fragility and antioxidant status in children with acute post streptococcal glomerulonephritis', *Clinica Chimica Acta*, vol. 308, 2001, pp. 155–161.

Devasena, T. (2010), *Enzymology*, Oxford University Press.

Devasena, T. (2010). *Techniques in Biochemistry*, Ahuja Publishers, New Delhi.

Dov Zipori (2009), *Stem Cell Biology and Regenerative Medicine*, Humana Press, Springer, New York.

Dudek, R.W. (1999), *High Yield-Cell and Molecular Biology*, Lippincott Williams and Wilkins, Philadelphia.

Gall, J.G. and J. Richard McIntosh (2001), *Landmark Papers in Cell Biology*, Cold Spring Harbour Laboratory Press, USA.

Garret, R.H. and Charles M. Grisham (1995), *Biochemistry*, Saunder's College Publishing, New York.

Gartner, L.P., James L. Hiatt, and Judy M. Strum (2003), *Cell Biology and Histology*, 6th edition, Lippincott Williams and Wilkins, New York.

Gary S. Stein, Renato Baserga, Antonio Giordano, and David T. Denhardt (1999), *The Molecular Basis of Cell Cycle and Growth Control*, Wiley, USA.

George Studzinski (2000), *Cell Growth, Differentiation and Senescence: A Practical Approach (Practical Approach Series)*, Oxford University Press, USA.

Go¨thel S.F. and Marahiel M.A. (1999), 'Peptidyl-prolyl cis-trans isomerases, a superfamily of ubiquitous folding catalysts', *Cellular and Molecular Life Sciences*, vol. 55; pp. 423–436.

Gonzalo G. Garcia and Richard A. Miller , 'Age-related defects in the cytoskeleton signaling pathways of CD4 T cells', *Ageing Research Reviews*, vol. 10, 2011, pp. 26–34

Goodsell, D.S. (1997), *The Machinery of Life*, Springer Verlag, New York.

Hames, B.D. and Hooper, N.M (2003), *Instant Notes Biochemistry*, 2nd edition, Viva Books Private Limited, New Delhi.

Hoffman, Ronald, Edward J. Benz Jr, Stanford J. Shattil, Bruce Furie, Leslie E. Silberstain, Philip Mc Glave, and Helen Hslop (2009), *Haematology: Basic Principles and Practice*, Churchill Livingstone, Elsevier, Philadelphia.

James L. Funderburgh, 'Keratan sulfate: Structure, biosynthesis, and function', *Glycobiology*, vol. 10, no. 10, pp. 951–958.

Janet M. Trowbridge and Richard L. Gallow (2002), 'Dermatan sulfate: New functions from an old glycosaminoglycan', *Glycobiology*, vol. 12, no. 9, pp. 117R–125R.

Jeremy. M. Berg, John L. Tymoczko, and Lubert Stryer (2002), *Biochemistry*, 5th edition, W.H. Freeman and company, New York.

Julien Royet and Roman Dziarski, 'Peptidoglycan recognition proteins: Pleiotropic sensors and effectors of antimicrobial defences', *Nature Reviews Microbiology*, vol. 5, 2007, pp. 264–277.

Karl G. Sylvester, and Michael T. Longaker, 'Stem cells review and update', *Archives of Surgery*, vol. 139, 2004, pp. 93–99.

Kaushik Deb and Satish Totey (2009), *Stem Cell Technologies: Basics and Applications*, McGraw-Hill, New Delhi.

Kierszenbaum, A.L. (2002), *Histology and Cell Biology*, Mosby Inc, USA.

Kratz, René Fester (2009), *Molecular and Cell Biology for Dummies*, Wiley, Indiana.

Lauren Pecorino (2008), *Molecular Biology of Cancer: Mechanisms, Targets, and Therapeutics*, Oxford University Press, New York.

Levinson Warren (2008), *Review of Medical Microbiology and Immunology*, McGraw Hill-Lange, New York.

Lewin Benjamin (2008), *Genes IX*, Jones and Barttlet Publishers Inc, UK.

Lewis J. Kleinsmith (2005), *Principles of Cancer Biology*, Benjamin Cummings, USA.

Linda A. Amos and W. Bradshaw Amos (1991), *Molecules of the Cytoskeleton*, The Guilford Press, New York.

Lodish, H., Arnold B., S. Lawrence Zipursky, Paul Matsudaira, David Baltimore, and James Darnell (2008), *Molecular Cell Biology*, 4th edition, W.H. Freeman and Company, USA.

Mara H. Sherman, Craig H. Bassing, and Michael A. Teitell, 'Regulation of cell differentiation by the DNA damage response', *Trends in Cell Biology*, vol. 21, 2011, pp. 312–319.

Mark Gerstein and Werner Krebs (1998), 'A database of macromolecular motions', *Nucleic Acids Research*, vol. 26, no. 18, 1998, pp. 4280–4290.

Mark J. Daniels, Maarten J. Chrispeels and Mark Yeager, 'Projection structure of a plant vacuole membrane aquaporin by electron cryo-crystallography', *Molecular Biology*, vol. 294, 1999, pp. 1337–1349.

Martinoia, Enrico, Agnès Massonneau, and Nathalie Frangne, 'Transport processes of solutes across the vacuolar membrane of higher plants', *Plant and Cell Physiology*, vol. 41(11), 2000, pp. 1175–1186.

Mazzarello P, 'A unifying concept: The history of cell theory', *Nature Cell Biology*, vol. 1(1), 1999, pp. E13–5.

Meyers R.A. (Ed.), Eric Westhof, and Pascal Auffinger (2000), *RNA Tertiary Structure. Encyclopedia of Analytical Chemistry*, John Wiley & Sons Ltd, Chichester.

Murray, R.K., Daryl K. Garner, Peter A. Mayes, and Victor W. Rodwell (2000), *Harper's Biochemistry*, 25th edition, McGraw-Hill, New York.

N.Sushma and T. Devasena, 'Aqueous extract of *Trigonella foenum graecum* ameliorates cypermethrin induced circulatory oxidative stress in male Wistar rats', *International Journal of Biological and Chemical Sciences*, vol. 4(1), 2010, pp. 25-33.

Nava Segev (2009), *Trafficking Inside Cells: Pathways, Mechanisms and Regulations*, Lands bioscience and Springer Science, New York.

Noriko Yasuhara, Masahiro Oka, and Yoshihiro Yoneda (2009), 'The role of the nuclear transport system in cell differentiation', *Seminars in Cell & Developmental Biology*, vol. 20, 2009, pp. 590–599.

Olivia Ndozangue-Touriguine, Jocelyne Hamelin, and Jacqueline Bréard, 'Cytoskeleton and apoptosis', *Biochemical Pharmacology*, vol. 76, 2008, pp. 11–18.

Paolo Sassone-Corsi, Margaret T. Fuller, and Robert Braun (2011), *Germ Cells*, Cold Spring Harbor Laboratory, New York.

Patricia Armati (2007), *The Biology of Schwann Cells: Development, Differentiation and Immunomodulation*, Cambridge University Press, UK.

Paul B. Fisher (1990), *Mechanisms of Differentiation*, vols. 1 & 2, CRC Press, USA.

Paustian, Timothy and Gary Roberts (2006), *Through the Microscope: A Look at All Things Small*, Lulu. com

Price, N.C. and Lewis Stevens (1999), *Fundamentals of Enzymology. The Cell and Molecular Biology of Catalytic Proteins*, 3rd edition, Oxford University Press, UK and Europe.

Rao, C. Vaman (2006), *Immunology*, 2nd edition, Narosa Publishing House, New Delhi.

Reinhard Laubenbacher, Valerie Hower, Abdul Jarra, Suzy V. Torti, Vladimir Shulaev, Pedro Mendes, Frank M. Torti, and Steven Akman, 'A systems biology view of cancer', *Biochimica et Biophysica Acta (BBA)*, vol. 1796, 2009, pp. 129–139.

Robert A. Weinberg (2006), *The Biology of Cancer*, Garland Science, New York.

Robert Lanza, Irving Weissman, James Thomson, and Roger Pederse (2004), *Handbook of Stem Cells*, vols. 1 2, Elsevier Academic Press, USA.

Robert Lanza, John Gearhart, Brigid Hogan, and Douglas Melton (2009), *Essentials of Stem Cell Biology*, 2nd edition, Academic Press, Canada.

Robert William and Hegner (2010), *The Germ-cell Cycle in Animals*, Nabu Press, USA.

Roger, J.B. King and Mike W. Robins (2006), *Cancer Biology*, 3rd edition, Benjamin Cummings, USA.

Roit, Ivan, Jonathan Brostoff, and David Male (1995), *Immunology*, Mosby Inc., London.

Sebastien Lansing and Tristan Rousseau (2010), *Cytoskeleton: Cell Movement, Cytokinesis and Organelles Organization (Cell Biology Research Progress)*, Nova Science Publishing Inc, USA.

Shay J.W. (1986), *Cell and Molecular Biology of the Cytoskeleton*, Springer, UK.

Song Li, Nicolas L'Heureux, and Jennifer Elisseeff (2011), *Stem Cell and Tissue Engineering*, World Scientific Publishing Company, Singapore.

Stavrovskaya A.A. and Stromskaya T.P. , 'Transport proteins of the ABC family and multidrug resistance of tumor cells biochemistry (Moscow)', *Biochemistry*, vol. 73, no. 5, 2008, pp. 592–604.

Susana M. Chuva, de Sousa Lopes, and Bernard A.J. Roelen, 'On the formation of germ cells: The good, the bad and the ugly', *Differentiation*, vol. 79, 2010, pp. 131–140.

Thomas, Lewis (1978), *The Lives of a Cell*, Penguin Books Limited, USA.

Vance, D.E. and Vance, J.E (eds.) (2008), *Biochemistry of Lipids, Lipoproteins and Membranes*, Elsevier, UK.

Voet, Donald, and Judith G. Voet (2008), *Biochemistry*, 3rd edition, John Wiley Sons, New York.

Weissman, I.L., 'Stem cells: Units of development, units of regeneration, and units in evolution', *Cell*, Vol. 100, 2000, pp. 157–168.

Yawata, Yoshihito (2008), *Cell Membrane. Red Cell as a Model*, Wiley-VCH-Verlag.
Yeagle, P.L. (2005), *The Structure of Biological Membranes*, 2nd edition, CRC Press, London.
Yu-li Wang, Klaus M. Hahn, Robert F. Murphy, and Alan F. Horwitz, 'From imaging to understanding: Frontiers in live cell imaging', *The Journal of Cell Biology*, vol. 174, no. 4, 2006, pp. 481–484.
Zon, L.I. (2001), *Haematopoiesis–A Developmental Approach*, Oxford University Press, New York.

WEB REFERENCES

http://cellbio.utmb.edu/cellbio
http://falcon.jmu.edu/~ramseyil/cellbiology.htm
http://gened.emc.maricopa.edu/bio/bio181/BIOBK/BioBookTOC.html
http://ghr.nlm.nih.gov/handbook/hgp.pdf
http://golgi.harvard.edu/BioLinks.html
http://library thinkquest.org
http://lipidlibrary.aocs.org/Lipids/whatlip/index.htm
http://stemcells.nih.gov/info/basics/basics1.asp
http://web.mit.edu/esgbio/www/cb/cbdir.html
http://www.abgent.com/downloads/posters/neurosciences.pdf
http://www.assaydesigns.com/corp/images/literature/poster-hsp-complexes.pdf
http://www.biology.eku.edu/RITCHISO/301notes1.htm
http://www.biology.eku.edu/RITCHISO/301syl.htm
http://www.cellbio.com/
http://www.emc.maricopa.edu/faculty/farabee/BIOBK/BioBookTOC.html
http://www.gac.edu/~cellab
http://www.kensbiorefs.com/cellchem.html
http://www.kensbiorefs.com/cellstructure.html
http://www.kensbiorefs.com/index.html
http://www.ncsu.edu/sciencejunction/terminal/imse/lowres/1/cellbiology.htm
http://www.nih.gov/about/researchresultsforthepublic/HumanGenomeProject.pdf
http://www.ornl.gov/sci/techresources/Human_Genome/publicat/microbial/13doeproj.html
http://www.sanger.ac.uk/PostGenomics/epigenome
www.biologie.uni-hamburg.de/.../cells2.html
www.dnareplication.info/
www.genome.gov/pressDisplay.cfm?photoID=38
www.ncbi.nlm.nih.gov/bookshelf/br.fcgi?book=g..References n 649
www.rpi.edu/dept/bcbp/molbiochem/MBWeb/.../carriers.htm
www.scientificpsychic.com/fitness/carbohydrat..
www.sparknotes.com/.../section1.html
www.wiley.com/.../membrane_transport/membrane_transport.htm

Index

Related Titles

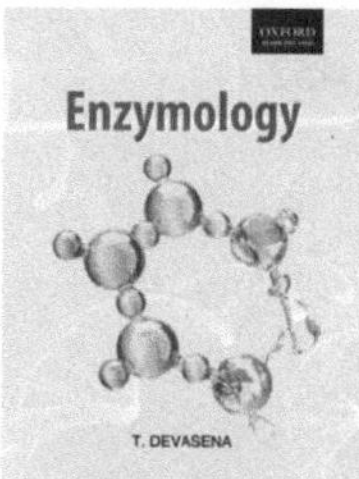

ENZYMOLOGY [9780198064435]

Devasena T., Centre for Nanoscience and Technology, A. C. Tech Campus, Anna University, Chennai

This book covers all the fundamental concepts of enzymology that include enzyme kinetics, enzyme inhibition, and enzyme regulation before proceeding to discuss the applications of enzymes in various domains such as molecular biology, cloning and genetic engineering. A separate chapter has been devoted to the study of enzyme engineering and technology. Comprehensive in its coverage of topics, the book is rich in illustrations supporting the theoretical discussion.

Key Features

- Addresses topics on mechanics of enzyme action, microbiological enzymes, purification and characterization of enzymes, enzyme reactors, industrial applications of enzymes, enzyme kinetics and synzymes
- Includes discussion of latest topics like biosensors, and biochips, and 'Enzyme Business in India'

GENETIC ENGINEERING [9780195696578]

Smita Rastogi (Department of Biotechnology, Delhi Technological University) and **Neelam Pathak** (Department of Biotechnology, Integral University, Lucknow)

Genetic Engineering is designed to serve as a textbook for students of biotechnology, life sciences, genetics, microbiology, biochemistry, and other related areas. Replete with comprehensive details, it will also serve as a reference for research students in these fields.

Key Features

- Gives an elaborative account of the basic as well as the applied aspects of genetic engineering through detailed illustrations, self-explanatory figures, and elaborate tables for easy understanding
- Provides an overview of recombinant DNA Biosafety Guidelines and National Regulatory Mechanism for their implementation

IMMUNOLOGY [9780198068266]

Raj Khanna, Former Professor, University of Lucknow

The book has been written primarily for undergraduate biotechnology, biochemistry and medical courses; it would also be useful to postgraduate students of biochemistry, bio-technology, and microbiology.

Key Features

- Presents numerous extensively-labelled colour illustrations, exhibits containing additional information and interesting facts accompanying the text
- Presents recent research areas such as regulatory T cells and their manipulation for clinical use and cancer immunotherapy including cancer vaccines

Other Related Titles

- 9780195697810 Pal and Ghaskadbi: Fundamentals of Molecular Biology
- 9780198073918 Sivasankar: Instrumental Methods of Analysis
- 9780195676884 Chakravarty: Immunology and Immunotechnology
- 9780195699609 Nallari: Medical Biotechnology
- 9780195687828 Bhattacharyya and Banerjee: Environmental Biotechnology
- 9780195692303 Ghosh and Mallick: Bioinformatics Principles and Applications
- 9780195676839 Bosu and Thukral: Bioinformatics Experiments, Databases, Tools, and Algorithms